The Laws of Belief

The Laws of Belief

Ranking Theory and its
Philosophical Applications

Wolfgang Spohn

OXFORD
UNIVERSITY PRESS

Great Clarendon Street, Oxford, OX2 6DP,
United Kingdom

Oxford University Press is a department of the University of Oxford.
It furthers the University's objective of excellence in research, scholarship,
and education by publishing worldwide. Oxford is a registered trade mark of
Oxford University Press in the UK and in certain other countries

First Edition published in 2012
First published in paperback 2014

Published in the United States of America by Oxford University Press
198 Madison Avenue, New York, NY 10016, United States of America

British Library Cataloguing in Publication Data
Data available

Library of Congress Cataloging in Publication Data
Data available

ISBN 978–0–19–969750–2 (Hbk)
ISBN 978–0–19–870585–7 (Pbk)

Time has told me, Ulli

Contents

Detailed Contents

Preface

So many prefaces went through my mind in the past few years. Now, when this book has miraculously come to an end and I am obliged to write this preface, my heart is silent. Relief overwhelms other thoughts and feelings.

Still I should say a few things here. I will be claiming that ranking theory delivers *the* dynamic laws of belief and is *the* legitimate sister of probability theory. The unfolding of these grand claims may well require so many pages. A substantial introduction to the contents of this book will be found in Chapter 1. It also offers a brief guide through the book and various options for partial reading, in case you do not want to read the book from start to end. Here, I just give a few remarks about the history of the book and a few reading instructions.

Around 1980 Peter Gärdenfors' papers on conditionals and belief change (1978, 1981) provided the best available account of conditional belief (besides Ellis 1979, which I was unaware of at that time), and this account appeared to offer more intelligible grounds for a theory of conditionals and causation than the similarity spheres used by David Lewis as a basis for those topics. However, I thought that Gärdenfors' account represented conditional beliefs and the dynamics of belief only in an incomplete way, too incomplete for the purposes of a theory of causation. For months I struggled to find a way of completing the account. Somehow the solution came to my mind in April 1982, and what was later called ranking theory was born.

Since then, I have written many papers on it. However, these papers appeared more and more absurd to me. They always started with the same short introduction to ranking theory and then proceeded to some new problem or application, or, still worse, they used ranking theory only as an implicit background. In this way, the power and beauty of the theory never came into full perspective.

Therefore I decided in 2004 to present ranking theory as completely as I could. I wanted to carefully explain all of the extensive details and applications at a moderate pace. So I knew I was starting a big enterprise. Still, I sometimes despaired at its size. In retrospect, however, I am utterly glad about my decision. I first thought that the book was simply a matter of writing up all the material I had collected so far. However, at almost all points I started to rethink matters, and in almost every case I clarified my ideas and improved my theorizing. So each chapter supersedes the previous papers on which it builds.

This process took more time than expected. In 2006 I essentially finished the first ten chapters, Chapter 14 was written in 2008, and the full first draft was sent to the publisher at the end of 2009. The slow progress made me nervous; I felt like the hare in the fairy tale "The Hare and the Hedgehog". As soon as I was finished with one chapter, I needed to start revising the others in order to account for the relevant books and papers that had appeared in the meantime. This was a race I could only lose. I plead for leniency in this respect.

The book proceeds in an almost textbook-like manner. It is, however, a research monograph through and through. And it is avowedly a piece of formal philosophy. I am aware that I am thereby excluding many readers and giving a hard time to many others. However, the far-reaching use of formal methods was entirely unavoidable. Without formal definitions, principles, theorems, and proofs, everything would have remained at the surface and nothing could have been seriously treated; the entire wit of the enterprise would have gone. Indeed, I did not want to avoid formal methods. They are very natural to me, and I am convinced of their broad utility in philosophy as one important set of tools among others (see also Spohn 2005c). So, one side goal of this book is to display the power of formal methods and to extend their scope in philosophy.

This was my excuse to the readers who are not yet full experts. I also have a request to the full experts. This book might be called delayed, since it takes up many topics that have been under discussion for decades and about which the experts will already have developed firm opinions. My request is: try to withhold those opinions as long as possible. Of course, you have to exert your critical expertise in the end. But the sooner you do, the more you obstruct your grasp of the various changes of perspective that drive this book.

At the outset I decided to write a book without footnotes. I integrated all potential footnotes into the main text, often as parenthetical remarks. Likewise, proofs are presented immediately after the theorems, if they are given at all. Sometimes I omit (difficult) proofs when they can be found in other places. I stuck to those decisions against some friendly advice and despite some textual hardships. For me at least, it is more convenient to learn to decide what to read and what to skip in a linear text than to have to continue reading at scattered places. I hope that you can share my attitude.

The text is ordered by 17 chapters and 91 sections and moreover by an extensive consecutive numbering of definitions, theorems, and examples, which is also my main tool for cross references. "$(m.n)$" refers to item n of chapter m, but I often omit parentheses in order to reduce their nesting. The same rule for parentheses holds for references to the bibliography. No ambiguity can thereby arise.

Over the many years I received a lot of help and good advice, much more than I could actually process. It is painful to finish a book and at the same time to know that so many parts are unfinished or could have been treated in better ways. Perhaps there is some truth in the soothing thought that such a book should not foreclose future research, but open a rich agenda of further inquiries. And this book really does, in every chapter.

My gratitude for all this help and advice is deep and comprehensive. Slightly shocked I realize that my bookkeeping has been quite insufficient for a long time. If I forget important names – and I certainly will – I sincerely apologize.

The first thanks go to my venerated teacher, Wolfgang Stegmüller, who was also the first supervisor of my Habilitationsschrift (1983), and to Godehard Link and

Kurt Weichselberger, whose comments on that thesis were most valuable. I first presented ranking theory to an international audience at an NSF conference at UCI in 1985. I am still grateful to Brian Skyrms and William Harper for this opportunity. Isaac Levi and Peter Gärdenfors immediately saw the importance of my contribution. Their work and the discussions with them provided a continuous and most fruitful challenge to me. Indeed, the rich literature on belief revision theory has provided an equally rich stimulus for me over many years. My largest debt here is to Hans Rott, my exchange with whom started in the early 80s and still continues. But I have richly profited also from discussions with André Fuhrmann, Wlodek Rabinowicz, Sven Ove Hansson, Erik Olsson, and Gordian Haas.

Daniel Hunter was about the first to constructively take up ranking theory and to make important contributions. I am still indebted to him for introducing me to the computer science community. Here, I am most grateful to Judea Pearl and his group, notably Dan Geiger, Moisés Goldszmidt, Adnan Darwiche, and Thomas Verma. They invented the label, and their use of ranking theory was most illuminating for me. Similarly, I have learnt a lot from the work of Prakash Shenoy, and later on from the work of Joseph Halpern. My gratefulness extends also to Emil Weydert and to Gabriele Kern-Isberner.

Causation is one of the most prominent topics in this book. Here, I have enormously profited from exchanges with Karel Lambert, Nancy Cartwright, Maria Carla Galavotti, Clark Glymour, Peter Spirtes, Richard Scheines, James Woodward, Chris Hitchcock, Luc Bovens, Max Kistler, and others already mentioned, as well as more recently from Michael Baumgartner and Luke Glynn. I am deeply grateful to all of them.

Very special thanks go to Franz Huber, Matthias Hild, and Arthur Merin. At different times they have developed the deepest understanding of the philosophical relevance of ranking theory and have helped me more than anyone else in unfolding this relevance in quite different directions.

As to the broader philosophical concerns that ultimately drive this book despite its rich formalism, Wolfgang Benkewitz has been a continuous partner to me since the early 80s; the discussions with him over matters of ontology, epistemology and philosophy of language and mind have been extremely valuable for me. The same holds for Wolfgang Freitag and Holger Sturm in recent years and for my wife Ulrike Haas-Spohn for all dealings with two-dimensional semantics – which loom large in this book, though mostly implicitly. I should also mention Martine Nida-Rümelin, Ansgar Beckermann, Volker Halbach, Andreas Kemmerling, and Franz von Kutschera who provided most stimulating input on various parts of this book.

Besides some of those mentioned, many persons have commented on parts of my manuscript: Horacio Arló-Costa, Arthur Pedersen, Stefano Bigliardi, Maryia Ramanava, Stefan Hohenadel, Benjamin Bewersdorf, Alexander Reutlinger, Ludwig Fahrbach, Jacob Rosenthal, Holger Leerhoff, and Christian Bender. In sum, their remarks considerably improved the manuscript. I also profited from my students and their stimulating questions and remarks when I occasionally presented parts of my stuff in courses and colloquia.

I am most grateful to two anonymous referees, not only for studying the entire manuscript, but also for providing still further valuable comments that I tried to integrate as well as I could.

In a way, I feel a very general and all the deeper gratitude towards all the authors referred to in the bibliography. Even if I do not actually talk to them, I live in a continuous conversation with them, which, I am sure, transcends soliloquy. Without this continuous conversation every part of the book would have been entirely impossible.

Thanks of a quite different sort go to Ruth Katzmarek, who typed the entire manuscript with unfailing care and patience for the many formulae; I am most grateful. I am deeply indebted to Iain Martel and also to Daniel Ranweiler, who checked the entire book regarding its imperfect English, indeed more imperfect than I had thought. Many thanks also to Christopher von Bülow for diligently reading the proofs and preparing the diagrams.

Finally, I am very grateful to Oxford University Press, to Peter Momtchiloff and all the staff behind and beside him, for accepting this book and so carefully giving the book its final shape.

Academic life in Germany has dramatically changed since around 2000, definitely to the detriment of such an enterprise as this one. Often I thought with indignation that it had become almost impossible to write such a comprehensive research monograph under such circumstances, unless one radically and irresponsibly shirked one's duties. I did not do this. No doubt this was one reason why this book took so long to complete. All the more, then, I am grateful for the institutional support I received from the DFG for a sabbatical term in Winter 2005/06 (Grant No. Sp 279/12-1), and from my university for a sabbatical year in 2008/09 (within the program "Free Spaces for Creativity"). During these sabbaticals my book advanced most fruitfully.

Another aspect that was continuously on my mind was the incompatibility of such an enterprise with family life. I was often absorbed to an irresponsible degree, I felt. My sons Lucas and Simon flourished nevertheless and knew well how to pursue their own ideas and interests. I am humbly grateful for the very happy times I had and still have with them.

I owe my deepest gratitude to my wife Ulli. She shared my ranking-theoretic existence from the beginning. I firmly recall the hike in April 1982 when I told her what I had discovered. And she has borne my ups and downs ever since. Well, she has not only borne them, she has been of the greatest understanding and support. And she finally convinced me of the actual title of this book. It is to her that it is dedicated.

The cover picture *Königsberg in Winter* was painted by my grandmother Gertrud Knopp (1879–1974) around 1920. It shows the view from the home office of my grandfather Konrad Knopp (1882–1957).

Wolfgang Spohn

1

Introduction

1.1 Some Reflections on Skepticism and the Problem of Induction

The skeptical challenge is the alpha and the omega of all our epistemological efforts, or so it has seemed for centuries. I agree that the skeptical challenge *is* important, though I find the emphasis placed on it to be excessive. In the end I will set it aside. The skeptic tells us that we can be certain of little, if anything, that we are hardly ever justified in holding any beliefs, and that we rarely, if ever, know anything. Certainty is corrupted by Cartesian methodical doubt, justifications dissolve in Agrippa's trilemma, and if one or both are lost, knowledge is lost as well. To be sure, the skeptic does not claim anything. He only asks nagging questions or suggests arguments that lead us, by our own lights, to skeptical conclusions.

I take it as evident that the falsity of such conclusions is much more credible than any premises apparently leading to them, and so I only feebly feel the force of the skeptical challenge. If the conclusion is that we can hardly be certain of anything, then absolute certainty is presumably taken as the standard of comparison. We may well agree; but then we would still claim to have sufficient certainty most of the time. If we should turn out to be hardly ever justified, we have apparently applied standards of justification that are too rigorous. And if it is said that we hardly know anything, it cannot be *our* notion of knowledge that is used.

Of course it is along these lines that the skeptical challenge has mostly been taken, namely as an analytic and a constructive challenge. What then *is* our notion of knowledge? What is the appropriate theory of degrees of certainty or belief pointed to by the vague notion of sufficient certainty? What might be a workable account of justification? Such are the questions the epistemologist must answer.

On the one hand, the various forms of the skeptical challenge have been helpful in approaching such questions because they save us from various naiveties and point to various deficiencies of ideas to which we might naturally tend. On the other hand, I sense that the preoccupation with the skeptical challenge is also detrimental. The constructive concern is distorted if the main intent is to overcome that challenge. A refutation of the skeptic should rather be a natural by-product of constructive theorizing. For instance, the mere antiskeptic tends to answer "how is it possible that x?" questions, which is to go only half way to answering the "how is it the case that x?" questions that are the focus of the constructively minded epistemologist.

Be this as it may, in this book I am not explicitly concerned with antiskeptical maneuvers, nor will I be engaged in an analysis of knowledge. Instead I am interested in the component concepts of knowledge. And so, throughout the entire book, I will be occupied with developing a detailed account of degrees of belief or certainty. It will become evident that I am deeply concerned with issues relating to reason and reasoning (throughout; but particularly in Chapters 6 and 7) and to justification (in Section 16.1). In Section 17.3 I will finally speculate about the relation between truth and belief and about the truth-conduciveness of our belief dispositions. What all this means in terms of knowledge does not, however, concern me here.

The background of this somewhat nonchalant attitude towards knowledge is that I find there to be a deep schism in contemporary epistemology. This schism, obscured by the English word "epistemology" as well as the German label "Erkenntnistheorie", is one between the theory of belief and the theory of knowledge. These are to a surprising extent pursued in diverging paradigms, by different methods, and by largely separate communities. (I have described this schism a bit more carefully, e.g., in Spohn 2001a.) The schism is not easily bridged. So it seems prudent to me that I should be cautious with claims concerning knowledge. The present book is instead intended solely as a contribution to the theory of belief, though it will open many perspectives on how to bridge the schism from that side.

So let us take a first step towards an account of belief formation. An important distinction is that between inferential and non-inferential beliefs. This distinction might derive from a misconceived notion of inference, an issue we will consider thoroughly. For the time being, though, we may take it for granted and observe that it entails a traditional and still effective basic division of epistemology. There is the problem of induction, which concerns what we do or should inferentially believe, and there is the problem of the belief base, which concerns the nature of our basic or non-inferential beliefs. If we answer both questions, then it seems that we will have delivered a full account of belief formation.

The problem of the belief base is perhaps the less perspicuous of the two insofar as there are several quite different candidates for basic beliefs: (i) a priori beliefs that involve us in intricate issues about the nature and extension of apriority, (ii) perceptual or observational beliefs that require an account of perception (clearly an entirely different matter), (iii) intuitions (not in a Kantian sense, but as ordinarily understood, though it is quite unclear that they are distinct from (i) and (ii)), and perhaps (iv) evident beliefs (though it is quite unclear that they extend beyond (i)–(iii)). However, I do not want to burden my investigation with these issues right from the beginning. We will rather see what we can say about apriority and about perception and perceptual beliefs from the vantage point of the theory to be developed in Section 6.5 and in the final Chapters 16 and 17.

So it is the problem of induction – the problem which David Hume so forcefully placed on our eternal philosophical agenda and which has provoked such multifarious and ingenious responses – that is in the center of the present investigation. This is the problem about which expectations to form regarding the future on the basis of our beliefs about the past, which general beliefs to entertain on the basis of our singular observations, which beliefs about the non-observed to infer from our beliefs about the observed – i.e., generally, which beliefs to infer from our basic, non-inferential beliefs. If inference could be reduced here to deductive inference, all would be fine, as fine as our very well developed accounts of deductive inference. But inference cannot be so reduced; deductive accounts will never lead us from perceptions to beliefs about laws, the future, or the unobserved. Non-deductive or inductive inference, as I will say (thus giving "inductive" its widest possible sense) must be accounted for as well, as Hume was acutely aware. We vitally depend on our inductively inferred beliefs.

At this point let me introduce a transformation of the problem of induction that will be of fundamental importance to my whole inquiry and that helps us see the problem in a more clear-headed way. (In Spohn (2000b) I discuss the subsequent thoughts somewhat more broadly.) I define an *inductive scheme* to be any function that projects from each possible sequence of data some set of beliefs or, more generally, some doxastic state. Clearly, there are a great many such inductive schemes. In a descriptive mood, we might want to determine the inductive scheme characterizing a given person. Since an inductive scheme is an infinite disposition, it is underdetermined by the finite set of data sequences and doxastic states the person goes through in the course of her life. In a normative mood, we want to know which inductive scheme is the right one to apply. This presupposes an answer to the question of how to find criteria for determining the right inductive scheme in the first place. Clearly this is the problem of induction as we all know it.

Normally, though, the issue of how to form one's beliefs is not put in this global way. More natural is the local question: how does a person, after having somehow acquired a rich body of beliefs and now receiving a new datum or piece of information, arrive at a new doxastic state? By following a *learning scheme*: that is, any function assigning to each (prior) set of beliefs or, more generally, doxastic state and each datum a new (posterior) set of beliefs or doxastic state. Thus a learning scheme is a dynamic law for doxastic states, about how doxastic states change under the influence of possible data. Again, there is the descriptive task of determining the actual learning scheme of a given person, which is again underdetermined by the person's actual doxastic career. And then there is the normative task regarding which learning scheme we should rationally adopt.

The point now is that inductive schemes and learning schemes roughly come to the same thing. Let us be a bit more precise: Each inductive scheme generates a learning scheme by assigning to any given data sequence a prior doxastic state and to this sequence, enriched by a new datum, another posterior doxastic state. From this we can abstract a well-defined learning scheme, albeit only if the following condition is satisfied: if the inductive scheme assigns the same doxastic state to two different data sequences, then it must also assign the same doxastic state to identical extensions of these data sequences. Otherwise, the posterior state would depend not only on the prior state and the new datum, but also on how the prior state was arrived at. I think the condition is plausible, but at this stage we need not debate it. Moreover, the inductive scheme does not generate a complete learning scheme in this way. So far only those doxastic states that are possible outputs of the inductive scheme are in the domain of the learning scheme; for all other doxastic states the learning scheme may be arbitrarily completed. But so much arbitrariness is harmless, for it affects only those doxastic states that are inaccessible from the point of view of the inductive scheme.

Conversely, each learning scheme generates an inductive scheme, provided it is accompanied by some initial or a priori doxastic state. Given such a starting point, iterated application of the learning scheme yields a doxastic output for each possible data sequence.

This near-perfect equivalence shows that the problem of induction can just as well be posed as the question of the appropriate learning scheme, i.e., the search for reasonable strategies of changing or revising one's beliefs in the light of new evidence. Still there are three differences emerging from this discussion, all of which will prove important.

(1) One point is that, as has just been seen, the dependence on an initial or a priori doxastic state is more salient when the problem is posed in terms of learning schemes. With respect to inductive schemes it comes into focus only by asking the somewhat artificial question about what to believe with regard to the zero input, the empty data sequence. For Bayesians, however, this dependence has always been obvious and crucial. Indeed, foundational discussions in statistics center precisely around the extent to which assumptions about a priori probabilities must be justified or can be avoided (e.g., see the professed Bayesians Howson, Urbach (1989, ch. 10), or the very representative assessment of Stuart, Ord (1991, ch. 31, in particular pp. 1225ff.)).

In traditional epistemology, by contrast, this dependence has been much less clearly recognized. One reason is, I suspect, the point just noted. Another reason – or rather another aspect of the skewed older picture – is the focus on the wrong kind of apriority. Traditionally, philosophers inquired into the *unrevisable a priori*, as one might call it, into the question about which beliefs do not depend on contingent facts of the empirical world and are hence invariably held whatever may come. This is of deep importance, but in the present context the concern is rather what to believe in the initial doxastic state upon which the learning schemes operate. These have an equal right to be called a priori beliefs, but they should be called *defeasibly a priori*, because there is no presumption that all of them are preserved in learning. Hence, the old inquiry was incomplete, since the features of initial doxastic states go beyond the unchangeable features.

This distinction between two kinds of apriority has been clear to Bayesian epistem- ology all along and seems to be receiving more acceptance from outside this circle as well. I will introduce and discuss this distinction more carefully in the Sections 6.5 and 17.1, as it is a major aim of this book to develop the notion of defeasible apriority and to arrive at substantial assertions about it.

(2) Another point is that the descriptive or normative questions relating to inductive schemes are global ones, whereas those relating to learning schemes are only local. Surely local questions are much easier to answer. For instance, in describing or evaluat- ing the transition from the prior to the posterior doxastic state one is exempted from describing or evaluating the provenance of the prior state; it can simply be taken as given. This facilitates matters enormously, even if, as we have seen, the many local answers add up to a global answer.

The point goes beyond a merely tactical level. As is well known, our apparent inability to answer Hume's problem of induction led to a thoroughgoing inductive skepticism which even deepened, for instance, in Goodman's new riddle of induction (cf. Goodman 1946) and in the meaning-skepticism of Kripke's Wittgenstein inter- pretation (cf. Kripke 1982). The latter is, in a way, a consequence of applying inductive skepticism to Quine's remark (Quine 1960, §17, and elsewhere) that language learning always requires a double induction concerning the reference and concerning the sign (although Quine there spots the original confusion of use and mention). However, the skeptical issues are also the result of posing global questions. They generate an all-or- nothing attitude: either we are able to positively answer the global questions, or the case is entirely forlorn. Hence, one lesson of the history of inductive skepticism is to turn away from the global questions – for instance, not to aspire to full justification of our beliefs, but to inquire only into the extent to which our beliefs or belief changes can be justified. Then the skeptic can no longer conclude with the skeptical verdict upon observing that complete justifications are missing; rather he must grapple with the specific local justifications proposed, and then the skeptical strategy is of no avail. All in all, the shift from the global to the local issues corresponds to the move from the argument over skepticism to constructive epistemology, which I have so emphatically endorsed. (Though our terminologies differ, here I feel very close to the exposition of the problem of induction by Lipton (1991, ch. 1) and his distinction between the problem of justification and the problem of description.)

(3) There is yet another, quite different point. In terms of inductive schemes the problem of induction is usually presented as one about inductive inference: here I am with my total evidence, and now what should I conclude from all this? This way of posing the question is misleading. It creates a false analogy between inductive and deductive inference. Deductive arguments are valid or invalid, and there is little dis- agreement about this. So one might expect the same to hold for inductive arguments, and if this expectation is frustrated, inductive skepticism ensues.

In terms of learning schemes, by contrast, the static perspective of unfolding the inferences from the premises given at a specific time is replaced by a dynamic perspective of changing one's beliefs under the pressure of new evidence. Prima facie, at least, this is an entirely different perspective. Exposing the false analogy between inductive and deductive inference is certainly liberating. Indeed, the point of the whole essay is, in a way, to display the vast superiority of the dynamic over the static perspective.

Let me thus sum up my short remarks on inductive and learning schemes by saying that the appropriate up-to-date phrasing of the problem of induction is this: (i) *Which laws or rules hold for the dynamics of doxastic states, and* (ii) *how are the initial doxastic states from which the dynamics starts to be characterized*? My reflection should have made clear that this *is* the perennial problem of induction, full stop.

This formulation makes clear why the so neatly posed problem of induction is in fact so inexhaustible. The search for the initial doxastic states seems to get lost either in early infancy on an individual level or in immemorial stages of mankind on an evolutionary or historical scale, though both also seem to be the wrong directions to inquire. So far, the label "initial" just indicates a problem area. And the question concerning the dynamic laws of doxastic states addresses so many different parties: neurophysiological mechanisms are relevant in answering this question; so are the cognitive sciences with their theories and findings concerning perception, learning, thought, memory, speech, and agency; so are the models and simulations provided by AI; so is the developmental and social psychology of cognitive functions; so are speculations about the evolutionary selection of certain cognitive mechanisms; and so, last but not least, are all studies concerning the sociology and the history of science and the general history of ideas. Virtually everybody can contribute to the study of the change and development of our notions, beliefs, and doxastic states in general. All this is at stake in the problem of induction.

Is there anything left for the philosopher to do? Yes. I am expressly not pleading for naturalized epistemology. The philosopher's field hides in the indeterminacy of asking for the laws *holding* for the dynamics of belief. Clearly this is ambiguous between a descriptive and a normative reading. One might say, then, that the philosopher's task is to fill the normative perspective. This might indeed be right, but the truth is more complicated; the philosopher's responsibility is both wider and narrower.

Probably the best thing to say is that the philosopher inquires into the rational laws of the dynamics of belief. However, the term "rational" is equally ambiguous between a normative and a descriptive perspective. Not that the term has two meanings. It has one meaning with two facets that are inextricably intertwined. The normative discussion about what is rational must not ignore the descriptive perspective. It can be shaken or even be disproved by empirical findings, simply because the results of the normative discussion must serve at least as an empirical idealization. Conversely the descriptive discussion cannot simply side-step the normative findings. Since we want to understand ourselves to some extent as rational, empirical theories about us must, however tacitly, appeal to an ideal standard of rationality, even if only for the purposes of explaining the extent to which and the way in which we miss it. This ideal standard is provided by the normative discussion. Hence, neither side is exclusively for the philosopher or for the empirical scientist. (I have made three attempts to describe the intricate relation between the normative and the descriptive perspective on rationality in Spohn 1993b, 2002b, and 2007. However, the issue was already paradigmatically set out by Hempel 1961/62.)

What can the philosopher contribute to the descriptive perspective? He might have a special competence in some particular fields of the manifold of empirical phenomena related to the dynamics of belief (e.g. in the history of ideas). What I would like to emphasize more, though, is that those occupied with this vast empirical manifold are likely to get lost in details. The philosopher is freer to speculate about general, or the most general, laws of belief, to specify general formal models that are at best loosely connected with empirical facts. One might, of course, doubt that there are any general laws in this realm and conclude that such speculation is entirely futile. This doubt is neither easily refuted nor easily confirmed. But it is certainly desirable to conceive the manifold of relevant phenomena under some general laws, even if only as idealizations that apply to reality only with massive help from correcting theories. Given this desire, the burden of proof is on the side of the skeptic; he must show the desire to be misguided.

Admittedly, the inspiration for general formal models derives from the normative perspective, which is not exclusively, but predominantly that of the philosopher. Here, however, we face another objection, drawn from naturalized epistemology. Quine once famously said that "normative epistemology is a branch of engineering", "the technology of truth-seeking". "The normative here, as elsewhere in engineering, becomes descriptive when the terminal parameter is expressed" (Quine 1986, pp. 663f.). If this were so, there would be no point in a genuinely normative discussion, and the philosopher could join the cognitive and other scientists in engineering.

Is Quine right? It would certainly be most gratifying if he were. If all normative conceptions in epistemology could be subsumed under the ultimate goal, the terminal parameter, of truth-seeking, then the result would be a perfect systematization of the normative point of view. Quine seems to take this for granted and thus creates the impression that the normative perspective is insignificant. My point, by contrast, is that such a perfect systematization is not predetermined and can at best be the outcome of a thorough and principally open-ended normative discussion. When I look at the vast literature of the last decades in scientific methodology, statistics, inductive logic, philosophy of science, and epistemology, I see an enormously rich and lively discussion about symmetry and relevance principles, simplicity criteria, enumerative and eliminative induction, inference to the best explanation, truth approximation, a mixed bunch assembled under the label "coherence", a lot of pragmatic procedures and justifications in the colloquial and in the philosophical sense of "pragmatic", various kinds of conditionalization rules, the principle of minimizing relative entropy and other principles of conservativity and minimal change, epistemic decision theory, profound limit and reflection principles, and so forth. It is certainly not (as it were) a technological issue, a matter of trial and error, to find out about the truth-conduciveness of all these rules and principles. It is rather a matter of normative theory construction, of fitting and trimming the principles so as to reach a reflective equilibrium among them and with our normative intuitions and practices. Indeed, as I will lay out in some more detail in Section 17.3, I see truth entangled in that equilibrium; truth and (theoretical) rationality help explaining each other. (This view will presuppose a strict distinction between two notions of truth, viz., a pragmatic or internal notion of truth thus entangled and the familiar correspondence notion. Quine has different views on truth; however, in his (1995, pp. 49f.) he gives much more space to normative epistemology than before and thus seems to soften his naturalistic attitude.) By all means, there are deep necessities and rich opportunities to discuss normative questions in epistemology, and I primarily see this essay as engaged in this perspective.

A lot of important and difficult issues have been raised in these preliminary reflections, all of which deserve closer scrutiny. However, I do not wish to continue the discussion in this way. In my view, philosophy tends to do far too much metatheorizing, it tends to discuss shapes, statuses, and relations of merely imagined theories. In the absence of actual theories there may be nothing better to do. However, my emphatic preference is to engage in actual theory construction and to discuss the virtues of actual theories. Actuality is the best proof of possibility. This is what I want to do in this book. The aim of the introductory remarks was only to partially unfold the epistemological setting within which my account is located. We will see in the second half of this book, i.e. in Chapters 12–17, how much this account can contribute to such philosophical issues.

1.2 The Agenda

These reflections clearly indicate the way before us. What I aim to provide is a perfectly general, formally rigorous normative theory of doxastic states and their laws of change. I have explained why this is equivalent to accounting for the problem of induction. In view of the multifarious importance of that problem, rich epistemological returns should be expected, and will indeed be forthcoming from such a theory.

But we have such a theory, haven't we? There is probability theory and the associated Bayesian epistemology, which have been thoroughly developed for centuries. Is that not good enough? Probability theory is indeed my paradigm; by all means, we must not fall below its standards. But it is *not* good enough, as we shall see in Section 3.3, and we must work on amendments and alternatives.

In fact, the historic development is not so clear-cut. Cohen (1977) has introduced a vivid opposition between what he calls *Pascalian* and *Baconian* probability. (There he uses the term "non-Pascalian"; the label "Baconian" is first used in Cohen (1980).) I like this opposition. Pascalian probability is probability *simpliciter* as we know it today. It emerged in the middle of the 17th century. (cf., e.g., Hacking 1975). Certainly, the origins date back much farther and one might well argue about just how far back. Still, the characteristic formal structure becomes discernible only then. Baconian probability signifies a different tradition of inductive reasoning, which is traced back by Cohen to Bacon (1620) at the latest. It was carried on by Jacob Bernoulli and Johann Heinrich Lambert in the 18th century (according to Shafer 1978) and by John Herschel, William Whewell, and John Stuart Mill in the 19th century (according to Cohen 1980), to mention just a few prominent names. And of course, Cohen sees himself working in this tradition since his (1970).

I do not think it is very clear what Baconian probability really is (even though the label is chosen as aptly as possible), for the course of history is quite ambiguous. However, I do not want to engage here in historic argument; it would be hard to agree, and it is of no particular use at the moment. My very rough and noncommittal historic picture is as follows:

The idea that there are degrees of belief expressing subjective uncertainty is very old (as well as the attempt to externalize this uncertainty by postulating more objective tendencies in the world or its parts). How could it fail to be? With the rise of modern science and philosophy, however, this idea gained particular importance in the 16th and 17th century. In a way, it was nothing less than the forms of our thought (understood in a wider than merely deductive sense) that were at stake. Of course it could not be but a long and laborious process by which this idea gained a firmer shape and by which these forms of thought received a more precise statement (e.g., modern set theory and its rigorous art of stating things is itself a very recent achievement). There might be many ways to try, and there is no reason why only one way should work in the end. Yet (Pascalian) probability was the first clear structure to emerge, and it advanced triumphantly, accruing ever more power and beauty. Whatever the alternatives that were considered, probability was always far ahead. The extent to which it interweaves with our cognitive enterprise has become nearly total (cf. the rich essays in Krüger et al. 1987).

However, alternative ideas existed all along (though even the term "alternative" might actually be too strong as alternatives to and variants of probability were not clearly distinguishable), and the philosophers mentioned above exemplify them. These alternatives led a meager existence in the shadow of probability theory and cannot even be said to have taken a clear and definite shape until the middle of the 20th century.

Still the alternatives did not simply vanish. Their mere existence expressed a continued dissatisfaction with the probabilistic paradigm and a feeling that there is more to inductive reasoning than probability. Indeed the methodology of many empirical disciplines is prima facie not probabilistic at all, even though Bayesians are quite successful in explaining away this negative impression. Around 1950 things finally started changing. This development was considerably accelerated from the 1970s onwards by the rise of artificial intelligence with its interest in implementing inductive reasoning in the computer.

Again, this is not the place for historic details. I will discuss some of the theories emerging in the last sixty years in the comparative Chapter 11. One might say, though, that the historic project of Baconian probability finally took a definite shape. Of course the shaping was not a matter of historically adequate interpretation. It is a matter of explication, and it is a commonplace that an explication need not capture the original intention exactly. It is only required to be similar with respect to its origin and to be precise and fruitful. Therefore, it was always clear that the label "Baconian probability" embodied a promise and not a historic description (this is why I feel absolved from giving accurate historic details).

I will propose my own explication of Baconian probability called *ranking theory*. This is the aim of my book. As I will explain in Section 5.2, its basic structure is essentially Cohen's. To this extent it can be conceived as a variant of Baconian probability. However, ranking theory also goes much further, as Chapter 5 will show. My aim will be to present a particularly rich and fruitful extension and development of Baconian probability, indeed the only one that is on a par with Pascalian probability.

Ironically the reason why the extension is so fruitful is that I did not consciously develop ranking theory in this tradition. I first developed it in my Habilitationsschrift (1983b, Sect. 5.3), and it first appeared in English in my (1988). There I called my subject matter ordinal conditional functions, in order to choose an artificial and clearly distinguishable term. This was a bad idea, and I happily accepted the proposal of Goldszmidt, Pearl (1992a,b) to call them ranking functions, a most suitable and grammatically felicitous label.

My focus in my (1983b, Sect. 5.3) was the problem of induction as restated in the previous section as the dynamic problem of belief change. My intention was to improve upon the theory of belief revision as initiated by Gärdenfors (1978 and 1981). This intention is valid even in the light of the more recent developments of AGM belief revision theory, so-called according to Alchourrón, Gärdenfors, Makinson (1985) (see also, for instance, Gärdenfors 1988 and Rott 2001), and it will again be my leading idea in Chapters 4 and 5 when motivating and introducing ranking theory. Being aware of the close relation between induction and causation that has been common ground since at least Hume, my deeper intention in my (1983b) was to develop an account of deterministic causation that is possibly superior to that of Lewis (1973b) and closely parallels existing accounts of probabilistic causation, in particular my own in Spohn (1980). Here causation will be discussed as an application of ranking theory only in Chapter 14.

The reason why I think ranking theory is a particularly fruitful variant of Baconian probability is precisely that it approaches induction via the dynamics of belief, an approach that, as I discovered only later on, was not well respected in the other attempts to explicate Baconian probability. This remark will be fully explained in Chapter 5. In any case, this is a further reason why the problem of belief revision is put at the center of my considerations.

I said at the beginning of this section that (Pascalian) probability theory is not a good enough response to the problem of induction – it needs amendments and alternatives. Ranking theory will turn out to deliver both. The close relation between ranks and probabilities will be obvious, rich, and complex. In a way, it is the task of the whole book to unfold this relation. Chapter 10, however, will be particularly devoted to it.

Thus, this book is about ranking theory, and the plan for the book is straightforward: develop the theory, compare it, and put it to epistemological use. However, let me display this plan here in a bit more detail:

The first major part of this book, comprising Chapters 2–9, will introduce ranking theory and develop various basic features in formal detail. Ranking theory is a theory of belief (and disbelief). Hence, *Chapter 2* will start explaining how the objects or contents of belief are to be understood. Sections 2.1–2.2 present the noncommittal story that will carry us through most of the book, and in Section 2.3 I commit myself to a specific understanding of belief. After all, I cannot write a book about belief and back out from saying what that is! *Chapter 3* briefly presents the big sister, probability theory, and its problems with representing belief. These are particularly well reflected in the lottery paradox. It will motivate the subsequent independent study of belief. *Chapter 4* will explain what can be taken as the standard account of belief and belief revision. This will be the starting point of my investigation.

The observation in Section 5.1 that the standard account states an incomplete dynamics of belief leads me, in the rest of *Chapter 5* (the central chapter of the book), to introduce ranking functions and their dynamic laws and thus to a full static and dynamic account of belief. *Chapter 6* will explicate the notion of a reason in ranking terms, start inquiring into its structure, and proceed to explicate the aforementioned two notions of apriority. I trust that these explications will hold and not collapse under their tremendous historic weight. This chapter will thus be my entrance to the philosophical applications in the second major part of this book.

Chapter 7 will elaborate on the complementary notion of (conditional) doxastic independence and its structure, and will lead to an investigation of the theory of Bayesian nets and its algorithmic virtues that apply to ranks just as well as to probabilities (however, I shall only begin the exploration of the very large field thereby opened). Since the numeric ranks assumed in ranking theory have often appeared arbitrary and insufficiently determined, *Chapter 8* will provide various rigorous measurement procedures for ranks and thus further justification for their basic features. The main method presented in Sections 8.3–8.4 will measure ranks via iterated belief contractions and thus arrive at a set of laws of iterated contraction that is correct and complete from the point of view of ranking theory. *Chapter 9* is a sort of appendix. The notion of a conditional rank is the crucial one doing most of the work in the previous chapters. It gives rise, however, to similar confusions about conditioning, supposing, and updating as does the notion of conditional probability. It is embarrassing how such a clearly defined notion can be so obscure at the same time. Chapter 9 will be my attempt to clear up the issue.

Chapters 10 and 11 present a presumably impatiently expected comparative interlude. *Chapter 10* is devoted to the most important comparison, namely, the comparison with probability theory. The parallel between ranking and probability theory will have been seen to be the most far-reaching, and it will indeed motivate the entire research program laid out in the epistemological applications of ranking theory in the second major part of the book. Therefore, some more thorough notes on the possibilities of formal unification and on the philosophical convergences and divergences are called for. Section 10.4 presents the most intelligible extension of ranking theory to a qualitative decision theory that I have found in the literature. *Chapter 11* presents a more varied bunch of comparisons: with the predecessor theories of George L. S. Shackle, Nicholas Rescher, and L. Jonathan Cohen; with the epistemological enterprises of Isaac Levi, Keith Lehrer, and John Pollock (the latter giving rise to deeper reflections on the nature of normative epistemology in Section 11.6, which thus further elucidate the basics of my project); and with AGM belief revision theory for a final time. Lastly, the chapter will end with a brief look into some formal alternatives such as formal learning theory, possibility theory, the Dempster–Shafer theory of belief functions, and non-monotonic reasoning in general.

The second major part of this book, comprising Chapters 12–17, will present and discuss various philosophical applications of ranking theory. Chapters 12–15 will focus on applications in philosophy of science and Chapters 16–17 on applications in epistemology. There is, however, no sharp boundary.

Indeed Chapters 12–15 set out to fulfill an all-encompassing program of explaining natural modalities as covertly epistemological modalities, a program designed to replace David Lewis' program of Humean supervenience and to carry out Simon Blackburn's strategy of quasi-realism or Humean projection in constructive detail. The start is made in *Chapter 12* with the notion of a (deterministic) law, the appertaining natural modality being lawlikeness or nomic necessity. Ranking theory seems ideally suited to capture the soft side of laws, their explanatory power, their support of counterfactuals, and most of all, their ability to be confirmed and projected. In the end I will arrive at a novel account of laws that is at the same time very familiar. It simply consists in a deterministic duplication of Bruno de Finetti's account of statistical laws and their confirmation, the interesting point being that such an insightful story is indeed available in terms of ranking theory. The natural continuation is a treatment of the quite obscure topic of ceteris paribus laws or conditions in *Chapter 13*. This widespread phenomenon has often been conjectured to reflect practices of defeasible reasoning in the natural and social sciences, the theoretical grasp of which has been hampered by an insufficient understanding of this form of reasoning. Ranking theory makes progress here as well. This in turn helps with an understanding of dispositions, since, as is well known, the reduction sentences explaining dispositions require a qualification by a ceteris paribus clause. In this vein the second half of the chapter illuminates the epistemology and the metaphysics of dispositions.

Chapter 14, by far the longest of the book, deals with causation, perhaps the most important natural modality. It summarizes more than 30 years of my thinking about this topic. The account I develop resembles the existing accounts vigorously discussed in the literature and is at the same time at odds with them in so many respects that it makes no sense to even start enumerating them here (see, however, the introduction of Chapter 14 and Spohn 2010b). Let me just note that the most important message of the chapter is that there is *one* notion of causation that can be developed into a probabilistic and a deterministic theory of causation in perfect parallel. The latter is, of course, stated in terms of ranking theory and argued to improve upon counterfactual analyses and their recent interventionist variations. Explanation is a topic in the vicinity of causation, which was originally intended to have its own chapter. Instead it has condensed into two appendices of Chapter 14, where a ranking-theoretic home is also provided for the so-called inference to the best explanation.

In some sense *Chapter 15* is the philosophically most important of this book and perhaps the most tentative. As just mentioned, I treat natural modalities as covertly epistemological. In the foregoing chapters I did this by explaining these modalities relative to a ranking function that can only be interpreted as an epistemic or doxastic state. This seems a forbiddingly subjectivist account of those modalities, of lawlikeness and causation. Chapter 15 attempts to take back this extreme subject-relativity by inquiring into the extent to which ranking functions can be objectively true or false. This includes, in a way, a sophisticated justification of the regularity theory of causation. The inquiry will be neither a complete success nor a complete failure and is thus all the more interesting. I myself am surprised how strictly one can still adhere to David Hume's doctrines; for this is what this chapter does in principle.

Chapters 16–17 turn, finally, to the more epistemological applications of ranking theory. Chapters 6–9 and 12–15 will have dealt, in their ways, with the inferential aspect of belief formation. Section 16.1 does so as well by proposing to identify degrees of belief or ranks with degrees of justification, a notion I will have deliberately avoided so far in this book. The rest of *Chapter 16*, however, is occupied with perception and thus with the basis of belief formation. By arguing for what I will call the Schein–Sein Principle and the Conscious Essence Principle I strictly derive a peculiar mixture of foundationalism, coherentism, and (minimal) externalism, which has considerable potential to clarify the vexed philosophical debate delineated by these labels.

The concluding *Chapter 17* deals with the other origin of our beliefs, the unrevisably a priori and the defeasibly a priori beliefs. Section 17.1 attempts to characterize these notions more specifically than before. Then, starting in Section 17.2 from a rationality principle that I will call the Basic Empiricist Principle, I will add and partially derive various further principles: a weaker and a stronger coherence principle, a very weak and a less weak principle of causality, and principles about the fundamental connection between reason and truth. I argue that all these principles are unrevisably a priori. The ultimate goal of this final investigation is to establish grounds for apriority other than mere conceptual grounds. It is indeed, in a very precise sense, the conditions of the possibility of experience or of learning that provide such stronger grounds and help determine the a priori forms of our thought, as Kant always claimed.

This brief overview indicates that we are heading for a very long, very demanding journey. I have thought very hard, but I cannot imagine any briefer, more informal, or less demanding story that could develop the full potential of ranking theory. My only hope is that the reader who really works through chapter by chapter, section by section, will have found the journey worthwhile in the end.

In truth one need not read through every chapter and section. The overview suggested that there is only a partial order among the chapters. For the reader's orientation I have displayed the partial order in a diagram that shows which chapters presuppose which and which chapters may be skipped. This is a rough sketch and may be taken as a first exercise for getting acquainted with directed acyclic graphs that will play a big role in this book! Within the chapters I will give more hints regarding the centrality or marginality of the various sections:

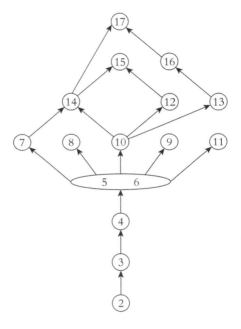

Figure 1.1

In reality the book is not quite as well ordered as this picture suggests. There are various topics that would have deserved a fuller or more systematic treatment, but I am simply unsure how to do it better. Let me highlight just a few important deficiencies: Section 11.6 on the nature of normative epistemology should have been more centrally placed and not amidst all the comparative discussions of Chapter 11, but since it is intertwined with my discussion of John Pollock's epistemology, I decided to leave it there. My discussion of apriority is quite scattered in the book; only Chapter 17 is entirely devoted to the topic. Issues of frame-relativity come up at various central parts of the book, but only obliquely; they would have deserved a general discussion in their own right. Section 14.9 treats them at least with respect to causation.

Many questions are opened and left open; this is how it should be. However, what makes me feel uncomfortable is that I sometimes proceed on open questions, for even such a foundational book apparently cannot start entirely from scratch. For instance, such key words as truth, truth conditions, propositions, and concepts are treated only on a small scale so as to be able to proceed, and hence in a painfully insufficient way. A symptom is that I will seem to deny the conceptual structure of propositions and to use it at the same time, an apparent inconsistency I did not take time to clear up. Another important neglect is that I simply did not manage to complete my comparative discussions (in Chapter 11 and elsewhere) even within my central concerns, i.e., regarding defeasible reasoning. Major omissions are, for instance, the so-called systems Z and Z^+ (cf. Pearl 1990, Goldszmidt, Pearl 1992a,b, 1996) and their further generalization by Weydert (1996, 2003) as well as the theory of inhibitory nets developed by Leitgeb (2004); they simply find no mention in my book. The reader will find many further instances.

Two important omissions, however, are intentional. First, I will have a lot to say about belief, about reasons and justification, and even about truth. Nevertheless, even though all ingredients are provided, I will nowhere systematically discuss *knowledge*. Secondly, I will treat many topics, such as laws, dispositions, and causation, that are most prominently treated in terms of *counterfactual* or subjunctive *conditionals*. Still, even though it would advance my competitive claims, I will nowhere try to give an analysis or a semantics of conditionals. The reasons for my abstinence are twofold: The negative reason is that in both cases I fear drowning in the mess and not being able to return to my proper business. There is, however, a positive reason, which is to make clear that, unless one has an intrinsic interest in knowledge and conditionals (which is, of course, fully legitimate) one might well do without them; the philosophical purposes to which these notions are put can well be served without engaging in them. I will not explicitly argue that point, for this would be a further issue. However, by the end of the book I will have provided everything necessary for assessing this claim.

Enough of announcements and apologies! You have the plan and intent of this book before your eyes, so at long last let us set to work!

2

Belief and its Objects

Any account of the statics and dynamics of belief must build on an account of the objects of belief – i.e., the entities which we are related to by our doxastic attitudes. In this chapter let us lay these foundations for the rest of the book.

The precise nature of the objects of belief is a deep and difficult issue most heatedly discussed in epistemology and in the philosophy of mind and language. It is even unclear whether we may at all assume a uniform nature of those objects. The literature has grown tremendously in the last 30 years: just search the terms "wide vs. narrow content", "proposition" or "propositional vs. non-propositional content", and "concepts" or "conceptual vs. perceptual content", etc. As it would be hopeless to engage at length in any of these discussions here – indeed that would result in another book – at the present the only reasonable strategy is to stay as neutral as possible so that most or many accounts concerning the objects of belief are compatible with our epistemological considerations. For this reason Section 2.1 will introduce the formal representation of these objects that I will use in this book and Section 2.2 will explain the extent to which this representation actually manages to take the recommendable neutral attitude.

Still, it is clear that any conception of the objects of belief is closely bound up with what one takes belief itself to be. The exact nature of belief is no less profound an issue than the nature of its objects. Indeed it is surprising how great the difficulties are in adequately capturing such a familiar phenomenon. As I will explain, relative neutrality with respect to the former goes with relative neutrality with respect to the latter. This is an attitude that will fortunately carry us through most of the book. The reason is that our investigations mostly concern the structure of belief and its dynamic laws and will thus be relatively independent from a more precise stance on the nature of belief and of its objects. Hence, we can largely avoid getting entangled in these issues. However, I do have a stance on them and so feel obliged to at least sketch it out in Section 2.3, even if it will play no crucial role until the final Chapters 16 and 17.

2.1 The Formal Representation

Our representation of the objects of belief is entirely standard. They are *propositions*, and propositions are represented as *sets of possibilities* from a given space of possibilities. Basically, the neutrality is achieved by being totally silent on the issue of what a possibility is. The underlying *space of possibilities* is primitive and always denoted by W (with subscripts where there is a need to distinguish various spaces of possibilities). Propositions, then, are subsets of W and denoted by A, B, C, D, and E, with or without subscripts. Hence, propositions are what the probability theorist calls events, a usage that I avoid here since it is colloquially and philosophically inappropriate.

For instance, if we simply consider a single throw of a die, we may assume $\{1, 2, 3, 4, 5, 6\}$ to be the underlying space of possibilities, and the proposition that an even number appears is represented by the set $\{2, 4, 6\}$. Alternatively we may consider a die being rolled infinitely many times. Then the underlying space of possibilities consists of all infinite sequences of the numbers 1, 2, 3, 4, 5, and 6. The proposition that a six appears in the 7th throw is represented by the (uncountable) set of all sequences, the 7th member of which is a six; the proposition that the average of the first 100 throws is at least 4 is given by the set of all sequences the first 100 members of which sum up to at least 400; and the proposition that a six is thrown all the time is just the singleton $\{\langle 6, 6, \ldots \rangle\}$. We will meet many more examples in the course of the text.

Is *any* subset of W a proposition? Usually we may assume this or simply not worry about the issue, particularly if W is finite. However, we will need more explicit flexibility here, and we should take the next standard move. If not all subsets of W are propositions, then we should at least assume that the set of propositions is closed under (finite, and possibly infinite) logical operations, i.e., forms an algebra. Set–theoretically, the negation of a proposition A is represented by its complement $\bar{A}$ (relative to W), and the conjunction and disjunction of two propositions A and B is, respectively, represented by the intersection $A \cap B$ and the union $A \cup B$. Hence, let us define:

Definition 2.1: $\mathcal{A}$ is an *algebra* over W iff $\mathcal{A} \subseteq \mathcal{P}(W)$, the power set of W, such that:

(a) $W \in \mathcal{A}$,
(b)   if $A \in \mathcal{A}$, then $\bar{A} \in \mathcal{A}$,
(c)   if $A, B \in \mathcal{A}$, then $A \cup B \in \mathcal{A}$.

$\mathcal{A}$ is a σ-*algebra* over W iff moreover

(d)   for each countable $\mathcal{B} \subseteq \mathcal{A}, \bigcup \mathcal{B} \in \mathcal{A}$,

and a *complete algebra* over W iff moreover

(e)   for each (uncountable) $\mathcal{B} \subseteq \mathcal{A}, \bigcup \mathcal{B} \in \mathcal{A}$.

An algebra is defined to be closed under negations and disjunctions, but this entails, of course, that it is also closed under conjunctions. Since we can distinguish closure under finite, countable, or uncountable disjunctions (and conjunctions), we have to distinguish different kinds of algebras. Clearly, each complete algebra is a σ-algebra, and each σ-algebra is an algebra, but the converses do not necessarily hold. However, if $\mathcal{A}$ is finite, and hence in particular if W is finite, then these distinctions collapse. Often we need not worry about which kind of algebra to consider – I will be specific if necessary. In any case, from now on only members of the implicitly or explicitly assumed algebra over W will be propositions. W and $\varnothing$ are propositions in each algebra. I call them, respectively, the tautological and the contradictory proposition, or simply the *tautology* and the *contradiction*.

Singletons of possibilities need not be propositions. But there might be propositions having a similar role:

Definition 2.2: The proposition A is an *atom* of the algebra $\mathcal{A}$ over W iff $A \neq \varnothing$ and there is no proposition B in $\mathcal{A}$ with $\varnothing \subset B \subset A$ (where "$\subset$" means "proper subset"). The algebra $\mathcal{A}$ is *atomic* iff each proposition is the union of its atoms, i.e., of the atoms contained in it.

Clearly, if singletons of possibilities are members of an algebra, then they are atoms, but the converse need not hold. Often, though, these singletons are not needed as propositions, and the atoms are good enough; an algebra may even be atomless. The standard example is given by the set of all finite unions of half-open intervals over the real line of the form $[a, b) = \{x \mid a \leq x < b\}$. This set is indeed an algebra, but it has no atoms because for each such interval there is a smaller one contained in it. However, each complete, and a fortiori each finite, algebra is atomic. (Simply consider for each $w \in W$ the intersection of all propositions containing w; this intersection is another proposition of the complete algebra, and it must be an atom.)

There is a final notion we will need throughout the book, namely the notion of a *variable* (this is what probability theorists call a 'random variable'). The term is standard, even though it is possibly confusing (for we are not talking about variables in a logical sense). For now, though, let us simply take it as a nominal definition:

Definition 2.3: A *variable* X is any *measurable* function from W into some space of values W_X. That X is *measurable* means that some algebra $\mathcal{A}$ is associated with W and another algebra $\mathcal{B}$ with W_X such that for any $X^{-1}[B]$ element $\mathcal{A}$, for any B element $\mathcal{B}$, i.e., for any set $B \in \mathcal{B}$ of values the set of possibilities w receiving a value $X(w)$ in B is a proposition in $\mathcal{A}$.

If we want to make the dependence on the associated algebras explicit, we speak of $\mathcal{A}$-$\mathcal{B}$-measurability and an $\mathcal{A}$-$\mathcal{B}$-variable, but usually we can omit this because the associated algebras will be clear from the context. Informally, measurability means that we can talk of the proposition that X takes some value in B, provided that B itself is a proposition over the values.

For illustration, let us return to our infinite series of throws of a die. Here, W was the set of all infinite sequences of the numbers 1, 2, 3, 4, 5, and 6. One variable, for instance, is X_7, the result of the 7th throw; it is the function assigning to each such sequence its 7th number, i.e., $X_7((a_n)_{n \in \mathbf{N}}) = a_7$ (where $\mathbf{N}$ is the set of positive integers). Another variable is Z_{100}, the average of the first 100 throws, which assigns to each sequence just the average of its first 100 numbers, i.e., $Z_{100}((a_n)_{n \in \mathbf{N}}) = \frac{1}{100} \cdot \sum_{n=1}^{100} a_n$.

The values of a variable can be anything, but usually we can consider numbers as values. A *binary* or *yes–no* variable, for instance, can be taken as a function from W into $\{0, 1\}$ (1 for yes, 0 for no). If the variable concerns the price of some good, its values are natural numbers (expressing the price in cents), possibly up to some maximal price. Or if the variable concerns location in space (of the center of gravity of some body), then its values are triples of real numbers (relative to some coordinate system). In all these cases the algebra associated with the set W_X of values of a variable X will be assumed to be the natural one, i.e., the power set $\mathcal{P}(W_X)$ of W_X, if W_X is finite, the σ-algebra of Borel sets, if W_X is the real line, etc. So, as far as the values are concerned, we may take the associated algebra for granted.

Indeed very often here, as well as in many textbooks, the order of explanation is reversed. We do not start with a given algebra of propositions over W and then confirm whether or not a certain function on W is a variable, i.e., measurable, but rather we start with some set of functions on W and turn them into variables, i.e., construct the algebra of propositions in such a way that all these functions are variables.

Let me explain this with an example. Let W be the set of all possible global weather courses in 2009, i.e., continuous sequences of global weather states from the beginning to the end of the year 2009. Formally, this is just some very rich set. Now, meteorologists are not interested in each and every detail of these weather courses. Suppose rather they have established a fine grid of weather stations over the earth and are only interested in the values measured at these stations every hour. So for each station s and every hour t in 2009 they consider the possible temperatures at s and t, the possible air pressures at s and t, the possible humidity at s and t, the possible amount of precipitation at s and t, the possible velocities and directions of the wind at s and t, etc. This makes for six or more variables at each s and t. Now all these variables should indeed be variables in the formal sense; for instance, the proposition that the temperature in Konstanz at 12pm on 1 August 2009 is 32°C should be a proposition over W. Hence, we simply assume that the set of weather courses to which the relevant function assigns 32°C is indeed a member of the relevant algebra over W. This we do for all the variables considered and their possible values, and in this way all the propositions the meteorologists are dealing with are generated.

Let me describe this in formal generality:

Definition 2.4: Let W be some space of possibilities and let U be some set of functions on W. With each function $X \in U$ a range W_X of possible values of X and an algebra $\mathcal{B}_X$ over W_X is associated. Let for any $X \in U$ and any $B \in \mathcal{B}_X A_{X,B} = X^{-1}[B]$ $= \{w \mid X(w) \in B\}$ be the proposition that X takes some value in B. Then define $\mathcal{A}(U)$, *the* (σ-, *complete*) *algebra generated by* U, to be the smallest (σ-, complete) algebra containing all these $A_{X,B}$. More generally, for each $U' \subseteq U$, $\mathcal{A}(U')$ is to be the (σ-, complete) subalgebra of $\mathcal{A}(U)$ generated by U', i.e., by all $A_{X,B}$ for $X \in U'$.

It is easy to see that this smallest algebra is well-defined, since the intersection of any number of algebras over W is again an algebra over W. Indeed, $\mathcal{A}(U)$ is the smallest algebra relative to which all functions in U are variables.

And so we always have two options when describing examples or specifying applications of our theory. We can start with the relevant space W of possibilities and directly assume the relevant algebra (for instance $\mathcal{P}(W)$). Alternatively we can start with a set U of variables – which I will often refer to as a (*conceptual*) *frame* or a *space of variables* – where W is implicitly contained as the domain of those variables and where the algebra associated with the values of the variables is tacitly understood; from here we can go on generating the algebra $\mathcal{A}(U)$ of propositions over W. Either way is legitimate, but often the second is more convenient. For the probability theorist, all this should be familiar.

One might wonder why the second option should be a genuine alternative. After all, starting with the functions in U *is* starting also with their domain W, isn't it? Yes, but we can go one step further. Starting with U should really be taken as starting with all the possible ranges W_X of the $X \in U$. Then we may *define* the space W of possibilities as the Cartesian product $\times \{W_X \mid X \in U\}$, conceive of each variable as the projection from this Cartesian product to the relevant component, and finally generate the propositions over the product space via these projections. Thus, $\mathcal{A}(U)$ is the (σ-, complete) algebra over W generated by all these projections. Taken in this way, the alternative is indeed a genuine one.

Let me stop here; the set theoretic constructions are already going on too long. I need only add one final piece of notation. We will often have to refer also to sets of variables. Hence, members as well as subsets of the frame or space U of variables will be indiscriminately denoted by X, Y, Z, or V (with or without subscripts), thus identifying, as it were, a variable with its singleton. This convention will be useful, but perhaps confusing as well, and I will minimize the confusion by always making explicit whether I refer to a single variable or to a set of variables.

In a way, we may conceive of a set $X = \{X_1, \ldots, X_n\}$ of variables as one compound variable taking compound values $x = \langle x_1, \ldots, x_n \rangle$ (this is why there is, in a way, no difference between variables and sets of them). Generally I will adopt the convention to use x, y, z as denoting possible values of the corresponding sets X, Y, Z of variables. Finally, the proposition that the set (or the compound variable) X takes the value x is symbolized by $\{X = x\} = \{w \mid X(w) = x\}$, where, to be precise, both are short for $\{w \mid X_1(w) = x_1, \ldots, X_n(w) = x_n\}$. Clearly the notation can be generalized to $\{X \in B\}$ as denoting the proposition that X takes some compound value from the set B of compound values.

Having thus explicitly introduced the formal framework of propositions and variables, what does it mean then to believe such a proposition? There is a very simple and commonly accepted explanation:

Explanation 2.5: Belief is the exclusion of possibilities. That is, the proposition A is *believed* (by s at t) – symbolically $\mathbf{B}(A)$, or $\mathbf{B}_{s,t}(A)$, in case the subject and the time of belief are to be made explicit – if and only if all possibilities in $\bar{A}$ are excluded (by s at t).

I distinguish the notion of belief, which is basic for this book, by symbolizing it by a bold $\mathbf{B}$. When I introduce notions in a formally completely explicit way as in (2.1) I speak of a definition. Sometimes, however, I have definitions, the definiens of which contains unclear or problematic notions, and then I only call them explanations; (2.5) is such a case, as is fairly obvious. Much later, first in Section 14.9, I will also make a similar distinction between theorems and assertions.

If I believe, e.g., that the temperature in Konstanz at 12pm on 1 August 2009 is 32°C, then I exclude all possibilities in which that temperature is different; all those possibilities, I take it, are not the actual one. To believe a proposition thus is to assume that the actual possibility is a member of the proposition, i.e., to take that proposition to be true.

These are apt phrases, helpful to relate belief to its set-theoretically constructed objects. However, they are only mere paraphrases. Of course believing something is taking it to be true, and what kind of attitude it is to exclude a possibility is in no way clearer: no further insight is thereby provided (this is why I call (2.5) an explanation). These paraphrases are just as neutral as our description of the objects of belief, as they were indeed intended to be.

So far the chapter has served its primary preparatory purposes. If you like, you may proceed immediately to Chapter 3, without much loss. The truth, however, is that there is a multitude of problems underneath the clear constructions of this section, problems that lead us into ever more obscure and speculative philosophical fields. I want at least to indicate what they are and that I am aware of them. The subsequent reflections are most likely misplaced in a mainly introductory chapter, but it would be irresponsible to omit them, and the most fitting place for them is in the present context.

2.2 Interpretation: The Noncommittal Way

I said at the beginning that it would be hopeless to engage into the vast philosophical discussion about the nature of the objects of belief and then proceeded on a conventional path by conceiving these objects as propositions, i.e., as sets of possibilities. Still, we should check the extent to which we have thereby achieved our aim of a neutral representation, so that, whatever the results of that philosophical discussion may be, they can be plugged into the theory that I am going to construct. We were successful on two counts, but not successful on two others, and it is important to be entirely clear about this. Let me explain:

The first aspect of neutrality lies in the variability of the underlying space of possibilities. To be sure, within each application this space must be fixed, and then all further variation of possibility is restricted to that space. However, we are free to consider different spaces in different applications; we do not need to assume one all-purpose space of possibilities that is good for all applications. Indeed, the question of what to say about the relation between the doxastic states pertaining to more or less fine-grained spaces of possibilities will open up very fruitful fields of theoretical investigation. It will come up at several places in Chapters 12–17, most explicitly in Section 14.9. The question is not likely to arise when we start, and stick to, working with the all-purpose space of maximally fine-grained possibilities.

This is not to disrespect, though, the search for this all-purpose space. Certainly we do not have beliefs-relative-to-an-application, but we do have beliefs (to various degrees) and their totality sums up to an overall doxastic state. The philosophical interest is to provide principal means for describing that overall state. Then, however, it seems that we have to go for the maximal space of possibilities. It is certainly not my intention to reject the idea of such a maximal space, for in Section 2.3 I too will endorse it. But it is also clear that the modest start with smaller and variable possibility spaces does at first neither burden us with, nor bar us from, difficult metaphysical speculations about the nature of the all-purpose space of maximal possibilities (perhaps the most profound contemporary speculation of this kind is offered by Lewis 1986a). As I said, our approach is neutral so far.

Still, let us tentatively stay for a while with these speculations, for then the second aspect of neutrality comes to the fore: Even the maximal possibilities, which philosophers like to call possible worlds, are still entirely open to interpretation. They can be taken in a noncommittal way as, say, in modal logic where they figure simply as uninterpreted items relative to which the terms of the language in question receive their extensions. They can be taken as full interpretations of a given formal language or as complete and consistent (infinite) stories in a given natural language. They can be any kind of ersatz worlds, as Lewis (1986a, pp. 137ff.) calls them. Or they can be taken seriously. Then we enter another big discussion: Are they Lewisian possible worlds, i.e., in some specific sense maximal possible objects? Or Wittgensteinian possible worlds, i.e., in some specific sense maximal collections of states of affairs? Recall Wittgenstein (1922), Proposition 1.1: "Die Welt ist die Gesamtheit der Tatsachen, nicht der Dinge." ("The world is the totality of facts, not of things.") Are these ideas at all suited for our epistemological purposes, i.e., to capture epistemic possibilities? Perhaps the latter should rather be construed as scenarios, as conceived by Chalmers (2006, pp. 75ff.), that are in some way uniquely related to metaphysical possibilities (in terms of canonical descriptions, as he calls them). And so forth. (Cf. the thorough discussions in Chalmers 2006.)

There is, moreover, the idea that, whatever the precise notion of a possible world, it will not do. Rather we need centered worlds as doxastic possibilities. Here, a *centered* possible world may be understood as a triple $\langle w, s, t \rangle$ consisting of a possible world w, a subject s, and a time t, both existing in w. This idea has been pushed by Lewis (1979a) among others, and is in my view quite convincing, but certainly not beyond dispute (cf. the discussions in Haas-Spohn 1995, Ch. 2). In fact, I believe, as I have argued in

Spohn (1998) and (2008a, Ch. 16), that it needs further refinement, that a doxastic possibility must be rather conceived as a sequence $\langle w, s, t, a_1, a_2, \ldots \rangle$ consisting of a centered world $\langle w, s, t \rangle$ and a potentially infinite sequence $\langle a_1, a_2, \ldots \rangle$ of objects all existing in w. These remarks already indicate how much there is to settle. But we need not settle it here; wherever we settle, it falls under our neutral start with a space W of possibilities.

Of course the issue of how precisely to conceive of doxastic possibilities is bound up with the issue of belief itself and with the issue of what the belief relation **B** really is that relates a doxastic subject at a given time to a proposition. The former issue cannot seriously be answered without taking a stance on the latter. So what does it mean for a subject to believe, to a certain degree, a proposition or a set of doxastic possibilities to obtain? I said it means to exclude all other possibilities, but this was not yet an informative response.

Again, various substantial positions are conceivable and indeed actually discussed in the literature. Certainly the most natural answer is that believing one out of a set of possibilities to obtain just means assenting to a given description of this set of possibilities; speech is the foremost manifestation of belief. This approach is taken already by Carnap (1947, §13), and it certainly pervades Quine's writings. However, assent possibly varies with different descriptions of the same set of possibilities, and so this approach does not obviously describe an attitude towards propositions, a point I will return to below. To believe one out of a set of possibilities to obtain might mean to stand in appropriate perceptual and inferential relations to this set or a suitable description thereof, where "appropriate" has a normative content unfolding in certain communicative commitments and entitlements. Such views are associated with the philosophy of Sellars (1963, in particular Ch. 11) and Brandom (1994). It might mean that this belief is part of a system of propositional attitudes, beliefs, desires, etc., that best systematizes the behavior and behavioral dispositions of the subject at hand. This is how Lewis (1986a, Sect. 1.4) explains it. It might consist in a certain judgmental disposition counterfactually manifested by a person upon being presented with this set of possibilities itself (and not a description or some other proxy of this set), a view endorsed in Haas-Spohn, Spohn (2001) and explained a bit more fully below. All of these views might need qualification by relevant normal or ideal conditions. In such summary fashion, all of this is terribly cryptic, and philosophers work hard to give substance to those views. Again, though, we need not take issue. However the dispute resolves, it should disprove neither our neutral conception of the objects of belief nor our neutral characterization of belief as the exclusion of possibilities.

Yet the neutrality is not perfect on two counts (by my accounting). One point is that we have apparently presupposed a uniform conception of the objects of belief. They are always sets of possibilities. While one might agree on this, one might think that there are fundamentally different kinds of possibilities involved. Indeed, I have indicated that there are different ideas as to the nature of possibilities, and then one might argue that there is not *one* correct idea or fundamental kind of possibilities to which the others reduce. That is, one might argue that there are different kinds of belief of equal right relating to different kinds of possibilities. For instance, one might argue that there are perceptual beliefs described in terms of perceptual possibilities and *de dicto* beliefs described in terms of linguistically given possibilities. Then one might further think about the locus of conceptual possibilities – whether percepts are already concepts, whether there are only linguistic concepts, or whether conceptual possibilities open up a third kind.

This would not really disprove our approach, though. It would only restrict it to each individual kind of belief, leaving the task of saying how the different kinds of belief relate or interact. This task would be of pressing concern to anyone defending such a doxastic plurality.

There is a second and more important bias in our approach. It would be appropriate to say that we have assumed outright belief to be propositional despite all our tolerance concerning the possibilities that make up propositions. This assumption has been attacked from various sides. Let me mention three main directions.

One criticism is based on a narrow understanding of propositions that equates them with sets of possible worlds. In this view, sets of centered worlds, for instance, are not propositions, but properties. This is, e.g., how Lewis (1979a) puts it. Hence, any account that favors such other objects of beliefs in order to cope with such things as indexical or egocentric beliefs is, by definition, not one of propositional belief. Clearly, there is no conflict here, for we may simply use a wider notion of proposition according to which any set of appropriate possibilities represents a proposition.

Another line of attack starts from the assumption that propositional belief is in any case conceptual or conceptually structured and then points out that there are various kinds of belief that are apparently non-conceptual. The most conspicuous example is perceptual belief. Perception is often thought to be conceptually unstructured (certainly also a repercussion of the strict Kantian separation between intuition and concept), and so one might likewise think that perceptual beliefs (surely an un-Kantian notion) have purely perceptual, non-conceptual content. This is a difficult issue that we ought not engage with. The only remark I wish to make at present is that *prima facie* our characterization of propositions as sets of possibilities has nothing to do with concepts. A possibility as such is *prima facie* unstructured, whereas an adequate description of the possibility (which may serve as a representation of it) is most richly conceptually structured. A detailed story would hence be needed to establish a connection. Therefore I am inclined to think that the allegedly non-conceptual perceptual beliefs may still be understood in terms of a space of possibilities. Let us defer the issue to Chapter 16, where we discuss perception and the basis of belief in more detail.

The third line of attack is just the reverse in a way. It insists that conceptual belief is at least the paradigm case of belief and that this is, as I just pointed out, not yet reflected in conceiving the objects of belief as sets of possibilities. Again we enter a large field of debate. What are concepts? The traditional answer "meanings of predicates" has proved insufficient and ambiguous, in particular under the influence of Putnam (1975).

Meanings tend to be associated with so-called wide contents and concepts with narrow contents. Since these terms mark problem areas rather than prospects for a solution, this issue has become one of the most contested areas in theoretical philosophy. One can only state that the discussion is thus far inconclusive; cf., e.g., Peacocke (1992) or the collections Margolis, Laurence (1999) and Villanueva (1998) or Fodor (1998) as a kind of conclusion of his long struggle with the topic; this is just the tip of the iceberg. Haas-Spohn, Spohn (2001) contains my view about concepts, which is well in line with the sets-of-possibilities approach. Indeed I believe that the nature of concepts is best understood within the framework of two-dimensional semantics as displayed, e.g., in Haas-Spohn (1995), Chalmers (1996, Sect. 2.4), Davies, Stoljar (2004), and García-Carpintero, Macia (2006); but this is definitely not the place for engaging with the topic. The only point I want to make is that despite the lack of clarity we can identify a serious problem here:

If we conceive of objects of belief as sets of possibilities, then we really conceive of them as pure contents. A pure content is nothing but a truth condition: a set of possibilities is true if and only if the one and only actual possibility is a member of it. The problem now is that such contents are not given directly to us. Having a belief is somehow having a mental representation in the belief mode (and not, say, in the mode of a desire or a supposition), which will usually be a conceptual representation or, if that is too unclear, a linguistic representation; this is, finally, something quite determinate. The belief is then endowed with content only because this representation is somehow related to the content or because the sentence representing the belief has a truth condition. But then it seems that it is rather the representation that should be the object of belief and not the content; different representations are different beliefs, even if they have the same content. As Quine has continuously insisted since his assault on meaning and analyticity, belief is, it seems, not a propositional, but rather a sentential attitude.

(Similar words might be used for explaining the distinction between wide and narrow contents, since different narrow contents can represent the same wide content. However, this is not what I am up to at the moment, all the more as the same narrow content can also represent different wide contents. The two distinctions would coincide only when narrow contents are equated with sentences or sentence-like entities. Usually, though, the narrow and wide contents are both located at the semantic and not at the syntactic level; they are both contents.)

So, to reiterate: The problem is that contents seem accessible only through representations, that in principle infinitely many representations represent the same content, and that the issue of whether two representations are equivalent (i.e., have the same content) can be computationally arbitrarily complex and is indeed undecidable above a certain level of complexity. Nobody knows or can be expected to know about all these equivalences. Restricting the discussion to logical equivalences, this is called the problem of logical omniscience. In any case, that there is something to know or not to know is entirely neglected by taking contents or truth conditions as the objects of belief. It can be accounted for only by taking the representations themselves as these objects. So, the latter is the option to be preferred.

This line of thought is tempting, but not cogent, and it has its problems as well. A standard problem is that we would like to be able to say that people speaking different languages and hence having different representations can have the same belief. If one inquires, then, what the representations have in common and only finds their content, we are back at our original account.

Without doubt the problem of logical omniscience and the more general problem it represents are profound ones. (For a thorough discussion see, e.g., Stalnaker 1999, chs. 13 and 14.) We cannot solve it now in a reasoned way, and we will indeed encounter it again and again. I am convinced that it indicates a schism in epistemology that is much more pervasive than usually thought. For the time being I have simply decided to take a certain stance on it, namely to resist the temptation and to insist on contents as objects of belief. This is the most important respect in which our approach is not neutral. Let me simply mention, not argue, three reasons for my decision:

One reason is that we are in good company; in that respect, I think, the majority of epistemologists are still with us. Of course, the argument from authority is dubious, but it is reassuring not to feel like an outsider. My main reason is that our stance is by far the more fruitful one. Within this stance, epistemological theory construction is incomparably richer than within the alternative. This reason should become clear in the course of the text. In fact most theories apparently operating on a sentential level assume the substitutability of logical equivalences and differ thus only superficially from our procedure. Without this substitutability we are left with hardly any theory at all. Although this appeal to theory construction sounds technological, it provides, I believe, a very strong reason that weighs heavily on my personal balance. My third reason is more philosophical. It relates to the fact that we are here engaged into normative epistemological theorizing. In my view this entails that the objects of belief should be conceived as contents or truth conditions or sets of possibilities and not as sentences or representations of contents. This is an essential point, although it presently is but a cryptic remark. Let me defer the argument to Section 11.6, where I specifically discuss John Pollock's views on epistemic normativity.

2.3 A More Determinate View

I suppose it would be wise to simply leave it at that relatively noncommittal, catholic view. It would be improper, however, to write a book on belief without trying to say what belief is and what it is about. So in this section I intend to do just that, in a rather descriptive manner, while avoiding deep discussions about the nature of belief, in which we might well get lost.

Belief is not just a disposition to assent or dissent, or only denoted by a theoretical term of belief–desire psychology, or simply a functional role, whether understood in causal or in normative terms, whether located within a narrow net reaching from perceptions to actions or movements or a wide net embracing the environment, the linguistic community, and all the transactions of the subject within that wide net. To be sure belief is all that, but it is also more. I take belief to be a vastly more counter-factual and normative disposition, only contingently related to all smaller-scale dis-positions of a linguistic, behavioral, or social kind. It is a judgmental disposition of maximal totality. Let me explain these dramatic words:

First, I take the objects of belief – belief in an absolute and not a somehow relative sense, relative to the application or conceptual framework – to be propositions built, indeed, from Lewisian possible worlds. I have no quarrel with them: if we accept pos-sible objects at all, their size does not matter. In other words, epistemic or doxastic possibilities are Lewisian possible worlds. They are like our actual universe that con-fronts us as an unstructured whole and as a challenge to our cognitive powers (though endowing it with structure starts at birth or earlier). They are Kantian noumenal worlds, though not unknowable or even unconceivable, but simply initially unknown and unconceived.

Such a Lewisian possible world, and only such a world, presents a complete mani-fold of experience to us; such a world contains everything that can be experienced. Conversely, our field of experience cannot be cut smaller. In principle – i.e., counter-factually – we can inquire into every part and corner of such a world. Indeed on a dark night we can already see, in a way, the entire spread of the universe with the naked eye, though we miss most details. Our actual field of experience is the actual universe, of course. However, for all we know, that universe might be very different from what it actually is; very different totalities of experience might confront us. Therefore, we have to envisage other totalities as well. So, as I say, doxastic possibilities are Lewisian possible worlds.

(There is an irony in that claim. Lewis (1986a) took his possible worlds to be metaphysically possible worlds accounting for metaphysical possibility, and since meta-physical possibility is, in an important sense, the most admissible kind of possibility, he also used his possible worlds for accounting for epistemic possibility. Chalmers (2006, Sect. 3) disagrees and thinks that epistemic possibilities must be construed in a different way. The irony is that I disagree in the opposite way. Lewisian worlds *only* serve as doxastic possibilities; Wittgensteinian possible worlds, totalities of states of affairs, are rather those serving as metaphysical possibilities. This is so because only Wittgensteinian possible worlds are fully analyzed into objects, properties, and relations that make up states of affairs, and because metaphysical necessity and possibility are constituted by the nature of objects, properties, and relations. However, since metaphysical necessity is not our present business, I will not further argue the point.)

I should hasten to add that a doxastic possibility is not only a Lewisian possible world, for at the very least it needs to be centered. It is an a priori truth that I am here now, somewhere in this world, and so I must presently find myself in each doxastic possibility, i.e., in its possible 'center'. Therefore, a doxastic possibility is at least a triple $\langle w, s, t \rangle$, where w is a Lewisian possible world, s an object in w, and t a time in w. (To be precise, we should avoid speaking of an object s here. w is just a maximal object, maximal in its spatiotemporal extension or, as Lewis (1986, pp. 72ff.) explains, in its space-time-analogical relations, and as such has only spatiotemporal or space-time-like parts. If an object is just such a mereological part of a world, and hence exists only in that world, as Lewis assumes, then such a world also contains objects. However, it does so only in a meager sense and not in our ordinary, modally loaded sense of "object". Hence, s should just be taken as some spatial extension in w at time t.)

Now what does it mean to believe a proposition construed in terms of such possi-bilities? That is, what does it mean to exclude such a possibility? It means undergoing a vastly counterfactual test. Here is the actual subject s^* at time t^* with her actual doxastic state, i.e., totality of beliefs. For the moment I want to neglect all indetermina-tion, uncertainty, degrees of belief, etc. Does this doxastic state of s^* at t^* exclude the doxastic possibility $\langle w, s, t \rangle$? In order to find out we fix her doxastic state, and then we let her do what she would actually do as well in order to check her beliefs. We only let her do so in an extreme, exhaustive way. That is, we first let her take the place of s at t. If s is insensible, or if s is sensible but not sensing exactly the way s^* is aware of at t^*, she knows that this is not the right center, and she can exclude the possibility $\langle w, s, t \rangle$. She can move around in w, look at and inquire into every place at every time. Most worlds will be black and barren and not yielding to her senses; these can be excluded right away. From the remaining ones, most worlds will be strange, if not alien, and not yield-ing to her concepts; they can be excluded as well. Still, there will be many worlds left that come very close to her beliefs, and it might require an arbitrary amount of inquiry to detect some deviations. Her inquiry includes scrutinizing every place and every time in w and subjecting any place of w to any kind of test or analysis without affecting w in its fixity. That is, we make the ideal counterfactual assumption that we can observe situations in w as if unobserved, an ideal we can actually often, but not universally realize (elementary particles change through observation; living matter is destroyed by too invasive an inquiry; persons do not behave naturally under detected observation; etc.). Her inquiry also includes taking the perspective of every sensible being in w, learning all languages in w, talking to everybody there is to talk to in w, etc. Mostly, this inquiry will reveal details she has no opinions about, but some details of w might be unexpected. If so, w is to be excluded. In all this, she has to check all the time whether the center s at t in w fits to her entire self-conception embodied in her actual doxastic state. Being in the right center is not simply a matter of presently having the right sensations. Her inquiry does not only result in a huge collection of singular observations, for she is required to form concepts that are suitable for w, to develop generalizations and systematizations, etc. If these are just expansions of the conceptual and theoretical machinery embodied in her actual doxastic state at t^*, all is fine, but if w enforces a revision, it is to be excluded.

Let us say that the subject has to come to a *maximally experienced and considered judgment* about $\langle w, s, t \rangle$. Indeed the subject has to do this for all doxastic possibilities. This is, in many respects, a vastly counterfactual supposition (even if, as indicated, most possibilities can be excluded long before completing the judgment). Still it should be clear that, in principle, no less will do. Then we may define:

> *Explanation 2.6*: A doxastic possibility $\langle s, t \rangle$ is a *doxastic alternative* of s^* at t^* iff the maximally experienced and considered judgment about $\langle w, s, t \rangle$ is compatible with the doxastic state of s^* at t^*, i.e., iff that judgment is only an expansion, but not a revision of this doxastic state. The possibility is *excluded* by s^* at t^* iff it is not a doxastic alternative of s^* at t^*. And the proposition A is *believed* by s^* at t^*, i.e., $\mathbf{B}_{s^*, t^*}(A)$, iff it contains all doxastic alternatives of s^* at t^*, i.e., iff all possibilities in $\bar{A}$ are excluded.

Again, I call this an explanation and not a definition, because it is full of ill-defined terms. I am well aware that my preliminary characterization of that maximally experienced and considered judgment is quite insufficient. I could also say that that judgment is something like an ideal theory about such a doxastic possibility in the sense explained and used by Putnam (1980) and elsewhere, for his descriptions are of a similar character.

On the one hand I have now put my cards on the table and sketched what I take belief and its objects to be. On the other hand it is clear that this should be the start of a thorough-going discussion – one that I am not going to lead, however, as my interests in this book lie elsewhere. Let me confine myself to five remarks:

First, despite the somehow overdrawn explanation 2.6, its basic idea appears entirely natural. Believing is taking to be true, and just that. This allows that beliefs (or their objects) may indeed be characterized by their truth conditions. But when is a belief true? When it corresponds to the facts. Yes, of course – if only we knew what the facts were! There is another answer: A belief is true iff it survives all further experience and belief formation (in the world we are living in). All (or most) of our beliefs are somehow premature or risky or fallible. We claim them to be true; this is what it means to have them. And that claim consists in the expectation that we ultimately do not receive any reasons to withdraw that belief. "Ultimately" can here be understood only in the vastly counterfactual way sketched above; no less would do, and no more is possible. This is the simple idea behind (2.6).

Second, as already indicated in the first remark, there are indeed two notions of truth – or so I claim. There is not only the correspondence notion of truth and its contemporary descendants (the semantic conception, the redundancy view, minimalist and disquotationalist accounts, etc., all of which look quite similar from some distance, even though they can of course be sharply distinguished) along with a somehow subordinate criterion of truth, as it has often been called. No, there are two notions of truth, the correspondence notion and its family and the one implicit in (2.6) that is often called the pragmatic notion of truth or described as ideal ascertainability or ideal justifiability, and which characterizes Putnam's internal realism (cf. his collection of papers 1983a). Approaches to a coherence theory of truth (cf. Rescher 1973 and 1985) or to an evaluative notion of truth (cf. Ellis 1990) are also attempting to grasp this notion. I granted already that these characterizations are still unsatisfactory, but this does not mean that there is nothing to characterize. The basic point is simply that the correspondence notion of truth is appropriate only with respect to Wittgensteinian possible worlds or metaphysical possibilities that contain all the facts to which our sentences, utterances, or beliefs possibly correspond. But that notion of truth does not apply to Lewisian possible worlds or epistemic possibilities that are very rich – but as such unstructured – objects; the correspondence finds no correspondent in a mere object. If there is a notion of truth for Lewisian possible worlds, it must be a different one, and the suggestion is that it is the pragmatic or internal notion sketched above (see also Spohn 2008b). I will say more on that internal notion in Chapter 17.

Of course, if we could establish a (one–one?) mapping between Lewisian and Wittgensteinian worlds, we could thereby transfer the correspondence notion of truth to Lewisian worlds. Yet this mapping hides all the problems of the relations between epistemology and ontology, and certainly it cannot proceed without using that other, pragmatic notion of truth.

If this point is correct, then it is of deep significance to two-dimensional semantics that intends to determine the extension of phrases and the truth of sentences relative to both epistemic and metaphysical possibilities. The diagonal or primary or A-intension and the secondary or C-intension of a sentence really provide its truth condition in two different senses of truth. The deep, if not deepest question, then, is how the two notions of truth, or how Lewisian and Wittgensteinian possible worlds, relate. This is not our present task, but I think it is a genuine problem and not a mere philosophical fiction – and it is a problem two-dimensional semantics must solve in order to work. (In Spohn 2008a, pp. 12–15, I have slightly expanded on these difficult issues.)

Third, I have taken great care to keep (2.6) free from any linguistic entanglements. Those entanglements are large and indeed almost inextricable (in Spohn 2008a, pp. 337–346, I more thoroughly discuss no less than six forms of linguistic entanglement). As a consequence, epistemology and philosophy of language have become nearly inseparable. In stark contrast to this, I vigorously wish to regain and maintain the idea of pure epistemology – indeed this whole book is an exercise in pure epistemology – and (2.6) is a cornerstone for doing so.

Thus, for instance, belief, **B**, as explained in (2.6) is neither *de dicto* nor *de re*. A basic idea behind (2.6) is to construe belief in an individualistic or internalistic way, i.e., as a psychological state in the narrow sense, as Putnam (1975, pp. 219ff.) has put it. Therefore (2.6) describes belief as an intrinsic disposition of a subject, though a vastly counterfactual one. I am convinced despite considerable attempts at alleviating arguments (such as Burge 1986) that to do otherwise would import causal disorder into psychology. Linguistic entanglements suspend internalism; hence, *de dicto* beliefs are as non-individualistic as *de re* beliefs, as Burge (1979) has convincingly argued. It is thus all the more important to secure an individualistic notion of belief and to describe how it lies at the base of those entanglements and the many forms of belief ascriptions.

By the way, (2.6) is my background for rejecting belief as a sentential attitude, as I did at the end of Section 2.2 (2.6) directly realizes the basic idea (2.5) of belief as the exclusion of possibilities without linguistic detour and hence directly establishes belief as propositional. It thus proposes a trivial solution, or a trivial avoidance, of the problem of logical omniscience. I will not discuss, though, whether this speaks against or in favor of (2.6). (Again, see also Section 11.6.)

Fourth, a remark about amending (2.6): According to (2.6) belief is a terribly holistic affair. Ascertaining whether a possibility is a doxastic alternative in principle requires looking at the entire possibility, and ascertaining whether a proposition is believed requires checking whether it contains all doxastic alternatives. What a roundabout way! As indicated in Section 2.2, one should rather think that propositions are somehow composed of concepts and that belief somehow is affirmation of propositions thus composed. In Haas-Spohn, Spohn (2001) we have attempted to do justice to this idea, not by adjusting belief, but by adjusting concepts to that account of belief. This looks even more roundabout, as a stubborn continuation of the odd strategy. Note, however, that we simply have no other choice when we want to relate concepts, propositions, and belief to the pragmatic or internal notion of truth appertaining to Lewisian possible worlds, a notion that is wedded to the idea of a maximally experienced and considered judgment.

Fifth, and last, another remark about amending (2.6): I do not think, in the end, that doxastic possibilities are merely centered Lewisian possible worlds; they need to be extended by a potentially infinite sequence $\langle a_1, a_2, \ldots \rangle$ of possible objects, so that a doxastic possibility takes the form $\langle w, s, t, a_1, a_2, \ldots \rangle$. I have thoroughly argued for this extension in Spohn (2008a, Ch. 16), where I call this, with some justification, the intentional conception of (belief) contents as opposed to the propositional conception embodied in (2.6). Interestingly, though, the argument – suggested by linguistic considerations, but painstakingly disentangled from them, and thus carried out within pure epistemology – is again driven by considerations on the dynamics of belief, just as with the above pragmatic notion of truth. Very roughly, the basic point is that, a priori or with doxastic necessity, the world we live in is a world of objects, and that this a priori truth is not adequately captured by (2.6), but only by the extension indicated. Ultimately this point leads us into deep issues regarding what an object *is* and how we conceive of objects. The philosophical importance of these issues cannot be overestimated, but they are simply not the focus of this book; our agenda is big enough without them.

I am well aware that this section was much too brief and forbiddingly speculative. I have sufficiently emphasized that, for the purposes of this book, I am much happier with the noncommittal view as explained in Sections 2.1–2.2; it should at least carry us up to Chapter 15. Yet I felt I should at the very least indicate that there are deep concerns about the notion of belief not addressed in this book, though they may – or may not – be treated along the lines suggested and referred to.

3

The Probabilistic Way

Up to this point I have tried to take an approach concerning belief and its objects that is formally clear and, regarding its interpretation, relatively neutral (in its official part) or relatively determinate (in its unofficial part). Let us now return to the problem of induction, or rather, as we concluded in Section 1.1, to the statics and dynamics of doxastic states. There are not all that many fundamentally different accounts, and one is by far the best established: probability theory (subjectively interpreted). We should start by examining at least some of its virtues, so as to preserve them, and some of its shortcomings, that we might improve upon.

Probability theory is by far the best established on various counts. First, there was always the idea that belief inevitably comes in degrees of (un)certainty, and that the ideal of maximal certainty, if it exists, is rarely achieved. It was unclear, though, just how to measure uncertainty. As I have mentioned in Section 1.2, historically probability theory emerged as the first clear model that gained increasing power and superseded possible alternative developments. Only recently are such alternatives being taken seriously and investigated in detail. For this reason probability theory simply has an overwhelming historical lead.

Second, probability theory – together with its various applications in statistics, statistical physics, etc. – is by far the most developed in terms of mathematics. Whenever the thought occurs to me that there are more persons professionally occupied with mathematical probability than with philosophy, I am filled with envy.

Moreover, if we are to tell in more detail what exactly belief has to do with action – one of the most important things to say about belief – then probability theory naturally expands to the most general theoretical account of this connection, i.e., decision theory, which compounds subjective probabilities and subjective utilities or values in order to determine rational action. Here the alternatives are weak and tentative, and without historic precedent (see, however, Section 10.4).

One very important point is that probability is firmly grounded in reality through relative frequencies. The precise nature of this grounding is not entirely clear, and the uncertainty is manifested in the ongoing discussion of frequentist and alternative understandings of objective probabilities (e.g., as propensities). In any case, it is evident that subjective probabilities are guided by observed relative frequencies, and statistics has developed into a tremendously rich and ramified discipline telling us just exactly how this guidance works (cf. Section 10.3 for more on this topic).

Finally, probability theory contains the most sophisticated ideas concerning the dynamics of doxastic states. This aspect, of course, is of particular interest to us.

For all these reasons it seems reasonable to have at least one chapter-long look at probability theory or the 'Bayesian' point of view. As I have emphasized, this is my paradigm theory, and even though I will present ranking theory as a genuine and fruitful alternative, I will develop it in close parallel to this paradigm.

3.1 The Basic Laws of Subjective Probability

The probabilistic account of the statics of belief is very simple. The doxastic state of a person at a given time concerning a certain algebra $\mathcal{A}$ of propositions over W is given by nothing but a probability measure:

Definition 3.1: Let $\mathcal{A}$ be an algebra over W. Then P is a *probability measure* on $\mathcal{A}$ iff P is a non-negative, normalized, and (finitely) additive function from $\mathcal{A}$ into $\mathbf{R}$, the set of reals, i.e., iff for all $A, B \in \mathcal{A}$:

(a)　$0 \le P(A) \le 1$　　　　　　　　　　　　　　　　　　　　[*non-negativity*],

(b)　$P(W) = 1$　　　　　　　　　　　　　　　　　　　　　　[*normedness*],

(c)　if $A \cap B = \varnothing$, then $P(A \cup B) = P(A) + P(B)$　　　　[*additivity*].

If $\mathcal{A}$ is a σ-algebra over W, then P is a σ-*additive probability measure* on $\mathcal{A}$ iff moreover for any sequence $\langle A_1, A_2, \ldots \rangle$ of pairwise disjoint propositions in $\mathcal{A}$ (i.e., such that $A_i \cap A_j = \varnothing$ for $i \ne j$)

(d)　$P\left(\bigcup_{i \in \mathbf{N}} A_i\right) = \sum_{i \in \mathbf{N}} P(A_i)$　　　　　　　　　　　　　　　　[σ-*additivity*].

Concerning the rational justification of the three basic properties (a)–(c), there are essentially four much-discussed lines of reasoning. The most famous line is the so-called Dutch-book argument, which starts from the normative premise that it is silly to engage in an incoherent system of bets, i.e., in a betting system with logically necessary loss, and which deduces (a)–(c) from this via some auxiliary assumptions which might be debatable. (The Dutch book argument goes back to Frank Ramsey and Bruno de Finetti. The label is due to Lehman 1955. For critical discussions see, e.g., Earman 1992, Sect. 2.3–2.4, and Maher 1993, Sect. 4.6.)

A second line of reasoning is provided by measurement theory. If certain structural properties of comparative judgments concerning what is likelier than what can be shown to be normatively convincing, then measurement theory can prove that there is only one way to quantify these comparisons which satisfies (a)–(c). (See, e.g., Fine 1973, chs. II and III, or Krantz et al. 1971, Ch. 5.) And if preferences over alternative options satisfy certain normatively highly plausible postulates, then measurement theory can even prove that there is a unique probability measure satisfying (a)–(c) and a utility function unique up to positively linear transformations such that expected utilities as calculated from these probabilities and utilities mirror these preferences. (This idea goes back to Leonard J. Savage; see Fishburn 1970, Ch. 14, and Maher 1993, Ch. 8.)

Another idea consists in Cox's theorem, which starts from some very innocent and convincing properties of any quantitative functions measuring uncertainty, and shows – again, with the help of some auxiliary premises – that any such function must be a probability measure. (See Cox 1946 and Aczél 1966, Sect. 7.1.4, for the argument, and Earman 1992, Sect. 2.5, for a critical discussion; it seems to be still unclear what the argument actually shows at all.)

A still different line of reasoning is offered by the so-called calibration theorem (cf. van Fraassen 1983), according to which a degree of belief has to agree with the expectation for the frequency of truth among the propositions assigned that degree of belief. The conclusion, then, is that degrees of belief, must conform to (a)–(c) just like relative frequencies.

A criticism and improvement of this argument is offered by Joyce (1998) via what he calls the norm of gradational accuracy: If a rational person strives to have credences that are as close to the truth as possible, and if the measure of closeness satisfies some plausible axioms, then the credences must be probabilities. (See, however, the objections of Maher 2002.)

These are ingenious and intricate arguments. Of course hardly anything in philosophy is beyond dispute, but even though one might find fault with the various lines of justification, the weight of normative argument in favor of (a)–(c) is tremendous. Personally, I find these arguments fascinating and worth pondering. I have slight doubts, however, concerning their efficacy; they seem to be justifying the unrivaled. It is a bit like arguing that space must be Euclidean when there is no other conception of space around.

Mathematically, σ-additivity is overwhelmingly useful, though it is interesting to see how far one can get by assuming only finite additivity. Epistemologically, it is quite contested whether σ-additivity can be justified for subjective probabilities. Some of the above justifications are extended to σ-additivity, but with less conviction (cf. the references given; see moreover Seidenfeld 2001 for six reasons why one might rather stick to finite additivity). However, we need not go more deeply into this issue.

A particularly crucial notion in view of our dynamic purposes is conditional probability:

Definition 3.2: Let P be a probability measure on $\mathcal{A}$, let $A, B \in \mathcal{A}$ and $P(A) > 0$. Then the (conditional) *probability of B given* or *conditional on A* is defined as

$$P(B \mid A) = \frac{P(A \cap B)}{P(A)}.$$

If $P(A) = 0$, $P(B \mid A)$ is undefined.

This definition is clear as can be, and so it is surprising to see that as soon as one starts applying this notion of conditional probability it is clouded by puzzles and problems. Despite the clear definition, there are great uncertainties about its meaning (see, e.g., Hájek 2003a). We will see that ranking theory is confronted with similar uncertainties. Hence, I will more fully discuss the topic only in Chapter 9. Moreover, as defined, it is perfectly clear that $P(\cdot \mid \cdot)$ is a two-place function applying to two propositions. Still, there has always been the strong temptation to understand "|" as a propositional function combining two propositions into one conditional proposition, so that P generally applies to single propositions, conditional and unconditional ones. However, the temptation should be resisted; it leads to awkward problems without clear solution (see in particular the trivialization results in Lewis 1976).

Let us state some simple consequences of the basic postulates. A first noteworthy observation is:

Theorem 3.3: Let $W = \{w_1, \ldots, w_n\}$ be finite, $\mathcal{A} = \mathcal{P}(W)$, and P a probability measure on $\mathcal{A}$. Then there are numbers $x_1, \ldots, x_n \geq 0$ such that $\sum_{i=1}^{n} x_i = 1$, $P(\{w_i\}) = x_i$ for $i = 1, \ldots, n$, and for each $A \in \mathcal{A}$, $P(A) = \sum \{x_i \mid w_i \in A\}$.

In other words, in the finite case we can reduce the probability measure, which is a set function, to a point function defined for all the possibilities in W. This reducibility is often convenient. (Note: I allow myself to state theorems without proof when they are well known or trivial.) Some further simple laws are stated in

Theorem 3.4: Let P be a probability measure on $\mathcal{A}$ and $A, B, C, A_1, \ldots, A_n \in \mathcal{A}$. Then we have:

(a) $P(\bar{A}) = 1 - P(A)$ [*law of negation*],

(b) $P(A \cup B) = P(A) + P(B) - P(A \cap B)$ [*general additivity*],

(c) $P(A \cap B) = P(B \mid A)P(A)$ (provided $P(A) > 0$) [*multiplication theorem*],

(d) $P(A_1 \cap \ldots \cap A_n) = P(A_n \mid A_1 \cap \ldots \cap A_{n-1}) \cdot \ldots \cdot P(A_2 \mid A_1)$,
     provided $P(A_1 \cap \ldots \cap A_{n-1}) > 0$) [*general multiplication theorem*],

(e) if $A_1, \ldots, A_n$ partition W (i.e., $A_i \cap A_j = \emptyset$ for $i \neq j$ and $\bigcup_{i=1}^{n} A_i = W$) and if $P(A_i) > 0$ $(i = 1, \ldots, n)$, then $P(B) = \sum_{i=1}^{n} P(B \mid A_i)P(A_i)$
     [*formula of total probability*],

(f) if $A \cap B = \emptyset$ and $P(C \mid A) \leq P(C \mid B)$, then $P(C \mid A) \leq P(C \mid A \cup B) \leq P(C \mid B)$
     [*law of disjunctive conditions*].

The law of negation tells that the probabilities of a proposition and its negation are functionally related. With general additivity we can calculate the probability of a disjunction even if the disjuncts are not disjoint. The multiplication theorems are most useful to calculate the probabilities of (longer) conjunctions because the conditional probabilities on the right hand side are often more easily accessible. This advantage is also exploited by the formula of total probability: one might be unsure about the absolute probability of B, but if one knows the probability of B conditional on alternative hypotheses A_i and the absolute probabilities of these hypotheses, then the probability of B results as a weighted mixture according to the formula. The law of disjunctive conditions, finally, spells out an obvious consequence of this weighing procedure.

At this point, each standard textbook continues with

Theorem 3.5: [*Bayes' Theorem*] Let $A_1, \ldots, A_n \in \mathcal{A}$ partition W, let $B \in \mathcal{A}$, $P(A_i) > 0$ ($i = 1, \ldots, n$), and $P(B) > 0$. Then for each k in $\{1, \ldots, n\}$,

(a) $P(A_k \mid B) = \dfrac{P(B \mid A_k) \cdot P(A_k)}{P(B)}$, and

(b) $P(A_k \mid B) = \dfrac{P(B \mid A_k) \cdot P(A_k)}{\sum\limits_{i=1}^{n} P(B \mid A_i) \cdot P(A_i)}$.

These are two versions. Version (a) follows by twice applying the definition of conditional probability; version (b) is (a) with the denominator having been expanded according to the formula of total probability (3.4e). The importance of this theorem lies in the fact that since its discovery it received a dynamic interpretation and thus served as the first and dominant model of belief change. The dynamic interpretation is this: The *posterior* probability $P(A_k \mid B)$, as it was called, of some hypothesis A_k given some evidence B is proportional to its *prior* probability $P(A_k)$ and to the *likelihood* $P(B \mid A_k)$ of the evidence B under the hypothesis A_k. The proportionality factor is given by the prior probability $P(B)$ of the evidence (see version (a)), which may be analyzed in terms of likelihoods and prior probabilities of alternative hypotheses (see version (b)). Since the latter are usually the ones more easily accessible, version (b) is the more useful one.

This interpretation is still not fully dynamical, since there is only one probability measure mentioned in the theorem and no points of time. But this is a pedantic remark; we will turn fully dynamic soon. First we need to continue with the question as to whether Definition 3.1 exhausts what can be said about the statics of belief in probabilistic terms. The answer is a clear "no"; the axioms of probability theory (3.1a–c) define only an agreed minimum of static rationality.

How to strengthen the minimum is, however, less agreed. The only fairly uncontroversial strengthening is that one should not assign the extremal probabilities 0 and 1 thoughtlessly and without necessity. This is the idea of regularity:

Definition 3.6: Let the algebra $\mathcal{A}$ over W be finite. Then the probability measure P on $\mathcal{A}$ is *regular* iff $P(A) = 0$ only for $A = \varnothing$.

Thus regularity means that all non-contradictory propositions receive some positive credibility, however small. This seems reasonable because there is at least the logical possibility of such propositions turning out true. There is also a slightly more sophisticated justification of the regularity postulate in terms of a Dutch book argument. Definition 3.6 does not give a general definition of regularity. If the algebra of propositions is infinite, it might be mathematically unavoidable that some non-empty propositions receive probability 0, and then Definition 3.6 is inapplicable. Carnap (1971b, Sect. 7) gives a workable explication of the idea of regularity also for the infinite case, but there is no need at present to go into these technical details.

Concerning additional postulates, Carnap, in his relentless efforts to construct an inductive logic, has set the agenda. He offered precise statements of symmetry postulates that are well known to have a much longer history. He proposed a priori measures for accounting for analogy influences, as he called it, and so on. (For all this see Carnap 1971 and 1980.) There are some efforts to continue on his program (e.g., Skyrms 1991 and Festa 1997), but we need neither go into further details, nor discuss the reasons why the program has not found many followers. Of course one reason – but by no means the only one – is that all the further rationality postulates are somehow problematic and debatable. On the other hand, I find it perfectly obvious that there must be further a priori rationality postulates; it is utterly incredible that the basic probability axioms and the regularity postulate should exhaust what to say about our degrees of belief from a rational point of view. But the issue is intricate, and we will return to it in terms of ranking theory at several places and more extensively in the final Chapter 17.

3.2 The Dynamics of Subjective Probability

Let us proceed from the statics to the dynamics. What can be said about belief dynamics in probabilistic terms? The standard answer was almost explicit in the interpretation of Bayes' theorem. If, for some hypothesis A_k, $P(A_k)$ is called the prior and $P(A_k \mid E)$ the posterior probability of A_k after some evidence E, this amounts to the claim that probability dynamics simply works by conditionalization.

Definition 3.7: Let P and P' be probability measures on $\mathcal{A}$ and $P(E) > 0$. Then P' is the *conditionalization* of P with respect to E iff for all $A \in \mathcal{A}$, $P'(A) = P(A \mid E)$.

Then the dynamic *law of (simple) conditionalization* states:

(3.8)   If P characterizes the doxastic state of some subject s at time t and E is the total evidence or information s receives between t and t', then the doxastic state P' of s at t' is the conditionalization of P w.r.t. E.

The wide use of Bayes' theorem shows that this law has been taken for granted since the rise of probability theory. However, why should the nominal definition of conditional probabilities suddenly acquire this dynamic meaning? This may appear surprising. A first strong, though not entirely cogent answer is ready at hand: The information or evidence E reduces the space of remaining possibilities from W to E. By excluding the possibilities outside E, however, the relative weights of the possibilities within E are not changed. This is exactly what conditionalization amounts to. If the premises of this argument are reasonable, then the law of conditionalization is indeed a law of rationality. The main attempt to fix this reasoning is in terms of a dynamic version of the Dutch book argument that is ascribed to David Lewis (cf. Teller 1976). The most general dynamic Dutch book argument is found in Skyrms (1990, Ch. 5). On the whole, however, the literature is more ambiguous on the dynamic version than it is on the static version (cf. Earman 1992, Sect.2.6, and Maher 1993, Ch. 5).

The ambiguity reflects the fact that a big discussion about probability dynamics was started with, and only with, Chapter 11 of Jeffrey (1965). The market has since become variegated, and there is no point in establishing a survey in this introductory chapter. We should pursue some ideas, though, in order to understand why conditionalization is only the beginning, not the end of the topic, and in order to prepare some of the later comparisons with ranking theory.

The best starting point is to recognize the conflict between regularity and conditionalization. Clearly, if P' is the conditionalization of P w.r.t. E, then P' is not regular; for instance $P'(\bar{E}) = 0$. Hence if one is really attached to the idea of regularity – that only the tautology can have maximal certainty – then even the firm evidence E cannot acquire this status; the possibilities outside E cannot be simply excluded. Conversely, if one finds conditionalization acceptable, one has to give up regularity. However, this has the further consequence of depriving the law of conditionalization of its full generality; it works only for evidence with positive probability and not for all possible evidence. The conflict, hence, is a substantial one. There are two obvious ways to go: give up conditionalization or give up regularity. Let us follow at least their beginnings.

Jeffrey (1965, Ch. 11) ingeniously showed a way to preserve regularity. The basic idea is straightforward: In simple conditionalization, the evidence E comes with maximal certainty; its posterior probability is $P'(E) = P(E \mid E) = 1$. It is this assumption which violates regularity. And it may appear unreasonably strong, for there seems to be always a slight chance of error in evidence.

Jeffrey's original example is the observation under candle light. Suppose your prior information about your friend's coat is that it is dark-colored, i.e., either blue (Bl), or green (Gr) or brown (Br), and that these alternatives seem equally likely to you. Hence $P(Bl) = P(Gr) = P(Br) = 1/3$. (If you think the example violates regularity from the outset, assume that the possibility space contains just these alternatives.) Now you see the coat for the first time, but only under candle light too dim to see clearly. So the coat rather seems to be blue to you, but you are quite unsure. What then is the result of your observation? Just some new degrees of uncertainty, say $P'(Bl) = 0.6$ and $P'(Gr) = P'(Br) = 0.2$. The details of the observational process might be hard to describe, but if doxastic states are represented as probability measures, then the outcome must be of that sort.

Jeffrey's point of view is that this example is not an exceptional one, but the normal case. If the light had been better, you might have ended up with $P'(Bl) = 0.9$. But even given bright day light there is a tiny chance of error; so still $P'(Bl) < 1$. There is no need to assume that observation or evidence should ever lead to absolute certainty. (I will plead only one exception in Section 16.4.)

But how then to describe probabilistic belief change? The general model is this. First, we assume an observational or evidential algebra $\mathcal{E} \subseteq \mathcal{A}$ of propositions. $\mathcal{E}$ is supposed to be finite with atoms $E_1, \ldots, E_n$. The idea is that $\mathcal{E}$ contains all those propositions, the probabilities of which get redistributed directly as an effect of the observation. These are, as it were, the newly acquired basic beliefs. How are the probabilities on $\mathcal{E}$ redistributed? That is specified in some probability measure Q on $\mathcal{E}$, which is a primitive element of the belief change. There is no deeper explanation how the doxastic subject arrives at Q.

The probabilities for propositions outside $\mathcal{E}$ must change, too. How? Here, Jeffrey makes the basic assumption that observation might affect the probabilities of $E_1, \ldots, E_n$, but not the probabilities conditional on $E_1, \ldots, E_n$. In traditional terms one might say that the inferential relations of $E_1, \ldots, E_n$ are not changed by making an observation on $E_1, \ldots, E_n$ themselves. According to the formula of total probability, we thereby have everything we need for determining the posterior probabilities:

Definition 3.9: Let P be a probability measure on $\mathcal{A}$ and Q be a probability measure on a finite subalgebra $\mathcal{E} \subseteq \mathcal{A}$ with atoms $E_1, \ldots, E_n$. Then P' is the (*generalized*) *conditionalization* of P with respect to Q iff for each $A \in \mathcal{A}$,

$$P'(A) = \sum_{i=1}^{n} P(A \mid E_i) \cdot Q(E_i).$$

The associated dynamic *law of generalized conditionalization* then says:

(3.10)   If P is the prior probability measure of s at t and if Q gives s's new probability for the propositions about which s acquired evidence between t and t', then s's posterior probability measure P' at t' is the conditionalization of P w.r.t. Q.

It immediately follows

Theorem 3.11: If P and Q are regular, so is the conditionalization of P w.r.t. Q.

This observation is essential for Jeffrey; it says that he succeeded with generalized conditionalization in defining a doxastic dynamics that moves entirely within the realm of regular probability measures. Is it a rational dynamics? The crucial assumption about the constancy of conditional probabilities is at least plausible, and there are more rigorous arguments, again of the Dutch-book type, establishing its rationality (see again Teller 1976 and Skyrms 1990, Ch. 5).

There are two prima facie shortcomings. First, one might object to making Q a primitive element of generalized conditionalization, since Q seems to be already the result of an interaction between the prior probabilities and the evidential process. For instance, if the coat's being blue were initially a remote possibility, so that, say, $P(Bl) = 0.01$, then that observation under candle light would not have raised the posterior probability $P'(Bl)$ to 0.6. But the objection can be met: there is a way of characterizing the observational import and its interaction with the prior probabilities that yields for any prior P a posterior probability measure Q on $\mathcal{E}$ (cf. Field 1978; see also the discussions of Garber 1980 and Jeffrey 1983, pp. 181–183). Second, our definition of generalized conditionalization was restricted to a finite evidential algebra $\mathcal{E}$. However, this restriction was purely technical and there are more advanced formal means to overcome this restriction and to define the conditionalization of P with respect to any Q continuous with P on any subalgebra $\mathcal{E}$ of $\mathcal{A}$. (There is a generalized notion of probability conditional on a σ-algebra that can be put to use here. See, e.g., Breiman 1968, Ch. 4, or Jeffrey 1971, Sect. 9.)

However, the last observation brings up another worry. Apparently, any P' can come from any regular P by conditionalization w.r.t. a suitable Q; simply choose $\mathcal{E} = \mathcal{A}$ and $P' = Q$. Hence, Jeffrey's generalized conditionalization (3.10) does not seem to impose any constraint at all on possible probabilistic dynamics.

There is a two-step reply. First, we may observe

Theorem 3.12: If P' is the conditionalization of P w.r.t. Q_1 on $\mathcal{E}_1$ and w.r.t. Q_2 on $\mathcal{E}_2$, then Q_1 and Q_2 agree on $\mathcal{E} = \mathcal{E}_1 \cap \mathcal{E}_2$, and if Q is the restriction of Q_1 (or Q_2) to $\mathcal{E}$, then P' is the conditionalization of P w.r.t. Q.

Proof: Let $E_{11}, \ldots, E_{1k}$ be the atoms of $\mathcal{E}_1$, let $E_{21}, \ldots, E_{2l}$ the atoms of $\mathcal{E}_2$, and let $E_1, \ldots, E_n$ be the atoms of $\mathcal{E}$. Our assumption is that for each $A \in \mathcal{A}$:

$$(*) \quad P'(A) = \sum_{i=1}^{k} P(A \mid E_{1i}) \cdot Q_1(E_{1i}) = \sum_{j=1}^{l} P(A \mid E_{2j}) \cdot Q_2(E_{2j}).$$

For $A = E_m$ $(m = 1, \ldots, n)$ $(*)$ immediately yields $Q_1(E_m) = Q_2(E_m)$. So, Q_1 and Q_2 agree on $\mathcal{E}$.

Now, take any E_m and focus on the $E_{1i} \subseteq E_m$ and $E_{2j} \subseteq E_m$; suppose this is the case for $i = 1, \ldots, k_m$ and $j = 1, \ldots, l_m$. For $A = E_{1i} \cap E_{2j} \neq \varnothing$ $(*)$ yields: $P'(E_{1i} \cap E_{2j}) = P(E_{2j} \mid E_{1i}) \cdot Q_1(E_{1i}) = P(E_{1i} \mid E_{2j}) \cdot Q_2(E_{2j})$. Hence, $Q_1(E_{1i}) / Q_2(E_{2j}) = P(E_{1i} \mid E_{2j}) / P(E_{2j} \mid E_{1i}) = P(E_{1i}) / P(E_{2j})$. Since this holds for all $j = 1, \ldots, l_m$, it entails in turn that $Q_1(E_{1i}) / Q(E_m) = P(E_{1i}) / P(E_m)$. Thus, finally, for any $A \in \mathcal{A}$:

$$P'(A \cap E_m) = \sum_{i=1}^{k_m} P(A \mid E_{1i}) \cdot Q_1(E_{1i}) = \sum_{i=1}^{k_m} P(A \mid E_{1i}) \cdot \frac{P(E_{1i})}{P(E_m)} \cdot Q(E_m)$$
$$= P(A \mid E_m) \cdot Q(E_m), \text{ and so}$$
$$P'(A) = \sum_{n}^{m=1} P(A \mid E_m) \cdot Q(E_m).$$

The theorem says, in effect, that whenever P' comes from P by generalized conditionalization there is a unique smallest algebra $\mathcal{E}$ and some unique probability measure Q on $\mathcal{E}$ such that P' is the conditionalization of P w.r.t. Q (Q is indeed simply the restriction of P' to $\mathcal{E}$):

Definition 3.13: For any two probability measures P and P' such that P' comes from P by generalized conditionalization, the $\mathcal{E}$ or rather the $\langle \mathcal{E}, Q \rangle$ that is unique according to (3.12) is called the (*evidential*) *source* of the transition from P to P'.

This was the rigorous step.

The second step is only partially clear. So much is clear: not every proposition can be an element of the evidential source driving a probabilistic belief change. There must be severe constraints restricting membership in this source. What is unclear is what these constraints might be. If we still believed in the empiricist separation of an observational and a theoretical language, then we could require that the propositions in the evidential source must be expressible in the observational language. But today we can no longer believe in this separation, and, as I said, it is unclear what to say instead. Certainly, further constraints would be needed to characterize a specific transition at a specific time. Maybe the evidential propositions are always indexical ones, referring to the contextually given time, place, and observer.

Levi (1967b) moreover attacked the idea that the evidential source contains a subjective distribution Q over the propositions in the source without further justification and argued that, if evidence is to justify belief change, it must be certain and hence constrained to be of a kind that can be certain. In effect he pleaded to return to simple conditionalization. Jeffrey (1983, p. 181) disagreed, and rightly so in my view. Section 5.1 will put forward what I take to be the essential argument. However, it is not the time to delve into, let alone attempt to solve this issue. It is part of the problem of the belief base that will be seriously addressed only in Chapter 16. Here, the point was only to show how in principle Jeffrey's generalized conditionalization must be supplemented in order to acquire bite.

This brief explanation of some basic aspects of Jeffrey's generalized conditionalization indicates that this is only the beginning of a big discussion. In a way, Jeffrey's revolutionary point was that the input driving doxastic change is not a single evidential proposition, but is itself probabilistic information. Now one might wonder why this probabilistic information should take the form of a distribution over an evidential partition $\{E_1, \ldots, E_n\}$ (or the algebra $\mathcal{E}$ generated by it). Maybe the input consists in probabilistically reassessing an arbitrary set of propositions not necessarily forming a partition or an algebra. This case is discussed by Wagner (1992), who proposes a generalization of Jeffrey conditionalization. Or maybe the input comes as any kind of probabilistic information, not necessarily specifying the probabilities of some propositions. It appears, then, that we have arrived at the most general case for which the rule of minimizing relative entropy has been proposed (cf., e.g., Shore, Johnson 1980 and Hunter 1991b). Surprisingly, this further generalization is not compatible with Wagner's generalization, as Wagner (1992, p. 255) points out, and so on. All this is just namedropping, but it does point to a substantial and fascinating discussion.

So much for one of the two ways of solving the conflict between conditionalization and regularity. The alternative is to give up regularity. Then we may stick to simple conditionalization (3.8), as mentioned. But this comes at the cost of no longer having a general dynamic law, for it works only for evidence having positive prior probability. One might agree that this is not really a restriction. Even if one gives up regularity and allows maximal certainty not only to the tautology, but at least to past evidence as well, then future possible evidence – so the argument goes – can never have been established or rejected with maximal certainty beforehand. This would be a kind of restricted regularity. But the argument is delicate. If one goes so far as to grant absolute certainty to past evidence – something emphatically denied by Jeffrey – then it is not so clear how much farther absolute certainty extends. Why should recollection be safer than prediction? I do not want to claim that there is no way to make the argument cogent. The point is only that there is no obvious way. Therefore it might be worthwhile to attempt to save conditionalization without relying on such an argument.

Is this possible? Yes. The idea must be to explain conditional probabilities even in the case where the condition has probability 0. Definition 3.2 cannot be used for this purpose. Hence, the idea has been to take conditional probabilities as a primitive notion such that they embrace absolute probabilities and the conditional probabilities derived from them. This idea can be traced back at least to John Maynard Keynes and Bruno de Finetti, but it seems to have gained power only in the 1950s, by Rényi (1955), Popper (1955), and others. Rényi (1962) is, as far as I know, the only mathematical textbook that develops probability theory in terms of this primitive notion. So, let us introduce:

Definition 3.14: Let $\mathcal{A}$ be an algebra over W and $\mathcal{B}$ a non-empty subset of $\mathcal{A}$. Then $\tilde{P}$ is a (σ-*additive*) *Popper measure* on $\mathcal{A} \times \mathcal{B}$ iff $\tilde{P}$ is a two-place function from $\mathcal{A} \times \mathcal{B}$ into $\mathbf{R}$ such that for all $A, B, C \in \mathcal{A}$ with $A, A \cap B \in \mathcal{B}$:

(a)   if $\tilde{P}(B \mid A) > 0$, then $B \in \mathcal{B}$,

(b) $\tilde{P}(\cdot \mid A)$ is a (σ-additive) probability measure on $\mathcal{A}$ with $\tilde{P}(A \mid A) = 1$,

(c) $\tilde{P}(B \cap C \mid A) = \tilde{P}(B \mid A) \cdot \tilde{P}(C \mid A \cap B)$.

Clause (b) says that, given any fixed condition, $\tilde{P}$ is just a probability measure in the first argument which gives that condition a probability of 1. Clause (c) is a conditional version of the law of multiplication (3.4c) and thus relates the probabilities under different conditions. Finally, (a) is a closure property of the set $\mathcal{B}$ of conditions. If B has positive probability under any condition, it qualifies as a condition as well. This is obviously a copy of Definition 3.2, according to which any B having positive absolute probability serves as a condition. Clause (a) entails that $W \in \mathcal{B}$, that the tautology is also a condition, and so we may define the absolute probability of a proposition A relative to a Popper measure $\tilde{P}$ simply as $\tilde{P}(A \mid W)$.

The closure property (a) is relatively liberal (and it should be mentioned that Rényi favors a still weaker version). It is, for instance, compatible with setting $\mathcal{B} = \{A \mid \tilde{P}(A \mid W) > 0\}$. In this case, a Popper measure would consist of nothing but absolute probabilities plus the entailed conditional probabilities, and we would not have made any progress at all. This is only the minimal case, though. The maximal case is where $\mathcal{B} = \mathcal{A} - \{\varnothing\}$ (since $\varnothing$ can never be among the conditions). In this case, I prefer to call $\tilde{P}$ a *full conditional measure*. Clearly this is the case we intend, for we then have conditional probabilities with respect to any non-contradictory condition whatsoever. So why not fix this maximal case in Definition 3.14 outright? The reason lies only in technical generality and flexibility.

What, then, is the law of belief change in this setting? It is again the *law of (simple) conditionalization*. This was the point of the whole exercise:

(3.15)   If the Popper measure $\tilde{P}$ characterizes the doxastic state of some subject s at t and if E is the total evidence s receives between t and t', then the subjective probabilities of s at t' are given by $P'(A) = \tilde{P}(A \mid E)$ for all $A \in \mathcal{A}$.

However, now the law is as general as the set B of conditions for which $\tilde{P}$ is defined and hence completely general in case $\tilde{P}$ is a full conditional measure.

Inquiry might continue here in various directions. For instance, the structure of Popper measures is quite complicated in ways that do not surface in Definition 3.14. Though it appears to resolve our concern, the law of conditionalization w.r.t. Popper measures will turn out to be clearly unsatisfactory, as we will see in Section 5.1. We will return to these issues more thoroughly in Section 10.2.

We may also look at other attempts at resolving the conflict between regularity and conditionalization. Levi proposed that learning some A initially having probability 0 is done by first contracting the set of full beliefs, which have probability 1, and then probabilistically assessing A within the thus-extended set of serious possibilities (cf., e.g., Levi 1984, pp. 148f. and p. 201). However, I will comparatively discuss Levi's rich epistemological doctrine only in Section 11.2.

We may take pause for the time being. The opposition between preserving regularity and preserving simple conditionalization is only one way to enter the discussion about the dynamics of subjective probabilities, but it served well for preparing a foil for motivating and elaborating ranking theory as constructed in the next chapters.

3.3 Probability and Belief

We should rather ask ourselves: why look at all for alternatives, why not be content with probabilistic epistemology? We have seen some of its problems, but we have seen no reasons why these problems should not be internally solvable. So, again, why not acquiesce?

This question invokes a large discussion. In a summarizing way it is not unfair to say that there is a deep schism between traditional and probabilistic epistemology that is essentially characterized by mutual misunderstanding. In fact, it is not the least of the aims of this book to overcome this schism.

Let me mention a paradigmatic case: Plantinga (1993a,b) is a major epistemological work of two rich and thoughtful volumes. Plantinga (1993a, Chs. 6 and 7) even contains a discussion of Bayesian epistemology, but he soon ends up dismissing it entirely, apparently frustrated. One cannot say that he reads Bayesianism with charity. But one must also say that Bayesian epistemology is not obliging at all. Indeed, as it presents itself, he has every reason to despair.

Plantinga's aim is to explicate the nature of justification, yet he finds that this notion is completely avoided by Bayesianism; the entire Bayesian literature makes no attempt to characterize justified belief. Plantinga then tries to extract a characterization by himself, but soon gives up. Or take truth, another notion central to traditional epistemology. It seems important that we are able to characterize our doxastic states, and not only propositions themselves, as at least to some extent true. However, it makes no sense at all to say that subjective probabilities are true. They can only be more or less adequate, well-informed, reasonable, etc. Even if a proposition has an objective chance and my subjective probability agrees with the chance, the subjective probability cannot be called true, but only my belief in that objective chance could be. Traditionally, we are interested in knowledge and we think that knowledge is something like justified true belief, but Bayesianism has apparently nothing at all to say about the key notions of justification and truth.

As I have presented the story, Bayesianism is an important, or the most important, contribution to the problem of induction. But there are other ways to set up what epistemology is about, and then the contribution of Bayesianism suddenly appears negligible. This is odd, but it is simply a symptom of the schism I am talking about. The schism might be seen as part of the schism between a theory of knowledge and the theory of belief that I already deplored in Section 1.1. From this perspective probability theory would have to be taken as part of the theory of belief.

However, the present schism is even deeper. Let me try to pin-point the source of trouble. It is, I believe, that Bayesianism does not even have the notion of belief, $\mathbf{B}$, but merely has the notion of degrees of belief. However, only beliefs are doxastic attitudes capable of being true, and we are used to speaking of the justification of beliefs rather than of the justification of subjective probabilities. Hence, if Bayesianism misses beliefs, it misses all the things related to belief and, worst of all, knowledge. Small wonder that the traditional epistemologist finds Bayesianism useless.

But is it really the case that belief cannot be accounted for in terms of subjective probabilities? Yes, this is at least the lesson taught by the famous lottery paradox (contemporarily first presented by Kyburg 1961, p. 197, but see also Hempel 1962, pp. 163–166). We might be tempted to say that belief is sufficient certainty, where "sufficient" is certainly vague and context-dependent. So we might try the translation:

(3.16) $\mathbf{B}(A)$ iff $P(A) \geq 1 - \varepsilon$ for some small $\varepsilon > 0$.

However, we may then devise a lottery with $n \geq 1/\varepsilon$ tickets of which exactly one wins. Let A_i ($i = 1, \ldots, n$) be the proposition that the i-th ticket wins and $A = \bigcup_{i=1}^{n} A_i$ the proposition that at least one ticket wins. Hence, $P(A) = 1$ and $P(\bar{A}_i) \geq 1 - \varepsilon$ for $i = 1, \ldots, n$. If the translation (3.16) were acceptable, this would entail $\mathbf{B}(A)$ and $\mathbf{B}(\bar{A}_i)$ for $i = 1, \ldots, n$. And if belief is closed under conjunction, this further entails $\mathbf{B}(\varnothing)$, the belief in the contradiction. Thus, (3.16) inevitably leads to absurd results, for every $\varepsilon > 0$.

(In personal communication Rudolf Schüßler informs me that this fact was already known and discussed in scholastic times. Terillus (1669, p. 60) states: "Assertio tertia: Consequens ex duabus propositionibus probabilibus rite illatum saepe non est probabile,

si aliunde non probetur quam ab illis praemissis." In English: "Third assertion: A conclusion that is correctly derived from two probable premises is often not probable, if
it cannot be derived from other premises than those." And then Terillus goes on to
demonstrate this assertion with a clear example. So one might well say that the lottery
paradox was known already at that time.)

A desperate way out, taken by Kyburg (1961, p. 197) and affirmed in Kyburg (1963,
1970), is to deny the closure of belief under conjunction. Believing is taking to be true,
and if I take A to be true and B to be true, how could I fail to take A-and-B to be true?
This seems to be the last resort.

More promising is perhaps the translation:

(3.17) $\mathbf{B}(A)$ iff $P(A) = 1$.

Thereby we would avoid the previous conclusion, because the set of propositions
having the probability 1 is closed under finite (and, in the σ-additive case, countable)
conjunctions. (It is not closed under uncountable conjunctions; perhaps, though, we
have no clear intuitions here.) However, the translation is implausible since it equates
belief with maximal certainty, where sufficient certainty, in all its vagueness, would
be enough. The problem is not only an intuitive one. According to the law of simple
conditionalization, every proposition having the prior probability 1 keeps the posterior probability 1. That is, given this law and (3.17), one is always stuck with one's old
beliefs; beliefs could only be accumulated, but never genuinely revised. However, this
is clearly not our notion of belief. We may, and indeed might be forced to, give up
some of our prior beliefs.

Perhaps, the law of conditionalization is the wrong one. But, so far, the only alternative we know is Jeffrey's generalized conditionalization, and this is even worse in this
respect. If we accept regularity, there is no belief at all according to (3.17), except in
the tautology.

Perhaps the law of simple conditionalization as well as the translation (3.17) should
be understood in terms of Popper measures. This is much better, for then one need not
be stuck with previously held beliefs. Indeed, at present I cannot argue against this idea.
In Section 10.2, though, we will see that it must be rejected as well, and with it the
whole idea of extending the probabilistic point of view with the help of Popper
measures.

If this argument holds good, then there is no way left to maintain the second translation. Basically, this situation has been clear for a long time, and hence the lottery paradox
kept on nagging. And it has found more elaborate responses. Somehow, Lehrer's epistemology, last formulated in Lehrer (2000), is based on a more complicated explanation of acceptance in terms of some kind of doxastic decision theory which includes
subjective probability. I will comment on it in Section 11.4. Levi (2004), the most
recent presentation of his epistemology, has a most sophisticated account of acceptance
that is ultimately based on subjective probabilities (and several other measures and
parameters). I will defer comparative remarks to Section 11.2. More approaches might

be mentioned, but let me just resume by saying that all accounts of belief in probabilistic terms, beyond the two just discussed and rejected, are not really convincing; they do not seem the way to go, as I will explain a bit more thoroughly in Section 10.2.

What is my conclusion, then? We do want an account of belief, **B**, of acceptance, of taking a proposition to be true, or whatever may express the same notion, which applies not only to tautologies and other maximal certainties, but also to contingent propositions. And as Chapter 1 urged us, we do want a static and a dynamic account of belief, for otherwise we will not connect up with the topics of traditional epistemology. Probabilistic epistemology is unable to provide such an account, as the lottery paradox forces us to recognize. Hence, the only choice left is to develop such an account independent of probability theory. This is what we will do in the next chapters, and this is what ranking theory is all about. So although we will at first entirely turn away from Bayesianism, it will soon be obvious that the similarities and relations of our account to probability theory are very close, indeed so close that one might see this book as presenting an extension of the Bayesian viewpoint. However, it will become clear in the course of the book that this perspective would not be fair, and in Chapter 10 I will argue that probability and ranking theory are on a par and symmetrically supplement one another. Still, these relations will pervade the entire book.

4

The Representation of Belief:
Some Standard Moves

4.1 The Basic Laws of Belief

To start, then, let us look for a representation of belief, independent of probability theory. This independence is psychologically difficult. In fact, already the term "belief" presents difficulties. If I believe a proposition, I take it to be true – I accept it; I shall use all of these phrases interchangeably, but there is inevitable vagueness. Suppose I say: "I believe that George Bush lied to the American Congress concerning the Iraqi threat." You ask: "Are you sure?" I reply: "No, I am not." You insist: "So, it's only likely, or likelier than not, that Bush lied?" I repeat: "No, it's not only likely, I really believe Bush lied." "How much would you bet then?" "I don't want to bet, it's hard to decide, anyway." And so the dialogue might continue for a while.

The essential point here is that, on the one hand, there is a deeply rooted tendency to project belief onto some scale of (un)certainty that is naturally interpreted in terms of probabilities. On the other hand, the scale does not really fit; belief is clearly not maximal certainty, but also not any probabilistic degree below the maximum. This is why you and I are talking at cross-purposes in that dialogue. Hence, we must firmly banish any probabilistic connotations in this chapter. We start out with belief being a yes-or-no affair, and indeed must do so, though we will be led again to degrees of belief, albeit not probabilistic ones.

In order to relieve the notion of belief of some of its vagueness, philosophers have applied qualifications like "full belief", "strong belief", "weak belief", etc. In order to dissociate myself from such specifications, I have often used the term "plain belief", but this tended to be understood as a further and possibly mystifying qualification. And so I prefer to return to plain talk of belief.

The static representation of belief seems to be an entirely straightforward matter: the doxastic state of a subject s at time t is characterized by her *belief set* $\mathcal{B}_{s,t} = \{A \in \mathcal{A} \mid \mathbf{B}_{s,t}\}$. This characterization is trivial. We may omit subscripts in the following.

The immediate question is whether any set $\mathcal{B}$ of propositions will do as a belief set, or whether there are static laws characterizing rational belief sets. There are two laws that were perhaps taken for granted in earlier times and that have been explicitly stated in doxastic logic since its onset in Hintikka (1962):

(4.1)   Belief sets are consistent.

(4.2)   Belief sets are deductively closed.

These two assumptions are still very common, but have also come under strong attack. Let us first examine the latter. Assuming (4.2) seems to require an endless process of reasoning, at least in cases where the underlying algebra $\mathcal{A}$ of propositions is infinite. However, we need not assume that every inferred belief is arrived at by explicit reasoning. Still, (4.2) requires a subject to have an infinity of beliefs. Certainly, but there is no problem. Beliefs need not be explicit or occurrent beliefs. They might also be dispositional beliefs that one has even though they are never actually thought – i.e., never occurrent in the course of one's life. For example, the belief that there are no wild polar bears in Africa may be such a belief (of course now you are thinking of it). And dispositional beliefs may well be infinite in number, even for finite brains.

Still, logical consequence is undecidable (in the technical sense). Hence, (4.2) assumes a belief set to be objectively and, a fortiori, subjectively undecidable. What should it mean then that a believer is disposed to believe such an undecidable set? This is the real problem behind (4.2), and it concerns (4.1) as well, since logical consistency is also undecidable. How, then, can it be required of rational persons? Was Frege irrational because he firmly believed in an inconsistent axiomatization of set theory? Clearly not. He would only have been irrational if he had stuck to the axiomatization after Russell's discovery of the inconsistency. But of course he immediately saw the impact of Russell's antinomy. For this reason one might want to conclude that (4.1–4.2) should be weakened to:

(4.3)   Belief sets are believed to be consistent.

(4.4)   Belief sets are closed under believed logical consequence.

These laws, however, appear quite empty. This is not to say that (4.4) is simply a weakening of (4.2). The subject may well believe certain logical consequences to obtain, though they do not. In that case (4.4) forces him to believe more than (4.2) does. But of course his belief set would then contain logical impossibilities. Paraconsistent logic may offer an account how this is possible. I doubt, though, that we can really make sense of such a situation. Anyway, in view of such problematic substitutes, (4.1–4.2) have been vigorously defended. All this is also known as the problem of logical omniscience (cf., e.g., Stalnaker 1999, Chs. 13 and 14).

One defense says that laws of rationality are always idealizations, and so are (4.1–4.2) in particular. This is probably true, but it is unsatisfying so long as the nature of the particular idealization is left unspecified. A better defense is given by Levi (1991, Sect. 2.1 and 2.7), who argues that (4.1–4.2) are not about belief, but rather doxastic commitments. These commitments need not be transparent and one might have them without realizing it. In this sense, it seems fully appropriate to say that the set of beliefs one is committed to is deductively closed. One may even accept the picture of belief that I laid out in Section 2.3, according to which (4.1–4.2) are trivially satisfied. Or one might also recommend a more pragmatic attitude, in which one inquires about the theory resulting from (4.1–4.2) – which is strong and well-behaved – and then compares it to alternative theories that try to get along without these assumptions.

However, we are not sufficiently prepared for a fruitful discussion of this difficult issue. For the time being, let me simply explain my preference. First, I take the pragmatic attitude very seriously. If I compare the theorizing based on (4.1–4.2) with the theorizing based on alternatives like (4.3–4.4), then I find the former incomparably richer. The theoretical epistemologist can *do* something given (4.1–4.2), but next to nothing without them. How much he can do is the subject of this book, and to present serious rivals is not my business (though I heavily doubt that there are any).

The second and less pragmatic response is equally important to me. In my view, we have decided the issue already in Section 2.1 with the almost non-committal view taken there, which opted for propositions as objects of belief. Recall that propositions are pure truth conditions and that my vote for propositions was thus a vote against taking any kind of mental or linguistic representations of truth conditions as objects of belief. In other words, we do not speak of a subject believing a certain truth condition under one representation and not believing it under another. Or to put it yet another way, if we start talking about representations at all, we are assuming that equivalent representations of the same truth condition are always recognized as such. Without this assumption it would have been from the outset illegitimate to choose propositions as objects of belief.

Let me repeat at this point what I have already emphasized in Chapter 2, namely that truth conditions need not be wide contents. Truth conditions in themselves are at the very least ambiguous. Two-dimensional semantics is able to clarify the ambiguity and also provide a narrow understanding of truth conditions. Likewise, representations need not be the same as modes of presentations. Both may be understood on a syntactic or computational level. Modes of presentation may, however, be understood purely on a semantic level that can be well accounted for by two-dimensional semantics. One and the same mode of presentation in the semantic sense might still be represented in different ways on the syntactic level. This is how my talk of representations in the previous paragraph should be understood.

In view of this, (4.1–4.2) appear rationally necessary. Consistency requires the subject to recognize that the actual possibility can never be in the contradictory, empty proposition. And deductive closure requires the subject to recognize that (i) if the actual possibility is in each of two propositions, then it is also in their intersection, and (ii) if the actual possibility is contained in some proposition, then it is so also in any superset. Or referring once more to representations of truth conditions, we might say that consistency requires the subject to know that sentences of the form $p \wedge \neg p$ cannot be true, and deductive closure requires the subject to know that $p_1, \ldots, p_n$ are true if and only if $p_1 \wedge \ldots \wedge p_n$ is true. Given knowledge of logical equivalences, consistency and deductive closure reduce to these primitive cases.

Of course, logical equivalence between sentences is, in turn, undecidable. And so one might remain as skeptical as before when working at the sentential level. But the above doubts do not apply, because we are starting on a different, i.e., the propositional level. Putting it in another way: There are various deductive or – more generally – computational relations between sentences or – more generally – between representations of truth conditions, and there are highly involved theories dealing with these relations. By taking belief to be a relation between persons and propositions, we disregard all of these computational relations. In focusing on the problem of induction, we take deduction for granted. Therefore, I will firmly stick to (4.1–4.2) as static rationality postulates.

One may protest now that, had this strong consequence already been clear in Section 2.1, one would never have agreed to take propositions as objects of belief. Now we are back at that dispute, and I won't repeat it here. Let me only reiterate that, from the point of view of normative theory construction, the decision to stay on the propositional level seems necessary to me, as I am going to explain further in Section 11.6.

Let us resume this brief discussion with

Definition 4.5: Let $\mathcal{A}$ be an algebra of propositions. $\mathcal{B} \subseteq \mathcal{A}$ is a *belief set* iff for all $A, B \in \mathcal{A}$:

(a) $\varnothing \notin \mathcal{B}$,
(b)   if $A, B \in \mathcal{B}$, then $A \cap B \in \mathcal{B}$,
(c)   if $A \in \mathcal{B}$ and $A \subseteq B$, then $B \in \mathcal{B}$.

If $\mathcal{A}$ is a complete algebra, then $\mathcal{B}$ is a *complete belief set* iff moreover

(d)   for any $\mathcal{B}' \subseteq \mathcal{B}, \bigcap \mathcal{B}' \in \mathcal{B}$.

For a complete belief set $\mathcal{B}, \bigcap \mathcal{B}$ is called its *core*.

Mathematicians call belief sets filters, but in this context our terminology is preferable. If $\mathcal{A}$ is finite, then any belief set is complete. A difference emerges only for infinite algebras.

Is (d) rationally required in the infinite case? One may doubt this, but, once again, for the wrong reasons. In connection with (c), (d) seems to postulate closure of belief sets under logical consequence from infinitely many premises, and this is clearly even more unworkable for human minds. But from the propositional point of view, the case is less clear. On the contrary, if you believe the actual possibility to be in all of an arbitrary number of propositions, then it should be clear to you that it cannot fail to be in their intersection as well. So I tend to accept (d) as well. I will indicate where my conclusions depend on it and where they do not.

For instance, an immediate consequence is

(4.6)   If $\mathcal{B}$ is a complete belief set and C its core, then $C \neq \varnothing$ and $\mathcal{B} = \{A \in \mathcal{A} \mid C \subseteq A\}$.

Thus, if a doxastic state is represented by a complete belief set, it may be represented by its core as well, which is a single proposition – i.e., the intersection of all propositions believed in this state. This makes for a very simple and convenient static representation of belief.

Is static rationality exhausted by Definition 4.5? By no means. The situation here is similar to the probabilistic case. Definition 4.5 provides only a minimum of rationality, but strengthenings of it are less clear. One obvious way to continue would be to ask which unrevisably a priori beliefs must be held in any state of belief. But then we are in the midst of a very contested area in philosophy which we are better to avoid for now (cf. Section 6.5 and Chapter 17). In any case, we will see that (4.1) and (4.2) take us very far. In a sense more fully explained in Chapter 5 and 8, the theory to be developed hardly goes beyond assuming (4.1) and (4.2).

4.2 The Dynamics of Belief

We have now come to the key question of this book. What can we say about the dynamics of doxastic states represented as (complete) belief sets or their cores? Let us make the question precise: Suppose that s's state at t is characterized by the prior core C and that s changes to the posterior core C' at t'. What drives the change? This may have many causes: forgetfulness and recollections, thoughtlessness, drugs, wishful thinking, exhaustion, etc. However, it was clear all along that we do not want to examine these kinds of changes, but only changes that are clearly rationally assessable. This is why we are always considering evidence as driving doxastic change. Let us start with the simple assumption that evidence comes in propositional form. In the previous chapter we began the same way, though we came to think of evidence in a different way with Jeffrey's generalized conditionalization. We will do so here as well, but there is still a long way to go.

Suppose, then, that s's total evidence between t and t' is captured by the proposition E. How is s's posterior core C' at t' rationally constrained? We have to distinguish two cases: (i) the case where the evidence E is consistent with the prior core C, i.e., $C \cap E \neq \varnothing$, and (ii) the case where it is not. The latter is surely possible and not even unusual. Indeed it happens all the time that evidence is contrary to our prior expectations.

Consider the consistent case first. It is governed by two highly plausible conditions. The first one is:

(4.7)   If $C \cap E \neq \varnothing$, then $C' \subseteq C \cap E$.

This says that the posterior state preserves all of the prior beliefs, accepts the evidence as well, and draws all the logical conclusions from combining evidence and prior beliefs. (4.7) thus sets a minimum for the posterior belief set and still allows it to be exceeded.

The rationale for (4.7) is very simple: Clearly, the evidence E must be believed in the posterior state. If E were not accepted in the posterior state, there would be no propositional evidence driving any change. Thus, the cases we are presently dealing with are characterized by the fact that the evidence E is accepted. (4.7) moreover says that unless the evidence forces the subject to reject any of her prior beliefs, she keeps them. It would be unmotivated and irrational to skip any of them. However, both claims have come under attack. In belief revision theory the acceptance of evidence is also called the success postulate, and it soon turned out not to be sacrosanct. The issue is discussed under the heading of "non-prioritized belief change" (cf., e.g., Hansson 1999). Concerning the second claim, Haas (2005, Sect. 2.9) has made a strong case that there may be tensions between prior beliefs and evidence which do not amount to inconsistency, but which may still be strong enough to prevent preservation of all prior beliefs. I will deal with these attacks when comparing my theory with the belief revision literature in Section 11.3. For the time being, I find the simple rationale of (4.7) good enough to proceed.

The other condition for the consistent case is this:

(4.8)   If $C \cap E \neq \varnothing$, then $C \cap E \subseteq C'$.

This is the converse of (4.7). It says that it would be irrational to accept in the posterior state more than the prior beliefs, the evidence, and their joint consequences. If the posterior state were to contain an excess belief A, how might it be justified? To my knowledge, no one has proposed a good idea. It seems that, if $C' \subseteq A$, but not $C \subseteq A$, then at least the material conditional "if E, then A" must be previously accepted, i.e., $C \subseteq \bar{E} \cup A$. But then A is entailed by C and E and is not an excess belief. I am not aware of any protest against (4.8) in the literature.

Clearly, (4.7) and (4.8) add up to:

(4.9)   If $C \cap E \neq \varnothing$, then $C' = C \cap E$.

In other words, (4.9) fully determines the posterior state in the case of consistent evidence. This case is governed by *belief expansion*, as it is usually called. There is presently nothing more to say about it.

What about the inconsistent case where $C \cap E = \varnothing$? Here, we can say depressingly little. We can safely claim:

(4.10)   If $C \cap E = \varnothing$, then $\varnothing \neq C' \subseteq E$.

That is, whatever the posterior core might be, it must be consistent again. The static rationality postulates must be observed in any case. And the evidence E has to be accepted in the posterior state, on the same grounds as in (4.7). Otherwise, there would not be any evidence driving any doxastic change (I said that I would discuss possible doubts in Section 11.3). So much seems clear.

The crucial point is that no more than (4.10) can be said about the inconsistent case. This is the lesson I draw from the continuous failure of various attempts that may be understood as at least implicitly addressing this issue. The famous metaphor of Quine (1951) suggests that we give up only the more peripheral beliefs in the case of con-travening evidence and stick to the more central ones, although in the worst cases we may have to revise even our most central beliefs. But which beliefs are central and which are peripheral? How are we to measure centrality? The irresistible suspicion is that it is precisely the revision behavior that defines what is more or less central. This is indeed the conclusion I will arrive at. But the suspicion is presently not helpful, for if revision defines centrality, centrality is no guide to revision.

Of course there have been various attempts to explain centrality without recourse to revision. For instance, a priori beliefs (in the strong sense) seem maximally central and hence unrevisable. So it would be useful to have a substantial theory of apriority, though Quine's intention was, of course, to criticize apriority or maximal centrality (which he called analyticity). Moreover, in order to get beyond (4.10), we need not only maximal, but graded centrality, since we want to know not only what to keep in any case, but also whether we should keep one thing rather than another. Here the observation may help that belief in laws seems to be more central than belief in accidental fact. True, but then we would have to know more precisely what lawlikeness is. Belief in simple hypotheses is more central than belief in complicated hypotheses. Maybe so, but how are we to define simplicity? Natural assumptions are more central than ad hoc assumptions. Certainly, but ad-hoc-ness is a very obscure matter. Beliefs expressible with projectible predicates are more central than those expressed with gruesome predicates. Yet this is a most delicate distinction, and − as is well known − Goodman (1955, Ch. 4) roughly arrived at the conclusion that the projectible pre-dicates basically are the ones successfully projected in the past. Perhaps revisions of beliefs in factual matters are guided by beliefs in counterfactual matters. To specify this relation would, however, only help if we had a firmer grip on counterfactuals. In any case Gärdenfors' (1978, 1981) strategy was the exact opposite of this, explaining counter-factuals in terms of belief revision and the so-called Ramsey test.

Philosophers have worked hard to make such ideas viable, but I find this to be a history of failures. We should modestly acknowledge that any move beyond (4.10) is highly problematic. (4.10) is the only clear and incontestable assumption we have about the case of contravening evidence. Yet this leaves us with an incomplete and therefore unsatisfactory dynamics of belief. It seems we must do more, but cannot do more. Is this an unsolvable dilemma?

We need not do more! This point is perhaps the most crucial one in the dialectics of the whole book. As should have been clear, all of the above-mentioned ideas for mak-ing progress on centrality and thus on belief revision were also ideas for somehow making objective progress on the problem of induction. If we give up on all these ideas and so yield to inductive skepticism, then the only choice left to us, one might think, would be to take belief revision, or dispositions to belief revision, as an unexplained primitive. This *is* the only choice, but it is not total defeat that condemns us to say only the poorest things about revision and induction. Rather it is the seed which will bear rich fruit, as we will see throughout the rest of the book. In my view, this revolutionary turn was initiated in the early writings of Gärdenfors (1978, 1981) on belief revision theory, but it did not receive proper emphasis. This is why I may be exaggerating here.

So what does it mean that we do not need to say more than (4.9–4.10) about the dynamics of belief, despite its glaring incompleteness? The simple, but essential point is that, whatever the doxastic transition from the prior to the posterior state, the disposition for that transition is already contained in the prior state. This disposition is not something over and above, but part of the prior doxastic state. Such a disposition may be formally represented by a selection function as defined in

Definition 4.11: Let $\mathcal{A}$ be an algebra over W. Then g is a *selection function* for $\mathcal{A}$ iff g is a function from $\mathcal{A} - \{\emptyset\}$ into $\mathcal{A}$ such that for all $A, B \in \mathcal{A} - \{\emptyset\}$:

(a) $\emptyset \neq g(A) \subseteq A$,
(b)   if $g(A) \cap B \neq \emptyset$, then $g(A \cap B) = g(A) \cap B$.

I will soon say where the term "selection function" comes from. But first we should understand what a selection function g is: $g(A)$ is to be interpreted as the core of the posterior belief state after receiving and accepting information A. As we saw above, this core fixes all of the posterior beliefs. $g(W)$ may be understood as the prior core, because the tautological information W does not effect any belief change at all. Since g yields a posterior core for any possible information, g indeed represents a *revision scheme*, a complete disposition for revising beliefs.

Why must such a scheme satisfy conditions (a) and (b)? Condition (a) is nothing but constraint (4.10) above, which holds generally and not only in the case of contravening evidence. And condition (b) is nothing but the postulate (4.9), slightly generalized: If we take $g(A)$, the core reached after information A, as the prior core and now consider information B consistent with that core, what is the posterior core generated by the second information? On the one hand, we may identify it with $g(A \cap B)$, since we may assume that in this consistent case the effect of receiving first information A and then information B is the same as that of receiving information $A \cap B$. On the other hand, postulate (4.9) tells us that the posterior core is just the conjunction of the prior core $g(A)$ and the new evidence B. And this is precisely condition (b).

I find this to be the most plausible way of explaining condition (b). Rott (1999) is right, however, in pointing out that (b) as such does not say anything about iterated revision. This is why he refers to (b) as a postulate of dispositional coherence as opposed to a genuine postulate of dynamic coherence. We will return to this point in Section 5.6, where we will see that dispositional and dynamic coherence are closely related.

With the help of this notion we can state a completely general dynamic law for belief, which I will also call the *law of simple conditionalization* because we will see in Theorem 10.4 that it corresponds to that law with respect to Popper measures. It says:

(4.12)   If the selection function g characterizes the doxastic state of the subject s at time t and if E is the total evidence s receives and accepts between t and t', then $g(E)$ is s's core at t', so that s believes A at t' iff $g(E) \subseteq A$.

How is it possible that this law is perfectly general although it does not go beyond the postulates (4.9–4.10) that were manifestly incomplete? To repeat: The trick was only to surrender the dynamic incompleteness on the level of beliefs to the prior disposition of the subject, to his prior revision scheme, whatever it may be, as long as it satisfies the minimal conditions (a) and (b) of Definition 4.11. Since the dynamic behavior is completely encapsulated in the prior revision scheme, it is no feat to state a complete dynamic law.

Is that to say that we are already finished with our business? Certainly not. We have only taken the first important step, and we will soon see why it is not as satisfying as it may appear. But it is useful first to get more familiar with the notion of a selection function by noting various equivalences and comparisons.

4.3 Equivalent Formulations

The first equivalence is very important, especially for the further development of my argument. We will introduce the formal concept first:

> *Definition 4.13*: Let $\mathcal{A}$ be an algebra over W. Then $\leq$ is an $\mathcal{A}$-*measurable weak well-ordering* of W, or an $\mathcal{A}$-*WWO* for short, iff $\leq$ is a weak order on W, i.e., a transitive and complete relation on W, such that for each $w' \in W$, $\{w \in W \mid w \leq w'\} \in \mathcal{A}$ and for each non-empty $A \in \mathcal{A}$ there is a $w \in A$ with $w \leq w'$ for all $w' \in A$, i.e., each non-empty $A \in \mathcal{A}$ has a minimum w.r.t. $\leq$. For any non-empty $A \in \mathcal{A}$, $\min_\leq(A) = \{w \in A \mid w \leq w'$ for all $w' \in A\}$ is the set of minimal members of A w.r.t. $\leq$. For a WWO $\leq$, $<$ is to denote the corresponding strict order and $\approx$ the corresponding equivalence relation.
>
> Equivalently we may conceive of an $\mathcal{A}$-WWO $\leq$ as an $\mathcal{A}$-*measurable well-ordered partition* $\langle D_\alpha \rangle_{\alpha < \zeta}$ for some ordinal number $\zeta \in \Omega$, or an $\mathcal{A}$-*WOP* for short, consisting of the well-ordered sequence of equivalence classes generated by $\approx$.

In this definition, Ω is the class of all ordinal numbers. The reader not familiar with them may simply assume that $\approx$ generates only a finite number of equivalence classes and that the corresponding $\mathcal{A}$-WOP is a finite sequence. Nothing really important is missed thereby. I should also mention that "weak well-ordering" is not an established term – as far as I know – because the characteristic property of well-orderings, the existence of a minimum, is usually stated only for linear orders (i.e., for transitive, complete, and antisymmetric relations).

WWOs (or WOPs) and selection functions are intertranslatable, as shown by

> *Theorem 4.14*: Let $\leq$ be an $\mathcal{A}$-WWO and define $g(A) = \min_\leq(A)$ for each non-empty A. Then g is a selection function for $\mathcal{A}$. Conversely, let g be a selection function

for $\mathcal{A}$ and define recursively $D_0 = g(W)$, $D_\beta = g(W - \bigcup_{\alpha < \beta} D_\alpha)$, and $\zeta = \min \{\alpha \mid D_\alpha = \varnothing\}$. Then $\langle D_\alpha \rangle_{\alpha < \zeta}$ is an $\mathcal{A}$-WOP. The selection function defined by the corresponding $\mathcal{A}$-WWO is g in turn.

Proof: g thus defined from the $\mathcal{A}$-WWO $\leq$ obviously satisfies (4.11a). Next suppose that $g(A) \cap B \neq \varnothing$, i.e., $\min_\leq(A) \cap B \neq \varnothing$. This directly implies that $\min_\leq(A \cap B) = \min_\leq(A) \cap B$. Thus, g satisfies (4.11b), too; i.e., it is a selection function.

Conversely, let $\langle D_\alpha \rangle_{\alpha < \zeta}$ be defined from the selection function g as above. $\langle D_\alpha \rangle_{\alpha < \zeta}$ is obviously an $\mathcal{A}$-WOP. Let $\leq$ be the corresponding $\mathcal{A}$-WWO. Does it define g? Yes, since for any $A \in \mathcal{A}$ with $\beta = \min \{\alpha \mid D_\alpha \cap A \neq \varnothing\}$ we have in virtue of (4.11b):

$$g(A) = g((W - \bigcup_{\alpha < \beta} D_\alpha) \cap A) = g(W - \bigcup_{\alpha < \beta} D_\alpha) \cap A = D_\beta \cap A = \min_\leq(A).$$

In view of Theorem 4.14 we may speak of *the* selection function *generated* by a WWO or a WOP and of *the* WWO or WOP *representing* a selection function.

The proof shows that we can interpret a WWO $\leq$ or the corresponding WOP $\langle D_\alpha \rangle_{\alpha < \zeta}$ as a *well-ordering of disbelief*. If I start with a belief set having the core $g(W)$ $= \min_\leq(W) = D_0$, and receive just the information that D_0 is wrong, then the core of my new belief set is $g(W - D_0) = \min_\leq(W - D_0) = D_1$. If D_1 turns out wrong as well, I change to the core $g(W - (D_0 \cup D_1)) = D_2$. And so on. Thus, D_0 contains the possibilities not disbelieved at all by me. D_1 contains the least disbelieved possibilities, among which I would locate the actual possibility if it is supposed not to be in D_0. D_2 contains the second least disbelieved possibilities, to which I would turn next. And so on. So we might generally say that a selection function g selects from a non-empty proposition A the proposition $g(A)$, which is the set of the possibilities that are most preferred or most plausible or – still better – least disbelieved within A. WWOs thus offer a way to understand more vividly how belief revision works according to a selection function. Note again that no claim is thereby made about specific preferences among possibilities. The only claim is that some such well-ordering of disbelief must underlie any rational revision scheme, whatever the ordering may be.

The formal account developed so far is standard in belief revision theory since at least 1985, as I will explain in the next section. It is much older, though, and well known from an apparently quite different context, viz. the theory of revealed preferences, developed by Samuelson (1938; 1947, Ch. V). This point has been urged by Rott (2001, Ch. 6–8). There the issue was how to choose from a given set W of options. The answer is straightforward: choose one of the optimal options, whatever your optimality criteria might be. But now the issue was slightly generalized: how to choose from any non-empty subset S of W? These choices must be coherent. The choice function f which selects from any $S \subseteq W$ the set $f(S)$ of optimal options in S must satisfy some rationality constraints. Two constraints were usually accepted:

(4.15)   if $s \in S \subseteq S' \subseteq W$ and $s \in f(S')$, then $s \in f(S)$; and
(4.16)   if $s, s' \in f(S)$ and $S \subseteq S' \subseteq W$, then $s \in f(S')$ iff $s' \in f(S')$.

These are properties (α) and (β) of Sen (1970, p. 17). (4.15) says that an option which is optimal in a larger set, must also be optimal in a smaller set, where it has less rivals. And (4.16) says that two options which are both optimal in a smaller set – and thus on equal footing – must not be treated differently within a larger set, where they may or may not be optimal. These are very plausible conditions, and hence any function f satisfying (4.15–4.16) was called a *choice function* for W.

Why were economists so interested in this notion? Because they could show that all and only such choice functions are governed by preference relations. More precisely, they could show that for each choice function f for W there is exactly one weak order $\geq$ over W – i.e., a transitive and complete relation over W, which may be interpreted as a preference relation over W – such that for any $S \subseteq W$, $f(S)$ is the set of maximal or most preferred options in S according to $\geq$, and moreover that each preference relation over W thus generates a choice function (cf. Sen 1970, pp. 17ff., Lemma 1*m and 1*q). In short, it is precisely choice behavior according to (4.15–4.16) that is rationalizable by preferences.

Why were economists so excited about this formal result? Because it showed how to make preferences behaviorally accessible. Preferences were clearly a basic notion of all microeconomic theory, but at the same time a hypothetical construct, a latent variable, as psychologists said, or a theoretical notion, as philosophers said, which is not directly observable. Choice behavior, however, *is* directly observable, and hence the formal result showed how coherent choice behavior makes the hypothetical construct of preferences observationally definable.

The relation of these familiar findings to our context is clear: it is identity. Selection functions are the same as choice functions. Condition (4.11a) states the basic fact that a choice function picks a non-empty subset from any non-empty set of options, and it is an easy exercise to show that condition (4.11b) is equivalent to (4.15–4.16). Moreover, the kinds of orderings involved are clearly the same. Hence, the equivalence between selection functions and WWOs is the same as that between choice functions and preference relations. This is quite revealing. The original motivation of (4.9–4.10) was not in terms of choices at all, but rather strictly in terms of belief dynamics. That belief revision is a problem of choice – a matter of choosing the most plausible or the least disbelieved possibilities – was suggested by the metaphors accompanying WWOs and has now been established by the observed straightforward parallel between revealed preference and belief revision.

The very same facts are known from conditional logic. There the issue was to specify truth conditions for subjunctives "if φ were the case, ψ would be the case", symbolically: $\varphi \mathbin{\square\!\!\rightarrow} \psi$. Stalnaker (1968) specified the truth conditions in terms of selection functions. (I have borrowed the term from him.) Roughly the idea was that $\varphi \mathbin{\square\!\!\rightarrow} \psi$ is true iff ψ is true in all worlds which are closest to the actual world among all the worlds in which φ is true. This presupposes a selection function that selects from φ-worlds all those closest to actuality. Stalnaker (1968) still assumed his selection functions to select exactly one φ-world closest to actuality. This stricter selection entailed the contested law of conditional excluded middle, as it was called, and was therefore generally given up later on. Lewis (1973a) preferred a semantics in terms of similarity spheres, which constitute nothing but a weak order among possible worlds, and he rigorously established the equivalence of (and subtle difference between) the two approaches (in Lewis, 1973a, pp. 58f.). So this is another area, in fact much closer to our context, where the equivalence of selection functions and WWOs is well known.

4.4 AGM Belief Revision Theory

Finally we should note that the account given so far is nothing but AGM belief revision theory, that received its standard form in Alchourrón, Gärdenfors, Makinson (1985). Since belief revision takes a different guise there, the equivalence is a bit more difficult to see. Let us follow the presentations of Gärdenfors (1988, Ch. 3) and Rott (2001, Ch. 4).

First of all, there is the difference that standard belief revision theory deals with sentences, not with propositions. It starts with a language $\mathbf{L}$ taken as the set of its well-formed sentences. Usually $\mathbf{L}$ is just propositional logic. The language $\mathbf{L}$ is accompanied by a logic as specified in the consequence relation Cn. For any set of sentences $K \subseteq \mathbf{L}$, $Cn(K)$ is the deductive closure of K. $\perp$ is a special sentence, the *falsum* (or contradiction). Next, K is a *belief set* iff $K = Cn(K)$, i.e., if K is deductively closed. This carries our notion of a belief set (Definition 4.5) onto the sentential level, with the small exception that the inconsistent belief set K, for which $\perp \in Cn(K)$, is allowed here as well. $\mathbf{K}$ is the set of all belief sets.

Now the goal of belief revision theory is to characterize revision functions that are functions from $\mathbf{K} \times \mathbf{L}$ into $\mathbf{K}$. A revision function $*$ carries any belief set K and any sentence $\varphi \in \mathbf{L}$ into a belief set $K * \varphi$, which is to be the result of revising K by accepting φ. This diverges from our framework in a significant way. Sentential belief sets may be understood as doxastic states just as well as our propositional belief sets. But our selection function g has quite a different format than an AGM revision function $*$. A selection function g was a scheme for revising a given belief set by any possible information and could hence be understood as a momentary doxastic disposition or state as well. By contrast, a revision function $*$ is a scheme for revising any belief set by any (sentential) information. This is much stronger, and not easily interpreted as a doxastic feature. It seems the only way to take it is as a permanent disposition to revise belief sets, whatever they may be. If this is so, then $*$ is not a momentary state, but a feature that stays fixed throughout one's entire doxastic career. That such a feature exists at all is, however, a substantial assumption that may be justly doubted and that we will reject in the next chapter.

Is there an explanation of how this difference has come about? I would contend that the reason is that AGM belief revision theory originated in philosophical logic and never fully changed sides from logic to epistemology. Logic is governed by somewhat different interests. The first interest is to define logical truth (or logical validity). Usually logical truth is truth in all models or valuations. Ellis (1976, 1979) and Gärdenfors (1978) had the path-breaking idea to define logical truth as membership in all belief sets. The approach is particularly fruitful when logical constants are to be explained in relation to operations on belief sets. For instance, it is most natural and suggestive in the case of various conditionals like "if, then", "even if", etc., which are closely connected to belief revision. The principal idea here is to postulate the so-called *Ramsey test* as providing a semantics for the subjunctive conditional $\Box\!\!\rightarrow$:

(4.17) $\varphi \Box\!\!\rightarrow \psi \in K$ iff $\psi \in K * \varphi$.

It is not so clear, by the way, whether the conditional explained by the Ramsey test is, or is intended to be, the same as the conditional as explained by Stalnaker (1968) and Lewis (1973a); rather, Lewis (1973a, pp. 70f.), when criticizing "the method of thought experiments", may be taken as rejecting the Ramsey test for subjunctives. So one should use different kinds of arrows cautiously. Yet the intricacies of the conditional are not our concern. In fact, we need not further pursue these ideas, because our interest does not lie in logic and semantics. My point is only that it is intelligible on this approach that Ellis and Gärdenfors started referring to *all* belief sets. For instance, if the Ramsey test is to provide a semantics of the subjunctive conditional $\Box\!\!\rightarrow$, * must be defined for all belief sets. And my further point is that this means thinking in logical rather than epistemological terms. (For a fuller discussion of the issue see Rott 2001, Sect. 3.6.)

For the moment, we may take the issue as a mere formality that should not hamper the further comparison of our selection and AGM revision functions. Still I suspect that it is the source of all my later dissatisfaction with standard belief revision theory.

Be this as it may, what does the AGM theory say about revision functions? They are characterized by eight axioms, as specified in

Definition 4.18: * is a *revision function* iff it maps $\mathbf{K} \times \mathbf{L}$ into $\mathbf{K}$ such that for all $K \in \mathbf{K}$ and $\varphi, \psi \in \mathbf{L}$:

(K*1)	$K * \varphi = Cn(K * \varphi)$	[*closure*],
(K*2)	$\varphi \in K * \varphi$	[*success*],
(K*3)	$K * \varphi \subseteq Cn(K \cup \{\varphi\})$	[*expansion 1*],
(K*4)	if $\neg\varphi \notin K$, then $Cn(K \cup \{\varphi\}) \subseteq K * \varphi$	[*expansion 2*],
(K*5)	if $\bot \notin Cn(K)$ and $\bot \notin Cn(\varphi)$, then $\bot \notin K * \varphi$	[*consistency preservation*],
(K*6)	if $Cn(\varphi) = Cn(\psi)$, then $K * \varphi = K * \psi$	[*extensionality*],
(K*7)	$K * (\varphi \wedge \psi) \subseteq Cn((K * \varphi) \cup \{\psi\})$	[*superexpansion*],
(K*8)	if $\neg\psi \notin K * \varphi$, then $Cn((K * \varphi) \cup \{\psi\}) \subseteq K * (\varphi \wedge \psi)$	[*subexpansion*].

These axioms look a bit circumstantial, but it is easily verified that they say no more and no less than our axioms (4.11a–b) for selection functions. One point is that deductive closure is presupposed by our dealing with belief cores only, so *closure* (K*1) is superfluous in our framework. It is mentioned in Definition 4.18 only to reinforce the point that a revision function takes belief sets as values. Similarly, *extensionality* (K*6) is superfluous in our framework dealing directly with propositions instead of sentences. *Success* (K*2) says that the information is accepted in any case – that is part of our axiom (4.11a). The other part of this axiom corresponds to *consistency preservation* (K*5).

Now it is obvious that, if φ is a tautology, then (K*7) and (K*8) specialize to (K*3) and (K*4). This makes the latter two redundant in Definition 4.18 (given extensionality and the fact that revision by a tautology runs empty, which is provable, though, only with (K*3) and (K*4)). The reason why they are always explicitly stated is that there is a vigorous debate about (K*7) and (K*8), as is indicated by the fact that Rott (2001, p. 109f.) discusses two partially equivalent alternatives to (K*7) and eight partially equivalent alternatives to (K*8). Therefore (K*3) and (K*4) are separately stated, whatever the fate of the stronger conditions.

Still, the comparison continues neatly. *Expansion 1* (K*3) is nothing but our preliminary postulate (4.8) and *expansion 2* (K*4) is the same as our postulate (4.7). (Note that set inclusion reverses by passing from belief sets to their cores.) Moreover, *superexpansion* and *subexpansion*, (K*7) and (K*8), exactly correspond to our axiom (4.11b). Turning our postulates (4.7–4.8) into our axiom (4.11b) is hence the same as strengthening (K*3) and (K*4) to (K*7) and (K*8).

All this shows that our considerations leading us to the definition of selection functions were, to wit, nothing but a brief introduction into standard belief revision theory. I have indicated that there is a lot of discussion about the axioms (K*1)–(K*8) within belief revision theory; hardly any axiom gets off uncriticized. But I will not take issue with these axioms. On the contrary, I have already explained why I endorse them. My point of criticism will be, as we will soon see, a different and conclusive one. Hence, I am obliged to comment on the discussion about (K*1)–(K*8) from my point of view. However, the commentary has to wait until Section 11.3, after the development of my positive theory.

For the moment, we should finish our comparison. It is most important to observe that revision functions constitute only one half of belief revision theory. The other half is given by so-called contraction functions, which are equally prominent. Belief contractions are assumed to be another basic type of belief change. They occur when you have somehow lost faith into one of your beliefs without replacing it by a new one. The problem is that you cannot simply eliminate one belief, for due to deductive closure the belief would have to return immediately. If one belief has to go, many others have to go as well in order to arrive at a belief set again. The further problem is that there is no unique elimination procedure. You have to make a lot of choices: Whenever you believe φ, you believe φ ∨ ψ and φ ∨ ¬ψ as well. So, if you give up φ, you have to give up φ ∨ ψ or φ ∨ ¬ψ or both. This holds for any ψ! Therefore, the more modest aim is only to axiomatically characterize contraction by rationality postulates. The standard postulates are specified in:

Definition 4.19: ÷ is a *contraction function* iff it maps $\mathbf{K} \times \mathbf{L}$ into $\mathbf{K}$ such that for all $K \in \mathbf{K}$ and $\varphi, \psi \in \mathbf{L}$:

(K÷1)	$K \div \varphi = Cn(K \div \varphi)$	[*closure*],
(K÷2)	$K \div \varphi \subseteq K$	[*inclusion*],
(K÷3)	if $\varphi \notin K$, then $K \div \varphi = K$	[*vacuity*],
(K÷4)	$\varphi \notin K \div \varphi$, unless $\varphi \in Cn(\varnothing)$	[*success*],
(K÷5)	$K \subseteq Cn((K \div \varphi) \cup \{\varphi\})$	[*recovery*],
(K÷6)	if $Cn(\varphi) = Cn(\psi)$, then $K \div \varphi = K \div \psi$	[*extensionality*],
(K÷7)	$(K \div \varphi) \cap (K \div \psi) \subseteq K \div (\varphi \wedge \psi)$	[*intersection*],
(K÷8)	if $\varphi \notin K \div (\varphi \wedge \psi)$, then $K \div (\varphi \wedge \psi) \subseteq K \div \varphi$	[*conjunction*].

(K÷1)–(K÷6) are very plausible and need no further explanation, though *recovery* (K÷5) has come under strong attack. (K÷7) and (K÷8), *intersection* and *conjunction*, are prima facie imperspicuous. Rott (2001, pp. 103f.) mentions five partially equivalent alternatives to (K÷7) and nine partially equivalent alternatives to (K÷8). The clearest way to summarize the content of (K÷7) and (K÷8) is by observing that, in the presence of the other conditions, they are jointly equivalent to the following "factoring" condition (cf. Alchourrón et al. 1985, observation 6.5):

(4.20) $K \div (\varphi \wedge \psi)$ is equal either to $K \div \varphi$ or to $K \div \psi$ or to $(K \div \varphi) \cap (K \div \psi)$.

There is no need to look at the ramified discussion. It suffices to have stated the standard axioms for future comparison.

However, it will be useful to introduce the propositional counterpart to (K÷1)–(K÷8). If (K∗1)–(K∗8) melt down to selection functions in the transition from the sentential to the propositional framework, a similar simplification might be forthcoming here as well. This is indeed the case. Here is the pertinent

Definition 4.21: Let $\mathcal{A}$ be an algebra over W. Then c is a *(propositional) contraction function* for $\mathcal{A}$ iff c is a function from $\mathcal{A} - \{W\}$ into $\mathcal{A}$ such that for all $A, B \in \mathcal{A} - \{W\}$:

(a) $\varnothing \neq c(\varnothing) \subseteq c(A) \subseteq c(\varnothing) \cup \bar{A}$ and $c(A) \not\subseteq A$ ("$\not\subseteq$" means "is not a subset of"),

(b) if $c(A \cap B) \not\subseteq A$, then $c(A) \subseteq c(A \cap B) \subseteq c(A) \cup c(B)$.

It is quite obvious how this corresponds to (K÷1)–(K÷8). As in the case of revisions, (K÷1) and (K÷6) are superfluous within the propositional framework. W cannot be contracted, as per (K÷4). $c(\varnothing)$ is the core of the belief set from which contraction starts (since contraction by $\varnothing$ runs empty in any case). Thus, clause (a) first restricts contraction to consistent belief sets. Its second claim amounts to *inclusion* (K÷2). Its third claim is equivalent to *recovery* (K÷5). Finally, its last claim is *success* (K÷4). Clause (b) is equivalent to *intersection* and *conjunction*, (K÷7) and (K÷8). It looks weaker since it restricts (K÷7) to the condition of (K÷8). However, if $c(A \cap B) \subseteq A$, then $c(A \cap B) \not\subseteq B$ must obtain because of *success*, and in this case (b) again claims $c(A \cap B) \subseteq c(A) \cup c(B)$. In fact, we must have $c(A \cap B) = c(B)$ in this case, for this follows from $c(A \cap B) \subseteq A$ and the fact that (a) implies $c(A) \subseteq c(\varnothing) \cup \bar{A}$ and $c(B) \subseteq c(\varnothing) \cup \bar{B}$. This, by the way, also explains how (4.20) comes about. Finally, *vacuity* (K÷3) is a consequence of clause (b) for $B = \varnothing$.

The most important reason for belief revision theorists to consider both revision and contraction was that they appeared to be interdefinable. Levi has indeed forcefully argued (first in 1977 and then continuously, most extensively perhaps in 1991) that belief revision should always be considered as a two-step procedure. If the evidence contravenes the prior beliefs, these have first to be appropriately contracted. In a second step, then, the contracted belief set should simply be expanded by the evidence. Thus he proposed to define:

(4.22) $K * \varphi = Cn((K \div \neg\varphi) \cup \{\varphi\})$ [*Levi identity*].

He then developed detailed accounts of contraction and expansion, which he kept strictly separate because in his view they form two entirely different doxastic activities. I will more thoroughly discuss his views in Section 11.2.

Conversely one may also reduce contraction to revision. This way is suggested by Harper (1977):

(4.23) $K \div \varphi = K \cap (K * \neg\varphi)$ [*Harper identity*].

The idea here is that to contract by φ is to imagine a revision by $\neg\varphi$ and then to check which of the prior beliefs would be maintained in that imagined revision. These are precisely the beliefs that survive contraction by φ.

Both identities are widely accepted, though they are not without criticism. The crucial observation, well known (cf., e.g., Gärdenfors 1988, Sect. 3.6) and thus cited here without proof, is

Theorem 4.24: If * is a revision function and if ÷ is defined by the Harper identity (4.23), then ÷ is a contraction function, namely that which would define * according to (4.22). Conversely, if ÷ is a contraction function and if * is defined by the Levi identity (4.22), then * is a revision function, namely that which would define ÷ according to (4.23).

The propositional counterpart is:

Theorem 4.25: If g is a selection function for $\mathcal{A}$, then c defined by $c(A) = g(W) \cup g(\bar{A})$ is a propositional contraction function for $\mathcal{A}$, namely that which would define g by the equation $g(A) = c(\bar{A}) \cap A$. And if c is a propositional contraction function for $\mathcal{A}$, then g thus defined is a selection function for $\mathcal{A}$, namely that which would define c according to the first equation.

This establishes a strong connection. If one finds fault with the postulates governing revision and sticks to (4.22–4.23), one has to find fault with the postulates for contraction as well, and vice versa. Or positively put: via the identities (4.22–4.23) there is strong mutual support among the revision and the contraction principles.

From the dynamic point of view developed thus far we must grant that belief contraction is not easily grasped. From this point of view we always need some force
affecting our doxastic state in order to change it. And so far the only force we know
or have considered is evidence. Some such force is required for contraction as well. We
do not contract out of the blue, but because something or somebody has raised our
doubt. Thus, it seems we must have also acquired some evidence in such cases. Perhaps
that evidence should not be understood as propositional evidence, as getting informed
about some proposition being true. But again, the only understanding of evidence we
have so far is propositional evidence.

The point of these remarks is not to criticize contraction. The point is to note
a tension. Contraction has found a lot of plausible interest, but I have no room for it
so far. This is a defect on my part, and so we must attend to this issue in our future
theorizing. Fortunately, everything will turn out well. Contraction will find a perfect
account in ranking theory. We will reproduce the postulates (K÷1)–(K÷8), or rather
Definition 4.21, and we will even be able to solve an unsolved problem in contraction
theory in Section 5.7. Indeed contraction will play a most crucial role for ranking
theory that will be thoroughly explained in Sections 8.3 and 8.4.

The final point to observe in this still introductory chapter is that belief revision
theorists have certainly been aware of the close relation of revision and contraction to
some kind of doxastic preferences. That there are choices to be made was particularly
obvious in the case of contraction. Here one had to hold on to the more preferred,
more central, or – according to the established usage – the more deeply entrenched
beliefs and give up the less preferred, less central, or less deeply entrenched ones.
Rationally, contractions must reflect some such entrenchment order. The same then
holds for revisions as well, because they also entail giving up previous beliefs.

As I have presented it, the ordering of disbelief was an ordering of possibilities, and
its relation to selection or revision was straightforward. The same was true of choice
and its relation to preference among options. In belief revision theory, matters are
obscured by the fact that there are orderings on three different levels. There is an
entrenchment order among sentences. This was the first and most natural to be considered in relation to contraction. Alchourrón et al. (1985), however, considered also a
preference relation between belief sets. In contracting K by φ one should keep all those
beliefs which are members of all of the most preferred maximal belief sets which are
subsets of K not containing φ. And there is yet a third ordering on the level of models
or possible worlds or possibilities which may serve as a semantic foundation of the
other orderings. This is the one we have used. It would go much too far to present all
this in detail. I refer the interested reader to Rott (2001, Chs. 7–8).

There, the relations between the various types of ordering and their connection to
various types of choice or selection or contraction are fully investigated. This is an
instructive, but complex inquiry. Because of the complexity, it then appears as a rediscovery that the venerable theory of choice and revealed preferences is underlying all
these epistemological developments. To a good extent, though, the complexity and the

rediscovery are home-made. The basic facts may certainly be summarized in the suc-cinct way presented above. One reason for the complexity is that belief revision theory is always carried out on the syntactic or sentential level. Things become much more perspicuous, though, when proceeding purely on the propositional or semantic level. In any case this is my preference, that I will continue to follow.

5

Ranking Functions

So far I have been presenting what may be considered to be more or less the standard representation of belief and its dynamics. I have indicated that there is a lot of discussion and a lot of superficial and substantial variation, and I know that I need to comment on this discussion. But first I must develop my own stance, for my comments on other positions can only flow from there. It is thus time to cross the threshold from the preparatory chapters to the pursuit of the constructive theory announced in the introduction. This will take place over the next four chapters.

5.1 The Problem of Iterated Belief Change

The first step is to fully understand why the account of the dynamics of belief given so far is deeply deficient and why there is no easy repair. A thorough analysis of the problem will then naturally lead us to a solution and to a satisfying dynamics of belief. However, this will not be a mere variation, but a clear extension of the present account in several crucial aspects.

So what is so bad about the dynamic account given so far? The problem lies in the law of simple conditionalization (4.12), for an entirely trivial reason: In this law, the prior state is characterized as a selection function and the posterior state as a belief set. This will not do. We saw that there is no complete dynamics if doxastic states are represented as belief sets. Our way out was to represent doxastic states as selection functions, but then we have to state a dynamic law for selection functions and not such a hybrid as (4.12). Gärdenfors and Rott have coined a name for this platitude:

(5.1)   *The Principle of Categorical Matching*: "The representation of a belief state after a belief change has taken place should be of the same format as the representation of the belief state before the change" (Gärdenfors, Rott, 1995, p. 37).

The reason is that the dynamics remains incomplete when this principle is violated. The law of simple conditionalization (4.12) might offer a complete account of the first change. But after the first step, it totters since the result of the first step is characterized as a belief set and we have no law taking this belief set as a start. In other words, (4.12) is unable to give an account of *iterated belief change*. Yet this is what any acceptable account of belief dynamics must give us.

It is important to note that the very same problem besets the probabilistic law of simple conditionalization (3.15) with respect to Popper measures. There the prior state was characterized as a Popper measure, but the posterior state was characterized only as a probability measure. And so we are stuck again with the first step and have no account of iterated probabilistic change. In this central aspect Popper measures are definitely unsatisfactory. The point has been still insufficiently grasped in the growing epistemological literature on Popper measures. We must either return to Jeffrey's radical probabilism and his generalized conditionalization or somehow improve upon Popper measures. I will do the latter in Section 10.2.

This problem with Popper measures was first noted by Harper (1976), but his solution was complicated and not well received. I urged the problem with respect to belief revision in Spohn (1983b, Sect. 5.3, and 1988, Sect. 3), and I will explain in Section 5.6 why I think that, in a way, the problem of iterated belief change has still not received proper attention in the belief revision literature. First, however, we must analyze much more closely what the problem really is.

Once again, the problem is to say which posterior selection function g' results from the prior selection function g and the evidence E. Thus put, the problem is not particularly perspicuous. It becomes much more vivid in terms of well-orderings: Suppose that your prior doxastic state is characterized by a WWO $\leq$ of disbelief or the corresponding WOP $\langle D_0, D_1, \ldots, D_n \rangle$ (which we may assume to be finite for expository purposes) and that you receive evidence E. What can we say then about your posterior WWO $\leq_E$? Let me repeat here my reasoning from Spohn (1983b, Sect. 5.3, and 1988, Sect. 3):

A first proposal might be that the evidence E has the effect of making the possibilities in E less disbelieved than those outside E, but does not touch the order in any further way. This already defines the posterior WWO $\leq_E$:

(5.2) $w \leq_E w'$ iff (a) $w \in E$ and $w' \in \bar{E}$ or (b) either $w, w' \in E$ or $w, w' \in \bar{E}$ and $w \leq w'$. The corresponding WOP is $\langle D_0 \cap E, \ldots, D_n \cap E, D_0 - E, \ldots, D_n - E \rangle$, *with all empty terms eliminated.*

The clause about empty terms is necessary: E need not overlap with all levels of disbelief. So some of the terms of that sequence can be empty and must be removed in order to turn this sequence into a WOP. This proposal looks attractive, but I have two formal objections and an intuitive one:

First, according to the proposal (5.2) belief changes almost never commute. This is most graphic in terms of WOPs. Receiving first evidence A and then evidence B results in the sequence

$$\langle D_i \cap A \cap B \ (i = 0, \ldots, n), D_i \cap \bar{A} \cap B \ (i = 0, \ldots, n),$$
$$D_i \cap A \cap \bar{B} \ (i = 0, \ldots, n), D_i \cap \bar{A} \cap \bar{B} \ (i = 0, \ldots, n) \rangle,$$

whereas receiving first evidence B and then evidence A yields

$$\langle D_i \cap A \cap B\ (i = 0, \ldots, n), D_i \cap A \cap \bar{B}\ (i = 0, \ldots, n),$$
$$D_i \cap \bar{A} \cap B\ (i = 0, \ldots, n), D_i \cap \bar{A} \cap \bar{B}\ (i = 0, \ldots, n) \rangle.$$

Hence, we end up with different WOPs iff both $A \cap \bar{B}$ and $\bar{A} \cap B$ are non-empty, i.e. iff neither A implies B nor B implies A. However, this appears to be far too little commutativity. Sometimes the order of evidence might play a role. Two pieces of evidence might somehow conflict, and then the later piece might supersede the earlier one. Yet the normal case is surely that logically independent pieces of evidence simply accumulate, and in that case one should expect the temporal order to be irrelevant. But this is exactly what is denied by the proposal (5.2).

One might defend (5.2) by pointing out that the belief set resulting from the two changes is always the same irrespective of their order, namely, the first non-empty $D_i \cap A \cap B$ (provided A and B are logically compatible). But this defense is insufficient. Our expectation is stronger and requires even sameness of the further disposition to change beliefs (at least in normal cases).

The second objection is that the proposal (5.2) does not allow any reversals of belief change. If we look at the above transition from the prior to the posterior WOP, then we obviously do not find any transition of the *same kind* that returns to the prior WOP. This would correspond to accepting "Sorry, the alleged evidence E was simply misinformation and is to be ignored". This objection might seem unfair insofar as it is not specifically directed against the present proposal, but instead against the whole approach of describing belief change through evidence. It is roughly the same point as the one noted above: that there is thus far no place for contractions in our dynamic picture (contractions being a kind of reversal of revisions). However, now when we want to do things right, we may not dismiss this objection as being too general.

So we have to ask ourselves: is general reversibility of belief change really wanted? The same point seems to apply as with regard to contraction: it is always some kind of new input that drives belief change, there is no mere cancellation of old input. Of course there might also be internal forces changing beliefs, e.g. forgetting or drugs. But such kinds of belief change not subsumable under categories of rationality are clearly irrelevant to the present point. So I can well agree that there is no such mere cancellation. However, I would like to refer to my emphasis in Section 2.2 that it is wise to avoid presupposing an all-purpose space of possibilities, but rather to choose the underlying space of possibilities according to the purposes at hand. So, even if the input were always propositional evidence relative to some very rich or even the all-purpose space of possibilities, it need not be so describable relative to the possibility space or the propositional algebra considered. We should still be able to describe the belief change affected by this input.

Let me expand the example above. First, I tell you: "E is the case." You change your beliefs accordingly. The evidential proposition need not be "Wolfgang says: 'E is the case'." Perhaps propositions about what I say, whether I am trustworthy and so on, are not represented within the possibility space or the propositional algebra at hand. But suppose E is, and hence the effect of my assertion just is that you revise your beliefs by E. Next I tell you: "Sorry, I have made a mistake. Please, simply ignore what I have told you before." Relative to a rich propositional algebra, the second evidential proposition is "Wolfgang says: 'Sorry, . . .'." But relative to the coarse algebra considered we do not find a proposition on which to conditionalize. On the coarse level the best we can say is that my second utterance undoes the first belief revision. And so our doxastic theory should provide means for describing such reversals and hence contractions as well.

(One might say that you simply recall what you believed before and thereby know how to return to your previous beliefs. However, this means invoking present beliefs about your past beliefs and hence again enriching the propositional algebra. I shall discuss such auto-epistemic expansions of the conceptual framework in Sections 9.2–9.3.)

In fact I believe there is a deep and powerful principle behind my argument. In Spohn (2008a, p. 357) I called it

(5.3)   *The Invariance Principle*: The propositional attitudes, their contents, and their static and dynamic laws must be so conceived as to be invariant under coarse- and fine-graining of the underlying conceptual and propositional framework.

It is clear how the principle applies to the present case: At a certain level of coarseness some belief changes can only be described as reversals, and since the principle requires our theory to work also on that level, we must be able to describe reversals. The principle also resolves the dispute between Jeffrey (1965/83) and Levi (1967b) mentioned in Section 3.2, in favor of Jeffrey: It may well be that a simple conditionalization (3.8) with respect to a sufficiently fine-grained propositional algebra is representable on a coarser level only by a generalized conditionalization (3.10). However, what can be described by a generalized conditionalization on a fine-grained level is still representable in this way on a coarser level. Hence, (3.10) satisfies the invariance principle (5.3), but (3.8) does not. In Spohn (2008a, Section 16.5) I mention four other applications in decision theory and epistemology, among them the argument in favor of the intentional conception of belief contents as opposed to the propositional conception, which I mentioned in the last remark of Section 2.3. Through these applications, which I take to be instructive and successful, I provide a kind of inductive argument for the invariance principle. By all means, the theory to be developed here will satisfy it also. But we need not deepen this discussion now; for the time being I have said enough about the desirability of a reversible dynamics of belief.

Let us turn, then, to the third intuitive objection against the proposal (5.2) for a dynamic law for WWOs or WOPs. It is simply that, by reshuffling the WWO or WOP upon evidence E in the way indicated, even the most far-fetched possibility within E gets less disbelieved than the most plausible possibility outside E. This is not precisely to return to the idea that evidence comes with absolute certainty, but it comes close to that. Maybe evidence sometimes is of that type, but normally it seems crazy to accept any kind of preposterous possibility rather than to give up the evidence again. This is particularly so in view of my remarks on reversibility – i.e., in view of the fact that the evidence E may be taken from some coarser possibility space. If E is, as in my above example, what I claim and not that I claim something, then this nearly maximal certainty for E would be inappropriate. In any case, (5.2) therefore appears unsuitable as a *general* model of belief change.

This criticism suggests another construction of the posterior WWO $\leq_E$ from the prior WWO $\leq$ or WOP $\langle D_0, \ldots, D_n \rangle$. Perhaps we should shift only the top possibilities within the evidence E to the top of the whole ordering. More explicitly, the proposal is this:

(5.4) $w \leq_E w'$ iff (a) $w \in \min_{\leq}(E)$ or (b) $w, w' \notin \min_{\leq}(E)$ and $w \leq w'$. If $\min_{\leq}(E) \subseteq D_k$, the corresponding WOP is $\langle D_k \cap E, D_0, \ldots, D_{k-1}, D_k - E, D_{k+1}, \ldots, D_n \rangle$, again with empty terms eliminated ($D_k - E$ might be empty).

My criticism of this second proposal is predictable:

First, we have not made any progress on commutativity. Again, two successive operations of the envisaged type commute only in trivial and never in substantial cases. But we would expect at least some substantial commutativity. Second, we have not made any progress on reversibility. Again, there is no change of that type undoing a previous change of that type. Third, we have now fallen into the other extreme of treating evidence as utterly weak. As soon as the slightest detail of the most plausible possibilities within E, i.e., of $\min_{\leq}(E)$, turns out wrong, (5.4) tells us to give up on E as well and to put our trust in possibilities excluding E. Sometimes this timidity might be appropriate, but it does not seem suitable as a general model.

5.2 Ranking Functions

What to do? From a constructive point of view, two lessons suggest themselves. First, if neither of the extremes (5.2) and (5.4) in handling the evidence will do, then no fixed middle course will do either. Evidence can be more or less firm, and the dynamic law must be flexible enough to account for this, as Jeffrey's generalized conditionalization in the probabilistic case did. Second, if we really want to define a reversible dynamics, then we must not eliminate the empty terms in the above proposals. This elimination was required for arriving at a posterior WOP, but it destroyed reversibility. For it is impossible, once the empty terms are eliminated, to reconstruct where their positions were. (In order to bring home this point I have couched my explanations mainly in terms of WOPs. In terms of WWOs the point would have been much less clear. That the reshuffling of a WWO creates empty positions, as it were, is not transparent and indeed formal nonsense.)

So we should keep the empty terms. This is all I want to propose, but we will see that it makes all the difference in the world. It will take several chapters to develop all this. Let me start with the basic notion, which I would like to explain a bit more extensively than I could in my previous papers on the subject matter.

We may keep empty terms by giving up on the equivalence between WWOs and WOPs and generalizing WOPs so as to contain empty members. But this would result in a somewhat clumsy notion. It is more elegant to define a function that maps each possibility to the position of the partition member containing it. This is done in

Definition 5.5: Let $\mathcal{A}$ be a complete algebra over W. Then κ is an $\mathcal{A}$-*measurable completely minimitive natural negative ranking function* iff κ is a function from W into $\mathbf{N}^+ = \mathbf{N} \cup \{\infty\}$ such that $\kappa^{-1}(0) \neq \varnothing$ and $\kappa^{-1}(n) \in \mathcal{A}$ for each $n \in \mathbf{N}^+$. κ is extended to propositions by defining $\kappa(\varnothing) = \infty$ and $\kappa(A) = \min \{\kappa(w) \mid w \in A\}$ for each non-empty $A \in \mathcal{A}$; $\kappa(A)$ is called *the negative rank* of A.

In Spohn (1988) I called this a natural conditional function. Later on I called it a ranking function, following a suggestion of Goldszmidt, Pearl (1992a,b). The reason for my present more sophisticated terminology will become clear in a moment. The condition of $\mathcal{A}$-measurability is introduced for the sake of formal correctness and for the most part will be left implicit.

We have lost a bit of generality in Definition 5.5 by restricting the range of a ranking function to the extended natural members in $\mathbf{N}^+$, whereas WOPs were allowed to arbitrarily climb up the ordinal hierarchy. I will discuss the restriction below. Mainly, however, we have achieved the desired generalization. $\kappa^{-1}(n)$ corresponds to the n-th member of a WOP, but $\kappa^{-1}(n)$ might be empty, whereas members of a WOP might be so only under the generalized understanding.

This has, however, a most important consequence: *numbers matter!* Two negative ranking functions might differ only in giving the empty terms different positions and might thus represent the same WWO, the same ordering of disbelief. Still, they have to represent different doxastic states if our notion is to make epistemological sense. I will have to explain the difference and will do so in due course. In any case what this means is that a negative ranking function is a *grading of disbelief*, and negative ranks are grades of *dis*belief. This is why I call them *negative*, even though they are non-negative integers.

As things developed, these grades appeared to be the main obstacle to a wider acceptance of such ranking functions. Numbers matter, but somehow the numbers seemed to fall from the sky and their epistemological meaning appeared obscure. For many years I thought that the introduction of grades of disbelief is sufficiently motivated by the foregoing argument and by the failure of all attempts to state a full dynamics of belief in terms of WWOs of disbelief. But sufficiency is relative to the critical standards applied. So I grant that the present introduction of these grades might not be entirely cogent. Matters will further improve in this regard in Chapter 8, when various ways of measuring grades of disbelief on a ratio scale will be presented.

Understanding a negative ranking function κ as a grading of disbelief is intuitively most helpful. Thus, $\kappa(A) = 0$ means that A is not disbelieved at all, $\kappa(A) = 1$ means that A is disbelieved to the least degree, $\kappa(A) = 2$ means that A is disbelieved to the second least degree; and so on. Hence, A is *believed* iff $\bar{A}$ is disbelieved to some positive degree, i.e.:

(5.6) $\mathbf{B}(A)$ iff $\kappa(\bar{A}) > 0$.

In the first place, negative ranking functions are a theory about disbelief, and belief is represented only in this clumsy doubly negative way. (Shackle (1949, 1961) is the only predecessor who also proceeded in such negative terms with his functions of potential surprise.)

This definition of belief is equivalent to saying that $\mathbf{B}(A)$ iff $\kappa^{-1}(0) \subseteq A$. Thus, $\kappa^{-1}(0)$ is the *core* of the doxastic state κ determining everything that is believed in that state, and deductive closure of the set of beliefs is automatically preserved. This finally explains the condition in Definition 5.5 that $\kappa^{-1}(0) \neq \varnothing$. The core must be consistent, as was argued in Section 4.1. It might well happen that both $\kappa(A) = 0$ and $\kappa(\bar{A}) = 0$; in that case, A is *neutral*, i.e., the subject is unopinionated or has no judgment concerning A.

Example 5.7, Tweety: A quick example might be helpful here. Look at Tweety, an entity which has acquired some fame in the nonmonotonic reasoning literature. Tweety has, or fails to have, each of the three properties: being a bird (B), being a penguin (P), and being able to fly (F). This makes for eight possibilities. Suppose you have no idea who or what Tweety is (for all you know, it might even be a car). Then your negative ranking function might be the following one (I am choosing the ranks in an arbitrary, though intuitively plausible way, just as I would have to arbitrarily choose plausible subjective probabilities if the example were a probabilistic one):

κ	$B \cap \bar{P}$	$B \cap P$	$\bar{B} \cap \bar{P}$	$\bar{B} \cap P$
F	0	4	0	11
$\bar{F}$	2	1	0	8

In this case, the strongest proposition you believe is that Tweety is *either* not a penguin and not a bird ($\bar{B} \cap \bar{P}$) *or* a flying bird and not a penguin ($F \cap B \cap \bar{P}$) – all other possibilities are disbelieved. So you neither believe that Tweety is a bird nor that it is not a bird. You are also neutral concerning its ability to fly. But you believe, for instance: (i) if Tweety is a bird, then it is not a penguin and can fly ($\bar{B} \cup (\bar{P} \cap F)$), and (ii) if Tweety is not a bird, then it is not a penguin ($B \cup \bar{P}$) (each conditional taken as material implication). Likewise you also believe: (iii) if Tweety is a penguin, it can fly ($\bar{P} \cup F$), and (iv) if Tweety is a penguin, it cannot fly ($\bar{P} \cup \bar{F}$). But you do so only because you believe that it is not a penguin in the first place. (Some like to call this a paradox of material implication.) If we understand the conditional differently, as we will later on, the picture will change. The large ranks in the last column indicate that you strongly disbelieve that penguins are not birds. And so we might discover even more features of this example. This may suffice for a first intuitive grasp of negative ranking functions.

Two (or three) laws which are too simple to require proof follow immediately:

Theorem 5.8: If κ is an $\mathcal{A}$-measurable complete negative ranking function, then we have for all propositions $A, B \in \mathcal{A}$ and all subsets $\mathcal{B}$ of $\mathcal{A}$:

(a)   either $\kappa(A) = 0$ or $\kappa(\bar{A}) = 0$ or both   [*the law of negation*],

(b) $\kappa(A \cup B) = \min \{\kappa(A), \kappa(B)\}$ [*finite minimitivity* or *the law of disjunction*],

(c) $\kappa(\cup\mathcal{B}) = \min \{\kappa(B) \mid B \in \mathcal{B}\}$

[*complete minimitivity* or *the law of infinite disjunction*].

(If necessary, the laws will be qualified as holding for negative ranking functions.)

I borrow the term "minimitivity" from Huber (2006). My term "law of disjunction" denotes the same property. The rationale behind (a) is simply consistency: A and $\bar{A}$ cannot both be disbelieved (just as $\bar{A}$ and A cannot both be believed). We will soon see that the rationale behind (b) and (c) is simply conditional consistency. But from Section 4.3 it is already clear that at least (b) is implicit in selection functions or the equivalent WOPs and the idea of an ordering of disbelief embodied in them.

Having arrived at Definition 5.5 as one, if not the only natural response to the problem raised in Section 5.1, it is quite obvious that there are a number of formal variants that should be considered. The variations unfold along different, though inter-related dimensions. The first concerns *point* versus *set functions*. In (5.5) we started with a point function – ranks were first defined for possibilities in W – and then derived the set function, i.e., the ranks for propositions in $\mathcal{A}$. However, we might as well directly start with the latter. The second dimension concerns the underlying *algebra* of proposi-tions. (5.5) assumed it to be complete, but this might be weakened. This corresponds to third dimension, the kind of *minimitivity* assumed. (5.5) entailed (5.8c) (complete minimitivity), but we might be content with the weaker finite minimitivity (5.8b). Finally, we might vary the range of ranking functions. Why not take real numbers as ranks instead of natural numbers as in (5.5), or ordinal numbers as previously suggested?

To summarize the various possibilities before discussing how they might reasonably combine:

Definition 5.9: Let $\mathcal{A}$ be either (a) an algebra, or (b) a σ-algebra, or (c) a complete algebra of propositions. Then κ is an $\mathcal{A}$-*measurable negative ranking function* iff κ is a function from $\mathcal{A}$ into either (d) $\mathbf{N}^+ = \mathbf{N} \cup \{\infty\}$, or (e) $\mathbf{R}^+ = \mathbf{R} \cup \{\infty\}$, or (f) Ω (the class of ordinals) such that

(g) $\kappa(W) = 0$,

(h) $\kappa(\varnothing) = \infty$ in cases (d) and (e) or $\kappa(\varnothing) = \Omega$ in case (f), and either

(i) $\kappa(A \cup B) = \min \{\kappa(A), \kappa(B)\}$ for all $A, B \in \mathcal{A}$, or

(j) $\kappa(\cup\mathcal{B}) = \min \{\kappa(B) \mid B \in \mathcal{B}\}$ for all countable $\mathcal{B} \subseteq \mathcal{A}$, with $\cup\mathcal{B} \in \mathcal{A}$, or

(k) $\kappa(\cup\mathcal{B}) = \min \{\kappa(B) \mid B \in \mathcal{B}\}$ for all $\mathcal{B} \subseteq \mathcal{A}$ with $\cup\mathcal{B} \in \mathcal{A}$.

κ is called *natural* in case (d), *real* in case (e), *ordinal* in case (f), (*finitely*) *minimitive* in case (i), σ-*minimitive* in case (j), and *completely minimitive* in case (k).

My original notion defined in Spohn (1983b, 1988) combined (c), (f), and (k); this is why I called them ordinal conditional functions. My reason was that this is the only notion that truly generalizes selection functions or WOPs. But it is hard to find applications where this generality is really needed. In Spohn (1988, footnote 16) I indicated a case where such an extended range might be required, namely in order to account for "the stubbornness with which some beliefs are held in the face of seemingly arbitrarily augmentable counter-evidence". Yet I have never elaborated on this remark and do not know whether it can be done. So in the absence of good applications we are better to stay content with the simpler versions (d) or (e), for there would be a price to pay for the generality. We will soon start calculating with ranks, and then we would need to use ordinal arithmetic, which is less well-behaved than ordinary arithmetic. For instance, ordinal addition and multiplication are not commutative. For this reason I will not use ordinal ranking functions in this book.

What else should be our choice? We need not exclude any combination. Clearly, though, condition (a) is naturally combined with (i), (b) with (j), and (c) with (k). Still there might be a point in considering other combinations of algebraic and minimitivity properties. For instance, Huber (2006, p. 464) defines pre-rankings combining (a) and (j) in order to inquire, with partially positive results, as to the extendibility of pre-rankings to σ-minimitive ranking functions on the induced σ-algebra, in analogy to Carathéodory's extension theorem in the probabilistic case. Here, however, I will not further attend to such exotic combinations.

The most important observation for us is that all these distinctions collapse when the algebra $\mathcal{A}$ is finite, and a fortiori when the space W of possibilities is finite. This will be so in many applications in this book, and in all these cases we are absolved from taking a stance.

Nonetheless we should discuss whether (i), finite, (j), σ-, or (k), complete minimitivity should be accepted. Let us look at a violation of complete minimitivity: suppose $\kappa (\bigcup \mathcal{B}) = 0$ for an infinite set $\mathcal{B}$ of propositions, even though $\kappa(B) > 0$ for each $B \in \mathcal{B}$. According to the intended interpretation this means that each $B \in \mathcal{B}$ is disbelieved, even though $\bigcup \mathcal{B}$ is not disbelieved. But this is inconsistent. If you exclude the actual possibility from all $B \in \mathcal{B}$, then you cannot say that it might still be in $\bigcup \mathcal{B}$. This is the same discussion as the one about the closure of belief sets under infinite conjunctions. And since we are still operating on the level of propositions, not sentences, the result is the same.

The first time this or a similar issue was discussed is in Lewis (1973a, Sect. 1.4) where he introduced the so-called Limit Assumption. I am speaking here in favor of this assumption, whereas he finds reason against it. In his context the issue is not really important because it does not affect the logic of counterfactuals. Logic is ordinarily about reasoning from a finite number of premises. But from a semantic or propositional point of view, I do not understand his reason. He requests us to counterfactually suppose that a certain line, which is actually shorter than an inch, were longer than an inch. Then he asks how long it would or might be. He argues in effect that for each

$\varepsilon > 0$ we should accept it as true that "if the line were more than an inch long, it would not be more than $1 + \varepsilon$ inches long". This strikes me as blatantly inconsistent even if we cannot prove this within a finitary logic. Therefore, I am strongly inclined to accept the Limit Assumption and the law of infinite disjunction or complete minimitivity.

We have not yet discussed the choice between (d) and (e), natural and real negative ranking functions. This issue is clearly connected with the previous one. If we assume σ- or complete minimitivity, then at least each countable set of ranks must have a minimum. (And recall my above argument with Lewis: it would not do to replace the minimum by the infimum.) In other words, the range of negative ranking functions would have to be well-ordered. Of course there are well-ordered sets of reals. Still, under that assumption it would be most natural to consider natural negative ranking functions – or ordinal ones (if we had not dismissed them for other reasons).

In probability theory, σ-additivity is essential for reasons of continuity, whereas complete additivity would be fatal. In ranking theory, this motive fails, for accepting σ-minimitivity already entails discontinuity or well-orderedness. And so I do not see any ground for discriminating the σ-versions (b) and (j) and the complete versions (c) and (k). Since we have found good reasons for the complete versions, we may finally do away with the σ-versions.

This leaves us with two main versions: the completely minimitive natural version, as already defined in (5.5), and the finitely minimitive real version. In order to reduce adjectives I will usually speak simply of *natural* and *real* (*negative*) *ranking functions*. It might be strategically wiser to start with the weaker notion of real ranking functions, which is beautifully developed in Hild (2001). However, as I am attached to the law of infinite disjunction, I have been accustomed to natural ranking functions for more than twenty years and will mostly work with this type of functions in this book. The only noticeable exception will be the measurement theory in Chapter 8. In fact, I might be somewhat sloppy in my usage. Since it often does not make much of a difference which version of a ranking function we are using, I will often leave it open – one need not be principled about this issue. Whenever necessary, I will be precise, and whenever I am not precise, the default reading should be that of natural ranking functions.

What about the issue of point versus set functions? If a ranking function is defined on a complete algebra and completely minimitive, it can be reduced to, or generated by, a point function on possibilities. If not, there is no guarantee. Huber (2006, pp. 465f.) gives an example in which a σ-minimitive ranking function κ on an algebra is not reducible to a point function. But he does prove that, if the algebra is countably generated, then κ is induced by a unique minimal point function (for details see Huber 2006, Theorem 1).

So much for negative ranking functions. It is clear how we came to think of them and hence of ranks as degrees of disbelief. This followed from our dynamic perspective that led us first to selection functions and WWOs. Apart from that, it would have been more natural to directly study positive belief and its degrees. This is indeed the more usual approach which we find, e.g., in the inductive support gradings of Cohen (1977), and in the possibility measures of Dubois, Prade (1988). Since his (1967a) Levi, develops both, the *b*-values (= belief) and the *d*-values (= disbelief), in tandem. Let me defer, though, more extensive comparative remarks to Chapter 11. The basic structure is always the same:

> *Definition 5.10*: Let $\mathcal{A}$ be an algebra over W. Then β is a *natural* or *real, finitely* or *completely minimitive positive ranking function* for $\mathcal{A}$ iff, respectively, β is a function from W into $\mathbf{N}^+$ or $\mathbf{R}^+$ such that:
>
> (a) $\beta(A) \geq 0, \beta(\emptyset) = 0$, and $\beta(W) = \infty$, and
> (b) $\beta(A \cap B) = \min \{\beta(A), \beta(B)\}$ for all $A, B \in \mathcal{A}$
> $\qquad\qquad\qquad\qquad\qquad\qquad$ [*finite minimitivity* or *the law of conjunction*], or
> (c) $\beta(\cap\mathcal{B}) = \min \{\beta(B) \mid B \in \mathcal{B}\}$ for all $\mathcal{B} \subseteq \mathcal{A}$ with $\cap\mathcal{B} \in \mathcal{A}$
> $\qquad\qquad\qquad\qquad$ [*complete minimitivity* or *the law of infinite conjunction*].

$\beta(A)$ is called the *positive rank* of A.
(If necessary, the laws will be qualified as holding for positive ranking functions.)

In the references given above, positive ranking functions are normalized so that $\beta(W) = 1$. We can, however, neglect this difference for the time being.

A positive ranking function β is to be understood as a *grading of belief*. If $\beta(A) > 0$, then A is believed (and $\bar{A}$ disbelieved). If $\beta(A) = 0$, then A is not believed. Indeed the law of negation for positive ranks reads the same as the law (5.8a) for negative ranks. According to (b), conjunctions are as strongly believed as the more weakly believed conjunct. The latter property was obvious and attractive to all proponents of this structure, but there is no present need to study their reasons. After all, positive ranks are just negative ranks in a new guise, and we know already why negative ranks have to have the corresponding property. The equivalence is stated in the trivial assertion:

> (5.11)   If κ is a negative ranking function for $\mathcal{A}$ and if $\beta(A) = \kappa(\bar{A})$ for each $A \in \mathcal{A}$, then β is a positive ranking function for $\mathcal{A}$, and vice versa.

So positive ranks are simply another version of the negative ranks discussed thus far. I introduce them here because they have, as indicated, rich historical precedent and because they appear to be intuitively more manageable. This is why they are often useful. Yet one disadvantage is clear: A positive ranking function β is a set function and never reducible to a point function. Usually we have $\beta(\{w\}) = 0$ for each $w \in W$ (if $\{w\}$ is a proposition). Instead the positive ranks of propositions can be reduced to the positive ranks of propositions of the form $W - \{w\}$, i.e., of the weakest non-tautological propositions in the algebra. But this does not seem to be an illuminating reduction. We will come across a more important disadvantage in the next section.

One might still complain that we have thus far not provided a notion that expresses belief and disbelief at the same time. But this desire is easy to satisfy:

Definition 5.12: Let $\mathcal{A}$ be an algebra of propositions. Then τ is a *two-sided ranking function* for $\mathcal{A}$ iff there is a negative ranking function κ for $\mathcal{A}$ such that $\tau(A) = \kappa(\bar{A}) - \kappa(A)$, or a positive ranking function β for $\mathcal{A}$ such that $\tau(A) = \beta(A) - \beta(\bar{A})$, for all $A \in \mathcal{A}$. $\tau(A)$ is called the *two-sided rank* of A.

A two-sided ranking function thus takes positive as well as negative values. The intended interpretation is, of course, that a proposition A is believed if $\tau(A) > 0$, disbelieved if $\tau(A) < 0$, and neutral or undecided if $\tau(A) = 0$. Indeed, for any $A \in \mathcal{A}$ we have:

$$(5.13) \quad \tau(\bar{A}) = -\tau(A) \qquad\qquad\qquad [\textit{law of negation (for two-sided ranks)}]$$

In earlier papers, I called two-sided ranking functions belief functions. But the term is used in multiple ways and hence unhappy. For instance, it would be equally apt for positive ranking functions (this is why I now use "β" for the latter, as suggested by Arthur Merin, and not for the two-sided ranks, as I did in earlier publications). Two-sided ranking functions often appeared to me to be the most handsome way to present ranking theory, and they often help to obtain the shortest description of some formal state of affairs. On the other hand there is no simple law of conjunction or disjunction for two-sided ranks, for belief and disbelief behave differently in this respect. Therefore there is no simple axiomatics of two-sided ranking functions; their simplest characterization is, as in Definition 5.12, via positive or negative ranking functions. This shows that two-sided ranking functions are a derived notion. But of course all of these notions are interdefinable, and so I will often take the liberty, if required for easy presentation, to switch between a negative ranking function κ and the corresponding positive ranking function β or the corresponding two-sided ranking function τ.

But let me say this once and for all: *My preferred tool will be (natural) negative ranking functions.* Again, it would be nice to reduce adjectives. So I will often talk of ranks and ranking functions *simpliciter*, and unless specified otherwise, the default interpretation will always be that I am referring to negative ranking functions.

So far I have presented the standard interpretation of ranking functions in their various forms. Before proceeding to the next step, I should mention that this interpretation is not obligatory. In fact, there is an important interpretational degree of freedom, which Matthias Hild has made clear to me. It is best explained in terms of a two-sided ranking function τ:

It may seem unfair that the range of belief extends to all positive reals (or integers) and the range of disbelief to all negative reals (or integers), whereas there is only one way to be neutral, namely by assigning rank 0. Why should neutrality not comprise a larger range of ranks? We could just as well distinguish some positive rank (or some positive number) z and define the closed interval $[-z, z]$ as the range of neutrality. So, $\tau(A) > z$ expresses belief in A, $\tau(A) < -z$ expresses disbelief in A, and everything in between expresses suspense of judgment:

$$(5.14) \quad B(A) \text{ iff } \kappa(\bar{A}) > z \text{ iff } \beta(A) > z \text{ iff } \tau(A) > z, \text{ where } z > 0.$$

This is a viable interpretation. In particular, consistency and deductive closure of the set of beliefs is preserved and, given completeness, the core would be the smallest or logically strongest proposition A such that $\tau(A) > z$. As far as I can see this interpretation could indeed be maintained throughout the book.

The interpretational freedom appears quite natural. After all, it is a familiar fact that the notion of belief is vague and can be taken more or less strict. Indeed in Section 3.3, when discussing the lottery paradox, we tried to capture this fact in probabilistic terms and found probability theory unsuited. I opened Section 4.1 with the same point. Now we have found a new means to do justice to this vagueness: how exactly the parameter z is fixed depends on how strictly we want to understand belief in the given context. The crucial point is that, however we fix the parameter z, we always get the formal structure of belief we want to have.

The principal lesson of Hild's observation therefore seems to be that it is not the notion of belief that is of basic importance. Rather it is the formal structure of ranks. *The study of belief is the study of that ranking structure.* Whether we say in the end that belief is positive or two-sided rank > 0 or > z is of minor importance.

Nevertheless it would be fatal to simply give up talking of belief in favor of the more abstract talk of ranks. *Ranks express belief* even if there is a degree of freedom. It is crucial not to lose the intuitive connection, and so I will usually stick to my standard interpretation and equate belief in A with $\tau(A) > 0$. But we should always keep in mind that this is a matter of choice; we could resolve semantic indecision just as well in another way.

5.3 Conditional Ranks

In the introductory Section 1.2 I have put great emphasis on the opposition between Pascalian and Baconian probability as introduced by Cohen (1977). However, I also mentioned that the label "Baconian probability" stands for a family of ideas that is historically not very well defined. If there is something unifying this family, I would like to suggest that it is the structure identified so far – i.e., the law of negation and the law of conjunction for positive ranks or something equivalent like the law of negation and the law of disjunction for negative ranks. In any case, these are also the basic laws emphasized by Cohen (1977, Ch. 14), but found at many other places as well. If this suggestion is fair, then ranking theory as stated thus far is nothing but a codification of Baconian probability.

If ranking theory merely consisted of these laws, then there would be no reason to get excited about it; after all, this structure has been around for a while. But there is a fundamental point in which ranking theory goes beyond this loose-knit family of Baconian probability. This is the issue of *conditional ranks*, to which we turn next. As we will observe more thoroughly in the comparative Chapter 11, none of the earlier proposals have – as far as I can see – an adequate equivalent to the notion of conditional ranks to be introduced here. My only explanation for this observation is that none of these proposals were explicitly guided by our dynamic perspective, aiming at an account of belief dynamics. Their motives were different ones that did not require seriously considering conditional ranks. So whatever is novel in this book ultimately derives from the explanation of conditional ranks.

Let us discuss the issue first in terms of a negative ranking function κ. For any two propositions A and B we want to explain the conditional negative rank $\kappa(B \mid A)$ of B given A. First, it is clear that $\kappa(\bar{A} \mid A) = \infty$ and hence $\kappa(A \mid A) = 0$. What logically contradicts the supposition is maximally firmly disbelieved; we might also say that it is disregarded. Second, we should recall how we argued above about changing WWOs of disbelief as proposed in (5.2) and (5.4). We argued that the evidence A should not change the ordering within A, i.e., the ordering conditional on A. Similarly, we should now say that conditional on A the possibilities in A do not only keep their relative ordering, but also their relative ranks, i.e., their ranking differences. The two points already entail

> *Definition 5.15*: Let κ be an $\mathcal{A}$-measurable natural negative ranking function and $A \in \mathcal{A}$ with $\kappa(A) < \infty$. Then, for any $w \in W$ the *conditional negative rank of w given A* is
> defined as $\kappa(w \mid A) = \begin{cases} \kappa(w) - \kappa(A) \text{ for } w \in A, \\ \infty \text{ for } w \notin A. \end{cases}$ For any $B \in \mathcal{A}$ the *conditional negative rank of B given A* is defined as $\kappa(B \mid A) = \min \{\kappa(w \mid A) \mid w \in B\} = \kappa(A \cap B) - \kappa(A)$.
> The latter equation is also the defining one in case κ is a real ranking function. (If κ were an ordinal ranking function, then we would define $\kappa(B \mid A) = -\kappa(A) + \kappa(A \cap B)$, thereby invoking ordinal subtraction.) The function $\kappa_A : B \mapsto \kappa(B \mid A)$ is called the *conditionalization of κ by A*. It clearly is a negative ranking function in turn.

Rewriting the definition, we get

(5.16) $\kappa(A \cap B) = \kappa(A) + \kappa(B \mid A)$ [*the law of conjunction* (for negative ranks)].

This says that the disbelief in A and the disbelief in B given A add up to the disbelief in $A \cap B$, a highly intuitive assertion. Moreover, the law of negation holds in a conditional form as well, since we obviously have for all $A, B \in \mathcal{A}$ with $\kappa(A) < \infty$:

(5.17) $\kappa(B \mid A) = 0$ or $\kappa(\bar{B} \mid A) = 0$ or both   [*conditional law of negation*]

Example 5.18, Tweety continued: For illustrating (negative) conditional ranks it might be helpful to briefly return to our example 5.7 of Tweety. There, I mentioned various examples of conditional propositions, some held vacuously true and others non-vacuously. Now we can see that precisely the conditional propositions non-vacuously held true correspond to conditional beliefs. For example, according to the κ specified there you believe: (i) that Tweety can fly given it is a bird (since $\kappa(\bar{F} \mid B) = 1$) and also (ii) given it is a bird, but not a penguin (since $\kappa(\bar{F} \mid B \& \bar{P}) = 2$), (iii) that Tweety cannot fly given it is a penguin (since $\kappa(F \mid P) = 3$) and (iv) even given it is a penguin, but not a bird (since $\kappa(F \mid \bar{B} \& P) = 3$). You also believe (v) that it is not a penguin given it is a bird (since $\kappa(P \mid B) = 1$) and (vi) that it is a bird given it is a penguin (since $\kappa(\bar{B} \mid P) = 7$). And so forth. These samples show that we have chosen an example that is at least qualitatively plausible.

It might be tempting to copy the idea of Popper measures (Definition 3.14) and take conditional ranks as the primitive notion. (5.17) and a conditional version of (5.16) then turn into defining characteristics:

Definition 5.19: Let $\mathcal{A}$ be an algebra over W and $\mathcal{B}$ a non-empty subset of $\mathcal{A}$. Then $\tilde{\kappa}$ is a *conditional negative ranking function* on $\mathcal{A} \times \mathcal{B}$ iff $\tilde{\kappa}$ is a function from $\mathcal{A} \times \mathcal{B}$ into $\mathbf{N}^+$ or $\mathbf{R}^+$ such that for all $A, B, C \in \mathcal{A}$ with $A, A \cap B \in \mathcal{B}$:

(a) $\tilde{\kappa}(\bar{A} \mid A) = \infty$,
(b)   if $\tilde{\kappa}(B \mid A) < \infty$, then $B \in \mathcal{B}$,
(c) $\tilde{\kappa}(B \mid A) = 0$ or $\tilde{\kappa}(\bar{B} \mid A) = 0$,
(d) $\tilde{\kappa}(B \cap C \mid A) = \tilde{\kappa}(B \mid A) + \tilde{\kappa}(C \mid A \cap B)$.

The *unconditional* rank of A may then be defined as $\tilde{\kappa}(A) = \tilde{\kappa}(A \mid W)$.

This is the conditional variant of negative ranking functions (Definition 5.9). It can be strengthened to a completely minimitive conditional ranking function by generalizing condition (c) from the finite partition $\{B, \bar{B}\}$ to any finite or infinite partition.

Conditional ranking functions generalize ranking functions in a similar way as do Popper measures with respect to standard probability measures. I do not want to discuss here whether this generalization can be made sense of and put to use, especially because the generalization might be spurious. Interestingly, the generalization vanishes in the conditional counterpart of ordinal ranking functions, which were the most general variant of completely minimitive negative ranking functions. The reason is roughly that the hierarchy of ordinal numbers is so vast that it can always encompass all levels of conditionality contained in a conditional ranking function. I will briefly return to this notion of a conditional ranking function in Section 8.2; otherwise it will play no role in this book.

But there is a reason for introducing conditional ranking functions here (besides potential intrinsic interest). The reason is that only the conditional law of negation is used here as the defining characteristic (c), whereas the law of disjunction (5.8b) was required for the unconditional notion. That law follows here: Given (5.19d), the conditional law of negation is equivalent to min $\{\tilde{\kappa}(A \cap B), \tilde{\kappa}(A \cap \bar{B})\} = \tilde{\kappa}(A)$, and this is clearly just a variant formulation of the law of disjunction. It is interesting to see that this law − or its positive counterpart, i.e., the law of conjunction for positive ranks, which appeared as the ranking-theoretic analogue of the requirement of deductive closure − is nothing but a *conditional consistency requirement*. So Definition 5.19 provides, in a way, the most illuminating axiomatization of ranking functions: they just encode conditional consistency and the Definition 5.15 of conditional ranks or its generalization (5.19d). We will deepen this observation in Chapter 8.

Let us now turn to the positive variant of conditional ranks. It is no less appealing. In view of the equation $\beta(A) = \kappa(\bar{A})$ Definition 5.15 turns into

Definition 5.20: Let β be a positive ranking function for $\mathcal{A}$ and $A \in \mathcal{A}$ with $\beta(\bar{A}) < \infty$. Then for any $B \in \mathcal{A}$ the *conditional positive rank of B given A* is defined as $\beta(B \mid A)$ $= \beta(\bar{A} \cup B) - \beta(\bar{A})$. The function $\beta_A: B \mapsto \beta(B \mid A)$ is called the *conditionalization of* β *by A* and is clearly a positive ranking function.

Again, this is tantamount to

(5.21) $\beta(\bar{A} \cup B) = \beta(B \mid A) + \beta(\bar{A})$ [*the law of material implication*].

Definition 5.20 says that the conditional positive rank of B given A results from the positive rank of the material implication $A \to B$ ($:= \bar{A} \cup B$) by subtracting from it the positive rank of the falsity of its antecedent, its trivial and uninteresting truth condition. In particular, if $\bar{A}$ is not believed, then the conditional rank and the rank of the material implication coincide.

It is instructive to compare this with the Ramsey test (4.17). It said that a conditional $\varphi \,\Box\!\!\to \psi$ is accepted in a belief set K iff ψ is accepted in $K * \varphi$. Given the Levi identity this is the same as the material implication $\varphi \to \psi$ being accepted in $K \div \neg\varphi$. This is strengthened to a quantitative equality by (5.20) and (5.21). However, the latter is more cautious since it talks only about conditional belief and not about the acceptance of conditionals.

The topic will come up again and I will always proceed with the same caution. As an epistemologist I am happy to describe conditional belief. I also find it entirely clear that our countless overt or covert conditional constructions somehow express conditional belief. How they exactly do so, however, is utterly complex. The expression relation here, as for belief in general, is superimposed by so many grammatical interactions and pragmatic features that I feel well advised to stay away from this semantic business. The suspicion that the Ramsey test is much too simple was indeed already raised by the impossibility theorem in Gärdenfors' (1988, Sect. 7.4), which shows that conditional belief cannot simply translate into belief in a possibly iterable conditional via the Ramsey test. Let us stick to epistemology.

Finally there are two-sided conditional ranks definable in a straightforward way:

Definition 5.22: Let τ be a two-sided ranking function for $\mathcal{A}$, and κ and β be the corresponding negative and positive ranking function for $\mathcal{A}$. Then for any $A, B \in \mathcal{A}$ with $\tau(A) > -\infty$ $\tau(B \mid A) = \kappa(\bar{B} \mid A) - \kappa(B \mid A) = \beta(B \mid A) - \beta(\bar{B} \mid A)$ is the *conditional two-sided rank of B given A*.

All of the positive and negative remarks about two-sided ranks apply here as well.

In Chapter 3 I noted some basic probabilistic theorems. Let us briefly state some similar ranking-theoretic facts.

Theorem 5.23: Let κ be a negative, β a positive, and τ the corresponding two-sided ranking function for $\mathcal{A}$. Let $A, B, C \in \mathcal{A}$ and $\{A_1, \ldots, A_n\} \subseteq \mathcal{A}$ be a partition of W. Then we have:

(a)   if $\tau(C \mid A) \leq \tau(C \mid B)$, then $\tau(C \mid A) \leq \tau(C \mid A \cup B) \leq \tau(C \mid B)$

[law of disjunctive conditions],

(b)   if $\tau(C \mid A) < \tau(C \mid B)$, then $\tau(C \mid A \cup B) = \tau(C \mid A)$ iff $\kappa(A \cap C) \leq \kappa(B \cap C)$, and $\tau(C \mid A \cup B) = \tau(C \mid B)$ iff $\kappa(A \cap \bar{C}) \geq \kappa(B \cap \bar{C})$,

(c) $\kappa(B) = \min\limits_{i \leq n} \, [\kappa(B \mid A_i) + \kappa(A_i)]$ [formula of the total negative rank],

(d) $\kappa(A_k \mid B) = \kappa(B \mid A_k) + \kappa(A_k) - \kappa(B)$ [Bayes' Theorem for negative ranks],

(e) $\beta(A_k \mid B) = \beta(\bar{B} \mid \bar{A}_k) + \beta(A_k) - \beta(\bar{B})$ [Bayes' Theorem for positive ranks].

Proof: (c), (d), and (e) are obvious, (a) not quite. The assumption in (a) was that $\tau(C \mid A) \leq \tau(C \mid B)$, i.e., $\kappa(\bar{C} \mid A) - \kappa(C \mid A) \leq \kappa(\bar{C} \mid B) - \kappa(C \mid B)$, i.e., $\kappa(A \cap \bar{C}) - \kappa(A \cap C) \leq \kappa(B \cap \bar{C}) - \kappa(B \cap C)$, or $c - a \leq d - b$, with obvious abbreviations. We have to show that $\tau(C \mid A) \leq \tau(C \mid A \cup B) \leq \tau(C \mid B)$, i.e., that

$c - a \leq \kappa(\bar{C} \mid A \cup B) - \kappa(C \mid A \cup B) \leq d - b$, i.e., that
$c - a \leq \kappa((A \cup B) \cap \bar{C}) - \kappa((A \cup B) \cap C) \leq d - b$, i.e., that
$c - a \leq \min \, [\kappa(A \cap \bar{C}), \kappa(B \cap \bar{C})] - \min \, [\kappa(A \cap C), \kappa(B \cap C)] \leq d - b$, or
$c - a \leq \min \, [c, d] - \min \, [a, b] \leq d - b$.

This holds (i) when $\min \, [c, d] = c$ and $\min \, [a, b] = a$, (ii) when $\min \, [c, d] = d$ and $\min \, [a, b] = b$, and (iii) when $\min \, [c, d] = c$ and $\min \, [a, b] = b \neq a$. The fourth case that $\min \, [c, d] = d \neq c$ and $\min \, [a, b] = a$ is excluded by our assumption.

This proves (b) as well, since it makes clear that the first equation of (b) holds iff $\min \, [a, b] = a$ and that the second equation of (b) holds iff $\min \, [c, d] = d$.

The law of disjunctive conditions says that the rank of some proposition conditional on some disjunction is always between the ranks conditional on the disjuncts. It thus corresponds to the probabilistic Theorem 3.4f. Note, however, that in the ranking-theoretic case it is not required that the disjuncts are disjoint. There is a further slight disanalogy in (5.23b). In the probabilistic counterpart, betweenness can fail to be proper, i.e., reduce to equality only if one of the disjuncts has probability 0 given the disjunction, whereas (5.23b) says it can hold even if the negative rank of the disjuncts given the disjunction is finite. We will repeatedly apply the law of disjunctive conditions.

The formula of the total negative rank (5.23c) will be useful as well. It tells how to derive the degree of disbelief in some proposition B from its conditional degrees of disbelief under alternative hypotheses, namely by adding to them the negative ranks of the alternatives (thus weighing, as it were, the conditional degrees with the negative ranks) and by looking for the least disbelieved combination. Similarly we might express (5.23d) by taking over the terminology from the probabilistic Bayes' Theorem in the following way: the posterior degree of disbelief in some hypothesis A_k is just the normalized sum of its prior degree of disbelief and the "negative ranking likelihood" of the evidence under that hypothesis (where normalization means subtraction of $\kappa(B)$). Thus, as in probability theory, the posterior rank is a simple function of the prior rank and the likelihood, and we may expect that the likelihood is the dominating factor in the long run. Indeed this will be proved in Section 12.5. (5.23e) is a translation of (5.23d) into positive ranks, but it is *prima facie* less perspicuous.

For the time being I will not go deeper into the significance of these observations. However, in Section 14.15 I will more thoroughly return to the ranking-theoretic version (5.23d) of Bayes' Theorem. I would like to add, though, that Hild (2001, Chs. 12 and 13) has developed a detailed line of thought showing deep interconnections between Bayes' Theorem for ranks and statistical methodology.

Here I want to add only one important remark. The reader might already have suspected a kind of algorithm for translating probabilistic theorems into ranking theorems. Theorem 5.23 displays this algorithm: adding probabilities translates into taking the minimum of negative ranks, multiplication of probabilities into addition of negative ranks, and division of probabilities into subtraction of negative ranks. With this algorithm we can generate the whole axiomatics of negative ranking functions from probability theory, and probabilistic theorems are almost guaranteed to thus translate into ranking theorems. This is no accident at all! We will uncover the deep truth behind this in Section 10.2. Indeed, the pervasive analogy between probability and ranking theory will be a *leitmotiv* for many of the subsequent chapters, which will develop the analogy in all its beauty. However, the analogy pertains only to negative ranks. For positive ranks, additional steps of translation are needed, making the analogy less direct and perspicuous, as is shown, for instance, by a comparison of Bayes' Theorem for negative and for positive ranks. This is the irresistible reason why I am bound to proceed in terms of negative ranking functions.

We also saw that the translation did not work perfectly in the case of the law of disjunctive conditions. This indicates that the analogy is not fully reliable and has its subtleties. I will return to the point in Section 10.2.

5.4 The Dynamics of Ranking Functions

We are now sufficiently prepared to take the final step and state a general dynamics for ranking functions. Of course, the idea is not that evidence E moves the prior state κ into the posterior state κ_E. Since $\kappa_E(\bar{E}) = \infty$, the evidence E would have to be maximally and unrevisably certain. But this is what we generally wanted to avoid. We may entirely disregard the possibilities outside E only on the *supposition* that E, but not on the *evidence* that E.

So what to do? Our discussion of the WWO or WOP change rules (5.2) and (5.4) already showed the way. Both rules proposed not to change the ordering of possibilities within the evidence E and within its negation $\bar{E}$. This was their plausible part, and it translates here into keeping the conditional ranks $\kappa(w \mid E)$ and $\kappa(w' \mid \bar{E})$ for $w \in E$ and $w' \in \bar{E}$ unchanged. Yet (5.2) made the evidence E maximally certain and (5.4) made it minimally certain in terms of WWOs, and we concluded that no strategy is the right one for all cases. But if we want to be flexible, then we have to assume the firmness with which the evidence is accepted as a further parameter characterizing belief change. The evidence is, then, not simply a proposition, but a proposition plus its posterior rank. However, this, together with the constancy of conditional ranks, determines the entire posterior ranking function.

Definition 5.24: Let κ be a natural negative ranking function for $\mathcal{A}$ and $A \in \mathcal{A}$ such that $\kappa(A), \kappa(\bar{A}) < \infty$, and $n \in \mathbf{N}^+$. Then the $A \rightarrow n$-*conditionalization* $\kappa_{A \rightarrow n}$ of κ is defined by $\kappa_{A \rightarrow n}(w) = \begin{cases} \kappa(w \mid A) \text{ for } w \in A, \\ \kappa(w \mid \bar{A}) + n \text{ for } w \in \bar{A} \end{cases}$. The $A \rightarrow n$-conditionalization will also be called *result-oriented* (for reasons soon to be clear).

Thus, the effect of the $A \rightarrow n$-conditionalization is to shift the possibilities in A upwards so that $\kappa_{A \rightarrow n}(A) = 0$ and the possibilities in $\bar{A}$ downwards so that $\kappa_{A \rightarrow n}(\bar{A}) = n$. Definition 5.24 exploits the reducibility of completely minimitive negative ranking functions to point functions. But this is not essential, and it is clear how to carry over (5.24) to finitely minimitive negative and even to positive ranking functions.

If one is attached to the idea that evidence consists in nothing but a proposition, then this additional parameter is a mystery. However, this idea was seen to be questionable already in our discussion of Jeffrey's generalized conditionalization (3.9). And so it is in our framework. The processing of evidence might be so automatic that one hardly becomes aware of that parameter. But I find it entirely natural that the evidence comes in varying degrees of certainty. Take, for instance, the proposition: "There are tigers in the Amazon jungle," and consider six scenarios: (a) I read somewhat sensationalist coverage in the tabloid press claiming this, (b) I read a serious article in a serious newspaper claiming this, (c) I hear the Brazilian government officially announcing that tigers have been discovered in the Amazon area, (d) I see a documentary on TV claiming to show tigers in the Amazon jungle, (e) I read an article in *Nature* by a famous zoologist reporting of tigers there, (f) I travel there myself and see the tigers. In all six cases I receive the information that there are tigers in the Amazon jungle, but with varying and, I find, increasing certainty. One might object that the proposition and evidence is clearly a different one in each scenario. But recall my insistence in Chapter 2 on the variability of the underlying space of possibilities or of the fine-grainedness of the algebra of propositions considered, and my brief discussion of the invariance principle (5.3). In all six cases the net information is that about the tigers. The net information is clearly of varying certainty, and perhaps only the net information and none of the more fine-grained pieces of evidence is considered in the algebra.

Definition 5.24 enables us to state a dynamic *law of conditionalization* for ranking functions:

(5.25)   If the prior doxastic state of the subject s at time t is characterized by the ranking function κ and if s receives evidence E with firmness n between t and t', then the posterior state of s at t' is represented by $\kappa_{E \rightarrow n}$.

Example 5.26, Tweety continued: For illustration, suppose you first learn and accept with firmness 2 that Tweety is a bird, say, because you heard Tweety chirping like a bird. Thus you shift $\bar{B}$ up by 2 and change to the $B{\to}2$-conditionalization κ' of the κ specified in (5.7):

κ'	$B \cap \bar{P}$	$B \cap P$	$\bar{B} \cap \bar{P}$	$\bar{B} \cap P$
F	0	4	2	13
$\bar{F}$	2	1	2	10

In κ' you believe that Tweety is a bird able to fly, but not a penguin. So, in κ' you believe more than in κ, and we might also call κ' an expansion of κ.

Next, to your surprise and, suppose, with firmness 1, you tentatively learn and accept that Tweety is indeed a penguin, perhaps because you get a glimpse of its shape, thus shifting P down by 1 and $\bar{P}$ up by 1 and moving to the $P{\to}1$-conditionalization κ'' of κ':

κ''	$B \cap \bar{P}$	$B \cap P$	$\bar{B} \cap \bar{P}$	$\bar{B} \cap P$
F	1	3	3	12
$\bar{F}$	3	0	3	9

Now you believe in κ'' that Tweety is a penguin bird that cannot fly. So you have changed your mind, and we may also call κ'' a revision of κ'.

Unlike the dynamic law (4.12) for selection functions, the law (5.25) is clearly iterable. Is it already fully general? Almost. We have a problem with the rank ∞. If $\kappa(A) = \infty$, then conditional ranks given A are undefined, and then the subsequent notions do not work. Therefore, I propose to make the same move as Jeffrey did in the probabilistic case. Let us define in analogy to (3.6):

Definition 5.27: A negative ranking function κ for $\mathcal{A}$ is called *regular* iff $\kappa(A) < \infty$ for all non-empty $A \in \mathcal{A}$.

(We saw already that the second strategy discussed in Section 3.2, i.e., the move to Popper measures, can also be carried over to the ranking-theoretic case, namely with the notion of a conditional ranking function (5.19). There I also noticed, however, that the conditional notion does not generalize the unconditional notion in the case of ordinal ranking functions, relative to which regularity would mean $\kappa(A) < \Omega$ for all non-empty $A \in \mathcal{A}$.)

Huber (2006) calls a negative ranking function κ on some algebra $\mathcal{A}$ *natural* iff it is induced by a pointwise negative ranking function taking only natural numbers as values. Clearly, for $\mathcal{A}$-measurable completely minimitive negative ranking functions regularity and naturalness coincide. Huber observes, however, that the two notions might fall apart for quite interesting reasons if $\mathcal{A}$ is not complete. For similar reasons

I will qualify the notion of regularity in Definition 12.4. For the time being, though, these niceties are irrelevant, and we may observe the theorem corresponding to (3.11):

Theorem 5.28: If κ is regular, $A \in \mathcal{A}$, $\varnothing \neq A \neq W$, and $n < \infty$, then $\kappa_{A \to n}$ is regular.

And so (5.25) defines a fully general and iterable dynamics at least within the space of regular ranking functions. This is so far my solution to the problem of iterated belief change and, in view of all the preliminary explanations, my contribution to the problem of induction. One essential point has been the step from WWOs to ranks and conditional ranks, but the other essential point was Jeffrey's insight about the nature of evidence: that evidence is not merely an evidential proposition, but a doxastic attitude towards evidential propositions. We will return to the issue from a more philosophical side in Chapter 16.

For the moment let me just add two variants of (5.24) and (5.25). Shenoy (1991) and several others after him pointed out that the parameter n as conceived in Definition 5.24 does not characterize the evidence as such, but rather the result of the interaction between the prior doxastic state and the evidence. And he proposed a reformulation of (5.24) with a parameter exclusively pertaining to the evidence:

Definition 5.29: Let κ be a natural negative ranking function for $\mathcal{A}$, $A \in \mathcal{A}$ such that $\kappa(A)$, $\kappa(\bar{A}) < \infty$, and $n \in \mathbf{N}^+$. Then the $A{\uparrow}n$-*conditionalization* $\kappa_{A{\uparrow}n}$ of κ is defined by $\kappa_{A{\uparrow}n}(w) = \begin{cases} \kappa(w) - m \text{ for } w \in A, \\ \kappa(w) + n - m \text{ for } w \in \bar{A} \end{cases}$, where $m = \min\{\kappa(A), n\}$. The $A{\uparrow}n$-conditionalization will also be called *evidence-oriented*.

The effect of this conditionalization is that, whatever the prior ranks of A and $\bar{A}$ are, the possibilities within A improve by exactly n ranks as opposed to the possibilities outside A. This is more perspicuous for the corresponding two-sided ranking function τ; we always have

(5.30) $\tau_{A{\uparrow}n}(A) - \tau(A) = n$.

It is thus fair to say that in $A{\uparrow}n$-conditionalization the parameter n exclusively characterizes the evidential impact. This is why I call it evidence-oriented. Correspondingly, $A{\to}n$-conditionalization is called result-oriented, because its parameter tells at which rank the evidence arrives after conditionalization. And so (5.28) relates to (5.24) just as the probabilistic variant of Field (1978), which was mentioned after Theorem 3.11, is related to Jeffrey's original proposal. Of course the two kinds of conditionalization are easily interdefinable:

(5.31) $\kappa_{A{\to}n} = \kappa_{A{\uparrow}m}$, where $m = \tau(\bar{A}) + n$.

Thus, translating example 5.26 into evidence-oriented terms, we might as well describe κ' as the $B{\uparrow}2$-conditionalization of the κ specified in (5.7), and κ'' as the $P{\uparrow}2$-conditionalization of κ'. Perhaps this description is even the more natural one. In the course of the book I will use both notions of conditionalization. Often things are more easily explained in terms of the one than in terms of the other.

We should perhaps follow Jeffrey all the way through and not only consider an evidential proposition E and its negation, but rather a whole evidential partition or algebra $\mathcal{E}$. Jeffrey's idea was that evidence somehow rearranges the doxastic state concerning the whole of $\mathcal{E}$. We can easily translate this generalization into our terms:

Definition 5.32: Let κ be a natural negative ranking function for $\mathcal{A}$, $\mathcal{E}$ some complete subalgebra of $\mathcal{A}$, and ε a natural negative ranking function for $\mathcal{E}$. Then the result-oriented $\mathcal{E} \rightarrow \varepsilon$-conditionalization $\kappa_{\mathcal{E} \rightarrow \varepsilon}$ of κ is defined by $\kappa_{\mathcal{E} \rightarrow \varepsilon}(w) = \kappa(w \mid E) + \varepsilon(E)$ for each $w \in W$, where E is the unique atom of $\mathcal{E}$ with $w \in E$.

This allows us to state our most general law of belief change, the *law of generalized conditionalization*:

(5.33)   If the prior doxastic state of the subject s at t is represented by the ranking function κ and if the evidence between t and t' produces the ranking function ε on the evidential algebra $\mathcal{E}$, then the posterior state of s at t' is given by $\kappa_{\mathcal{E} \rightarrow \varepsilon}$.

It is clear that I have described *result-oriented* $\mathcal{E} \rightarrow \varepsilon$-conditionalization in (5.33). But it is straightforward to formulate the *evidence-oriented* version of generalized conditionalization:

Definition 5.34: Let κ be a natural negative ranking function for $\mathcal{A}$, $\mathcal{E}$ some complete subalgebra of $\mathcal{A}$, and ε a natural negative ranking function for $\mathcal{E}$. Then the *evidence-oriented* $\mathcal{E} \uparrow \varepsilon$-conditionalization $\kappa_{\mathcal{E} \uparrow \varepsilon}$ of κ is defined by $\kappa_{\mathcal{E} \uparrow \varepsilon}(w) = \kappa(w) + \varepsilon(E) - m$ for each $w \in W$, where E is the unique atom of $\mathcal{E}$ with $w \in E$ and $m = \min \{\kappa(E') + \varepsilon(E') \mid E'$ is an atom of $\mathcal{E}\}$.

As compared with the probabilistic counterpart of Field (1978) I find this an elegant version of evidence-oriented conditionalization. However, the point is not merely elegance, but rather that in the probabilistic case the parameters characterizing the firmness of evidence are any positive or negative real numbers and so cannot be understood as probabilities. The case is remarkably different with (5.34), where the strength of evidence itself can be understood as a ranking function.

As in the probabilistic case, it appears that the law (5.33) is empty: any ranking function κ' can come from any regular ranking function κ by a suitable $\mathcal{E} \rightarrow \varepsilon$-conditionalization. My response is the same as in Section 3.2. We have

Theorem 5.35: If κ' is the $\mathcal{E}_1 \rightarrow \varepsilon_1$-conditionalization and the $\mathcal{E}_2 \rightarrow \varepsilon_2$-conditionalization of κ, and $\mathcal{E} = \mathcal{E}_1 \cap \mathcal{E}_2$, then $\varepsilon = \varepsilon_1 = \varepsilon_2$ is an $\mathcal{E}$-measurable ranking function and κ' is the $\mathcal{E} \rightarrow \varepsilon$-conditionalization of κ.

Proof: We may copy the proof of Theorem 3.12. Let again $E_{11}, \ldots, E_{1k}$ be the atoms of $\mathcal{E}_1$, $E_{21}, \ldots, E_{2l}$ the atoms of $\mathcal{E}_2$, and $E_1, \ldots, E_n$ be the atoms of $\mathcal{E}$. Our assumption now is that for each $A \in \mathcal{A}$:

(*) $\kappa'(A) = \min_{i=1,\ldots,k} [\kappa(A \mid E_{1i}) + \varepsilon_1(E_{1i})] = \min_{j=1,\ldots,l} [\kappa(A \mid E_{2j}) + \varepsilon_2(E_{2j})]$.

For $A = E_m$ $(m = 1, \ldots, n)$ (*) immediately yields $\varepsilon_1(E_m) = \varepsilon_2(E_m)$. So, ε_1 and ε_2 agree on $\mathcal{E}$.

Now, take any E_m and focus on the $E_{1i} \subseteq E_m$ and $E_{2j} \subseteq E_m$; suppose this is the case for $i = 1, \ldots, k_m$ and $j = 1, \ldots, l_m$. For $A = E_{1i} \cap E_{2j} \neq \varnothing$ (*) yields: $\kappa'(E_{1i} \cap E_{2j}) = \kappa(E_{2j} \mid E_{1i}) + \varepsilon_1(E_{1i}) = \kappa(E_{1i} \mid E_{2j}) + \varepsilon_2(E_{2j})$. Hence, $\varepsilon_1(E_{1i}) - \varepsilon_2(E_{2j}) = \kappa(E_{1i} \mid E_{2j}) - \kappa(E_{2j} \mid E_{1i})$ $= \kappa(E_{1i}) - \kappa(E_{2j})$. Since this holds for all $j = 1, \ldots, l_m$, it entails in turn that $\varepsilon_1(E_{1i}) - \varepsilon(E_m) = \kappa(E_{1i}) - \kappa(E_m)$. Thus, finally, for any $A \in \mathcal{A}$:

$$\kappa'(A \cap E_m) = \min_{i=1,\ldots,k_m} [\kappa(A \mid E_{1i}) + \varepsilon_1(E_{1i})]$$
$$= \min_{i=1,\ldots,k_m} [\kappa(A \mid E_{1i}) + \kappa(E_{1i}) - \kappa(E_m)] + \varepsilon(E_m) = \kappa(A \mid E_m) + \varepsilon(E_m).$$

So, finally, $\kappa'(A) = \min_{i=1,\ldots,m} [\kappa(A \mid E_m) + \varepsilon(E_m)]$.

In other words, as in the probabilistic case (3.13) we are justified to accept

Definition 5.36: For any two ranking functions κ and κ' such that κ' comes from κ by generalized conditionalization, the algebra $\mathcal{E}$ or rather the $\langle \mathcal{E}, \varepsilon \rangle$ that is unique according to (5.35) is called the *(evidential) source* of the transition from κ to κ'.

Again the further task would be to specify informative constraints on suitable evidential sources.

This consideration shows, by the way, that the law (5.33) of generalized conditionalization is indeed independent of the coarse- or fine-grainedness, i.e., of the granularity of the underlying algebra $\mathcal{A}$ of propositions. We may describe a change from κ to κ' on the basis of a fine-grained algebra $\mathcal{A}$ as a conditionalization (5.25) or a generalized conditionalization (5.33). Now we might be interested only in the subalgebra $\mathcal{B}$ of $\mathcal{A}$ and only in the restrictions of κ and κ' to $\mathcal{B}$. The point, then, is that we can still describe the change from κ restricted to $\mathcal{B}$ to κ' restricted to $\mathcal{B}$ as an instance of generalized conditionalization (5.33), though not necessarily in terms of (5.25). In this sense (5.33) is general enough to not require a specific level of fine-grainedness of the propositions in order to make belief change describable. It thus fully satisfies the invariance principle (5.3). I consider this to be an essential virtue.

Let us look at generalized conditionalization from a still more abstract perspective. What the $\mathcal{E} \rightarrow \varepsilon$-conditionalization $\kappa_{\mathcal{E} \rightarrow \varepsilon}$ of κ in effect does is to reshuffle the conditionalizations κ_E of κ by the various atoms E of $\mathcal{E}$. If we combine these according to their original "weights" $\kappa(E)$, then of course we return to the original κ. This is what the formula (5.18a) of the total negative rank says. And if we combine them with their new "weights" $\varepsilon(E)$, then we arrive at $\kappa_{\mathcal{E} \rightarrow \varepsilon}$. The general mathematical idea behind this is that it makes sense to mix ranking functions by a ranking function, just as we can produce weighted mixtures of probability measures with probabilities as weights. The general notion is this:

Definition 5.37: Let $\mathcal{A}$ be an algebra over W, Λ a set of ranking functions for $\mathcal{A}$, and ρ a complete ranking function for Λ. Define κ by $\kappa(A) = \min \{\lambda(A) + \rho(\lambda) \mid \lambda \in \Lambda\}$

for all $A \in \mathcal{A}$. Then, κ is a ranking function for $\mathcal{A}$ in turn (the proof is trivial) and is called the *mixture* of Λ by ρ.

For the time being, this is just a mathematical observation. But we will return to it. The idea of mixing ranking functions will acquire great importance in Chapter 12.

I started this chapter by criticizing the preliminary proposals (5.2) and (5.4) for defining a dynamics that is neither commutative nor reversible. Have we done any better now? Yes. In terms of evidence-oriented conditionalization we have indeed reached full success. It is reversible:

(5.38) $(\kappa_{A \uparrow n})_{\bar{A} \uparrow n} = \kappa$.

In other words, $\bar{A} \uparrow n$-conditionalization undoes $A \uparrow n$-conditionalization. Moreover, it is commutative without restriction, as is easily verified:

(5.39) $(\kappa_{A \uparrow m})_{B \uparrow n} = (\kappa_{B \uparrow n})_{A \uparrow m}$.

However, an $A \uparrow m$ and a $B \uparrow n$-conditionalization usually do not add up to an $A \cap B \uparrow m + n$-conditionalization, but rather the joint effect of two simple conditionalizations can in general only be summarized as one step of generalized conditionalization (5.33).

In terms of result-oriented conditionalization, the results are not quite as nice. As is easily checked, reversibility takes a less elegant form and commutativity holds only restrictedly, i.e., only when the second result does not destroy the first result. (In Spohn 1988, pp. 118f., I have stated the restriction rigorously; see also (5.58) below for a special case. See, moreover, the discussion in Jeffrey 1983, pp. 182f., who is somehow attached to result-oriented conditionalization and so thinks he has to defend failure of commutativity.)

5.5 Conditionalization Generalizes AGM Belief Revision and Contraction

A further important reason why the account in the previous section is satisfactory is that result-oriented conditionalization (5.19) preserves and generalizes belief revision theory as presented in Chapter 4. This point was already indicated in the continuation (5.26) of the Tweety example. The present section will flesh out the point and will thus make our conditionalization a bit more vivid. Let us take up belief revision first.

How can we describe the revision $K * \varphi$ of a belief set K by φ with the present means? If φ expresses the proposition A and if the core $\kappa^{-1}(0)$ of the ranking function κ represents the belief set K, then this revision corresponds to an $A \rightarrow n$-conditionalization of κ for some $n > 0$. Because $\kappa_{A \rightarrow n}(\bar{A}) = n > 0$, A is accepted in $\kappa_{A \rightarrow n}$ as revision requires. But which n? This does not matter as far as the core of the resulting ranking function is concerned. $\kappa_{A \rightarrow m}$ and $\kappa_{A \rightarrow n}$ have the same core given $m, n > 0$, namely the set of possibilities of minimal rank within A. Of course for $m \neq n$ $\kappa_{A \rightarrow m}$ and $\kappa_{A \rightarrow n}$ are different doxastic states. It is only that the difference does not show up in the accepted beliefs.

If $\kappa(A) > 0$, then A is initially disbelieved and the $A \rightarrow n$-conditionalization $(n > 0)$ is a genuine revision, as in the second change in example 5.26. If $\kappa(A) = \kappa(\bar{A}) = 0$, then neither A nor $\bar{A}$ is initially believed and the $A{\rightarrow}n$-conditionalization ($n > 0$) is an expansion, as in the first change in example 5.26. If $\kappa(\bar{A}) > 0$, then A is initially believed, and the revision does not show any change. However, the $A{\rightarrow}n$-conditionalization still makes sense. If $\kappa(\bar{A}) < n$, the $A{\rightarrow}n$-conditionalization describes a process where one receives further reason for A strengthening the already existing belief in A. This is a kind of belief change that, to my knowledge, until now has never been considered. It is even possible that initially $\kappa(\bar{A}) > n$. Then the $A{\rightarrow}n$-conditionalization describes the reception of counter-evidence that weakens, but does not destroy the belief in A. Thus the possibilities of change provided by ranking conditionalization are much richer than those considered in belief revision theory. This is no surprise given our additional parameter.

Belief revision as defined by $A{\rightarrow}n$-conditionalization for some $n > 0$ satisfies all the revision postulates $(K{*}1)$–$(K{*}8)$ (see Definition 4.18). To see this let us temporarily define the *revision* $*_\kappa$ w.r.t. κ as the function such that for each $A \in \mathcal{A} *_\kappa(A)$ is the core of $\kappa_{A{\rightarrow}n}$ for some $n > 0$ (it does not matter which). Then it is easily verified that:

(5.40)   If κ is regular, $*_\kappa$ is a selection function.

(If κ is not regular, then we would have to slightly generalize the notion of a selection function.) Recalling the equivalence between the axioms for selection functions (4.11a–b) and the revision postulates $(K{*}1)$–$(K{*}8)$, this shows that AGM belief revision theory is preserved in ranking theory. This is nice, but it also burdens us with the obligation to defend the standard postulates against the large critical discussion they have been subjected to. I will do so in Section 11.3. This brief discussion has also shown that, in a way, belief revision is ambiguous from the ranking-theoretic point of view. It is not a uniquely defined change of doxastic states. The ambiguity does just not surface at the level of belief sets.

So far we have left out the $A{\rightarrow}0$-conditionalization of κ, a significant neglect, for $A{\rightarrow}0$-conditionalization is an interesting case. It might mean a downgrading or an upgrading of A, depending on A's initial two-sided rank. In any case, it ends up with $\kappa_{A{\rightarrow}0}(A) = \kappa_{A{\rightarrow}0}(\bar{A}) = 0$, i.e., with being neutral or unopinionated about A. Thus it does the same as a contraction. To be precise, it does more – it is a kind of two-sided contraction: If A is believed in κ, then the $A{\rightarrow}0$-conditionalization effects a contraction by A. If $\bar{A}$ is believed in κ, then it effects a contraction by $\bar{A}$ (and if κ is neutral about A, neither contraction changes anything). So if we want to capture contraction as it was intended in Chapter 4, we should accept the following

Definition 5.41: Let κ be a negative ranking function for $\mathcal{A}$ and $A \in \mathcal{A}$ such that $\kappa(\bar{A}) < \infty$. Then the *contraction* $\kappa_{\div A}$ of κ by A is defined as

$$\kappa_{\div A} = \begin{cases} \kappa, \text{ if } \kappa(\bar{A}) = 0, \\ \kappa_{A{\rightarrow}0}, \text{ if } \kappa(\bar{A}) > 0 \end{cases}$$

The (*propositional*) *contraction function* $\div_\kappa$ w.r.t. κ is defined as the function assigning to each $A \in \mathcal{A}$ such that $\kappa(\bar{A}) < \infty$ the proposition $\div_\kappa(A) = \kappa_{\div A}^{-1}(0)$ (= the core of $\kappa_{\div A}$).

Of course the definition works for any kind of ranking function. Note that $\kappa_{\div A}$ is a ranking function while $\div_\kappa(A)$ is a proposition. At the moment we only need the latter notion, but the contraction of ranking functions will become important later on. Indeed, in contrast to revision, contraction is a uniquely defined operation from the ranking-theoretic point of view, which is why I have given it an official definition (a point that will acquire great significance in Section 8.3). Note, however, that the uniqueness holds only if belief is represented as in (5.6). Within the representation (5.14) liberating neutrality, contraction loses its uniqueness, because giving A any rank between $-z$ and z counts as giving up the belief in A. However, in that case we might conceive of (5.41) as defining what may be called the *central contraction* which reestablishes uniqueness because it is the only one after which A and $\bar{A}$ are equally neutral, i.e., where $\tau_{\div A}(A) = \tau_{\div A}(\bar{A}) = 0$. Note also that the contraction (5.41) is clearly iterable, just like conditionalization.

The first line of Definition 5.41 covers the trivial case in which contraction does not change anything, and the second line describes the interesting case in which A is initially believed and in which single contraction by A generates a weaker core or belief set. This is indeed contraction as conceived in Chapter 4:

(5.42)   If κ is regular, $\div_\kappa$ is a propositional contraction function as defined in (4.21).

Proof: It is obvious from (5.41) that $\varnothing \neq \div_\kappa(\varnothing) \subseteq \div_\kappa(A) \subseteq \div_\kappa(\varnothing) \cup \bar{A} \not\subseteq A$, i.e., that (4.21a) holds for $\div_\kappa$.

As to (4.21b), we have first to show that $\div_\kappa(A \cap B) \subseteq \div_\kappa(A) \cup \div_\kappa(B)$ (this is the propositional (K÷7)). If either contraction, $\div_\kappa(A)$ or $\div_\kappa(B)$, is vacuous, then the contraction $\div_\kappa(A \cap B)$ is so too. Hence the postulate trivially holds in this case. If both $\div_\kappa(A)$ and $\div_\kappa(B)$ are genuine contractions, so is $\div_\kappa(A \cap B)$. In this case, (5.41) tells us that $\div_\kappa(A) = C \cup \{w \in \bar{A} \mid \kappa(w) = \kappa(\bar{A})\}$, $\div_\kappa(B) = C \cup \{w \in \bar{B} \mid \kappa(w) = \kappa(\bar{B})\}$, and $\div_\kappa(A \cap B) = C \cup \{w \in \bar{A} \cup \bar{B} \mid \kappa(w) = \kappa(\bar{A} \cup \bar{B})\}$, which is hence a subset of $\div_\kappa(A) \cup \div_\kappa(B)$ due to the law of disjunction for negative ranks.

Secondly, we have to show that $\div_\kappa(A) \subseteq \div_\kappa(A \cap B)$ provided that $\div_\kappa(A \cap B) \cap \bar{A} \neq \varnothing$ (this is the propositional (K÷8)). Again, there is nothing to prove in the trivial case. If $\div_\kappa(A)$ is a genuine contraction, so is $\div_\kappa(A \cap B)$, and the equations just stated concerning (K÷7) apply. Hence, the proviso in effect ensures that $\kappa(\bar{A}) = \kappa(\bar{A} \cup \bar{B})$ and thus that the required subset relation holds.

The final step is to check whether revision and contraction as explained in ranking-theoretic terms are related by the Levi and the Harper identity (4.22) and (4.23). This is indeed the case:

(5.43) $*_\kappa(A) = \div_\kappa(A) \cap A$, and $\div_\kappa(A) = *_\kappa(W) \cup *_\kappa(\bar{A})$,

as we know already from Theorem 4.25. And so we may conclude that belief revision theory is completely preserved in ranking theory, as I noted in Spohn (1988, footnote 20). Gärdenfors (1988, Sect. 3.7) has emphasized this as well.

5.6 An Appendix on Iterated AGM Belief Revision and Contraction

What I have presented in this chapter is essentially contained already in my (1983b, 1988). So far I have added only some variant notions and some alternative explanations. Since my account arose in response to the problem of iterated belief change and the problem was posed as a challenge to belief revision theory, one should expect to see noticeable efforts from that side to solve the problem. Therefore, as a kind of appendix to this chapter, I will briefly review these efforts. I can state, though, my conclusion in advance: We will find an important reason why the problem has not been taken as really pressing in the belief revision literature, and maybe because of this the contributions to the problem can only be called disappointing from our vantage point. For this reason I will feel justified to continue developing ranking theory in this book.

When we consider the problem of iterated belief change from the point of view of belief revision theory, its challenge is hard to perceive. I explained in Section 4.4 that a selection function g and an AGM revision function $*$ have different formats: $*$ tells us how to revise any belief set, whereas g tells only how to revise a given belief set having the core $g(W)$. And I said that this is the root of all further differences. This is now apparent, for a revision function $*$ trivially defines iterated belief change. Starting with any belief set K, it delivers $(\ldots (K * \varphi_1) * \ldots) * \varphi_n$, for any $\varphi_1, \ldots, \varphi_n$. So there is no problem here at all. But this is so only at the expense of presupposing a permanent disposition $*$ to change whatever belief set one may happen to have. I noted in the previous chapter that this presupposition is highly problematic, and we indeed rejected it in this chapter. There is nothing impossible about having the same belief set at two different times, but with different dispositions to change it.

In any case, my picture is quite different. I am looking for a suitable notion of a momentary doxastic state, whatever it may be, which allows us to state a general dynamic law for the whole of these states, in agreement with the principle of categorical matching (5.1), and not for some parts of them while keeping other parts constant. We observed that selection functions do not offer such a notion precisely because of the problem of iterated belief change. And if we see a ranking function κ as consisting of two components, a belief set or a core $\kappa^{-1}(0)$ and a disposition to change the belief set somehow embodied in the rest of the ranking function – not a helpful separation in my view – then the rule of $A{\to}n$-conditionalization tells us how to change both components at once. I suspect that this difference of perspective is responsible for a lot of misunderstanding. Darwiche, Pearl (1997, Sect. 3) have already emphasized the point, and Rott (1999) and (2001, Sect. 3.6) has also recognized its importance and commented on it.

This is not to say that there is no problem at all from the perspective of belief revision theory. It only takes a different and apparently less pressing form. The problem is not to define iterated belief change, the problem is rather to say more definite things about it – to investigate more fully into the rationality postulates governing it. In the meantime the discussion has become almost confusingly rich. There is no point in resuming it here, especially because Rott (2009) has given an excellent up-to-date survey and because I am hoping to essentially settle the issue in Section 8.4. So I will confine myself to some classificatory remarks and to some first glimpses.

A first confusing aspect is that even single belief change is not at all a settled affair and that the issues there radiate to iterated belief change. For instance, one fundamental issue is whether belief change should be discussed in terms of belief bases (that we neglected) or in terms of deductively closed belief sets (as was done here). Iterated belief change in terms of belief bases seems to be an even more difficult topic (see, e.g., the remarks in Rott 2001, Sect. 5.2.3); Williams (1995) attempts to cope with it by applying ranking-theoretic means in her theory of so-called transmutations. However, Rott (2009) carries out his comparative discussion by starting from belief bases and thus bridges the difference. I will not pursue the point; it is too remote from the approach taken here.

Even then, though, single belief change is contested. The full AGM postulates of belief revision and contraction have been vigorously criticized, mainly as too strong (see, e.g., the impressive list of alternative postulates in Rott 2001, Sec. 4.2 and 4.3). I will discuss some criticisms in Section 11.3. Of course this debate heavily affects iterated belief change, although there seems to have been little effort put into working out the consequences. A noticeable exception is Bochman (1999) who provides a general framework for representing iterated contraction without presupposing the full AGM postulates. Still, his emphasis is on the representation of weaker postulates for single contractions and not on iterations.

Once again, I will not pursue this line of investigation. As explained in (5.40) and (5.42), ranking theory embraces the full AGM postulates, and so I agree with the main discussion of iterated belief change that proceeds on the basis of belief sets and the full AGM postulates. Let us take a brief look at what has been achieved there.

A first point is this: Rott (1999) is right in insisting that the postulates $(K*7)$ and $(K*8)$, as they stand, are only postulates of dispositional coherence, as he says, relating various possible one-step revisions. Still, the intuition behind them is rather about two-step revisions; this is how I explained them in Section 4.2. If we make this intuition explicit, they say:

$$(5.44*) \quad \text{if } \neg\psi \notin K * \varphi, \text{ then } (K * \varphi) * \psi = Cn((K * \varphi) \cup \{\psi\}) = K * (\varphi \wedge \psi).$$

That is, if the second piece of information is consistent with the belief set reached after the first piece of information, then the second revision is nothing but an expansion.

Yet (5.44*) is not a genuine law of iterated revision. The second equation is nothing but $(K*7)$–$(K*8)$ or, equivalently, the law (4.11b) for selection functions, and the first equation simply applies $(K*3)$–$(K*4)$ to $K * \varphi$. Thus (5.44*) does not go beyond what was already contained in our assumptions about one-step revisions. Accordingly, (5.44*) is satisfied by (5.2) and (5.4) as well as by (5.25).

It is useful to consider what this and the further conditions on iterated revisions mean in terms of WWOs and what they say about how to change the prior WWO $\leq$ guiding the first revision upon receiving evidence E into the posterior WWO $\leq_E$ guiding possible second revisions. We will be going back and forth between the sentential and the propositional framework in what follows. So it is perhaps helpful to explicitly state the relation between sentential revision and WWOs:

(5.45)   Let $\mathbf{L}$ be a set of sentences closed under propositional logic. For $\varphi \in \mathbf{L}$, let $T(\varphi) \subseteq W$ be the truth-condition of φ or the proposition expressed by φ. So, $\{ T(\varphi) \mid \varphi \in \mathbf{L} \}$ forms an algebra $\mathcal{A}$ over W. Finally, let $\leq$ be an $\mathcal{A}$-WWO and $K = \{ \varphi \mid \min_\leq(W) \subseteq T(\varphi) \}$. Then revision relative to $\leq$ is defined as $K * \varphi = \{ \psi \mid \min_\leq(T(\varphi)) \subseteq T(\psi) \}$.

This is the connection we had assumed all along. In the sequel φ will stand for the sentential evidence and E for the propositional evidence (hence $E = T(\varphi)$). In terms of WWOs, then, (5.44*) is equivalent to:

(5.44$\leq$) $w \leq_E w'$ for all $w' \in W$ iff $w \in \min_\leq(E)$.

The crucial question is how to go beyond (5.44). Rott (1991b, Sect. 6) was about the first within the belief revision paradigm to make a proposal about how to change epistemic entrenchment orderings (which correspond to our WWO $\leq$), later called *irrevocable belief revision* by Segerberg (1998) (who restricted it to suppositional reasoning) or *radical belief revision* by Rott (2003, Sect. 8.2) (who, however, was dissatisfied by it). In Rott (1991b, Sect. 6), he still conjectures that the difficulties are due to limitations of proceeding purely in terms of entrenchment relations.

Later on in Rott (1999, 2003), he is more optimistic and defines and discusses approvingly what he calls *moderate belief revision*. Within the belief revision paradigm, it was first introduced by Nayak (1994), but many others have discussed it as well. However, a little inspection shows, as Rott (2003, p. 131) notes as well, that moderate belief revision is equivalent to the idea (5.2) discussed in Section 5.1. Thus, I cannot see the grounds for positively considering this proposal.

Boutilier (1993) proposes to amend (5.44) in an opposite way. This is not immediately perspicuous because he is concerned with a superficially different problem. He is attached to the Ramsey test and to the issue of explaining belief in conditionals. He rightly observes that one-step belief revision as explained in (K*1)–(K*8) does not say anything specific about revising belief in conditionals. This is what he wants to supplement (as do others engaged with the same problem). But, of course, if belief in conditionals is explained by belief revision (via the Ramsey test), then the revision of beliefs in conditionals becomes very similar to iterated belief change. Still I find that this means mixing up the problems, and I prefer to stay away from beliefs in conditionals.

If only the net effect of Boutilier's constructions for iterated belief revision is observed, then his proposal amounts to what is called *conservative belief revision* by Rott (2003) (since it is conservative insofar as it preserves a maximal number of beliefs in conditionals). To define such a revision method was Boutilier's original goal, and this is why it is still under discussion. Again, though, a little inspection shows (see Darwiche, Pearl 1997, Theorem 11) that conservative revision is equivalent to our old proposal (5.4). And so from my point of view conservative revision is as unsuitable as moderate revision. I am surprised to see that these kinds of revision still appear to be live options in the belief revision literature.

By contrast, Darwiche, Pearl (1997) make real progress, indeed (almost) as much as one can within the confines of the belief revision paradigm. They supply a lot of intuitive support for the following four postulates:

(5.46*)   if $\varphi \in Cn(\psi)$, then $(K * \varphi) * \psi = K * \psi$,
(5.47*)   if $\neg\varphi \in Cn(\psi)$, then $(K * \varphi) * \psi = K * \psi$ (= 5.40*),
(5.48*)   if $\varphi \in K * \psi$, then $\varphi \in (K * \varphi) * \psi$,
(5.49*)   if $\neg\varphi \notin K * \psi$, then $\neg\varphi \notin (K * \varphi) * \psi$.

But I find their theoretical support, which consists in explaining what (5.46*–5.49*) mean in terms of WWOs, is more important. In this way these postulates become fully perspicuous. If $\leq$ is the prior WWO associated with the belief set K, $\leq_E$ the posterior WWO brought about by revision by φ (E being the truth condition of φ), and $<$ and $<_E$ the corresponding strict relations, then the following conditions on the posterior WWO appear plausible:

(5.46≤)   if $w, w' \in E$, then $w \leq_E w'$ iff $w \leq w'$,
(5.47≤)   if $w, w' \in \bar{E}$, then $w \leq_E w'$ iff $w \leq w'$ (= 5.41≤),
(5.48≤)   if $w \in E, w' \in \bar{E}$, and $w < w'$, then $w <_E w'$,
(5.49≤)   if $w \in E, w' \in \bar{E}$, and $w \leq w'$, then $w \leq_E w'$.

(5.46≤) and (5.47≤) postulate that the ordering of disbelief is preserved within E and within $\bar{E}$, an idea we have maintained from the beginning of this chapter. And (5.48≤) and (5.49≤) require that a possibility in E which is initially less (or at most as) disbelieved than a possibility in $\bar{E}$ remains so after the change. Again this is a requirement that is satisfied by (5.2), (5.4), and (5.25), the proposal we finally maintained. Darwiche, Pearl (1997) now prove in their Theorem 13:

(5.50)   Given the relation (5.45), (5.46*–5.49*) are, respectively, equivalent to (5.46≤–5.49≤).

Thus (5.46≤–5.49≤) confer their plausibility on (5.46*–5.49*).

At the same time, the theorem makes clear why these conditions can hardly be strengthened within the confines of the belief revision paradigm. The only case left open by (5.46≤–5.49≤) is the case where $w \in E, w' \in \bar{E}$, and $w' < w$. It is clear that the revision by E must upgrade the position of w as opposed to w'. But unless one sticks to proposals as rigid as (5.2) and (5.4), there is no general way of saying whether or not w is sufficiently upgraded to end up with $w \leq_E w'$. Hence that case must remain open as long as we conceive of evidence as merely propositional. It can be settled only by some such idea as our $E{\rightarrow}n$-conditionalization.

In fact a slight improvement is feasible. (5.48≤) and (5.49≤) only say that $w \in E$ does not worsen its position as compared to $w' \in \bar{E}$ in a revision by E. However, if the revision is not running empty (because E is already believed), then $w \in E$ should even improve its position in relation to $w' \in \bar{E}$; that is, if $w \leq w'$, then $w <_E w'$. When turning to contraction, I will state this improvement in a more precise way.

This seems to exhaust iterated belief revision in terms of entrenchment orders and propositional evidence. So it might appear that Nayak (1994) has drawn the right conclusion. He is still attached to the ordinal approach to belief revision theory; however, he conceives of evidence not as a mere proposition, but as a whole entrenchment relation or ordering of disbelief. If that evidential ordering is more complex, then this idea is even reminiscent of our general evidence-oriented $\mathcal{E}{\uparrow}\varepsilon$-conditionalization (5.34). However, if one takes a closer look at his conditions (EE9*) and (EE10*) (p. 366) and Definition 1 (p. 372), one sees that the evidential ordering dominates the prior ordering. Thus, in the simple case where the evidential ordering distinguishes only two levels, Nayak's idea reduces to what we above called moderate revision, i.e., to our old (5.2). Again, we find no genuine progress.

So we may conclude that all attempts to state a doxastic dynamics on the level of selection functions, WWOs, or WOPs either fall into one of the overly rigid extremes of (5.2) and (5.4) (or slight variants thereof) or pay for their more adequate flexibility with incompleteness, as did Darwiche, Pearl (1997). Of course my brief discussion here did not, and could not, strive for completeness. There are many more models of belief revision, illuminating in various aspects and contributing also to the problem of iterated belief revision. In particular, there are various attempts to deal with iterations which, roughly, work by somehow relativizing selection functions and entrenchment orders to further parameters, by turning 'two-dimensional' (this has nothing to do with the two-dimensional semantics alluded to in Chapter 2). This is done, for instance, by the two-place selection functions envisaged by Hansson (1993), by global belief revision proposed by Rabinowicz (1995), by revision by comparison introduced by Cantwell (1997) and studied by Fermé, Rott (2004), by restrained revision considered by Booth, Meyer (2006), by bounded revision invented by Rott (2007), and so on; see the most useful overview in Rott (2009). However, I could not see that there are attempts that do radically better than the ideas already discussed. This is, of course, not a proof that it is impossible to do so with such means, but the fact that almost twenty years of thorough-going research got rather bogged down in ramifications, far from resounding success, are at least strong evidence. So far there is every reason to favor our ranking-theoretic approach and its straightforward, general, and complete account of the dynamics of belief.

Despite the negative overall conclusion, our discussion has gathered some positive findings concerning iterated revision. In view of the close correlation between revision and contraction it is useful to translate these findings into postulates about iterated contractions. They will be equally incomplete. However, we will return to the topic in Section 8.4 from a different perspective. There we will find a complete set of iterative postulates, though only in terms of contractions. So it will be interesting to compare how far we have got with our present considerations.

The positive findings about iterated revisions consisted in the postulates (5.46)–(5.49) (and the slight improvement observed after 5.50). Let us denote the posterior WWO generated from the prior WWO $\leq$ by contraction with respect to the proposition E (as expressed by the sentence φ) by $\leq_{+E}$. $\leq$ governs the first contraction, $\leq_{+E}$ governs the second contraction, and therefore assertions about $\leq_{+E}$ are assertions about iterated contractions. The postulates for $\leq_{+E}$ corresponding to (5.46$\leq$)–(5.49$\leq$), including their improvement, are straightforward to state:

(5.51$\leq$)   if $w, w' \in E$, then $w \leq_{+E} w'$ iff $w \leq w'$,

(5.52$\leq$)   if $w, w' \in \bar{E}$, then $w \leq_{+E} w'$ iff $w \leq w'$,

(5.53$\leq$)   if $w \in \bar{E}, w' \in E$, and $w \leq w'$, then $w <_{+E} w'$, provided there is a $v \in E$ with $v < v'$ for all $v' \in \bar{E}$,

(5.54$\leq$)   if there is a $v' \in \bar{E}$ with $v' \leq v$ for all $v \in E$, then $w \leq_{+E} w'$ iff $w \leq w'$.

(5.51$\leq$)–(5.52$\leq$) again state that the ordering of disbelief does not change within E and $\bar{E}$. And (5.53$\leq$) corresponds to the slight improvement of (5.48$\leq$)–(5.49$\leq$): the proviso states the contraction to be a genuine one, since it says in effect that E is initially believed; and then (5.53$\leq$) states that the possibilities in $\bar{E}$ get *less* disbelieved as compared to the possibilities in E. Finally, (5.54$\leq$) treats the case of vacuous contraction, since satisfying its antecedent amounts to not believing E, so that there is nothing to contract.

What do these postulates mean in terms of iterated contractions? Let us first explicitly state how contractions are related to WWOs:

(5.55)   Under the same suppositions as in (5.45) let contraction relative to $\leq$ be defined as $K \div \varphi = \{\psi \mid \min_\leq(W) \cup \min_\leq(T(\neg\varphi)) \subseteq T(\psi)\}$.

Again, this is the connection we had assumed all along; cf. (5.41). Under (5.55) the corresponding contraction postulates run as follows:

(5.51$\div$)   if $\psi, \chi \in Cn(\neg\varphi)$, then $\chi \in (K \div \varphi) \div \psi$ iff $\chi \in K \div \psi$,

(5.52$\div$)   if $\psi, \chi \in Cn(\varphi)$ and $\psi, \chi \in K \div \varphi$, then $\chi \in (K \div \varphi) \div \psi$ iff $\chi \in K \div \psi$,

(5.53$\div$)   if $\varphi \in K$ and $\varphi \wedge \neg\psi \notin K \div \psi$, then $\varphi \rightarrow \psi \in (K \div \varphi) \div \psi$,

(5.54$\div$)   if $\varphi \notin K$, then $(K \div \varphi) \div \psi = K \div \psi$.

Perhaps the best way to understand these postulates is to prove them to be equivalent to the postulates about orderings:

(5.56)   Given the relation (5.55), (5.51$\div$)–(5.54$\div$) are, respectively, equivalent to (5.51$\leq$)–(5.54$\leq$).

Proof: As to (5.54): The suppositions of (5.45) entail that $\varphi \notin K$ iff there is a $v' \in \bar{E}$ with $v' \leq v$ for all $v \in E$. The consequent of (5.54$\leq$) entails that of (5.54$\div$). Conversely, if there were w, w' with $w \leq_{+E} w'$ and $w' < w$, then we could choose $\psi \in L$ such that $T(\psi) = \{v \mid v \approx w'\}$ (since $\leq$ is an $\mathcal{A}$-WWO), and then $(K \div \varphi) \div \psi \neq K \div \psi$.

As to (5.53): Again the provisos of (5.53$\leq$) and (5.53$\div$) are equivalent. From (5.53$\leq$) we may proceed to (5.53$\div$) as follows: According to (5.55), $\varphi \rightarrow \psi \in (K \div \varphi) \div \psi$ translates into $\min_{\leq_{+E}}(W) \cup \min_{\leq_{+E}}(T(\neg\psi)) \subseteq T(\varphi \rightarrow \psi)$, i.e., $\min_{\leq}(W) \cup \min_{\leq}(T(\neg\varphi)) \cup \min_{\leq_{+E}}(T(\neg\psi)) \subseteq T(\neg\varphi) \cup T(\psi)$. Clearly, $\min_{\leq}(W) \subseteq T(\psi)$, since $\psi \in K$. Trivially, $\min_{\leq}(T(\neg\varphi)) \subseteq T(\neg\varphi)$. Hence, it remains to show that $\min_{\leq_{+E}}(T(\neg\psi)) \subseteq T(\neg\varphi)$. We have supposed that $\varphi \notin K \div \psi$, i.e., $\min_{\leq}(W) \cup \min_{\leq}(T(\neg\psi)) \not\subseteq T(\varphi)$, i.e., $T(\neg\varphi) \cap \min_{\leq}(T(\neg\psi)) \neq \varnothing$ (since $T(\neg\varphi) \cap \min_{\leq}(W) = \varnothing$ due to the proviso $\varphi \in K$). Hence, for all $w \in T(\neg\varphi) \cap \min_{\leq}(T(\neg\psi))$ and $w' \in T(\varphi) \cap T(\neg\psi)$ $w \leq w'$ and, according to (5.53$\leq$), $w <_{+E} w'$. Therefore, we indeed have $\min_{\leq_{+E}}(T(\neg\psi)) \subseteq T(\neg\varphi)$.

Conversely, for any $w \in \bar{E}$ and $w' \in E$ with $w \leq w'$ there is a $\psi \in L$ such that $T(\psi) = \{v \in \bar{E} \mid v \approx w\} \cup \{v \in E \mid v \approx w'\}$ (since $\leq$ is an $\mathcal{A}$-WWO). This ψ satisfies the suppositions of (5.53$\div$). We have just seen that (5.53$\div$) then amounts to $\min_{\leq_{+E}}(T(\neg\psi)) \subseteq T(\neg\varphi) = \bar{E}$, i.e., $w <_{+E} w'$.

As to (5.51): With (5.55), (5.51$\div$) says that if $\bar{E} \subseteq T(\psi), T(\chi)$, then $\min_{\leq_{+E}}(W) \cup \min_{\leq_{+E}}(T(\neg\psi)) = \min_{\leq}(W) \cup \min_{\leq}(T(\neg\varphi)) \cup \min_{\leq_{+E}}(T(\neg\psi)) \subseteq T(\chi)$ iff $\min_{\leq}(W) \cup \min_{\leq}(T(\neg\psi)) \subseteq T(\chi)$. Since $T(\neg\psi) \subseteq E$, (5.51$\leq$) ensures that $\min_{\leq_{+E}}(T(\neg\psi)) = \min_{\leq}(T(\neg\psi))$. Hence, (5.50$\div$) holds true. Conversely, for any $w, w' \in E$ we may choose ψ such that $(T(\neg\psi)) = \{v \in E \mid v \approx w$ or $v \approx w'\}$. We have just seen that (5.51$\div$) entails that $\min_{\leq_{+E}}(T(\neg\psi)) = \min_{\leq}(T(\neg\psi))$. With (5.55) this translates into $w \leq_{+E} w'$ iff $w \leq w'$.

As to (5.52): With (5.55), (5.52$\div$) say that if $E \subseteq T(\psi), T(\chi)$ and $\min_{\leq}(W) \cup \min_{\leq}(T(\neg\varphi)) \subseteq T(\psi), T(\chi)$, then $\min_{\leq}(W) \cup \min_{\leq}(T(\neg\varphi)) \cup \min_{\leq_{+E}}(T(\neg\psi)) \subseteq T(\chi)$ iff $\min_{\leq}(W) \cup \min_{\leq}(T(\neg\psi)) \subseteq T(\chi)$. Since $T(\neg\psi) \subseteq \bar{E}$ and $T(\neg\psi) \cap \min_{\leq}(T(\neg\varphi)) = \varnothing$, (5.52$\leq$) ensures that $\min_{\leq_{+E}}(T(\neg\psi)) = \min_{\leq}(T(\neg\psi))$. Hence, (5.52$\div$) holds true. Conversely, for any $w, w' \in \bar{E}$ we may choose ψ such that $(T(\neg\psi)) = \{v \mid v \in \bar{E}, v \notin \min_{\leq}\bar{E},$ and $v \approx w$ or $v \approx w'\}$. We have just seen that (5.52$\div$) entails that $\min_{\leq_{+E}}(T(\neg\psi)) = \min_{\leq}(T(\neg\psi))$. With (5.55) this translates into $w \leq_{+E} w'$ iff $w \leq w'$.

Our findings in Section 8.4 will confirm all postulates (5.51)–(5.54); but they will also go considerably beyond them.

Nayak et al. (2007) have further strengthened the postulates for iterated contraction. They convincingly argue for what they call "Principled Factored Insertion" (pp. 2570f.). It says:

(5.57$\div$) If $\varphi \vee \psi \in (K \div \varphi) \div \psi$, then $(K \div \varphi) \div \psi = K \div \varphi \cap K \div (\neg\varphi \vee \psi)$;
if $\neg\varphi \vee \psi \in (K \div \varphi) \div \psi$, then $(K \div \varphi) \div \psi = K \div \varphi \cap K \div (\varphi \vee \psi)$;
and if $\varphi \vee \psi, \neg\varphi \vee \psi \notin (K \div \varphi) \div \psi$, then $(K \div \varphi) \div \psi = K \div \varphi \cap K \div (\varphi \vee \psi) \cap K \div (\neg\varphi \vee \psi)$.

Contrary to appearances, (5.57$\div$) is not a reduction of iterated contraction to single contraction, since the single contractions $K \div \varphi$, $K \div (\varphi \vee \psi)$, and $K \div (\neg\varphi \vee \psi)$ alone do not tell which of the three cases of (5.57÷) applies. Once again, (5.57÷) will be entailed by the postulates derived in Section 8.4. So I fully endorse (5.57÷). However, the disagreement would start again if (5.57÷) were considered to be half way to a complete characterization of so-called lexicographic contraction, as Nayak et al. (2007) do, since lexicographic contraction is not compatible with ranking-theoretic contraction (5.41) when iterated.

5.7 A Further Appendix on Multiple Contraction

There is an unsolved issue in belief revision theory that is, I think, best understood as an application of generalized conditionalization (5.33) – no wonder that it is unsolved. This is the issue of so-called multiple contraction, which I would like to address now. This section may thus be understood either as a contribution to belief revision theory or as an illustration of the power of ranking conditionalization. Actually, though, this section is merely a digression that fits in here better than elsewhere. This is why I have inserted it here, but you might as well skip it.

Let us return for a moment to the linguistic framework of AGM belief revision theory where we deal with sentences φ of a language $\mathbf{L}$ (= the set of well-formed formulae) and deductively closed belief sets K in $\mathbf{K}$ (= the set of all such belief sets); cf. Section 4.4. We have seen that the contraction $K \div \varphi$ of some belief set K by some sentence φ is an epistemologically most important operation which we could adequately account for in ranking theory (cf. Definition 5.41). Now it appears quite natural to think also of contracting a belief set K by a whole set of sentences $L \subseteq \mathbf{L}$, resulting in a belief set $K \div L$. This is *multiple contraction*, a term introduced by Fuhrmann (1988). One might well wonder how it behaves or should behave.

The first thing to note about it is that it is ambiguous: $K \div L$ might mean that all sentences in L are to be contracted from K, or that at least one sentence in L is to be contracted from K. The latter is called *choice contraction* by Fuhrmann, Hansson (1994) and denoted by $K \div \{L\}$. (Fuhrmann, Hansson (1994) is my main reference in the sequel, but Fuhrmann (1997, Ch. 3) should be consulted as well. Their notation for choice contraction is $K \div \langle L \rangle$, but I would like to reserve the angle brackets for denoting iterated contraction.) Choice contraction is a familiar case. It is a standard observation in philosophy of science that a failed prediction does not necessarily falsify the scientific theory used for the prediction. The fault might be anywhere: in the theory, in the auxiliary hypotheses, in the assumptions about initial and boundary conditions, etc. So at least one of all these claims must be given up in order to block the failed prediction. This is a case of choice contraction. And it is obvious how to account for it, at least if L is finite, i.e., $L = \{\varphi_1, \ldots, \varphi_n\}$. If $L \subseteq K$, then $K \div \{L\}$ is tantamount to $K \div \varphi_1 \wedge \ldots \wedge \varphi_n$, since that conjunction is deleted from K if and only if at least one of the conjuncts has to go as well. And if not $L \subseteq K$, then choice contraction runs empty, and we may trivially equate $K \div \{L\}$ with $K \div \varphi_1 \wedge \ldots \wedge \varphi_n$. Hence, as Fuhrmann, Hansson (1994, p. 72) have already observed, choice contraction reduces to ordinary contraction by single sentences in this finite case (and within our propositional framework we could even skip the finiteness constraint).

The interesting case is, therefore, the first understanding of multiple contraction, the contraction of all members of L from K, which is called *package contraction* by Fuhrmann, Hansson (1994) and denoted by $K \div [L]$. Package contraction seems to be a natural generalization of ordinary contraction, and equally important. Imagine I tell you a lot of stories one day and you believe all of them, since I always appeared to be a well-informed and trustworthy guy. The next day I meet you again and apologetically say: "I am awfully sorry. I don't know what happened to me yesterday, somehow it was as if I were drugged. In any case, please forget everything I told you yesterday. It was all lies." What you face, then, is a problem of package contraction. And it is a difficult problem even if you know everything about ordinary contraction. Fuhrmann, Hansson (1994) argue more extensively for the ubiquity of package contraction.

Why is it difficult? Let us look at the simplest genuine case, where $L = \{\varphi, \psi\}$, i.e., at the contraction $K \div [\varphi, \psi]$ of two sentences φ and ψ from K. It clearly cannot be explained as $K \div \varphi \wedge \psi$. To contract by the conjunction guarantees only that at least one of the conjuncts has to go, but the other may be retained, so that the package reduction would be unsuccessful. Success, i.e., $(K \div [L]) \cap L = \varnothing$ is a basic requirement. In other words, it would be wrong to equate package contraction with choice contraction. They agree only in the degenerate case of contraction by a singleton.

Nor may package contraction $K \div [\varphi, \psi]$ be explained as $K \div \varphi \vee \psi$. This would indeed guarantee success, for if the disjunction has to give way, then the disjuncts have to do so as well. However, the proposal is clearly too strong. One might well give up both disjuncts while retaining the disjunction. In any case, this should not be excluded.

Package contraction must also be distinguished from iterated contraction. The easiest way to see this is that iterated contractions need not commute. Possibly $(K \div \varphi) \div \psi \neq (K \div \psi) \div \varphi$. When one asks in such a case with which of the two terms $K \div [\varphi, \psi]$ should be identified, the obvious answer is 'none'. There is no such asymmetry in the idea of package contraction.

Hansson (1993) gives a nice example in which commutativity of iterated contractions intuitively fails.

In the ongoing conflict between India and Pakistan, troops have been sent to the border from both sides. A friend has told me that an agreement has been reached between the two countries to withdraw the troops. I believe in this. I also believe that each of the two governments will withdraw its troops if there is such an agreement, but for some reason my belief in the compliance of the Pakistan government is stronger than my belief in the compliance of the Indian government.

Let s denote that there is an agreement to withdraw troops on both sides, p that the Pakistan government is going to withdraw its troops and q that the Indian government is going to withdraw its troops. Then (the relevant part of) my belief base is $\{s, s \to p, s \to q\}$.

Case 1: The morning news on the radio contains one single sentence on the conflict: "The Indian Prime Minister has told journalists that India has not yet decided whether or not to withdraw its troops from the Pakistan border." When contracting q, I have to choose between retaining s and $s \rightarrow q$. Since my belief in the latter is weaker, I let it go, and the resulting belief base is $\{s, s \rightarrow p\}$.

The evening news also contains one single sentence on the conflict, namely: "The Pakistan government has officially denied that any decision has been taken on the possible withdrawal of Pakistan troops from the Indian border." I now have to contract p. This involves a choice between retaining s and retaining $s \rightarrow p$. Because of my strong belief in the latter, I keep it, and the resulting belief base is $\{s \rightarrow p\}$.

Case 2: The contents of the morning and evening news are interchanged.

In the morning, when contracting p from the original belief base $\{s, s \rightarrow p, s \rightarrow q\}$, I retain $s \rightarrow p$ rather than s, because of the strength of my belief in the former. The resulting belief base is $\{s \rightarrow p, s \rightarrow q\}$. The contraction by q that takes place in the evening leaves this set unchanged. (Hansson 1993, p. 648)

If we accept the ranking-theoretic account of contraction (5.41), then we may state the conditions of commutativity in a precise way. Let us return to the propositional framework. Recall that $\kappa_{\div A}$ was the contraction of the negative ranking function κ by the proposition A and that $\div_\kappa(A)$ was the core of $\kappa_{\div A}$, which was the contraction of the core of κ on the basis of κ, i.e., the kind of contraction dealt with in AGM belief revision theory. On the basis of κ we are clearly able to iterate contraction: we may simply set $\div_\kappa\langle A, B \rangle = \div_{\kappa_{\div A}}(B)$. After this reminder we may state

> *Theorem 5.58*: Let C be the core of κ. Then $\div_\kappa\langle A, B \rangle \neq \div_\kappa\langle B, A \rangle$ if and only if $C \subseteq A, B, \kappa(B \mid \bar{A}) = 0$ or $\kappa(A \mid \bar{B}) = 0$, and $\kappa(\bar{B} \mid \bar{A}) < \kappa(\bar{B} \mid A)$ (which is equivalent to $\kappa(\bar{A} \mid \bar{B}) < \kappa(\bar{A} \mid B)$).

Instead of stating the easy proof, let me rather explain the gist of the matter. It is, roughly, that the last condition requires the positive relevance of A to B (and vice versa) and that the first conditions then have the effect either that $A \cap \bar{B}$ is disbelieved (or, "if A, then B" believed) after contracting first by A and then by B, but not after the reverse contraction, or that $\bar{A} \cap B$ is disbelieved (or "if B, then A" believed) after contracting first by B and then by A, but not after the reverse contraction (or that both is the case). This is exactly how non-commutativity of contraction can come about. Indeed the survival of material implication in iterated contractions will play a crucial role in the measurement procedures of Chapter 8. Note, by the way, how nicely Hansson's example fits the conditions of Theorem 5.58. In any case, we must conclude that package contraction is not iterated contraction.

Fuhrmann, Hansson (1994) discuss a final option, namely that $K \div [\varphi, \psi] = K \div \varphi \cap K \div \psi$. Then, even though package contraction is not an ordinary contraction, it might be explained in the latter terms and thus could be reduced away as an independent phenomenon. However, they are not happy with that option, either, because they believe it to be incompatible with the approach they prefer (cf. p. 62, *ibid.*). I believe they erred, as we will soon see. On the other hand, it is intuitively not so clear that this is the right explanation. So one must look for another approach anyway.

The only approach left for Fuhrmann, Hansson (1994) is the axiomatic one: if we cannot define package contraction, then we can at least try to characterize it. And so they start appropriately generalizing the postulates $(K\div 1)$–$(K\div 8)$ (cf. 4.19). This works convincingly for $(K\div 1)$–$(K\div 6)$, and they even produce representation results for their generalization (see their Theorem 9 on p. 59). But they are not sure what to do with $(K\div 7)$ and $(K\div 8)$. Sven Ove Hansson told me that he no longer believes in the proposals made there on p. 56. Research seems to have reached a dead end at this point.

Let us now look at package contraction from the vantage point of ranking theory. We start with a negative ranking function κ for $\mathcal{A}$ and a finite set $\mathcal{B} \subseteq \mathcal{A}$ of propositions, and then we ask how to change κ and its core C so that none of the propositions in $\mathcal{B}$ are still believed in κ. We have seen above that we may restrict attention to $\mathcal{B}' = \{A \in \mathcal{B} \mid C \subseteq A\}$, the propositions in $\mathcal{B}$ believed in κ, since contraction is vacuous for the other propositions in $\mathcal{B}$. We have seen, moreover, that we may have to deal with logical combinations of propositions in $\mathcal{B}'$. Let us focus hence on the algebra $\mathcal{B}^*$ generated by $\mathcal{B}'$. There is no reason why propositions outside $\mathcal{B}^*$ should become relevant.

Now we should proceed as follows: We should start by contracting $\bigcap \mathcal{B}'$, the strongest believed proposition in $\mathcal{B}^*$. This is the same as choice contraction by $\mathcal{B}'$, and we have noted that it removes some beliefs in $\mathcal{B}'$. If we are lucky, it even removes all beliefs in $\mathcal{B}'$. If so, we are done with the package reduction. Yet this would be exceptional. Normally, we will have moved from the prior core $C = C_0$ to a larger core $C_1 \supseteq C_0$ which still includes belief in some propositions in $\mathcal{B}'$. So in a further step we again proceed as cautiously as possible and contract the strongest proposition in $\mathcal{B}^*$ still believed, i.e., $\bigcap\{A \in \mathcal{B}^* \mid C_1 \subseteq A\}$. Package contraction might now be completed. If not, then we have arrived at a core $C_2 \supseteq C_1$ that still holds on to some other beliefs in $\mathcal{B}'$. In this case, we add a third step, and so on, until all beliefs in $\mathcal{B}'$ are removed. This procedure must stop after finitely many steps.

The conception behind this procedure is the same as in ordinary contraction of a single proposition: change as little as possible till the contraction is successful. Minimal change translates here as giving up the weakest beliefs, i.e., those with the lowest positive rank.

Let us cast this into formal definition: Let $\mathcal{E} = \{E_0, \ldots, E_k\}$ be the set of atoms of $\mathcal{B}^*$. Let $E_0 = \bigcap \mathcal{B}'$. So, $\kappa(E_0) = 0$ and $\kappa(E_i) > 0$ for $i = 1, \ldots, k$. Thus, the first contraction informally described above is an $E_0 \to 0$-conditionalization. Thereby some further atoms receive rank 0, say E_1 and E_2, so that $E_3 \cup \ldots \cup E_k$ is still disbelieved. The second contraction outlined above then is an $E_1 \cup E_2 \to 0$-conditionalization, and so on. Let $R = \{\kappa(E) \mid E \in \mathcal{E}\}$ be the set of ranks occupied by the atoms of $\mathcal{E}$. Let $m = \min \{n \in R \mid \bigcup\{E \in \mathcal{E} \mid \kappa(E) > n\} \subseteq \bar{A}$ for all $A \in \mathcal{B}'\}$. If we set only all $E \in \mathcal{E}$ with $\kappa(E) < m$ to 0, contraction of the whole of $\mathcal{B}'$ is not yet completed. If we set all $E \in \mathcal{E}$ with $\kappa(E) \leq m$ to 0, contraction of $\mathcal{B}'$ is successful. And if we set more atoms to 0, we have contracted more than necessary. So m is the margin where our contraction procedure stops. Hence define the ranking function ε on $\mathcal{E}$ by $\varepsilon(E) = 0$ if $\kappa(E) \leq m$ and

$\varepsilon(E) = \kappa(E) - m$ if $\kappa(E) > m$. My proposal for explicating package contraction thus results in the following

> *Definition 5.59*: Let κ, $\mathcal{B}$, $\mathcal{E}$, and ε be as just explained. Then, the *package contraction* $\kappa_{+[\mathcal{B}]}$ of κ by $\mathcal{B}$ is the $\mathcal{E}{\rightarrow}\varepsilon$-conditionalization of κ. And the *package contraction* $\dotdiv_\kappa[\mathcal{B}]$ *of the core* of κ by $\mathcal{B}$ is the core of $\kappa_{+[\mathcal{B}]}$.

In this way package contraction turns out as a special case of generalized conditionalization (5.33). Note that my intuitive explanation of package contraction was in terms of successive contractions. But in order to describe the result in one step we require the expressive power of generalized conditionalization. Note also that this account can be immediately transferred to complete algebras and contractions of infinite sets.

It easily checked that this model of package contraction satisfies all the postulates endorsed by Fuhrmann, Hansson (1994, pp. 51–4). If we accept this explication, we can immediately complete the theory of package contraction. First, it is obvious from the construction above that:

(5.60)   if $\mathcal{B} \subseteq C$, then $\dotdiv_\kappa[\mathcal{B}] \subseteq \dotdiv_\kappa[C]$.

A fortiori, we have

(5.61) $\dotdiv_\kappa[\mathcal{B} \cap C] \subseteq \dotdiv_\kappa[\mathcal{B}] \cup \dotdiv_\kappa[C]$,

which translates into the ranking-theoretic framework what Fuhrmann, Hansson (1994, p. 56) propose as generalization of (K÷7) (see 4.19). Moreover, it is obvious from our construction that:

(5.62)   if for no $B \in \mathcal{B} \dotdiv_\kappa[C] \subseteq B$, then $\dotdiv_\kappa[\mathcal{B} \cup C] \subseteq \dotdiv_\kappa[C]$.

If by contracting C the whole of $\mathcal{B}$ is contracted as well, then our iterative procedure for contracting $\mathcal{B} \cup C$ must stop at the same point as that for contracting C. (5.62) is what Fuhrmann, Hansson (1994, p. 56) offer as generalization of (K÷8) (see 4.19). Thus their tentative proposals are in fact confirmed by our model.

Indeed the most illuminating result concerning our explication (5.59) is:

> *Theorem 5.63*: $\dotdiv_\kappa[A_1, \ldots, A_n] = \dotdiv_\kappa(A_1) \cup \ldots \cup \dotdiv_\kappa(A_n)$. (Recall that the union of cores is equivalent to the intersection of the corresponding belief sets.)

> *Proof*: (5.60) entails that $\dotdiv_\kappa(A_i) \subseteq \dotdiv_\kappa[A_1, \ldots, A_n]$ for $i = 1, \ldots, n$. This proves one direction.
>
>   Conversely, assume that $\dotdiv_\kappa[A_1, \ldots, A_{i-1}] \subseteq \dotdiv_\kappa(A_1) \cup \ldots \cup \dotdiv_\kappa(A_{i-1})$. If $\dotdiv_\kappa[A_1, \ldots, A_{i-1}] \cap \bar{A}_i \neq \varnothing$, then $\dotdiv_\kappa[A_1, \ldots, A_i] = \dotdiv_\kappa[A_1, \ldots, A_{i-1}]$ and there is nothing more to show. If $\dotdiv_\kappa[A_1, \ldots, A_{i-1}] \cap \bar{A}_i = \varnothing$, then $\kappa_{+[A_1,\ldots,A_{i-1}]}(\bar{A}_i) > 0$, and hence $\dotdiv_\kappa[A_1, \ldots, A_i]$ $= \dotdiv_\kappa[A_1, \ldots, A_{i-1}] \cup \{w \in \bar{A}_i \mid \kappa(w) \leq \kappa(w')$ for all $w' \in \bar{A}_i\} = \dotdiv_\kappa[A_1, \ldots, A_{i-1}] \cup \dotdiv_\kappa(A_i)$. This inductively proves the reverse direction.

I take this to be a desirable theorem. It might have been difficult to motivate it as a definition of package contraction, but if it is a consequence of a plausible explication, then this establishes mutual support for the explication and the theorem. In some sense, the theorem is perhaps disappointing. It says that package contraction is reducible to ordinary contraction, after all, and is not an independent general issue.

I am not sure whether I am thereby contradicting Fuhrmann, Hansson (1994). They have doubts about (5.60) (see p. 62, *ibid.*) and hence about the ensuing assertions. But they have these doubts because of their weaker axiomatic basis, which might indeed leave room for denying (5.60–5.63). The only disagreement we may have is that I find my explication (5.59) utterly plausible and hence see no need to retreat to a weaker axiomatic characterization.

6

Reasons and Apriority

So far, our predominant concern was to account for the statics and in particular the dynamics of belief. This led us to introduce ranking theory, and my suggestion was that this theory is the only one that fully meets this concern. However, the foregoing chapter only scratched at the surface of ranking theory. Many things are implicitly contained in the structure so simply defined, and we have to unfold them. This will occupy us for a while. There is no better starting point than the ranking-theoretic explication of the concept of a reason. This will be the subject of the next section. We have to study its formal properties in Section 6.2 and some further central aspects in Sections 6.3–6.4. It will occupy us, though, throughout the entire book. The final Section 6.5 is devoted to the complementary, but not exactly congruent notions of maximal certainty, unrevisability, and apriority.

6.1 Explicating Reasons

A very important philosophical use of probability theory consists in confirmation theory. As such it is most successful; Bayesian confirmation theory seems to dominate the scene. Earman (1992) is quite symptomatic in this respect. It intends to be critical of Bayesianism, but at the same time underscores its central position. Carnap (1962, pp. xv ff.) paradigmatically distinguished two probabilistic notions of confirmation: firmness and increase in firmness (after realizing that previous discussions were confusing the two notions). According to the first, a hypothesis is confirmed by the evidence if the (subjective or logical) probability of the hypothesis given the evidence is high, where the vague "high" at the very least means "greater than .5", but usually something stronger. According to the second, a hypothesis is confirmed by the evidence if the evidence is positively relevant to, or raises the probability of, the hypothesis. This second notion has become much better entrenched, and it has received various forms which are argued to more or less adequately measure the strength of confirmation (for a recent survey see Fitelson 2001).

The high probability criterion was not well received, at least as a notion of probabilistic confirmation. How should one translate it into ranking-theoretic terms? Here the vagueness of "high" becomes even worse. Which positive rank should a hypothesis have, i.e., how firmly should it be believed in order to count as confirmed? The weakest translation apparently is that A confirms B iff B is believed given A, i.e., $\beta(B \mid A) > 0$. In analogy, we might call this the *positive rank* notion of confirmation. In fact, though, this is a misnomer; what is defined thereby is simply not confirmation, it is conditional belief, which we have already studied to some extent. So this is not something we should pursue further under the label "confirmation".

The alternative notion is all the more exciting. I find the notion of positive relevance extremely well anchored in ordinary language. A confirms B, A supports B, A speaks in favor of B, A is a reason for B, etc.: all of these locutions express that A somehow strengthens the belief in B, that B is more firmly believed with A than without A. Of course there is also the opposite notion of negative relevance: that A disconfirms B, A defeats B, A speaks against B, A is a reason against B, etc. Without an account of conditional degrees of belief you cannot express this idea of positive or negative relevance. As soon as you have such an account, however, you can, and then it is interesting to see how it behaves. In any case, ranking theory is able to express this idea. Let us turn this observation into

> *Definition 6.1*: Let κ be a negative and τ the corresponding two-sided ranking function for $\mathcal{A}$, and $A, B \in \mathcal{A}$. Then A is a *reason for* B or *positively relevant to* B w.r.t. κ iff $\tau(B \mid A) > \tau(B \mid \bar{A})$, i.e., iff $\kappa(\bar{B} \mid A) > \kappa(\bar{B} \mid \bar{A})$ or $\kappa(B \mid A) < \kappa(B \mid \bar{A})$. A is a *reason against* B or *negatively relevant to* B w.r.t. κ iff $\tau(B \mid A) < \tau(B \mid \bar{A})$. Finally, A is *relevant to* B or *dependent on* B w.r.t. κ iff A is a reason for or against B w.r.t. κ.

Since my (1983b, Sect. 6.1) I have offered this explication again and again. I emphatically use here the philosophically ennobled term "reason", simply because Definition 6.1 captures one ordinary meaning of that term well. In German the point is even a bit clearer. Here "reason" disambiguates into "Grund" and "Vernunft"; and it is just the notion of a *Grund* or *Begründung* that is explicated in Definition 6.1. It is good to have a precise account of the ordinary meaning before philosophically elevating it, for my claim will be that this account fits well into philosophical contexts. Indeed if one tracks the various philosophical uses of "reason" and approaches them with the present explication, one finds oneself on a journey full of insights and surprises. My discussion of the philosophical applications of ranking theory in the Chapters 12–17 will essentially take us on this journey. Part of it will be a more extensive comparative discussion of other philosophical notions of a reason in Section 16.1, which will fit better in that later context.

Definition 6.1 requires various comments. First, it should be emphasized that being a reason is a relation between propositions, according to (6.1). This might appear doubtful. Colloquially, it is assumptions, conjectures, beliefs, etc., that are thus related. However, these entities are subject to a familiar type/token ambiguity. As types they are essentially contents, and thus it should be clear from our discussion in Chapter 2 that propositions are our uniform substitute for all these entities.

Second, it is obvious from (6.1) that the reason relation is relativized to a subjective doxastic state, i.e., a ranking function. People need not agree in their reasons. You might take the argument I put forward in just the opposite way. – I: "Mick Jagger is really the greatest pop star on earth. He even pleases the Queen." You (ironically): "Yes, precisely." – This relativity may be disappointing. One may have hoped for a more objective notion. I will say a bit more about this point in Section 6.4. In a way, though, this issue will occupy us throughout the book, in particular in Chapter 15 and again in Section 16.1. For the moment, we should simply accept that this relativity is an unavoidable consequence of our key explication that reasons strengthen the belief in what they are a reason for; what strengthens what is up to the subject in the first place.

Next we should always observe the distinction between *being* a reason and *having* a reason. Saddam's alleged mobile underground laboratories *are* reason to believe that he is (on the verge of) possessing weapons of mass destruction. Everyone agreed on this. But only George W. Bush and some other statesmen *had* a reason to believe in the latter; they believed in the laboratories and hence in the weapons. Moreover, we should distinguish a factive and a non-factive sense of having a reason. In the factive sense, the reason must obtain or be true, in the non-factive sense the reason need only be believed. So Bush had a reason to believe in the weapons only in the non-factive sense, because he believed in the laboratories. But he did not have it in the factive sense, because there were no such laboratories and hence no such factual reason to be had. (As is well known, the same ambiguities can be found in the notion of (scientific) explanation.)

It is important to keep in mind that Definition 6.1 is only about what *is* a reason for what. It is neutral as to whether the reason is actually given to the subject. But the latter could be defined. If A is a reason for B w.r.t. κ, κ *has* A as a reason for B if moreover $\tau(A) > 0$ (i.e., $\kappa(\bar{A}) > 0$); for the factive sense, the truth of A is additionally required. People differ very much in having reasons; that is why we exchange relevant information all the time. Note, however, that the relativity stated above is not confined to the having of reasons. It may run deep and even concern what *is* a reason.

My initial explanation that A is a reason for B if B is more firmly believed with A than without A was intentionally ambiguous: "without A" might mean "not given A" or "given not-A". The issue is familiar from Bayesian confirmation theory, where the defining clause might be: (a) $P(B \mid A) > P(B \mid \bar{A})$, or (b) $P(B \mid A) > P(B)$. There is not much to choose though. If all three terms are defined, (a) and (b) are equivalent. However, there is a tiny difference between (a) and (b), namely in the case that $P(A) = 1$. In that case, (a) is undefined and (b) is defined, but false (whatever B is). (This, by the way, is the problem of old evidence: when evidence A is given, it acquires probability 1 and is hence no longer positively relevant to whatever it should confirm; cf., e.g. Earman 1992, Ch. 5). So, condition (b) is slightly more general than (a) and therefore the received option.

Definition 6.1 disambiguates in the opposite way; it prefers (a) $\tau(B \mid A) > \tau(B \mid \bar{A})$ over (b) $\tau(B \mid A) > \tau(B)$. Why? In the ranking-theoretic case, (a) and (b) make a real difference. Look at the following simple example of a negative ranking function κ and its corresponding two-sided version τ:

κ	B	$\bar{B}$
A	0	1
$\bar{A}$	1	0

$\Rightarrow$

$\tau(B \mid A) = 1,$

$\tau(B) = 0,$

$\tau(B \mid \bar{A}) = -1.$

κ just believes that A if and only if B; so imagine for illustration A = "it's dark" and B = "it's night". τ shows that A is a reason for B in the sense of (a) as well as (b). Now suppose that κ gets informed about A and thus changes into the $A{\to}1$-conditionalization κ' (with the corresponding τ'). The following figures result:

κ'	B	$\bar{B}$
A	0	1
$\bar{A}$	2	1

$\Rightarrow$

$\tau'(B \mid A) = 1,$

$\tau'(B) = 1,$

$\tau'(B \mid \bar{A}) = -1.$

Of course, we should have expected that if A gets believed to some degree, B gets believed to the very same degree. This entails, however, that according to alternative (b) the status of A as a reason for B depends on how firmly A and hence B get believed, and, as I have chosen the example, (b) tells us that A is no longer a reason for B w.r.t. κ'. This seems unacceptable. By learning that A obtains, A must not change its status as a reason; being a reason is independent of its being had. Alternative (a) precisely conforms to this intuition; the two-sided ranks conditional on A and on $\bar{A}$ do not change through $A{\to}n$-conditionalization for any n. That is how I have explained $A{\to}n$-conditionalization. I take this to be decisive reason for preferring option (a) over (b) and to state Definition 6.1 as I did.

(By the way, the probabilistic version (a) also turns out to be superior to (b), if both are stated in terms of Popper measures. This is the solution of the problem of old evidence proposed by Joyce (1999, Sect. 6.4). Of course, I tend to agree; my dissatisfaction with Popper measures and its resolution will be discussed only in Section 10.2.)

The final essential observation about Definition 6.1 is that we can distinguish different kinds of reasons. This is due to the special role of rank 0 (this is why we do not find the corresponding differentiation in probabilistic reasons).

Definition 6.2: Let $\kappa, \tau, A,$ and B as in Definition 6.1. Then A is a

$$\left.\begin{array}{l} \textit{supererogatory} \\ \textit{sufficient} \\ \textit{necessary} \\ \textit{insufficient} \end{array}\right\} \textit{reason for } B \text{ w.r.t. } \kappa \text{ iff} \left.\begin{array}{l} \tau(B \mid A) > \tau(B \mid \bar{A}) > 0 \\ \tau(B \mid A) > 0 \geq \tau(B \mid \bar{A}) \\ \tau(B \mid A) \geq 0 > \tau(B \mid \bar{A}) \\ 0 > \tau(B \mid A) > \tau(B \mid \bar{A}) \end{array}\right\} . \text{ Moreover, } A \text{ is a}$$

supererogatory, sufficient, necessary, or *insufficient reason against* B w.r.t. κ iff, respectively, A is a supererogatory, sufficient, necessary, or insufficient reason for $\bar{B}$ w.r.t. κ.

The hopefully suggestive qualifications "supererogatory" and "insufficient" are novel: a supererogatory reason strengthens the belief in something already believed, whereas an insufficient reason weakens, but does not eliminate the disbelief in something still disbelieved. A reason that is not sufficient might still be necessary, whereas an insufficient reason is not even necessary; "insufficient" is a stronger opposite to "sufficient" than "not sufficient". The qualifications "sufficient" and "necessary" are familiar and fitting. A sufficient reason for B suffices to believe B, whereas a necessary reason for B is necessary to give up disbelief in B. Clearly there is only one way to belong to two kinds of reasons, namely by being a necessary *and* sufficient reason. Otherwise the categories are disjoint. Note that the terminology proposed in (6.2) depends on my ruling adopted in Section 5.2 that positive two-sided ranks already express belief. If we use the interpretational freedom suggested in (5.14) and take belief to be two-sided rank larger than some $z \geq 0$, then (6.2) should be adapted accordingly. We would then have a fifth type of reasons raising the rank within the range of neutrality.

Definition 6.2 contains two terminological changes. In previous publications I spoke of weak instead of insufficient reasons. This appears now infelicitous to me, since one often and naturally speaks of strong and weak reasons, where the latter are quite different from insufficient reasons. Also I spoke of additional instead of supererogatory reasons. However, "additional" suggests that there are yet other reasons, and this connotation is to be avoided. (The new terms "insufficient" and "supererogatory" were suggested to me by Arthur Merin; see Merin 2008.)

There are two important messages contained in this classification. The one is that my remark about carrying the present explication to the traditional philosophical contexts of the notion of a reason applies all the more to the present explication of the notion of a necessary and/or sufficient reason. That traditional notion seems to have a fixed meaning; all the more surprising is, as we will see, its reinterpretation through (6.2). It would indeed be a most revealing exercise to interpret the countless central appeals to something like inferential relations within epistemology, philosophy of language, and elsewhere (provided they are not obviously restricted to deductive relations) as referring to the (sufficient) reason relation as explicated in (6.2). The other message is that there are the further categories of supererogatory and insufficient reasons that are, as far as I can see, entirely neglected in the tradition, but will get important roles to play in our theoretical development.

The previous chapter has already shown the centrality of conditional ranks; without them we could not have explained a dynamics of belief. Now they have also helped in gaining a notion of a reason. The two points can be combined: there is not only a dynamics of belief, but there is also a dynamics of reasons. Of course, the reason relation might change. The proposition that Tweety is a bird is a reason to assume that it can fly, but given that it lives in the Antarctic the proposition is a reason against this assumption. Such situations may be accounted for by the following straightforward notion:

Definition 6.3: Let κ be a negative ranking function for $\mathcal{A}$, and $A, B, C \in \mathcal{A}$. Then A is a (*supererogatory, sufficient, necessary*, or *insufficient*) *reason for* or *against B conditional on* or *given C* w.r.t. κ iff, respectively, A is a (supererogatory, sufficient, necessary, or insufficient) reason for or against B w.r.t. $κ_C$.

The notion of a conditional reason will be used throughout. It will acquire particular importance in Chapter 14 on causation, where I will maintain the thesis that causes are nothing but a special kind of conditional reasons.

6.2 The Formal Structure of Reasons

So much for conceptual clarifications appertaining to the basic explication (6.1). Now we should study how reasons formally behave. This is not a trivial matter, and we will find expected as well as unexpected features.

We should perhaps start our investigation by introducing the notion of a deductive (or demonstrative, logical, cogent) reason necessarily entailing its conclusion. This is an interesting (and well-studied) relation only on the sentential level; on the propositional level on which we are operating it trivially reduces to the subset relation.

Definition 6.4: $A \in \mathcal{A}$ is a *deductive reason for* $B \in \mathcal{A}$ iff $A \subseteq B$.

If the definiens is not satisfied, there remains a possibility of A being true and B being false. Note that deductive reasons are fully objective and not relativized to doxastic states. Their behavior is obvious:

Theorem 6.5: For all propositions A, B, C in $\mathcal{A}$ the following assertions hold:

(a) A is a deductive reason for A [*reflexivity*],
(b)   if A is a deductive reason for B and B for C, so is A for C [*transitivity*],
(c)   if A is a deductive reason for B and B a deductive reason for A, then $A = B$ [*antisymmetry*],
(d)   if A is a deductive reason for B, so is $\bar{B}$ for $\bar{A}$ [*contraposition*],
(e)   if A is a deductive reason for C, so is $A \cap B$ [*monotony* or *strengthening of the antecedent*],
(f)   if A and B are deductive reasons for C, so is $A \cup B$,
(g)   if A is a deductive reason for B and for C, it is so for $B \cap C$ and for $B \cup C$,
(h)   the impossible proposition $\varnothing$ is a deductive reason for A,
(i) A is a deductive reason for the necessary proposition W.

(Of course, antisymmetry (c) holds only on the propositional level.) The theorem is as trivial as it is important, since we tend to think of reasons on the model of those deductive reasons, and hence our intuitions are in the firm grip of that theorem, a grip we will have to fight.

We should note, as an aside, that what I called the positive rank notion of confirmation, i.e., conditional belief, is one way of loosening deductive inference or reasoning, simply because $A \subseteq B$ entails $\beta(B \mid A) = \infty > 0$. As (4.18) or (5.40) reveal, this notion satisfies clauses (a), (f), (g), (h), (i), but it does not conform to the clauses (b), (c), (d), and (e). In particular, monotony holds only in the weakened form of *rational monotony*: if B is not disbelieved given A, then if C is believed given A, C is also believed given $A \cap B$. Indeed, as Rott (2001, Ch. 4) explains in full, non-monotonic reasoning and belief revision (i.e., conditional belief) can be fully parallelized.

Here, of course, we are instead interested in the observation that our ranking-theoretic reason relation is another way of loosening deductive reasons:

Theorem 6.6: Let κ be a negative ranking function for $\mathcal{A}$, and $A, B \in \mathcal{A}$ such that $\kappa(A)$, $\kappa(\bar{B}) < \infty$. Then, if A is a deductive reason for B, A is a sufficient or supererogatory reason for B w.r.t. κ.

Proof: Since $A \subseteq B$, we have $\tau(B \mid A) = \kappa(\bar{B} \mid A) - \kappa(B \mid A) = \kappa(A \cap \bar{B}) - \kappa(A \cap B)$ $= \kappa(\varnothing) - \kappa(A) = \infty$ and $\tau(B \mid \bar{A}) = \kappa(\bar{B} \mid \bar{A}) - \kappa(B \mid \bar{A}) = \kappa(\bar{A} \cap \bar{B}) - \kappa(\bar{A} \cap B) = \kappa(\bar{B}) - \kappa(\bar{A} \cap B) < \infty$.

Hence, our general reason relation embraces the deductive reason relation (with the exception of the extreme cases mentioned in (6.6)). In order to emphasize this point, I often say that reasons in the sense of (6.1) are *deductive or inductive* reasons, where "inductive" just means "non-deductive" and thus assumes its widest sense. Indeed, the inductive part is essential, and this is why the general behavior is quite different.

Theorem 6.7: Let κ be a negative ranking function for $\mathcal{A}$, and $A, B, C \in \mathcal{A}$. Then the following assertions hold:

(a)   if $\kappa(A), \kappa(\bar{A}) < \infty$, then A is a reason for A w.r.t. κ [*reflexivity*],

(b)   if A is a reason for B w.r.t. κ, B is a reason for A w.r.t. κ [*symmetry*],

(c)   if A is a reason for B w.r.t. κ, then A is a reason against $\bar{B}$, $\bar{A}$ a reason against B, and $\bar{A}$ a reason for $\bar{B}$ w.r.t. κ,

(d)   if $A \cap B = \varnothing$ and if A and B are reasons for C w.r.t. κ, then $A \cup B$ is a reason for C w.r.t. κ,

(e)   if A is reason for B and for C given B w.r.t. κ, then A is a reason for $B \cap C$ w.r.t. κ,

(f) $\varnothing$ and W are not reasons for A w.r.t. κ.

Each assertion holds also with respect to κ_D for any $D \in \mathcal{A}$ with $\kappa(D) < \infty$, i.e., if all reason relations are conditioned to a fixed condition D.

Proof: (a), (c) and (f) are trivial. (b) is almost trivial: A is a reason for B iff $\tau(B \mid A) > \tau(B \mid \bar{A})$ iff $\kappa(A \cap \bar{B}) + \kappa(\bar{A} \cap B) > \kappa(A \cap B) + \kappa(\bar{A} \cap \bar{B})$ iff $\tau(A \mid B) > \tau(A \mid \bar{B})$ iff B is a reason for A.

(d) is a bit harder. For abbreviation, put $\tau(C \mid A) = a$, $\tau(C \mid \bar{A}) = c$, $\tau(C \mid B) = b$, $\tau(C \mid \bar{B}) = d$, $\tau(C) = e$, $\tau(C \mid A \cup B) = x$, and $\tau(C \mid \bar{A} \cap \bar{B}) = y$. We want to show that $x > y$. Our premises are $a > c$ and $b > d$. The law of disjunctive conditions (5.23a) tells us that $e \in [a, c]$ (the closed interval between a and c), $e \in [b, d]$, $e \in [x, y]$, and also $x \in [a, b]$. Hence, with the premises, $a \geq e \geq c$, and $b \geq e \geq d$, and $x \geq e \geq y$. Now suppose $x = y$. This would entail $x = y = e$ and $a = e$ or $b = e$. However, the law of disjunctive conditions also tells us that $c \in [b, y]$ and $d \in [a, y]$. Hence, the supposition also entails $c = d = e$. But this and $a = e$ or $b = e$ contradicts our premises. Hence $x > y$.

(e) Again, let us abbreviate notation by putting $\kappa(A \cap B \cap C) = a$, $\kappa(A \cap B \cap \bar{C}) = b$, $\kappa(\bar{A} \cap B \cap C) = c$, $\kappa(\bar{A} \cap B \cap \bar{C}) = d$, $\kappa(A \cap \bar{B}) = e$, and $\kappa(\bar{A} \cap \bar{B}) = f$. Now the first premise $\tau(B \mid A) > \tau(B \mid \bar{A})$ translates into

(i) $e - \min(a, b) > f - \min(c, d)$,

the second premise $\tau(C \mid A \cap B) > \tau(C \mid \bar{A} \cap B)$ translates into

(ii) $b - a > d - c$,

and the intended conclusion $\tau(B \cap C \mid A) > \tau(B \cap C \mid \bar{A})$ translates into

(iii) $\min (b, e) - a > \min (d, f) - c$.

Does (iii) follow from (i) and (ii)? Yes. To see this, distinguish three cases. First, it may be that $b - a > d - c \geq 0$. In this case, (i) reduces to $e - a > f - c$, and then (i) and (ii) clearly entail (iii). Second, it may be that $0 \geq b - e > c - d$. In this case, (i) reduces to $e - b > f - d$; adding this to (ii) yields $e - a > f - c$; and thus the second case reduces to the first. Finally, it may be that $b - a > 0 > d - c$. Now (i) reduces to $e - a > f - d$. Since $0 > d - c$, this entails $e - a > f - c$. And so the last case also reduces to the first.

It is remarkable both what is in the theorem and what is not. It will take some paragraphs to unfold its content.

Reflexivity (6.7a) is unsurprising and acceptable as a limiting case. A is trivially a reason for A, but not an illuminating one. (6.7f) is unexciting as well. It will have consequences, though. W is epistemically necessary or a priori true and $\varnothing$ a priori false – notions I will slightly more carefully introduce in Section 6.5. However, we may observe already now that (6.7f) says that the a priori is not a reason for anything and that there are conversely no reasons for the a priori. I mention this just in order to indicate that even such minor assertions as (6.7f) have a rich philosophical background worth discussing.

(6.7c) describes the expected dialectics of reasons and counter-reasons. Indeed Merin (1997, 2002) proposes to characterize negation formally by way of that dialectics rather than by way of its truth table. (Semantics is not our business here, but I should mention that there is a semantic research program – decision theoretic semantics – which admits utilities instead of truth values as basic semantic values, and hence in particular relevance values (which can behave like utilities). See Merin 1999, 2005, 2006a,b) for the foundations and the development of this program.)

Symmetry (6.7b) is perhaps the most conspicuous feature of our reason relation. It is familiar from probabilistic positive relevance and is now seen to hold in the ranking-theoretic framework as well. In a way, it is intuitively appealing. Positive relevance can also be graphically described as support, which clearly appears as something mutual. (However, there is no presumption that mutual support is equally strong both ways; moreover, symmetry does not hold within the subcategories of sufficient, necessary, supererogatory, or insufficient reasons.) Why do we nevertheless tend to think that reasons have a direction?

The basic point is that deductive reasons are our prevalent conception of reasons and deductive reasons have a direction; antisymmetry (6.5c) is the only way to revert them. The point can also be expressed more neutrally. Reasons are closely related with (deductive *or* inductive) inference, and inference appears to be asymmetric. We infer the general from the singular, the future from the past, the unobserved from the observed, we confirm the hypothesis by the evidence. Listing these alleged asymmetries makes clear, though, that they do not hold. Inferences run in any direction; we infer the singular prediction from the general hypothesis, the observable from the unobserved, etc. The asymmetry is not in what reasons *are*, but rather in the reasons we have or get. Of course, evidence, i.e., the reasons we get, rather concerns singular past observations and not the future, the general, or the unobserved. I will more carefully discuss observation and the basis of belief later in Chapter 16. However, already now it is clear that this asymmetry is not one of reasons as such and thus not an objection to (6.7b). On the contrary, it is within the scope of our indicated account of having and getting reasons.

A further possible confusion might come into play at this point. We noted after Theorem 6.5 that the positive rank notion of confirmation or simply conditional belief might also claim to provide a weakened notion of inference, a kind of non-monotonic inference. It indeed satisfies the intuition of non-symmetry. This claim might be legitimate, but we then have to distinguish that notion of inference from our deductive or inductive reasons. The crucial difference is that, in contrast to our reasons, the positive rank notion of confirmation or inference does not require the relevance of the premise to the conclusion. This is why the two notions conform to different laws. If we add positive relevance to conditional belief, we of course arrive at the notion of a sufficient or supererogatory reason.

Symmetry (6.7b) should be combined with Theorem 6.5. Then it becomes clear that a proposition is a reason not only for what it logically entails, but also for what logically entails it. In qualitative confirmation theory (cf. Niiniluoto 1972, p. 28), these postulates were called the entailment condition and the converse entailment condition. They met a lot of suspicion because together with the transitivity of the confirmation or reason relation they immediately led to paradox, because $A \cup B$ would then be confirmed by A and B by $A \cup B$ and hence by A; everything would confirm everything. However, transitivity does not hold for our reason relation, and so no paradox threatens.

Indeed, transitivity is perhaps the most conspicuous omission from Theorem 6.7. Consider the following example:

Example 6.8:

κ	$B \cap C$	$B \cap \bar{C}$	$\bar{B} \cap C$	$\bar{B} \cap \bar{C}$
A	0	1	2	1
$\bar{A}$	3	4	2	4

Here we have $\kappa(\bar{B} \mid A) = 1 = \kappa(B \mid \bar{A})$; so A is a necessary and sufficient reason for B. Moreover, we have $\kappa(\bar{C} \mid B) = 1 = \kappa(C \mid \bar{B})$; so B is a necessary and sufficient reason for C. Nevertheless, we have $\kappa(\bar{C} \mid A) = 1$ and $\kappa(\bar{C} \mid \bar{A}) = 2$. This means that A is an (insufficient) reason *against* C. This example does not only refute transitivity, but also destroys all hope that at least sufficient and/or necessary reasons might be transitive. The example is chosen in a minimal way; the critical situation cannot be realized with smaller natural ranks.

Again, the failure of transitivity is familiar from probabilistic relevance. It is counter-intuitive only if reasons are understood as deductive reasons. Apart from this, the failure should be unsurprising. It can also be observed in the case of counterfactuals, which is much closer to our concern than the probabilistic situation. Lewis (1973a) emphatic-ally warns against the fallacy of transitivity. His counterexample is this:

If Otto had gone to the party, then Anna would have gone.
If Anna had gone, then Waldo would have gone.

∴ If Otto had gone, then Waldo would have gone.

The fact is that Otto is Waldo's successful rival for Anna's affections. Waldo still tags around after Anna, but never runs the risk of meeting Otto. Otto was locked up at the time of the party, so that his going to it is a far-fetched supposition; but Anna almost did go. Then the premises are true and the conclusion false. (Lewis 1973a, pp. 32f.)

The case serves our purposes as well: replace throughout the counterfactual "if, then" by "is a reason for", and you have a nice intuitive counter-example to the transitivity of the reason relation.

These structural observations are fundamental, also philosophically. Look at the Agrippan justification trilemma, which drives the epistemological discussions of the last decades more fiercely than ever. It proceeds from the assumptions that every belief must be justified and that reasons can justify only if they are justified in turn. This is what I will call the trilemma generator (16.1). If this is right, then beliefs are (i) ultimately grounded in some base that is self-justifying or the only case not in need of justification (foundationalism), or (ii) justified in a circular way (coherentism), or (iii) backed up by infinite trees of reasons (infinite regress). Hardly anyone has dared to endorse the infinite regress (for an exception see Klein 1999). Coherentism looks unacceptable because of the structural intuition that the reason relation is acyclic and a fortiori asymmetric. The exceptional base of foundationalism either violates the first assumption of the trilemma or violates the structural intuition that the reason relation is irreflexive. And even if the base can be made acceptable, foundationalism works only if the base justifies the rest at one fell swoop or if justification is transitive.

I will more carefully discuss the issue in Chapter 16. The present remarks are only intended to emphasize how crucial it is to get clear about the structure of the reason relation. Structural intuitions are not good enough; theory is required. My remarks already indicate where my sympathies concerning the Agrippan trilemma will come to lie in the end.

We are not finished yet with the discussion of Theorem 6.7. The set of deductive reasons for a given proposition C has a very simple structure, as is underscored by monotony (6.5e). By contrast (6.7d + e) and their symmetric reversals display quite complicated behavior exhibited by the set of reasons for a given proposition C. According to (6.7d) this set is closed under disjoint unions, and (6.7e) (or rather its symmetric reversal) makes things even more complicated by stating only a relation between conditional and unconditional reasons.

That's how it is; there is no simplification in sight. For instance, the proviso of (6.7d) is essential. If A and B are reasons for C, neither $A \cap B$ nor $A \cup B$ need be reasons for C; even with one of the latter two added, the other does not follow. This is easily seen. That $A \cup B$, e.g., is a reason for C w.r.t. κ is equivalent to $\kappa((A \cup B) \cap \bar{C}) - \kappa((A \cup B) \cap C) > \kappa(\bar{A} \cap \bar{B} \cap \bar{C}) - \kappa(\bar{A} \cap \bar{B} \cap C)$, but there is nothing in the premises that would constrain the latter two terms, and one can choose them such that the consequence is refuted. Again, it does not help to restrict consideration to necessary and/or sufficient reasons. Salmon (1975) emphatically warns against related fallacies in the probabilistic case. These warnings apply here no less.

Does Theorem 6.7 completely characterize the (conditional) reason relation? Certainly not. I only hope to have stated the most salient laws. But surely the missing laws will complicate, not simplify the picture. Would it be so difficult to complete Theorem 6.7? I fear this to be the case. That no complete characterization is known in the unfamiliar ranking-theoretic case is no surprise. The support for my pessimism lies in the fact that, to my knowledge, no complete axiomatization is available for the parallel notion of probabilistic positive relevance that has been investigated for many years. Why should ranking theory be any better off?

On the other hand, I conjecture that there is a (hitherto unknown) complete characterization. The validity of each finite assertion about reason relations, i.e., its truth in all models, should be decidable: the size of the modeling algebra should have an effectively specifiable upper limit, and the ranks used by the modeling ranking function should also have an effectively specifiable upper limit. If so, the validity of such an assertion is decided by finitely many possible ranking functions, and hence decidable. And if validity is decidable, it can be completely axiomatized. However, even if this argument can be made rigorous, it is quite another thing to actually find an adequate axiomatization.

Having thus seen the features of the ranking-theoretic reason relation, we can take three different attitudes. First, we might find the properties of the reason relation too complicated and unintuitive and therefore reject the model on which it is based. Then we could try to directly explicate our intuitions about confirmation, an approach that has led nowhere; this is at least how I view the history of qualitative confirmation theory starting with Hempel (1945) and given up in the 70s (cf. the survey by Niiniluoto 1972).

A second way would be to try to find other models with a better-behaved logic of confirmation. Surprisingly, this attempt has hardly been made; I do not know of any model explicitly designed for this purpose. Of course, there are quite a number of alternative models of doxastic states. However, in our comparative discussion in Chapter 11 we will find that these models do not pay special attention to positive relevance. All this is not promising.

The third approach is to trust in our model, to appreciate the straightforwardness of our explication of the reason relation, and hence to confidently accept its consequences, even if they consist in a complicated formal structure. This is of course the option I will adopt. We have to acknowledge that the logic of confirmation is complicated, and we have to work on it positively, not grudgingly. At the moment, though, I have no more to offer than the theorems of this section. Still, we now have at least the prospects of a qualitative confirmation theory, a genuine one that is not simply the qualitative coarsening of a quantitative, i.e., probabilistic, confirmation theory. In the next chapter we will elaborate more, and more intelligible, structure.

6.3 The Balance of Reasons

There is another aspect of reasons that I would like to discuss separately. Intuitively, we weigh reasons. This is a most important activity of our mind. We not only weigh practical reasons in order to find out what to do, but also theoretical reasons. We wonder whether or not we should believe B, we survey the reasons speaking in favor of and against B as completely as possible, we weigh these reasons, and we hopefully reach a conclusion. In emphasizing this I do not want to deny the phenomenon of inference proper. How could I? However, what is represented as an inference often takes the form of such a weighing procedure. 'Reflective equilibrium' is a familiar and somewhat more pompous metaphor for the same thing.

If the balance of reasons is such a central phenomenon, the question arises: how can epistemological theories account for it? The question is less well addressed than one should think; the metaphor is hardly taken up in formal epistemology. What can we say about it from our point of view?

A first trivial, but interesting observation is that deductive reasons cannot be weighed at all. On pain of irrationality, one cannot have deductive reasons and counter-reasons for one and the same assertion. Having these reasons means believing them, and such reasons would have to be contradictory. We have to turn away from deductive reasoning as the dominant model of reasoning.

Bayesians, by contrast, have a perfectly natural representation of the balance of reasons, which I take to be a very strong argument in favor of Bayesianism, even though it is rarely made explicit. Let us take a brief look at this representation:

Let P be a (subjective) probability measure over $\mathcal{A}$ and let B be the focal proposition. Let us consider only the simplest case, consisting of one reason A for B and the automatic counter-reason $\bar{A}$ against B. Thus, in analogy to Definition 6.1, $P(B \mid A) > P(B \mid \bar{A})$. How does P balance these reasons and thus fit in B? The answer is straightforward, we have:

(6.9) $P(B) = P(B \mid A) \cdot P(A) + P(B \mid \bar{A}) \cdot P(\bar{A})$.

This means that the probabilistic balance of reason is a *beam balance* in the literal sense, as shown in figure 6.10 below. The length of the beam is $P(B \mid A) - P(B \mid \bar{A})$; its two ends are loaded with the *weights* $P(A)$ and $P(\bar{A})$ of the reasons; $P(B)$ divides the lever into two parts of length $P(B \mid A) - P(B)$ and $P(B) - P(B \mid \bar{A})$ representing the *strength* of the reasons; and then $P(B)$ must be chosen so that the beam is in balance. Thus interpreted (6.9) is nothing but the law of levers. This is a beautiful picture.

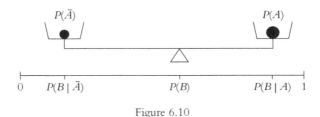

Figure 6.10

Ranking theory has a representation, too, and I am wondering what else is on offer. According to ranking theory, the balance of reasons works like a *spring balance*. Let κ be a negative ranking function for $\mathcal{A}$, τ the corresponding two-sided ranking function, B the focal proposition, and A a reason for B. So, $\tau(B \mid A) > \tau(B \mid \bar{A})$. We know from the law of disjunctive conditions (5.23a) that $\tau(B \mid A) \geq \tau(B) \geq \tau(B \mid \bar{A})$. But where in between is $\tau(B)$ located? A little calculation proves the following specification:

(6.11)   Let $x = \kappa(B \mid \bar{A}) - \kappa(B \mid A)$ and $y = \kappa(\bar{B} \mid A) - \kappa(\bar{B} \mid \bar{A})$. Then

    (a) $x, y \geq 0$ and $\tau(B \mid A) - \tau(B \mid \bar{A}) = x + y$,

    (b) $\tau(B) = \tau(B \mid \bar{A})$, if $\tau(A) \leq -x$,

    (c) $\tau(B) = \tau(B \mid A)$, if $\tau(A) \geq y$,

    (d) $\tau(B) = \tau(A)$, if $-x < \tau(A) < y$ and $x, y > 0$,

    (e) $\tau(B) = \tau(B \mid \bar{A}) + \tau(A)$, if $-x < \tau(A) < y$ and $x = 0$,

    (f) $\tau(B) = \tau(B \mid A) + \tau(A)$, if $-x < \tau(A) < y$ and $y = 0$.

This does not look as natural and straightforward as the probabilistic beam balance. Still, it is not so complicated to interpret (6.11) as a spring balance. The idea is that you hook in the spring at a certain point, that you extend it by the force of reasons, and that $\tau(B)$ is the point to which the spring extends. Consider first the case where $x, y > 0$, which is depicted in figure (6.12) below. There you hook in the spring at point 0 and exert the force $\tau(A)$ on the spring. This force might transcend the lower stopping point $-x$ or the upper stopping point y. In this case, the spring extends exactly to the stopping point, as (6.11b) and (6.11c) say. Or the force $\tau(A)$ is smaller. In that case, the spring extends exactly by $\tau(A)$, according to (6.11d).

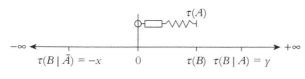

Figure 6.12

The second case is that $x = 0$ and $y > 0$. Then you fix the spring at $\tau(B \mid \bar{A})$, the lower point of the interval within which $\tau(B)$ can move. The spring cannot extend below that point, says (6.11b). But according to (6.11c + e) it can extend above, by the force $\tau(A)$, but not beyond the upper stopping point. For the third case $x > 0$ and $y = 0$ just reverse the second picture and apply (6.11f). In this way, the force of the reason A, represented by its two-sided rank, pulls the two-sided rank of the focal proposition B to its proper place within the interval fixed by the relevant conditional ranks. In this way the rank of each proposition is fixed by the forces exerted on it by all the other propositions. Of course, this is only a figurative way of saying that one could not change the rank of a single proposition in a ranking function without violating the ranking-theoretic laws.

What is the analogy between the probabilistic beam balance and the ranking spring balance? The counterpart to the probabilistic weights of reasons $P(A)$ and $P(\bar{A})$ is given by the ranking force $\tau(A)$ of the reason pulling into one or the other direction. And the counterparts to the probabilistic strengths of reasons $P(B \mid A) - P(B)$ and $P(B) - P(B \mid \bar{A})$ is given by the lower bound $\tau(B \mid \bar{A})$ and the upper bound $\tau(B \mid A)$ to which the string can at most extend.

The ranking-theoretic picture is, I admit, not as smooth as the probabilistic one. One might therefore prefer the probabilistic balance of reasons. However, it is good to have an alternative and theory-induced model of that balance allowing for genuine non-trivial balancing. As I said, I am wondering whether there are any further alternatives. In any case, one must not ignore the issue; epistemology is incomplete without a plausible stance on it. And to repeat, such a stance requires a plausible notion of a reason, which in turn requires a well-working account of conditional degrees of belief, at least if positive relevance is a basic characteristic of reasons. This is why probability and ranking theory can offer weighing models at all.

6.4 The Force of Reasons

Jeffrey (1965) deliberately gave his revolutionary Chapter 11 the title "probability kinematics". Kinematics describes how things move. Dynamics, by contrast, explains why things move as they do and thus has to appeal to such notions as force and inertia. In our case, the moving things are changing beliefs, and the forces are the causes of belief change. Jeffrey felt he had at best quite incomplete things to say about these causes, hence the modest term "kinematics".

So far I used the term "dynamics" instead, because it is more familiar and because I did not want to discuss this fine distinction. However, I was not just sloppy; I think I am using the term properly. This is what I want to briefly explain in this section.

Belief change has many causes: experience and many others, among them, above all, limited and failing memory. It is hard to develop theories about these other causes, and they do not seem to be the philosopher's business anyway (though the issue of rational mnemonic strategies has fascinating philosophical depth; cf. Spohn 1978, Sect. 4.4, or van Fraassen 1995, Sect. 7). Putting these other causes aside, philosophers tend to draw a very rationalistic picture of the human mind, and this is fortunate: somebody must do so, and in a responsible and reflective manner. One of the basic tenets of this rationalistic picture is that beliefs are held for reasons and that changes in belief must not occur without reason. The former is familiar from the theory of knowledge, but some emphasize the latter instead. For instance, Peirce (1877) insisted that doubt must be more than mere paper doubt, it must be real and living, nourished by reasons, and that the goal of inquiry then is the removal of that doubt; Levi (2004, Sect. 1.1–1.2) infers that it is primarily belief changes that are in need of justification. The issue looks idle. At least each acquired belief has been acquired at some point through a belief change, and so if the belief change has a reason, so has the belief, and vice versa.

Let me focus here on belief change. The rationalistic picture is, I said, that it is driven by reasons. However, in Section 3.2 and Chapters 4 and 5 we discussed – successfully in my view – various laws of doxastic change without mentioning the notion of a reason, and I referred to a bulk of literature that is similarly silent in this respect. Indeed, I was very surprised about this lack of discussion when surveying that literature.

To be a bit more precise: My surprise refers to the main bulk of the AGM belief revision literature. Elsewhere the situation is not quite as bad. For instance, an important paradigm in AI is provided by so-called truth or reason maintenance systems (RMS) initiated by Doyle (1979). There, a doxastic state consists not only of a set of beliefs, but also of a reason relation (specifying which assumptions support and undermine which conclusions); and the beliefs held in such a state must conform to those reasons. I shall not further comparatively discuss this approach, though closer inspection would certainly show that reasons behave very differently in RMS and in ranking theory. Similarly, John Pollock assumes a number of conclusive and defeasible reason-schemata that govern belief formation (cf., e.g., Pollock 1995). In both of these accounts the reason relation or reason-schemata just come in the form of an extensional list and are not derived from an underlying explication or theory; rigorous theorizing then refers to the determination of beliefs through the extensionally given reasons. In Section 11.5 I will thoroughly discuss Pollock's work in relation to my approach.

However, in the AGM literature, which has so far been my main reference point, the topic is mostly lacking. Quite late and in response to RMS, Gärdenfors (1990) claimed, with reference to Spohn (1983a), that reasons are implicit in the AGM picture, i.e., can be reconstructed from the entrenchment relation. He is not happy with the circular structure of reasons thus entailed, and Doyle (1992), responding to Gärdenfors (1990), concurs (though both missed the criticism of that reconstruction in footnote 18 of Spohn 1983a, which is the more disturbing one in my view). I will not continue with that discussion; the point rather is that even the exchange between Gärdenfors and Doyle did not have further repercussions (though one might view the attempts at belief-base revisions and contractions as being guided by such concerns; cf., e.g., Fuhrmann 1991 or Hansson 1999). Within the AGM paradigm it was most explicitly Haas (2005) who attempted to constructively build belief revisions on justificational structure (of a different kind inspired by Keith Lehrer).

In any case, there seems to be a tension here. How were we able to state a belief dynamics without explicitly referring to reasons? What is the relation between the former and the latter? Given the present explication of reasons, the answer is obvious. According to the conditionalization rules (5.25) and (5.33), I change my rank for a proposition only if I receive reasons for or against it. If the evidential propositions are irrelevant for a proposition, its rank remains unchanged. This is how we have defined reasons! Reasons thus drive doxastic change. This is, in a nutshell, my utterly trivial account of the connection between reasons and doxastic change. But we must look at this answer a little bit more carefully. I cannot leave its triviality without comment.

There might first be a general question about causation. I had emphasized that I am developing a normative account or laws of rationality for doxastic states. Where do factual causes enter a normative account? This is an easy question, though. Surely, we are not directly discussing actual people moved by actual causes. We model rational subjects, and this model has a normative thrust. But when subjects exactly conform to this model, then, no doubt, the conditionalization rules are causal laws of belief change (whatever the subvenient microcausal story may be).

Still, we have to say more carefully what are here the causes and what the effects. According to the conditionalization rules, the causes are the changed degrees of beliefs in the propositions in what I have defined in (3.13) and (5.36) as the unique evidential source of the change, and the effects are the changed degrees of belief in the other propositions. This change is reason driven. Reasons are causes in exactly this sense. The rules seem to state here a one-step causal process, but this appears so only because the rules are still coarse. One should expect that this process may be analyzed as consisting of many steps of inference. With the methods explained in Chapter 8 we will be able to give a more detailed representation.

Of course, the question remains: where does the change in that evidential source come from? So far we have been silent on this issue. It seems obvious, though, that now we would have to go into the details of the perceptual process that is somehow ultimately rooted in the external world. This appears to be a matter of the psychology, or even the physiology, of perception and hence not directly the philosopher's task. Therefore Jeffrey felt himself to have at best a very incomplete causal story about belief change. Maybe, though, the philosophy of perception can also contribute a little bit to this issue. However, we should not now engage in this tricky field; I will return to it in Sections 16.2–16.4, where we will discuss foundationalist issues concerning the observational base of belief formation in more detail.

In any case, it should be clear by now that I am as modest as Jeffrey and that I have no argument with him. I have explained how reasons are the causes of belief change and hence I am using the more pretentious term "dynamics". At the same time I have emphasized that, like Jeffrey, I have a causal story only about the subsequent and not about the initial doxastic change. So "dynamics" only has a restricted intent.

You will nevertheless be disappointed by this section. I have opened it with a grand philosophical claim: reasons drive belief change. And then I have made it true by definition. That's cheating, isn't it? Yes and no. I admit that I have not yet given a substantial account of reasons. Yet I have given an explication of the notion of a reason, one I find to be very natural, I have started the theoretical development of this explication, and I felt that I should point out this philosophical consequence of the explication at this early stage. It is better to have a trivial account of the grand claim instead of not redeeming it at all; depth and triviality have always been close neighbors in the history of philosophy.

In fact, we are standing here at the same dialectical point as in Section 4.2 after the observed vacuity of (4.10), which I commented on with strong words. There the crucial issue was what to say about belief change in the face of belief-contravening evidence, and my recommendation was to leave the issue unanswered, i.e., to leave it to the rational subject and to simply describe what she does (initially in terms of selection functions and then in a more sophisticated way). Likewise here. How the relevancies and the reasons are distributed is so far entirely left to the subject and her ranking function. The notion of a reason is basically subject-relative and therefore still empty. If we had directly headed for a more substantial and objective account of reasons, we would have got lost in difficult and possibly futile philosophical discussions. By side-stepping all this, we have moved, I hope, to neutral ground on which things can be substantially advanced as well, though in a different sense. The further hope is to thus build a sound base from which these philosophical discussions can be more fruitfully addressed. This is my view of the constructive heritage of inductive skepticism. For the moment, though, we are still engaged in the formal development of ranking theory. So let us continue with this line of investigation in Chapters 15–17.

6.5 Maximal Certainty, Unrevisability, and Apriority

Having thus proposed an explanation for the intimate relation between reasons and belief change, we should also take a look at which kinds of belief cannot be changed. They seem to be captured by the notions in the section title, which, however, are subtly different. The purpose of this section is simply to define these notions and explain their relation in order to fix their meanings for the rest of this book. A more serious philosophical discussion – in particular of apriority, which is so central to epistemology – will have to wait until the final Chapter 17.

Maximal certainty is the simplest of the notions:

Definition 6.13: The proposition A is *maximally certain* w.r.t. the negative ranking function κ iff $\kappa(\bar{A})$ is maximal, i.e., $= \infty$ in the case of natural or real ranking functions.

The explication of unrevisability is also without alternative:

Definition 6.14: The proposition A is *strictly unrevisable* w.r.t. the negative ranking function κ iff $\kappa(A) = \kappa_{\mathcal{E} \to \varepsilon}(A)$ for all $\mathcal{E} \to \varepsilon$-conditionalizations $\kappa_{\mathcal{E} \to \varepsilon}$ of κ.

That is, a proposition A is strictly unrevisable iff there is no possible doxastic change that can change the degree of belief in A, where, in our context, we have to describe possible doxastic changes with our most general law of conditionalization (5.33).

Are the two notions coextensive? Yes, trivially:

(6.15) A is strictly unrevisable w.r.t. κ iff A is maximally certain w.r.t. κ.

This is so, since whatever has rank ∞ keeps it after any conditionalization and whatever does not have rank ∞ can be changed through conditionalization.

But there is a gap. (6.14) refers to all possible $\mathcal{E} \to \varepsilon$-conditionalizations. Yet in Sections 3.2 and 5.4 I have intimated that not all algebras $\mathcal{E}$ and $\mathcal{E}$-measurable ranking functions ε might be suitable as evidential sources of belief change and have deferred the difficult question of what the required restrictions on evidential sources might be. This leads us to a weaker notion of unrevisability:

Explanation 6.16: The proposition A is *unrevisable* w.r.t. the negative ranking function κ iff $\kappa(A) = \kappa_{\mathcal{E} \to \varepsilon}(A)$ for all possible evidential sources $\mathcal{E}$ and all $\mathcal{E} \to \varepsilon$-conditionalizations $\kappa_{\mathcal{E} \to \varepsilon}$ of κ.

The precise content of (6.16), however, is unclear, so long as we do not know what possible evidential sources might be (recall the remarks on my distinction between definitions and explanations after (2.5)). This issue will be more thoroughly discussed only in Sections 16.2–16.4. Clearly what is strictly unrevisable or maximally certain is also unrevisable. However, a proposition might be unrevisable and still less than maximally certain simply by being independent of all possible evidential sources. In Chapter 17 we will have a closer look at this strange case.

Due to the equipollence of belief change or conditionalization and reasons, as explained in the previous section, the notions introduced are closely connected to the absence of reasons:

(6.17)   If the proposition A is maximally certain or strictly unrevisable, then there are no reasons for or against A, and if A is unrevisable, then there is no possible evidential source that is a reason for or against A.

So much about certainty and unrevisability and their relation to reasons. Apriority needs a bit more circumspection. After all, it is one, if not the central notion of philosophical epistemology, even though (or perhaps because) its status was always contested. Kant, in his attempt to establish synthetic principles a priori, argued that it is importantly different from analyticity, whereas those unable to follow Kant held apriority and analyticity to be extensionally equivalent (even if the two notions should be conceptually distinct). This equivalence has finally been overcome with Kripke (1972). At present, I do not want to discuss such issues (although it will become clear in Chapter 17 that I rather side with Kant, though perhaps not in a way he would really agree with). In particular, I do not wish to engage in the inquiry as to what is a priori. The present interest is only to define the term. The traditional discussion focused on judgments or principles or sentences or propositions as a priori. Thus it seemed to be something about their contents that is characteristic of their apriority. This appears slightly misleading to me. I think the dynamic perspective of ranking theory is particularly suited for grasping this notion.

To begin with, apriority is often described as epistemic necessity, where the latter term is then taken in the sense of modal logic as truth in all epistemically possible worlds. This is not wrong, of course, but I think the description should be taken more literally. Kant, with his constant formula of the necessary conditions of the possibility of experience, has made clear that apriority is about the doxastic constitution of our minds, about our ability to at all grasp, learn, or experience anything. This suggests that it not only concerns judgments or propositions:

Explanation 6.18: A feature of a doxastic state is *unrevisably a priori* iff it is rationally necessary, i.e., if all possible rational doxastic states have it.

The adverb "unrevisably" indicates that, as will soon become clear, I intend to introduce another notion of apriority as well. The restriction to rational doxastic states is as obvious as important; we should not try to speculate about irrationalities in actual doxastic states and their behavior. This implies something important:

(6.19)   All principles of doxastic rationality are unrevisably a priori.

Thus, if we should rationally satisfy the basic axioms of ranking theory or, e.g., the principle (5.19c) of conditional consistency, these are a priori principles. If we should rationally satisfy the axioms of mathematical probability, they are a priori features. Of course, we heatedly discuss the principles of rationality; they emerge in normative discourse and are not fixed in advance. Still, it should be clear that these discussions are about the a priori constitution of our minds or our a priori forms of thought in the required sense.

(6.18) is more general than the usual explication insofar as it allows any doxastic feature to be a priori. This generality was exploited in (6.19) and it will prove useful throughout. However, (6.18) is easily specialized to the usual notion, since the belief in a certain proposition is also a feature of doxastic states:

Explanation 6.20: A proposition is *unrevisably a priori* iff it is believed in all possible rational doxastic states.

Again, we will not discuss right now which propositions are unrevisably a priori.

Of course, we have:

(6.21)   Unrevisably a priori propositions are strictly unrevisable in the sense of (6.14) and hence maximally certain.

This explains the label. One should note, however, that (6.20) is stronger than (6.14). A proposition A need not be maximally certain in all doxastic states, but it might become so in some ranking function κ. In this case, A would be strictly unrevisable w.r.t. κ, but not unrevisably a priori. (6.17) and (6.21) entail that there are no reasons for or against unrevisably a priori propositions. This agrees well with my commentary to (6.7f) after the proof of (6.7). (Of course, this is not to say that philosophical discussions about which propositions are a priori would be unreasonable, but simply that they appeal to standards of reasonableness outside the scope of the investigations in this book.)

Explanation 6.20 well fits the traditional explanation that the a priori is independent of all experience (so that it must belong to the forms of our thought or to the constitution of our mind). Unrevisably a priori propositions are characterized precisely by this independence. However, the traditional formulation is ambiguous, and our dynamic perspective is particularly apt for uncovering this ambiguity. That a proposition is believed independently of all experience might mean that it is believed given *any* experience whatsoever or that it is believed given *no* experience whatsoever, i.e., *before* all experience. (6.20) only grasps the first notion. It is certainly the traditional notion and the one Kant intends. It is also the one Quine (1951) attacks when he attacks analyticity (this is salient in his second line of argument, when he proposes to replace unrevisability by centrality).

However, the second notion is no less important. It is most prominent in the probabilistic context. When Bayesians fight over the shape of a priori probabilities and when Carnap makes specific, though ever more modest proposals for them in his program of inductive logic from his (1950) to his (1971a,b, 1980), apriority takes on this second sense. Such apriority need not be unrevisable. For example, attempting to satisfy some principle of insufficient reason, we might start with an equal distribution over the six possible results of a throw of a die. Experience, however, might suggest that the die is loaded and hence make us deviate from the a priori distribution. In other words, a priori probabilities are defeasible.

In non-probabilistic contexts the second notion is not so salient, perhaps because the dynamic perspective is not so salient. Another reason is perhaps that the issue of apriority is usually presented as being about a priori *knowledge*. However, only the unrevisably a priori propositions may be assumed to be known, whereas what is a priori in a defeasible sense need not be knowledge; it might even turn out false. Still, even in non-probabilistic contexts the second notion seems to be fairly well acknowledged by now. Pollock certainly cut a lot of ice; for example, his (1990) and (1995) are full of defeasibly a priori rules and principles. Field (1996) started distinguishing weak apriority, which is believability justifiable on a basis other than empirical evidence, and strong apriority, which adds empirical indefeasibility to the weak form. Quite a different example is provided by rules of presumption in the theory of interpretation; cf. Scholz (1999). And whenever philosophers speak of such things as prima facie assumptions, these might well be defeasibly a priori in the present sense:

> *Explanation 6.22*: A feature of a doxastic state is *defeasibly a priori* iff all possible rational initial doxastic states have it; and a proposition is *defeasibly a priori* iff it is believed in all possible initial doxastic states.

The notion of defeasible apriority will unfold its great importance from Chapter 12 onwards.

Of course, the puzzling notion here is that of an *initial* doxastic state, as a result of which (6.22) is rather a working program than a completed explication. Somehow we have to start with the defeasibly a priori features, but we need not stick to them. Where or when to place the start is, however, quite unclear. Saying that an initial doxastic state is prior to all experience is not so helpful either, for it cannot be taken literally unless we want to revive the notion of innate ideas and to inquire into the doxastic fixations of babies in the womb of their mothers, or even in their genetic endowment. Clearly only some relative start can be intended with the notion of initiality, e.g., the start of some specific inquiry. For instance, at first I know nothing about a die except that it is a die, and now I start collecting data about it by repeatedly throwing it. This can at least serve as a rough guideline.

For the time being, the above explications and observations may suffice. We will discuss all the apriority issues of this section more thoroughly in the final Chapter 17. My present interest was only to introduce these notions in a fairly clear and representative way and to explain their relation to the notion of belief change that was examined in the previous chapters, and thus to the notion of reason that was looked at in the previous sections. Having thus fixed their meaning is important, because the notions of apriority will be of significant use again and again, even though their systematic discussion will be confined to the last chapter.

7

Conditional Independence and Ranking Nets

There are not only reasons, but also lack of reasons. Positive and negative relevance go hand in hand with irrelevance or independence. In a way, the latter is an even more important notion. When everything is connected with everything, when reasons run in all directions, when their balance gets excessively complex, then one needs to departmentalize, to draw boundaries, in order to be able to confine oneself to what is inside the boundary, establish order for distinguishing direct and indirect reasons, and so on. This is what epistemological economy urges and what common wisdom tells us. One economical method is just neglect and disregard. This is the method of the dull. Or it can be a great skill acquired through rich experience with complex matters defying systematic treatment. In any case it is untheoretical. If one wants to proceed on a theoretical basis, then one has to turn to the main tool for accomplishing all this: the notion of conditional independence. Its centrality is made clear by probability theory. Many textbooks are organized around independence assumptions: After the basics are done, the first topic covers sequences of independent random variables with no dependence at all (a rich topic). Next come Markov processes characterized by a very strong conditional independence assumption (a still richer topic). And finally, perhaps some space is left for studying even more complex stochastic processes with weaker independence properties. This chapter will study the possible benefits the notion of conditional independence delivers to ranking theory.

I have to make, though, a precautionary remark in advance. It was obvious that the basic development of ranking theory in Chapter 5 was at the same time a commentary – partially explicit, partially implicit, and partially still to be made explicit – regarding the literature on belief revision theory. Given the vastness of this literature, it is also clear that this commentary is bound to remain highly incomplete. This chapter will fare still much worse with respect to the Bayesian net literature, which has indeed become a huge industry, even in the literal sense.

In a way, my German (1976/78) and its English by-product (1980) were among the predecessors of all this literature. They approached the theory of probabilistic causation from the philosophical side – Suppes (1970) was my most important reference work –, proved the essential laws of probabilistic conditional independence between sets of variables (apparently for the first time), explicitly gave them the elaborate causal interpretation that they are usually given nowadays, pointed out some computational benefits, and in fact did all this in a decision-theoretic context which took action or intervention variables into account. Only afterwards did I become aware of the highly relevant and richly developed statistical methods of multivariate analysis and path analysis in particular (which originated with the work of Sewell Wright; see Wright (1921, 1934)) and of the contemporary work in mathematical statistics promoted by Philip Dawid, Harri Kiiveri, Steffen L. Lauritzen, Terry P. Speed, David J. Spiegelhalter, Nanny Wermuth and many others (see, e.g., Dawid (1979) who independently proved the same laws of conditional independence, Lauritzen (1979), Darroch et al. (1980), Wermuth (1980), Isham (1981), Wermuth, Lauritzen (1983), Kiiveri et al. (1984)). The same applies to the most relevant work in econometrics (see in particular Haavelmo (1943) and Granger (1969, 1980)). Through Fishburn (1964, Ch. 12) I was made aware of the theory of sequential and Markovian decision processes, but had not followed its important contemporary development into the theory of influence diagrams (cf., e.g., Howard, Matheson (1981)). In any case, it was crucial for my original development of ranking theory in my (1983b) to demonstrate that this kind of theorizing carries over in a neat and complete way.

Moreover, I only later noticed the rising interest and contribution of AI. Indeed, the topic exploded in the 1980s with the work of Judea Pearl and his many collaborators. One reason was that his work was located directly in the context of AI and combined theory and application in a most detailed and elaborate way. But there was a more substantial reason, namely, that Pearl and his group firmly wedded the topic of conditional independence with graph-theoretic methods. The work of the 80s is beautifully summarized in Pearl (1988), a true milestone. Around the same time, Clark Glymour and his collaborators ingeniously employed these theoretical means for causal analysis, a topic not yet systematically treated in Pearl (1988) (see Glymour et al. (1987) and Spirtes et al. (1993) with particular emphasis on the applications in statistics). Since then these theoretical developments have been firmly established in AI and statistics with overwhelming success.

In comparison to all this, I can pursue here only very modest aims, namely to show that the basic facts allowing for these developments hold in ranking theory as well and thus to indicate that there is a ranking-theoretic duplicate for at least large parts of this field. For this purpose I will need to explain the probabilistic original as well. At the same time, I will try to convey that these results are not only of technical or applied interest, but also philosophically important in that they provide a convincing model of inductive inference. For a more advanced and comparative treatment of these issues see Halpern (2003). Recommendable recent textbooks on the probabilistic theory of Bayesian nets are Cowell et al. (1999) and Jensen (2001). And, despite the fast development of the field, Pearl (1988) is still a must. The results presented here will again prove most important in Chapter 14, when I turn to causation and causal inference.

7.1 Independence Among Propositions and Among Variables

My study will start with a somewhat formalistic investigation of conditional independence in the ranking-theoretic and probabilistic sense. It will only become more vivid with an example and the introduction of the graph-theoretic methods in the next section. The first step is to introduce the basic notion.

> *Definition 7.1*: Let κ be a negative ranking function for $\mathcal{A}$, τ the corresponding two-sided ranking function for $\mathcal{A}$, and $A, B, C \in \mathcal{A}$ with $\kappa(C) < \infty$. Then A is *independent of B conditional on* or *given C* w.r.t. κ iff $\tau(A \mid C) = \pm\infty$ or $\tau(B \mid A \cap C) = \tau(B \mid \bar{A} \cap C)$ or iff, equivalently, $\kappa(A \cap B \mid C) + \kappa(\bar{A} \cap \bar{B} \mid C) = \kappa(A \cap \bar{B} \mid C) + \kappa(\bar{A} \cap B \mid C)$. This situation will be symbolized by $A \perp_{\kappa} B \mid C$, possibly without the subscript. A is (unconditionally) *independent of B* w.r.t. κ, symbolically: $A \perp_{\kappa} B$, iff $A \perp_{\kappa} B \mid W$.

As is to be expected, the behavior of independence is not smoother than that of positive relevance. Here is the pertinent theorem:

> *Theorem 7.2*: Let κ be a negative ranking function for $\mathcal{A}$ and $A, B, C, D \in \mathcal{A}$. Then we have
>
> (a) if $\kappa(C) < \infty$, then $A \perp_{\kappa} B \mid C$ for all $B \in \mathcal{A}$ iff $\kappa(A \mid C) = \infty$ or $\kappa(\bar{A} \mid C) = \infty$; a fortiori, $W \perp_{\kappa} B \mid C$ *and* $\varnothing \perp_{\kappa} B \mid C$ *for all* $B \in \mathcal{A}$;
>
> (b) if $\kappa(C) < \infty$ and $A \cap C \subseteq B \cap C$, then $A \perp_{\kappa} B \mid C$ iff $\kappa(A) = \infty$ or $\kappa(\bar{B}) = \infty$; a fortiori, if $\kappa(A \mid C), \kappa(\bar{A} \mid C) < \infty$, then not $A \perp_{\kappa} A \mid C$ [*irreflexivity*];
>
> (c) if $A \perp_{\kappa} B \mid C$, then $B \perp_{\kappa} A \mid C$ [*symmetry*];
>
> (d) if $A \perp_{\kappa} B \mid C$, then $A \perp_{\kappa} \bar{B} \mid C$, $\bar{A} \perp_{\kappa} B \mid C$, and $\bar{A} \perp_{\kappa} \bar{B} \mid C$;
>
> (e) if $\mathcal{B}$ is a partition of W and $A \perp_{\kappa} B \mid C$ for all $B \in \mathcal{B}$, then $A \perp_{\kappa} \bigcup \mathcal{B}' \mid C$ for all $\mathcal{B}' \subseteq \mathcal{B}$;
>
> (f) if $A \perp_{\kappa} B \mid D$ and $A \perp_{\kappa} C \mid B \cap D$, then $A \perp_{\kappa} B \cap C \mid D$.

Proof: (a), (b), (c), and (d) are trivial.

For the proof of (e), let $\mathcal{B} = \{B_1, \ldots, B_n\}$, $\kappa(A \cap B_i \mid C) = a_i$ and $\kappa(\bar{A} \cap B_i \mid C) = b_i$ $(i = 1, \ldots, n)$. The premise then says that for all $i = 1, \ldots, n$ $a_i + \min \{b_j \mid j \neq i\} = b_i + \min \{a_j \mid j \neq i\}$. If all $a_i = \infty$, then $\kappa(A \mid C) = \infty$, and (e) follows trivially from (a); likewise in case all $b_i = \infty$. Therefore we may assume without loss of generality that $a_1 = \min \{a_i \mid i = 1, \ldots, n\} < \infty$. Hence $a_1 - \min \{a_j \mid j \neq 1\} = b_1 - \min \{b_j \mid j \neq 1\} \leq 0$, and so $b_1 = \min \{b_i \mid i \neq 1, \ldots, n\}$. Our premise thus says that for all $i \neq 1$ $a_i - b_i = \min \{a_j \mid j \neq i\} - \min \{b_j \mid j \neq i\} = a_1 - b_1$ or $a_i = b_i = \infty$. This entails for any $R \subseteq \{1, \ldots, n\}$ that $\min \{a_i \mid i \in R\} - \min \{b_i \mid i \in R\} = \min \{a_i \mid i \notin R\} - \min \{b_i \mid i \notin R\}$ or that both terms on the left or on the right side are ∞. So, the latter says $A \perp_{\kappa} B \mid C$ for $B = \bigcup\{B_i \mid i \in R\}$, either by definition or by (a), as was to be shown.

For the proof of (f), finally, simply replace all ">" by "=" in the proof of Theorem 6.7(e).

As mentioned, we have to consider probabilistic independence as well. Let me repeat some textbook knowledge:

Definition 7.3: Let P be a probability measure on the σ-algebra $\mathcal{A}$, and $A, B, C \in \mathcal{A}$ with $P(C) > 0$. Then A is *independent of B given C* w.r.t. P, symbolically: $A \perp_P B \mid C$, iff $P(A \mid C) = 0$ or $P(\bar{A} \mid C) = 0$ or $P(B \mid A \cap C) = P(B \mid \bar{A} \cap C)$ or iff, equivalently, $P(A \cap B \mid C) = P(A \mid C) \cdot P(B \mid C)$. A is *independent of B* w.r.t. P, symbolically: $A \perp_P B$, iff $A \perp_P B \mid W$.

Probabilistic independence is governed by the following laws:

Theorem 7.4: Let P be a probability measure for $\mathcal{A}$, and $A, B, C, D \in \mathcal{A}$. Then:

(a) $A \perp_P B \mid C$ for all $B \in \mathcal{A}$ iff $P(A \mid C) = 0$ or $P(A \mid C) = 1$; a fortiori, $W \perp_P B \mid C$ and $\varnothing \perp_P B \mid C$ for all $B \in \mathcal{A}$;

(b) if $P(C) > 0$ and $A \cap C \subseteq B \cap C$, then $A \perp_P B \mid C$ iff $P(A) = 0$ or $P(B) = 1$; a fortiori, if $0 < P(A \mid C) < 1$, then not $A \perp_P A \mid C$ [*irreflexivity*];

(c) if $A \perp_P B \mid C$, then $B \perp_P A \mid C$ [*symmetry*];

(d) if $A \perp_P B \mid C$, then $A \perp_P \bar{B} \mid C$, $\bar{A} \perp_P B \mid C$, and $\bar{A} \perp_P \bar{B} \mid C$;

(e) if $B \cap C = \varnothing$ and $A \perp_P B \mid D$, then $A \perp_P C \mid D$ iff $A \perp_P B \cup C \mid D$;

(f) if $A \perp_P B \mid D$, then $A \perp_P C \mid B \cap D$ iff $A \perp_P B \cap C \mid D$.

Theorem 7.4 reveals quite a complex structure of conditional probabilistic independence. The ranking-theoretic analogue is structurally a bit poorer. The laws (a) − (d) + (f) of (7.2) and (7.4) correspond. Note, however, the "iff" in (7.4f) that is not present in (7.2f). Moreover, note that (7.2e) is substantially weaker than (7.4e). It is also weaker than the corresponding law (6.7d) for positive relevance. These differences derive from Theorem 5.23a–b, the law of disjunctive conditions, and the ranking-theoretic possibility that $\tau(B \mid A) = \tau(B) \neq \tau(B \mid \bar{A})$, which is not a probabilistic possibility.

The fact that (conditional) independence in the ranking-theoretic or in the probabilistic sense behaves in this and no other way cannot be sufficiently emphasized. Independence (like necessity) has so many interpretations; my counting usually runs up to a dozen, each with its own theory. According to the more salient ones (e.g., logical or causal independence) the laws of independence are different ones. Hence, intuitive confusion is almost impossible to avoid. The only help is to get clear about the interpretation, the theory underlying it, and the laws of independence entailed thereby (a maxim, by the way, deplorably neglected by all the idealists claiming the world to be dependent on the mind). So, in the case of ranking and probabilistic independence, one must be clear that, if A is independent of B and of C, it may or may not be independent of $B \cap C$ or $B \cup C$. Also, if A and B are independent given C and given D, then they may or may not be independent given $C \cap D$ or $C \cup D$. Such strengthenings of Theorems 7.2 and 7.4 simply do not hold!

Do Theorems 7.2 and 7.4 completely characterize conditional independence in this ranking-theoretic and the probabilistic sense? Certainly not. I do not know of any relevant theorems in the probabilistic case and a fortiori in the ranking-theoretic case. I should, however, avoid the impression that nothing beyond Theorem 7.4 is known about probabilistic independence. On the contrary, the literature on the measurement of probability is highly revealing. There the task is to specify conditions on a relation $\geq$ (= "is at least as likely as") which are necessary and sufficient for proving that there is a unique probability measure P agreeing with $\geq$ (in the sense that $P(A) \geq P(B)$ iff $A \geq B$). There are many variants of this type of result. In particular, it is tempting to jointly axiomatize the order relation $\geq$ and a qualitative independence relation $\perp$. Such an axiomatization contains axioms solely about $\geq$, solely about $\perp$, and mixed ones. The pure axioms for $\perp$ do not go beyond Theorem 7.4 (or its consequences for unconditional independence). Hence, the relevant representation results establish in a way the completeness of these axioms, though only in the context of the further pure and mixed axioms for $\geq$. Classic sources for such results are Domotor (1969, Ch. 2), Fine (1973, Ch. 2), and Krantz et al. (1971, Sect. 5.8). Beyond that context, matters are not as well explored. Still, my abstract conjecture concerning the decidability and hence complete characterizability of positive relevance at the end of Section 6.2 carries over to conditional independence claims; they should be decidable as well. Moreover, the next consideration will allow a more positive picture.

This is the consideration that many independencies among propositions arise accidentally, as it were, without deeper significance. For instance, concerning the toss of a die, the propositions $\{2, 3, 5\}$, to throw a prime number, and $\{2, 4\}$, to throw a positive power of 2, are probabilistically independent. Why should one attempt to account for significant and insignificant independencies alike?

For this reason it seems to be a good idea to search for more stable or even for essential independence, whatever this might mean. Here I am in effect touching on deep issues. The innocent word "stable" will acquire greater significance later on, and we will see in Sections 14.4 and 17.4 that the request for stability is finally answered only by causal independence. However, this is still a long way to go. Presently, we are just taking conventional steps. The next step, and indeed an important one, is to focus on (conditional) independence among sets of variables, not propositions. A glance at probability and statistics textbooks shows that this is in fact the only notion of interest, even though it builds on independence concerning propositions. In any case, we have to study this notion in some detail.

In order to do this, we must make use of the second way of representing possibilities introduced in Chapter 2, in particular in Definitions 2.3 and 2.4. Recall that U is the set of all variables considered, and W is the set of possibilities defined from U as explained there. X, Y, Z, and V will be used to denote members or subsets of U, and for $X \subseteq U$, $\mathcal{A}(X)$ is the subalgebra of $\mathcal{A}(U)$ over W generated by X. Moreover, I will assume W to be finite, i.e., the set U of variables as well as the range of each variable $X \in U$ is finite. I will restrict myself to a few remarks about the generalization to the infinite case, which is feasible, but complicated. The finiteness assumption entails that all algebras considered are atomic.

Independence among variables is defined in the following way for ranks and probabilities alike:

Definition 7.5: Let κ be a negative ranking function, and P a probability measure over $\mathcal{A}(U)$, and $X, Y, Z \subseteq U$. Then X is *independent of Y given* or *conditional on Z* w.r.t. κ or P, symbolically: $X \perp_\kappa Y \mid Z$ or $X \perp_P Y \mid Z$, iff for all $A \in \mathcal{A}(X)$, $B \in \mathcal{A}(Y)$, and atoms C of $\mathcal{A}(Z)$ $A \perp_\kappa B \mid C$ or, respectively, $A \perp_P B \mid C$. And X is (unconditionally) *independent of Y* w.r.t. κ or P, symbolically: $X \perp_\kappa Y$ or $X \perp_P Y$, iff $X \perp_\kappa Y \mid \varnothing$ or, respectively, $X \perp_P Y \mid \varnothing$.

In other words, X and Y are independent given Z if and only if, given any complete realization of the variables in Z, propositions about X and propositions about Y are independent. Note that the reference to the atoms of $\mathcal{A}(Z)$ is essential here. It is quite open whether or not the independence of X and Y still holds if given less information about Z, i.e. disjunctions of atoms of $\mathcal{A}(Z)$. Theorems 7.2 and 7.4 (or rather what they omit) show this. Because of this reference, the generalization to the infinite case and non-atomic algebras $\mathcal{A}(Z)$ is not trivial. It is straightforward only for complete ranking functions. In the probabilistic case one has to resort to a generalized notion of conditional probabilities that is standard in mathematics, but not needed here (cf., e.g., Breiman 1968, Ch. 4, or Jeffrey 1971, Sect. 9).

Unconditional independence among variables is governed by

Theorem 7.6: Let κ be a negative ranking function and P a probability measure over $\mathcal{A}(U)$. Then we have for all $X, Y, Z \subseteq U$:

(a) $X \perp_\kappa \varnothing$, and if $X \perp_\kappa X$, then $X \perp_\kappa U$ *[trivial independence]*,

(b) if $X \perp_\kappa Y$, then $Y \perp_\kappa X$ *[symmetry]*,

(c) if $X \perp_\kappa Y \cup Z$, then $X \perp_\kappa Y$ *[decomposition]*,

(d) if $X \perp_\kappa Y$ and $X \cup Y \perp_\kappa Z$, then $X \perp_\kappa Y \cup Z$ *[mixing]*.

The same laws hold for $\perp_P$ in place of $\perp_\kappa$.

Proof: (a), (b), and (c) are trivial. And (d) immediately follows from (7.8d–e), *weak union* and *contraction*, shown in (7.9) and (7.10) to hold for P and κ. Note how the strange case of self-independence $X \perp_\kappa X$ can come about. It obtains exactly when there is one atom A of $\mathcal{A}(X)$ with $\kappa(A) = 0$ and $\kappa(B) = \infty$ for all other atoms B of $\mathcal{A}(X)$.

Theorem 7.6 is remarkable, since it is known to completely characterize unconditional independence among variables:

Theorem 7.7: Let $\perp$ be any binary relation for subsets of U satisfying (7.6a–d), *trivial independence, symmetry, decomposition*, and *mixing*. Then there exist a probability measure P and a negative ranking function κ over $\mathcal{A}(U)$ such that $\perp_P = \perp_\kappa = \perp$.

I conjectured the probabilistic part of this theorem in my (1976, Sect. 3.2), but it was proved only by Geiger, Paz, Pearl (1988, Theorem 3) and independently by Matús

(1988) (cf. Matús 1994). As far as I can tell, this proof may be directly translated into ranking theory.

Conditional independence among variables is more complicated and more interesting; it will occupy us for the rest of this chapter. I apologize that the next pages continue in the formalistic spirit of the previous ones. As mentioned, though, the next sections will make a lot of intuitive sense of the subsequent formal observations. Instead of looking directly at $\perp_\kappa$ and $\perp_P$, let me first define the abstract notions of a semi-graphoid and a graphoid, originally introduced by Pearl, Paz (1985):

> *Definition 7.8*: A ternary relation $\perp$ among subsets for some set U is a *semi-graphoid* (*over* U) iff for all $X, Y, Z, V \subseteq U$:
>
> (a) $X \perp \varnothing \mid V$, and if $X \perp X \mid V$, then $X \perp U \mid V$ [*trivial independence*],
> (b)   if $X \perp Y \mid V$, then $Y \perp X \mid V$ [*symmetry*],
> (c)   if $X \perp Y \cup Z \mid V$, then $X \perp Y \mid V$ [*decomposition*],
> (d)   if $X \perp Y \cup Z \mid V$, then $X \perp Y \mid Z \cup V$ [*weak union*],
> (e)   if $X \perp Y \mid V$ and $X \perp Z \mid Y \cup V$, then $X \perp Y \cup Z \mid V$ [*contraction*].
>
> $\perp$ is a *graphoid* (*over* U) iff additionally:
>
> (f)   if $X \perp Y \mid Z \cup V$ and $X \perp Z \mid Y \cup V$, then $X \perp Y \cup Z \mid V$ [*intersection*].

The labels of the axioms are also due to Pearl, Paz (1985). Of course, *contraction* (e) has nothing to do with contraction in belief revision. Though the axioms might at first appear confusing, a little reflection shows them to be quite intuitive. Let me quote Pearl (1988, p. 85) (with the variables appropriately exchanged):

The *symmetry* axiom states that in any state of knowledge V, if Y tells us nothing new about X, then X tells us nothing new about Y. The *decomposition* axiom asserts that if two combined items of information are judged irrelevant to X, then each separate item is irrelevant as well. The *weak union* axiom states that learning irrelevant information Z cannot help the irrelevant information Y to become relevant to X. The *contraction* axiom states that if we judge Z irrelevant to X after learning some irrelevant information Y, then Z must have been irrelevant before we learned Y. Together, the *weak union* and *contraction* properties mean that irrelevant information should not alter the relevance of other propositions in the system; what was relevant remains relevant, and what was irrelevant remains irrelevant. The *intersection* axiom states that unless Y affects X when Z is held constant or Z affects X when Y is held constant, neither Z nor Y nor their combination can affect X.

The next sections will provide further insight into these axioms.

With Definition 7.8 we can succinctly state how conditional probabilistic independence behaves:

> *Theorem 7.9*: For each probability measure P over $\mathcal{A}(U)$ $\perp_P$ is a semi-graphoid. If P is regular, i.e., if $P(A) = 0$ only for $A = \varnothing$, then $\perp_P$ is a graphoid.

As far as I know, this theorem was first proved in Spohn (1976/78, Sect. 3.2). The proof is extended to the case of infinitely, even uncountably many variables in Spohn (1980).

It was independently established by Dawid (1979). Pearl, Paz (1985) hoped that (7.9) completely characterizes conditional probabilistic independence among variables, as I had hoped before in my (1976/78). Alas, we were proved wrong, as I am about to outline.

Let me skip the proof for Theorem 7.9. However, I should repeat the proof in my (1983b, Sect. 5.3; 1988, Sect. 6) for the analogous ranking-theoretic claim:

Theorem 7.10: For each negative ranking function κ over $\mathcal{A}(U)$ $\perp_\kappa$ is a graphoid.

Proof: It is quite trivial to show that (7.8a–d) hold for $\perp_\kappa$.

Concerning (e), (7.2f) states the relevant implications of independence statements for all atoms A of $\mathcal{A}(X)$, B of $\mathcal{A}(Y)$, C of $\mathcal{A}(Z)$, and D of $\mathcal{A}(V)$; (7.2e) then extends the relevant conclusion to all $A \in \mathcal{A}(X)$ and all $B \in \mathcal{A}(Y \cup Z)$.

Concerning (f), take any $A \in \mathcal{A}(X)$ and any atom D of $\mathcal{A}(V)$. Let $B_1, \ldots, B_m$ be the atoms of $\mathcal{A}(Y)$ and $C_1, \ldots, C_n$ those of $\mathcal{A}(Z)$. Set $a_{ij} = \kappa(A \cap B_i \cap C_j \cap D)$ and $b_{ij} = \kappa(\bar{A} \cap B_i \cap C_j \cap D)$. The proof of (7.2e) has shown that our premises entail that for each $i = 1, \ldots, m$ there is a c_i such that $c_i = a_{ij} - b_{ij}$ for all $j = 1, \ldots, n$ for which not $a_{ij} = b_{ij} = \infty$, and similarly that for each $j = 1, \ldots, n$ there is a d_j such that $d_j = a_{ij} - b_{ij}$ for all $i = 1, \ldots, m$ for which not $a_{ij} = b_{ij} = \infty$. Hence, there must be an e such that $e = c_i = d_j$ for all i and j. So, we finally have for all i and j either $a_{ij} = b_{ij} = \infty$ or $a_{ij} - b_{ij} = \min \{a_{kl} \mid k \neq i, l \neq j\} - \min \{b_{kl} \mid k \neq i, l \neq j\}$, i.e., for all i and j $A \perp_\kappa B_i \cap C_j \mid D$ in either case. Applying (7.2e) once more yields the desired conclusion.

Note that regularity needs to be presupposed in the probabilistic case for proving $\perp_P$ to be a graphoid, whereas $\perp_\kappa$ is a graphoid even if κ is not regular.

We will see in the next section that in a sense the graphoid structure captures everything important about conditional probabilistic and ranking independence. However, it certainly does not capture everything there is. For instance, the following fact was well known:

(7.11) Let P be a probability measure over $\mathcal{A}(U)$ and $Z \in U$ a binary variable. Then, if for $X, Y, V \subseteq U$ $X \perp_P Y \mid V$ and $X \perp_P Y \mid V \cup \{Z\}$, so $X \cup \{Z\}$ $\perp_P Y \mid V$ or $X \perp_P Y \cup \{Z\} \mid V$.

The ranking-theoretic analogue of (7.11) does not hold. (7.11) was not a serious threat of the possible completeness of (7.9) because of its special assumption of the two-valuedness of Z.

However, Studeny (1989) was able to show that $\perp_P$ has the following complex general property:

(7.12) Let P be a probability measure over $\mathcal{A}(U)$ and $X, X', Y, Y' \subseteq U$. Then $X \perp_P$ $X', Y \perp_P Y' \mid X, Y \perp_P Y' \mid X'$, and $X \perp_P X' \mid Y \cup Y'$ if and only if $Y \perp_P Y'$, $X \perp_P X' \mid Y, X \perp_P X' \mid Y'$, and $Y \perp_P Y' \mid X \cup X'$. The same holds given any $Z \subseteq U$.

This was the fatal blow to the completeness conjecture concerning (7.9). Again (7.12) does not carry over to ranking theory; in Spohn (1994, p. 186) I presented a pertinent counter-example. Moreover, it can be shown (cf. Spohn 1994, p. 188):

(7.13)   Let P be a regular probability measure over $\mathcal{A}(U)$ and $X, X', Y, Y' \subseteq U$. Then, if $Y \perp_P Y' \mid X \cup X'$, and if three of the four independencies $X \perp_P X'$, $X \perp_P X' \mid Y$, $X \perp_P X' \mid Y'$ and $X \perp_P X' \mid Y \cup Y'$ hold, the fourth holds as well. The same holds given any $Z \subseteq U$. The ranking-theoretic analogue of this theorem is true of each negative ranking function κ over $\mathcal{A}(U)$.

Neither last nor least, Studeny (1992, p. 382) presents a whole family of properties:

(7.14)   Let P be a probability measure over $\mathcal{A}(U)$ and $X_0, \ldots, X_n \in U$ n variables ($n \geq 3$). Then $X_0 \perp_P X_i \mid X_{i+1}$ for $i = 1, \ldots, n - 1$ and $X_0 \perp_P X_n \mid X_1$ if and only if $X_0 \perp_P X_{i+1} \mid X_i$ for $i = 1, \ldots, n - 1$ and $X_0 \perp_P X_1 \mid X_n$. The corresponding assertion holds relative to a negative ranking function κ.

Again, the ranking part of (7.14) was shown to hold for a special case in Spohn (1994, pp. 187f.); the general proof is due to Studeny (1995, Sect. 5). However, Studeny (1992) proved not only the probabilistic part of (7.14), but even the much stronger theorem that for each $n \geq 3$ the assertion of (7.14) is logically independent of all statements of the form "if $X_1 \perp_P Y_1 \mid Z_1, \ldots, X_{r-1} \perp_P Y_{r-1} \mid Z_{r-1}$, then $X_r \perp_P Y_r \mid Z_r$," for all $r \leq n$. In other words, Studeny has provided us with an infinite family of logically independent axioms for conditional probabilistic independence among variables. In Studeny (1995, Theorem 1) he showed that this result carries over to ranking theory.

There is no point in deepening this study here. By citing (7.12–7.14) I just wanted to give a flavor of what is on the market. After the first discovery of independence properties beyond the graphoid axioms, vigorous research started continuously producing interesting and ramified results. If you want to know how complicated matters become even for only four variables, see Matús (1999). Probabilistic independence was the main subject, but Spohn (1994) and Studeny (1995) also inquired into ranking independence in a more thorough-going way. The interested reader should study Studeny (2005), an excellent and most comprehensive summary in book length.

However, as far as I know, and as Studeny (2005, pp. 16f.) confirms, there is still no complete axiomatization of conditional independence. The result by Studeny (1992 and 1995) to which I just referred shows that there is no finite complete axiomatization. However, he mentions in Studeny (1992, p. 394) that there might nevertheless be a countable recursive axiomatization of conditional independence. This would well agree with my conjecture after (7.4) that conditional independence statements are decidable. I should moreover mention that Studeny has achieved a complete characterization of probabilistic conditional independence in terms of what he calls supermodular functions and structural imsets (see Studeny 2005, proposition 5.3, p. 89, and Theorem 5.2, p. 101). Let us leave the topic at this point. The next sections will make clear why we may entirely focus on the graphoid axioms. They will be the subject of our further study.

7.2 An Example: Mr. Holmes and his Alarm

So far, so good. We might well ask, though: what is the point of all these abstract observations about conditional independence? This will become much clearer when we combine these observations with graph-theoretic concepts and results. Before proceeding in the abstract way, let me first discuss an elaborate, perhaps tedious, but definitely rewarding example – I have proceeded without exemplification for too long! Pearl (1988, Sect. 2.2) developed his basic observations with an example about Sherlock Holmes, in which Holmes has been extraordinarily displaced by over 80 years and 5000 miles to Los Angeles in the 1980s. This example is devoid of any statistical data that would bias us towards probabilistic thinking from the outset. Hence, it seems to me to be particularly suited for an adaptation in ranking theory. The example will involve six variables in the end, which help to explain various kinds of reasoning.

Example 7.15: Mr. Holmes, a distinguished detective in LA, has acquired a villa in Beverly Hills. Naturally, he has installed a burglar alarm. While in his office downtown, he suddenly receives a call from his neighbor Dr. Watson who informs him that the alarm is apparently sounding. What should Holmes infer from this call?

So far three yes–no variables are involved. B (= burglary) with the two possible values or realizations b_1 = yes = "Holmes' villa fell victim to a burglary" and b_0 = no = non-b_1, S (= sounding) with the values s_1 = yes = "the alarm is sounding" and s_0 = no, and W (= Watson's testimony) manifesting as w_1, Watson's reporting of the alarm's sounding, or as w_0 = non-w_1. There is a temporal order in these variables which is certainly also a reasoning order and perhaps even a causal order and which may be depicted thus:

Figure 7.16

Accordingly, we may specify Mr. Holmes' initial ranking function κ_0 or its two-sided version τ_0. In a probabilistic story there would be no statistics backing up Holmes' subjective probabilities; they would just have to be chosen in some plausible way. Likewise, I will just specify Holmes' ranks in some plausible way throughout the example. For instance, Holmes initially takes a burglar attack to be quite unlikely. Hence:

$(7.15a)$ $\tau_0(b_1) = -3,$

where b_1 is a short label for the proposition $\{B = b_1\}$. Moreover, he trusts in the proper functioning of his alarm, but not very firmly. Burglars might be clever and avoid the alarm and it is not so remote a possibility that the alarm has been activated for other reasons. Hence:

(b) $\tau_0(s_1 \mid b_1) = 2$ and $\tau_0(s_1 \mid b_0) = -2$.

Finally, Holmes has not so high an opinion of his neighbor, Watson. If the alarm sounds, Watson may or may not be so attentive as to call him. And if the alarm does not sound, Watson would presumably not call him. But this is not certain. He has learned that Watson is always good for bad jokes. Hence:

(c) $\tau_0(w_1 \mid s_1) = 0$ and $\tau_0(w_1 \mid s_0) = -1$.

Do these assumptions determine Holmes' ranking function for these three variables? Not as such. However, it is plausible to assume that $W \perp_{\kappa_0} B \mid S$, i.e., that Watson's testimony and the burglary are independent given the alarm's (non-) activity; this is also suggested by figure 7.16. Under this additional assumption τ_0 is completely determined by the above figures, simply through the repeated application of the law of conjunction for negative ranks. In particular, the assumption entails that

(d) $\tau_0(s_1) = -2$ and $\tau_0(w_1) = -1$;

i.e., initially Holmes believes his alarm to be mute and is a bit surprised about Watson's call.

Should he worry about that call? Through this call, his doxastic state shifts from τ_0 to τ_1. Suppose that

(e) $\tau_1(w_1) = 10$,

i.e., that Holmes is very sure that it was really Watson who called him and that he has not misunderstood what Watson was saying. Of course, he is not absolutely sure about this, which is why I chose rank 10. In other words, κ_1 is the result-oriented $w_1 \rightarrow 10$-conditionalization of κ_0. What does κ_1 imply about S and B? We can reason backwards in our diagram. First, we can infer from (c) and (d) that

(f) $\tau_i(s_1 \mid w_1) = -1$ and $\tau_i(s_1 \mid w_0) = -2$ for $i = 0, 1$;

or rather, we can infer this for $i = 0$ and know that this holds for $i = 1$ as well, since information about W does not change the conditional ranks in (f). So we have with (e):

(g) $\tau_1(s_1) = -1$.

Thus Holmes has received a reason for the proposition that his alarm is sounding, but only an insufficient one in the sense of Definition 6.2; he still disbelieves it. This is so because Watson is such an untrustworthy guy. Note that because of this, i.e., because the ranks in (f) are so small, other assumptions in (e), $\tau_1(w_1) = 100$, for instance, or $\tau_1(w_1) = 5$, would not have changed the conclusion (g). The reason can come with great force (= its posterior rank). If its strength is small (= the ranks conditional on it), it cannot move much. This agrees well with what we have observed about the (spring) balance of reasons in (6.11–6.12).

In the same way, we can reason further backwards to B, which is the variable of ultimate interest. We first infer from (a) and (b) that

(h) $\tau_i(b_1 \mid s_1) = -1$ and $\tau_i(b_1 \mid s_0) = -5$ for $i = 0, 1$;

and then with (g) that

(i) $\tau_1(b_1) = -2.$

Because the burglary is initially so unlikely and because the alarm is not a particularly strong sign, Holmes is still adverse to the burglar hypothesis. Note that in this second step we have implicitly assumed that the conditional independence of B and W given S is preserved in the change from κ_0 to κ_1. Otherwise, it would not have been correct to infer in (h) that, if it holds for $i = 0$, then it does so for $i = 1$ as well. We will see only in Section 7.4 why this preservation may be generally assumed. Conditionalization as such guarantees only that the ranks conditional on the evidence are preserved, but $W \perp_{\kappa_1} B \mid S$ secures that the ranks for the indirectly inferred (B) conditional on the directly inferred (S) are also preserved.

The whole procedure very much resembles the reasoning with *inverse probabilities*. If one interprets the arrows in figure 7.16 causally, as one might be inclined to do, the exercise may be called *inference from effects to causes*; it may as well be called *abduction*. Pearl (1988, p. 42) also speaks of cascaded inference.

So Holmes does not yet think that he should do something about his alarm or that he should even worry about a burglary. Still, he is a bit disturbed. Maybe he should gain a bit more certainty? He decides to call his other neighbor, Mrs. Gibbon, an elderly lady usually at home and certainly more reliable than Dr. Watson, despite her occasional drinking problems. So a new variable enters the picture, G (= Mrs. Gibbon's testimony) having two values, g_1 = "yes, Mrs. Gibbon confirms that the alarm is ringing", and g_0 = "no, she doesn't". It fits in into the reasoning (and causal) order in the following way:

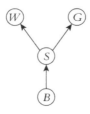

Figure 7.17

Let us again specify Holmes' prior beliefs about Mrs. Gibbon:

(7.15j) $\tau_0(g_1 \mid s_1) = 2$ and $\tau_0(g_1 \mid s_0) = -4$.

Thus Holmes is confident that, when the alarm is sounding, she will tell him so, and he is even more confident that, when the alarm is off, she will not erroneously tell him it is on. Together with (a), (b), and (c), (j) seems sufficient to calculate κ_0 for the whole algebra generated by B, S, W, and G, since the ranks for G depend only on S. This is indeed correct. It indicates, however, that we have made (or read off from the diagram) a further independence assumption, namely $G \perp_{\kappa_0} \{B, W\} \mid S$.

Hence, this is a case of *independent testimonies*. If Holmes suspected that Watson and Gibbon had somehow coordinated their statements, then this independence assumption would not be justified. Note that "independent testimonies" does not mean that G and W would be independent. This would entail a most unwanted random distribution of testimonies. No, it rather means that $G \perp_{\kappa_0} W \mid S$, i.e., that the testimonies are independent given the testified. In this case it is maximally useful to have several testimonies. This kind of situation has recently aroused much interest from confirmation theorists; see Bovens, Hartmann (2004) and Olsson (2005). We will briefly study the general ranking-theoretic situation in Section 7.5. Under a causal reading, this independence assumption is nothing but Reichenbach's *common cause principle*: two effect variables, both causally dependent on a third one (the "common cause"), are thereby rendered independent. However, as I have just slipped into causal talk for the third time in this section, let me emphasize that this is so far only a natural and figurative way of speaking. What all this really has to do with causation is a difficult issue requiring careful philosophical analysis. This issue has to wait till Chapter 14. Then, however, my slips will turn out to be fully justified.

Let us see how Holmes is helped by Mrs. Gibbon's independent testimony. Again, we can calculate, from (d), (g), and (j), his initial expectation about Mrs. Gibbon:

(k) $\tau_0(g_1) = -2$ and $\tau_1(g_1) = -1$.

Of course, since he initially believes that his alarm is mute, he expects Mrs. Gibbon to confirm this – to an equally firm degree, as the comparison of (d) and (k) shows. As (k) displays, this expectation has weakened through Watson's call, but not vanished. Unavoidably, the call entangles him in Mrs. Gibbon's endless stories, and when he manages to hang up after fifteen minutes, he feels she has only vaguely answered his question. Still, she had casually indicated something like "the alarm was sounding". Hence, let us suppose that the conversation changes his τ_1 into τ_2 such that:

(l) $\tau_2(g_1) = 1$.

That is, κ_2 is generated by a $g_1 \rightarrow 1$-conditionalization of κ_1.

What is the effect of this change on the other variables? Concerning S, the target of the independent testimonies, we can first infer from (d) and (j) that

(m) $\tau_i(s_1 \mid g_1) = 2$ and $\tau_i(s_1 \mid g_0) = -4$ for $i = 0, 1, 2$.

Again, this follows directly only for $i = 0$; due to the relevant conditional independency it holds also for $i = 1$ after learning about W; and for $i = 2$ it is built in into conditionalization w.r.t. propositions about G. Hence, we have after the conversation with Mrs. Gibbon:

(n) $\tau_2(s_1) = 1$.

This time, the rank of the hypothesis that the alarm is sounding has increased by 2 and not only by 1, as in the case of Watson. In the latter case, this was due to the weakness of his testimony. Mrs. Gibbon's testimony, by contrast, could have been still more effective. However, since it was so hazy and came with so little force, it was not. In Section 7.5, we will derive a general formula for the very simple interplay of independent testimonies (at least for the case of certain evidence).

Since (h) obtains for $i = 2$ as well (another effect of the independence assumptions) we can then conclude that

(o) $\tau_2(b_1) = -1$.

Again, we can see that a more reliable testimony from Mrs. Gibbon would not have allowed firmer conclusions about the burglary. Since the alarm is not very significant (as we chose the example) the belief that there has been a burglary cannot rise above (o).

The case of Mrs. Gibbon demonstrates another aspect that Pearl (1988, pp. 47f.) illustrates by introducing a further variable, Mr. Holmes' daughter, namely the fact that our model also accounts for *prediction*. In (k) we had computed Holmes' prediction about how Mrs. Gibbon would respond when phoned. Of course, this prediction has changed through Watson's call (and, as we chose the example, it was disproved by his actual call). In causal terms, this prediction proceeded by an *inference from causes to effects*. To be precise, it did so in the initial state. After Watson's testimony the revised prediction was generated by reasoning first from effects (W) to causes (S) and then from causes (S) to effects (G).

Holmes does not feel relieved from his uncertainty. He also realizes – see the comment on (o) – that further information about the alarm does not teach him more about the burglary. How else could he gain further insight? Here is a new idea (and the last one we will discuss). Maybe the burglary is not the only possible explanation of the alarm's sounding. There might be other causes, and a verification of such an alternative cause might dispel the worry about the burglary. Thus Holmes starts wondering whether the alarm might have been triggered by an earthquake. To be sure, it might have, and if there had been an earthquake, then it would certainly be on the news. So he starts listening to the news on the radio. (Let us forget about the slight incoherencies of the story, e.g., that he should have noticed the earthquake directly, and that an earthquake should be even more reason for concern about his home).

In order to represent this new consideration, we have to draw a somewhat more complex picture:

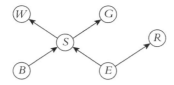

Figure 7.18

Let us build in a further complication the point of which will become clear at the end of our discussion: Suppose that the variables E and R have three possible realizations: there might have been no earthquake (e_0), or a minor one (e_1), or a major one (e_2). Accordingly, there might be three different news on the radio: no earthquake is mentioned (r_0), or a minor one worth reporting (r_1), or a major one (r_2).

Again, we have first to specify Holmes' prior beliefs. Fortunately, earthquakes are quite exceptional, and major ones may not even occur within a lifetime. So we may assume:

(7.15p) $\tau_0(e_0) = 5$, $\tau_0(e_1) = -5$, and $\tau_0(e_2) = -20$.

(Note that if $\{A_1, \ldots, A_n\}$ is a propositional partition, then there is at most one A_i such that the two-sided rank $\tau(A_i)$ of A_i is positive, and if $\tau(A_i) > 0$, then $\max_{j \neq i} \tau(A_j) = -\tau(A_i)$.)

We also have to specify the prior beliefs about the news given the actual happenings: Holmes has great confidence in public news. Hence:

(q) $\tau_0(r_i \mid e_j)$	r_0	r_1	r_2
e_0	10	−10	−20
e_1	−4	4	−8
e_2	−20	−10	10

The somewhat smaller figures in the second row reflect Holmes' slight doubts that a minor earthquake would be found to be worth reporting.

Finally, we have to specify Holmes' prior beliefs about the earthquake's effect on his alarm. However, we cannot do so separately. Rather we have to make assumptions about how the burglary and the earthquake jointly affect the alarm. For instance:

(r) $\tau_0(s_1 \mid b_i, e_j)$	b_0	b_1
e_0	−2	2
e_1	1	5
e_2	4	8

The assumptions (p), (q), and (r), together with (a), (c), and (j) suffice to determine the ranks for all propositions generated by the six variables in figure 7.18. However, they do so only under additional independence assumptions: $B \perp_{\kappa_0} E$, burglaries and earthquakes occur independently; and $R \perp_{\kappa_0} \{B, S, W, G\} \mid E$, the news reports are only about the earthquake, and, given that, they are irrelevant to the alarm sound and the rest of the story. Note that we were not free to specify (r) as we like; we had to agree with the initial stipulation (b).

Indeed, (a), (p), and (r) entail the following joint distribution for B, E, and S.

(s)	$\kappa_0(e_i, b_j, s_u)$	(b_0, s_0)	(b_0, s_1)	(b_1, s_0)	(b_1, s_1)
	e_0	0	2	5	3
	e_1	6	5	13	8
	e_2	24	20	31	23

This table clearly entails (b). Is the table affected by the testimonies of Dr. Watson and Mrs. Gibbon? Yes, but only through S. $\tau_0(s_1) = -2$ shifts to $\tau_2(s_1) = 1$, as we have seen, and hence the s_1-columns in (s) have to be decreased by 2 and the s_0-columns increased by 1 in order to obtain the corresponding table for κ_2.

From (p) and (q) we may moreover infer that

(t) $\tau_0(r_0) = 5, \tau_0(r_1) = -5$, and $\tau_0(r_2) = -13$,

i.e., that Holmes firmly expects not to hear any disastrous news. Tracking the influence of the testimonies on the expectations concerning the radio news we find:

(u) $\tau_2(r_0) = 3, \tau_2(r_1) = -3$, and $\tau_2(r_2) = -11$.

This is as it should be. If it gets more plausible that the alarm is sounding, then the possible causes – burglary or earthquake – get more plausible as well, and so do the relevant news reports.

Now we find Holmes shocked to hear a report that a minor earthquake has indeed occurred in the LA area. Holmes does not trust his ears. Maybe he has overheard something? In any case, let us suppose that

(v) $\tau_3(r_0) = -3, \tau_3(r_1) = 3$, and $\tau_0(r_2) = -14$,

i.e., that κ_3 is the $r_1 \to 3$-conditionalization of κ_2.

What is the impact of this news on Holmes' assessment of the burglary? We already have some practice in how to compute this. First, the beliefs concerning the earthquake get redistributed according to (v) and the unaltered (q):

(w) $\tau_3(e_0) = -3, \tau_3(e_1) = 3$, and $\tau_3(e_2) = -21$.

So, Holmes now believes that a minor earthquake has occurred.

Recall how we have modified (s) in order to represent κ_2. Now the table for κ_2 needs to be changed again in order to account for (w). This results in:

(x) $\kappa_3(e_i, b_j, s_u)$	(b_0, s_0)	(b_0, s_1)	(b_1, s_0)	(b_1, s_1)
e_0	4	3	9	4
e_1	4	0	11	3
e_2	28	21	35	24

From this table we can immediately read off that

(y) $\tau_3(s_1) = 4$ and $\tau_3(b_1) = -3$.

And so the news has a dramatic effect on Holmes' opinions. On the one hand, he is now quite sure that his alarm is sounding, but on the other hand, the burglary hypothesis has been clearly disconfirmed. Whether this is to his comfort may remain open. In any case, it is easily checked that if the news had confirmed his expectations (u) about the absence of an earthquake, he would have been stuck with his weak opinions (n) and (o). All this is as we should have intuitively expected.

Pearl (1988, pp. 49f.) calls this story a case of *explaining away*, and indeed this is what happened to the burglary hypothesis. Philosophers might also call this an *inference to the best explanation*. Given the data, the minor earthquake better explains the presumed sounding of the alarm than a possible burglary. In view of the large philosophical literature on IBE I should be careful with this label. As we will see in Section 14.15, however, its use here is fully justifiable.

There is a final point we can learn from the story. Everybody who has followed through all the calculations will have noticed that our assumptions about the major earthquake played no role whatsoever. We could have cancelled the row for e_2 throughout and the calculations and results would have been just the same. The reason is that by assuming $\tau_0(e_2) = -20$ in (p) we have put the hypothesis from the outset in such a bad position that it could not gain any relevance as far as we have developed the story. Of course, it might have become relevant in other scenarios. The point of including the major earthquake was just to demonstrate this fact and not some superfluous complication.

In fact, the major earthquake was a dummy for any very unlikely cause of the alarm's sounding. It was what philosophers of science call a *catch-all hypothesis*. The Bayesians were always uneasy with catch-all hypotheses. If they get probability 0, they are excluded entirely, they can never enter the picture. But how to give them a positive probability? How to assess the data's likelihood according to them, how to apply Bayes' Theorem to them? One might say that the probability of the catch-all hypothesis is approximately 0 and the probabilities of the genuine hypotheses are only approximately what they are, but this means giving up theory.

For ranking theory this problem does not arise at all. This is just what the major earthquake intended to illustrate. The ranks of the genuine hypotheses are exactly what they are, undisturbed by the low two-sided rank of the catch-all hypothesis. This is why there is no need whatsoever to speculate about possible specifications of the catch-all hypothesis; this can wait until the epistemic situation has become so critical that one can no longer avoid seriously considering it. However, this situation may never arise. Until this critical point, not thinking of any catch-all hypothesis at all and thinking of it, but giving it a very low two-sided rank, are indistinguishable. Again, the significance of this observation will be more thoroughly discussed in Section 14.15.

Allow me a bit of pathos at the end of this section: When one studies David Hume's epistemology in his (1748) or also in his (1739) and his laws of association, which are in fact laws of thought or reasoning, then his view is unmistakeable: "All reasonings concerning matter of fact seem to be founded on the relation of *Cause and Effect*" (Hume (1748, Sect. IV, p. 22). And his examples are constantly of this kind. Hume might have overstated his case; indeed, many contemporary philosophers of science think so. However, Hume is certainly right to a large extent, maybe even to a larger one than is usually acknowledged. In a way, this will be the upshot of Chapter 14. In any case, the crucial question is then: how exactly does causal reasoning work? Well, our example (7.15) *shows exactly how it works*. This is its deep significance.

Once again I have slipped into causal talk; it is too tempting. In Chapter 14 we will see, however, that the causal talk is most appropriate. There, the importance of the present chapter will become fully intelligible.

7.3 Ranking Nets and Bayesian Nets

The reader has perhaps discovered some patterns in the computations of my example. There must be a logic or a theory behind them! There is indeed. It is well-developed through many ramifications for the probabilistic side, its transferability to the ranking-theoretic side is clear in principle, but as far as I know only the basic points have been actually carried out for ranking theory. The pioneering ranking-theoretic work is Hunter (1990, 1991a). Let me explain the basic theory in the rest of this chapter, including the probabilistic side so ingeniously developed mainly by Judea Pearl and his group.

We have to start with the graph-theoretic notions that somehow played a crucial, but obscure role in the previous section. A graph consists of a set U of *nodes*, which in our case is the set U of all variables considered, and a set of connections between nodes called *edges*. For various purposes it might be useful to distinguish several kinds of edges, as we will do in Section 14.10. Here we will consider only one kind, the *arrows* or *directed* edges (only these kinds occurred in the figures (7.16–7.18) of our example). If all edges of a graph are directed, then the graph is also called *directed*. Finally, the arrows cannot run any way they like, but must follow some order. The crucial feature, exemplified in (7.16–7.18), is their *acyclicity*; the arrows must not run in cycles. So our core notion is this:

Definition 7.19: $\langle U, \rightarrow \rangle$ is a *directed acyclic graph*, or a *DAG* for short, iff U is a non-empty finite set and $\rightarrow \subseteq U \times U$ is a set of ordered pairs of members of U, i.e., a relation on U, such that there are no $X_1, \ldots, X_n \in U$ $(n \geq 1)$ with $X_1 \rightarrow X_2, \ldots,$

$X_{n-1} \to X_n$, and $X_n \to X_1$ (where $X_i \to X_j$ means $\langle X_i, X_j \rangle \in \to$). If $<$ is some strict linear order on U (i.e., an irreflexive, transitive, and trichotomous relation on U, where the latter means: either $X < Y$ or $Y < X$ or $X = Y$), we say that the graph $\langle U, \to \rangle$ is *consonant with* $<$ iff $\to \subseteq <$, i.e., iff $X \to Y$ entails $X < Y$. The set $\{X \mid X < Y\}$ of nodes *preceding* Y according to $<$ is denoted by $pr_<(Y)$.

Studeny (2005, p. 46) mentions that the term "directed acyclic graph" is misleading. DAGs should rather be called acyclic directed graphs. He is right, but I will stick to the customary usage. Clearly a directed graph is acyclic if and only if it is consonant with some strict linear order of its nodes. It is natural to interpret an edge $X \to Y$ of a DAG as expressing that Y *directly causally depends on* X. (The emphasis is on "direct"; causal dependence as such may be conceived as transitive, whereas the edges $\to$ of a DAG are generally not transitive). In that case we call the DAG a *causal graph*. We certainly understood our figures (7.16–7.18) in this way. However, we do not yet know what a direct causal dependency is. This causal understanding will not be required for the subsequent theory in any way.

Likewise, in the causal interpretation the strict linear order $<$ consonant with a DAG is naturally understood as the temporal order of the nodes, i.e., of the times at which the variables are realized. If we exclude simultaneous variables, then the temporal order is indeed strictly linear, and the causal order agrees with the temporal one, since we think of causation as directed in time. Again, though, the temporal connotation might be helpful, but it is not required. For the theory to be developed, the order $<$ may be any strict linear order whatsoever.

Let me collect some more or less self-explaining, continuously used graph-theoretic notions in the next

Definition 7.20: Let $\langle U, \to \rangle$ be a DAG. Then we say:

(a) $X \in U$ is a *parent of* $Y \in U$ iff $X \to Y$, the set of parents of Y is denoted by $pa(Y)$.

(b) X is a *child of* Y iff Y is a parent of X.

(c) X is an *ancestor of* Y iff there are $X_1, \ldots, X_n \in U$ ($n \geq 2$) such that $X = X_1 \to \ldots \to X_n = Y$. The set of ancestors of Y is denoted by $an(Y)$.

(d) X is a *descendant of* Y iff Y is an ancestor of X. X is a *non-descendant of* Y iff X is not a descendant of Y. The set of non-descendants of Y will be denoted by $nd(Y)$.

(e) A *path from* X *to* Y is any sequence $\langle X_1, \ldots, X_n \rangle$ of n different nodes in U ($n \geq 1$) such that $X = X_1$ and $Y = X_n$ and for each $i = 1, \ldots, n-1$ either $X_i \to X_{i+1}$ or $X_{i+1} \to X_i$.

(f) Two nodes X and Y are *connected* iff there is a path from X to Y. Being connected is obviously an equivalence relation (i.e. reflexive, symmetric, and transitive). The equivalence classes of that relation are called the (*connectivity*) *components* of $\langle U, \to \rangle$.

(g)   The DAG $\langle U, \rightarrow \rangle$ is *connected* iff U is its single component, i.e., iff there is a graph from each $X \in U$ to each $Y \in U$.

(h)   The DAG $\langle U, \rightarrow \rangle$ is *singly connected* iff for all $X, Y \in U$ there is at most one path from X to Y.

Note that a singly connected graph need not be connected. Our figures (7.16–7.18) were all singly connected. We will see in the next section that this feature is not necessary, but it will considerably facilitate subsequent theorizing. Clearly, this feature is not mandatory under a causal interpretation. One variable might causally depend on another in more than one way, two variables might have common causes as well as common effects, and so on. For this reason, singly connected graphs will turn out to be only a useful special case.

The next important point is that there are also graph theoretic independence notions, indeed conditional ones. Intuitively, the more appropriate term is *separation*: two sets X and Y of nodes are separated by a third one Z. Separation is what we *see* in the graph. However, in view of the close relation to the independence notions considered so far, we may well speak here also of X and Y being independent given Z.

The most obvious notion of separation would be this: $X \subseteq U$ and $Y \subseteq U$ are separated by $Z \subseteq U$ iff each path from some member of X to some member of Y leads through Z, i.e., contains a member of Z. However, paths are undirected and the direction of edges does not play any role for them. So, this notion of separation neglects the direction of edges and is suitable only for undirected graphs not considered here.

There is a less obvious notion of separation, so-called *d-separation* (directed separation), which does all the theoretical work for DAGs. Its definition looks complicated, but it is in fact intuitively and computationally very manageable. Let us consider a DAG $\langle U, \rightarrow \rangle$, a path $\langle X_1, \ldots, X_n \rangle$ from the node $X = X_1 \in U$ to the node $Y = X_n \in U$, and a set of nodes $Z \subseteq U$ not containing X and Y. And let us ask the intuitive question whether information about X is transmitted through the path to Y and whether Z allows or blocks this transmission. This depends on the shape of the path. There are three different types of nodes it can contain.

The node X_i may be a *chain node* (or a *serial connection*). It is so if and only if the path takes the shape $X_{i-1} \rightarrow X_i \rightarrow X_{i+1}$ or the shape $X_{i+1} \rightarrow X_i \rightarrow X_{i-1}$ at the node X_i. Then X_i clearly transmits information, and Z can block this transmission only by containing X_i itself. If the state of X_i is given, then X_{i-1} cannot inform about X_{i+1} or vice versa.

Alternatively, the node X_i may be a *fork node* (or a *diverging connection*). This is so if and only if the path takes the shape $X_{i-1} \leftarrow X_i \rightarrow X_{i+1}$ at X_i. This case is not different from that of a chain node. The fork node X_i transmits information from X_{i-1} to X_{i+1} or the other way around, and Z can block this transmission only by containing X_i. If the state of X_i is given, X_{i-1} and X_{i+1} are thereby rendered independent.

The final case is a bit more complicated. The node X_i may also be a *collider* (or a *converging connection*). It is so if and only if the path takes the shape $X_{i-1} \rightarrow X_i \leftarrow X_{i+1}$ at X_i. The idea then is that X_i does *not* transmit information. The neighbors X_{i-1} and X_{i+1}

contribute separately, i.e., independently to X_i, and so learning about X_{i-1} does not teach us anything about X_{i+1} via that path. Therefore, the path is blocked at a collider right away. Now, though, Z might suspend the blocking. Z might do so by containing X_i itself. If the state of X_i is given, then information is transmitted; learning about the contribution of X_{i-1} to the state of X_i tells us about what the contribution of X_{i+1} might have been. Z might also have that effect more indirectly by containing merely some descendant of X_i in the graph $\langle U, \rightarrow \rangle$. In other words, the path is or remains blocked by Z at a collider X_i iff Z does *not* contain X_i or any descendant of X_i.

In this discussion I have continually used epistemological terminology, "transmission of information", etc., but so far that use was not backed up by any epistemological theory (this back-up is a later part of the story). It was purely an appeal to intuitions in order to make the graph-theoretic notion of d-separation intuitively intelligible. Let us summarize our discussion in

Definition 7.21: Let $\langle U, \rightarrow \rangle$ be a DAG. Then we say:

(a)   A path from $X \in U$ to $Y \in U$ is *blocked by* $Z \subseteq U$ iff X or Y or some chain node or some fork node of the path is in Z or if there is at least one collider of the path such that neither it nor any of its descendants are in Z.

(b)   A path from $X \in U$ to $Y \in U$ is *allowed by* $Z \subseteq U$ iff neither X nor Y nor any chain or fork node of the path is in Z and if each colliders of the path are such that it itself or some of its descendants are in Z.

(c)   If X, Y, and Z are three subsets of U, then X and Y are *d-separated by* Z (in $\langle U, \rightarrow \rangle$), symbolically: $X \perp_d Y \mid Z$, iff each path from some node in X to some node in Y is blocked by Z or, equivalently, iff no path from some node in X to some node in Y is allowed by Z.

For an illustration return to figure (7.18). There, $\{B\}$ and $\{R\}$ are d-separated by $\varnothing$, by $\{E\}$, also by $\{E\}$ plus any of $\{S, W, G\}$, but not by some of $\{S, W, G\}$ alone. $\{B, W\}$ and $\{G, R\}$ are d-separated by $\{E, S\}$, but not by $\{S\}$ or by $\{E\}$ alone. And so on.

I should mention that Lauritzen et al. (1990) have developed another method for reading off conditional independencies from a DAG, the so-called moralization criterion. They also proved it to be equivalent to the d-separation criterion. The criteria have complementary virtues. For us, it suffices to be familiar with one method, and I have chosen the one that appears intuitively more easily accessible to me.

What we are dealing with now are in fact four conditional independence relations. There is the graph-theoretic notion $\perp_d$ of d-separation, there is the abstract notion $\perp$ of a graphoid (and of a semi-graphoid), and there are two epistemological notions $\perp_P$ and $\perp_\kappa$, conditional independence w.r.t. a probability measure P or a ranking function κ. We have substantially, though not completely clarified the relation between the latter three in Section 7.1. But how does d-separation relate to them? This is useful to know for the simple reason that d-separation is the intuitively and computationally best manageable notion. The answer to this question will also provide a theoretical foundation for our so far somewhat opaque practice in the example in Section 7.2 of reading off epistemological independencies from graph-theoretic ones.

The answer is difficult to prove, but easy to state. Let us first discuss d-separation and graphoids. A first answer is

Theorem 7.22: For each DAG $\langle U, \rightarrow \rangle \perp_d$ is a graphoid.

This is part of Theorem 11 of Pearl (1988, p. 128) and indeed easily shown. However, d-separation has more than the graphoid properties. The most conspicuous one is perhaps:

(7.23)   If $X \perp_d Y \mid W$ and $X \perp_d Z \mid W$, then $X \perp_d Y \cup Z \mid W$ [*composition*].

This property is a direct consequence of Definition 7.21. Natural as it may be, though, I hope to have sufficiently emphasized that doxastic independence does not conform to (7.23).

In fact, d-separation has a complex behavior. Theorem 11 of Pearl (1988, p. 128) mentions two further laws, and on p. 131 Pearl conjectures that there is indeed no finite complete axiomatic characterization of d-separation in DAGs. Studeny (2005, pp. 50f.) reinforces the point by outlining an involved argument to the effect that d-separation is not axiomatizable at all by axioms of the form "if $X_1 \perp_d Y_1 \mid Z_1, \ldots$, $X_m \perp_d Y_m \mid Z_m$, then $X_{m+1} \perp_d Y_{m+1} \mid Z_{m+1}$ or $\ldots$ or $X_n \perp_d Y_n \mid Z_n$".

These observations do not appear encouraging. The graphoid structure is a common core, but otherwise doxastic and graph-theoretic independence are disparate and complicated, each in its own way. However, we will next see that under natural assumptions the common core is complete and all that we need. Let us discuss this first in relation to d-separation, where Verma, Pearl (1990) first proved the relevant theorems.

The crucial point is that certain dependencies and independencies, construed as d-separation, are inherent in the DAG structure. To begin with, a child can never be d-separated from its parents. We can indeed observe something stronger: for each $X \in U$ we have $X \perp_d an(X) - pa(X) \mid pa(X)$. That is, each node is d-separated by its parents from its further ancestors. If the DAG is just a chain, i.e., a connected graph without diverging or converging nodes, this is the classical Markov property. The set of parents is even the minimal set doing this. That is, if $Y \subseteq an(X)$, then $X \perp_d an(X) - Y \mid Y$ if and only if $pa(X) \subseteq Y$. This reflects the point that a node cannot be d-separated from any of its parents. As I said, these independencies are built in into the DAG structure. Hence, if we are to compare d-separation with the abstract graphoid notion of independence, the comparison may include only graphoids that respect these basic independencies.

We can improve upon the observations just made. Let $<$ be any strict linear ordering of the variables in U consonant with the DAG $\langle U, \rightarrow \rangle$ (there is always at least one, and usually there are many). Then our Definition 7.21c of d-separation clearly entails that for any $X \in U$ even $X \perp_d pr_<(X) - pa(X) \mid pa(X)$, where $pr_<(X)$ was defined to be

$\{Y \in U \mid Y < X\}$. Note that this is considerably stronger than the above observation. The latter referred only to our stipulation about chain nodes in our motivation of d-separation, whereas now we have also made use of our stipulation about fork nodes. Again, the parents of X form the minimal set with this property; if $Y \subseteq pr_<(X)$, then $X \perp_d pr_<(X) - Y \mid Y$ if and only if $pa(X) \subseteq Y$.

Under a causal reading, with $<$ being the temporal order, we may call $pa(X)$ the *total cause* of X. (Here I am slipping into sloppy language that is common in this field. Of course variables do not cause each other, they causally depend on one another. When causation will be our proper topic, I will avoid such sloppiness). And we may call any member of $pa(X)$ a *direct cause* of X. The causal terminology has somehow persisted in the literature even where $<$ is not the temporal order. So now we can understand content and relevance of the following

Definition 7.24: Let U be any non-empty set and $<$ a strict linear order on U.

(a)   Then L is a *list of total causes* (for U) iff L is a set of pairs containing for each $X \in U$ exactly one pair $\langle X, Y \rangle$ such that $Y \subseteq pr_<(X)$ (Y being the total cause of X according to L). And $\perp_L$ is the corresponding set of independence statements $\{X \perp pr_<(X) - Y \mid Y$, for all $\langle X, Y \rangle \in L\}$ or the corresponding set of triples.

(b)   Likewise, L is a *list of direct causes* (for U) iff $L \subseteq U \times U$ is a set of pairs such that if $\langle X, Y \rangle \in L$, then $X \in U$ and $Y \in pr_<(X)$. $\perp_L$ then is the corresponding set of independence statements $\{X \perp Y \mid pr_<(X) - Y$, for all $Y \in pr_<(X)$ and *not* $\langle X, Y \rangle \in L\}$ or the corresponding set of triples.

In these terms, we had arrived at the conclusion that some list L of total or direct causes is inherently associated with any DAG, that d-separation w.r.t. this DAG automatically satisfies $\perp_L$, and that the comparison of d-separation with (semi-)graphoids may therefore consider only graphoids including $\perp_L$. The comparison then leads to a most satisfying result.

Theorem 7.25: Let U be a non-empty set and $<$ a strict linear order on U.

(a)   To each list L of total or direct causes for U there corresponds exactly one DAG $\langle U, \rightarrow \rangle$, and vice versa, via the correspondence: $Z \rightarrow X$ iff $Z \in Y$ for $\langle X, Y \rangle \in L$, if L is a list of total causes, or, respectively, $Z \rightarrow X$ iff $\langle X, Z \rangle \in L$, if L is a list of direct causes.

(b)   If L is a list of total causes for U, then for all $X, Y, Z \subseteq U$ $X \perp Y \mid Z$ according to the semi-graphoid $\perp$ generated by $\perp_L$, i.e., the smallest semi-graphoid containing $\perp_L$, if and only if $X \perp_d Y \mid Z$ w.r.t. the DAG $\langle U, \rightarrow \rangle$ corresponding to L.

(c)   If L is a list of direct causes for U, then for all $X, Y, Z \subseteq U$ $X \perp Y \mid Z$ according to the graphoid $\perp$ generated by $\perp_L$ if and only if $X \perp_d Y \mid Z$ w.r.t. the DAG $\langle U, \rightarrow \rangle$ corresponding to L.

Remarks (instead of a proof): Part (a) of the theorem is obvious. We had seen above that $\perp_d$ is a graphoid and satisfies $\perp_L$. Hence the "if"-part of (b) and (c) obtains. The "only if"-part is the hard one. Here I simply refer the reader to Theorem 2 and corollary 1 of Verma, Pearl (1990).

What is the difference between (b) and (c)? Applying the *intersection* axiom (7.8f) characteristic of graphoids to a list of direct causes or the associated independence statements clearly yields the corresponding list of total causes or the associated independence statements. The literature usually refers to total causes (or simply to a causal list because the distinction of (7.24) is not made), since semi-graphoids are more relevant in the probabilistic context (cf. Theorem 7.9). For this reason, the interesting point in comparing (b) and (c) is that the semi-graphoid generated by a list of total causes is the same as the graphoid generated by the corresponding list of direct causes. Thus the only contribution of the *intersection* axiom for graphoids lies in getting from the one to the other list.

The upshot of Theorem 7.25 is that (semi-)graphoids as such may be insufficient to characterize d-separation, but together with lists of total or direct causes they are sufficient. This is a perfect answer to our query about the relation between d-separation and (semi-)graphoids. Is there a similar result concerning the epistemological independence relations $\perp_p$ and $\perp_\kappa$? Yes, there is. This is the next point I am going to explain.

Let us deal with both the probabilistic and the ranking-theoretic side at one go. We are now considering a finite set U of variables with a strict linear order $<$, a DAG $\langle U, \rightarrow \rangle$ consonant with this order, and a probability measure P or, respectively, a negative ranking function κ over the propositional algebra $\mathcal{A}(U)$ generated by U. The DAG is there to somehow represent the measure or the ranking function. But when might it be said to do so? Exactly when P or κ respects the basic independencies in the form of a causal list. Let us work out what this means.

It first means that P or κ satisfies the *Markov condition* w.r.t. the DAG (the term used by Spirtes et al. 1993, p. 33) or that the DAG is an *I-map* (*independence map*) of P or κ (the term introduced by Pearl 1988, p. 92 and p. 119).

(7.26) P satisfies the *Markov condition* w.r.t. $\langle U, \rightarrow \rangle$ and $<$ iff for each node $X \in U$ $X \perp_p pr_<(X) - pa(X) \mid pa(X)$. Likewise for κ. This entails that $\langle U, \rightarrow \rangle$ is an *I-map* of P, or, respectively, κ in the sense that all independencies represented in the graph actually hold w.r.t. P or κ, i.e., that for all $X, Y, Z \subseteq U$, if $X \perp_d Y \mid Z$ w.r.t. $\langle U, \rightarrow \rangle$, then $X \perp_p Y \mid Z$ or $X \perp_\kappa Y \mid Z$.

The entailment holds because Theorem 7.25 says that $\perp_d$ *is* the semi-graphoid generated by the independencies defining the Markov condition and because Theorems 7.9 and 7.10 say that this semi-graphoid is a subrelation of $\perp_p$ and of $\perp_\kappa$.

Note that the Markov condition entails that the parents screen off their child not only from the other preceding variables, but from all other non-descendants:

(7.27) If P satisfies the Markov condition w.r.t. $\langle U, \rightarrow \rangle$ and $<$, then for each $X \in U$ $X \perp_p nd(X) - pa(X) \mid pa(X)$. Likewise for κ.

Proof: Let $Y_1, \ldots, Y_n$ be the non-descendants of X not preceding X arranged in the order $<$. Suppose we have already shown that $X \perp_P \{Y_1, \ldots, Y_{i-1}\} \cup pr_<(X) - pa(X) \mid pa(X)$. Y_i is also independent from all its other predecessors given its parents. According to (7.8d) *(weak union)* this entails $X \perp_P Y_i \mid pr_<(Y_i) - \{X\}$. This and the supposition together entail, according to (7.8e) *(contraction)*, that $X \perp_P \{Y_1, \ldots, Y_i\} \cup pr_<(X) - pa(X) \mid pa(X)$.

Spirtes et al. (1993, p. 33) prefer to specify the Markov condition in the form (7.27) rather than (7.26). This is more elegant since it avoids reference to an underlying strict linear order. But it is instructive to make the consequence (7.27) explicit. Moreover, in Chapter 14 on causation we will have to refer to some such order, i.e., the temporal one, anyway.

The relation between measure or ranking function and representing graph is yet tighter. It also means that P or κ satisfies the *minimality condition* w.r.t. the DAG (the term introduced by Spirtes et al. 1993, p. 34) or that the DAG is a minimal *I*-map of P and κ (Pearl's term in his 1988, p. 119):

(7.28) P satisfies the *minimality condition* w.r.t. $\langle U, \to \rangle$ and $<$ iff for each $X \in U$ $pa(X)$ is the smallest set Y such that $X \perp_p pr_<(X) - Y \mid Y$. Likewise for κ. This entails that $\langle U, \to \rangle$ is a *minimal I-map* of P or, respectively, κ in the sense that $\langle U, \to \rangle$ is an *I*-map of P or κ and no proper subgraph of $\langle U, \to \rangle$ (containing less edges) is an *I*-map of P or κ. We also say in this case that $\langle U, \to \rangle$ *agrees with* P or, respectively, κ.

The entailment holds for the same reason as in (7.26). Moreover, it is clear that for each probability measure P or ranking function κ over $\mathcal{A}(U)$ there is such a minimal *I*-map and possibly many; different underlying strict linear orders may produce different causal lists and different *I*-maps.

With (7.28) we have in fact reached the celebrated notion of a *Bayesian net* and the twin notion of a *ranking net* which deserve a separate

Definition 7.29: A triple $\langle U, \to, P \rangle$ consisting of a DAG $\langle U, \to \rangle$ and a probability measure P over $\mathcal{A}(U)$ is a *Bayesian net* iff $\langle U, \to \rangle$ is a minimal *I*-map of, or agrees with, P (or if P satisfies the minimality condition and, a fortiori, the Markov condition w.r.t. $\langle U, \to \rangle$ and some consonant strict linear order). Similarly, $\langle U, \to, \kappa \rangle$ is a *ranking net* iff the DAG $\langle U, \to \rangle$ agrees with the ranking function κ.

So our example in the previous section was in fact an extensive illustration of ranking nets. Why is this a celebrated notion? In AI, because of its algorithmic or computational features (which I am going to sketch in the next section), and in statistics because it promises important progress on the relation between cause and correlation which was obscure and extremely disturbing from the onset of the discipline (this issue will be discussed only in Chapter 14 on causation).

In the present line of thought, though, we must realize that we still have not knitted the relation between measure or ranking function and representing graph tightly enough. If $\langle U, \to \rangle$ agrees with P, then all independencies read off from the graph do obtain w.r.t. P. And the graph does so in an optimal way in the sense that any graph containing fewer arrows would sanction too many independencies. We can do no better. But have we done well enough? This is still the central question of this section. And the answer is apparently in the negative. The independencies w.r.t. P may well go beyond those of a (minimal) I-map of P. Indeed, we had seen in (7.12–7.14) that the laws of probabilistic and ranking independence go beyond the graphoid properties. So what we are looking for are *perfect* I-maps (Pearl 1988, p. 92) or *faithful* measures (Spirtes et al. 1993, p. 35):

> (7.30) P satisfies the *faithfulness condition* w.r.t. $\langle U, \to \rangle$, or P and $\langle U, \to \rangle$ are *faithful* to one another, or $\langle U, \to \rangle$ is a *perfect I-map* of P, iff $\perp_P = \perp_d$ w.r.t. $\langle U, \to \rangle$. Likewise for κ.

The important news is that the faithfulness condition is satisfiable:

> *Theorem 7.31*: For each DAG $\langle U, \to \rangle$ there is a probability measure P and a ranking function κ faithful to it.

This is a deep insight and difficult to prove. The probabilistic part was first proved by Geiger, Pearl (1988, Theorem 11), who specified a suitable discrete measure P taking only finitely many values. Spirtes et al. (1993, p. 390, Lemma 3.5.7) gave a different proof by constructing a suitable member of the class of multivariate normal distributions, which are of particular importance to statistics and which have the nice feature that a vanishing partial correlation already amounts to conditional independence. Meek (1995) has shown that the probabilistic part of (7.31) holds even if the size of the range of the variables in U is given in advance. The transfer of this result to ranking theory is the achievement of Hunter (1991). His construction of a suitable ranking function also works for any size of the range of variables. In any case, the proofs are involved and extend over several pages. I will not attempt to reproduce them here.

In fact, Spirtes et al. (1993, pp. 68f., Theorem 3.2) present a much stronger assertion:

> *Theorem 7.32*: Let $\langle U, \to \rangle$ be a DAG and $\mathcal{P}$ be the set of multivariate normal distributions satisfying the minimality condition w.r.t. $\langle U, \to \rangle$. Then the subset $\mathcal{P}'$ of $\mathcal{P}$ of distributions *not* being faithful to $\langle U, \to \rangle$ has Lebesgue measure 0 within $\mathcal{P}$.

This theorem shows that, though there are unfaithful measures, almost all measures (of a certain type at least) are faithful. Hence, it suggests that one is justified in ignoring unfaithful measures, though this significance is contested. In Section 14.9 we will return to the possible role of faithfulness. I suppose that the theorem carries over to real-valued, but maybe not to integer-valued ranking functions. However, I do not know of any positive results in this direction.

The upshot of all these considerations is the same as with the comparison of the (semi-)graphoid structure with d-separation. d-separation was richer than graphoids, but as far as inference from a list of total or direct causes is concerned, it was not. This situation now obtains as well.

Definition 7.33: Let I be any set of conditional independence statements of the form $X' \perp Y' \mid Z'$. Then $X \perp Y \mid Z$ is (a) *probabilistically*, (b) *regularly probabilistically*, (c) *ranking-theoretically implied* by I iff for (a) all probability measures P, (b) all regular probability measures P, (c) all negative ranking functions κ satisfying I, $X \perp_P Y \mid Z$ or, respectively, $X \perp_\kappa Y \mid Z$ obtains.

Thus, we have finally reached the main conclusion of this section:

Theorem 7.34: Let U be a non-empty set of variables and $<$ a strict linear order of U.

(a) If L is a list of total causes for U, then for all $X, Y, Z \subseteq U$ $X \perp_d Y \mid Z$ w.r.t. the DAG $\langle U, \rightarrow \rangle$ corresponding to L if and only if $X \perp Y \mid Z$ is probabilistically implied by $\perp_L$, i.e., if $X \perp_P Y \mid Z$ for all P satisfying the Markov condition w.r.t. $\langle U, \rightarrow \rangle$ and $<$.

(b) If L is a list of direct causes for U, then for all $X, Y, Z \subseteq U$ $X \perp_d Y \mid Z$ w.r.t. the DAG $\langle U, \rightarrow \rangle$ corresponding to L if and only if $X \perp Y \mid Z$ is regularly probabilistically or ranking-theoretically implied by $\perp_L$, i.e., if $X \perp_P Y \mid Z$ for all regular P or, respectively, $X \perp_\kappa Y \mid Z$ for all κ satisfying the Markov condition w.r.t. $\langle U, \rightarrow \rangle$ and $<$.

Proof: The "only if"-part of (a) and (b) is an immediate consequence of Theorems 7.9, 7.10, and 7.25. And the "if"-part of (a) and (b) is an immediate consequence of Theorem 7.31. Since there are faithful measures or ranking functions, probabilistic and ranking-theoretic implication cannot go beyond d-separation.

As I mentioned after (7.21) we had accumulated four different independence relations. To our satisfaction we have now seen that all four are equivalent and need not be distinguished in the presence of lists of direct or total causes. Our original discoveries (7.9) and (7.10) thus turned out to offer *complete* characterizations of probabilistic and ranking independence among sets of variables under a restriction that is quite comfortable.

DAGs and the associated d-separation played a threefold role here. One role is simple: to provide a graphic understanding of conditional independence (a point that can hardly be underestimated). A second role is to provide a standard of comparison: if d-separation is seen to be equivalent to the one and to the other, these must be equivalent too. The third is a heuristic role: by studying what is built into the DAG structure, one has detected the appropriate restriction under which the equivalence holds. Thus, there can be no doubt about how graph-theoretic methods have enriched the topic.

Let me emphasize at the end of this section that I have scratched here only at the surface. We have not looked at the hard proofs. We have not looked at the tremendous variations of the themes, concerning the kinds of graphs, concerning separation criteria, concerning further theorems, concerning families of probability measures, concerning representations of doxastic states, etc. Let me again simply refer to Studeny (2005), the most recent comprehensive reference about these matters.

7.4 Propagation in Local Ranking and Bayesian Nets

I have just mentioned three important roles of DAGs and d-separation in our inquiry into conditional independence. However, I have still neglected their most important general point. After all, one must not forget that it was mainly AI researchers who developed these concepts and theories, and they have one aim dominating all others: ease of computation. The structures we have studied will be of tremendous help here. This section will explain some basic points that will also enable a better understanding of what was going on in our example in Section 7.2.

The most basic point is this: Subjective probability delivers a beautiful model of inductive reasoning, the only one indeed for a long time. Hence it is natural to try to implement it in a computer. This is hard, though, and therefore AI researchers were at first reluctant to engage in probability theory and preferred other kinds of uncertainty algorithms like MYCIN and other so-called expert systems. The origin of the difficulties is obvious. Imagine just a small application, an algebra of propositions generated by 100 binary variables $X_1, \ldots, X_{100}$. We are thus dealing with 2^{100} possibilities, and in order to specify a probability measure for this algebra, we have to specify $2^{100} - 1$ non-negative values for these possibilities (which together with the last one must add up to 1). One can see that even the biggest and fastest computers soon reach their limit.

Usually we would not directly specify the probabilities of the possibilities. One would rather exploit the law of multiplication (3.4d) according to which:

$$P(x_1, \ldots, x_{100}) = P(x_1) \cdot P(x_2 \mid x_1) \cdot \ldots \cdot P(x_{100}) \mid (x_1, \ldots, x_{99}).$$

By itself this does not reduce the number of degrees of freedom. For X_1 we have to specify 1 probability value, for X_2 conditional on X_1 2 probabilities (one for each of the two values X_1 may take), ..., for X_{100} conditional on $X_1, \ldots, X_{99}$ 2^{99} probabilities, and again $1 + 2 + \ldots + 2^{99} = 2^{100} - 1$.

Imagine, though, that each of the 96 variables $X_5, \ldots, X_{100}$ depend on only four of the preceding variables so that, for instance, $P(x_{100} \mid x_1, \ldots, x_{99}) = P(x_{100}) \mid (x_{80}, x_{85}, x_{90}, x_{95})$. Then we have to specify 1 probability for X_1, 2 for X_2, 4 for X_3, 8 for X_4, and 16 for each of the remaining 96 variables, all in all 1551 probabilities. What a terrific reduction! The story would be exactly the same in ranking-theoretic terms.

In general, let $\langle U, \rightarrow, P \rangle$ be a Bayesian net, let u be any tuple of values the variables in U may take and let for $Y \subseteq U$ u_Y be the tuple of values the variables in Y take according to u. Then we have the factorization:

$$(7.35) \quad P(u) = \prod_{X \in U} P(u_X \mid u_{pa(X)}).$$

Likewise, if $\langle U, \rightarrow, \kappa \rangle$ is a ranking net, we have:

$$(7.36) \quad \kappa(u) = \sum_{X \in U} \kappa(u_X \mid u_{pa(X)}).$$

The computational advantage offered by these factorizations depends, of course, on the density of the underlying graph. In many applications, though, the density will be moderate. Thereby, probabilistic thinking becomes at all feasible for (electronic and biological) computers. In particular, the number of the degrees of freedom increases only linearly and not exponentially with the number of variables.

We have thus seen that, given a Bayesian net $\langle U, \rightarrow, P \rangle$, the probabilistic input required for inferring the whole of P can be considerably reduced. A natural question then is: where is this input stored? The answer is obvious from (7.35): we can store it in the graph itself, piecewise at the single nodes. The first idea suggested by (7.35) is to store at each node the distribution for this node conditional on its parents. This means that we have unconditional information only at the initial nodes without parents. The idea is impractical, though, since in order to know the absolute probabilities for a variable we have to work through the net from the initial nodes up to this variable.

A better idea is to associate with each node unconditional information about the node and its parents. This includes the conditional information, but goes beyond it in containing the unconditional information about the parents. This means, of course, that the pieces stored at each node can no longer be chosen independently; the pieces must fit together. This might appear to be a disadvantage as compared with the first idea that offered this kind of independence. We will soon see, though, the benefits of the present approach.

Before explaining it let me, however, restrict our further considerations to singly connected DAGs (7.20h) that contain at most one path between any two nodes. Most applications will not conform to this constraint. I am assuming it here only for expository reasons, since the account to be developed works in a particularly impressive way for singly connected DAGs. The general theory that becomes increasingly involved is beyond our scope. Let us conclude the discussion so far with

Definition 7.37: Let $\langle U, \rightarrow \rangle$ be a singly connected DAG, and let for each $X \in U$ X^* denote $\{X\} \cup pa(X)$, i.e., the set consisting of X and its parents. Then $\langle U, \rightarrow, (P_X)_{X \in U} \rangle$ is a *local Bayesian net* iff for each $X \in U$ P_X is a probability measure over $\mathcal{A}(X^*)$ such that for all $X, Y \in U$ P_X and P_Y agree on $\mathcal{A}(X^* \cap Y^*)$. P is the associated *global measure* over $\mathcal{A}(U)$ iff, borrowing the notation of (7.35–7.36), for all realizations u of U

$$P(u) = \prod_{X \in U} P_X(u_{X^*}) \bigg/ \prod_{X \neq Y \in U} P_X(u_{X^* \cap Y^*}).$$

(It does not matter whether we write P_X or P_Y in the denominator.) Likewise, $\langle U, \rightarrow, (\kappa_X)_{X \in U} \rangle$ is a *local ranking net* iff for each $X \in U$ κ_X is a negative ranking function for $\mathcal{A}(X^*)$ such that for all $X, Y \in U \kappa_X$ and κ_Y agree on $\mathcal{A}(X^* \cap Y^*)$. κ is the associated *global ranking function* for $\mathcal{A}(U)$ iff for all u

$$\kappa(u) = \sum_{X \in U} \kappa_X(u_{X^*}) - \sum_{X \neq Y \in U} \kappa_X(u_{X^* \cap Y^*}).$$

Note that the global measure is well defined in this way only because the graph is singly connected. For more general DAGs, stricter consistency requirements are required. (Cf. Jensen 1996, pp. 76f.)

If P is the global measure associated with the local Bayesian net $\langle U, \rightarrow, (P_X)_{X \in U} \rangle$, does this entail that $\langle U, \rightarrow, P \rangle$ is a Bayesian net? No, look at the simple DAG:

Figure 7.38

If this DAG and P form a Bayesian net, then the factorization (7.35) applies and automatically ensures that $X \perp_P Y$. A local Bayesian net, by contrast, contains P_X on $\mathcal{A}(X)$, P_Y on $\mathcal{A}(Y)$, and P_Z on $\mathcal{A}(\{X, Y, Z\})$, and in the latter any dependency between X and Y can be built in. So a local Bayesian net can contain fewer conditional independencies than dictated by d–separation w.r.t. its DAG.

Have we made a mistake? No, we will soon see that local nets are the ones we need. We can do two things, however, in order to reduce the difference. We can first ask which independencies are respected by a local Bayesian net $\langle U, \rightarrow, (P_X)_{X \in U} \rangle$ in case there are fewer of them than d–separation yields for $\langle U, \rightarrow \rangle$. An answer is provided by what I call d*-separation:

Theorem 7.39: Let $\langle U, \rightarrow, (P_X)_{X \in U} \rangle$ be a local Bayesian net, where $\langle U, \rightarrow \rangle$ is singly connected. Say that a path from $X \in U$ to $Y \in U$ is blocked* by $Z \subseteq U$ iff Z contains some chain or fork node of the path or X or Y itself (thus, paths cannot be blocked* at colliders at all). And say that $X \subseteq U$ and $Y \subseteq U$ are *d*-separated by* $Z \subseteq U$, symbolically $X \perp_{d^*} Y \mid Z$, iff all paths from some node in X to some node in Y are blocked* by Z. Then the associated global measure P satisfies the Markov condition w.r.t. d*-separation; that is, if for $X, Y, Z \subseteq U$ $X \perp_{d^*} Y \mid Z$, then $X \perp_P Y \mid Z$. The ranking-theoretic counterpart of this assertion holds as well.

Proof: Any chain or fork node $T \in U$ divides the graph $\langle U, \rightarrow \rangle$, since it is singly connected, in two halves V and V' such that $\{V, \{T\}, V'\}$ is a partition of U and any path from V to V' contains T as a chain or fork node. Indeed, T may allow several partitions of this kind. Then it is easily seen that

$$P(u) = \prod_{X \in U} P_X(u_{X^*}) \bigg/ \prod_{X \neq Y \in U} P_X(u_{X^* \cap Y^*})$$

factors out so that

$$P(u) = P(u_V, u_T) \cdot P(u_{V'}, u_T) \, / \, P(u_T).$$

Hence, due to singly-connectedness, we have for each such partition $V \perp_P V' \mid T$. Let us call each such conditional independency a *covering* one (since it covers the whole of U).

Now suppose that $X \perp_{d^*} Y \mid Z$. It is not difficult to see then that $X \perp_P Y \mid Z$ is entailed by the covering independencies: Let $X = \{X_1, \ldots, X_m\}$ and $Y = \{Y_1, \ldots, Y_n\}$. Because the path from X_1 to Y_1 is blocked* by Z, there must be covering independency which entails $X_1 \perp_P Y_1 \mid Z \cup \{X_2, \ldots, X_m, Y_2, \ldots, Y_n\}$ by *weak union* (7.8d). Suppose, we have already shown that $\{X_1, \ldots, X_k\} \perp_P \{Y_1, \ldots, Y_l\} \mid Z \cup \{X_{k+1}, \ldots, X_m, Y_{l+1}, \ldots, Y_n\}$. Then there must be further covering independencies entailing for each $i = 1, \ldots, k$ that $X_i \perp_P Y_{l+1} \mid Z \cup \{X_{i+1}, \ldots, X_m, Y_{l+2}, \ldots, Y_n\}$. Applying *contraction* (7.8e) several times, we first get $\{X_1, \ldots, X_k\} \perp_P Y_{l+1} \mid Z \cup \{X_{k+1}, \ldots, X_m, Y_{l+2}, \ldots, Y_n\}$ and then $\{X_1, \ldots, X_k\} \perp_P \{Y_1, \ldots, Y_{l+1}\} \mid Z \cup \{X_{k+1}, \ldots, X_m, Y_{l+2}, \ldots, Y_n\}$. Completing induction in this way we may finally conclude that $X \perp_P Y \mid Z$.

Of course, we cannot expect that d*-separation w.r.t. $\langle U, \to \rangle$ delivers a minimal *I*-map of P, because the local net might entail some independencies among the parents of a node which are not represented by d*-separation.

The second thing we can do is to define local nets for which the difference does not arise:

Definition 7.40: $\langle U, \to, (P_X)_{X \in U} \rangle$ is a *sound local Bayesian net* iff it is a local Bayesian net and if, P being the associated global measure, $\langle U, \to, P \rangle$ is a Bayesian net. Likewise for *sound local ranking nets*.

We will soon see, however, that soundness is not a stable property.

Now that we have stored local information at single nodes (and added the required explanations), we should consider the dynamic question: How does such a local net process information? How does it learn? We gave the global answers with our conditionalization rules (3.8, 3.10) and (5.25, 5.33). The crucial issue, therefore, is whether we can turn the global answers into local ones, whether we can update or conditionalize the local net without transcending the local structure. Can we simply propagate the evidence through the local net? Yes, this is the ultimate purpose of all the formal developments in this and the previous section. Obviously, one can invest unlimited computational wit in this issue. This has indeed been done; cf., e.g., Neapolitan (1990, Chs. 6–7) and Jensen (1996, Ch. 4, and 2001, Ch. 5) for deeper inquiries. Pioneering work for the probabilistic case is due to Pearl (1982) and Kim, Pearl (1983); Pearl (1988, p. 232) mentions some preceding work, interestingly also from judicial reasoning. Hunter (1990) was the first to carry this over to the ranking-theoretic case. In what follows, I stick to his paper.

So, let us see how propagation works, and let us primarily look at the ranking-theoretic side (after all, that is our primary interest). Hence, we are considering a singly connected local ranking net $\langle U, \rightarrow, (\kappa_X)_{X \in U} \rangle$ building up the global ranking function κ for $\mathcal{A}(U)$. Suppose that the net receives uncertain evidence on a variable V somewhere in the net, reflected in a ranking function ε on $\mathcal{A}(V)$. How is the whole net to be updated by this evidence? How is the $\mathcal{A}(V) \rightarrow \varepsilon$-conditionalization of κ, which we will henceforth denote by κ', to be computed?

The procedure is entirely straightforward. However, it is so only because we are focusing on singly connected graphs; otherwise computations might become quite messy. Let $U_1 = \{V\} \subseteq U$, let $U_m = \{X \mid$ the path from V to X contains exactly m variables (including initial and terminal node)$\}$, and let the longest path starting at V have length n. So $\{U_1, \ldots, U_n\}$ is a partition of V. This partition is well defined only because $\langle U, \rightarrow \rangle$ is singly connected. Now, evidence has directly induced the new ranking function ε on $\mathcal{A}(V)$. This means that we have to set $\kappa' = \varepsilon$ for $\mathcal{A}(V)$.

In a first immediate step we can update the local information κ_V at V. Let us still follow the convention that a lower case letter, say, z denotes possible values (or the corresponding propositions) from the range W_Z of the (possibly singleton) set Z of variables denoted by the corresponding upper case letter. Then the first updating step is defined by the equation:

(7.41) $\kappa'_V(v^*) = \kappa_V(v^* \mid v) + \varepsilon(v)$.

This is what $\mathcal{A}(V) \rightarrow \varepsilon$-conditionalization requires.

Now suppose we have already advanced m steps and updated local information at all nodes Y in $U_1 \cup \ldots \cup U_m$ to which, hence, a new κ'_Y is attached, and consider any $Z \in U_{m+1}$. Since $\langle U, \rightarrow \rangle$ is singly connected, there is exactly one $Y \in U_m$ immediately preceding Z on the path from V to Z. Two cases must now be distinguished: (i) Z may be a parent of Y. In that case $Z \in Y^*$, and κ'_Y specifies new ranks for $\mathcal{A}(Z)$. So we may derive new ranks for the whole of $\mathcal{A}(Z^*)$ by setting

(7.42) $\kappa'_Z(z^*) = \kappa_Z(z^* \mid z) + \kappa'_Y(z)$.

(ii) Z may be a child of Y, so that $Y \in Z^*$. Then κ'_Y specifies new ranks for $\mathcal{A}(Y)$ and now we may derive new ranks for $\mathcal{A}(Z^*)$ by the formula

(7.43) $\kappa'_Z(z^*) = \kappa_Z(z^* \mid y) + \kappa'_Y(y)$.

Note that we can undertake the $m + 1$th step for all $Z \in U_{m+1}$ at once; there is no interdependence of the computations requiring a more stepwise procedure. This is why Hunter (1990) speaks of *parallel* belief revision. The evidence really spreads through the net like a shock wave.

Clearly the procedure stops after n steps. Then we have attached new local information κ'_X at each node $X \in U$, and we have thereby arrived at the updated local ranking net:

Definition 7.44: Let us say that the posterior local ranking net $\langle U, \rightarrow, (\kappa'_X)_{X \in U} \rangle$ is generated from the prior local net $\langle U, \rightarrow, (\kappa_X)_{X \in U} \rangle$ and ε by *simple propagation* iff it is generated according to the procedure defined by (7.41–7.43).

As I said, nothing could be easier.

In fact, what we did in our extended example in Section 7.2 was nothing but to repeatedly conduct the various steps of simple propagation. Because I was more interested in demonstrating the effects of learning, I specified the posterior ranks concerning single variables, the burglary, the sounding of the alarm, etc., and did not really make explicit the entire local information at each node. Only when computations became a little bit more complex did I specify the local information in (7.15s + x). Yet the local information was always the base of my calculations, and each reader duplicating them will have done so by calculating it as well. So simple propagation is the logic, the inference method, underlying example (7.15) and causal inference in general (at least as long as we are dealing only with singly connected DAGs).

Indeed, simple propagation is not only a model of causal inference, it is at the same time a model for the causation of beliefs. Recall that when discussing this topic in Section 6.4 I presented belief change as a one-step causal process, with a new distribution (of ranks or probabilities) over the unique evidential source as cause, and an entire posterior doxastic state as effect. This was a very coarse description. With simple propagation, we now also have a model for the details of the process. The spreading of the experiential shock wave through our epistemic fabric has found many metaphorical expressions in the philosophical literature. Here we have a precise model on a sound theoretical basis (deductive inference would provide one as well, but I had emphasized that it is not well suited for our purposes).

However, we have not yet answered the most important question. Does simple propagation deliver what we want? What guarantees that the global measure associated with the updated local net is really the κ' we want, the $\mathcal{A}(V) \rightarrow \varepsilon$-conditionalization of κ? This is guaranteed only by the graph structure and the independencies entailed by it, as the proof of the following theorem shows (this is Theorem 3 of Hunter 1990):

Theorem 7.45: If the posterior local net $\langle U, \rightarrow, (\kappa'_X)_{X \in U} \rangle$ is generated by simple propagation from the prior local net $\langle U, \rightarrow, (\kappa_X)_{X \in U} \rangle$ and the ranking function ε on $\mathcal{A}(V)$, then the $\mathcal{A}(V) \rightarrow \varepsilon$-conditionalization κ' of the global ranking function κ associated with the prior net is the global ranking function associated with the posterior net.

Proof: Let $\{U_1, \ldots, U_n\}$ be as before. The restriction of κ' to $\mathcal{A}(U_1)$ *is* ε. Hence, thus restricted the assertion of the theorem is trivially correct. Now suppose the assertion is correct when restricted to $\mathcal{A}(U_1 \cup \ldots \cup U_m)$; that is, for all $Y \in U_1 \cup \ldots \cup U_m$ the restriction of κ' to $\mathcal{A}(Y^*)$ is κ'_Y. And take any $Z \in U_{m+1}$. Again, we have to distinguish two cases. Let us first look at the case where Z is a child of some unique $Y \in U_m$. Then we have:

$$\kappa'(Z^*) = \min_v \kappa'(z^*, v) = \min_v [\kappa'(z^* \mid v) + \kappa'(v)]$$
$$= \min_v [\kappa(z^* \mid v) + \varepsilon(v)] \qquad \text{(since } \kappa' \text{ is the } \mathcal{A}(V) {\to} \varepsilon\text{-conditionalization of } \kappa)$$
$$= \min_v [\min_y \{\kappa(z^* \mid y, v) + \kappa(y \mid v)\} + \varepsilon(v)] \qquad \text{(law of conjunction and disjunction)}$$
$$= \min_v [\min_y \{\kappa(z^* \mid y) + \kappa(y \mid v)\} + \varepsilon(v)] \qquad \text{(because } Z^* \perp_{d^*} V \mid Y)$$
$$= \min_y \{\kappa(z^* \mid y) + \min_v [\kappa(y \mid v) + \varepsilon(v)]\}$$
$$= \min_y \{\kappa(z^* \mid y) + \kappa'(y)\}$$
$$= \min_y \{\kappa_Z(z^* \mid y) + \kappa'_Y(y)\} \qquad \text{(by induction hypothesis)}$$
$$= \kappa'_Z(z^*) \qquad \text{(because of } Y \in Z^* \text{ and (7.43)).}$$

Note the crucial premise concerning d*-separation. However the path runs from V through Y to Z, Y is either a chain or a fork node of that path and thus blocks* it.

In the other case where Z is a parent of Y, the same chain of equations goes through. Simply replace "y" by "z" throughout. In the final line we of course appeal to (7.42) instead of (7.43). And the premise $Z^* \perp_{d^*} V \mid Z$ now holds because Z is a chain node on the path from V through Y and Z to the parents of Z.

Needless to say, all this works for the probabilistic side in the same way. Let me add two observations. First, soundness of local nets is not preserved by simple propagation. This is already clear from our toy graph (7.38); evidence on the variable Z might well produce a dependence between X and Y. This was in fact the point of the manner in which d-separation behaves at colliders. Hence, we would be ill-advised to focus only on sound local nets. Second, we may positively observe that $\langle U, \to, (\kappa'_x)_{x \in U} \rangle$ is again a local ranking net; that is how we defined it. Thus, Theorem 7.39 applies to it as well; that is, simple propagation preserves all the independencies entailed by d*-separation w.r.t. $\langle U, \to \rangle$ in the posterior net. Hence we can repeat the procedure as often we like.

So far, we have studied a case that was special in various ways. One way in which it was special was that the evidence directly affects only one variable. It seems desirable, though, to generally account for evidence that hits a local net at various nodes at once. Unfortunately, to quote Hunter (1990, p. 248) with adapted notation, "if X and Y are dependent variables, then updating first on X and then on Y will not in general give the same result as updating in the reverse order. Thus simultaneous updating on dependent evidence is problematic, since it is not clear what the answer should be. One can approach the problem of simultaneous multiple updatings by first considering a special case in which the above problem does not arise. This is the case in which all the evidence events are learned with certainty." Thus he sets out on p. 249 to explain a modification of simple propagation that works for multiple certain evidence, i.e., in the case that for some $V \subseteq U \varepsilon(v_0) = 0$ for exactly one tuple of values for V and $\varepsilon(v) = \infty$ for all $v \neq v_0$. He then goes on "to argue that this case it not so special as it may appear" (p. 248). The trick he applies on p. 250 for generalizing that technique to the case of multiple uncertain evidence consists in (i) enlarging the graph while preserving its singly-connectedness, by adding to each variable X about which uncertain evidence is received a binary dummy variable as a child (roughly saying that a specific kind of evidence about X is or is not received), (ii) giving the values of X just those degrees of belief conditional on the dummy variable being positive that they have according to the uncertain evidence, (iii) thus identifying the case of multiple uncertain evidence with a case of multiple certain evidence about all the dummy variables being positive, and (iv) finally applying the mentioned technique suitable for multiple certain evidence. All this mirrors a much larger discussion on the probabilistic side and purports to show that these problems can be satisfactorily dealt with in ranking theory as well. (We will be able take up this scenario from a different perspective in Section 9.3. It might also be interesting to relate this account of multiple uncertain evidence to my account of multiple contraction in Section 5.7.)

The major restriction was that we considered only singly connected DAGs. It was clear that simple propagation works only for such graphs; only then does the shock wave of evidence expand so regularly. However, various algorithms have been proposed that work in a perfectly general way. Lauritzen, Spiegelhalter (1988) proposed what was later transformed into the so-called HUGIN method. Shafer, Shenoy (1990) invented another algorithm. Lazy propagation, due to Madsen, Jensen (1999), integrated both proposals. And so forth. All these methods are technically quite involved. There would be no point in delving into these matters in a philosophy book. It is nice to see that these methods work without restriction. However, it is no surprise that they work well only if the underlying graphs are not too dense. These remarks make clear that there is no end here to algorithmic craftsmanship. For recent textbooks I refer the reader to Cowell et al. (1999) and Jensen (2001).

To my knowledge, these advanced methods have not been carried over from the probabilistic to the ranking-theoretic side, although I do not expect principled obstacles. However, we have observed on several occasions that the analogy between the two sides is not perfect. Hence, one can be sure only if the translation is actually carried out. On the other hand, we have seen that ranking theory is computationally much simpler than probability theory. So there might be clever algorithms that work for ranks, but which are not translatable into probabilities. All this might be worth exploring.

The overarching restriction surely was that we only considered directed acyclic graphs. I mentioned that many other kinds of graphs are considered in the literature. Moreover, we have only dealt with probabilities and ranks. Alternative modellings of doxastic states are eager to prove their representability in graph structures; see the overview in Halpern (2003, Ch. 4). In this way, questions, possibilities, and answers increase exponentially, and the relevant literature does so as well. I confess I am intimidated by these perspectives. Still, I think I have reached the main goal of this chapter. I have shown that the basic theory, comprising conditional independence, graphoids, DAGs, d-separation, and the appertaining algorithms, holds for ranking theory as well. Therefore, ranking theory can be combined with all of the theoretical developments from this base.

7.5 Appendix on Independent Testimonies

In Section 7.2 I gave examples of various interesting situations. It might thus be instructive to study the general ranking logic of these special situations. As an appendix to this chapter let us do this for at least one type of situation, namely, the case of independent testimonies as exemplified by Dr. Watson's and Mrs. Gibbon's calls. If we still restrict ourselves to binary variables, the general situation is this: there is a binary variable X with realizations x^+ and x^-; we are interested in X, and we have some prior assessment $\tau(x^+)$ of X; if we do not know anything about X, then $\tau(x^+) = 0$. There are n testimonies, symptoms, signs, or however you may call them; these are the binary variables $Y_1, \ldots, Y_n$ taking values y_i^+ and y_i^- ($i = 1, \ldots, n$). We have an initial assessment of the reliability of these signs, which need not be the same for each; so suppose $\tau(y_i^+ \mid x^+) = a_i \geq 0$ and $\tau(y_j^- \mid x^-) = b_j \geq 0$. (So, if there is a correlation between X and Y_i, it is always y_i^+ that is positively correlated with x^+ and hence also y_i^- with x^-.)

Additionally, we assume the testimonies to be independent; that is $Y_i \perp_\kappa \{Y_1, \ldots, Y_{i-1}, Y_{i+1}, \ldots, Y_n\} \mid X$ for all $i = 1, \ldots, n$. Hence, the situation is represented by the following DAG:

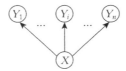

Figure 7.46: n independent testimonies

Let us use $\mathbf{y}$ as a variable for the possible realizations of all signs, i.e., for sequences $\langle y_1, \ldots, y_n \rangle$ with + and − somehow distributed as superscripts, and let $\mathbf{y}^+ = \{i \mid y_i^+$ is a member of $\mathbf{y}\}$ and $\mathbf{y}^- = \{j \mid y_j^-$ is a member of $\mathbf{y}\}$. So conditional independence (together with the initial ignorance about X) says that for each $\mathbf{y}$

$$\kappa(x^+, \mathbf{y}) = \kappa(x^+) + \kappa(\mathbf{y} \mid x^+) = \kappa(x^+) + \sum_{i \in \mathbf{y}^-} b_{j \in \mathbf{y}^-} \text{ and}$$

$$\kappa(x^-, \mathbf{y}) = \kappa(x^-) + \kappa(\mathbf{y} \mid x^-) = \kappa(x^-) + \sum_{i \in \mathbf{y}^-} a_i.$$

Now we observe all symptoms, listen to all testimonies, and thus receive the evidence $\mathbf{y}$, which need not be consonant in the sense that it contains only pluses or only minuses; the evidence might be conflicting. What should we infer from $\mathbf{y}$ about X? Very simple, we have $\tau(x^+ \mid \mathbf{y}) = \kappa(x^- \mid \mathbf{y}) - \kappa(x^+ \mid \mathbf{y}) = \kappa(x^-, \mathbf{y}) - \kappa(x^+, \mathbf{y})$, that is:

$$(7.47) \quad \tau(x^+ \mid \mathbf{y}) = \tau(x^+) + \sum_{i \in \mathbf{y}^-} a_i - \sum_{i \in \mathbf{y}^-} b_{j \in \mathbf{y}^-}.$$

So we have only to add the strengths of the positive signs to our prior assessment $\tau(x^+)$ and to subtract the strengths of the negative signs; that's all. In other words, independent testimonies can be simply summed up. The formula simplifies further when all signs point into the same direction. This is an intuitively most satisfying result.

I hasten to add that (7.47) assumes that the evidence is certain. If evidence is uncertain, i.e., if the members of **y** receive some finite posterior rank, then the strength of the signs might be hampered by their uncertainty, and we have to replace (7.47) by something more complicated. In fact, I had assumed in Section 7.2 that Mrs. Gibbon's testimony is relevantly uncertain; this is why (7.15d + g + n) do not conform to (7.47). However, if the evidence is not maximally, but just sufficiently certain, (7.47) will stand.

What might be the corresponding probabilistic formula? It is less simple. Again, we proceed from some prior assessment $P(x^+)$; ignorance about X is probably best expressed by $P(x^+) = P(x^-) = 1/2$. Again, the signs might have varying reliability. So assume $P(y_i^+ \mid x^+) = a_i \geq 1/2$ and $P(y_j^- \mid x^-) = b_j \geq 1/2$. With the above notation we hence have

$$P(x^+, \mathbf{y}) = P(x^+) \cdot \prod_{i \in y^+} a_i \cdot \prod_{j \in y^-} (1 - a_j) \text{ and}$$

$$P(x^-, y) = P(x^-) \cdot \prod_{i \in y^+} (1 - b_i) \cdot \prod_{j \in y^-} b_j.$$

Therefore, $P(x^+ \mid \mathbf{y}) = P(x^+, \mathbf{y}) / P(\mathbf{y}) = P(x^+, \mathbf{y}) / [P(x^+, \mathbf{y}) + P(x^-, \mathbf{y})] = 1 / [1 + P(x^-, y) / P(x^+, y)]$, or, to resume:

$$(7.48) \quad P(x^+ \mid y) = \frac{1}{1 + q}, \text{ where } q = \frac{P(x^-)}{P(x^+)} \cdot \prod_{i \in y^+} \frac{a_i}{1 - b_i} \cdot \prod_{j \in y^-} \frac{1 - a_j}{b_j}.$$

Bovens, Hartmann (2004, formula (3.19) in Sect. 3.3, and Sect. 5.2) derive the same formula. It is qualitatively correct and plausible, but quantitatively not very transparent. I am much more satisfied with (7.47), and I would like to see the result of casting such texts as Bovens, Hartmann (2004) into ranking-theoretic terms. In fact we will see some unexpected use of (7.47) right in the next chapter, use which (7.48) could not have in probability theory.

8

The Measurement of Ranks

Numbers matter! This was my conclusion after Definition 5.5, our first encounter with ranking functions. In the meantime we have seen substantial pieces of theorizing on the basis of this definition. I used to think that the argument in Section 5.1 concerning iterated belief revision, the fact that ranking theory thus generalizes AGM belief revision theory, and the prospects for substantial theorizing were sufficient grounds for accepting ranking theory. They were so within Artificial Intelligence, where researchers are perfectly happy with numerical theories, if they work in a plausible and useful way. Not so in philosophy, and hence neither in belief revision theory, which was originally a philosophers' product. There, the uneasiness about what the numbers really mean persisted. What exactly is the difference between two numerically different, but ordinally equivalent ranking functions? Just vague feelings concerning strength of belief? This would be a poor answer, and as long as there is no better one, ranks appear arbitrary.

Uneasiness is one thing, but is there really an objection? Yes, to some extent. Cardinal utility became acceptable only after von Neumann, Morgenstern (1944, Ch. 3) proved that preferences conforming to certain axioms determine cardinal utilities on an interval scale. Thus, the cardinal concept turned out to be definable by, or reducible to, the ordinal concept; one cannot accept the one and reject the other. Ranks likewise are psychological magnitudes, and hence it appears legitimate to demand a measurement theory for them, too.

Whether this is really so depends, however, on one's opinions about the legitimacy of theoretical terms in general. On that issue, opinions have dramatically changed in the last 60 years, and the operationalism that supported this demand became outmoded long ago. It is fair to say that the dominant view now is that a rigorous measurement theory is nice to have, but not required. Theoretical terms need not be strictly definable in observational terms in order to be meaningful; and sometimes it appears that they must not be. Of course, theoretical terms need some operational definition in a much vaguer sense; one must be able to say how theoretical and empirical claims are related. How to describe this connection, though, is quite unclear and contested, despite long-standing efforts in philosophy of science. One of the philosophical by-products of ranking theory will consist in a constructive proposal on this matter (see Sections 13.3–13.4 on dispositional terms that were often taken to be paradigmatic theoretical terms). In any case, definability or reducibility in a logical sense, as a measurement theory would provide, seems too strong a demand.

Presumably, though, the uneasiness does not draw on outmoded operationalism, but is rather inspired by logical theory. Customarily, any logical calculus is ennobled by a correctness and completeness – i.e., soundness – theorem. There are calculi that live well without such a theorem. Still, I need not rehearse here the historic examples of the tremendous insight delivered by such soundness theorems. If the calculus looks reasonable, if the semantics is intelligible, and if a soundness theorem proves them to be equivalent, then mutual support makes for a nearly unassailable theory.

AGM belief revision theory has these virtues. Originally, it came in a logical guise (cf. Section 4.4). There, it was proved that the revision axioms (K∗1–8) (4.18), and the contraction axioms (K÷1–8) (4.19), are exactly those justified by an underlying entrenchment relation; this was AGM's completeness theorem. In our propositional framework this assertion boiled down to Theorem 4.14. By contrast, ranking theory did not offer a comparable result, and that is why ranks appeared somehow arbitrary.

Now I do not want to further argue the case in a defensive way, by pointing to other virtues, by referring to contemporary philosophy of science, etc. Rather, I want to face the challenge right away. If I am asked for a measurement theory for ranks, I will give one. In fact, I will be offering an overkill of no less than three different measurement methods.

The first is what I call the method of upgradings, which I briefly envisaged in my (1999b, Sect. 3). I do not want to rely on it, but I reproduce it here for introductory purposes and for the sake of completeness. The second and the third method are the main ones; they are essentially based on the theory of difference measurement as propounded in Krantz et al. (1971, Ch. 4). The second will simply copy the measurement of conditional probabilities. The third method is my official one, as it were. I will show that ranks are uniquely determined by iterated contractions up to a ratio scale; it will essentially suffice for this purpose to provide rules for two steps of contraction. For the logically minded, this is a kind of completeness result similar to the one given in belief revision theory; for the operationalistically minded, this is a definition of (theoretical) ranks in terms of (empirical) belief. Thereby, we can also close the gap left open at the end of Section 5.6; that is, on the base of ranking theory we can completely specify the laws of iterated contraction.

This official method was first invented by Matthias Hild around 1997 (see Hild 1997), but not published. I discovered it independently and presented an awkward version in my (1999b). Finally, we joined efforts in Hild, Spohn (2008), and Sections 8.3–8.4 will reproduce essential parts of this paper. In any case, I think that these measurement results give strong additional support to ranking theory.

In order to round up the topic, an appendix, Section 8.5, will sketch two further arguments offering a possible justification of the structure of ranking functions.

8.1 The Method of Upgradings

When attempting to measure ranks, we may take belief as the behavioral or observational base. Of course, belief is in turn a propositional attitude, a disposition, indeed a very wild one, if I was right in Section 2.3. Still, belief is introspectively easily accessible and has clear and direct behavioral manifestations, particularly in sincere speech. Degrees of belief are much worse in both respects. In fact, I need not emphasize here how close and dear the relation between speech and belief is to philosophers; it always has been. So, I will not try to base the measurement of ranks on something still more directly observable.

Hence, I will assume that what we can observe about a negative ranking function κ is its core $\kappa^{-1}(0)$, i.e., whether $\kappa(A) = 0$ or > 0. The task is to use such information in order to somehow infer the whole of κ. What we have seen in Chapter 4 is that revisions (or contractions) w.r.t. various propositions do not provide the required cardinal information; selection functions are equivalent only to WOP's or WWO's; cf. (4.14).

It is not hard to see, though, how to do better. Let us focus on natural negative ranking functions, and let us define:

Definition 8.1: κ' is an *upgrading* of the natural negative ranking function κ w.r.t. $A \neq \varnothing$ iff there is an m with $0 < m \leq \kappa(A)$ such that $\kappa' = \kappa_{A \uparrow m}$, i.e., κ' is the (evidence-oriented) $A \uparrow m$-conditionalization of κ. If $\kappa(A) > 0$, $\kappa_{A \uparrow 1}$ is the *minimal upgrading* of κ w.r.t. A and denoted by $\hat{\kappa}_A$. Upgradings can be iterated like all conditionalizations. So, we can continue to define $\hat{\kappa}_A^n = (\ldots (\kappa_{A \uparrow 1}) \ldots)_{A \uparrow 1}$ (n times) to be the *n-fold minimal upgrading* of κ w.r.t. A.

Only disbelieved propositions can be upgraded according to (8.1), possibly until they are no longer disbelieved. In fact, we have the entirely trivial

Theorem 8.2: $\kappa(A) = n > 0$ iff $\hat{\kappa}_A^{n-1}(A) > 0$ and $\hat{\kappa}_A^n(A) = 0$.

In other words: Simply by observing what is (dis-)believed after iterated minimal upgradings and what is not, we can uniquely determine any integer-valued rank. This is already my first method of measuring ranks; let us call it the *method of upgradings*. It is quite a shallow method, I agree; but it is not without interest and deserves two pages of further consideration.

What I want to explain is that this is indeed a genuine measurement method in the sense that it does not presuppose any cardinality; the cardinal ranks are generated by mere counting, clearly a legitimate method. So, what is presupposed on the right-hand side of (8.2)? First, that we can determine what is (dis-)believed according to a possible ranking function and what not, and second, that we can ascertain upgradings. We have already assumed the first. What about the second?

Upgrading is an ordinal notion. We might directly rely on the judgment of the subject himself. For any doxastic change the subject should be able to tell whether a given proposition is less disbelieved or less incredible after the change than before, i.e., whether the change is an upgrading w.r.t. that proposition. Clearly, only a comparative judgment is thereby required from the subject.

We can, however, tell a more detailed story at this point. Being a reason is also an ordinal notion, and we may assume that the subject is able to tell what is a reason for what according to his present state and also what is an insufficient reason for what; cf. (6.2). Now, we can provide him with an insufficient reason for a given proposition; what he undergoes then is an upgrading w.r.t. that proposition.

In this way we can also make sense of iterated upgradings. We can utilize here (7.47) and our observations about independent testimonies. If the (weak) reasons are independent in the sense specified there, then we can give one reason after the other to the subject, and what results according to (7.47) is a simple addition of reasons and hence an iterated upgrading. If we then simply count how many reasons we have to give him until he gives up his disbelief, we can thus infer its original negative rank.

We can cast this procedure into a somewhat schematic scenario. When seemingly honest and serious people tell us something, this is a reason for that thing. Suppose you disbelieve that Nixon was a democrat. Now your neighbor comes by and tells you that Nixon was a democrat. You do not trust your ears, but the first seeds of doubt are sown. Another neighbor comes by, or maybe you ask him, and he tells you the same thing. And so on. You do not believe in a conspiracy; you believe all the reports to be independent. After twenty statements, but not before, you give in and think: "Maybe, my memory fails me, and they are right!" If so, then your original negative rank for Nixon's being a democrat was 20. This is an artificial story, granted; but it is a behavioral story with exclusively qualitative and comparative assumptions.

In explaining this I have somehow slipped from upgradings in general to minimal upgradings, and one might suspect that this is a hidden cardinal assumption in the right-hand side of (8.2). Not so. If the strengths of reasons take discrete levels, as they do according to natural ranking functions, then there is also a minimal strength, and it is determined by purely ordinal judgments.

At this point, you might revive your doubts about completeness. Perhaps, real ranking functions are more realistic or appropriate, at least in this context. Then the notion of a minimal upgrading makes no sense, and our method seems to fail. However, the method does not really depend on minimal upgradings. We can as well work with arbitrary upgradings. When we weigh a body, we do not take a large number of one-gram weights until we have outweighed the body. Rather, we take heavier weights that are calibrated by lighter weights. Likewise here. We need not amass minimal reasons generating iterated minimal upgradings. We can observe that one stronger reason has the same effect as two independent and equally weak reasons and infer that the former is exactly twice as strong as the latter. And so forth.

To state the issue more theoretically: We have seen in (7.47) that independent reasons, signs, or testimonies are additive. This entails that we can apply the whole theory of extensive measurement (cf., e.g., Krantz et al. 1971, Ch. 3) to the present case. And this theory is known to work even if there are no minimal weights, lengths, reasons, or whatever.

This idea of measuring the strength of reasons and of disbelief is quite illuminating, I find; it might come as a surprise that we do not need more than the most basic part of measurement theory, namely, the extensive part. Still, I do not want to dwell further on this idea.

One reason is that my hint about extensive measurement may suffice; it is a routine matter to elaborate on it. The other reason is that I am not so sure how appealing the method really is. One point is that the vagueness of the notion of belief we noted in (5.14) is painfully apparent in the method of upgradings. When exactly did you lose your disbelief in a series of gradual upgradings? When the 18th neighbor told you that Nixon was a democrat? Or with the testimony of the 19th? Or really only after the 20th neighbor? It would not make much sense to press you on that issue. Another concern is the required independence of the reasons (conditional on what they are a reason for). Our world does indeed branch in such complex ways that for most items there will be a considerable number of independent reasons. However, above a certain limit we are reluctant to assume that independence. If very many signs point into the same direction, then the congruence is presumably not only produced exclusively by what they indicate. If this is so, then the independence in question cannot be postulated to an arbitrary extent.

However, the ideas we have preliminarily discussed here will return in modified form in the next sections. So, whatever the ultimate judgment is on the measurement method of upgradings, it serves us as an intuitively and formally simple introduction to our present topic.

8.2 A Difference Measurement of Conditional Ranks

Let us approach our topic in a, as it were, more professional or standard way. The right thing seems to be to look at the measurement of subjective probabilities; after all, we are attempting here something similar. My reference text will be Krantz et al. (1971), in particular Chapter 4 on difference measurement and Chapter 5 on probability representations; it contains everything we will need.

There, the starting point is a comparative relation ≤ to be interpreted as "is at most as credible or plausible as". Indeed, each measurement starts with such a relation comparing objects in the respect to be measured: length, temperature, utility, or whatever. ≤ is usually assumed to be a weak order, i.e., a transitive and complete relation. One can do also without completeness, but then measurement becomes much harder.

Clearly, the starting point is not enough. If the order ≤ is not too rich, we can map its domain into the real numbers in an order-preserving way. However, this yields only an ordinal measurement; any monotone transformation of the mapping will do as well. This is not a genuine measurement; it does not mean using the real numbers as numbers. The mapping must be determined more uniquely.

The basic way to achieve a more unique mapping is always roughly the same. It consists in identifying within the order $\leq$ so-called standard sequences that are equally spaced, i.e., for which the assertion that the distance between adjacent numbers of the sequence is equal is endowed with empirical meaning. The simplest example is extensive measurement for which some empirical operation of addition is defined. In the case of length, this operation is putting lengthy objects in a row; in the case of weight, it is putting weighty objects together on a scale; etc. The most elementary explanation of this procedure I know of is still Hempel (1952, Ch. III). Measurement theory has shown great ingenuity in constructing most diverse kinds of standard sequences.

In the case of subjective probability, it is the additivity (3.1a) of probabilities and its reflection in the order $\leq$ that allow the construction of the pertinent standard sequences. Thus, subjective probability is amenable to ordinary extensive measurement (cf. Krantz et al. 1971, Sect. 5.2). This measurement indeed yields an absolute scale, i.e., a unique probability measure, since the extreme values must be 0 and 1.

Clearly, we cannot carry over this method to ranking theory. The ranking-theoretic counterpart to probabilistic additivity is minimitivity or the law of disjunction (5.8b). This law refers only to the order and not to the arithmetical properties of ranks, and though many proponents of the law held this to be a virtue, it is of no avail in our present context.

It seems clear how we can make progress. I emphasized in Section 5.3 that it is only conditional ranks that make use of the arithmetical structure of ranks. Thus, if we compare conditional ranks $\kappa(B \mid A)$, we actually compare differences $\kappa(B \cap A) - \kappa(A)$ of ranks, and if we judge such differences to be equal, a purely ordinal judgment, we might be able to construct suitable standard sequences for ranks as well. Let me return to example 5.7 about Tweety; it enables me to illustrate the core of the measurement procedures to come:

Example 8.3, Tweety for a final time: Looking at the table in example 5.7, we find there specified a ranking function κ for the eight propositional atoms, entailing ranks for all 256 propositions involved. Focusing on the atoms, we are thus dealing with a realm $X = \{x_1, \ldots x_8\}$ and a numerical function f such that

$$f(x_1) = 0, \ f(x_2) = 4, \ f(x_3) = 0, \ f(x_4) = 11,$$
$$f(x_5) = 2, \ f(x_6) = 1, \ f(x_7) = 0, \ f(x_8) = 8.$$

This induces a lot of difference comparisons. Denoting pairs by $(x - y)$ for mnemonic reasons, we have, e.g., $(x_5 - x_6) \leq (x_2 - x_1)$ or $(x_8 - x_4) \leq (x_4 - x_8)$. At present we can think of such differences as expressing conditional ranks; for instance, the last difference $(x_4 - x_8)$ simply represents the conditional rank $\kappa(F \mid \bar{B} \cap P)$. Thus, comparing conditional ranks *is* comparing such differences. In the next section we will exploit the fact that such differences express reason or relevance relations.

The crucial question now is: Do these comparisons help to determine f? Yes, the example was so constructed: First, we have $(x_1 - x_3) \approx (x_3 - x_1) \approx (x_1 - x_7)$. This entails $f(x_1) = f(x_3) = f(x_7)$. Let us choose this as the zero point of our scale; i.e., $f(x_1) = 0$.

Next, we have $(x_5 - x_6) \approx (x_6 - x_1)$. If we choose $f(x_6) = 1$ as our unit, this entails $f(x_5) = 2$. Then, we have $(x_2 - x_5) \approx (x_5 - x_1)$, entailing $f(x_2) = 4$, and $(x_8 - x_2) \approx (x_2 - x_1)$, entailing $f(x_8) = 8$. Finally, we have $(x_4 - x_8) \approx (x_2 - x_6)$ so that $f(x_4) = 11$. In this way, the difference comparisons determine f uniquely up to a unit and a zero point and hence determine the ranking function κ uniquely up to a unit.

Of course, the example worked only because the numbers were luckily chosen. Still, the case is instructive; the whole point of the theory of difference measurement is to find conditions under which such a construction is guaranteed to be successful.

If we continue to focus on the comparison of conditional ranks, we find an exact model for this construction on the probabilistic side. Krantz et al. (1971, Sect. 5.6) discuss the idea of measuring conditional probabilities. It involves the comparison of ratios $P(B \mid A) = P(A \cap B) / P(A)$; but this does not make a difference, since taking the logarithm of such ratios transforms them into differences. Interestingly, Krantz et al. (1971, Sect. 5.6.4) develop, for the first time, the possibility of measuring non-additive conditional probability, for which additivity (3.1a) and the corresponding comparative features are dropped and only the law of multiplication (3.4c) and its comparative features are kept; of course, the term "probability" should then be put into scare quotes, as they do. This is a result we can directly and routinely translate into our ranking-theoretic framework. I will do so in the rest of this section, by focusing on the simplest case and only indicating some possible generalizations at the end.

So, our starting point is a four-place relation $A \mid B \leq C \mid D$, to be interpreted as "A given B is at most as *disbelieved* as C given D"; note the reversal of the order from greater credibility to greater incredibility. What we want to show is that if that relation satisfies certain axioms, then there is a negative ranking function κ such that $\kappa(A \mid B) \leq \kappa(C \mid D)$ iff $A \mid B \leq C \mid D$, and this ranking function is to some extent unique, i.e., measured on a ratio scale, since we have a natural zero-point. Let me first introduce the crucial axioms and discuss them afterwards:

Definition 8.4: Let $\mathcal{A}$ be an algebra over W, and $\leq$ a relation over $\mathcal{A} \times (\mathcal{A} - \{\varnothing\})$, i.e., $A \mid B \leq C \mid D$ only if $A, C \in \mathcal{A}$ and $B, D \in \mathcal{A} - \{\varnothing\}$. Let $<$ be the corresponding strict relation, i.e., $A \mid B < C \mid D$ iff $A \mid B \leq C \mid D$, but not $C \mid D \leq A \mid B$, and $\approx$ the corresponding indifference, i.e., $A \mid B \approx C \mid D$ iff $A \mid B \leq C \mid D$ and $C \mid D \leq A \mid B$. Then $\leq$ is a *conditional disbelief comparison* for $\mathcal{A}$ iff for all, A, A', B, B', C, C':

(a) $\leq$ is a weak order, i.e., a transitive and complete relation on $\mathcal{A} \times (\mathcal{A} - \{\varnothing\})$
    [*weak order*],

(b) $W \mid W \approx A \mid A \leq B \mid A < \varnothing \mid A$, if $B \neq \varnothing$ [*minimum and maximum*],

(c) $B \mid A \approx A \cap B \mid A$ [*conditionality*],

(d) $B \mid A \approx W \mid W$ or $\bar{B} \mid A \approx W \mid W$ [*conditional consistency*],

(e) if $C \subseteq B \subseteq A$, $C' \subseteq B' \subseteq A'$, $B \mid A \leq B' \mid A'$, and $C \mid B \leq C' \mid B'$, then $C \mid A \leq C' \mid A'$, and if either hypothesis is $<$, the conclusion is also $<$
    [*monotonicity*].

The comparison $\leq$ is *Archimedean* iff it moreover satisfies:

(f)   every standard sequence is finite, where $(A_1, A_2, \ldots)$ is a *standard sequence* iff for
all: $A_i \in \mathcal{A} - \{\varnothing\}$, $A_i \subseteq A_{i+1}$, and $W \mid W < A_1 \mid A_2 \approx A_i \mid A_{i+1}$.

Finally, the comparison $\leq$ is *full* iff for all $A, B, C, D \in \mathcal{A} - \{\varnothing\}$:

(g)   if $A \mid B \leq C \mid D$, there exist C' with $C \cap D \subseteq C' \subseteq C'$ and D' with $C \cap D \subseteq D'$
$\subseteq D$ such that $A \mid B \approx C' \mid D \approx C \mid D'$.

Let me explain what these axioms mean and why most of them are highly plausible. A first crucial assumption is that the items to be compared by $\leq$ are in $\mathcal{A} \times (\mathcal{A} - \{\varnothing\})$. This says that $\varnothing$ is the only proposition not allowed as a condition. This entails that $\leq$ can only be represented by regular ranking functions (5.27), since only they have conditional ranks so widely defined. We will see that this assumption is not required; I make it for the sake of simplicity.

The core of (8.4) is formed by (a)–(e). These are necessary axioms in the sense that they are entailed by the intended numerical representation. Axiom (a), weak order, goes without saying. So does (b); $A \mid A$ is minimally and $\varnothing \mid A$ maximally disbelieved. Conditionality (c) says that, given A, B and $A \cap B$ are equivalent; that is how conditioning works. (c) has the effect that one can restrict the comparison $\leq$ to pairs $B \mid A$ with $B \subseteq A$; all the other pairs are taken care of by (c). The conditional consistency axiom (d) is characteristic of ranking theory; it does not appear in the axiomatization of "non-additive probability" of Krantz et al. (1971, p. 228). It would be doubtful, if $\leq$ were just a plausibility or credibility ordering. However, $\leq$ orders conditional disbelief; and there is no lesser disbelief than no disbelief at all. Hence, (d) simply says that, given any condition, B and $\bar{B}$ cannot both be disbelieved. Monotonicity (e) is a necessary axiom as well; it follows from the law of conjunction (5.16). It is also intuitively plausible: the interval from C to A is, as it were, the sum of the interval from C to B and from B to A; likewise for C', B', and A'; and hence the comparisons should behave as stated.

These axioms are also testable. Empirical subjects can be asked to make such comparisons of conditional disbelief. And since the axioms are plausible, subjects might well be expected to conform to these axioms.

This is different with the Archimedean axiom (f). It is a necessary axiom, too. In fact, every representation of some order in the real numbers requires such an axiom; this is due to the Archimedean structure of the field of real numbers. It is hardly testable, though. It is, however, trivially satisfied if the algebra $\mathcal{A}$ is finite; and there are indeed finite models of the axioms (a)–(g).

The only condition that is not necessary is the fullness axiom (g). So, we have not arrived at a set of conditions necessary and sufficient for numerical representation. Such sets are rarely attainable. Usually, the necessary axioms will leave some leeway for representation, and only if the domain ordered by $\leq$ is rich enough, this leeway vanishes. For our purposes it is axiom (g) that requires the appropriate kind of richness.

Now we are prepared to state the intended

Theorem 8.5: Let ≤ be a full Archimedean conditional disbelief comparison for $\mathcal{A}$. Then there is a regular real negative ranking function κ for $\mathcal{A}$ such that for all $A, C \in \mathcal{A}$ and $B, D \in \mathcal{A} - \{\varnothing\}$

$$\kappa(A \mid B) \leq \kappa(C \mid D) \text{ iff } A \mid B \leq C \mid D.$$

And for each negative ranking function κ′ also fulfilling this equivalence there is a real $x > 0$ such that $\kappa' = x \cdot \kappa$.

This is a most satisfying representation theorem saying that numerical ranks are measured on a ratio scale by appropriate comparative data.

The proof requires involved constructions that I will not attempt to reproduce here. This is indeed unnecessary, since Theorem 8.5 is almost exactly Theorem 8 of Krantz et al. (1971, p. 228) representing ≤ by a "non-additive probability" Q. Let me instead quickly explain the small differences.

First, their theorem represents $A \mid B$ as a numerical ratio $Q(A \cap B) / Q(B)$. As mentioned, this is insignificant; in fact, their proof first arrives at a difference representation on p. 237 that they turn into a ratio representation by exponentiation.

Second, the most conspicuous difference is that I have added axiom (8.3d) that is not found in their theorem. This axiom entails that their numerical representation Q must additionally satisfy the conditional law (5.17) of negation and is hence a negative ranking function.

Third – now we are already entering possible generalizations of (8.5) – Krantz et al. assume a set $\mathcal{N}$ of propositions excluded from the conditioning position, so that ≤ only orders $\mathcal{A} \times (\mathcal{A} - \mathcal{N})$. I have simply assumed that $\mathcal{N} = \{\varnothing\}$. Some of their axioms fix the structure of $\mathcal{N}$; this is indeed the only further difference between their axioms and mine. With this generalization, Theorem 8.5 would not be restricted to regular ranking functions; we will realize it in the next section.

Finally, we might give up the Archimedean axiom (8.4f). Then it turns out that conditional ranking functions as defined in (5.19) are suitable for representing also non-Archimedean conditional disbelief comparisons; I have not inquired, though, as to whether one could thereby measure ordinal ranking functions (5.9f). In any case, a ramified measurement theory for ranking functions might be developed along these lines. Again, I am content here with demonstrating the basic opportunities.

So, this is how far standard measurement theory carries us with respect to ranking functions. Without doubt, this theory is an impressive theoretical achievement; it is pleasing to see that it extends to ranking theory as well. Still, I am not fully satisfied. We have not only improved the range and power of measurement as compared with the method of upgradings, we have also lost some of the latter's virtues. One virtue was that that method promised to start from a more basic level, the level of belief and disbelief, whereas the present methods always start from the comparative level. We may observe the analogous difference in utility theory when it is based on preferences or on choice behavior, a difference only bridged by the theory of revealed preference. Another virtue of the method of upgradings is even more important. It attempted, though still in an imperfect way, to measure ranks by the reason-driven dynamics of belief, i.e., with the help of the notions so central to ranking theory. This connection is not at all perspicuous in the methods of this section. These virtues should be saved. So there is space for further improvement. We will fill the space in the next section.

8.3 Measuring Ranks by Iterated Contractions

When attempting to combine the advantages of the previous methods we should turn to contractions. The reason is that conditionalizations in general are also characterized by the input parameter expressing the strength with which the evidence is received. Basing the measurement of ranks on belief change in general hence means assuming this numerical input. We had this problem with the method of upgradings, which I tried to dissolve by referring to minimal upgradings (or some more complicated substitute). Whatever the plausibility of this solution, it would be better to entirely avoid the problem. This is what contractions will allow us to do. Contraction by A is essentially $A \to 0$-conditionalization (cf. 5.41), i.e., unlike in revision, the numerical input parameter is fixed in contraction. Hence, we do not rely on hidden numerical information when proceeding from intuitively well understood contractions.

Recall, though, that in (5.14) I proposed to account for the vagueness of the notion of belief by pointing out that the range of unopinionatedness may extend to the interval $[-z, z]$ of two-sided ranks instead of shrinking to the point $\{0\}$. This was an attractive idea. However, we thereby lose the uniqueness of contractions. Is this a problem in the present context? No. As I noted after (5.41) we may then refer to what I there called the central contraction that reestablishes uniqueness. Still, it is easier in this section to neglect (5.14), to interpret only rank 0 as absence of disbelief, and to thus deal with a unique contraction.

In (5.41) I defined single contractions w.r.t. ranking functions and mentioned that this definition is iterable; in fact, in Section 5.7 we preliminarily envisaged iterations. Let me make this explicit:

> *Definition 8.6*: Let κ be a negative ranking function for $\mathcal{A}$ and $A_1, \ldots, A_n \in \mathcal{A}$ such that $\kappa(\bar{A}_i) < \infty$ $(i = 1, \ldots, n)$. Then the *iterated contraction* $\kappa_{\div\langle A_1, \ldots, A_n\rangle}$ of κ by $\langle A_1, \ldots, A_n\rangle$ is defined as $\kappa_{\div\langle A_1, \ldots, A_n\rangle} = (\ldots (\kappa_{\div A_1}) \ldots)_{\div A_n}$; this includes the iterated contraction $\kappa_{\div\langle\rangle} = \kappa$ by the empty sequence $\langle\rangle$. And the *iterated contraction* $\div_\kappa$ w.r.t. κ is defined as that function which assign to any finite sequence $\langle A_1, \ldots, A_n\rangle$ of propositions with $\kappa(\bar{A}_i) < \infty$ the proposition $\div_\kappa\langle A_1, \ldots, A_n\rangle = \kappa^{-1}_{\div\langle A_1, \ldots, A_n\rangle}$ (= the core of $\kappa_{\div\langle A_1, \ldots, A_n\rangle}$). Hence, $\div_\kappa\langle\rangle$ is the core of κ.

And just in order to fix the format of our discussion, let me add

> *Definition 8.7*: Let $\mathcal{A}$ be an algebra of propositions over W and $\mathcal{N} \subseteq \mathcal{A}$. (The intended interpretation of $\mathcal{N}$ is the set of propositions with infinite rank the complements of

which cannot be contracted.) Let $\mathcal{N}^c = \{\bar{A} \mid A \in \mathcal{N}\}$, and let $\mathcal{A}^+_{\mathcal{N}}$ denote the set of all finite (possibly empty) sequences of propositions from $\mathcal{A} - \mathcal{N}^c$. Then $\div$ is a *potential iterated contraction*, a *potential IC*, for $(\mathcal{A}, \mathcal{N})$ iff $\div$ is a function from $\mathcal{A}^+_{\mathcal{N}}$ into $\mathcal{A}$. A potential IC $\div$ is an *iterated ranking contraction*, an *IRC*, for $(\mathcal{A}, \mathcal{N})$ iff there is a negative ranking function κ such that $\mathcal{N} = \{A \in \mathcal{A} \mid \kappa(A) = \infty\}$ and $\div = \div_\kappa$.

Given this terminology, our principal aim is to measure ranks with the help of iterated contraction on a ratio scale (we know already that single contractions do not achieve this aim). This means reconstructing a ranking function κ from its iterated contraction $\div_\kappa$, and doing so uniquely up to a multiplicative constant. This is what we will do in this section. The further aim, completing the investigation in the next section, is to state which properties potential IC's must have in order to be IRC's, i.e., IC's suitable for measuring ranks. Of course, (8.7) does not count as an answer; we will be searching for informative properties that do not refer to ranking functions. My focus in these two sections will be on carefully explaining how these constructions work; for the proof that they indeed work I will essentially refer to Hild, Spohn (2008).

We will reach our principal aim in four simple steps. The first step is familiar; it consists in the old observation that the ordering of negative ranks, i.e., of disbelief, may be inferred from single contractions. In our terms, this means that we have for each negative ranking function κ and all $A, B \in \mathcal{A}$:

(8.8) $\kappa(A) \leq \kappa(B)$ iff $\div_\kappa$ is not defined for $\langle \bar{B} \rangle$ or $\div_\kappa \langle \bar{A} \cap \bar{B} \rangle \not\subseteq \bar{A}$.

That is, B is at least as disbelieved as A, or $\bar{B}$ is at least as firmly believed as $\bar{A}$, either if $\bar{B}$ is maximally believed or if giving up the belief in $\bar{A} \cap \bar{B}$ entails giving up the belief in $\bar{A}$. Let us fix this connection without reference to ranks:

Definition 8.9: Let $\div$ be a potential IC for $(\mathcal{A}, \mathcal{N})$. Then the *potential disbelief comparison* $\trianglelefteq_\div$ associated with $\div$ is the binary relation on $\mathcal{A}$ such that for all $A, B \in \mathcal{A}$: $A \trianglelefteq_\div B$ iff $B \in \mathcal{N}$ or $\div \langle \bar{A} \cap \bar{B} \rangle \not\subseteq \bar{A}$. The associated *disbelief equivalence* $\triangleq_\div$ and the *strict disbelief comparison* $\trianglelefteq_\div$ are defined in the usual way. The disbelief comparison associated with the IRC $\div_\kappa$ is denoted by $\trianglelefteq_\kappa$, so that (8.8) entails that $A \trianglelefteq_\kappa B$ iff $\kappa(A) \leq \kappa(B)$.

Of course, such a potential disbelief comparison $\trianglelefteq_\div$ is well-behaved and thus a proper disbelief comparison only if the associated potential IC $\div$ is well-behaved. For instance, $\trianglelefteq_\div$ is a weak order only if $\div$ satisfies Definition 4.21. Let us defer, though, the systematic inquiry as to what good behavior amounts to in the end. At present, the relevant observation is that single contractions do not carry us beyond this point.

Hence, we must take further steps. The second step is the crucial one. Hild (1997) first discovered it, and I had the same idea later on in Spohn (1999b). It consists in the observation that all the reason relations discussed in Chapter 6, i.e., positive relevance, negative relevance, and irrelevance, can also be expressed in terms of contractions, albeit only iterated ones. This is the content of

Theorem 8.10: Let κ be a negative ranking function, and let $A \to B = \bar{A} \cup B$ be the set-theoretical version of material implication. Then we have:

(a) A is a reason for B, i.e., positively relevant to B, w.r.t. κ iff $\dot{-}_\kappa\langle A, \bar{A}, B, \bar{B}\rangle \subseteq$ $A \to B$ or $\dot{-}_\kappa\langle A, \bar{A}, B, \bar{B}\rangle \subseteq \bar{A} \to \bar{B}$ or both.

(b) A is negatively relevant to B w.r.t. κ iff $\dot{-}_\kappa\langle A, \bar{A}, B, \bar{B}\rangle \subseteq \bar{A} \to B$ or $\dot{-}_\kappa\langle A, \bar{A}, B, \bar{B}\rangle \subseteq A \to \bar{B}$ or both.

(c) A is irrelevant to B w.r.t. κ iff none of $A \cup B, A \cup \bar{B}, \bar{A} \cup B$, and $\bar{A} \cup \bar{B}$ is a superset of $\dot{-}_\kappa\langle A, \bar{A}, B, \bar{B}\rangle$.

This theorem requires a few comments. The most important point, I think, is the tremendous intuitive appeal of these assertions. If A and B are irrelevant to each other, if they have nothing to do with one another, then after eliminating any belief or disbelief about A and about B, nothing weaker concerning A and B – no disjunction or material implication – should remain; this is what (c) says. Conversely, if A and B are somehow connected to each other, this connection should survive that elimination, and then it is obvious that the survivals in (a) express positive relevance and those in (b) negative relevance. Recall also the law of material implication (5.21) for positive ranks; its intuitive plausibility has the same source.

Formally, Theorem 8.10 refers to fourfold contraction. But it is obvious that at most two of them can be genuine contractions; if the contraction by A is genuine, that by $\bar{A}$ must be vacuous. However, the theorem requires eliminating any opinion about A; this is why it had to be stated as it is. Likewise for B.

One might suspect an incoherence in (8.10), since I have emphasized that these relevance notions are symmetric (cf. 6.7b) and, by contrast, that iterated contractions do not commute (cf. 5.58). However, it does not make a difference in (8.10) whether we refer to $\dot{-}_\kappa\langle A, \bar{A}, B, \bar{B}\rangle$ or to $\dot{-}_\kappa\langle B, \bar{B}, A, \bar{A}\rangle$. In (8.10a), for instance, it may happen that the two iterated contractions do not retain the same of the two specified material implications, but both retain at least one of the two.

I introduced Theorem 8.10 only because it most perspicuously shows how reasons are reflected in contractions. For our measurement purposes the most convenient notion is non-negative relevance, i.e., the complement of case (8.10b). Moreover, we have to more generally refer to conditional relevance. These two points are taken care of in the next theorem, which is immediately intelligible on the basis of (8.10); its proof at the same time demonstrates the claims of (8.10).

Theorem 8.11: Let κ be a negative ranking function for $\mathcal{A}$ and $A, B, C \in \mathcal{A}$ such that $\kappa(C) < \infty$. Then A is not a reason against B, or non-negatively relevant to B, given C w.r.t. κ iff $\kappa(A \mid C)$ or $\kappa(\bar{A} \mid C)$ or $\kappa(B \mid C)$ or $\kappa(\bar{B} \mid C)$ is infinite or $\kappa(A \cap B \cap C) - \kappa(A \cap \bar{B} \cap C) \leq \kappa(\bar{A} \cap B \cap C) - \kappa(\bar{A} \cap \bar{B} \cap C)$, i.e., iff neither $(C \cap A) \to \bar{B}$ nor $(C \cap \bar{A}) \to B$ is a superset of $\dot{-}_\kappa\langle C \to A, C \to \bar{A}, C \to B, C \to \bar{B}\rangle$ or the latter is undefined.

Proof: A is non-negatively relevant to B given C according to Definition 6.3 iff $\tau(B \mid A \cap C) \geq \tau(B \mid \bar{A} \cap C)$ iff $\kappa(\bar{B} \mid A \cap C) - \kappa(B \mid A \cap C) \geq \kappa(\bar{B} \mid \bar{A} \cap C) - \kappa(B \mid \bar{A} \cap C)$ iff $\kappa(A \mid C)$ or $\kappa(\bar{A} \mid C)$ or $\kappa(B \mid C)$ or $\kappa(\bar{B} \mid C)$ is infinite or $\kappa(A \cap$

$B \cap C) - \kappa(A \cap \bar{B} \cap C) \leq \kappa(\bar{A} \cap B \cap C) - \kappa(\bar{A} \cap \bar{B} \cap C)$. This proves the first equivalence.

As to the second equivalence, it is clear that the exceptional clauses are equivalent. So let us suppose that all four contractions are defined, let κ_i denote the ranking function resulting from performing the first i of the 4 contractions considered, and let us abbreviate $\kappa(A \cap B \cap C) = x$, $\kappa(A \cap \bar{B} \cap C) = y$, $\kappa(\bar{A} \cap B \cap C) = z$, and $\kappa(\bar{A} \cap \bar{B} \cap C) = w$. What we have to show then is that $x - y \leq z - w$ iff $\kappa_4(A \cap B \cap C) = \kappa_4(\bar{A} \cap \bar{B} \cap C) = 0$. We should observe that for any proposition D, if $\kappa_i(D) = 0$ and $j \geq i$, then $\kappa_j(D) = 0$; this is so because beliefs never increase by contractions. Now we have to distinguish four cases:

First, $x \leq y$ and $z \leq w$: Then, we have first $\kappa_1(\bar{A} \cap B \cap C) = 0$ and then $\kappa_2(A \cap B \cap C) = 0$, $\kappa_2(A \cap \bar{B} \cap C) = y - x$, and $\kappa_2(\bar{A} \cap \bar{B} \cap C) = w - z$. Hence, $\kappa_3(\bar{A} \cap \bar{B} \cap C) = 0$ iff $w - z \leq y - x$. Thus, our equivalence holds in this case.

Second, $x \leq y$ and $w < z$: This already entails $x - y \leq z - w$. However, we also have first $\kappa_1(\bar{A} \cap \bar{B} \cap C)$ and then $\kappa_2(A \cap B \cap C) = 0$. So, the equivalence holds in this case, too.

Third, $y < x$ and $z \leq w$: In this case, we cannot have $x - y \leq z - w$. However, we also have first $\kappa_1(\bar{A} \cap B \cap C) = 0$ and then $\kappa_2(A \cap \bar{B} \cap C) = 0 < \kappa_2(A \cap B \cap C)$. Since the third and the fourth contraction run empty in this case, we also have $\kappa_4(A \cap B \cap C) > 0$. Thus, again, the equivalence holds.

Fourth, $y < x$ and $w < z$: Then, we have first $\kappa_1(\bar{A} \cap \bar{B} \cap C) = 0$ and then $\kappa_2(A \cap \bar{B} \cap C) = 0$, $\kappa_2(A \cap B \cap C) = x - y$, and $\kappa_2(\bar{A} \cap B \cap C) = z - w$. The 3rd contraction is vacuous. The 4th is not. Rather, we have $\kappa_4(A \cap B \cap C) = 0$ iff $x - y \leq z - w$. Thus, the equivalence also holds in the final case.

Note that, in contrast to (8.10), the minimal number of vacuous contractions occurring in this proof was only one. Hence, in the general case conditional relevance shows up in up to three genuine contractions.

Note, moreover, that we can express the equivalence of (8.11) also in the following way:

(8.12)   For any four mutually disjoint propositions $A, B, C, D \in \mathcal{A}$ with finite ranks $\kappa(A) - \kappa(B) \leq \kappa(C) - \kappa(D)$ iff $A \cup B$ is non-negatively relevant to $A \cup C$ given $A \cup B \cup C \cup D$ w.r.t. κ iff neither $\bar{A}$ nor $\bar{D}$ is a superset of $\div_\kappa \langle \bar{A} \cap \bar{B}, \bar{C} \cap \bar{D}, \bar{A} \cap \bar{C}, \bar{B} \cap \bar{D} \rangle$.

(8.12) most conspicuously connects up with example 8.3. There I illustrated how difference comparisons help us to a measurement procedure and how these difference comparisons amount to comparing conditional ranks. (8.12) tells us, alternatively, how the comparisons represent reason or relevance relations and how they may be expressed in terms of contractions.

Let us follow this line, and let us make the same transition as we did from (8.8) to (8.9) and adopt the following

Definition 8.13: Let $\div$ be a potential IC for $(\mathcal{A}, \mathcal{N})$. Then the *potential disjoint differ-ence comparison* (*potential DisDC*) $\leq^d_{\div}$ associated with $\div$ is the four-place relation defined for all quadruples of mutually disjoint propositions in $\mathcal{A} - \mathcal{N}$ such that for all such propositions A, B, C, D $(A - B) \leq^d_{\div} (C - D)$ iff $\div\langle \bar{A} \cap \bar{B}, \bar{C} \cap \bar{D}, \bar{A} \cap \bar{C}, \bar{B} \cap \bar{D}\rangle \not\sqsubseteq \bar{A}, \bar{D}$ – where the ordered pair of A and B is denoted by $(A - B)$ simply for mnemonic reasons. The associated *disjoint difference equivalence* $\approx^d_{\div}$ and the *strict disjoint difference comparison* $<^d_{\div}$ are defined in the usual way. The potential disjoint difference comparison associated with the IRC $\div_\kappa$ is denoted by $\leq^d_\kappa$, so that (8.10) entails that $(A - B) \leq^d_\kappa (C - D)$ iff $\kappa(A) - \kappa(B) \leq \kappa(C) - \kappa(D)$.

So, it is clear where we are heading. On the one hand, I have shown how to derive such a difference comparison from iterated contractions. On the other hand, we know that if such a difference comparison behaves in the appropriate way, we can use it for a difference measurement of ranking functions. Put the pieces together and you have a measurement of ranks in terms of iterated contractions.

However, we are not yet fully prepared for this final step. If we want to apply the general theory of difference measurement to the present case, the difference comparison must hold for any four propositions, not only for any four mutually disjoint propositions. The required extension is the third step of our measurement procedure.

There are various options at this point. What is in any case required is a certain rich-ness of the set of propositions. So, we are about to state a structural condition with the help of which our measurement procedure will succeed, but not a necessary condition entailed by ranking theory and its definition of iterated contraction. Therefore, one might strive for as parsimonious a structural condition as possible. Hild (1997) pro-posed one way, I indicated another in Spohn (1999b); presumably they are not yet optimal. However, parsimony has technical costs. Therefore, I prefer to choose here an extremely simple structural condition that is moderately demanding and intuitively highly intelligible.

My idea is to straightforwardly require that for each proposition with finite rank there are at least n mutually disjoint equally ranked propositions for some $n \geq 4$. With this assumption we can extend any potential disjoint difference comparison to all quadruples of propositions. This is the content of the next two definitions:

Definition 8.14: A potential IC $\div$ for $(\mathcal{A}, \mathcal{N})$ is called *n-rich* iff for each $A \in \mathcal{A} - \mathcal{N}$ there are n mutually disjoint propositions $E_1, \ldots, E_n$ such that $A \triangleq_{\div} E_i$ for $i = 1, \ldots, n$. A ranking function κ is called *n-rich* iff the associated IRC $\div_\kappa$ is *n-rich*.

Definition 8.15: Let $\div$ be a 4-rich potential IC for $(\mathcal{A}, \mathcal{N})$. Then the *potential doxastic difference comparison* (*potential DC*) $\leq_{\div}$ associated with $\div$ is the quarternary relation defined for all propositions in $\mathcal{A} - \mathcal{N}$ such that for all $A, B, C, D \in \mathcal{A} - \mathcal{N}(A - B) \leq_{\div} (C - D)$ iff there are four mutually disjoint propositions $A', B', C', D' \in \mathcal{A} - \mathcal{N}$ such that $A \triangleq_{\div} A', B \triangleq_{\div} B', C \triangleq_{\div} C', D \triangleq_{\div} D'$, and $(A' - B') \leq^d_{\div} (C' - D')$, i.e., $\div\langle \bar{A}' \cap \bar{B}', \bar{C}' \cap \bar{D}', \bar{A}' \cap \bar{C}', \bar{B}' \cap \bar{D}'\rangle \not\sqsubseteq \bar{A}', \bar{D}'$. Again, the associated *doxastic difference equivalence* $\approx_+$ and the *strict doxastic difference comparison* $\prec_+$ are defined in the usual way.

In fact, it takes very little to be rich. Suppose the ranking function κ for $\mathcal{A}$ is not rich. How can we extend it to a rich one? Very easily: Just take k new propositions $N_1, \ldots, N_k$ that are entirely neutral, i.e., we have no beliefs about them so that $\tau(N_i) = 0$ for $i = 1, \ldots, k$ and such that $N_1, \ldots, N_k$ are independent from each other and from all the old propositions in $\mathcal{A}$. Let $\mathcal{A}^*$ be the algebra generated by $\mathcal{A} \cup \{N_1, \ldots, N_k\}$. The assumptions determine how to extend κ to $\mathcal{A}^*$; in particular, we have for any $A \in \mathcal{A}$ $\kappa(A) = \kappa(A \cap N_1' \cap \ldots \cap N_k')$ for all $N_i' \in \{N_i, \bar{N}_i\}$ $(i = 1, \ldots, k)$. Thus, the extended κ *is* 2^k-*rich*. And surely, it is easy to find k propositions that are neutral in this sense; take, for instance, k fair coins not occurring in the propositions of $\mathcal{A}$ and put $N_i =$ "coin i shows heads in the next throw". Richness thus appears to be a very modest structural condition. Observe also that my little ranking-theoretic story about the k neutral propositions $N_1, \ldots, N_k$ could as well be expressed in terms of iterated contractions; independence is the same as irrelevance. We will see that our representation result will require no more than the assumption of 6-richness.

After this auxiliary move, we can take the fourth step and complete our measurement procedure. If potential DC's have the right properties, we can construct a ratio scale from them. What this means and how this goes can be directly read off from the standard theory of difference measurement; cf. Krantz et al. (1971, Ch. 4). We have to do no more than copy Definition 3 and Theorem 2 from there, p. 151, and slightly adapt it for our purposes:

Definition 8.16: $\leq$ is a *doxastic difference comparison* (DC) for $(\mathcal{A}, \mathcal{N})$ (with $\approx$ being the associated equivalence and $<$ the associated strict comparison) iff $\leq$ is a quarternary relation on $\mathcal{A} - \mathcal{N}$ such that for all $A, B, C, D, E, F \in \mathcal{A}$:

(a) $W \notin \mathcal{N}$, and $\mathcal{N}$ is an ideal in $\mathcal{A}$, i.e.: $\varnothing \in \mathcal{N}$; if $A \in \mathcal{N}$ and $B \subseteq A$, then $B \in \mathcal{N}$; and if $A, B \in \mathcal{N}$, then $A \cup B \in \mathcal{N}$ [*undefinedness*],

(b) $\leq$ is a weak order on $(\mathcal{A} - \mathcal{N}) \times (\mathcal{A} - \mathcal{N})$ [*weak order*],

(c)   if $(A - B) \leq (C - D)$, then $(D - C) \leq (B - A)$ [*sign reversal*],

(d)   if $(A - B) \leq (D - E)$ and $(B - C) \leq (E - F)$, then $(A - C) \leq (D - F)$
            [*monotonicity*],

(e)   if $(A - W) \leq (B - W)$, then $(A - W) \approx (A \cup B - W)$ [*law of disjunction*].

The DC $\leq$ is *Archimedean* iff, moreover, for any sequence $A_1, A_2, \ldots$ in $\mathcal{A} - \mathcal{N}$:

(f)   if $A_1, A_2, \ldots$ is a strictly bounded standard sequence, i.e., if for all i $(A_1 - A_1) \prec (A_2 - A_1) \approx (A_{i+1} - A_i)$ and if there is a $D \in \mathcal{A} - \mathcal{N}$ such that for all i $(A_i - A_1)$ $\prec (D - W)$, then the sequence $A_1, A_2, \ldots$ is finite.

Finally, the DC $\leq$ is *full* iff for all $A, B, C, D \in \mathcal{A} - \mathcal{N}$:

(g)   if $(A - A) \leq (A - B) \leq (C - D)$, then there exist $C', D' \in \mathcal{A}$ such that $(A - B)$ $\approx (C' - D) \approx (C - D')$.

As in Section 8.2, it is obvious that (a)–(e) are axioms necessary for measurement; (a) and (e) contain the adaptation to ranking theory. The Archimedean axiom (f) is also necessary. Again, it might be worthwhile investigating whether by dropping the Archimedean axiom we could measure something like conditional or ordinal ranking functions. (g), finally, is a structural axiom which is not entailed by ranking theory, but which is required for the solvability of all the measurement inequalities.

I am not claiming that DC's are intuitively readily accessible. Indeed, they are not, I find. Disbelief comparisons or their positive counterparts, entrenchment relations, are highly accessible. Likewise, we have, I think, a good intuitive grasp of conditional disbelief comparisons. This is why I have proposed the measurement method in the previous section. By contrast, we have no safe intuitive assessment of doxastic differences between four arbitrary propositions, even if they are mutually disjoint. Therefore I did not start this section with Definition 8.16 that plays only a mediating role, but rather explained how difference judgments reduce to readily accessible relevance judgments and how all these assessments reduce to even more readily accessible iterated contractions.

The theorem appertaining to (8.16) finally establishes the measurability of ranking functions.

Theorem 8.17: Let ≤ be a full Archimedean DC for $(\mathcal{A}, \mathcal{N})$. Then there is a real negative ranking function κ for $\mathcal{A}$ such that for all $A \in \mathcal{A}$ $\kappa(A) = \infty$ iff $A \in \mathcal{N}$ and for all $A, B, C, D \in \mathcal{A} - \mathcal{N}$ $(A - B) \le (C - D)$ iff $\kappa(A) - \kappa(B) \le \kappa(C) - \kappa(D)$. If κ' is another negative ranking function with these properties, then there is an $x > 0$ such that $\kappa' = x \cdot \kappa$.

Proof: Since ≤ satisfies the conditions (b), (c), (d), (f), and (g) of (8.16), Theorem 2 of Krantz et al. (1971, p. 151) tells us that there is a real-valued function f on $\mathcal{A} - \mathcal{N}$ such that $(A - B) \le (C - D)$ iff $f(A) - f(B) \le f(C) - f(D)$, and that f is unique up to a positive linear transformation. Because of condition (e), this function must satisfy the law of disjunction for ranking functions. We can extend f to $\mathcal{N}$ by putting $f(A) = \infty$ for $A \in \mathcal{N}$; because of condition (a), f still satisfies the law of disjunction. Condition (a) also says that $f(W) < \infty$. Hence, we can choose $f(W) = 0$, and then f is a negative ranking function. As such it is thus unique up to a positive multiplicative constant.

To sum up: We have seen how potential IC's induce potential DisDC's, how rich potential IC's induce potential DC's, that the right kind of potential IC's, for instance, IRC's, indeed induce DC's, and that DC's determine ranking functions uniquely up to a multiplicative constant. In particular, this means that we may start from a rich ranking function κ, then consider only the rich IRC $\dot{+}_\kappa$ associated with κ, and finally reconstruct the whole of κ from $\dot{+}_\kappa$ in the unique way indicated. This appears to me to be a most satisfying representation result. The only data we need are the beliefs under various iterated contractions. These data reflect not only the comparative strengths of beliefs, they also reflect the comparative nature of reasons and relevance. And these inferred comparisons suffice to fix the cardinal structure of ranking functions. In this way our final measurement method combines the virtues of the first two.

Note, however, that all measurement methods proposed measure the ranks of each subject on her own ratio scale; they provide no theoretical ground whatsoever for comparing the scale units for different subjects. We thus face a problem of interpersonal rank comparison, similar to the problem of interpersonal utility comparison (that is worse because even the zero point of the utility scales is interpersonally incomparable). Subjective probabilities do not pose that problem, because they are usually measured on an absolute scale (for an exception see Jeffrey 1965/83, Ch. 8, and Bolker 1966). I think this problem is not an artifact of ranking theory; it rather is a real problem that often plagues us in everyday discourse. I will not further dwell on it, but it certainly deserves further attention.

8.4 The Laws of Iterated Contraction

We are not yet finished. I just said that the right kind of potential IC's measure ranks and that IRC's are the right kind. However, we are still missing a general characterization of what the right kind is that avoids reference to ranking functions. The required information is, of course, implicit in Definitions 8.16 and 8.15 and the auxiliary (8.9), (8.13), and (8.14). This is why I neglected the issue in Spohn (1999b). Yet, the implicit information needs to be explicitly elaborated in a perspicuous way, in order for the content and import of our measurement result to become intelligible, as Hild (1997) essentially demonstrated.

Note that we can thereby also close a gap left open in Section 5.6. There we despaired of finding stronger, let alone complete laws of iterated contraction. Now they will fall right into our lap as a consequence of our measurement theory; at least, Theorem 8.17 shows that the laws we will find are complete given the structural conditions of richness and fullness. Let me represent here the essential results of Hild, Spohn (2008).

We start with the required characterization of iterated contractions:

Definition 8.18: Let $\mathcal{A}$ be an algebra of propositions over W and $\mathcal{N} \subseteq \mathcal{A}$. Let again be $\mathcal{N}^c = \{\bar{A} \mid A \in \mathcal{N}\}$ and $\mathcal{A}_{\mathcal{N}}^+$ the set of all finite sequences of propositions from $\mathcal{A} - \mathcal{N}^c$. I will use S as a variable for members of $\mathcal{A}_{\mathcal{N}}^+$. Then $\div$ is an *iterated contraction* *(IC)* for $(\mathcal{A}, \mathcal{N})$ iff $\div$ is a potential IC for $(\mathcal{A}, \mathcal{N})$ such that for all $A, B, C \in \mathcal{A} - \mathcal{N}^c$, and $S \in \mathcal{A}_{\mathcal{N}}^+$:

(IC1)   the function $A \mapsto \div\langle A \rangle$ is a propositional contraction function (cf. 4.21)
[*single contraction*],

(IC2)   if $\div\langle \varnothing \rangle \not\subseteq A$, then $\div\langle A, S \rangle = \div\langle S \rangle$ [*strong vacuity*],

(IC3)   if $\bar{A} \cap \bar{B} = \varnothing$, then $\div\langle A, B, S \rangle = \div\langle B, A, S \rangle$ [*restricted commutativity*],

(IC4)   if $A \subseteq B$ and $\div\langle A \rangle \not\subseteq A \cup \bar{B}$, then $\div\langle A \cup \bar{B}, B, S \rangle = \div\langle A, B, S \rangle$
[*path independence*],

(IC5)   if $C \subseteq \bar{A}$ or $\bar{C} \subseteq \bar{A} \cap \bar{B}$ and $A \trianglelefteq_+ B$, then $A \trianglelefteq_{\dotplus(C)} B$, and if the inequality in
the antecedent is strict, that of the consequent is strict, too

[*order preservation*],

(IC6) $\mathcal{N}$ is an ideal, i.e., $\varnothing \in \mathcal{N}$, $A \in \mathcal{N}$ and $B \subseteq A$ entail $B \in \mathcal{N}$, and $A, B \in \mathcal{N}$
entails $A \cup B \in \mathcal{N}$ [*domain*],

(IC7) $\dotplus_{(S)}$ is an IC   [*iterability*].

A notational slip occurs in (IC5). Disbelief comparisons, difference comparisons, etc.,
are explained relative to potential IC's; thus, strictly taken, the notation "$\trianglelefteq_{\dotplus(C)}$" is nonsense. However, for typographic reasons I allow myself always to write "$\trianglelefteq_{\dotplus(S)}$" instead
of the more correct "$\trianglelefteq_{\dotplus(S)}$"; there is no danger of confusion. This understood, let me
immediately add a most useful consequence of (IC5):

(8.19)   for any IC $\dotplus$, if $C \subseteq \bar{A} \cap \bar{B}$ or $\bar{C} \subseteq \bar{A} \cap \bar{B}$, then $A \trianglelefteq_+ B$ iff $A \trianglelefteq_{\dotplus(C)} B$

[*order equivalence*].

Proof: If $A \trianglelefteq_+ B$, then $A \trianglelefteq_{\dotplus(C)} B$ due to (IC5). If $B \triangleleft_+ A$, then $B \triangleleft_{\dotplus(C)} A$, again due to
(IC5). Because both, $\trianglelefteq_+$ and $\trianglelefteq_{\dotplus(C)}$, are weak orders, as we will show in (8.20), the
latter says that if not $A \trianglelefteq_+ B$, then not $A \trianglelefteq_{\dotplus(C)} B$. This proves the reverse direction.

Before proceeding to the formal business, we should first look at the intuitive and
formal content of these axioms. At the same time, it is interesting to examine the
extent to which they go beyond our incomplete efforts in Section 5.6 to come to
grips with iterated contraction (and revision).

Iterated contractions must, of course, behave like single contractions at each single
step; therefore (IC1). (IC2) goes beyond vacuity for single contractions – whence
called strong – since it says that a vacuous contraction leaves not only the beliefs
unchanged, but indeed the entire doxastic state as reflected in possible further contractions. This is certainly how vacuous contractions were intended, even though it was
not expressible in terms of single contractions. (5.54) was our first attempt to realize
this intention. There is nothing much to say about (IC6). (IC7) goes without saying; it
lies at the heart of iteration that it can be carried out without limit. Of course, (IC7)
does not make Definition 8.18 circular; it only states succinctly what I could have
attained by stating all the other axioms more clumsily for all $\dotplus_{(S)}$.

Hence, (IC3)–(IC5) are the proper laws of iterated contraction. I find them intuitively convincing, though, of course, my intuitions are already shaped by ranking
theory. Let me briefly discuss them.

Order equivalence (8.19) was our very first idea when thinking about iterated belief
change; (5.2) and (5.4) already embody it. Darwiche, Pearl (1997) assumed it with
their postulates (5.46≤) and (5.47≤) for revisions, and our (5.51≤) and (5.52≤) were the
corresponding postulates for contraction. Indeed, this idea seems to be universally
maintained. Here, *order equivalence* (8.19) is only a consequence of (IC5), *order preservation*. (IC5) goes beyond (8.19) exactly in its asymmetric part, i.e., by maintaining order
preservation not only under the supposition $C \subseteq \bar{A} \cap \bar{B}$, but also under the weaker

supposition $C \subseteq \bar{A}$. It thus expresses nothing but the idea of the other Darwiche/Pearl postulates (5.48≤) and (5.49≤), translated into contraction terms (and slightly improved) in (5.53≤). In sum, (IC5) embodies nothing but the four well-accepted Darwiche/Pearl postulates transferred to contraction.

Hence, it is exactly with (IC3) and (IC4) that our axiomatization of iterated contraction goes beyond the present state of the art. As to (IC3), *restricted commutativity*, we had found plausible reasons in Section 5.7 for expecting that iterated contractions do not always commute, and (5.58) described the conditions under which they do not do so. One condition was that the two contracted propositions A and B are positively relevant to each other (and the point then was that, though at least one of the two material implications expressing the positive relevance must survive the two contractions, it may be that one of them survives after $\div\langle A, B\rangle$ and the other survives after $\div\langle B, A\rangle$). However, in (IC3) we assumed an extreme negative relevance, i.e., that "if A, then *not* B" is logically true. This deductive relation holds across all doxastic states whatsoever. So, (5.58) can never apply to such A and B, whatever the doxastic state, and we may accept *restricted commutativity* as an axiom. In other words, if two disbeliefs are logically incompatible, there can be no interaction between giving up these disbeliefs, and hence it seems also intuitively convincing that the order in which they are given up should not matter at all.

The intuitive content of (IC4), *path independence*, can be described as follows: Suppose you believe A. Then you also believe the logical consequences of A. Let B be one of them; $B \to A$ is another. Now you contract by A. This entails that you have to give up at least one of B and $B \to A$. Suppose you keep B and give up $B \to A$. What *path independence* claims is that it does not make a difference whether you give up A (= $(B \to A) \cap B$) and then B or whether you give up $B \to A$ right away and then B. The description is still simpler in terms of disbelief: Suppose you disbelieve two logically incompatible propositions, and you have to contract both of them. Then you can either contract one after the other, or you can first contract their disjunction, and if you still disbelieve one of them, you then contract it as well. (IC4) says that both ways result in the same doxastic state. This seems entirely right to me.

This may suffice as an explanation of the intuitive appeal of our iterated contraction axioms. A further point that I find remarkable is that these axioms make assertions only about one-step and two-step contractions. (IC1) and (IC2) characterize single contractions, (IC3) and (IC4) both say that two different two-step contractions come to the same thing, and (IC5) compares the result of a one-step and a two-step contraction (since claims about "$\trianglelefteq_+$" are claims about single contractions). What is remarkable about this is that we do not need any independent assumptions about the interaction or relation between three or more steps of contraction, even though we have been referring to such longer contraction chains all the time; recall that conditional relevance showed up only in threefold contractions.

The crucial point now is that the potential DC $\leq_+$ generated according to (8.15) by an IC $\div$ satisfying Definition 8.18 is indeed a DC according to Definition 8.16. The proof is not so difficult, but very tedious, carried out in the series of lemmata 6.1–6.19 in Hild, Spohn (2008); there is no point in here reproducing all this material. What we thereby derive are the properties (8.16a–e) of the DC $\leq_{\div}$; that it is, moreover, Archimedean and full (i.e., also satisfies properties 8.16f–g), does not follow from Definition 8.18. Hence, let us simply accept

Definition 8.20: Let $\div$ be an IC for $(\mathcal{A}, \mathcal{N})$. Then $\div$ is called *Archimedean* iff the DC $\leq_{\div}$ induced by $\div$ is Archimedean. And $\div$ is called *full* iff the DC $\leq_{\div}$ induced by $\div$ is full.

I have no ambition to express the Archimedean property purely in terms of iterated contraction; this appears an unilluminating exercise. Likewise, though fullness is easily translated into contractions, its original explanation in terms of difference comparisons is the most perspicuous.

When combined with Theorem 8.17, then, the upshot is

Theorem 8.21: For any IC $\div$ for $(\mathcal{A}, \mathcal{N})$, the potential DC $\leq_{\div}$ induced by $\div$ is a DC for $(\mathcal{A}, \mathcal{N})$. And any 6-rich full Archimedean IC $\div$ for $(\mathcal{A}, \mathcal{N})$ is an IRC for $(\mathcal{A}, \mathcal{N})$, i.e., there is a real negative ranking function κ for $\mathcal{A}$ with $\div = \div_{\kappa}$. Moreover, for each ranking function κ' with $\div = \div_{\kappa'}$ there is an $x > 0$ such that $\kappa' = x \cdot \kappa$.

(See also Theorem 6.21 in Hild, Spohn (2008), which, however, is an immediate consequence of that series of lemmata.)

Theorem 8.21 concludes my presentation of the method of measuring ranking functions through iterated contractions. It also shows that (IC1)–(IC7) are necessary properties of iterated contractions and (together with the Archimedean property) jointly sufficient ones in the presence of richness and fullness; under such structural assumptions they offer a complete characterization of iterated contractions – at least if iterated contraction is conceived as proposed in the ranking-theoretic way.

To summarize: We have seen that the laws of iterated contraction are intuitively appealing. The last two sections have shown that they are adequately embedded in a rich theoretical context relating them to disbelief comparisons or entrenchment orderings, relevance judgments, etc. Finally, we have seen that they provide a rigorous justification of using *numerical* ranks. This indeed was the main goal of this chapter.

8.5 Two Further Lines of Justification

There are two further ideas about how one might argue in favor of the ranking structure, which I would like to at least mention at the end of this chapter for the sake of completeness.

One idea is found in Hild (2001, Ch. 7). Hild explains how to carry over Cox's (1946, 1961) justification of the probabilistic structure of degrees of belief to ranking theory. This translation can indeed be easily summarized in

Theorem 8.22: Let $\mathcal{A}$ be a propositional algebra over W. Let β be a function from $\mathcal{A}$ into $\mathbf{R}^+ = \mathbf{R} \cup \{\infty\}$ and g be a function from $\mathbf{R}^+ \times \mathbf{R}^+$ into $\mathbf{R}^+$ satisfying the following conditions:

(a)   if $A \subseteq B$, then $\beta(A) \leq \beta(B)$ [*monotonicity*],

(b) $\beta(A \cap B) = g(\beta(A), \beta(B))$ [*conjunction modularity*],

(c)   for any x and y in the range of β with $x \leq y$ there are $A, B \in \mathcal{A}$ such that
      $\beta(A \cap B) = x$ and $\beta(A) = y$ [*fullness*].

Then $g(x, y) = \min (x, y)$ for all x and y in the range of β. And if $\beta(W) = \infty$ and $\beta(\varnothing) = 0$, then β is a real positive ranking function for $\mathcal{A}$.

Proof: This is Theorem 1 of Hild (2001, Sect. 7.2). I reproduce his proof: Conjunction modularity entails the symmetry of g, and monotonicity entails the idempotency of g over the range of β, i.e., $g(x, y) = g(y, x)$ and $g(x, x) = x$ for all x and y in the range of β. Applying conjunction modularity once more we find $g(\beta(A \cap B), \beta(A \cap B))$ $= g(\beta(A \cap B), \beta(A))$. Thus, fullness entails $g(x, x) = g(x, y)$ for all x and y in the range of β with $x \leq y$. Hence, by idempotency, $g(x, y) = x$ for $x \leq y$, and by symmetry also $g(x, y) = y$ for $y \leq x$. So, $g(x, y) = \min (x, y)$ for all x and y in the range of β.

Observe that the theorem is about degrees of belief and thus about positive, not negative ranking functions. The point of the theorem is clear. (c) is again a structural condition on the richness of the algebra and as such not to be criticized. (a) is a universally accepted condition for any kind of degrees of belief. So, the crucial condition characteristic of Cox's type of argument is (b). Indeed, Cox replaces (b) by the condition

(d) $\beta(A \cap B) = g'(\beta(A), \beta(B \mid A))$ for some function g',

in order to show with the help of additional assumptions that β must be (transformable into) a probability measure. Hild emphasizes that the crux of Cox's procedure lies in these additional and somewhat hidden assumptions, whereas he has stated his assumptions completely in (8.22).

If condition (b) were, or appeared to be, convincing, Theorem 8.22 would provide further support for ranking theory. However, Hild is careful in his conclusions. The problem is not that (b) does not appear sufficiently plausible. The problem is rather that similar conditions appear equally plausible. Why prefer (b) to (d)? Or (d) to (b)? Why not accept

(e) $\beta(A \cup B) = g''(\beta(A), \beta(B))$ for some function g'' [*disjunction modularity*] or

(f) $\beta(\bar{A}) = h(\beta(A))$ for some function h [*negation modularity*]?

The further problem is that one cannot accept all of the plausible conditions. As Hild (2001, Section 7.2, Theorem 3) shows, in the presence of *monotonicity* (a) and *fullness* (c) any two of (b), (e), and (f) together trivialize degrees of belief insofar as they allow for no more than two degrees. So we must choose, but we have no obvious guidance.

Hild's argument thus takes a critical turn against Cox's line of reasoning. Not only probability theory can be justified along this line, as many Bayesians fond of it seem to assume, for alternative accounts of degrees of belief can be so justified as well. However, nobody has provided any argument as to why one should prefer this rather than that kind of modularity. And it seems difficult to do so without question-begging.

Indeed the only way to argue for conjunction modularity seems to me by question-begging. If β is supposed not to be just any measure of uncertainty, but to express genuine degrees of belief (including belief simpliciter), then one can argue convincingly for the law of conjunction (5.10b) for positive ranks which entails *conjunction modularity*. However, then we know beforehand that $g = \min$ and do not require (8.22) in order to infer this. Hence, I tend to think that this line of argument is not particularly successful as a justification of ranking theory. Rather, (8.22) gives us another very weak set of premises from which we can deduce positive ranking theory that is already known to be desirable, and this is valuable information, too. (Of course, we could argue in the very same way about *disjunction modularity* (e) and negative ranking theory.)

Huber (2007) pursues a different line of reasoning. He picks up from my remarks in Chapter 4 and after (5.19) that ranking theory can be developed from the assumptions of consistency and deductive closure, and attempts to turn them into a rigorous argument. He does so by utilizing what I have called the method of upgradings in Section 8.1. Roughly, he proceeds as follows:

He uses the empirical method described in Section 8.1 for defining the degrees of entrenchment of a proposition A in someone's doxastic state; the method was simply to determine how many minimal upgradings of $\bar{A}$ are required for eliminating the belief in A. Ranking theory is not presupposed by that method; what is referred to is only the empirical process of repeatedly giving minimal reasons against A to someone from independent sources.

This process comprises many doxastic states: the initial one, as well as all the states resulting from these upgradings. What Huber (2007, Theorem 4.2) shows then is that, if the belief sets of all these states − not only the initial one, but also all the subsequent ones − conform to the normative requirement of consistency and deductive closure, then and only then these degrees of entrenchment behave like ranks.

And he extends his argument to the conditionalization rules (5.25) and (5.33). That is, he shows that, if in a change from prior to posterior degrees of entrenchment some specific entrenchment function ε on an evidential algebra $\mathcal{E}$ is adopted and if all the belief sets resulting from subjecting the posterior degrees of entrenchment to the method of upgradings are consistent and deductively closed − that is, not only the unconditional belief sets, but also those conditional on the atoms of $\mathcal{E}$ − then, and only then, the posterior degrees of entrenchment form a ranking function that is the $\mathcal{E} \rightarrow \varepsilon$-conditionalization of the prior degrees of entrenchment. For details see Huber (2007, Theorems 7.3 + 7.5).

Huber draws an illuminating parallel to the Dutch-book justification of subjective probabilities. There we start from empirically given betting quotients. These betting quotients are closely related to degrees of uncertainty; indeed, under certain assumptions the two may be equated. Moreover, the betting quotients must conform to the basic normative requirement that they do not allow a Dutch-book against them with sure loss. Then, the argument goes, the degrees of uncertainty must behave like probabilities. It is well known that this type of argument extends to the various conditionalization rules for changing subjective probabilities; this is the dynamic Dutch-book argument (see my references after (3.1) and (3.8)). The pragmatic nature of this argument, i.e., of its basic normative premise, is obvious, though, and gave rise to the search for purely epistemic justifications of Bayesianism.

Huber (2007) pursues an exact parallel. He starts with the empirically given degrees of entrenchment, which are closely related to degrees of (dis)belief and may be equated with them under certain assumptions. The degrees of entrenchment, initial and subsequent ones, must conform to the basic normative requirement of generating consistent and deductively closed belief sets. All this entails, then, that the degrees of (dis)belief must behave like ranks and must change according to the conditionalization rules for ranks. And, Huber insists, this is an epistemic, not a pragmatic justification of ranking theory.

I expressed some reservation concerning the method of upgradings in Section 8.1. Therefore I am unsure how fully I should endorse Huber's argument. Perhaps, though, this reservation corresponds in the present context to the (not so telling) doubts about defining fair betting quotients in the way required for the Dutch-book argument. In any case, Huber's argument definitely opens an illuminating perspective. If it holds good, then the statics and the dynamics of ranks embody nothing but the preservation of consistency (4.1) and deductive closure (4.2) throughout one's doxastic career. We started our discussion of the representation of belief with (4.1) and (4.2). Now the circle closes. All our rich and ramified theorizing may finally turn out to be based on no more than (4.1) and (4.2) in a dynamic perspective.

9

Supposing, Updating, and Reflection

9.1 Conditional Belief

After all these theoretical developments, let us pause, before continuing with the comparative part of this book, for a chapter-long while, with an issue I should have raised already in Section 5.3. There I could hardly overstate the importance of conditional ranks, representing (graded) conditional belief. So far, however, we have not properly reflected on the nature of conditional belief, clearly a neglect, in particular in view of the well-known perplexities surrounding the equally difficult notion of conditional probability. Let us use this chapter to make good for this neglect.

So, what is conditional belief? Simply a derived notion, we may reply as a first attempt. After all, we *defined* conditional ranks in (5.15). Hence, it seems, if we understand unconditional ranks or degrees of belief, there is no further mystery. The corresponding claim is that conditional probability is simply a defined notion according to (3.2).

The latter claim has often been criticized, most forcefully perhaps in twin papers by Hájek (2003a,b). Hájek (2003a) first insists that Definition 3.2 is really an *analysis* of conditional probability, which is somehow antecedently understood; so it raises a claim of adequacy. Secondly, he argues that (3.2) is not completely satisfactory as an analysis. The argument proceeds via so-called "trouble-spots" of a probability assignment, among them propositions of probability 0, conditional on which probabilities can be intuitively meaningful, even though they are not delivered by (3.2). So, he concludes that conditional probability should be taken as a primitive notion that is only constrained by (3.2) as far as it is applicable, or rather by clause (3.13c) of the definition of Popper measures.

In principle, Hájek's argument carries over to ranking functions and to Definition 5.15 of conditional ranks, since ranking functions can have "trouble spots", too; at least they can fail to be regular so that conditional ranks are not completely defined. Hence, one may again conclude that it is rather conditional ranking functions as defined in (5.19) that are to be considered as basic. Some caution is required, though, since the argument from failure of regularity does not apply to ordinal ranking functions (5.9f), which are always regular (according to the most natural generalization of regularity to these functions); I indicated this point after (5.19). In any case, Hájek's point that (5.15) only offers a less than complete analysis of conditional ranks, and thus of conditional belief, is doubtlessly correct. Therefore, the question remains what our antecedent understanding of these notions might be.

Well, what might it be? One idea is to say that conditional belief is nothing but belief in a conditional. However, the basic point is that we understand conditionals much more poorly than conditional belief; so the hope should be to analyze conditionals in terms of conditional belief, rather than the other way around. This was indeed the idea behind the Ramsey test (4.17).

We have learned by now that this idea is highly problematic. It was first ventured by Lewis (1976) with respect to probabilities, with devastating results; the identification of conditional probabilities with probabilities of conditionals entailed various trivialization theorems. Analogous impossibility results for conditional belief were first proved by Gärdenfors (1986); cf. also Gärdenfors (1988, Sect. 7.4–7.6).

One might try to find fault with the other premises of these theorems. However, in view of the extent to which these premises can be weakened (cf. Rott 1989), this strategy does not appear promising. Moreover, there are deep intuitive doubts about the Ramsey test; Gärdenfors (1988, pp. 159f. + 165f.) discusses some of them. A possible response might be that the Ramsey test applies only to so-called epistemic conditionals. Whether these form a natural or an artificial kind is, however, in turn open. For instance, it is doubtful whether epistemic conditionals can be identified with indicative conditionals. As I have already indicated, conditionals are a messy topic, and the closer one looks at linguistic details, the messier it gets. I prefer to stay away from them. (See, however, the excellent surveys of Edgington 1995 and Bennett 2003.)

In any case, I think it is fair to say that the vigorous discussion of these matters has settled with the view that conditional belief is not to be analyzed in terms of belief in conditionals and at best helps partially in analyzing conditionals. The Ramsey test is tempting, but deceptive.

There is a prima facie quite similar, though different, attempt to analyze conditionality. It is more suggestive with respect to conditional probability and refers to the rule (3.8) of simple conditionalization that tightly connects conditional probabilities with probability dynamics. Maybe $P(B \mid A)$ is the probability I would have for B if I learned A, as (3.8) says it is. In this way, we might again analyze conditional probability in terms of unconditional probability, though this time not as an unconditional probability of a conditional, but rather through a conditional about unconditional probability. (This sentence perfectly reveals how confusing the issue is.)

Some doubt is cast on this idea by the fact that it does not smoothly carry over to conditional ranks, since we did not state a rule of simple conditionalization for ranks. Rather, our basic type of $A \rightarrow n$-conditionalization was already a simple case of generalized conditionalization, and in terms of the latter the connection between conditional ranks or probabilities and the dynamics of ranks or probabilities is not quite as tight.

Indeed, the doubt is more than justified. Hacking (1967, pp. 314ff.) already argues that the rule of simple conditionalization is really a non-tautological dynamic assumption about the relation between conditional probability and the probability given facts or evidence (his terminology vacillates slightly) and hence does not offer an analysis of either concept. The case is more clearly expressed and more explicitly argued by Weirich (1983). He gives three different types of example in which conditional probabilities and probabilities given knowledge of a condition fall apart. Let us focus here on the first kind of example, which, according to van Fraassen (1980b, p. 503), were already suggested by Richmond Thomason (the other two kinds import quite different difficulties and potential confusions).

Imagine Ann pondering at t about Julius Caesar's last thought. One plausible possibility is A = "Caesar's last thought was of Cleopatra". So, if P_t denotes Ann's probabilities at t, we have $P_t(A) > 0$, so that probabilities conditional on A are defined. Thus, $P_t(A \mid A) = 1$. However, it would be false to claim that, if A were the case, then Ann would be certain of A. Whatever Caesar's last thought, Ann's opinions about it do not counterfactually depend on it. So, generally, we may not equate $P_t(B \mid A) = x$ with "if A were the case, $P_t(B)$ would be x".

So much for one variant. Weirich's argument concerns the more important variant. Although $P_t(A) > 0$, Ann is absolutely sure that she will never have a firm opinion about Caesar's last thought. That is, if B is the proposition "for some $t' > t$, $P_{t'}(A) = 1$", then $P_t(B) = 0$. Hence also $P_t(A \cap B) = 0$ and $P_t(B \mid A) = 0$. However, it is trivially true that, if Ann were to learn at $t' > t$ that A, then Ann would be certain of A at t', i.e., $P_{t'}(A) = 1$. Hence, B is counterfactually true, and since Ann would be aware of what she learns and would know her own doxastic state, we counterfactually even have $P_{t'}(B) = 1$. All in all, if Ann were to learn at t' that A, then $P_{t'}(B) = 1$. Therefore, $P_t(B \mid A) = 0$ cannot mean that, if Ann were to learn A at some t', then $P_{t'}(B)$ would be 0.

To summarize: However, we construe the counterfactual condition, $P_t(B \mid A) = x$ is not equivalent to $P_{t'}(B)$ counterfactually being x for some $t' > t$. Conditional probability cannot be understood in this way.

In the previous paragraphs I have deliberately spoken of some third person Ann. Things are even more confusing in the first-person perspective, since the expression and the assertion of beliefs get mixed up in this case. (The ensuing problems surface in Moore's paradox in the simplest way.) When I say: "If A, B is likely (or certain)", I seem to assert some kind of conditional relation between A and B, and we might try various distributions of doxastic operators in order to capture what I thereby express: "it is likely or certain for me that $A \, \Box\!\!\rightarrow B$", "$A \, \Box\!\!\rightarrow$ it is likely (or certain) for me that B", "I believe $A \, \Box\!\!\rightarrow$ it is likely or certain for me that B". If our discussion of the third-person perspective is correct, then none of these readings of what I express applies. (To my knowledge, Benkewitz (2006, Ch. 5) is the most thorough-going discussion of the many readings of the related assertion: "He or, respectively, I believe p *because q.*"

Rosefeldt (2000, Sect. 7.d) informs me that even Kant already struggled with these ambiguities at various places and used them to argue for the freedom of thought.)

So, we still have no analysis of conditional belief or conditional subjective probability. I think Hájek (2003a,b) is right; there is no analysis. They are primitive concepts. Or rather, they are theoretical concepts which characterize momentary doxastic states and thus help us to state dynamic laws for doxastic states such as the conditionalization rules without being definable through these laws. Animals, at least highly developed ones, certainly have doxastic states (see, however, Bennett (1976) for a still exemplary explanation of the various levels of complexity in the ascription of something like doxastic states). Hence, they have a dynamics of doxastic states and consequently some kind of conditional states, even if it is difficult to tell which.

The case is comparable to the notion of (instantaneous) velocity. Velocity is a theoretical notion characterizing the momentary state of motion of a body and helping us to describe its kinematics. Of course, instantaneous velocity is closely related to average velocity within brief intervals of time in the familiar way, but only given further assumptions. For instance, the body has to exist for while; it would, however, have a velocity even if it were to be destroyed in the next moment. And it has to change its velocity continuously; otherwise, the relation between instantaneous and average velocity would fail.

What distinguishes us from other highly developed animals is our ability to access our own doxastic states, an ability manifested in or, maybe, equivalent to our capacity to linguistically express our doxastic states. A dog might have conditional beliefs, but if so, it is very difficult to infer them from its behavior. In the human case, by contrast, we simply have to take notice of the utterances of our fellows.

However, the relation between conditional belief and probability and its linguistic expressions is highly complex in turn. It is clear that they are somehow related to our rich conditional vocabulary. We saw, though, when briefly discussing the Ramsey test, that this vocabulary is a messy field not furnishing reliable guidance. Perhaps the purest way to express conditional belief and probability is in terms of *hypothetical supposition*: Suppose that A. Would you believe B then? Or how likely would B be then for you? Your answers display your conditional beliefs and probabilities. Not that the language of supposition would be free of the ambiguities of our other conditional idioms; we can, e.g., make a counterfactual instead of a hypothetical supposition and ask for its counterfactual consequences. Still, the way I have just put it seems to be clearer than others. (Because talk of suppositions is ambiguous as well, the kind of supposition directly eliciting conditional beliefs and probabilities is often called matter-of-fact supposition; cf. Joyce 1999, p. 182, and Arló-Costa 2001b.)

Indeed, matters are often described in this way in the literature. There is the logic of hypothetical reasoning (pursued since Rescher 1964), i.e., of the way suppositions work; this is one thing. And there is the theory of updating, of processing information or evidence and changing doxastic states accordingly; that is another thing. Levi (1996) was perhaps the one to most extensively argue that these are two distinct theoretical projects. My main concern in the Chapters 3–5 was the theory of updating. What the discussion in this section suggests is that all the conditional doxastic notions – conditional beliefs, ranks, and probabilities – are directly related to the logic of hypothetical or matter-of-fact supposition; the condition operates as a supposition. They are only indirectly related to the theory of updating, insofar doxastic change conforms to conditionalization rules statable with the help of conditional doxastic notions. (And they are only indirectly related to the theory of counterfactual reasoning, which I do not discuss.)

These remarks also throw a different light on AGM belief revision theory. My main objection was that it does not account for iterated updating or belief change. However, this criticism does not apply if suppositions are our concern. For one might say that there are no genuine iterated suppositions, since making a further supposition under an already given supposition can be considered to be the same as making a single conjoint supposition. In this perspective AGM belief revision theory turns out to be a theory of supposition. This conclusion is endorsed and defended at length by Levi (1996, Ch. 4) and Haas (2005, Ch. 6). Haas' argument is a different one. He finds the postulates $(K*1)$–$(K*8)$ (4.18) defective in various ways when seen as postulates of belief revision, but perfectly acceptable as postulates governing supposition (or "coerced revision", as he says). I have deferred my argument with the criticisms of the postulates $(K*1)$–$(K*8)$ as postulates of belief revision to Section 11.3, but I share the conclusion. It is an irony, of course: it means that AGM belief revision theory can only be saved by denying that it is about belief revision.

Arló-Costa, Thomason (2001) analyze the situation still more carefully. They allow for iterated suppositions or hypothetical revisions, as they call them, and they make the above assumption explicit, i.e., the assumption that several iterated suppositions are the same as a single conjoint supposition; this is their postulate of cumulativity (Arló-Costa, Thomason 2001, p. 502). They then show that it is incompatible with the AGM belief revision postulates, and they conclude that "the AGM postulates capture only *some* features of supposition, and *some* features of rational inquiry" (p. 502) (where "rational inquiry" is their term for what we call "updating").

However, as their paper makes amply clear, the fault they find with the AGM theory is less with its postulates, but already with its simple conception of a doxastic state as a belief set. We had occasion to criticize this point in Section 5.6. In the suppositional context, Arló-Costa's and Thomason's critical point is that a supposition need not change the belief set, namely when the supposition is already contained in the belief set (this is the postulate of invariance (p. 501), which is entailed by the AGM postulates), while the supposition may, and does, change the entrenchment ordering; this is why invariance is inadequate in the suppositional context. (For all this, see also Arló-Costa 2001b.) So, it seems that AGM belief revision theory is inadequate even as a theory of supposing. However, this is due to its format, not due to its postulates. If all suppositions are considered as, or implicitly reduced to, one-step suppositions, then the deficient format does not show up and the AGM postulates remain adequate, as observed above.

9.2 Reflection

So far, I have emphasized the difference between supposing and updating. Even in simple conditionalization, where it seemed to vanish, it persists in problem cases as discussed by Weirich (1983). There is, however, a way – a very illuminating one, I find – to represent updating directly in suppositional terms; namely by engaging in so-called auto-epistemology. Hild (1998a,b) has thoroughly elaborated the issue for the probabilistic and even for the qualitative case, and we may smoothly copy it in ranking-theoretic terms. Let me briefly explain the point; for all details, though, see Hild (1998a,b).

Auto-epistemology deals with one's beliefs about one's own future, present and past beliefs, or, more generally with one's future, present and past doxastic states being considered in one's present doxastic state. We have beliefs about many parts of the world. We are also part of the world, indeed a part we are very well acquainted with. So, of course, our own mental and especially our doxastic states are an important topic of our theorizing. Auto-epistemology hence is an important part of epistemology. It has one central principle, the reflection principle; indeed I could not find any further independent principle of auto-epistemology in the literature. In order to explain it, we first have to introduce some suitable notation.

Since we will talk, as before, only about one doxastic subject, we need not make it explicit in the notation. However, we have to consider various points of time. So, let P_t or, respectively, κ_t denote the actual subjective probability measure or negative ranking function of our subject at time t. Moreover, there are not only actual doxastic states of our subject, there are also possible ones. So, P_{wt} or κ_{wt} is to denote our subject's doxastic state at time t in the possible world w. This allows us to express a lot of auto-epistemic propositions: $\{w \mid P_{wt}(A) = x\}$ is the proposition that the subject's probability for A at time t is x, $\{w \mid \kappa_{wt'} = \lambda\}$ is the proposition that the subject's ranking function at t' is λ, etc. As usual, I will slightly abbreviate these expressions as $\{P_{wt}(A) = x\}$, $\{\kappa_{wt'} = \lambda\}$, etc., thus giving the curly brackets a w-binding function. Auto-epistemology then requires that all these auto-epistemic propositions are in the domain of the subject's actual doxastic states.

At the risk of being pedantic, let us assume that all the possible doxastic states are not auto-epistemic; their common domain is to be some algebra $\mathcal{A}$ consisting of ordinary propositions. Let $\mathcal{A}^*$ denote the algebra generated by $\mathcal{A}$ and all the above auto-epistemic propositions. Only the actual states P_t and κ_t are then supposed to be about $\mathcal{A}^*$. Things would get unnecessarily complicated if we were to allow the reflected doxastic states to be auto-epistemic as well (although one would have to do this in a full treatment of auto-epistemology). Let us call the whole construction resulting in $\mathcal{A}^*$ and the measure P_t or the ranking function κ_t an *auto-epistemic set-up*.

Given a probabilistic auto-epistemic set-up, the version in which the *reflection principle* is usually discussed is the probabilistic principle:

(9.1)   For any $A \in \mathcal{A}$ and $t' > t$ $P_t(A \mid \{P_{wt'}(A) = x\}) = x$.

This says that, given your future probability of A is x, your present probability of A should be x, too. Or more colloquially, you should now trust your future judgment. So, in this perspective the principle is a dynamic one. Gaifman (1988) has introduced a more general perspective. He thinks of $P_{wt'}$ not as your own future probability measure, but of anybody's at any time. And he defines such a $P_{wt'}$ to be an *expert function* for you at t just in case your P_t satisfies (9.1). Of course, you might also consider experts that are so only for a restricted subject matter and not necessarily for the whole of $\mathcal{A}$. In this reading, (9.1) says that you should trust the experts, but it is true by definition, since an expert is defined as one you trust. Hájek (2003b, pp. 192–5) points to the difficulties arising when you accept several experts on the same issue.

In this terminology, the original reflection principle (9.1) says you should consider your future self as an expert. This makes clear that the reflection principle cannot be a universal maxim. Given that you will have forgotten things you should not trust in your forgetfulness; given that your usual clear-headedness is clouded by (too much) alcohol, you should get sober and not let your sober self be guided by your drunken self. This is a familiar objection (see, e.g., Maher 1993, Sect. 5.1). Van Fraassen (1995a) contemplates saving the generality of (9.1) by assuming that doxastic subjects lose their continuity and existence in the exceptional cases. This appears, though, to be an extravagant way of avoiding the problems.

Despite such restrictions one must realize the fundamental importance of the reflection principle. The usual case is that one's future doxastic state is better informed than the present one; one learns and improves one's point of view in the course of time. Unlike the conditionalization rules the reflection principle as such does not specify any details about how exactly that learning is to work; it only says that the improvements must be such as to conform to (9.1). This looks like a very abstract constraint, but we will see that it has specific consequences.

The reflection principle is sometimes called Miller's principle because Miller (1966) launched a surprising attack against it, long before it was positively acknowledged. The main effect of the attack, though, was to promote the insight that we have here something valuable. Spohn (1978, pp. 161f.) and Goldstein (1983) endorsed another principle called the *iteration principle* by Hild (1998a):

(9.2)   for any $A \in \mathcal{A}$ and $t' > t$:

$$P_t(A) = \sum_{w \in W} P_{wt'}(A) \cdot P_t(\{w\}) = \sum_{x \in [0,1]} x \cdot P_t(\{P_{wt'}(A) = x\}),$$

where the sum should be replaced by an integral in the likely case of W being uncountable. The iteration principle says that your present opinion is a weighted mixture of your possible future opinions, the weights being your present probabilities for arriving at the various possible future opinions. It is easily shown to be equivalent to the reflection principle (cf. Hild 1998a, Theorem 2.2) and thus subject to the same restrictions, as already noted in Spohn (1978, p. 166).

Skyrms (1980, appendix 2) discusses a slight generalization of the reflection principle (= constraint 2, p. 163) that he calls principle M in later writings:

(9.3)   for any $A \in \mathcal{A}$, $t' > t$ and any interval $I \subseteq [0, 1]$:

$$P_t(A \mid \{P_{wt'}(A) \in I\}) \in I.$$

Strictly taken, (9.1) will usually be undefined since the condition in (9.1) will usually have probability 0. This mishap does not plague (9.3), which is therefore preferable. Of course, to the extent to which (9.1) is defined it entails (9.3), basically due to the law of disjunctive conditions (3.4f). I will neglect this point in the sequel.

Skyrms (1980, p. 162) mentions other precursors. Actually, though, the reflection principle is quite obviously suggested by de Finetti's representation theorem and his philosophy of probability (see de Finetti 1937). Still, it received its name and its first extensive philosophical discussion only by van Fraassen (1984), which hence deserves to be used as the common reference text.

Hild (1998a, p. 328) uses a slightly stronger version of the *reflection principle*:

(9.4)   for any $t' \geq t$ $P_t(\cdot \mid \{P_{wt'} = Q\}) = Q$.

This is stronger since the condition specifies the totality of your future doxastic state w.r.t. $\mathcal{A}$. And it is stronger since it allows $t' = t$ as a limiting case; your present opinion is trivially an expert for your present opinion. Hild (1998a) then shows (in his Theorem 2.1) that the reflection principle (9.4) entails *auto-epistemic transparency*:

(9.5)   for any t $P_t(\{P_{wt'} = Q\}) = \begin{cases} 1, \text{if } Q = P_t \text{ restricted to } \mathcal{A}, \\ 0, \text{otherwise}, \end{cases}$

and *perfect memory*:

(9.6)   for any $t' < t$ $P_t(\{P_{wt'} = Q\}) = \begin{cases} 1, \text{if } Q = P_{t'} \text{ restricted to } \mathcal{A}, \\ 0, \text{otherwise}, \end{cases}$

Memory loss is simply among the circumstances under which the reflection principle fails.

So much for the probabilistic side. My point now is predictable: All the principles, restrictions, and arguments apply to ranking theory just as well. We should accept the *reflection principle* in ranking-theoretic form:

(9.7)   for any $t' \geq t$ $\kappa_t(\cdot \mid \{\kappa_{wt'} = \lambda\}) = \lambda$.

as a fundamental principle of rationality, subject to the restrictions already discussed. It is again equivalent to the *iteration principle*, as Hild calls it:

(9.8)   for any $A \in \mathcal{A}$ and $t' > t$:

$$\kappa_t(A) = \min_{w \in W} (\kappa_{wt'}(A) + \kappa_t(w)) = \min_{n \in N^+} (n + \kappa_t(\{\kappa_{wt'}(A) = n\})).$$

And it entails a ranking-theoretic version of *auto-epistemic transparency*:

$$(9.9) \quad \text{for any } t \; \kappa_t(\{\kappa_{wt'} = \lambda\}) = \begin{cases} 0, \text{ if } \lambda = \kappa_t \text{ restricted to } \mathcal{A}, \\ \infty, \text{ otherwise,} \end{cases}$$

and of *perfect memory*:

$$(9.10) \quad \text{for any } t' < t \; \kappa_t(\{\kappa_{wt'} = \lambda\}) = \begin{cases} 0, \text{ if } \lambda = \kappa_{t'} \text{ restricted to } \mathcal{A}, \\ \infty, \text{ otherwise,} \end{cases}$$

The credentials of these principles are those of the probabilistic counterparts. There is, however, independent support as well. Since its inception in Hintikka (1962) doxastic logic has had its own static reflection principle, namely the principle that I believe that I believe that *p* if and only if I believe that *p*. This principle is certainly not uncontested. I think, however, that all plausible counter-examples make use of different notions of "believe" in the iterated clause. Hence I fail to see a serious objection. Given the representation of belief in negative ranking-theoretic terms, the reflection principle of doxastic logic states:

$$(9.11) \quad \kappa_t(A) > 0 \text{ iff } \kappa_t(\{\kappa_{wt'}(A) = 0\}) > 0,$$

which, unsurprisingly, is an immediate consequence of auto-epistemic transparency (9.9).

One even finds a dynamic principle for belief, even though it has apparently not left deeper traces in the philosophical discussion. Binkley (1968) discussed the surprise examination paradox, and in the attempt to prove the teacher's announcement to be epistemically inconsistent, he used what was later called *Binkley's principle* (cf. Hild 1998a, pp. 348f.): namely that if you believe today that you will believe tomorrow that *p*, you should believe already today that *p*. Clearly, this is a qualitative counterpart to the probabilistic reflection principle (9.1), and it is subject to the same restriction. If, for instance, you believe today that by tomorrow you will come to believe that *p* by wishful thinking, the principle does not apply. Translated into ranking-theoretic terms, it says:

$$(9.12) \quad \text{for any } t' > t, \text{ if } \kappa_t(\{\kappa_{wt'}(A) = 0\}) > 0, \text{ then } \kappa_t(A) > 0.$$

$$(9.13) \quad (9.12) \text{ is entailed by the ranking-theoretic reflection principle (9.7).}$$

Proof: (9.7) says that $\kappa_t(A \mid \{\kappa_{wt'}(A) = 0\}) = 0$. Together with the premise of (9.12) this entails (i) $\kappa_t(A \cap \{\kappa_{wt'}(A) = 0\}) > 0$. (9.7) also says that $\kappa_t(A \mid \{\kappa_{wt'}(A) > 0\}) > 0$. (This corresponds to Skyrms' generalization (9.3) of (9.1).) Therefore (ii) $\kappa_t(A \cap \{\kappa_{wt'}(A) > 0\}) > 0$. (i) and (ii) together imply $\kappa_t(A) > 0$. (9.11) may be deduced from the reflection principle in the same way.

These are the two predecessors of ranking-theoretic auto-epistemology I am aware of (which are older than Hild (1998a, Sect. 5), the first full investigation of the auto-epistemology of belief sets).

9.3 Auto-Epistemic Updating

Auto-epistemology is a most interesting topic in itself. I have announced, though, that it helps in clarifying the relation between supposing and updating. How does it do so? Let us continue to follow Hild (1998a). He builds on the notion of a protocol, forcefully reintroduced by Shafer (1985), and puts it into an auto-epistemological perspective as already suggested by Shafer (1985, p. 266) in his notion of a subjective protocol.

They explained the need for protocols with Freund's puzzle (cf. Freund 1965). Let me instead use the somewhat simpler puzzle of the three prisoners (cf. Mosteller 1965, problem 13): Suppose you and your two fellows Jim and Tim travel abroad. Out of the blue you are imprisoned by a whimsical dictator and accused of an absurd crime. The sentence is that two of you are to be executed the next day and the third, perhaps to be determined by lot, to be let free. During the night you try to get some information from the guard as to whether you are the lucky one. The guard is reluctant, but after a bribe he tells you: "Oh, I am so sorry for your friend Jim; he will be executed at dawn." Does your chance of being let free thereby increase from 1/3 to 1/2? It might seem so. On the other hand, you knew from the outset that at least one of Jim and Tim will be executed; so you seem to have received no useful information at all. The point is that there is no good answer, the case is underdetermined. You need to know, one might say, the rules of the information game, i.e., the *protocol* of the situation. Maybe the guard is determined to inform you about Jim; in this case your hope of getting free indeed increases. (Balancing the positive news in this case, the guard might instead have had negative news for you, indicating that it is Jim who will be let free.) Maybe, though, the guard simply gives you the name of one of your fellows to be executed, and at random if both are; in this case your hope is as slim as before. Maybe the guard is somehow biased in naming Jim or Tim when both are to be executed; then the case is still different. And if you do not know the rules of the game, you at least need to have assumptions about the rules, i.e., a *subjective protocol*. Only relative to such assumptions you can know what to infer.

Let us give a general theoretical description of updating in an auto-epistemological perspective as suggested by this example. First, let us confine ourselves to considering just one stage from t to t'; this will do. Next, the auto-epistemic reasoner must envisage a set $\mathcal{TE}$ of possible pieces of total evidence (acquired between t and t'). $\mathcal{TE}$ may be a set of propositions from $\mathcal{A}$. Recall from Section 3.2, though, how Jeffrey has widened our conception of evidence. So $\mathcal{TE}$ may also consist of pairs $\langle \mathcal{E}, Q \rangle$, where $\mathcal{E}$ is some (evidential) subalgebra of $\mathcal{A}$ and Q a probability distribution over $\mathcal{E}$. Later on we will consider a still more general possibility. A *protocol* or *evidence function* π then tells us which total evidence you receive between t and t' in the world w, i.e., π is a function from W into $\mathcal{TE}$. The point of the prisoners' story was exactly that it may be completed by differing protocols or evidence functions. Given π and any $e \in \mathcal{TE}$, we can define $TE(e) = \{w \mid \pi(w) = e\}$ to be the proposition that the subject receives total evidence e between t and t'. For each $e \in \mathcal{TE}$, $TE(e)$ is an auto-epistemic, not an ordinary proposition.

Hence, we assume that the auto-epistemic expansion $\mathcal{A}^*$ of $\mathcal{A}$ also contains all propositions of the form $TE(e)$.

How should the subject respond to some piece of total evidence? The most abstract answer is that he should conform to some update rule u that assigns to each probability measure P on $\mathcal{A}$ and each piece $e \in TE$ a probability measure $u(P, e) = P^{u,e}$ on $\mathcal{A}$. So, if the subject obeys the update rule u and receives $e \in TE$, then $P_{t'} = P_t^{u,e}$ (restricted to $\mathcal{A}$). More generally, if the subject obeys the update rule u and expects or considers only probabilistic changes that are driven by evidence according to that rule, then

(9.14)   for all $w \in W$ $P_{wt'} = P_{wt}^{u,\pi(w)}$.

An auto-epistemic set-up is therefore called *evidence-driven* iff it satisfies (9.14).

Now one obvious update rule is the rule u according to which the subject simply conditionalizes on the auto-epistemic proposition that he has received total evidence e. Hild (1998a, p. 332) calls that *auto-epistemic conditionalization*:

(9.15) $P_{t'} = P_t^{u,e} = P_t(\cdot \mid TE(e))$.

It does not take much, then, to prove:

(9.16)   For evidence-driven auto-epistemic set-ups satisfying (9.14) auto-epistemic conditionalization (9.15) is equivalent to the reflection principle (9.4).

(Cf. Hild 1998a, Theorem 2.3.) In other words, when the only probabilistic changes envisaged are those driven by evidence according to a certain protocol or evidence function, the only update rule compatible with, and indeed tantamount to, the reflection principle is auto-epistemic conditionalization. Hence, if we accept the reflection principle within the auto-epistemic perspective, updating reduces to a special case of supposing according to (9.15). Suppose you were to learn $TE(e)$: that is exactly the doxastic state you arrive at when you actually learn $TE(e)$. This is the insight I promised at the beginning of this section. It is not particularly surprising once the construction of auto-epistemic set-ups is explicitly developed; the real gain of clarity lies in that development. Note also that the example from Weirich (1983) we discussed above is adequately treated by this construction.

Hild (1998a) hastens to add that auto-epistemic conditionalization (9.15) must by no means be confused with simple or generalized conditionalization (3.7) or (3.9), and in (1998b) he more carefully argues that they may even come into conflict, in which case (3.7) and (3.9) have to give way. They agree only under special conditions.

Consider simple conditionalization first. Here, the possible pieces of total evidence in TE simply consist of propositions in $\mathcal{A}$. Let E be such a proposition in TE. Clearly, auto-epistemic (9.15) turns into simple conditionalization (3.7) if and only if:

(9.17) $P_t(A \mid TE(E)) = P_t(A \mid E)$ for all $A \in \mathcal{A}$ and $E \in TE$ [*crux*].

Crux, as Hild calls it, might well fail to hold. The story of the three prisoners is a case in point. Under the one protocol where the guard merely informs you about the fate of your fellow Jim, (9.17) holds good; but it does not under the other protocol where the guard just gives you the name of one of those to be executed.

In the case of generalized conditionalization, the pertinent pieces of evidence e in $\mathcal{TE}$ are of the form $e = \langle \mathcal{E}, Q \rangle$ (or simply a distribution over the atoms of $\mathcal{E}$). Here, the agreement of auto-epistemic with generalized conditionalization requires two conditions:

(9.18) if $e = \langle \mathcal{E}, Q \rangle$ and $E \in \mathcal{E}$, then $P_t(E) \mid TE(e)) = Q(E)$ [*reliability*],

(9.19) if $e = \langle \mathcal{E}, Q \rangle$ and E is an atom of $\mathcal{E}$, then for all $A \in \mathcal{A}$
$\qquad P_t(A \mid E \cap TE(e)) = P_t(A \mid E)$ [*evidential independence*].

In fact, Hild (1998a, Theorem 3.5) proves that (9.18) and (9.19) are necessary and sufficient for (9.15) and (3.9), auto-epistemic and generalized conditionalization, to coincide. Note that (9.17) may also be split up into the two assertions $P_t(E \mid TE(E)) = 1$ (*reliability*) and $P_t(A \mid E \cap TE(E)) = P_t(A \mid E)$ (*evidential independence*). These conditions are so widely satisfied that their role is easily overlooked. It is therefore important to make them explicit.

Wagner (1992) has proposed a further generalization of Jeffrey's generalized conditionalization (3.9) that fits nicely into the present auto-epistemological perspective; this is why I want to briefly discuss it. His scenario is that "the total evidence relevant to revising some prior probability P_t on W allows us to assess on a related possibility set Ω a probability measure μ". And he assumes "that our understanding of the relationship between outcomes in Ω and those in W is expressed by a function $\Gamma : \Omega \to \mathcal{A}$ with $\Gamma(\omega)$ denoting the narrowest event in $\mathcal{A}$ presently understood to be entailed by the outcome $\omega \in \Omega$. How do μ and Γ constrain the possible revisions of the prior in question?" (Wagner 1992, pp. 246f.; I have slightly adapted the notation.)

μ and Γ first induce a function m on $\mathcal{A}$ defined by

(9.20) $m(E) = \mu(\{\omega \in \Omega \mid \Gamma(\omega) = E\})$.

Since $\sum_{E \in \mathcal{A}} m(E) = 1$, m is a *basic probability assignment* on $\mathcal{A}$ in the sense of Shafer (1976, p. 38) (and hence is in general *not* a probability measure on $\mathcal{A}$). We might restrict m to its *focal elements* E for which $m(E) > 0$, since only these focal elements are positively indicated by the total evidence μ via Γ; let $\mathcal{E}$ be the set of focal elements of m. Wagner then gives a sophisticated argument to the effect that the posterior $P_{t'}$ on $\mathcal{A}$ should be determined by:

(9.21) $P_{t'}(A) = \sum_{E \in \mathcal{E}} m(E) \cdot P_t(A \mid E)$.

We might call this *Wagner's more generalized conditionalization*. It looks deceptively like Jeffrey's generalized conditionalization (3.9). But, of course, (9.21) is more general; it reduces to (3.9) only when the focal elements in $\mathcal{E}$ form a partition of W.

Note that $m(E^*)$ is *not* the posterior probability of $E^* \in \mathcal{E}$. Rather, we have $P_{t'}(E^*)$ $\geq b(E^*) := \Sigma\{m(E) \mid E \in \mathcal{E}, E \subseteq E^*\}$, where b is the Dempster-Shafer belief function induced by the basic probability assignment m. $P_t(E^*)$, in effect, adds up all the support given to E^* by the pieces of evidence in $\mathcal{E}$, which is $m(E)$ for $E \subseteq E^*$, 0 for $E \cap E^* = \varnothing$, and some intermediate figure for the other $E \in \mathcal{E}$. This explains at the same time why (9.21) reduces to generalized conditionalization under Jeffrey's assumption that $\mathcal{E}$ forms a partition of W.

Wagner (1992, p. 252) gives a droll example in which the focal elements do not form a partition (notation again adapted):

> *The Linguist:* You encounter a native of a certain foreign country and wonder whether he is a Catholic northerner (w_1), a Catholic southerner (w_2), a Protestant northerner (w_3), or a Protestant southerner (w_4). Your prior probability P_t over these possibilities (based, say, on population stat-istics . . .) is given by $P_t(w_1) = 0.2$, $P_t(w_2) = 0.3$, $P_t(w_3) = 0.4$, and $P_t(w_4) = 0.1$. The individual now utters a phrase in his native tongue which, due to the aural similarity of the phrases in question, might be a traditional Catholic piety (ω_1), an epithet uncomplimentary to Protestants (ω_2), an innocuous southern regionalism (ω_3), or a slang expression used throughout the country in question (ω_4). After reflecting on the matter you assign subjective probabilities $\mu(\omega_1) = 0.4$, $\mu(\omega_2) = 0.3$, $\mu(\omega_3) = 0.2$, and $\mu(\omega_4) = 0.1$ to these alternatives. In the light of this new evidence how should you revise P_t?

The background assumption is that no Protestant would utter ω_1 or ω_2 and no north-erner ω_3. This determines the mapping Γ from $\Omega = \{\omega_1, \omega_2, \omega_3, \omega_4\}$ into $\mathcal{A}$ relevant to the example, and it is clear that the values of Γ do not form a partition. So, the case cannot be handled by Jeffrey's generalized conditionalization. Still, (9.21) is applicable and solves the problem.

Now, I do not want to engage in a debate of Wagner's proposal and argument, all the more as his interest apparently is not only to generalize conditionalization, but also to build a probabilistic bridge to the Dempster-Shafer theory of belief func-tions (cf. Shafer 1976), a topic we are not at all prepared to discuss. (I will offer a very brief comparison of Dempster-Shafer belief functions and ranking functions in Section 11.9.) As announced, the only point I want to make is that Wagner's scenario and his more generalized conditionalization naturally fit in into our auto-epistemic perspective:

Wagner can state his case only with the auxiliary possibility space Ω mapped by Γ into the original possibility space W. The content of this construction became intelligible in such examples as the quoted one. The natural translation into our framework is to view Ω simply as a part of the original possibility space W and to interpret Γ, which specifies the evidence received according to Wagner, as a kind of protocol mapping "real" possibilities in W into the auto-epistemic expansion $\mathcal{A}^*$ over W.

Thus translated, the space $\mathcal{TE}$ of possible pieces of total evidence in Wagner's scenario consists of $\mathcal{A}$ and (finite) mixtures of $\mathcal{A}$. That is, we have $\mathcal{A} \subseteq \mathcal{TE}$, and if $\mathcal{E}$ is a finite subset of $\mathcal{A}$ and m a probability distribution over $\mathcal{E}$ (so that $m(E) \geq 0$ for $E \in \mathcal{E}$ and $\sum_{E \in \mathcal{E}} m(E) = 1$), then $\langle \mathcal{E}, m \rangle \in \mathcal{TE}$. This includes the pieces of total evidence required for Jeffrey's conditionalization. Thereby, Wagner's crucial assumption (14) (1992, p. 250) translates into the condition *crux* (9.17) for any $E \in \mathcal{E} \subseteq \mathcal{A} \subseteq \mathcal{TE}$.

Let us make a further natural assumption, namely that upon receiving mixed evidence $e = \langle \mathcal{E}, m \rangle$ one is in the same doxastic state concerning $\mathcal{A}$ as when expecting to receive certain evidence $E \in \mathcal{E}$ to the degree $m(E)$; the degrees of uncertainty of the propositions in $\mathcal{E}$ are the same in both cases, and there is no further difference between the two states w.r.t. $\mathcal{A}$ (of course, they differ on $\mathcal{A}^*$). To state this formally: Let $e = \langle \mathcal{E}, m \rangle$, and $TE_{\mathcal{E}} = \bigcup_{E \in \mathcal{E}} TE(E)$ be the proposition that the total evidence will be (certain and) one of the $E \in \mathcal{E}$. Our assumption thus is:

(9.23)   If $P_t(TE(E) \mid TE_{\mathcal{E}}) = m(E)$ for all $E \in \mathcal{E}$, then $P_t(A \mid TE(e)) = P_t(A \mid TE_{\mathcal{E}})$
for all $A \in \mathcal{A}$ [*mixture*].

Then, we again have the result:

(9.24)   *Crux* (9.17) and *mixture* (9.23) are sufficient for (9.15) and (9.21), auto-epistemic and Wagner's more generalized conditionalization, to agree.

Proof: We have $P_{t'}(A) = P_t(A \mid TE(e))$ (by 9.15) $= \sum_{E \in \mathcal{E}} m(E) \cdot P_t(A \mid TE(E))$ (by 9.23) $= \sum_{E \in \mathcal{E}} m(E) \cdot P_t(A \mid E)$ (by 9.17) $= P_{t'}(A)$ (by 9.21).

The auto-epistemic reinterpretation of Wagner's more generalized conditionalization finally helps to understand what was going on at the end of Section 7.4. After explaining the update algorithm of simple propagation for information about a single variable, I mentioned there how Hunter (1990) extends this algorithm in order to deal with multiple uncertain evidence about several variables, namely by introducing for each evidential variable a binary dummy variable that is observed with certainty. This seemed to be a technical trick, but now we can understand the dummy variables as auto-epistemic variables so that the certain evidence that all these dummy variables are positive can be identified with the proposition $TE(E)$ in the auto-epistemic reconstruction of Wagner's proposal.

So far, I have explained auto-epistemic conditionalization only within a probabilistic setting; this is the only setting in which these matters are discussed in the literature. My continuous theme, of course, is that the same discussion can be led within a ranking-theoretic setting. This is essentially a matter of replacing P's by κ's; so, it would perhaps be too tedious to carry out this exercise. It is clear, though, that auto-epistemic conditionalization also works in ranking-theoretic terms, is equivalent to the ranking-theoretic reflection principle in the relevant set-up, and reduces to the usual forms of conditionalization under the corresponding conditions. All this is a relevant part of ranking theory.

In fact, this part is, in a way, already developed by Hild (1998a, Sect. 5). He did so not in ranking-theoretic terms, but rather in terms of AGM belief revision theory, with which the parallel considerations work relatively well. But since, as we have seen, ranking theory includes AGM belief revision, Hild's account is entailed by ranking-theoretic auto-epistemology, as he surely saw.

Important as it is, within the course of this book this presentation of auto-epistemology was rather a digression that will not be presupposed in the subsequent chapters. I will return to these topics from a somewhat different angle only Chapter 16 when further discussing evidence and the basis of belief.

10

Ranks and Probabilities

Despite my declared intention of writing a book on ranking theory I have developed relevant portions of probability theory in parallel, in particular in Chapters 6, 7, and 9. I plead a didactic excuse; my point was always to make ranking theory more vivid and intelligible. Still, I might have left the reader bewildered. I concluded in Section 3.3 that we must develop the theory of belief independently of probability theory, and then it might appear that I have done nothing but copy the Bayesian point of view, so perfectly in fact that one might wonder whether I am simply preaching a sophisticated form of Bayesianism, despite my disclaimer. So, we must try to get clear about the relation between ranks and probabilities. This will turn out to be a surprisingly rich and rewarding topic, which I want to tackle in this chapter in a slightly more systematic way. The topic cannot be exhausted now; the relation will pervade the rest of the book. Still, we have deferred the issue for too long already; moreover, the present discussion will be most useful as a background for the chapters in which this relation will unfold further.

10.1 Once More: Belief and Probability

We may, in fact, directly continue our discussion from Section 3.3. There, I pointed to the difficulties of explaining belief in terms of probabilities, as high-lighted by the lottery paradox. I was content with stating that there is no obvious solution and thus motivated an independent investigation of belief, the dynamics of which naturally led us to introduce ranking theory and to develop it in Chapters 5–8. So, the relation between ranks and probabilities is basically still an issue of the relation between probabilities and belief.

That relation is very puzzling, indeed. It appears highly incredible that there are two frames of mind, two different kinds of doxastic attitudes that coexist unrelated in one person. However, we can escape this conclusion only when we positively specify that relation; we must take a stance. Which one, though, is hard to say. Let us take an overview of the available options.

An intuitively and theoretically most satisfying option is epistemological *monism*: there is just one basic description of doxastic states. Monism can take various forms. One form is *eliminativism*: the view that, although we presently have various ways to describe our doxastic attitudes, only one will be needed and survive in the end; the other ones can simply be eliminated; their relation to the basic notion is of no further interest. The question is: which notion might be the lucky one? It cannot be the notion of belief (or its sophistication in form of ranks); we cannot forget about subjective probabilities and confine ourselves to talking about beliefs. It might be some third notion, who knows? However, it is idle to speculate; I cannot see any present candidate that would be able to play that role. Of course, it might be probability itself; this is *radical probabilism* so ably defended by Richard Jeffrey since his (1965). I feel strongly attracted by the unity and systematicity of this position; in comparison all other positions are a jumble. I hope the reader can sense my efforts to preserve these virtues. None the less, this position is also deeply unsatisfactory; I find it entirely incredible that our talk of belief should merely be a convenient device, yet a mistake ultimately to be eliminated from the theoretical point of view. We must tell a positive story about belief.

Monism may also take the form of *reductionism*; our multifarious doxastic idioms might ultimately be explained by, or reduced to, one basic notion. Again, the question is: which one might it be? Of course not the notion of belief; subjective probabilities cannot be reduced to it (in particular, subjective probabilities do not reduce to beliefs about objective probabilities). Neither can it be the notion of probability, at least the standard one. This was just the point of the lottery paradox. In Section 3.3 I reproduced only the familiar arguments refuting direct methods of reduction. There are also more sophisticated methods; but the proposals I have seen do not appear satisfactory to me. We shall briefly look at some of them in Section 10.2. There is a third option: both, belief and probability, might reduce to something else; perhaps some non-standard notion of probability, perhaps something that should not be called probability at all. This is a live option, which I will explore in Section 10.2. I will also present my own proposal for realizing this option. Still, I will remain skeptical about reductionism; in Section 10.3 I will rather conclude that the feasible forms of reductionism are not philosophically enlightening.

The alternative to monism is *dualism* (or pluralism, if there were even more doxastic conceptions at stake). It is no wonder, by the way, that we can duplicate here many positions familiar from the mind-body problem; the relation of two notions or fields always faces these kinds of options. (In these terms, we have already discussed the counterparts of Churchland's eliminativism, panpsychism, materialism, and Russell's neutral monism; I see no counterpart for Davidson's anomalous monism and none for epiphenomenalism. The analogy will continue.)

Dualism may take the form of *interactionism*, the view that both notions, belief and probability, are required to draw an adequate epistemological picture, and that the task is hence to give an account of their interplay without reducing one to the other. This might appear to be the most sensible form of dualism. It is, I find, championed in the work of Isaac Levi, which I will critically discuss more thoroughly in Section 11.2. Quite a different kind of interactionism is represented by Hild (2001) who has many interesting things to say about how ranking and probability theory mesh, about how heavily ranking-theoretic ideas are implicitly used in current statistical methodology. However, I am not prepared to assess Hild's type of interactionism; we would have to explore too deeply into the foundations of statistics, itself a delicate topic. I am not aware of any further plausibly elaborated forms of interactionism.

In fact, I am reluctant to opt for interactionism. My experience rather is that belief and probability are like oil and water; they do not mix easily. So, the dualistic alternative to interactionism is *separatism*, the view that there are indeed two distinct doxastic modes. You may be described as being in the one or as being in the other, perhaps on different occasions; but there is no good way to mediate between or combine the two modes. I sense the absurdity of this position; therefore I am not determinately promoting it. However, it is obvious that I have so far adopted *methodological separatism*, as one might call it. This was my conclusion in Section 3.3, and this will be my further strategy.

"Methodological" means that my separatism is only a temporary research strategy. If a convincing version of reductionism or interactionism were presented, I would happily accept it. But I do not see such versions, and so I do not know better than to follow my maxim. I am really unsure what the ultimate judgment on the matter might be. My non-interactionist dualism is all the more tolerable, I find, as it is backed up by a kind of preestablished harmony; my separatism is in fact a *parallelism*, as we already had plenty of occasions to observe. I will continue to argue for methodological separatism in Section 10.3.

Illuminating as it may be, we should not overdraw, though, the metaphorical loan from the philosophy of mind. Let us rather look at the matter itself. In the next section I will discuss formal possibilities of realizing reductionism. Whatever the philosophical merits of these possibilities, they will be highly instructive in any case. In Section 10.3, then, I will give a broader philosophical comparison of the uses and merits of ranks and probabilities that will speak in favor of my methodological separatism. There, I will also motivate the appendix 10.4 on ranking-theoretic decision theory.

10.2 Ways of Formal Unification

Reductionism has been proposed in quite a number of ways. This section has five parts. First, I shall discuss the relation between ranking functions and non-standard probabilities; perhaps the former, and thus also the notion of belief, reduce to the latter. Secondly, I shall discuss whether Popper measures offer an acceptable account of belief, as is strongly suggested in the literature. This will, thirdly, lead me to my own proposal for unifying ranks and probabilities within a hybrid notion, which I take to transcend even an extended probabilistic point of view. Then I shall return to some recent attempts to circumvent the lottery paradox and to account for belief in terms of standard probability. Finally, two quite different strategies for realizing reductionism will be mentioned.

The best starting point for inquiring into possible unifications is the striking similarity of ranking with probability theory, which might have made me appear to be a mere copycat. (To a large extent I *am* a copycat; probability theory is just too beautiful a model.) This similarity must have an explanation. It is indeed quite simple, and I indicated it already at the end of Section 5.3. I mentioned there that there is an algorithm – translate the sum of probabilities into the minimum of ranks, the product of probabilities into the sum of ranks, and the quotient of probabilities into the difference of ranks – that is almost guaranteed to translate probabilistic axioms, definitions, and theorems into ranking-theoretic ones. The "almost" raises flags, but we must first ask: why does the algorithm work so well? The first one to answer the question was Kurt Weichselberger in his report on my Habilitationsschrift (1983b), but many independently made the same observation:

The translation of products and quotients of probabilities suggests that negative ranks are simply the logarithm of probabilities with respect to some base $x \in (0, 1)$. This does not seem to agree with the translation of sums of probabilities. It can be made to fit, however, by making the logarithmic base smaller and smaller and turning it into an infinitesimal in the end. (Precisely because of this limit relation Goldszmidt, Pearl (1992a, 1996) elaborated the probabilistic ε-semantics of Adams (1975) into their system Z^+ based on ranking theory.)

Thus we come to study the relation between ranks and non-standard probabilities. Let us state it precisely: Let $\mathbf{R}$ be the set of standard real numbers and $\mathbf{R}^* \supseteq \mathbf{R}$ the set of *non-standard reals* according to some non-standard model of analysis. Then $x \in \mathbf{R}^*$ is infinite iff $|x| > r$ for all $r \in \mathbf{R}$, finite iff it is not infinite, and infinitesimal iff $|x| < r$ for all $r \in \mathbf{R}$ with $r > 0$. We know that for each finite non-standard real $x \in \mathbf{R}^*$ there is a unique $r \in \mathbf{R}$ and a unique infinitesimal i such that $x = r + i$; we may hence define $\mathrm{st}(x) = r$ as the *standard part* of x. Let us extend this notion for our temporary purposes by defining $\mathrm{st}(x) = \pm\infty$, respectively, for any positive or negative infinite $x \in \mathbf{R}^*$. Then we have the following simple theorem:

Theorem 10.1: Let $\mathcal{A}$ be an algebra over W, P be a finitely additive non-standard probability measure on $\mathcal{A}$ taking values in $\mathbf{R}^*$, and i some infinitesimal. Define for all $A \in \mathcal{A}$ $\kappa(A) = \mathrm{st}(\lg_i P(A))$. Then κ is a real negative ranking function on $\mathcal{A}$ according to (5.9e).

Proof: Obviously, $\kappa(A) \geq 0$, $\kappa(W) = \mathrm{st}(\lg_i 1) = 0$, and $\kappa(\varnothing) = \infty$ (with a slight extension of the logarithm). Now let $\kappa(A) = r \leq \kappa(B) = s$. Thus there are infinitesimals j and k such that $P(A) = i^{r+j}$ and $P(B) = i^{s+k}$. Now we have first $P(A \cup B) \geq P(A)$ and hence $\kappa(A \cup B) \leq \kappa(A) = r$. Second, we have $P(A \cup B) \leq P(A) + P(B) = i^{r+j} + i^{s+k} \leq 2i^{r+j}$ and hence $\kappa(A \cup B) \geq \mathrm{st}(\lg_i 2i^{r+j}) = \mathrm{st}(r + j + \lg_i 2) = r$, since $\lg_i 2$ is also an infinitesimal. To conclude, we have $\kappa(A \cup B) = \min \{\kappa(A), \kappa(B)\}$.

This theorem explains a lot. It first explains why our translation algorithm works so marvelously and why ranking and probability theory could be developed so largely in parallel in Chapters 5–7 and 9. Secondly, it can explain why ranks are measured on a ratio scale; instead of i we may as well use some other infinitesimal i' as logarithmic base in Theorem 10.1, and the resulting ranking function κ' is $\kappa' = \lg_i i' \cdot \kappa$.

Finally, it explains why the translation is not perfect. The point that a ranking function is only the *standard part* of an infinitesimal logarithm of a non-standard measure entails various modifications. Let me list all the deviations seen so far: we found them in (5.23a), the law of disjunctive conditions that entails, as far as I can see, most of the other deviations; in (6.7d), concerning the behavior of the reason relation, the probabilistic version of which could be strengthened; in (6.11), the weighing of reasons; in (7.2e), concerning the independence of propositions; in (7.9) as compared with (7.10), concerning the graphoid property (7.8f); and in (7.11) and (7.12), concerning the further properties of conditional independence. And we will find further deviations: after (12.7), concerning the so-called null confirmation of laws; in (12.9) and (12.19), concerning positive and non-negative instantial relevance; in the mixture Theorems 12.14, 12.18, and 12.21; in (14.14), the positive claim of which has no probabilistic counterpart and which spreads to (14.81); in (14.20–14.21), showing that the case of multi-valued variables is slightly more complicated in ranking-theoretic than in probabilistic terms; and finally in (14.82–14.83), concerning some aspects of the inference to the best explanation. If one compares the proofs of the probabilistic and ranking-theoretic counterparts in these cases of deviation, one sees that it is due to this standard part assumption that the proof works for one side, but not the other. However, I do not know of any systematic account of the deviations.

In view of Theorem 10.1 one might think there is an easy answer concerning the relation between belief and probability: Bayesianism in terms of non-standard probabilities is the common story behind both. Non-standard Bayesianism has in fact been suggested by David Lewis several times, for the first time in Lewis (1980, p. 268) (though his concern was to generalize regularity; he was silent on the issue between eliminativism and reductionism concerning belief). It might now appear as a kind of fraud that I am giving this answer only in Chapter 10. I could have written a standard textbook on Bayesian epistemology instead, a somewhat boring enterprise, and then I could have pointed out its extended interpretation and application in terms of non-standard probabilities. But I should not have. My methodological separatism is certainly preferable, I think, and part of it, of course, was to develop ranking theory in parallel, but independently.

For this way of unification meets several problems. The basic point is, of course, that non-standard Bayesianism may bolster eliminativism, but it is inadequate as an analysis of belief. In Section 3.3, we argued that the conjunction of beliefs and the dynamics of belief could not be simultaneously handled in probabilistic terms. This argument is invalidated by the non-standard extension. Still, the intuitive point remains: Believing is to be certain in some sense, but hardly ever in the extremely strong sense of having probability 1 or $1 - i$ for some infinitesimal i. You can safely bet your life against infinitesimal probabilities. Or think of Jorge Luis Borges' Book of Sand, with a continuum of pages; you can be sure never to find the same page again. Belief, however, is a much more mundane thing; you would not bet your life on most of your beliefs. The fact that David Lewis did not elaborate on his proposal might indicate that he also thought of it as a technical, but not a philosophical extension of Bayesianism. This objection will also apply to the method of unification via Popper measures discussed next, and it will be explored further in the next section.

Moreover, I would like to point to an algebraic incompatibility. After (5.9) I had argued that a negative ranking function might reasonably be defined on a complete algebra of propositions and should then satisfy the law (5.9k) of complete minimitivity. This cannot be duplicated in terms of Theorem 10.1; even a set of infinitesimal uncertainties can conjoin to a logical impossibility, if only the set is large enough. That is, we may also set up an infinite lottery paradox (cf. Smith 2010). Of course, this is an abundance problem not arising in finite applications. When we indulge in abundance, though, it is certainly an open issue how to resolve the problem.

Finally, I find non-standard probability theory simply ontologically incredible. Syntactically, the non-standard reals are inconspicuous, that is what they are made for. They are first-order indistinguishable from the real numbers and even second-order indistinguishable, as long as all the properties and relations of non-standard reals one refers to are internal; and one cannot refer to external relations in the object language. In the metalanguage, though, the picture changes. There, non-standard reals turn out to be model-theoretic or ontological monsters, as becomes clear when one studies the ultrapower construction of the non-standard reals modulo ultrafilters over infinite sets (cf., e.g., Keisler 1976). I am not a fan of Occam's razor; my purely ontological credo is: let a thousand flowers bloom! Usefulness, though, is a different matter, and before postulating such an incredibly rich field of degrees of belief, I would like to see stronger arguments than just the formal possibility of unification offered by Theorem 10.1. If the same epistemological goals can be served by less extravagant ontologies, this would surely be preferable.

For instance, one might consider whether pre-Robinson infinitesimals might do as well, a line of thinking that started with the attempts of Paul de Bois-Reymond and Giuseppe Veronese to conceive of geometry and the continuum in a non-Archimedean way, that led to the theory of real closed fields, and that was finally topped by Abraham Robinson's non-standard analysis. (An excellent survey of this development is offered by Ehrlich 1994.) There, we find more parsimonious infinitesimal extensions of the real numbers (such as quotients of polynomials) that are less suited for doing analysis, but might provide sufficiently many degrees of belief.

From an epistemological point of view we are indeed not committed to such infinitesimal extensions; their main point so far was to offer a formal explanation of why ranks behave so much like probabilities. So, let us put infinitesimal theorizing to one side and think about other possibilities of unification. We may take up the thread of our discussion of the lottery paradox in Section 3.3. There we observed that we might establish the closure of belief under conjunction by equating belief with probability 1 (as done in 3.17), and preserve the revisability of such beliefs with the help of Popper measures as defined in (3.14); this would be a solution of the lottery paradox that would also be ontologically much simpler than the one just discussed in terms of non-standard probability measures. So, maybe, Popper measures offer an attractive path to unification, and epistemology should ultimately proceed in terms of them.

This view, which might still be called an extended Bayesian one, is gaining, it seems, increasing popularity. Van Fraassen (1995b) is a sophisticated attempt to develop the logic of full belief in terms of Popper measures and is hence a major contribution to the topic of this section. Arló-Costa further elaborates this point of view in a series of papers (2000, 2001a,b). Joyce (1999, ch. 6) builds his account of conditional belief on Popper measures. Hájek 2003a provides further affirmation. And the trend continues.

However, I find the attempt misguided; Popper measures are unsuited for that uni-ficatory epistemological role. There is, again, the basic intuitive objection that belief is simply not absolute probability 1. One may therefore give an eliminativist rather than a reductionist twist to this view. This does not help, though; there is a more decisive theoretical objection; I mentioned it already at the beginning of Section 5.1. It is that Popper measures and their dynamics in terms of the associated law of simple conditionalization violate the principle of categorical matching (5.1), and hence founder at the problem of iterated probabilistic change. Indeed, it was there that this problem was first clearly perceived by Harper (1976). So, the situation of Popper measures is similar to that of selection functions (4.11) and AGM belief revision the-ory as a whole.

In fact, the situation is not only similar, it is exactly the same as I showed in Spohn (1986) where I stated probabilistic counterparts to Definition 4.13 and Theorem 4.14. Let me summarize my results for the finite case:

> *Definition 10.2*: Let $\mathcal{A}$ be a finite algebra over W. Then $\langle P_0, \ldots, P_n \rangle$ is a (*finite*) *hier-archy of probability measures* on $\mathcal{A}$ iff each P_j ($j = 0, \ldots, n$) is a probability measure on $\mathcal{A}$ such that there is a largest $C_j \in \mathcal{A}$ with $P_j(C_j) = 1$ and $P_i(C_j) = 0$ for all $i < j$. (Hence, $C_0 = W$, and if there is some such C_j, there is also a unique largest one for a finite $\mathcal{A}$.)

Then we have

> *Theorem 10.3*: Let $\langle P_0, \ldots, P_n \rangle$ be a hierarchy of probability measures on $\mathcal{A}$, $\langle C_0, \ldots, C_n \rangle$ being the appertaining sequence of propositions according to (10.2). Let $\mathcal{B} = \{ B \in \mathcal{A} \mid P_i(B) > 0$ for some $i = 0, \ldots, n \}$, and define $\tilde{P}$ for $A \in \mathcal{A}$ and $B \in \mathcal{B}$ by $\tilde{P}(A \mid B) = P_i(A \mid B)$, where $i = \min \{ j \mid P_j(B) > 0 \}$. Then $\tilde{P}$ is a Popper measure on $\mathcal{A} \times \mathcal{B}$ (in the sense of 3.13). Conversely, for each Popper measure $\tilde{P}$ on $\mathcal{A} \times \mathcal{B}$ there is exactly one hierarchy $\langle P_0, \ldots, P_n \rangle$ of probability measures on $\mathcal{A}$ that represents $\tilde{P}$ in the way just specified.

The complements $\bar{C}_j$ of those C_j ($j = 1, \ldots, n$) are just the belief cores as defined in van Fraassen (1995b) and Arló-Costa (2000); they satisfy $\tilde{P}(\bar{C}_j) = 1$ and the other defining conditions of belief cores.

In Spohn (1986) I called such hierarchies dimensionally well-ordered families (this is the traditional term) and extended them to arbitrary ordinal hierarchies; Theorem 3–5 there generalize Theorem 10.3 to any σ-algebra $\mathcal{A}$ and to any σ-additive Popper measures on $\mathcal{A}$ (and some $\mathcal{B} \subseteq \mathcal{A}$) (where σ-additivity turns out to be essential for implying the well-ordering of the family).

Now it is obvious that such hierarchies correspond to our well-ordered partitions (4.13) and Theorem 10.3 to Theorem 4.14. To state this a bit more precisely:

Theorem 10.4: Let $\langle P_0, \ldots, P_n \rangle$ be a hierarchy of probability measures on $\mathcal{A}$, such that P_n is regular (i.e., $P(A) = 0$ only if $A = \varnothing$), and let $\langle C_0, \ldots, C_n \rangle$ be the appertaining sequence of propositions according to (10.2). Then $W = C_0 \supseteq \ldots \supseteq C_n$, and the sequence $\langle D_1, \ldots, D_n \rangle$ with $D_i = C_{i-1}$ ($i = 1, \ldots, n$) is a well-ordered partition of W. Let $\tilde{P}$ be the Popper measure defined from $\langle P_0, \ldots, P_n \rangle$ as in (10.3) and g the selection function for $\mathcal{A}$ defined from $\langle D_1, \ldots, D_n \rangle$ according to (4.14). Then we have $g(A) \subseteq B$ iff $\tilde{P}(B \mid A) = 1$.

Proof: First, note that $\tilde{P}$ is indeed a full conditional measure in the sense explained after (3.14); that is, $\tilde{P}(B \mid A)$ is defined for all $A \neq \varnothing$, just as $g(A)$ is. This is so because P_n is assumed to be regular.

Then we have $P_j(C_{i+1}) = 0$ for $j < i + 1$, thus $P_i(C_i \cup C_{i+1}) = 1$ and $P_j(C_i \cup C_{i+1}) = 0$ for $j < i$, and hence $C_i \cup C_{i+1} \subseteq C_i$, because C_i is the largest proposition for which these equations hold. Therefore $C_n \subseteq \ldots \subseteq C_0$.

Moreover, D_i is the smallest proposition for which $P_i(D_i) = 1$. For suppose $E \subset D_i$ and $P_i(E) = 1$. Then $P_i(D_i - E) = 0$, hence $P_j(C_{i+1} \cup D_i - E) = 0$ for $j < i + 1$ and $P_{i+1}(C_{i+1} \cup D_i - E) = 1$, in contradiction to the assumption that C_{i+1} is the largest proposition for which these equations hold.

Now, for any $A, B \in \mathcal{A}$ with $A \neq \varnothing$ and $i = \min \{ j \mid D_j \cap A \neq \varnothing \}$, we have $g(A) \subseteq B$ iff $D_i \cap A \subseteq B$ (this is how g is defined) iff $P_i(B \mid D_i \cap A) = 1$ (because D_i is the smallest proposition with $P_i(D_i) = 1$) iff $P_i(B \mid A) = 1$ (because $P_i(D_i) = 1$) iff $\tilde{P}(B \mid A) = 1$ (according to 10.3).

In this way, selection functions mirror the 0–1-structure of Popper measures, as I already observed at the end of Spohn (1986). In the general σ-additive case the situation is more complicated, but not essentially different. (See also Arló-Costa, Thomason (2001, sect. 6 for a more elaborate analysis of the situation.)

The upshot of this correspondence is that the entire critical discussion of AGM belief revision theory in Chapter 5 applies equally to Popper measures. If there is no reasonable dynamics of WWO's or WOP's, there is none of hierarchies of probability measures and thus none of Popper measures. I do not find any real advance over Harper (1976), who first noticed the problem of iterated change for Popper measures. This was indeed my critical thrust in Spohn (1986) in connection with Spohn (1988), even though it went largely unnoticed: to show that Popper measures provide just as unsatisfactory an epistemological picture as AGM belief revision theory does.

Let me make the difficulties a bit more vivid. We might first undo the violation of the principle of categorical matching (5.1) by explaining not only conditional probabilities, but also conditional Popper measures. If $\tilde{P}$ is a Popper measure on $\mathcal{A} \times \mathcal{B}$, we can define the *conditionalization* $\tilde{P}_E$ of $\tilde{P}$ by E through:

(10.5) $\tilde{P}_E(A \mid B) = \tilde{P}(A \mid B \cap E)$.

$\tilde{P}_E$ is a Popper measure in turn, though it is defined only for $\mathcal{A} \times \mathcal{B}_E$, where $\mathcal{B}_E = \{B \mid B \cap E \in \mathcal{B}\}$. And then we might suggest that by learning evidence E the prior Popper measure $\tilde{P}$ moves to the posterior Popper measure $\tilde{P}_E$.

However, this is certainly inadequate as a general model. The evidence E thereby becomes maximally certain, not only in the standard probabilistic sense that $\tilde{P}_E(E) = 1$, but in the much stronger sense provided by Popper measures that says that $\tilde{P}_E(E \mid B) = 1$ for all conditions $B \in \mathcal{B}_E$ for which $\tilde{P}_E$ is defined. Nothing could cast any doubt on the evidence E even in the extended space of doubt provided by Popper measures.

We had the same problem with simple conditionalization (3.7) within standard probability theory, and generalized Jeffrey conditionalization (3.9) solved that problem. Hence, we might think of adopting the same solution for Popper measures. This seems straightforward. When the prior Popper measure $\tilde{P}$ learns uncertain evidence with probability x $(0 < x < 1)$, it moves to the posterior Popper measure $\tilde{P}'$ defined by:

$$
(10.6) \quad \tilde{P}'(A \mid B) = \begin{cases} x \cdot \tilde{P}_E(A \mid B) + (1 - x) \cdot \tilde{P}_{\bar{E}}(A \mid B), \text{ if } B \in \mathcal{B}_E \text{ and } B \in \mathcal{B}_{\bar{E}}, \\ \tilde{P}_E(A \mid B), \text{ if } B \in \mathcal{B}_E \text{ and } B \notin \mathcal{B}_{\bar{E}}, \\ \tilde{P}_{\bar{E}}(A \mid B), \text{ if } B \notin \mathcal{B}_E \text{ and } B \in \mathcal{B}_{\bar{E}}, \end{cases}
$$

(where I restrict (10.6) to the evidential partition $\{E, \bar{E}\}$ only for the sake of simplicity). Thus, $\tilde{P}'$ is again a Popper measure on $\mathcal{A} \times \mathcal{B}$.

However, this suggestion is no good, either. Let me point out only one problem. According to a Popper measure $\tilde{P}$ a proposition A can have three different statuses: (i) $\tilde{P}(A) > 0$, (ii) $\tilde{P}(A) = 0$, but $A \in \mathcal{B}$ (and hence $\tilde{P}(A \mid B) > 0$ for some $B \in \mathcal{B}$, or (iii) $A \notin \mathcal{B}$ (so that $\tilde{P}(A \mid B) = 0$ for all $B \in \mathcal{B}$). According to (10.5) many propositions keep their status. (10.5) may also send propositions from status (ii) to status (i), namely when the evidence E has prior status (ii). And it sends many propositions to status (iii), namely those in $\mathcal{B}$, but not in $\mathcal{B}_E$. However, it cannot send any proposition from status (i) to status (ii).

The same holds for (10.6). The propositions of status (iii) are the same for the prior $\tilde{P}$ and the posterior $\tilde{P}'$, since they are defined for the same set $\mathcal{B}$ of conditions. The uncertain evidence E as well as $\bar{E}$ has posterior status (i), whether their prior status is (i) or (ii). And so, many propositions may be sent from status (ii) to (i). However, again no proposition can acquire status (ii), if it did not have it before. There is no good reason for this asymmetry. If propositions can have those three statuses, then experience should be able to assign all three statuses to them.

This is only the beginning of problems. But it already suggests that we run here precisely in the issue discussed in Section 5.1: what is the effect of evidence on the entrenchment order or ordering of disbelief, or, here, on the hierarchy of probability measures embodied in a Popper measure? A more thorough-going study would analyze how the hierarchy representing the prior $\tilde{P}$ relates to the hierarchies representing the posterior $\tilde{P}_E$ (according to 10.5) or $\tilde{P}'$ (according to 10.6). We need not go through all this here. But this study would confirm the conclusion of Section 5.1: that we cannot solve the issue as long as we restrict consideration to orderings of disbelief or hierarchies of probability measures.

Arló-Costa, Thomason (2001) seem to disagree by explicitly proposing a revision procedure for Popper measures which they call hypothetical revision. However, I find that they confirm the stance taken here. Very roughly, they propose (10.5) as one step of revision; i.e., the revision $\tilde{P}*E$ of $\tilde{P}$ by E is simply $\tilde{P}_E$. They observe that this step of revision can be iterated, yielding, say, $(\tilde{P}*E)*F$ – provided that $F \in \mathcal{B}_E$. And they observe that this kind of revision, hypothetical revision, satisfies what they call the postulate of (conjunctive) cumulativity: $(\tilde{P}*E)*F = \tilde{P}*(E \cap F)$ (pp. 488f.). In all that, they are very explicit that hypothetical revision is to capture only the logic of supposition and not the logic of inquiry, as they say, or of updating, as I chose to say. (See also Arló-Costa 2001b.)

They are perfectly right. I rejected AGM belief revision theory as a theory of updating beliefs in Section 5.1 and concluded in Section 9.1, with many others, that it is suited, rather, as an account of suppositions (if iterated suppositions are tacitly reduced to one joint supposition by cumulativity). Theorem 10.4 suggests that this assessment holds for Popper measures, too. And it is precisely this assessment at which Arló-Costa and Thomason arrive as well. (I should mention, though, that their considerations are much more sophisticated, in particular insofar as they make the iteration of suppositions explicit and are thus able to distinguish several versions of the postulate of cumulativity and to observe an incompatibility of their hypothetical revision with the AGM postulates $(K*1)$–$(K*8)$ (4.18) in this more explicit context. Cf. my brief remarks at the end of Section 9.1.)

A different, though related point of criticism should still be mentioned; I find it entirely ignored in the literature. In Spohn (1983a, footnote 18) I made the well-hidden remark that AGM belief revision theory does not provide us with an adequate notion of a (conditional) reason and thus of (conditional) dependence and independence. This criticism carries over to Popper measures, I suppose. In any case, my philosophical applications of ranking theory in Chapters 12–17 do not directly refer to the solution of the problem of iterated change provided by ranking theory. Instead, they entirely depend on the adequacy of the ranking-theoretic account of reasons in Chapter 6 and of (conditional) dependence and independence in Chapter 7. Hence, my contention is that none of these applications would work in terms of AGM belief revision theory or in terms of Popper measures. I will not elaborate on this strong claim, but it is clear that it poses a big challenge to all those defending the epistemological utility of Popper measures.

Let me proceed to the third possible way of reduction, my own constructive one. If the only good solution of the problem of iterated belief change consists in advancing from selection to ranking functions, then the only good solution of the problem of iterated probabilistic change consists in analogously amending Popper measures. I proposed to do so in Spohn (1988, Sect. 7). However, I never actually carried out the proposal since I found the point so obvious. It was only in my discussion with Isaac Levi in Spohn (2006a) that I did so, in an appendix. Let me repeat it here, in a somewhat more general way, as a further proposal – indeed, as my proposal – for formally unifying belief and probability.

There are two ways to carry out the proposal; one may take the ranking-theoretic or the probabilistic side as a starting point. Let me do the latter first in order to connect up with the discussion of Popper measures.

Definition 10.7: Let $\mathcal{A}$ be an algebra over W. Then $\Pi = (P^r)_{r \in R}$ is a *ranked family of* (σ-*additive*) *probability measures* on $\mathcal{A}$ iff the following conditions hold:

(a)　R is a set of non-negative ranks, and $0 \in R$,

(b)　for each $r \in R$ P^r is a (σ-negative) probability measure on $\mathcal{A}$,

(c)　for each $r \in R$ there is a $C_r \in \mathcal{A}$ such that $P^r(C_r) = 1$ and $P^s(C_r) = 0$ for all $s \in R$ with $s < r$,

(d)　for each $A \in \mathcal{A}$ either $P^r(A) = 0$ for all $r \in R$ or $\{r \in R \mid P^r(A) > 0\}$ has a minimum.

The (*negative*) *rank* $\kappa_\Pi(A)$ of $A \in \mathcal{A}$ according to Π is defined as $\kappa_\Pi(A) = \begin{cases} \min\{r \in R \mid P^r(A) > 0\}, \text{if this set is non-empty,} \\ \infty, \text{otherwise} \end{cases}$. If $A \in \mathcal{A}$ and $\kappa_\Pi(A) = s < \infty$, then *the conditionalization of* $\Pi = (P^r)_{r \in R}$ *by* A is the family $\Pi_A = (P^r_A)_{r \in R_A}$, where $R_A = \{r - s \mid P^r(A) > 0\}$ and for each $r \in R_A$ and $B \in \mathcal{A}$ $P^r_A(B) = P^{r+s}(B \mid A)$.

It is obvious how Definition 10.7 generalizes hierarchies of probability measures and hence Popper measures: hierarchies are indexed by consecutive non-negative integers, whereas ranked families are indexed by non-negative reals (which may also be non-consecutive non-negative integers), since (10.7) is intended to build on negative ranking functions that do not necessarily satisfy completeness (5.9k). Hence, in contrast to hierarchies, ranked families are not necessarily well-ordered, and so we had to include condition (d) in order to ensure that κ_Π really is a negative ranking function.

The crucial generalization, however, takes place in conditionalization. In this respect, (10.7) might look similar to (10.5). However, (10.7) preserves rank differences in conditionalization (if ranks do not become infinite), whereas index differences in hierarchies are not preserved by conditionalization (10.5) (this is the point about the elimination of empty terms or indices that became so crucial in Section 5.1). Of course, conditionalization of ranked families is not recommendable as their dynamic rule. However, because of the preservation of rank differences by conditionalization,

(10.7) can be developed into adequate dynamic rules in analogy to (5.24), (5.29), (5.32), and (5.34) — something that cannot be done with hierarchies of probability measures. This was my criticism of Popper measures above.

We might as well emphasize the ranking-theoretic aspect; this is the version presented in Spohn (2006a):

> *Definition 10.8*: Let $\mathcal{A}$ be an algebra over W. Then ρ is a *probabilified (negative) ranking function* for $\mathcal{A}$ iff ρ is a function from $\mathcal{A}$ into $\mathbf{R}^+ \times \mathbf{R}$ such that the following conditions hold (wherein ρ_1 is the first and ρ_2 the second component of ρ so that $\rho(A) = \langle \rho_1(A), \rho_2(A) \rangle$):
>
> (a) $\rho(\varnothing) = \langle \infty, 0 \rangle$, $\rho(W) = \langle 0, 1 \rangle$, and $\rho_1(A) < \infty$ iff $\rho_2(A) > 0$,
>
> (b)   if $\rho_1(A) = r < \infty$, then there is a $C_r \in \mathcal{A}$ with $\rho(C_r) = \langle r, 1 \rangle$,
>
> (c)   if I is a finite index set, if the $A_i \in \mathcal{A}$ ($i \in I$) are mutually disjoint, and if $A = \bigcup_{i \in I} A_i$, then $\rho_1(A) = \min_{i \in I} \rho_1(A_i)$, and $\rho_2(A) = \Sigma\{\rho_2(A_i) \mid \rho_1(A_i) = \rho_1(A)\}$.
>
> ρ is *σ-additive* iff condition (c) holds also for countable index sets I. If $\rho_1(A) = r < \infty$, $\rho_1(A \cap B) = s < \infty$, and $\rho(C_s) = \langle s, 1 \rangle$ according to condition (b), then the *conditional ranked probability of B given A* is defined $\rho(B \mid A) = \langle \rho_1(B \mid A), \rho_2(B \mid A) \rangle$, where $\rho_1(B \mid A) = \rho_1(A \cap B) - \rho_1(A) = s - r$ and $\rho_2(B \mid A) = \rho_2(A \cap B \cap C_s) / \rho_2(A \cap C_s)$; and if $\rho_1(A) = r < \infty$ and $\rho(A \cap B) = \langle \infty, 0 \rangle$, then $\rho(B \mid A) = \langle \infty, 0 \rangle$.

If ρ is a probabilified ranking function, ρ_1 is a ranking function; this is where the emphasis lies. ρ_2, by contrast, is not probability-like at all and does not make much sense by itself. Still, it is clear that we have defined the same thing twice over; only my sense of symmetry led me to clumsily introduce the two definitions. They entail

> *Theorem 10.9*: Let $\Pi = (P^r)_{r \in R}$ be a ranked family of (σ-additive) probability measures on $\mathcal{A}$. Define for each $A \in \mathcal{A}$ $\rho(A) = \langle \infty, 0 \rangle$ iff $P^r(A) = 0$ for all $r \in R$ and $\rho(A) = \langle r, P^r(A) \rangle$ iff $\kappa_\Pi(A) = r < \infty$. Then ρ is a (σ-additive) probabilified ranking function for A. Conversely, let ρ be such a function. Define $R = \{\rho_1(A) \mid \rho_1 < \infty\}$, let for each $r \in R$ C_r be some proposition with $\rho(C_r) = \langle r, 1 \rangle$ as guaranteed by (10.8b), and define $P^r(A) = \rho_2(A \cap C_r)$ if $\rho_1(A \cap C_r) = r$ and $P^r(A) = 0$ otherwise. Then $\Pi = (P^r)_{r \in R}$ is a ranked family of (σ-additive) probability measures on $\mathcal{A}$.

The theorem is too trivial to require a proof; these conceptual exercises are going on for too long, anyway. Still, there can be no doubt about the philosophical significance of this notion (or these notions). It is superior to the notion of Popper measures for the reasons explained. It offers at least a formal unification of the theory of belief and the theory of subjective probability, and as such it is as adequate as the component theories are for their domains. It postulates as few doxastic degrees of uncertainty as required for this unification and avoids an extravagant wealth of such degrees (in which non-standard probability theory indulges). And most importantly perhaps, despite its looking like a hybrid (in both representations I have given), it is not only a unified notion, but has a unified theory. This was the point of developing considerable portions of ranking and probability theory in parallel in Chapters 3, 5, 6, 7, and 9, an exercise I will continue. All the laws common to both sides govern probabilified ranking functions as well, certainly enough to speak of *one* natural theory.

The explanation for there being a substantial unified theory is, of course, the same as that for the substantial parallel between ranks and probabilities. If ρ is a probabilified ranking function and i an infinitesimal, then Q defined by $Q(A) = \rho_2(A) \cdot i^{\rho_1(A)}$ is not exactly a non-standard probability measure, but is infinitesimally close to one in the sense that there is a non-standard probability measure P such that $P(A) \, / \, Q(A)$ is infinitesimally close to 1 for all propositions A with $\rho_1(A) < \infty$.

This brings me to a side remark worth noting. There seems to be a tension between the results of McGee (1994) and my observations in this section. McGee (1994, pp. 181f.) states a correspondence between Popper measures and non-standard probability measures (in his Theorem 1); this is more carefully spelled out in Arló-Costa, Thomason (2001, sect. 3). Now I have suggested another correspondence between probabilified ranking functions and non-standard probability measures. At the same time I have claimed that probabilified ranking functions generalize Popper measures (just as ranking functions generalize selection functions). This seems to form an inharmonious triangle.

There is no contradiction, though. McGee's correspondence is one-many, designed to preserve conditional probabilities. My correspondence is also one-many, designed to preserve simple as well as generalized conditionalization. However, mine is, so to speak, one to less many. So, the two correspondences agree with the fact that there is also a one-many correspondence between Popper measures and probabilified ranking functions. In other words, when it is said Popper measures and non-standard probabilities "amount to the same thing" (McGee 1994, p. 181) or "are two sides of the same coin" (Arló-Costa, Thomason 2001, p. 513), this is correct insofar as the structure of conditional probabilities and of iterated suppositional conditionalizations (10.5) or hypothetical revisions is concerned. If the dynamic structure of simple and generalized conditionalization rules is at issue, then it is rather probabilified ranking functions and non-standard probabilities that amount to the same thing. Certainly, the latter respect is crucial.

To resume: We have seen two ways of formally unifying the theory of belief and the theory of probability, namely non-standard probability measures and my notion of ranked families of probability measures or probabilified ranking functions. I have expressed my ontological preferences for the latter. We have seen epistemological reasons why Popper measures are unsatisfactory and need to be developed into the more complicated ranked families of probability measures. If there were epistemological reasons requiring further amendments with a still larger ontology of doxastic degrees, or even the ontology of non-standard reals, I would be happy to accept it. However, presently I do not see any such reasons. Therefore, I prefer to stick to probabilified ranking functions.

We might accept this unification as establishing reductionism: belief and probability reduce to some third thing. I am reluctant; I prefer, as I am going to explain in Section 10.3, my methodological separatism. But even if we were to accept this reduction, it would certainly be inappropriate to say that this reduction merely amounts to an extended form of Bayesianism. An epistemology based on Popper measures might still be sub-sumed under the somewhat vague label of Bayesianism. However, ranked families of probability measures are a true hybrid. They inherit probabilistic ideas, of course, but to the same extent they also inherit ranking-theoretic ideas that were independently motivated and justified in Chapters 5 and 8.

A fourth reductionist strategy, which apparently avoids extravagancies and returns to standard probabilistic terms, has recently received attention. It starts by reconsidering the lottery paradox. The paradox resulted from the assumption (3.16) that belief is the same as some high probability $\geq 1 - \varepsilon$, which is incompatible with the conjunctive closure of belief. Kyburg (1961, 1963, 1970) initially gave up conjunctive closure, a proposal that understandably did not gain much support. So, we might weaken the high probability criterion. Assuming high probability to be only sufficient for belief does not yet avoid the lottery paradox. However, we might say that high probability is sufficient for belief *ceteris paribus*, i.e., in the absence of defeaters – where the lottery paradox situation is one of the potential defeaters. Of course, this idea entirely depends on exactly specifying those defeaters. Douven, Williamson (2006) have bad news for this idea, though; they showed that the lottery paradox generalizes at least to those defeaters that have been proposed so far. This does not completely refute the idea; but its prospects look dim.

Another weakening of the high probability criterion is to take it only as a necessary condition: beliefs must have high probability $\geq 1 - \varepsilon$, but not everything having prob-ability $\geq 1 - \varepsilon$ needs to be a belief. Leitgeb (2010) shows this condition to be satisfiable. Slightly more precisely, he proves: Let P be a standard probability measure on some algebra $\mathcal{A}$, and $C \in \mathcal{A}$ be the core of a belief set (as defined in 4.5) entailing a (still restricted) notion of conditional belief defined by $\mathbf{B}(B \mid A)$ iff $C \cap A \subseteq B$, provided that $C \cap A \neq \varnothing$. Then P and conditional belief are related by the high probability criterion (in the sense that $\mathbf{B}(B \mid A)$ entails $P(B \mid A) \geq 1 - \varepsilon$) if and only if C is a P-stable set – where C is a P-stable set iff C meets the high probability criterion under all conditions consistent with it, i.e., $P(C \mid D) \geq 1 - \varepsilon$ for all D with $C \cap D \neq \varnothing$. For many P, though not for all, there are P-stable sets. Leitgeb then goes on to generalize this result to full conditional belief, i.e. to selection functions (defined in 4.11). And he shows that, if there is a P-stable set, there may be many, depending on the threshold $1 - \varepsilon$. However, all the P-stable sets are nested (indeed, in a well-ordered way), thus providing more or less strict standards of belief. This probabilistic reduction of condi-tional belief is finally brought to bear on many topics where conditional belief is philosophically relevant.

Here, I only want to point to this interesting development. The nesting of P-stable sets closely resembles the belief cores of van Fraassen (1995b) relative to a Popper measure. Maybe a translation of Popper into non-standard probability measures would reveal that van Fraassen (or Arló-Costa 2001a) presents a non-standard version of Leitgeb's standard story. However, precisely because of this parallel, I suspect that the ranking structure will not be forthcoming from Leitgeb's constructions. We may have iterated probabilistic change, but the induced iterated change of stable sets is presumably very different from the ranking-theoretic conditionalizations (5.24) and (5.32). And these differences will spread to the philosophical applications (that I will pursue in Chapters 12–17). However, all this awaits further scrutiny.

Fifth and finally, I should point to two further quite different methods of formal unification, which transcend the probabilistic point of view. The first consists in the theory of Dempster-Shafer belief functions ingeniously elaborated by Shafer (1976). It purports to generalize probability theory, and it apparently embraces ranking theory; at least the consonant belief functions, as Shafer (1976, Ch. 10) calls them, look quite similar to positive ranking functions. Shafer thus promises an entirely independent and original way to realize reductionism. Let me defer my comparative discussion of the Dempster-Shafer theory to Section 11.9 (where we will discover that there are unbridgeable differences after all).

The other method of formal unification to be finally mentioned here is brilliantly executed by Halpern (2003). His approach may be called structuralistic. He first collects all or the main models for uncertain reasoning, such as probability measures and ranking functions, but also inner and outer measures, lower and upper probabilities, Dempster-Shafer belief functions, and possibility measures. Then he looks for their common structure that he subsequently captures by what he calls *plausibility measures*. These are functions from the algebra of propositions considered into some set D of plausibility values. At first, we know very little about this set D: only that it contains a unique minimum plausibility for the contradiction and a unique maximum plausibility for the tautology and that it is partially ordered so that we can state the law that a logically weaker proposition is at least as plausible as a logically stronger one (cf. Halpern 2003, pp. 50f.). However, in due course the set D of plausibility values receives an increasingly rich algebraic structure according to the epistemological needs. For instance, if we want to assume that the plausibility value of a disjunction functionally depends on the values for its disjoint disjuncts, then we have to introduce a suitable algebraic operation $\oplus$ on D (which we might call addition even if it is the minimum operation for numbers). In order to define conditional independence, we need further structure, and then we can even carry out the theory of Bayesian nets in terms of plausibility measures (cf. Halpern 2003, Sect. 3.9, 4.3, and 4.5.4). If we want to build a theory of expectation, we introduce still further structure (cf. Halpern 2003, Sect. 5.3). And so forth.

This approach is highly illuminating, and I recommend studying it. I have no argument with it at all. It is just not mine. My approach was the reverse one, starting with a firm interpretation and then formalizing it in a theory as adequate and fruitful as possible. Since we have a fairly determinate understanding of belief (and of subjective probability, anyway), this approach was feasible. And when we turn to a more philo-sophical comparison in the next section and to all the philosophical applications in subsequent chapters, I would not know how to do it without building on such a firm interpretation. However, having carried out the comparisons and applications it might well be worth returning to Halpern's structuralist approach (even though I will not do so), in order to check the extent to which these applications generalize to his plausibility measures.

10.3 Philosophical Convergences and Divergences

So far, my repeated argument for methodological separatism and against taking the formal unifications as substantial ones was the simple point that belief is not (infinitesi-mally close to) maximal probabilistic certainty. The point is telling, I find. Still, we should broaden our perspective and look at the philosophical issues involved in a com-parison of ranks and probabilities. We will find convergences and divergences, and we will thus better understand what would have to be achieved by a unification or reduc-tion and why there is as yet no account coming to terms with all the requirements.

What are these issues? There are, first, the three aspects relevant to any kind of internal state: the input, the internal processing, and the output – or in terms more pertinent to doxastic states: the *experiential*, the *inferential*, and the *behavioral* aspect. There are, secondly, three related aspects characteristic of doxastic states due to their representational nature; I call them the *truth*, the *reality*, and the *objectivity* aspects. These aspects are, I sense, distinct, though related; our doxastic states somehow aim at truth, by somehow being anchored in reality, and by therein claiming objectivity in some sense.

Let me take up these six issues more or less in this order, not as such, for this would take up far more than one section, but only with a view to comparing ranks and prob-abilities. Most of the issues will be discussed in great detail in Chapters 12–17, on the philosophical applications of ranking theory. Therefore this section will at the same time serve as a kind of preview on these chapters. Only the behavioral aspect will play no further role in this book; this is why I will deal with it in slightly more detail in this section and in the appendix 10.4.

The first aspect to consider is the *inferential* one, since it is the aspect that was at the center of our theoretical development. Recall my introductory remarks in Section 1.1 about the progress from a static view of inductive inference to a dynamic under-standing of the problem of induction; this is why our focus in Chapter 5 was on stating adequate conditionalization rules. Recall furthermore our generalization in Section 6.2 from deductive to inductive reasons and our investigation into the doxastic edifice in Chapter 7. Thereby, we gained a detailed account of how various beliefs hang together, how their degrees depend on one another, how association, in Hume's terms, works, and thus how inference works (even though I hardly used that word). At the same time I always emphasized that the ranking-theoretic account of these matters is essentially the same as the Bayesian account, and we have seen the reason for this in the previous section. So, even if belief and probability have to be kept separate, the theory of relevance, connectedness, and inference is structurally virtually the same for both. This is an important point of convergence of probabilistic and ranking-theoretic epistemology.

In fact, "important" will be seen to be a kind of understatement, when we look at the list of topics related to this inferential or relevance aspect. Confirmation and explanation are among these topics, and they are two of the few fundamental issues in the philosophy of science. Explanation is first and foremost (or no more than?) causal explanation, and hence I do not hesitate to subsume the topic of causation under the inferential part of epistemology as well; in fact, this was my original motive in Spohn (1983b) for developing ranking theory. The notion of law or lawlikeness, finally, is of the same kin. These issues require careful development, and they will receive it in Chapters 12–14. Basically, the parallel between probability and ranking theory will hold good throughout these issues; it will be the guiding line of our discussion. These fundamental and pervasive philosophical applications are indeed the main source of my enthusiasm for ranking theory.

To some extent we have already dealt with the *experiential* or *input* aspect, too, namely when introducing the various rules of conditionalization (3.8), (3.10), (5.25), (5.33), and (9.15). The rules differed in their understanding of evidence, and I explained the progress lying within these different understandings. However, all this does not yet amount to a systematic discussion of the nature of evidence; we will pursue this only in Chapter 16 on perception and the basis of belief. It is quite obvious, though, that the parallel between probabilistic and ranking-theoretic epistemology will be preserved in this aspect, too.

We have not at all dealt, though, with the *behavioral* or *output* aspect. Here we find a grave divergence, possibly to the detriment of ranking theory. Since the days of my dissertation (Spohn 1976/78) on decision theory, I admired probability theory for its smooth extension to a powerful behavioral theory, i.e., decision theory; probability and utility seemed to be made for each other, and actually are in their joint measurement, as invented by Savage (1954, chs. 3 and 5) (cf. Fishburn 1970, Ch. 14). So, when thinking of ranking theory I was painfully aware that comparable achievements were entirely out of sight for me. I could not think of an appropriate notion of utility and I was unable to construct a reasonable theory of integration or expectation in ranking-theoretic terms. Hence, the crucial notion of expected utility was doubly out of reach. I took comfort, then, from the other virtues of ranking theory.

There are two considerations, however, that brighten the picture. I am still not sure whether they put ranking theory on an equal footing with probability theory in this respect. Yet they open up the issue; they point to theoretical developments that deserve scrutiny and might be promising.

The first point is that quite a number of alternative representations of doxastic states have been invented in the last 30 years, all facing the same problem and succeeding at least formally. There is no point in reviewing here these efforts; let me only mention some of them. Economists were, as far as I know, the first to investigate the issue; they rediscovered the general Choquet theory of integration for defining expected utility for non-additive representations of degrees of belief; for early contributions see, e.g. Gilboa (1987), Schmeidler (1989), Jaffray (1989), or Sarin, Wakker (1992); a more recent one is Wakker (2005). The AI side followed; see, e.g., Dubois, Prade (1998), Brafman, Tennenholtz (2000), or Giang, Shenoy (2005). A kind of textbook presentation is offered by Walley (1991, Ch. 3). Halpern (2003, Ch. 5) gives a succinct recent survey, even defining a notion of expectation for his plausibility measures in the structuralistic spirit mentioned at the end of the previous section. His bibliographical notes on pp. 185−8 are particularly instructive. And most recently, almost encyclopedic information is provided by Wakker (2010).

However, none of the works referred to inquire into a ranking-theoretic version of decision theory − with the exception of Halpern (2003, Sect. 5.2.4) (after all, his plausibility measures are intended to generalize ranking functions as well); and his remarks are scant and not encouraging. All the more I should point to Giang, Shenoy (2000) who developed a specific constructive version of a ranking-theoretic decision theory. They take the standard axiomatic treatment of utility by Luce, Raiffa (1957, Sect. 2.5), and they show how to directly translate this treatment into ranking-theoretic terms. Conservative decision theorists might find this odd. However, since it belongs to a full picture of the potential of ranking theory, I will give a sketch of their treatment in the appendix 10.4 of this chapter.

There is a further important facet to the behavioral aspect, namely *linguistic behavior*. Listening and talking, reading and writing take half of our lives and are indeed unique to us humans. One might try to subsume linguistic behavior under a general theory of human behavior. In a sense, the first one to do so was Grice (1957), who explained linguistic behavior and thus linguistic meaning by a characteristic kind of practical syllogism. Grice's intentional semantics indeed deserved the attention it received. It was perfected by Lewis (1969), who gave an admirable analysis of conventions in game- and thus decision-theoretic terms and could then explain linguistic conventions as a specific kind of conventions of truthfulness and trust. In my view this is still the most advanced proposal for unifying truth-conditional semantics and a use theory of meaning.

It is clear, though, that these are sophisticated and debatable constructions for understanding linguistic behavior as a kind of intentional action. They must not blind us to the fact that there are a lot of laws specific to linguistic behavior, whatever their relation to a general theory of intentional action. The most primitive of these laws is the so-called disquotation principle: if a seriously and sincerely asserts "p" or assents to "p?", then a believes that p. This is, no doubt, our simplest and most powerful principle for the ascription of belief; all other methods for inferring beliefs or doxastic states are more indirect and burdened with uncertainty. The fundamentality of the principle is reflected in the large debate about its precise content, status, and significance (for a recent discussion and a strong argument that the disquotation principle is essentially to be qualified by a ceteris paribus clause, see Kemmerling 2006). There is no point here in studying the matter more carefully. The only observation I want to make is that the disquotation principle is phrased in terms of belief and not in terms of any other doxastic attitude. I have not seen any, say, probabilistic version of the principle, and there is apparently no good idea about how to state one. Belief is that doxastic attitude first and foremost accessible by serious and sincere speech; the notion of belief is, as it were, made for the disquotation principle.

If this is so, my methodological separatism might have a substantial core. If the behavioral bases for belief and probability fall apart, belief and probability themselves fall apart, at least initially. A supplementary unification must then be a matter of hypothesis and construction, which may or may not succeed. I am thereby also suggesting that a theory of linguistic behavior, a linguistic pragmatics, should be stated rather in ranking-theoretic than in probabilistic terms. This suggestion, however, is almost entirely unredeemed; there are as yet hardly any applications of ranking theory in the philosophy of language.

There are so far just the beginnings of an application. This is within the project of a decision-theoretic semantics developed by Merin (1997, 1999, 2005, 2006a,b). This approach to the recursive composition of meanings deviates starkly from the received view in semantics when, in its most radical version, it assumes a space of ultimate semantic values consisting not of truth values, but of utility and, as a special case, relevance values. The latter may be conceived in probabilistic terms or, as Merin (2006b, 2008) shows, in ranking-theoretic terms. This should be a testing ground for the explanatory potential of ranking theory and for the parallels between probability and ranking theory already noted under the inferential or relevance aspect.

Now to the other three aspects − those due to the representational nature of doxastic states − where we will again find a mixed balance. The first of them is the *truth* aspect, as I call it, and it is the one where ranking theory is clearly superior to probability theory. Ranks represent beliefs, and beliefs can be true or false, whereas subjective probabilities can be assessed in various ways, as bold or cautious, as well informed, as reasonable, etc., but never as true or false. Even if a subjective probability coincides with the relevant objective probability, we cannot call it true; only the belief in the objective probability could be. One might think of substitutes. Degrees of belief might conform to degrees of truthlikeness; however, it even seems to be unclear, after 30 years of inquiry into truthlikeness, whether truthlikeness behaves like probability (cf. Oddie 2001). Or one might, as Joyce (1998) does, replace accuracy (= truth) by gradational accuracy and then follow his interesting argument that degrees of belief confirming to the so-called norm of gradational accuracy must behave like probabilities. I am not going to discuss such substitutes. Clearly, though, they are substitutes, and their soundness can be established only by intricate argument. By contrast, nothing could be more straightforward than the truth or falsity of a belief.

This point is of the greatest significance. It seems that ranking theory is much better able to connect up with traditional epistemology. Already in Section 1.1, I mentioned the schism between the theory of knowledge and the theory of belief, where traditional epistemology is rather occupied with the former, and in Section 3.3 I mentioned that Bayesianism in turn forms a special camp within the theory of belief. Both divides led to a complete non-understanding between traditional and Bayesian epistemology, as manifested, for instance, by Plantinga (1993a, chs. 6–7), who despairs of finding insights in Bayesianism he can use. Ranking theory has the potential to overcome both divides. It shows that the theory of belief (in the broad sense only distinguishing it from the theory of knowledge) can be developed in an equally sophisticated way in other than probabilistic terms, and it purports to deal with the central notions of a theory of knowledge: with belief, truth, and justification.

Connectibility and actual connection are two different things, though. I do not want to explicitly discuss the notion of knowledge in this book. To do so in a responsible way would require more care than I have space for, and would perhaps lead us astray from my central intention. So, in particular, I will remain silent on Gettier's problem and the discussion it has triggered. Still, the Chapters 16–17 will, I believe, amply show how well traditional epistemological concerns can be addressed with the help of ranking theory (and Section 11.4 will critically comment on Keith Lehrer's account of knowledge by proving a kind of trivialization result).

Despite my great emphasis, one might object that the point I just made was superficial (even in a literal sense). What is believed according to a negative ranking function only depends on its core, the set of possibilities receiving rank 0. There are, however, many ranking functions having the same core, but differing, possibly wildly, in their distribution of ranks larger than 0 (i.e. below the surface) and hence in their inductive or learning strategies. Calling any such ranking function true if only its core is true appears shallow. This would make substantial sense only if such things like inductive strategies could also be called true. This appears doubtful. Did I not emphasize the fundamentally subjective character of learning strategies in Sections 4.2 and 6.4? The objection is well taken. Still, I would like to insist on my emphasis. Talk about beliefs is our direct entry ticket to truth talk, and we must indeed inquire how far it takes us. By contrast, probabilistic talk bars even that entry.

However, this does not exonerate me from attempting to overcome this objection; I will do so amply. In fact, the objection might be strengthened in two ways. First, one might say that the crucial aspect is less the truth aspect, but rather the *reality* aspect, the point that our doxastic states must not be free floating, but somehow be grounded in, or guided by, reality. As long as a doxastic state consists solely of beliefs, the anchor is provided by the truth of the beliefs; the reality aspect does not then go beyond the truth aspect. The objection above was that this anchoring of beliefs does not extend to ranks; the non-zero ranks seem indeed to be free-floating. However, the objection continues, subjective probabilities *are* grounded in reality, even if they cannot be called true; they are grounded in relative frequencies. Observed relative frequencies guide our subjective probabilities and are ever more approximated by them. Whatever the precise relation, the vague talk of anchoring or guidance definitely applies. Thus, the apparent advantage of ranking functions over subjective probabilities seems to flip over into a clear disadvantage.

For many years I saw the objection and did not know how to circumvent it. It proves nothing of what we have done so far to be wrong, but ranking theory should try to close the open flank. Well, let us ask: Is there a ranking-theoretic counterpart to the notion of relative frequency and its role for probability theory? This is an odd question; it is hard to see how at all to start thinking about it. For now, I will only say that I believe that in ranking theory this role is played by the notion of an *exception*. *Just as subjective probabilities are guided by observed relative frequencies, negative ranks are guided by the number of observed exceptions.* For the time being this remark must remain obscure and unexplained. However, when we study laws, ceteris paribus conditions, and causation in Chapters 12–14, and inquire into their objective counterparts in Chapter 15, this remark will unfold its great significance; see my conclusion in Section 15.7. So far, I only wanted to indicate that there are indeed deep problems and substantial answers, so that the balance of ranking theory with regard to the reality aspect is quite positive, after all.

The above objection might be further deepened into what I call the *objectivity* aspect. The question is not only how our subjective attitudes find guidance; it might also be whether they have some objective counterparts they should strive to match. Subjective probability has an objective counterpart, namely objective probability or chance. What this is is still highly contested, and how it relates to subjective probability is again highly contested. Are chances limiting relative frequencies? Or propensities? But what are the latter? Graded dispositions? What might these be? Perhaps chances are just projections of subjective probabilities? But what does projection mean here? Perhaps, more vaguely, they can only be understood through their relation to subjective probabilities? Which relation, though? Unending questions! Yet, at least, chances do not seem to be mere fictions. And if the world is governed by objective chances, our subjective probabilities can do no better at each given time than precisely mirror the chances at that time. (This is an informal version of the Principal Principle of Lewis 1980.) So, even though it is highly unclear what objectivity amounts to in the case of probabilities, there is a positive intuition and a constructive debate elaborating on it. (In Spohn (2010a) I presented my own, basically projectivistic attempt at coming to grips with the notion of objective probability and its problems.)

Hence, the question is: Is there an analogous objective counterpart to subjective ranking functions? Yes, there is. In Chapter 15, I will construe this question as the question that arose above, namely to what extent not only beliefs, but an entire ranking function – and thus the learning strategies embodied in it – can be called true or false. And I will argue, or rather constructively prove, that natural laws and, in particular, causal laws provide that counterpart. In this way, we can recapture objectivity after our decidedly subjectivistic turn in Section 4.2. I will not elaborate on the consequences of my reasoning for the notion of objective probability; this is not our topic. Still, I think there are lessons to be drawn. In any case, my reasoning will be very much in line with my (2010a) attempt to analyze objective probability; the analogy to probability theory will continue in a highly illuminating way.

This response to the objectivity issue will also be seen to provide a response to the issue of grounding ranks in reality. I said before that beliefs are grounded in reality, insofar as they are true or false. Chapter 8 has shown us in a precise way how ranks are grounded in, i.e., uniquely determined by, the dynamics of beliefs (in the most convincing version of Section 8.3–8.4, by the dynamics of contractions). When we put the two pieces together, we should also see how our learning strategies find their grounds in reality. Indeed they do: insofar their objective counterparts are actually true, they will lead us only from true to true beliefs. Again, this point will be elaborated in Chapter 15, which, as should be obvious by now, will carry a heavy philosophical burden.

So, as I stated earlier, the comparison on the various scores rather took the shape of a preview; the issues involved are too broad to admit of a short treatment. What are we to conclude from the preliminary discussion? I feel confirmed in my methodological separatism. Insofar as we have found substantial convergences between ranks and probabilities, they might well be preserved in formal unifications; but I cannot find much sense in thus pasting over the divergences.

In my view, the two most important issues that should, but will not be clarified further in this book are the following. One issue is whether or not the linguistic expression of beliefs is to be taken as something special. Maybe we get clearer about the form of our doxastic states only by further clarifying the epistemological role of language.

The other issue was indicated in my discussion of the reality aspect. Often we presume a deterministic picture of laws and regularities, which is usually disappointed by exceptions that provoke us to improve our picture. Equally often, we approach reality with a statistical attitude and start observing relative frequencies. What, however, determines which attitude we take? There seems to be a grey area in which rare events can be taken as exceptions or as events of very low frequency. When do we take them as this and when as that? I have no good idea and cannot recall having read anything illuminating on the issue. I suspect a good answer would help us to a more substantial understanding of the relation of beliefs and ranks on the one side and probabilities on the other. In this book, though, I do not intend to make progress on these two issues.

Hence, I will maintain my methodological separatism in this book, unsatisfactory as it may be. Whether it can be resolved is an open issue. And whether it would resolve then into reductionism or into a well-elaborated interactionism is an open issue, too.

10.4 Appendix on Ranking-Theoretic Decision Theory

Since I will not return to decision- and action-theoretic matters, let me use this appendix to represent the account of Giang, Shenoy (2000), which is, as far as I know, the best elaborated attempt to state a decision theory in ranking-theoretic terms. Giang and Shenoy simply take the axiomatic treatment of expected utility of Luce, Raiffa (1957, Sect. 2.5) and translate it into ranking theory; this works in a surprisingly straightforward way:

We start with a (finite) set $A = \{a_1, \ldots a_n\}$ of basic alternatives or outcomes. From these Luce and Raiffa build up simple and compound lotteries as probabilistic mixtures of basic alternatives or less complex lotteries. In a similar way we may define a *simple ranking lottery* $L = [r_1 a_1, \ldots r_n a_n]$ to be an n-tuple of pairs $\langle r_i, a_i \rangle$ such that $\min_{i \leq n} r_i = 0$, where $a_i \in A$ and $r_i \in \mathbf{N}^+ = \mathbf{N} \cup \{\infty\}$ is the negative rank expressing the disbelief with which the alternative a_i is expected. "Lottery" may sound inappropriate; however, a lottery is nothing but an issue with uncertain outcomes, and in principle the uncertainty can be modeled by any kind of uncertainty measure.

Let $\mathcal{L}_0$ be the set of simple lotteries. Define recursively that if $L_1, \ldots, L_k$ are lotteries, then $L = [r_1 L_1, \ldots, r_k L_k]$ is a lottery in turn, where $r_i \in \mathbf{N}^+$ and $\min_{i \leq k} r_i = 0$. Let $\mathcal{L} \supseteq \mathcal{L}_0$ be the set of all lotteries. Now the agent is supposed to have a preference ordering $\geq$ not only over the basic alternatives, but over all lotteries; as usual, $>$ is correspondingly to denote strict preference and $\approx$ indifference. This preference order should satisfy six assumptions or axioms, directly translatable from Luce, Raiffa (1957, Sect. 2.5):

(10.10) $\geq$ is s weak order, i.e., a complete and transitive relation over A
[*ordering of alternatives*].

(10.10) goes without saying; without it, we could not get started.

(10.11)   Let $L = [s_1 L_1, \ldots, s_m L_m]$ be a compound lottery, where each $L_i = [r_{i1} a_1, \ldots, r_{in} a_n]$ is a simple lottery. Define the reduced lottery $L' = [s_1' a_1, \ldots, s_n' a_n]$ such that $s_j' = \min_{i \leq m}(r_{ij} + s_i)$ for all $j = 1, \ldots, n$. Then $L' \approx L$
[*reduction of compound lotteries*].

This axiom is as familiar as it is crucial. In order to define the s_j' we have, of course, to apply the notion of a ranking mixture, as introduced already in (5.37).

The next axiom says that indifferent lotteries are substitutable:

(10.12)   If $L_i \approx L_i'$, then $[r_1 L_1, \ldots, r_k L_k] \approx [r_1 L_1, \ldots, r_{i-1} L_{i-1}, r_i L_i', r_{i+1} L_{i+1}, \ldots, r_k L_k]$
[*substitutability*].

Let us assume that $a_1 > a_2 \geq \ldots > a_n$, so that a_1 is the best and a_n the worst alternative. A *standard lottery* then is any lottery of the form $[r_1 a_1, r_n a_n] = [r_1 a_1, \infty a_2, \ldots, \infty a_{n-1}, r_n a_n]$. With this notion we can state:

(10.13)   For each alternative $a_i \in A$ there is a standard lottery L such that $a_i \approx L$
[*quasi-continuity*].

Of course, we should assume:

(10.14) $\geq$ is a transitive relation over the whole of $\mathcal{L} \cup A$ [*transitivity*].

The final assumption concerns the comparison of standard lotteries:

(10.15)   Let $L_1 = [r_1 a_1, r_n a_n]$ and $L_2 = [s_1 a_1, s_n a_n]$ be two standard lotteries. Then $L_1 \succeq L_2$ iff either (i) $r_1 = s_1 = 0$ and $r_n \geq s_n$, or (ii) $r_1 = 0 < s_1$, or (iii) $r_n = s_n = 0$ and $0 < r_1 \leq s_1$ [*qualitative monotonicity*].

This just says that among two lotteries between the same two alternatives, the preference goes to that lottery in which the better alternative is more credible or less disbelieved.

Within the probabilistic setting, these axioms allow us to assign a numerical utility to each alternative measured on an interval scale so that the expected utilities of simple and compound lotteries derived from those utilities agree with the preferences among them. Giang, Shenoy (2000) show how to duplicate this. They first prove (Lemma 1, p. 223):

(10.16)   If the preference relation $\succeq$ on $\mathcal{L} \cup A$ satisfies (10.10)–(10.15), then for each lottery or basic alternative there exists exactly one standard lottery indifferent to it.

Thus, we only need to assign utilities to the standard lotteries; this assignment may then directly be extended to all lotteries and basic alternatives via (10.16). In the probabilistic case, the best alternative gets utility 1, for instance, and the worst utility 0, so that the utility of a standard lottery is just the probability with which it results in the best alternative. Similarly, we might define here the qualitative utility $U^*([ra_1, sa_n])$ of a ranking standard lottery, as the degree of belief or two-sided rank with which the best alternative is expected according to the lottery. Thus:

(10.17) $U^*([ra_1, sa_n]) = s - r.$

Hence, $U^*(a_1) = \infty$ and $U^*(a_n) = -\infty$, and the utilities of the other alternatives are integers in between. One must not understand this in a probabilistic way (in which it makes no sense); it just artificially expresses that a_1 is the best and a_n the worst alternative. All genuine standard lotteries, and lotteries avoiding infinite negative ranks ∞, receive a finite utility.

U^* is not only a natural choice, it is the only choice, as Giang and Shenoy prove in their Theorem 1 (p. 224):

(10.18)   Suppose the preference relation $\succeq$ on $\mathcal{L} \cup A$ satisfies (10.10)–(10.15). Then there exists a (qualitative) utility function U from $\mathcal{L} \cup A$ into $\mathbf{Z}^+ = \mathbf{Z} \cup \{\infty, -\infty\}$ such that for all lotteries $L, L' \in \mathcal{L}$ $U(L') \geq U(L)$ iff $L' \succeq L$ and for any (simple or compound) lottery: $U([r_1 L_1, \ldots, r_n L_n]) = \min_{i \leq k} [r_i + \max (U(L_i), 0)] - \min_{i \leq k} [r_i + \max (-U(L_i), 0)]$. In fact, U^* as defined in (10.17) is the only utility function U of this kind.

(10.18) looks awkward, in the same way that two-sided ranks are computationally awkward. Giang and Shenoy choose a more elegant presentation in which the analogy to probabilistic expected utility is more perspicuous, however, at the price of assuming a more peculiar space of utilities instead of Z^+ (which I wanted to avoid introducing here).

This may suffice as a sketch of the account of Giang, Shenoy (2000); I am content with having acknowledged here this important theoretical option. Let me close with a few remarks.

First, it should be recalled that (10.18) is a sufficient basis for a full-blown decision theory. As is well known, even the most complex strategy choice reduces to the choice between compound lotteries (as long as strategies only prepare for experienced external circumstances). So, (10.18) indeed provides a powerful formal option.

Whether it also makes intuitive and philosophical sense needs to be carefully checked. Giang, Shenoy (2000) offer a detailed example. They also offer a comparative discussion, in the course of which they point out that their account might also be understood as an order-of-magnitude abstraction from non-standard decision theory, as already envisaged by Goldszmidt, Pearl (1992b, 1996) and Tan, Pearl (1994). I am pleased, though, that Giang and Shenoy did not choose this approach, but rather the axiomatic one conforming with my methodological separatism and parallelism.

My remark about the grey area of overlap between probability and ranking theory at the end of the previous section possibly extends to their decision theoretic expansions. There I said that there is a grey area in which rare events can be taken as exceptions or as events of very low frequency. It is well known that standard decision theory is plagued by various paradoxes that originated from the famous Allais paradox, and that arise from the certainty or near-certainty of prospects (cf., e.g., Kahneman, Tversky 1979). Psychologists report here an overestimation as well as an underestimation of small probabilities (or rare events) by agents. However, one might as well see such cases as sliding into different doxastic states, i.e. states that are instead captured by ranking theory and accompanied by a different decision rule such as the one sketched here. Maybe this observation throws a new light on those paradoxes.

Such examples and remarks may be promising. Still, ranking-theoretic decision theory should be much more severely tested in practice, though in an unprejudiced way. (The latter is not easy to satisfy since all practitioners are, I assume, theoretically biased to formal or informal decision theory as standardly conceived.) Only then we can know whether we are dealing here with an idle play or a serious and useful theoretical option. I for my part have no clear prediction as to the outcome of such a test; the issue seems thoroughly open.

11

Comparisons

The previous chapter was devoted to a large-scale comparison of ranking theory with probability theory, surely the most honorable standard of comparison. This resulted in significant claims, the truth of which awaits proof. Before entirely focusing on this proof and hence on the many philosophical applications of ranking theory, let me pause for another larger comparative chapter. There are so many further formal representations of doxastic states in the field, and we are well advised to locate ranking theory therein. I certainly cannot strive for completeness; the ramifications are too hard to survey. Still, I should say how ranking theory relates to a number of important approaches; this should be an informative enterprise on its own. I do this pairwise; this entails that this chapter will be a mixed bag, as indicated by the section headings. So, you may also skip it. On the other hand, I will use the occasion for various relevant observations mainly of a methodological kind that would not find a natural place elsewhere, e.g. on the nature of normative epistemology in Section 11.6, on the methodology of relating examples and theory in Section 11.3, on the deep methodological divide between Isaac Levi and me in Section 11.2, or on the interestingly variant approach of formal learning theory in Section 11.7. So, let me just take you along on my comparative round trip.

11.1 Predecessors: Rescher, Cohen, Shackle

In Section 1.2 I already boldly claimed to be a successor of Francis Bacon. Of course, this was a dramatic device, although Cohen's opposition "Baconian vs. Pascalian Probability" is well taken. In the main, the historic texts are so informal that it does not appear fruitful to search for clear predecessors. I should mention, though, Shafer (1978), who identifies early attempts by Jacob Bernoulli and Johann Heinrich Lambert at developing alternatives to what has now become standard probability theory. Still, for me the history of formal alternatives only starts in the 20th century; only then was the level of formalization sufficiently advanced to determinately discern distinct developments. In this section, I will focus on three accounts that clearly display essential ingredients of ranking theory – i.e., the Baconian structure of Section 5.2 – namely those of George Shackle, Nicholas Rescher, and Jonathan Cohen. This is their temporal order. However, I will discuss the work of Shackle last, since it naturally leads over to the next section on Isaac Levi. I should also emphasize that my narrow comparative purpose prevents me from doing justice to the wider goals and achievements of these authors.

I mentioned already in Section 4.4 that belief revision theory is guided by principles very similar to those of counterfactual logic as developed by Thomason, Stalnaker (1968), Stalnaker, Thomason (1970), and Lewis (1973), the latter having become established as the standard account. This was no accident; Gärdenfors' work, starting with his (1978), grew out of a disagreement with that account of counterfactuals. However, this development of counterfactuals was preceded by Rescher (1964), which was the first formal attempt, as far as I know, to come to grips with this recalcitrant topic. No wonder that we there find ideas similar to those in ranking theory. Let us briefly take a closer look.

In Chapter 4, Rescher (1964) reduces the problem of counterfactual conditionals to that of belief-contravening hypotheses, and this is what he wants to solve in his account of hypothetical reasoning. A belief corpus becomes inconsistent when enriched by a belief-contravening hypothesis. The problem, then, is how to reason from an inconsistent set of beliefs. Rescher solves the problem by grouping all the possible beliefs (or sentences of a certain formal language) into a series of modal categories $M_0 \subseteq \ldots \subseteq M_n$. Here, M_0 contains exactly the logical truths or maximal certainties; the accepted statements are found in the categories $M_0, \ldots, M_{n-1}$ (acceptance need not be deductively closed according to Rescher); and M_n contains all sentences of the given language. Hypothetical reasoning then proceeds from the largest modal category with which the belief-contravening hypothesis is compatible; that is Rescher's compatibility-restricted entailment described in his Chapter 6. So these modal categories closely correspond to the entrenchment orderings found later in belief revision theory, and are indeed directly translatable into our framework of positive ranking functions.

If β is a positive ranking function taking only finitely many values $0, x_1, \ldots, x_m$, ∞, then $\beta^{-1}(\infty), \beta^{-1}(\{x \mid x \geq x_m\}), \ldots, \beta^{-1}(\{x \mid x \geq x_1\}), \beta^{-1}(\{x \mid x \geq 0\})$ is just a family of modal categories $M_0, \ldots, M_n$ ($n = m + 2$) (if we neglect the distinction between propositions and sentences). And they obey the same laws. Rescher's procedure for generating modal categories, as described in (1964, pp. 49f.), makes them closed under conjunction; this is our law (5.10b) of conjunction for positive ranks. And he observes on p. 47 that all the negations of sentences in modal categories up to M_{n-1} must be in $M_n = \beta^{-1}(0)$, i.e., not believed; this is our law of negation for positive ranks. So, we clearly find here ranking theory as developed in Section 5.2.

Rescher used this formal apparatus in his coherence theory of truth (1973) and further developed it into an account of plausible reasoning in his (1976). In the latter book he elaborates on a variety of applications of his theory, still worthy of study and clearly displaying its wider relevance. However, Rescher's plausibility indexing, though numerical, remains intuitive; theoretically, only an ordinal scale is established. Moreover, there is no sign that Rescher considered something like conditional plausibility indices; this was foreign to his approach. Thus, he laid remarkable foundations, but remained stuck in their beginnings.

Similar remarks in effect apply to the work of Jonathan Cohen, essentially contained in his two important books (1970) and (1977). Rescher's and Cohen's starting points, however, are quite different. As mentioned, Rescher (1964) was first interested in counterfactuals and realized only later on in his (1976) that his theory applies to inductive inference. By contrast, Cohen's supreme goal was to study inductive reasoning. Moreover, whereas Rescher (1976, Ch. IV) was very brief in his comparison with probability, Cohen clearly aimed at developing a full-fledged alternative to probability theory, as reflected in his opposition of Pascalian probability and inductive support (as he called it in his 1970) or Baconian probability (a term he coined only in his 1980). In fact, his (1977) is an impressive document of dualism and even separatism; this work is, as far as I know, the first explicit and powerful articulation of the attitude I have taken here as well. Furthermore, whereas I will cover here most of the applications Rescher had in mind, Cohen (1977) extends his account to quite different fields not at all touched here, in particular to legal reasoning; these applications are further developed by Schum (2001) (for a broader textbook see Anderson et al. 2005, in particular chs. 9–10). So, Cohen's book still is a most valuable complement to the present work considerably widening its applications.

Yet Cohen's formal core, though it looks quite different, does not get any further than Rescher's. Cohen's functions of inductive support again correspond to my positive ranking functions. Cohen clearly endorsed the law (5.10b) of conjunction for positive ranks; see his (1970, pp. 21f. and p. 63). He also endorsed the law of negation, but he noticed its importance only in his (1977, pp. 177ff.), whereas in (1970) it is well hidden in Theorem 306 on p. 226. All this is a bit unperspicuous, since Cohen's presentation is formally burdened. He is somehow attached to the idea that the modal operators $\le^i$ expressing inductive support $\ge i$ behave like iterable S4-necessities, and he even intertwines his many modal operators with first-order predicate calculus.

There is apparent formal progress in Cohen's account. He is explicit on the relationality of inductive support: it is a two-place function relating evidence and hypothesis. Hence, one might expect him to present a genuine account of conditionality. He does not, however. His conditionals behave like strict implication, as is particularly obvious from his (1970, p. 219, def. 5), a feature conclusively criticized already by Lewis (1973a, Sect. 1.2–1.3). Moreover, the relationality is not a genuine one, since Cohen discusses only laws of inductive support with fixed evidence. There is one exception, though, the consequence principle, as he calls it in (1970, p. 62). Translated into my notation it says for a positive ranking function β that

(11.1) if $A \subseteq B$, then $\beta(C \mid A) \ge \beta(C \mid B)$,

which clearly must not hold in ranking theory. These remarks sufficiently indicate that the aspect so crucial for ranking functions is ill-conceived in Cohen's work. He again did not get beyond the formal foundations as presented in Section 5.2. I wonder whether Rescher's and Cohen's limitations are also due to the fact that their accounts were phrased in (counterparts to) positive ranking functions, in terms of which it is much harder to come up with an account of conditional ranks (as we could already observe in Section 5.3 when comparing (5.16) and (5.21)).

Their work was preceded by that of George Shackle (1949, 1961/69), an economist, whose concern was not counterfactuals or inductive inference, but rather decision in the face of uncertainty. One essential point of his theoretical development was the idea that there are two kinds of uncertainty, there are "distributional uncertainty variables" governed by probability, and there are "non-distributional uncertainty variables" to be accounted for in some other way, as possibilities having a degree of "potential surprise". These degrees indeed behave according to the basic Baconian structure. It is really surprising from how many different angles one can arrive at more or less the same result. Thus, Shackle also ends up with an antagonistic structure, reminiscent of my separatism, although his intention was to develop two special cases of a general theory.

Shackle recognizes the duality of positive and negative ranks, he is explicit that potential surprise expresses certainty of wrongness, i.e., disbelief, and that conversely there is certainty of rightness (1969, p. 74). Thus, his functions of potential surprise correspond to my negative ranking functions; and we explicitly find the law of negation, this is axiom (9) in his (1969, p. 81), and the law of disjunction for negative ranks, this is his axiom (4) and/or (6) in his (1969, p. 90).

It is most interesting to see how hard Shackle struggles with an appropriate law of conjunction for negative ranks. The first version of his axiom 7 (1969, p. 80) claims, translated into our notation, that

(11.2) $\kappa(A \cap B) = \max \{\kappa(A), \kappa(B)\}$.

He accepts the criticism this axiom has met, and changes it into a second version (1969, p. 83), which I would argue must be translated as

(11.3) $\kappa(B) = \max \{\kappa(A), \kappa(B \mid A)\}$

(and is hence no law of conjunction at all). He goes on to state that it would be fallacious to infer that

(11.4) $\kappa(A \cap B) = \min [\max \{\kappa(A), \kappa(B \mid A)\}, \max \{\kappa(B), \kappa(A \mid B)\}]$.

All this does not lead very far. However, he has more to say. In (1969, Ch. 24) he is remarkably modern in discussing "expectation of change of own expectations", that is, in our terms, he is engaging in auto-epistemology (cf. Section 9.3). I interpret his formula (i) on p. 199 as slightly deviating from the second version (11.3) of his axiom 7 in claiming that

(11.5) $\kappa(A \cap B) = \max \{\kappa(A), \kappa(B \mid A)\}$, which could count as a law of conjunction. It is remarkable that Shackle explicitly considers the notion of conditional degrees of potential surprise and gives it an important role; however, it remains informal and in fact is formally a primitive notion; (11.5) would be unsuitable for defining it. On pp. 204ff. Shackle even considers the equation

$$(11.6) \quad \kappa(A \cap B) = \kappa(B) + \kappa(B \mid A),$$

i.e., our law (5.16) of conjunction for negative ranks. But he rejects it. Let me fully quote the decisive passage in his (1969, p. 205):

Our rule [i.e. (11.5)] thus bounds a range of diverse conceivable rules of which the other extreme might perhaps consist in *adding* the degrees of potential surprise respectively assigned ... [as in (11.6)]. But this other extreme rule, opposite to ours, would surely be unrealistic, and much more so than our own rule. The boxer who wonders whether he can become champion of the world does not *add together* the prowess of this opponent and that to reckon the strength of the opposition. He may have many rivals, but it is the most formidable of these who measures, for him, the difficulty of attaining the championship. We claim three things for our rule: first, that of the two extremes of the range of diverse conceivable rules, it is the less remote from reality; secondly, that it has greater simplicity than any other rule; and thirdly, that this simplicity is not bought at too high a sacrifice of realism.

So, Shackle was on the verge of getting things right here, but decided to go the wrong way. The example of the boxer is not very clear; I would have said that, given that I will beat the strongest opponent, my degree of disbelief that I will beat the others is 0, and hence my degree of disbelief that I will beat the strongest is the same as that of beating all (also according to (11.6)). What is more realistic is hard to assess, anyway. I guess what really moves him is simplicity, which, for him, is governed by his ordinalist preconception; already in the preface (p. xv) he emphasizes that he has "avoided entirely the assumption that possibility, or potential surprise, can be treated as a cardinal variable". However, cardinality was the key in Section 5.3, and we were able to defend it in Chapter 8. All in all, it seems fair to say that Shackle's struggle with conjunction and conditional degrees of potential surprise has not led to a clear result, so that in the end he did not really get beyond the basic Baconian structure in Section 5.2. Still, Isaac Levi is right in emphasizing the importance of Shackle's work, which was remarkably early and clear-sighted, and in several respects more sophisticated than that of other predecessors.

11.2 Isaac Levi's Incomparable Epistemological Picture

Isaac Levi was indeed the first to realize the importance of Shackle's functions of potential surprise for epistemology in general, and through his insistence they have become a familiar part of the philosophical discussion. Therefore it seems appropriate for my comparisons to turn now to Levi's work. Thereby, I also roughly preserve the temporal order of my comparisons. Later comparisons come later.

Beware, though! On the one hand, Levi does not formally develop Shackle's functions. Hence my criticism of the previous section will persist. On the other hand, he integrates them into a proliferating epistemological edifice that is entirely of his own making; they play there quite a different role. For instance, Levi is free of the ordinalist reservations so crucial for Shackle. More importantly, Levi has no use for Shackle's distinction between distributional and non-distributional uncertainty variables. Therefore, the functions of potential surprise take on a more general meaning in Levi's work, and do not support separatistic tendencies as they did for Shackle.

Indeed, for me Isaac Levi is the champion of a dualist, but interactionist picture of belief and probability, as opposed to my separatism. This was the guiding thread of my comparative paper, Spohn (2006). However, this is not our only and perhaps not our most important difference. Possibly the best summary is to say that we disagree on nearly every issue of strategy and detail, even though we are moved by the same interests and concerns. I find this most disturbing.

Therefore I will not start an argument in this section; each attempt to do so would be caught in the proliferating differences. I will rather confine myself to explaining the differences in a descriptive mood as well as I can. Ultimately, the argument is the whole book; this would be the appropriate standard of comparison. What to conclude according to this standard, though, I will leave to the reader.

In order to explain the differences let me briefly sketch the basics of Levi's epistemology without attempting to do justice to the richness of his work. Levi's epistemology has indeed been highly consistent over the decades. Many essential elements were laid out already in Levi (1967a) and are still found today. A lot has been added since, but only a few errors, in his view, had to be corrected. Here, I mainly refer to Levi (2004) that, among other things, beautifully summarizes large parts of his view and its development.

The first cornerstone of Levi's epistemology is his notion of *full belief*. Full beliefs are free of any doubt; this is Levi's Peircean heritage. They are not in need of justification; they rather form the current base for further inquiry. They are, as Levi often says, maximally and equally certain or infallible (from their own point of view). Still, they are corrigible, they can change, the changes need to be justified, and Levi pays a great deal of attention to this justification. This is Levi's peculiar combination of infallibilism and corrigibilism.

Before turning to the dynamics of full beliefs, the first thing to note is the place they assign to subjective or, to use Levi's term, *credal probabilities*. Full beliefs exclude certain possibilities and leave open others: they thus provide a standard of possibility, *serious possibility*, as Levi calls it. Serious possibilities define the space of further inquiry, which at the same time is the space of probability; only serious possibilities (and sets of them) genuinely receive credal probabilities. Full beliefs thus have probability 1, not in the sense that I would bet my life on them, but by default, as it were, since they define the frame of probabilistic judgment. The possibilities excluded by the full beliefs have no proper probabilistic role since they are bound to have probability 0. In this way, belief and probability have different roles not reducible to each other. Therein lies Levi's dualism (in the sense of Section 10.1).

In order to see how this dualism develops into interactionism we have to look at Levi's sophisticated ideas about how to change full beliefs and how to correlatively change credal probabilities. There are two basic ways in which full beliefs change: expansion and contraction. Levi provides detailed justificatory accounts for them, while all other changes can only be indirectly justified by getting decomposed into such basic moves. The decomposition is also required because the two basic kinds of change are guided by entirely different principles. Let us therefore look at them separately.

Expansion of full belief is an epistemic decision problem, according to Levi. On the one hand, one seeks valuable information; on the other hand, one wants to avoid error. Thus, one's acceptance is torn between stronger and weaker propositions – in my terminology. Levi prefers to talk of conjectures, hypotheses, etc., since in his analysis doxastic attitudes are relations to sentences and sets of sentences. Let us stick here, though, to our convention adapted in Chapter 2. Levi solves the decision problem posed by acceptance by determining the *expected epistemic utility $EV^*(h)$* of each proposition or hypothesis h. The simplest expression he derives for EV^* (cf. Levi 2004, p. 86) is:

(11.7) $EV^*(h) = Q(h) - q \cdot M(h)$.

Here, $Q(h)$ is the subject's *credal probability* for h. M is the subject's *informational value determining function* (the *undamped* version, to be precise), which, Levi argues, behaves like mathematical probability, although its interpretation must be sharply distinguished from that of credal probability. h's informational value itself is then given by $1 - M(h)$; thus, the stronger a proposition, the greater its informational value. q, finally, is an *index of boldness* between 0 and 1.

This label finds its explanation in Levi's *decision rule*. He argues that one should reject all and only those strongest propositions or serious possibilities that are not yet excluded by one's full beliefs and that carry negative expected epistemic utility. Equivalently, one should accept precisely the negations of these propositions (and their logical consequences). Hence, the greater one's q, the bolder one is in rejecting (and accepting) propositions.

Levi's interactionism is now evident. On the one hand, full beliefs delimit a space for subjective or credal probabilities; on the other hand, credal probabilities are needed to solve the decision problem of expanding full beliefs. In this way, belief and probability depend on each other, though neither reduces to the other.

Now we can also understand how Shackle's functions of potential surprise find a place in Levi's picture. Instead of fixing the index of boldness in advance, one may as well consider the maximal index at which the proposition h is still unrejected and define it as its *degree of unrejectibility*

(11.8) $q(h) = \max \{q \mid EV^*(h) \geq 0\} = \min \{Q(h) \mathbin{/} M(h), 1\}$.

Thus, $q(h) = 1$, if h is not rejected for any q. The more easily a proposition is rejected, the more surprising it is to learn that it obtains; hence, the *degree of potential surprise* of h may be defined as $d(h) = 1 - q(h)$. And the more easily a proposition is accepted, the more plausible it is; hence, the dual notion of the *degree of plausibility* of h may be defined as $b(h) = d(\text{non-}h)$.

These definitions entail (cf. Levi 2004, p. 90): All and only propositions excluded by full beliefs are maximally surprising, and all and only full beliefs themselves are maximally plausible. If h is surprising or plausible to some positive degree, non-h is not; that is, if $d(h) > 0$, then $d(\text{non-}h) = 0$, and if $b(h) > 0$, then $b(\text{non-}h) = 0$. The potential surprise of a disjunction is the minimum of the surprise values of its disjuncts; that is, $d(g \text{ or } h) = \min \{d(g), d(h)\}$. And the plausibility of a conjunction is the minimum of the plausibility degrees of the conjuncts; that is, $b(g \text{ and } h) = \min \{b(g), b(h)\}$. In other words, d and b are negative and positive ranking functions; they satisfy the Baconian structure as described in Section 5.2 and as proposed by Shackle (1949).

There is no saying whether credal probabilities or degrees of plausibility have more claims on being called degrees of belief. They coexist, they are different, but they are related (via the presented chain of definitions). Credal probabilities are relevant for assessing risk and for determining expected values or utilities in practical and epistemic decision making. By contrast, the set of propositions h with $b(h) > x$ is consistent and deductively closed for each $x > 0$, just like the set of full beliefs; hence, degrees of plausibility are suited for assessing changes of full beliefs through expansion.

How, finally, do probabilities get rearranged after expansion? Here, Levi always adhered firmly to simple conditionalization (3.8). This may suffice as a very brief description of Levi's account of expansion.

Contraction is also an epistemic decision problem, according to Levi, but quite a different one. There is no truth guarantee for our full beliefs; they must be conceived as corrigible. Sometimes we indeed find reasons to give up some of them. How? In contrast to expansion this problem is one-dimensional, as it were; there is no risk of error in contraction, since one thereby acquires no beliefs and hence no false beliefs. Therefore, the only parameter guiding contraction is informational value. One is losing information in contraction, and this loss needs to be minimized.

Thus put, contraction appears much simpler than expansion. There is a complication, though. At first, one might think one can apply the same informational value determining function M that was used in expansion. However, if contraction were required to minimize informational loss in this sense, the resulting contraction behavior would be heavily defective. This is the starting point of quite an involved discussion (cf. Levi 2004, Ch. 4) in which Levi arrives at the conclusion that informational loss must be measured not by the undamped version of M mentioned above and behaving like mathematical probability, but rather by a *damped* version M^* (version 2, to be precise).

It is useful to briefly summarize Levi's results in our terms: Let the proposition $C \subseteq W$ be the core corresponding to some prior belief set or corpus K. A contraction is then just any movement from the prior core C to some larger core $C \cup D$ (entailing less beliefs); and what is called a *maxichoice contraction* of a belief set is here represented by $C \cup \{w\}$ for any $w \in \bar{C}$; these are minimal contractions amounting to giving up only the belief in $\overline{\{w\}}$ and all stronger beliefs. Each maxichoice contraction incurs an informational loss which, for each $w \in \bar{C}$, may be measured by some real $M^*(w) > 0$; and we may complete M^* by defining $M^*(w) = 0$ for $w \in C$ (in this case moving from C to $C \cup \{w\}$ is a vacuous contraction and thus no loss at all). How, then, should we extend M^* to all contractions? If M^* were undamped, informational loss would be additive. The rule of minimizing informational loss would then entail that, if forced to a contraction, we would have to choose a maxichoice contraction. This is universally agreed to be intuitively and theoretically absurd. Hence, Levi (2004, p. 141) explicitly extends M^* to arbitrary propositions D by stipulating $M^*(D) = \max \{M^*(w) \mid w \in D\}$ to be the damped informational loss incurred by moving from C to $C \cup D$.

Now, giving up the belief in some proposition A, i.e. contracting by A, means moving to some $C \cup D$ such that $C \cup D \not\subseteq A$, i.e. $D \cap \bar{A} \neq \varnothing$. The *minimal damped informational loss* incurred by a contraction by A may then be defined as

(11.9) $\beta(A) = \kappa(\bar{A}) = \min \{M^*(D) \mid D \cap \bar{A} \neq \varnothing) = \min \{M^*(w) \mid w \in \bar{A}\}.$

Obviously, κ and β thus defined are negative and positive ranking functions. With respect to damped informational loss the rule of minimizing informational loss does not yet fully determine contraction. A maxichoice contraction to $C \cup \{w\}$ with $w \in \bar{A}$ and $M^*(w) = \kappa(\bar{A})$ would conform to the rule, but would be inadequate. Standard AGM or ranking contraction $\div_\kappa(A) = C \cup \{w \in \bar{A} \mid M^*(w) = \kappa(\bar{A})\}$ (5.41) does so as well. Levi disagrees and argues that his rule of ties entails the most far-reaching contraction called *mild*, i.e., choosing the largest possible D minimizing damped informational loss. That is, he defines (cf. Levi 2004, pp. 142f.):

(11.10) $\div_{mild}(A) = C \cup \{w \in W \mid M^*(w) \leq \kappa(\bar{A})\}.$

The various contractions are interdefinable; we could clearly describe Levi's mild contraction as a package contraction, by the means of Section 5.7, or as an iterated contraction, and he could describe AGM contraction as a mild contraction followed by a suitable expansion. Hence, we may perhaps abstain from discussing Levi's argument in favor of mild contraction in (2004, Sect. 4.6–4.8 and Ch. 5), although I will cast doubt on Levi's intuitive motivation in the next section. My point in explaining all this is rather that we also find the Baconian structure of Section 5.2 on the contraction side of Levi's theory, namely in the form of minimal damped informational loss. This observation will be relevant in my subsequent comparison.

One must take great care, though, not to confuse the various uses of this structure by Levi. Minimal damped informational losses are not degrees of plausibility, since plausibility, like probability, is a doxastic attitude relating to serious possibilities, whereas damped informational loss is a valuational attitude relating to the possibilities excluded by full beliefs. One might be tempted to interpret minimal damped informational losses as prior degrees of plausibility, i.e., as degrees of plausibility one would have in a hypothetical state before acquiring substantial full beliefs. But this cannot fit, either. Prior plausibility, if it makes sense, would be a doxastic attitude governing expansion of that hypothetical ignorant state by bringing informational value and risk of error into a balance, but minimal damped informational loss cannot be so described.

A final point needs to be mentioned concerning contraction. Contraction widens the standard of serious possibility. So the question arises, what might be the probabilistic assessment of the extended serious possibilities? Prior credal probability provides no answer, since the serious possibilities previously excluded have prior probability 0. Here, Levi's theory furnishes a further element called *confirmational commitments*, which need not, but usually do form a stable epistemic ingredient that also provides belief-contravening conditional probabilities and thus a way of rearranging credal probabilities after contraction (cf., e.g., Levi 1984, pp. 148f. and p. 201). This underscores Levi's profound interactionism.

So much for contraction. Of course, there are other changes of full beliefs. In particular, there are revisions in which one is forced to accept a hitherto unexpected, i.e., rejected proposition h, and residual shifts, as Levi calls them, where one replaces a formerly accepted proposition f by another proposition g. Any such change can and, as Levi insists, *must* be decomposed into a chain of contractions and expansions. As explained, contractions and expansions have their own, but quite different justifications that can be applied successively, but cannot be mixed to yield direct justifications of other forms of changes. In particular, Levi vigorously rejects the apparent symmetry between the Levi identity (4.22) and the Harper identity (4.23) as displayed in Section 4.4. Only the Levi identity yields a proper analysis.

This may suffice as a sketch of Levi's epistemological picture. The richness of detail in his many writings simply had to be neglected here. Still, I hope that I have fairly represented some basic moves. Since I only strive for a descriptive comparison, we need not go more fully into the strengths, justifications, and consequences of his position.

A superficial look already displays many differences. Levi takes expansions and contractions as the only basic epistemic moves and conceives all other epistemic changes as composed by basic moves, whereas I do not make the distinction between basic and composite moves and rather propose $A{\rightarrow}n$-conditionalization (5.24) or, more generally, $\mathcal{E}{\rightarrow}\varepsilon$-conditionalization (5.32) as a general description of epistemic changes. I thereby comply with standard AGM single contraction, whereas Levi prefers his mild contraction obeying somewhat different laws. Levi is also dedicated to epistemic decision theory, as his account of expansion shows particularly well, whereas I do not find much sense in conceiving doxastic changes in such terms. My emphasis then was on iterated changes; this is why I insisted on the principle of categorical matching (cf. Section 5.1). By contrast, Levi (2004, Sect. 5.11) argues against this principle. The point is that he does not have the difficulties with iteration I found in AGM belief revision theory. For him, informational value is the only parameter governing contraction and an essential parameter governing expansion, and he assumes it to be stable, at least within a given inquiry. Thus, he does not need a dynamic account of it, and can still claim to explain iterated belief change. And so we may continue.

However, this list already indicates that our epistemological pictures are at cross purposes in nearly every respect. Given the similarity of our epistemological concerns, I am really stunned by this fact. It also explains why it does not make much sense to argue about details as just listed; each such argument would inevitably come back to the incomparable frameworks in the end. Hence, let me rather continue by pointing out more important strategic differences.

The strategic difference that was perhaps most perspicuous in my exposition in Spohn (2006) concerns our views about the relation between belief and probability. We are both dualists, but Levi is an interactionist, whereas I outed myself as a separatist in Section 10.1. In (2006, p. 348) Levi says he still finds it difficult to take me at my word. The point, of course, as I more fully explained in Chapter 10, is that I am not a committed separatist, but only a methodological one. I have explained that my preferred way of realizing reductionism proceeds in terms of ranked probabilities in Section 10.2, but also why I find this reduction formally satisfying and yet not substantially illuminating.

Still, I find interactionism, of which Levi's is the best elaborated version, less attractive. The point is that Levi's interactionism entirely depends on his notion of full belief, that has no place at all in my picture. As Levi (2006, pp. 347f.) rightly notes, this is our most basic difference. (And he rightly observes there that this difference recalls an old dispute he had with Richard Jeffrey (cf. Jeffrey 1965/1983, Ch. 11, and Levi 1967), in which I take Jeffrey's side even though I am not a radical probabilist.) Let me explain:

I have described the central role of full beliefs in Levi's picture, which, for him, are both infallible (from their point of view) and corrigible. Levi finds only one way to locate his full beliefs in my picture and thus to adapt mine to his, namely by suggesting that I am a Parmenidean skeptic (cf. Levi 2004, p. 19 and p. 94). This is a person, according to Levi, for whom the only standard of serious possibility is logical or conceptual possibility and for whom full beliefs are restricted to unrevisably a priori beliefs. Such a person is called a skeptic because the full beliefs of a subject precisely set the undoubted frame within which she is conducting her inquiries. Thus, if the Parmenidean has such a narrow notion of full belief, she has a very large notion of doubt.

According to this suggestion, what I call beliefs can be understood as expansions of Parmenidean full beliefs. This seems fitting, since Levi's picture firmly adopts Shackle's functions of potential surprise in determining how to expand one's beliefs according to the chosen index of boldness and since my ranking functions behave in the same way. Indeed, I am a particularly bold expander according to Levi, since by holding that the minimal negative rank 1 already amounts to disbelief I am in fact applying the maximal index of boldness. By contrast, I have no contractions in Levi's sense since my Parmenidean full beliefs are in fact incorrigible.

Relatively speaking, this might be the best embedding of my views into Levi's, and it explains his persistent tendency to equate my negative ranking functions with

Shackle's functions of potential surprise. Still, it would be better to just acknowledge that the embedding does not fit at all. I grant that I have the notion of unrevisably a priori beliefs; A is such a belief only if $\kappa(\bar{A}) = \infty$ (cf. 6.20 and 6.21). And unlike Levi (2004, p. 11) I do not see this notion as a remnant of a "degenerating research program"; we will rather observe in Chapters 13, 16, and in particular 17 how many substantial things can be said about apriority. However, I would not call a priori beliefs full beliefs. There is simply no place and no use for full beliefs in my picture.

I have beliefs. What they are is vague. I admitted this in the discussion of (5.14), and tried to account for it by defining belief in A not as $\kappa(\bar{A}) > 0$, as I did initially, but as $\kappa(\bar{A}) > z$ for some z. Therefore, by the way, I need not be attached to being maximally bold in Levi's sense. The theory, however, is unaffected by the choice of z (which is why I returned to setting $z = 0$). (Levi 2004, p. 95, footnote, realizes the same point.) In Spohn (1988) I called my subject matter plain beliefs, thus trying to emphasize that I am plainly talking about belief. However, this invited the misunderstanding that I was talking about yet another specific kind of belief. This is not so. Thus, I have long since avoided all qualifying adjectives; they are all artificial and theory-induced. However my beliefs are fixed, then, I do not see why I should treat them as doubtful unless they are full. I need not justify them unless questioned, but I can justify them even if unquestioned. That is, each proposition receives the rank that is assigned to it according to my balance of reasons (as explained in Section 6.3), and if this balance assigns it a positive two-sided rank (or a rank $\geq z$), then I believe it and do not doubt it.

Still, my beliefs are revisable or corrigible. Doubts can be raised, though not by mere thinking; like Levi's account, ranking theory is not a computational theory about thought processes, but about doxastic states reached, as it were, when all possible thinking is done (more on this in Sections 11.5–11.6). Doubts can be raised through experience, e.g., through objections by interlocutors; conditionalization will then establish a new balance of reason; and so some prior beliefs may fall victim to doubt.

Hence, as I have always emphasized, ranking functions are good for treating both expansion and contraction. They are formally similar to Shackle's functions of potential surprise, though conditional ranks go formally beyond them. But they are not confined to the expansion role to which the latter are confined in Levi's theory. Indeed, they are not so confined in Levi's own picture; there they also play the contraction role under the label "minimal damped informational loss".

All this makes clear that Levi's picture and mine differ from the ground up. In the background, I see a difference of a methodological kind in our philosophical attitudes towards epistemology that might make our differences more intelligible. It is that Levi approaches epistemology in a *constructive-synthetic* way, I think, whereas my view is rather *structural-analytic*. (Isaac Levi shared this characterization in oral communication.) Levi has very detailed and instructive stories to tell about our epistemological machinery. There are small and big gears. A central gear is informational value, for instance, but it is driven by several smaller cogs, such as explanatory power, simplicity, generality, and special interests. And Levi carefully explains how the gears interact in order to produce judgment, belief, and probability.

I refuse to tell such stories, since I am not so clear about all these gears. What is simplicity? What is explanatory power? What is overall informational value? What is the measurement theory for all these magnitudes? None, I suppose. As long as such questions remain open, I am in doubt as to whether the machinery is a real, hypothetical, or imaginary one. Recall my remark after (4.10) in Section 4.2 that I do not want to appeal to these kinds of notions and that this would presumably be the most important dialectic move in this book. We saw the force of this remark in Section 6.4 when I reversed the relation between reasons and the belief changes driven by them. Here we see it again. In Chapter 4, I then decided to tell my structural story. Whatever the inner epistemological workings, the resulting structure must be such and such, in fact a ranking structure, if my arguments in Chapter 5 and the measurement theory for ranks in Chapter 8 hold good. My further procedure will continue to be analytic and thus the reverse of Levi's. Having established the structure my hope is to be able to analyze and lay bare parts of this machinery: lawhood, explanation, causal inference, justification, etc. (even though informational value will not belong to my categories). This will occupy us throughout the Chapters 12, 14, and 16.

One aspect of this methodological difference is, as we could see, our divergence over the epistemological locus of ranking functions. The inner workings of expansion and contraction are entirely different for Levi; hence, he can see only one place for ranking functions. By contrast, there is the same structure on both sides, for me as well as for Levi (see 11.9); hence, I apply ranking functions across the board.

Another aspect concerns justification in epistemology. Levi is modest in a way; he wants to justify only belief changes and not current beliefs themselves. This is part of his deeply rooted pragmatism. Concerning the justification of belief changes he is less modest; his inquiry into the epistemological machinery is to uncover that justification in detail. I do not say that Levi is pursuing an objectivist notion of justification here. He continuously emphasizes the relativity of his parameters to the inquiry at hand. His index of boldness is entirely subjective; there is no prescription of a wise choice. There is no need for people to agree on basically subjective informational value. And so on. Still, I sense the remnants of an objectivist notion in Levi's work. Even if a lot of wheels are (partially) subjective, his hope is to identify at least some objectively valid components (such as explanatory power) and at least some objectively valid connections between the components (such as expansion according to expected epistemic value).

This is not my view. I am immodest in thinking that I can also justify my current beliefs. Perhaps there is no real difference here, since all of my current beliefs except the a priori ones, result from belief changes. However, I apply a thoroughly subjectivist notion of justification; this is a consequence of the just mentioned dialectic move in Section 4.2 (see also Section 16.1). Like Levi, I take it as axiomatic for rational subjects that it is reasons that drive their doxastic dynamics. But I do not pretend to have an antecedent notion of a reason to be plugged in here. Rather, as I have explained in Section 6.4, I read the axiom inversely: for rational subjects reasons are whatever drives their doxastic dynamics. This is why I also reverse the methodological order. Of course, I will not forever stay on the subjectivist side. The investigation of objectification in Chapter 15 will explore the extent to which conditional belief and reason can be objectified, and the investigation of apriority in Chapter 17 will reveal that rationality enforces more substantial structure than hitherto found.

Clearly, both the structural-analytic and the constructive-synthetic method are legitimate and fruitful. Ideally, they are complementary and illuminate the same matter from opposite sides. Here, the ideal apparently does not obtain. I presently do not see how the tunnels Isaac Levi and I are drilling into the epistemological mountain could meet in the middle.

11.3 AGM Belief Revision Theory Once More, and the Problematic Force of Examples

The development of belief revision theory was driven to a considerable extent by the ongoing debate between Isaac Levi and the AGM theorists, a little bit of which surfaced when we observed that Levi's mild contraction (11.10) diverged from AGM contraction (4.19). Therefore it might be fitting to return to belief revision theory one last time.

When one compares the first book-length presentation of the AGM theory by Gärdenfors (1988) to such comprehensive overviews as Hansson (1999) and Rott (2001), one immediately sees how the straightforward beginnings have enormously ramified within a few years. I do not find these ramifications easy to follow: a lot of variant axioms, a lot of deductive connections among them, a lot of representation results, a lot of uncertain intuitions, and a lot of informal principles that turn out to be indeterminate and interpretable – all in all, many formal possibilities and much indecision. Perhaps this is an inevitable stage of maturation.

Ranking theory is full of implicit comments on these developments, but it would be too tedious to elaborate on them in explicit comparisons. I have sufficiently explained both my criticism of the AGM theory and the very large extent to which ranking theory builds on it. Still, there remains a dialectic burden I should relieve. On the one hand, ranking theory endorsed all the postulates of AGM revision and AGM contraction. On the other hand, the ramifications of belief revision theory were strongly motivated by intuitive doubts about these postulates. Hence, ranking theory apparently faces the very same doubts. I thus feel urged to do something to reduce the tension between the AGM theory and its critics.

Again, completeness is not my aim here. My discussion will be confined to a few important examples. This will suffice to show diverse ways in which the application of theory is indeterminate and indirect. The suggestion will be, of course, that there will always be enough room to reconcile ranking theory with specific examples.

Let me start with the postulate (K÷7), *intersection* (cf. 4.19), which stated (in the language of belief revision theory, with φ and ψ being sentences and K a deductively closed belief set):

(11.11) $(K \div \varphi) \cap (K \div \psi) \subseteq K \div (\varphi \wedge \psi)$.

Sven Ove Hansson is very active in producing (counter-)examples. In (1999, p. 79) he tells a story undermining the plausibility of (11.11):

I believe that Accra is a national capital (φ). I also believe that Bangui is a national capital (ψ) As a (logical) consequence of this, I also believe that either Accra or Bangui is a national capital (φ ∨ ψ).
Case 1: 'Give the name of an African capital' says my geography teacher.
'Accra' I say, confidently.
The teacher looks angrily at me without saying a word. I lose my belief in φ. However, I still retain my belief in ψ, and consequently in φ ∨ ψ.
Case 2: I answer 'Bangui' to the same question. The teacher gives me the same wordless response. In this case, I lose my belief in ψ, but I retain my belief in φ and consequently my belief in φ ∨ ψ.
Case 3: 'Give the names of two African capitals' says my geography teacher.
'Accra and Bangui' I say, confidently.
The teacher looks angrily at me without saying a word. I lose confidence in my answer, that is, I lose my belief in φ ∧ ψ. Since my beliefs in φ and in ψ were equally strong, I cannot choose between them, so I lose both of them.
After this, I no longer believe in φ ∨ ψ.

At first blush, Hansson's response to case 3 sounds plausible. I suspect, however, this is so because the teacher's angry look is interpreted as, respectively, φ and ψ being false. So, if case 1 is actually a revision by ¬φ, case 2 a revision by ¬ψ, and case 3 a revision by ¬φ ∧ ¬ψ, Hansson's intuitions concerning the retention of φ ∨ ψ come out right.

Still, let us assume that the teacher's angry look makes me only insecure so that we are dealing only with contractions. I think case 3 is quite ambiguous then. The look might make me uncertain about the whole of my answer. So I contract by φ ∧ ψ, thus give up φ as well as ψ (because I am indifferent between them) and retain φ ∨ ψ. It is more plausible, though, that the look makes me uncertain about both parts of my answer. So I contract by φ and by ψ. Either this is understood as a package contraction (5.59), in which case I still retain φ ∨ ψ. Or I contract by φ and *then* by ψ (or the other way around) – thus performing an iterated contraction – in which case φ ∨ ψ is in the end no longer believed (at least if φ and ψ are doxastically independent in the ranking-theoretic sense). This, I take it, is a doubly sufficient explanation of Hansson's intuition without the need to reject (11.11).

After this warm-up let me turn to the most contested of all contraction postulates, *recovery* (K÷5) (cf. 4.19), which asserts (in the language of belief sets):

(11.12) $K \subseteq Cn((K \div \varphi) \cup \{\varphi\})$

Hansson (1999, p. 73) presents the following example: Suppose I am convinced that George is a murderer (= ψ) and hence that George is a criminal (= φ); thus φ, ψ ∈ K. Now I hear the district attorney stating: "We have no evidence whatsoever that George is a criminal." I need not infer that George is innocent, but certainly I contract by φ and thus also lose the belief that ψ. Next, I learn that George has been arrested by the police (perhaps because of some minor crime). So, I accept that George is a criminal, after all, i.e., I expand by φ. Recovery then requires that ψ ∈ $Cn((K ÷ φ) ∪ \{φ\})$, i.e., that I also return to my belief that George is a murderer. I can do so only because I must have retained the belief in φ → ψ while giving up the belief in φ and thus in ψ. But this seems absurd, and hence we face a clear counter-example against (11.12).

This argument is indeed impressive – but not unassailable. First of all, it seems clear that the conditionalization rules (5.25) and (5.33) of ranking theory are extremely flexible; any doxastic movement you might want to describe can be described with them. The only issue is whether the description is natural. However, that is the second point: what is natural is quite unclear. Is the example really intended as a core example of contraction theory, such that one must find a characterization of contraction that directly fits the example? Or may we give more indirect accounts? Do we need, and would we approve of, various axiomatizations of contraction operations, each fitting at least one plausible example? There are no clear rules for this kind of discussion, and as long as this is so the relation between theory and application does not allow any definite conclusions.

Let us look closely at the example. Makinson (1997b) observes (with reference to the so-called filtering condition of Fuhrmann 1991, p. 184) that I believe φ (that George is a criminal) *only because* I believe ψ (that George is a murderer). Hence I believe φ → ψ, too, *only because* I believe ψ, so that by giving up φ and hence ψ the belief in φ → ψ should disappear as well. This implicit appeal to justificatory relations captures our intuition well and might explain the violation of recovery (11.12) (though the "only because" receives no further explication). However, I find the conclusion of Makinson (1997b, p. 478) not fully intelligible (in which "theory" means "belief corpus"):

Examples such as those above . . . show that even when a theory is taken as closed under consequence, recovery is still an inappropriate condition for the operation of contraction when the theory is seen as comprising not only statements but also a relation or other structural element indicating lines of justification, grounding, or reasons for belief. As soon as contraction makes use of the notion "*y* is believed only because of *x*", we run into counterexamples to recovery . . . But when a theory is taken as "naked", i.e. as a bare set of statements closed under consequence, then recovery appears to be free of intuitive counterexamples.

I would have thought that the conclusion is that it does not make much sense to consider "naked" theories, i.e., belief states represented simply as sets of sentences, in relation to contraction, since the example makes clear that contraction is governed by further parameters not contained in that simple representation. This is exactly the conclusion elaborated by Haas (2005, Sect. 2.10).

I now face a dialectical problem, though. A ranking function is clearly not a naked theory in Makinson's sense. It embodies justificatory relations; whether it does so in a generally acceptable way, and whether it can specifically explicate the "only because", does not really matter. Nevertheless, it is my task to defend *recovery*. Indeed, my explanation for our intuitions concerning George is a different one:

First, circumstances might be such that recovery is absolutely right. There might be only one crime under dispute, a murder, and the issue might be whether George has committed it, rather than whether George is a more or less dangerous criminal. Thus, I might firmly believe that he is either innocent or a murderer so that, when hearing that the police arrested him, my conclusion is that he is a murderer, after all.

These are special circumstances, though. The generic knowledge about criminals to which the example appeals is different. In my view, we are not dealing here with two sentences or propositions, φ and ψ, of which one, ψ, happens to entail the other, φ. We are rather dealing with a single scale or variable that, in this simple case, takes only three values: "not criminal", "criminal, but not a murderer", and "murderer". The default for such scales or variables is that a distribution of degrees of belief over the scale is *single-peaked*. In the case of negative ranks this means that the distribution of ranks over the scale has only one local minimum.

In my initial belief state in which I believe George to be a murderer the ranking minimum is at the value "murderer". Now, a standard AGM contraction by φ (or a $\varphi \to 0$-conditionalization) produces a two-peaked distribution: both "not criminal" and "murderer" receive rank 0 and only the middle value receives a higher rank. This just reflects the retention of $\varphi \to \psi$. Thus, AGM contraction violates the default of single-peakedness.

Precisely for this reason we do not understand the district attorney's message as an invitation for a standard contraction. Rather, I think the message "there is no evidence that George is a criminal" is tacitly supplemented by "let alone a murderer", in loose agreement with Grice's maxim of quantity. That is, we understand it as an invitation to contract by ψ (George is a murderer), and then, if still necessary, by φ or, what comes to the same, by $\varphi \land \neg\psi$. In other words, we understand it as an invitation to perform a mild contraction (11.10) by φ in Levi's sense. Given this reinterpretation there is no conflict between *recovery* and the example.

Levi (2004, pp. 65f.) finds another type of example to be absolutely telling against *recovery* (see also his discussion of still another example in Levi 1991, pp. 134ff.). Suppose you believe that a certain random experiment has been performed (= φ), say, a coin has been thrown, and furthermore you believe in a certain outcome of that experiment (= ψ), say, heads. Now, doubts are raised as to whether the experiment was at all performed. So, you contract by φ and thereby give up ψ as well. Suppose, finally, that your doubts are dispelled. So, you again believe in φ. Levi takes it to be obvious that it should then be entirely open to you whether or not the random ψ is the case – another violation of *recovery*.

I do not find this story so determinate. Again, circumstances might be such that *recovery* is appropriate. For instance, the doubt might concern the correct execution of the random experiment, it might have been a fake. Still, there is no doubt about its result, if the experiment is counted as valid. In that case *recovery* seems mandatory.

However, I agree with Levi that this is not the normal interpretation of the situation. But I have a different explanation of the normal interpretation. In my view, the point of the example is not randomization, but presupposition. ψ presupposes φ (in the formal linguistic sense); one cannot speak of the result of an experiment, unless the experiment has been performed. Now it seems to be a plausible rule that if the order is to withdraw a presupposition, then one has to withdraw the item depending on this presupposition explicitly, and not merely as an effect of giving up the presupposition. In the above case this means first to contract by ψ and then by φ. One may test my rule with non-random examples; take, e.g., φ = "Jim was a smoker" and ψ = "Jim quit smoking".

This pragmatic rule concerning presuppositions is quite different from my above observation about scales. The pragmatic effect, however, is the same. And again this effect agrees with Levi's mild contraction (11.10). Note, by the way, that what I described as special circumstances in both examples above can easily be reconciled with mild contraction; informational loss is plausibly distributed under these circumstances in such a way that mild contraction and AGM contraction arrive at the same result.

So, I entirely agree with Levi on the description of the examples. I disagree on their explanation. Levi feels urged to postulate another kind of contraction operation governed by different axioms, and Makinson has the hunch that taking account of justificatory relations will lead to such a different contraction operation. By contrast, I find AGM contraction sufficient on the theoretical level and invoke various pragmatic principles explaining why more complex things might be going on in certain situations than simple single AGM contractions. It is clear that I have invoked such principles only in an exemplary way. A systematic inquiry into the pragmatics of belief revision and contraction would be most fascinating, but is entirely out of the scope of the present investigation. (Cf. Merin 1999, 2003a,b, and 2006, who has made various interesting and relevant observations concerning presuppositions and scale phenomena.)

Let us also look at some of the revision postulates. A larger discussion originated from the undue rigidity of the *success* postulate (K÷4) (cf. 4.18) saying

(11.13) $\varphi \in K * \varphi$,

i.e., that the new evidence must be accepted. Many thought that "new information is often rejected if it contradicts more entrenched previous beliefs" (Hansson 1997, p. 2) or that if new information "conflicts with the old information in K, we may wish to weigh it against the old material, and if it is . . . incredible, we may not wish to accept it" (Makinson 1997a, p. 14). Thus, belief revision theorists tried to find accounts for what they called non-prioritized belief revision. Hansson (1997) is a whole journal issue devoted to this problem.

The idea is plausible, no doubt; the talk of weighing notoriously remains a mere metaphor in belief revision theory; and the proposals are too ramified to be discussed here. (I take them as another symptom of the fact that the whole framework is unfit for dealing with the dynamics of belief.) Is ranking theory able to deal with non-prioritized belief revision?

Yes. After all, ranking theory is made for the metaphor of weighing. So, how do we weigh new evidence against old beliefs? In Section 5.5 I explained revision by A as result-oriented $A{\rightarrow}n$-conditionalization (5.24) for some $n > 0$ (the result was the same for all $n > 0$). And I did so just in order to satisfy *success*. However, we noticed already in Section 5.4 that evidence-oriented $A{\uparrow}n$-conditionalization (5.29) is a more adequate characterization of belief dynamics insofar as its parameter n pertains only to the evidence. Now we can see that it is exactly suited for describing non-prioritized belief revision:

If we assume that evidence always comes with the same firmness $n > 0$, then $A{\uparrow}n$-conditionalization of a ranking function κ is sufficient for accepting A if $\kappa(A) < n$ and is not sufficient for accepting A otherwise. (One might object that the evidence A is here only weighed against the prior disbelief in A. But insofar the prior disbelief in A is a product of a weighing of reasons as described in Section 6.3, the evidence A is also weighed against these old reasons.) It should not be difficult to prove that $A{\uparrow}n$-conditionalization with a fixed n is a model of screened revision as defined by Makinson (1997a, p. 16). And if we let the parameter n sufficient for accepting the evidence vary with the evidence A, we should be able also to model relationally screened revision (Makinson 1997a, p. 19).

Was this a defense of *success* (11.13) and thus of AGM belief revision? Yes and no. What I have argued was that within ranking theory we may define belief revision in such a way as to satisfy *success* without loss, since ranking theory provides other kinds of belief change which comply with other intuitive desiderata and which we may, but need not, call belief revision. Ranking theory seems to cover all intuitive needs and thus renders superfluous the search for variant characterizations of belief revision.

For a final example note that (K*4), *expansion 2* (cf. 4.18), directly implies a postulate called *preservation*:

(11.14)   if $\neg\varphi \notin K$, then $K \subseteq K * \varphi$.

This has met intuitive doubts as well. Rabinowicz (1996) discusses the following simple story: Suppose that given all my evidence I believe that Paul committed a certain crime (= ψ); so $\psi \in K$. Now a new witness turns up producing an alibi for Paul (= φ). Rabinowicz assumes that φ, though surprising, might well be logically compatible with K; so $\neg\varphi \notin K$. However, after the testimony I no longer believe in Paul's guilt, so $\psi \notin K * \varphi$, in contradiction to *preservation* (11.14).

Prima facie, Rabinowicz's assumptions seem incoherent. If I believe Paul to be guilty, I thereby exclude the proposition that any such witness will turn up; the appearance of the witness is a surprise initially disbelieved. So, we have $\neg\varphi \in K$ after all, and *preservation* does not apply.

Look, however, at the following ranking function κ and its $\varphi\to 6$- or $\varphi\uparrow 9$-conditionalization κ':

κ	ψ	$\neg\psi$
φ	3	6
$\neg\varphi$	0	9

κ'	ψ	$\neg\psi$
φ	0	3
$\neg\varphi$	6	15

As it should be, the witness is negatively relevant to Paul's guilt according to κ (and vice versa); more precisely, Paul's being guilty is a necessary and sufficient reason for assuming that there is no alibi. Hence, we have $\kappa(\neg\psi) = 6$, i.e., I initially believe in Paul's guilt, and confirming our first impression, $\kappa(\varphi) = 3$, i.e., I initially disbelieve in the alibi.

However, I have just tacitly assumed that rank > 0 is the criterion of disbelief. We need not make this assumption. I emphasized in (5.14) that we might conceive disbelief more strictly, say, as rank > 5. Now note what happens in our numerical example: Since $\kappa(\neg\psi) = 6$ and $\kappa(\varphi) = 3$, I do initially believe in Paul's guilt, but not in the absence of an alibi (though one might say that I have positive inclinations toward the latter). Paul's guilt is still positively relevant to the absence of the alibi, but neither necessary nor sufficient for believing the latter. After getting firmly informed about the witness, I change to $\kappa'(\neg\varphi) = 6$ and $\kappa(\psi) = 3$, that is I believe afterwards that Paul has an alibi (even according to our stricter criterion of belief) and do not believe that he has committed the crime (though I am still suspicious).

By exploiting in this way the vagueness of the notion of belief, we have thus found a model that accounts for Rabinowicz's intuitions. Moreover, we have described an operation that may as well be called belief revision, even though it violates *preservation* (11.14). Still, this is not a refutation of *preservation*. If belief can be taken as more or less strict, belief revision might mean various things and might show varying behavior. And the example has in fact confirmed that, under our standard interpretation (with disbelief being rank > 0), belief revision should conform to preservation.

These four examples may suffice as a discussion of AGM revision and contraction postulates. I have deliberately responded to these examples in quite diverse ways, since I do not have, and do not believe in, a unified treatment of all the examples. As a theory, ranking theory is powerful and of a piece. But in applications a multitude of interpretations and supplementary considerations come into play. This is what I wanted to demonstrate, by alluding to the vagueness of belief, by relating revision to result-oriented and to evidence-oriented change, by understanding a given contraction as either a single, or a package, or an iterated contraction, and by invoking various pragmatic principles for the transition from evidential (linguistic) input to belief change. Hence, I certainly do not want to claim that my treatment of the examples is the only correct one. My intention was to display plausible possibilities – and thus to prove my conclusion that the relation between theory and application is most intricate and indirect and hardly ever allows for clear and determinate inferences.

A final remark: Haas (2005) is concerned about the same and similar examples, but attempts to respond with a unified account. His key is to explicitly build on justificatory structures, and since those structures are at most implicit in AGM theorizing (as observed in Section 6.4, and above in the discussion of the recovery postulate) I am sympathetic to Haas' attempt – at least in principle. A closer look reveals, however, that our approaches are hardly comparable. More precisely, his key idea is that at the root of all the problematic examples is the fact that in AGM belief revision theory belief sets can be simply expanded by *consistent* information. And he proposes to replace this by the Levi-inspired idea that a belief set can be expanded only if the expansion can be justified. That is, he basically distinguishes prima facie acceptance and justified acceptance that selects a subset from all prima facie accepted sentences or propositions. Thus, we start with a stable belief set K each part of which is justifiedly accepted, receive evidence φ, prima facie accept φ, and finally reach a new stable belief set K' that may or may not justifiedly accept the evidence and thus may or may not expand or revise the old K. Hence, the behavior of belief revision crucially depends on how the justified subset is selected from a set of prima facie acceptances.

This notion of justification has an explicit Lehrerian flavor; that is, it is shaped according to the notion of coherence as conceived, e.g., by Lehrer (2000, p. 126). However, this notion is foreign to my framework. I have a local relation of positive relevance or being a reason, as explained in Chapter 6. And I can call ranking functions, like probability measures, coherent, in the sense explained in Section 6.3 – namely that each proposition receives the rank it must receive according to the ranks of the other propositions and the internal reason relations. However, this is not Lehrer's coherence. We will have to see the extent to which ranking theory supports the talk of global justification in Section 16.1. In any case, as will become clearer in the next section, I have no place for the distinction between prima facie and justified acceptance.

Therefore, I do not see how to support Haas' account from the ranking-theoretic side. This agrees with his conclusion in Haas (2005, Sect. 9.2) that we approach our common topic from different angles. The relation between our accounts still needs to be clarified.

11.4 Undefeated Justification Defeated: A Criticism of Lehrer's Account of Knowledge

The last remarks suggest relating Lehrer's elaborate account of knowledge to ranking theory. This will be our next task. I just noted that I have no use for Lehrer's distinction between prima facie and justified acceptance; this will not be the only incomparability. In other respects, though, ranking theory seems to be ideally suited for amending Lehrer's account; so the comparison may well be fruitful. In any case, we should take a closer look. My discussion will be essentially based on Lehrer (2000) – the most recent full presentation of his theory adding important simplifications and clarifications to the first edition Lehrer (1990) – and on the exchange I had with Lehrer in Spohn (2003) and Lehrer (2003).

Lehrer's basic notions are concisely summarized in his definition of an *evaluation system* in Lehrer (2000, p. 170, D1), which consists of three components: (a) an *acceptance system*, i.e., a set of accepted statements or propositions; (b) a *preference system*, i.e., a four-place relation among all statements or propositions saying for all A, B, C, D whether A is *more reasonable to accept*, given or on the assumption that C than B is given or on the assumption that D – which I symbolize here as $A \mid C > B \mid D$ (although Lehrer mentions only unconditional preference in D1, it is clear that he requires conditional preference in his subsequent definitions); and (c) a *reasoning system*, i.e., a set of inferences each consisting of premises and a conclusion.

The idea behind (c) is that what a person accepts, and justifiedly accepts, depends also on the inferences she carries out. Here I understand Lehrer to be referring to *deductive* inferences, or rather to the inferences taken by the person to be deductively valid and sound. In any case, when introducing the reasoning system in (2000, p. 127), Lehrer refers only to "cogent" reasoning, and generally the notion of validity makes clear sense only relative to deductive inference. The idea behind (b), by contrast, is to take care of non-deductive reasoning in the widest sense, that is always a matter of weighing reasons and objections on the basis of some such preference system. Naturally, the preference system will be the focus of my comparison.

I have to adapt Lehrer's evaluation systems in three steps. They are required by my comparison; but then the comparison will proceed in a surprisingly smooth way. Of course, the issue will then be to what extent these adaptations are fair and justified.

The first step is that I will neglect Lehrer's distinction between belief and acceptance. I explained ranking theory as a theory of belief. But it could just as well be interpreted as a theory of acceptance. I do not see much of a difference and I am unable to make any difference within ranking theory. Moreover, I take it that any kinds of propositions are suited to be objects of acceptance. By contrast, it is important for Lehrer that the propositions to be accepted always have the special form "I accept that A", where A may be any kind of proposition. Again, I neglect the point on the basis of the one-one correspondence between Lehrer's and my objects of acceptance. This understood, we can directly translate Lehrer's acceptance system into ranking theory:

(11.15) A is *accepted* by the negative ranking function κ iff $\kappa(\bar{A}) > 0$, or iff $\kappa^{-1}(0) \subseteq A$, and $\{A \mid A$ is accepted by $\kappa\}$ is the *acceptance system* of κ.

The second step consists in fixing the reasoning system once and for all. The reference to propositions in the sense of Chapter 2 and the postulate of deductive closure (4.2) are so deeply entrenched in ranking theory that it can be compared only with similar theories. So, I impose these assumptions on Lehrer's account and thereby ignore all questions of syntactic structure and of logical equivalence or entailment. That is, I assume the acceptance system to be consistent and deductively closed. This entails in particular the assumption that the reasoning system is maximal and has no independent role to play. Since Lehrer repeatedly emphasizes that he is interested in acceptance only insofar as it is governed by the aim of truth, and since logically equivalent sentences serve that aim equally, one may think that the objects of acceptance are propositions. However, Lehrer is not explicit on this issue, and he certainly does not want to assume these rationality constraints. So, at this point I am deviating from Lehrer's specifications, and we will have to check whether the deviation is a fatal or a marginal one.

The third and for our purposes most important step is to find a model for Lehrer's second component of an evaluation system, the preference system. Here, the first suggestion is to appeal to a probability measure P and to define that $A \mid C > B \mid D$ iff $P(A \mid C) > P(B \mid D)$. However, the relation between probability and acceptance is problematic for the insuperable reasons explained in Section 3.3. Moreover, as I will point out below, there is a particular feature in Lehrer's notion of neutralizing an objection that prevents any probabilistic interpretation. Olsson (1998) discusses further difficulties of a purely probabilistic construal of justified acceptance.

For similar reasons Lehrer, too, has given up on finding purely probabilistic foundations, which he still hoped to build in Lehrer (1971, 1974). There he suggests, moreover, that the foundations may be construed as some kind of epistemic decision theory. The hint is still found in Lehrer (2000, pp. 145ff.) and also used for a solution of the lottery paradox. However, I have already expressed my doubts about epistemic decision theory in Section 11.2.

Hence, we should take the same move as at the end of Section 3.3, and put probability theory to one side. We might look instead at theories dealing directly with acceptance or belief. We know a large variety of such theories, such as default logic, AGM belief revision theory, various accounts of defeasible and non-monotonic reasoning, etc. However, none of them can model Lehrer's relation $A \mid C > B \mid D$. At best, one finds models for the special case $A \mid C > B \mid C$ of Lehrer's preference relation which refers twice to the same condition. But this won't do since, as we will see below, Lehrer requires the full conditional four-place relation.

The point I am heading for is clear. Ranking theory is the *only* theory I know that is capable of accounting for Lehrer's relation $A \mid C > B \mid D$. We may simply define:

(11.16) A is *more reasonable to accept given C* than B *given D* relative to κ, i.e., $A \mid C > B \mid D$, iff $\kappa(\bar{A} \mid C) > \kappa(\bar{B} \mid D)$. And A is more *reasonable to accept* than B, $A > B$, iff $A \mid W > B \mid W$, i.e., $\kappa(\bar{A}) > \kappa(\bar{B})$ (where W is still the set of all possibilities considered, or the tautological proposition).

Therefore it might seem that Lehrer's account of knowledge is wedded to ranking theory.

However, the marriage is unhappy. After having defined all three components of Lehrer's evaluation system in ranking-theoretic terms, it is an easy exercise to also translate Lehrer's theory of justification, and thus his account of knowledge, into ranking theory.

Let us see how this works:

(11.17) B is an *objection* to A iff $A \mid \bar{B} > A \mid B$.

This is D4 of Lehrer (2000, p. 170). Is being an objection a relation among accepted propositions or among propositions in general? The latter, according to (11.17), though Lehrer might intend the former. However, there is no issue here, since we will have to stipulate that justified acceptance entails acceptance and will thus consider only objections to accepted propositions. Otherwise, we would get nonsensical results.

(11.18)   The objection B to A is *answered* iff B is an objection to A, i.e., $A \mid \bar{B} > A \mid B$, *and* $A > B$.

This is literally D5 of Lehrer (2000, p. 170).

(11.19) C *neutralizes the objection B to A* iff B is an objection to A, i.e., $A \mid \bar{B} > A \mid B$, $B \cap C$ is not an objection to A, i.e., $A \mid \bar{B} \cup \bar{C} \leq A \mid B \cap C$, and $B \cap C$ is at least as acceptable as B, i.e., $B \cap C \geq B$.

This is D6 of Lehrer (2000, p. 170). Indeed, it is this last condition that prevents a probabilistic interpretation of Lehrer's preference system because it would be empty in this interpretation; no objection could then be neutralized. In ranking-theoretic terms, however, this consequence need not be feared.

Thus, we finally arrive at

(11.20) A is *justifiedly accepted*, or the acceptance of A is *personally justified*, relative to κ *in the strong sense* iff A is accepted and each objection to A is answered or neutralized by some C.

This is the literal translation of D3 and D7 of Lehrer (2000, pp. 170f.). My qualification "in the strong sense" indicates, however, that we will have to consider a weaker sense as well.

This chain of definitions yields weird results. What is justifiedly accepted according to (11.20) is an unintelligible selection from the accepted propositions. In order to define this selection let $E_m = \bigcup \{D \mid \kappa(D) \geq m\}$; hence, E_m is the logically weakest proposition having at least rank m. Then we have:

Theorem 11.21: A is justifiedly accepted relative to κ in the strong sense iff, given $\kappa(\bar{A}) = n > 0$, for all $m \geq n$ with $E_m - E_{m+1} \neq \varnothing$ $\bar{A} \cap E_m - E_{n+m+1} \neq \varnothing$ holds true, or, equivalently, iff for all $m \geq n$ with $\kappa(E_m) = m$ there is a $D \subseteq \bar{A}$ with $m \leq \kappa(D) \leq n + m$.

This is Theorem 1 of Spohn (2003); there is no sense in repeating its proof here. The only point in displaying (11.21) is that it makes no intuitive sense at all. I could not find any way to express the necessary and sufficient condition for justified acceptance in a more plausible or perspicuous form. The only conclusion is that if Lehrer's definitions force us to distinguish between justifiedly and unjustifiedly accepted propositions in *this* way, then there is something wrong either with Lehrer's definitions or with my reconstruction. We will have to discuss this at the end.

Indeed, there is cause for suspicion. For instance, objections may be restricted to inductive objections, where B is an inductive objection to A iff $\kappa(\bar{A} \mid B) < \kappa(\bar{A} \mid \bar{B}) < \infty$. It is, moreover, unclear whether Lehrer really meant (11.19) as stated. Perhaps, the idea of neutralization is better expressed by the condition that *given* the neutralizer C, the objection B to A is no longer an objection to A (i.e., $A \mid \bar{B} \cap C \le A \mid B \cap C$). However, as far as I have checked, this leads nowhere. Theorem 11.21 thereby changes considerably, but does not improve. Likewise, one may restrict (11.17) to accepted propositions; then; however, even the contradiction $\varnothing$ would turn out to be justifiedly accepted.

A better cure is suggested by the fact that (11.21) implies that justified acceptance is not even deductively closed: if A is justifiedly accepted and logically implies B, B still need not be justifiedly accepted. This seems to be a flaw in (11.20) as it stands. Indeed, thinking about what one is personally justified in accepting means also working through one's reasoning system. Given my specification of the reasoning system, this leads us to the following weaker and more adequate sense of justified acceptance:

(11.22) A is *justifiedly accepted* relative to κ *in the weak sense* iff A is logically implied by propositions justifiedly accepted in the strong sense.

In fact, Lehrer (1971, p. 221) took recourse to the same move when observing that the rule of induction he had proposed there is not deductively closed.

What is justifiedly accepted in this sense? This is answered by

Theorem 11.23: A is justifiedly accepted relative to κ in the weak sense iff A is accepted by κ.

(For a proof, see Spohn 2003, p. 137). In a way, this looks much nicer, but it is certainly also unwelcome, as it trivializes Lehrer's theory of justification. After all, the justification game was supposed to do some work. Theorem 11.23 would not change under the modifications to Lehrer's definitions mentioned above.

These results also affect Lehrer's theory of knowledge. In order to see how, we must first turn to undefeated justification and the ultrasystem. According to Lehrer (2000, p. 171, D8) a person's ultrasystem is generated from her evaluation system by deleting from the latter all acceptances of, preferences for, and reasonings from, falsehoods. This is immediately explicable in ranking-theoretic terms.

Let $@ \in W$ be the true or actual possibility in W, and let us consider any evaluation system, i.e., ranking function κ. If, exceptionally, everything accepted in κ should were true, then no falsehoods could intrude into the justifications with respect to κ. In this case, κ would be its own ultrasystem.

Usually, though, some false proposition will be believed in κ. This entails $\kappa(\{@\}) > 0$. Thus, the logically weakest falsehood accepted by κ is $\overline{\{@\}}$. Now, if we contract κ by $\overline{\{@\}}$, not only this falsehood, but also all other, stronger falsehoods must go (since the contracted acceptance system is again deductively closed). Hence, in the resulting ranking function κ^+ exactly the true among the propositions accepted by κ are accepted. Moreover, if A is true and B is false, then $\kappa^+(A) = 0 \leq \kappa^+(B)$, so that in the preference system provided by κ^+ no false proposition is ever preferred to a true one, as required by Lehrer. Finally, we need not worry about the reasoning system of κ^+, for the familiar reason. All in all, the explication is remarkably smooth, and we may translate Lehrer's D8 into:

(11.24)   The ultrafunction κ^+ of κ is $\kappa_{+\overline{\{@\}}}$, the contraction of κ by $\overline{\{@\}}$.

This covers also the exceptional case where no falsehood is accepted in κ, i.e., where $\kappa^+ = \kappa$.

On this basis, Lehrer's remaining definitions may be immediately translated. D9 of Lehrer (2000, p. 171) turns into:

(11.25)   The justification for accepting A in κ is *undefeated* (or *irrefutable*) iff A is justifiedly accepted in κ^+.

And the goal of the whole enterprise, Lehrer' explication of knowledge DK (2000, pp. 169f.) becomes:

(11.26)   *A is known* in κ iff (i) A is accepted in κ, (ii) A is true, i.e., $@ \in A$, (iii) A is justifiedly accepted in κ, and (iv) the justification for accepting A in κ is undefeated.

Lehrer's claim that knowledge reduces to undefeated justified acceptance, i.e., to condition (iv) is now easily confirmed. Alas, stronger results obtain. If knowledge is based on justified acceptance in the strong sense, just those propositions are known which satisfy the condition of (11.21) relative to κ^+. Unpalatable knowledge! If knowledge is based on justified acceptance in the more adequate weak sense, we get

Theorem 11.27: A is known in κ iff A is true and A is accepted in κ.

Proof: (11.23) reduces condition (iii) of (11.26) to (i). (11.23) applies to κ^+ as well. Hence, (iv) reduces to acceptance in κ^+ and thus to (i) and (ii).

So, again, the sophisticated considerations relating to the ultrasystem seem empty, and knowledge reduces to true belief, a conclusion Lehrer definitely wants to avoid.

In Spohn (2003, Sect. 4) I made various soothing remarks, but Lehrer (2003) took the above explications and theorems as what they plainly were, namely an unfriendly takeover; this was also my favored conclusion. Consequently Lehrer rejects my initial adaptive steps. He insists on the special form of the items accepted by the acceptance system. I granted this point, but still do not see how it can make a crucial difference. More importantly, Lehrer points out that the reasoning system is something much more subjective and incomplete. I assumed, as it were, the reasoning system of the logically omniscient, but we must not do this when dealing with real humans. Hence, we must not assume the acceptance system to be consistent and deductively closed. Presumably, the same remark applies to the preference system that need not be as orderly and completely specified as with a ranking function. Lehrer (2003, p. 331) even adds that the reasoning system need not be confined to deductive reasoning. In my view, though, this blurs his position even more; I no longer then see the division of labor between the reasoning and the preference system.

All in all, in his reply (2003) Lehrer takes a much more computational view than I found in his (2000); he wants to give an account of our actual cognitive structure, which defies rigorous theorizing, and not of any unrealistic idealizations. This would mean that Lehrer and I are talking on incomparable levels and that any attempt at formal reconstruction would be out of place, anyway.

I have two responses. One response is given only in the next section where the comparison with John Pollock's defeasible reasoning gives rise to the issue of whether epistemology should proceed on a syntactic/computational or a semantic/regulative level, a distinction that avowedly divides Pollock and me, but apparently also Lehrer and me, and that will be further discussed in Section 11.6.

The other response is that, while my reconstruction might in some sense not apply to real humans, Lehrer's account should apply as well to more perfected or idealized subjects. After fully working through one's reasoning and one's preference system, one should still, or all the more, have justified belief and knowledge in Lehrer's sense. And then we are bound to take my three adaptive steps. Working through one's (deductive) reasoning system means having a consistent and deductively closed acceptance system. And working through one's preference system means arranging the comparative reasonableness of conditional acceptance according to (11.16) and some ranking function; at least, this proposal lacks feasible alternatives, as was extensively defended in Chapters 5, 7, and 8. Hence, there is no way for idealized subjects to escape Theorems 11.21 and 11.23.

Focusing on (11.23) (and neglecting 11.21), one may say that this is exactly right. After having fully worked through one's reasoning and preference system, playing various justification games should not further alter the acceptance system; then acceptance should coincide with personally justified acceptance, as (11.23) asserts. My more general discussion in Section 16.1 of the notion of justification as used in traditional epistemology will arrive at the same conclusion (16.2). Thus far, a kind of agreement between Lehrer's notion of personal justification and ranking theory may be stated.

However, this sanguine view does not carry over to Theorem 11.27 saying that after fully working through one's reasoning and preference system all one's true belief or acceptances qualify as knowledge. Of course, this conclusion also depends on my ranking-theoretic translation (11.24) of Lehrer's ultrasystem, which, however, was perfectly literal, given the general translational scheme. Von Kutschera (1982, Sect. 1.3) and Sartwell (1991, 1992) might welcome this conclusion, but it is unacceptable for Lehrer and intuitively doubtful. Maybe ranking theory characterizes internal coherence well. Still, ever so much internal coherence of beliefs and ranks seems by itself unable to turn true beliefs into knowledge. Hence there remains a deep concern about Lehrer's account.

To be sure, I have not discussed the spirit of Lehrer's account of knowledge. To that end I should have engaged in a philosophical discussion of the various strategies and approaches towards knowledge, something I have declined to do in this book. Still, it is difficult to discuss spirits as such; one must look at their materialization. Indeed, Lehrer's account of knowledge is attractive not least because it offers a relatively specific materialization. However, if my ranking-theoretic reconstruction applies, this materialization seems hard to maintain.

11.5 John Pollock's Theory of Defeasible Reasoning: An Ally?

I presented ranking theory as a contribution to the problem of induction, and in Section 1.1 I already introduced my basic decision to approach this topic in terms of the dynamics of doxastic states rather than in terms of inductive reasoning or inference. This distinction is a strict one in principle, but in practice the difference is less clear. Indeed, (11.15)–(11.19) was a brief ranking-theoretic reconstruction of Lehrer's account of defeasible reasoning that he required for defining his notion (11.20) of personal justification – where "defeasible reasoning" is just another label for inductive or perhaps for non-probabilistic inductive reasoning.

This suggests extending our comparisons to the work of John Pollock who has presented the most elaborate, detailed, and philosophically minded account of defeasible reasoning. An in-depth comparison is out of place, all the more as my philosophical applications of ranking theory in the coming chapters will return to some ideas similar to Pollock's. On a strategic level, though, a comparison is feasible, and it will look at both the difference in principle and the similarity in practice. Pollock, Gillies (2000) compared defeasible reasoning with belief revision theory. The relation between belief revision and ranking theory was clear from the outset in Spohn (1988). So I attempted to complete the triangle of comparisons in Spohn (2002a), on which this and the next section essentially relies.

Since the 70s, John Pollock has developed his theory in many books and papers. Pollock (1990) is perhaps his most complete account of specific rules of inductive inference, and Pollock (1995) is his most elaborate presentation of his overall architecture, the so-called OSCAR project, an all-encompassing and most ambitious program for truly artificial intelligence which has occupied him since, and which, in particular, is intended to cover practical reasoning as well – an aspect I will neglect here. This book is a most suitable basis for my comparison. Let me start with a very rough sketch of this architecture.

In his theory Pollock draws a large and detailed picture of doxastic states as huge nets or *graphs of inferences*, reasons, justifications, or *arguments*. Each argument consists of a set of *premises*, a *conclusion*, and an inference rule or *reason-schema* applied in the argument. There are two kinds of argument. The first are *conclusive* or non-defeasible arguments, which we know well enough from deductive logic. The essence of his theory, though, lies in the second kind, the *defeasible* arguments instantiating defeasible reason-schemata that get our inductive machinery running.

That a reason-schema or the instantiating arguments are defeasible is to say that they have *defeaters*. Therefore, they need to be amended by a specification of their possible defeaters. They come in two kinds. There are *rebutting* defeaters; these are arguments arriving at the opposite conclusion. And there are *undercutting* defeaters; they undermine the connection between premises and conclusion of the defeated argument and hence imply that the conclusion is not established by the premises.

Of course, defeating arguments can in turn be defeated in both ways. And all this can mesh with conclusive arguments. In this way, a big and complicated architecture of more or less provisional justifications emerges. (Cf. Pollock 1995, Ch. 1 and Sect. 1.1–1.3.)

This picture needs to be filled in, and Pollock does so amply. The picture has a start, a substance, a form, and a goal:

The *start* is *perception*: Without perception there are no premises to start with and no conclusion to arrive at. Perceptions form the indispensable base of the whole wide web of reasons and beliefs. However, the base is defeasible, and it is governed by the following defeasible reason-schema: if p is a perceptible state of affairs, then infer p from "it looks as if p". This argument step moves us from a phenomenal premise to a conclusion about the external world (cf. Pollock, Cruz 1999, pp. 38ff., and Pollock 1995, Sect. 2.2). Having arrived there, the inductive net can start proliferating. (I will adopt and thoroughly discuss something quite similar that I will call the Schein-Sein principle in Section 16.3; what are defeasible inference rules for Pollock will turn into defeasibly a priori postulates within my framework.)

The *substance* is provided by the many specific defeasible reason-schemata, the fullest collection of which is found in Pollock (1990); cf. in particular pp. 339–44. Despite the general acknowledgment of the crucial importance of defeasible reasoning, the richness and specificity of Pollock's proposals by far outruns other accounts. We have already seen an example, the rule governing perception, for which Pollock states two potential undercutting defeaters: one for the case where the subject perceives something, but believes she is perceiving something else, and one for the case that the subject believes she is in unreliable perceptual circumstances.

There are quite a number of further defeasible inference rules. Just to give the flavor, the most important one for Pollock is the *statistical syllogism*: If G is projectible with respect to F, then infer Ga from Fa and "the probability of F's being G is greater than $1/2$" (where the strength of the inference depends on that probability). Again, this inference can be defeated, most importantly by a subproperty defeater: if G is projectible also with respect to H, then the above inference is undercut by Ha and "the probability of F-and-H's being G differs from that of F's being G". There are rules for enumerative and statistical induction arriving at universal and probabilistic generalizations. And so forth.

Then there is a *form* that provides rules for combining the many arguments entertained by the subject into an integrated inference graph. Individual arguments have strengths, and a formal theory is required for specifying the strengths within a complex argument structure. This is governed by the weakest link principle and the no-accrual-of-reasons principle, as Pollock calls them. Arguments can be defeated, and a formal theory is required for determining the defeat statuses of the nodes of the inference graph, that is, for each possible conclusion one must determine whether it is actually supported (with some strength) or defeated by the inferences considered (cf. Pollock 1995, Sect. 3.3–3.8). All in all, inference graphs easily become quite messy, as exemplified by Pollock (1995, p. 131) for the paradox of the preface.

Finally, all this reasoning has a *goal*, namely to arrive at a set of justified and warranted beliefs. Prima facie, it is not clear whether the goal can be reached at all. There are two issues. First, one wonders whether, given a certain stage of reasoning with its multiply entangled strengths and defeatings, the various conclusions may be unambiguously said to be undefeated or defeated. All kinds of crazy situations are imaginable, and Pollock has long struggled with them. However, his present theory of defeat statuses seems to get the issue under control (cf. Pollock 1995, Sect. 3.6–3.9).

A *justified* belief, then, is a conclusion that comes out undefeated under this theory. Justification is here still relative to a certain stage of reasoning. Rightly so, according to Pollock, because a subject is always at, and acts always from within, an unfinished state of reasoning. This suggests, though, that there is also a stronger notion of justification, which Pollock calls ideal warrant. The defeat statuses can change in unforeseen ways as soon as a further argument is considered. Hence, the subject should stepwise extend his inference graph, until all (possibly infinitely many) arguments in principle available to him are considered. The second issue, then, is whether this process is at all well-behaved. There is no guarantee. Pollock thus defines a conclusion to be *ideally warranted* if it is unambiguously undefeated in the maximal inference graph in which the subject, *per impossibile*, takes into account the arguments available to him all at once. And the stepwise extension of the inference graph is well-behaved if it eventually arrives at precisely the ideally warranted conclusions or beliefs. Pollock (1995, Sect. 3.10–3.11) specifies conditions under which this process is well-behaved, as so defined.

In this way, Pollock draws a detailed dynamic picture of reasoning and its goal. In another sense, the picture is still static; the whole edifice of reasoning and thus the set of ideally warranted beliefs rest on a given input of perceptions. By contrast, the focal question usually considered (and answered by ranking theory) is what happens to the beliefs when the input changes; only by answering this question do we acquire a fully dynamic picture. However, Pollock has no difficulties with this question, for he can enrich the stock of perceptions and set his reasoning machinery in motion again, so that new sets of justified and ideally warranted beliefs will result (cf. Pollock, Gillies 2000, Sect. 4).

This brief sketch of Pollock's theory may suffice as a base for the subsequent discussion. How is it related to ranking theory? The last paragraph gives a hint; it seems that Pollock's defeasible reasoning ends where ranking theory starts. Reasoning, as Pollock describes it, yields ideally warranted beliefs (if it could be completed), whereas ranking theory, or, for that matter, belief revision theory, describes the static and dynamic structure of these ideally warranted beliefs. A neat division of labor? No. There is no way at all of plugging the two theories together in order to yield a more comprehensive epistemology.

To begin with, one might again sense the opposition between a synthetic-constructive and a structural-analytic procedure that we already observed between Isaac Levi and me. Indeed, the opposition is more explicit here. Pollock, unlike Levi, offers a decidedly computational theory specifying every piece of reasoning from which our epistemological edifice is constructed. At the same time, Pollock's theory is avowedly normative; it is not an empirical theory about our actual reasoning. I believe that these two features do not fit together and display a deep misunderstanding of the nature of normative theorizing. This issue, though, is of independent interest; therefore I will defer it to the next section, in which I discuss more generally the normative status of ranking theory.

Even if we could ignore this fundamental difference, Pollock's defeasible reasoning and ranking theory could not be combined. I mentioned already that Pollock can deal with the dynamics of ideal warrant on his own, and his account diverges from that of AGM belief revision and ranking theory. For instance, Pollock (1995, p. 140) observes (in belief revision theoretic terms where K represents the set of ideally warranted beliefs given the input up to now and $K * A$ represent that set additionally given the input A) that ideal warrant as defined by him satisfies the law

(11.28)   if $A \in K$ and $B \in K$, then $B \in K * A$,

usually called restricted or *cumulative monotony* (cf. Rott 2001, pp. 112f.). By contrast, both AGM belief revision and ranking theory endorse the stronger law of *rational monotony* (cf. the postulate (K*4) in Definition 4.18):

(11.29)   if $\bar{A} \notin K$ and $B \in K$, then $B \in K * A$.

Pollock, Gillies (2000, pp. 88–90) note further discrepancies concerning expansion and contraction.

The incongruent overlap extends much farther, though, not just to the dynamics of ideal warrant, but to the whole field of defeasible reasoning. Instead of focusing on our difference in principle – that Pollock is dealing with inferences or arguments and I am not – it is worth emphasizing the sameness of our intentions, which is most obvious from the fact that we are both dealing with *reasons*. We do give different explications of reasons – Pollock a computational or procedural one, and I one in terms of positive relevance (cf. also my comparative discussion in Section 16.1) – but clearly we thereby intend to grasp the same subject matter. Thus, ranking theory should also be seen as a logic of defeasible reasoning.

What we say about reasons indeed agrees to a large extent. Of course, we both acknowledge deductive reasons as the only ones that can never be defeated. Ranking theory agrees that defeasible reasons can only be defeated in Pollock's two ways: If A is positively relevant and thus a reason for B, there can be a C that is negatively relevant and thus a reason against B; in this case C is a rebutting defeater. And if A is positively relevant for B, there can be a C such that, given C, A is no longer positively relevant for B; in this case C is an undercutting defeater (which can in turn be undercut by D, in case A is positively relevant for B given $C \cap D$, and so on). Moreover, ranking-theoretic reasons have strengths as well, since positive relevance is a matter of degree.

The fact that ranking theory offers a logic of defeasible reasoning is most evident from Chapter 7. If Pollock receives a new piece of evidence, he works through his graph of inferences in the way indicated; if I do so I run the propagation algorithm for local ranking nets as described in Section 7.4. These are competing descriptions.

From this perspective we find Pollock and I disagreeing on many details of what I above called the form of reasoning. Take, for instance, Pollock's weakest link principle. It has two components that need to be distinguished. On the one hand, it refers to the strength of premises. In deductive reasoning this is the only issue, and then the principle says: "The degree of support of the conclusion is the minimum of the degrees of support of its premises" (Pollock 1995, p. 99). This corresponds to the law of disjunction (5.8b) for negative ranks, which thus escapes Pollock's argument against probabilistic or similar construals of degrees of support. Insofar we agree.

On the other hand, the weakest link principle defines how degrees of support propagate through chains of reasoning. Pollock assumes that inferences are transitive. If one has arrived at a certain conclusion, even with less than maximal support, he thinks one can proceed from this conclusion neglecting how one has arrived at it. Here, I think, Pollock grossly underrates the holistic character of defeasible reasoning, which can be cut up in pieces only with the help of suitable (conditional) independence assumptions that may or may not hold, as shown in Chapter 7. In ranking theory, in any case, we saw that reasons need not be transitive (cf. Section 6.2).

The weakest link principle concerning chains of reasoning is obeyed by ranks only in special cases. Suppose that A is a reason for B, i.e., $\tau(B \mid A) - \tau(B \mid \bar{A}) = x > 0$, and B a reason for C, i.e., $\tau(C \mid B) - \tau(C \mid \bar{B}) = y > 0$. It is most natural to measure the strengths of the reasons or the degrees of support, respectively, by x and y (other ranking-theoretic explications of this notion would not be more favorable for Pollock). What can we say, then, about the strength of the reasoning chain, i.e., about $\tau(C \mid A) - \tau(C \mid \bar{A})$? The following partial observation will do:

(11.30)   If A is a necessary or sufficient reason for B (cf. 6.2) and B is so for C, so that $\tau(B \mid A) = x_1 \geq 0 \geq -x_2 = \tau(B \mid \bar{A})$ and $x = x_1 + x_2 > 0$ and $\tau(C \mid B) = y_1 \geq 0 \geq -y_2 = \tau(C \mid \bar{B})$ and $y = y_1 + y_2 > 0$ *and* if A and C are conditionally independent given B as well as given $\bar{B}$, then $\tau(C \mid A) = \min(x_1, y_1)$ and

$\tau(C \mid \bar{A}) = -\min(x_2, y_2)$. Thus, under these assumptions, $\tau(C \mid A) - \tau(C \mid \bar{A})$
$= \min(x, y)$ iff either $x_1 \leq y_1$ and $x_2 \leq y_2$ or $x_1 \geq y_1$ and $x_2 \geq y_2$.

Hence, even under these favorable circumstances the weakest link principle might, but need not hold according to ranking theory. Things would be more complicated if we were to consider supererogatory and insufficient reasons as well. And the conditional independence assumption is indispensable, of course. (Cf. also the discussion of the positive relevance condition (14.66) for indirect causation, and Theorems 14.68 and 14.74.)

Similar remarks apply to Pollock's no-accrual-of-reasons principle. The principle is right in denying that the support a conclusion receives jointly from two arguments is always the sum of the supports it receives from the individual arguments. But it makes a strong assumption instead, namely that the joint support is the maximum of the individual supports. This violates intuition as well as ranking theory. It suffices to refer here to Section 7.5 on independent testimonies, where it is shown that, given suitable conditional independence assumptions, the strengths of reasons do sum, and where it is suggested that things might get complicated without such assumptions. Contra Pollock's proposal, no simple principle will do.

These remarks show that what I have called the form of Pollock's inference graph is quite unsatisfactory from a ranking-theoretic perspective. We have a dispute. How could it be settled? Pollock cannot do much more than to adduce intuitive support for his principles. They might seem plausible, and alternatives might look less plausible or run more obviously into unacceptable consequences. Pollock lacks further normative guidance that could help him to answer such questions. This is my deeper point. I find his account normatively defective, and I will more generally discuss in the next section why this is so. Ranking theory is different; its normative depth is much greater. In the end, of course, it must appeal to normative intuitions as well. But, in between, it has longer stories to tell. For instance, it can precisely explain when the weakest link and the no-accrual-of-reasons principle hold and when they do not.

The situation reminds me of that in early modal logic. Many modal axioms had been proposed, and many of them were plausible. However, their consequences and their mutual fit were not well understood. There is no doubt that this situation has massively improved with the invention of possible world semantics for modal logics, even though its basic notions might be questioned in turn.

There is a much closer parallel. The basic criticism of early treatments of uncertainty in AI such as MYCIN and its successors was just this: that they distribute uncertainties according to some plausible and manageable rules, that implausibilities are somehow eliminated after discovery, and that all this leads to an ad hoc patchwork without principled guidance (cf. Pearl 1988, Sect. 1.2). By contrast, Pearl (1988) started from the best-entrenched normative theory we had, i.e., probability theory, and showed how to make it computationally manageable via the theory of Bayesian nets. This is why his work met with such enthusiasm, at least among the more theoretically minded

AI researchers. The lesson is pertinent here as well – even though ease of computation, which as such would be bad normative advice, is not among Pollock's primary objectives.

I have an entirely different attitude towards what I have called the substance of Pollock's account of defeasible reasoning. All the theories in the field – belief revision theory, non-monotonic logic, ranking theory as far as developed in Chapters 5–8, even Levi's epistemology, etc. – are concerned with the form of inductive reasoning. Pollock, by contrast, has made the most eminent efforts to state specific defeasible reason-schemata, i.e., rules of inductive reasoning. Without them, the form would be empty. Clearly, this is rather a philosophical than a formal task, and that may be why the more formally oriented approaches have not taken it on. Still, it needs to be tackled.

In this respect, Pollock's work is a shining example for me. Of course, we again have many points of disagreement in detail. For instance, Pollock attempts to integrate probability into his account of defeasible reasoning – among other things through his statistical syllogism mentioned above – whereas I have sufficiently explained my methodological separatism. And it will be clear that when tackling this philosophical task in the subsequent chapters, I will be following quite different paths. My motive, though, is the same as Pollock's. I learned the motive from Carnap's long-standing attempts from his (1950) to his (1980), to give substance to inductive logic by stating various principles that transcend pure probability theory. Pollock, though, has shown that something similar is feasible within the realm of qualitative inductive reasoning.

11.6 On the Nature of Normative Epistemology

I indicated in the previous section that I have a dispute with John Pollock about normativity in epistemology. This is a topic of general interest that I want to tackle in terms of an opposition between computational and regulative (or syntactic and semantic, or performance- and commitment-oriented) conceptions of epistemology. In the course of my discussion I will also explain, and thus close the gap I left in Section 2.2, why I think that within normative epistemology the objects of belief must be conceived as contents, i.e., as propositions or sets of possibilities. Because of this general interest I devote a separate section to this topic.

However, let us be clear about the status of this discussion. It is *not* a discussion of the normative foundations of ranking theory, which I hope I have made extensively clear in the course of this book. They consist in the postulates (4.1) and (4.2) of consistency and deductive closure, the principle of categorical matching and the argument in Section 5.1 motivating the introduction of ranks, the various measurement procedures in Chapter 8 including Section 8.5 and the assumptions they rest on. Finally, they are incorporated into a holistic argument for the overall normative adequacy of the explications, theorems, and applications of ranking theory. All this stands by itself and is independent of the present meta-level discussion.

And let me be modest. There is a big discussion in epistemology about epistemic norms. There is an even bigger discussion about the normative foundations of language, which certainly affects epistemology due to its close connection with philosophy of language. And there is a still broader discussion about the normative foundations of social institutions in general and the suggestion that these normative aspects pervade human affairs – including even matters of individual psychology such as concepts, contents, and propositional attitudes. No doubt, all these issues are involved in our present discussion; I even have sympathies for this suggestion (cf. Spohn 2011). Clearly, though, these issues are beyond my and this book's scope. So, compared with the range and importance of the topic, this section will be most incomplete. These remarks may sufficiently excuse my relatively marginal treatment of this topic.

To begin with, it seems clear from an empirical perspective that epistemology must proceed in computational terms in the final reckoning – which we might well never reach. Although naturalized epistemology has to cover the wide range from neurophysiology up to the history of ideas, each level being most interesting on its own, it seems that the basic level is a computational or syntactic one (cf. e.g., the powerful pleading of Fodor (1975) for a representational theory of the mind). We have to go below that level by studying the biochemical mechanisms of the brain's computations, we have to rise above it in order to find out what the computations mean in behavioral and semantic terms, and we might find that the computations are quite unlike those of our computers. Still, the metaphor of the brain's being a biological computer is the best one we presently have. And it might well be that we find computational or 'software' descriptions also on much higher levels of organization.

This is what Pollock thinks is also the case also from a normative point of view. For him (see Pollock, Cruz 1999, Ch. 5) epistemic norms must be internal and internally justified; this is his way of rejecting all forms of externalism in epistemology. They need not be explicitly or consciously applied. On the contrary, the ability to conform to them is usually an implicit know-how, just as we have with grammatical rules or riding a bicycle. And they are procedural norms very much like inference rules; this is why I say that Pollock locates them at a computational level.

Why does he do so? Because, for him, norms do not merely provide an external standard for evaluating performances. Rather, norms have to govern behavior, and epistemic norms have to govern reasoning, from an internal perspective. Subjects must be able to internalize and to apply them, even if only tacitly. This entails that norms have to be directly accessible in Pollock's sense. Pollock is thus driven, I think, by a very rigid interpretation of the familiar slogan "'ought' implies 'can'". Ideally, you simply do, and can do, what you ought to do; such must be the content of norms.

I see no reason for being so rigid. Norms are to be realized, in the first place, and then we have a normative argument, whatever its methodology, about what it is that is to be realized. Realiz*ability* is another issue. We might at first have no idea how to realize the norms we accept. We may realize them only roughly. Or we may only attempt to realize them and do something that at least moves us in the right direction.

Only if the realization of the norm is completely independent of whatever we do may the norm be said to reduce to a mere wish. Emission standards, for instance, are norms in this sense; how to fulfill them is a matter of technological ingenuity. The norm "be consistent!" is such a norm. Even if there are no algorithms to decide about consistency, we can improve our consistency. The same applies to the norm: "believe only truths and as many of them as possible!", which some take to be the most basic epistemological norm. (I find this norm to be most convincingly justified in a pragmatic, i.e., decision theoretic way by the theorem about the non-negative value of cost-free information – cf. Skyrms (1990, Ch. 4) for a most general version. A ranking-theoretic analogue of this theorem might be attempted via Section 10.4). This is a perfectly good norm even if we had no methodology for searching for truths and no criterion for distinguishing between truths and falsehoods. Clearly, though, we can act to conform ever closer to this norm, and this is good enough. " 'Ought' implies 'can' " should be understood most generously.

So far, I have argued that norms need not be conceived procedurally or computationally. My aim, though, is to argue that they must not be conceived merely in this way. The first point to observe here is that procedures, rules, or computations can be correct only in the sense of being admissible, not in the sense of being mandatory. As long as the result is correct, good, reasonable, or justifiable, the way it is reached does not matter (as long as it does not contain mistakes that cancel each other). What is, for instance, 23×29? – Pause for calculation. – 667, you are correct! How did you calculate it? $20 \times 29 + 3 \times 29$? Or $23 \times 30 - 23$? Or $26^2 - 3^2$, using the third binomial formula? That is my favorite method, if feasible. Perhaps you knew it by heart or you used a calculator, whatever its algorithm. There is no instruction how to proceed; any admissible way is fine.

Pollock could agree thus far. The point dividing us is that we have two different notions of admissibility. The only sense in which a rule can be admissible for Pollock is that it can be decomposed into several applications of basic rules. For me a rule or procedure is admissible if it is sanctioned by some semantic or regulative standard; no appeal to basic rules is thereby required.

This division entails entirely diverging characters of the normative discussion. On a purely computational level, the basic rules or procedural steps are also normatively basic. They provide the axiomatic standard of correctness. And we must follow where the basic rules lead us. We cannot protest, if they should lead us to unreasonable results, because this would mean appealing to further normative standards of reasonableness. One might reply that so far the basic rules are only conjectured and that the conjectures are, of course, under the control of such further standards. Still, these further normative standards cannot be taken as such, but merely as indicative of the true basic rules. This looks like a distorted picture of normative argument.

By contrast, the regulative notion of admissibility is free of such restrictions. This notion allows for the familiar weighing of normative intuitions, singular judgments, and general principles until some normative reflective equilibrium is reached. This is vague, but this is how it is (although we might, of course, search for some (formal) model of normative judgment formation). Certainly, such considerations have computational implications, and we should elaborate on them. If we are lucky, we might even find a correct and complete set of rules to satisfy our normative conception. The standard of correctness and completeness, though, lies on the regulative level of epistemic commitments. This seems to me to be a much truer picture of our normative life.

My brief discussion of Pollock's weakest link principle and non-accrual-of-reasons principle in the previous section nicely illustrates this opposition. This is why I called Pollock's conception normatively defective. To restrict epistemic norms to the computational level of performance is simply too narrow a conception.

In order to keep my argument general I have refrained from a positive specification of the regulative level. Anything that may be plausibly advanced in a normative discussion is fine; and if it is not a directly accessible and applicable rule or procedure, it belongs to the regulative level. My examples "be consistent!" and "believe only truths!" belong to it, as well as principles of minimal change so dear to belief revision theory (provided they have a precise content). Likewise, the normative assumptions on which ranking theory rests pertain to that level.

If this argument holds, it also extends to our discussion in Chapter 2 of the objects of belief. The point is simple. If one states computational or procedural rules, one must relate to the medium in which the computations are carried out, i.e., to sentences or other representational items; these sentences or items are then the appropriate attitudinal objects. By contrast, each wider normative conception on the regulative level will induce a notion of regulative equivalence, as one might call it. Two sentences or representational items are equivalent in this sense if they are treated equally by the regulative norms and cannot be distinguished by them. This sounds cryptic, but only because I am speaking of the regulative level so unspecifically. If, for instance, the basic epistemic norms refer only to the possible truth of the items believed, as my two examples do, then the regulative equivalence is simply truth-conditional and thus logical equivalence. Then we might still take sentences or other syntactic representations as objects of belief and subject them to the regulative norm that regulatively equivalent items should have equal doxastic status (whether or not that equivalence is computable or recognized). Equivalently, we might relate doxastic attitudes more directly to the objects abstracted from the representational items by regulative equivalence, for instance to truth-conditions or propositions if that equivalence is truth-conditional equivalence.

This is exactly the step we took in Section 2.2. There, at the end, I gave three reasons for this move. Now I have finally substantiated the last and philosophically most important of these reasons. It is the very nature of normative epistemology that forces us to conceive of the objects of belief in this way.

Basically the same point can be found already in Stalnaker (1976), who argues that propositional attitudes such as beliefs and desires are tied to rational explanation (which alludes to norms of rationality), and that rational explanation requires the attitudes to be strictly propositional, i.e., to take propositions in our sense as objects. (To be precise, Stalnaker takes propositions to be sets of worlds; however, Lewis (1979a, Section XI) already insists on a wider sense of propositions that is used here as well.) Levi (1991, Sect. 2.1 and 2.7) expresses basically the same point in a still different way. As mentioned already in Section 4.1, he thinks of belief rather in terms of doxastic commitments. This is a normative matter. If these commitments refer to truth conditions, then, so Levi argued, one is committed to the principles of consistency and deductive closure. Likewise, one is committed to treat logically equivalent items alike, i.e., to take the objects of belief to be something like propositions.

One might still have doubts about my argument for the normative deficiency of the computational level. The art of computation surely is a huge scientific enterprise, occupying mathematicians and statisticians, engineers and information scientists. It surely is a basically normative enterprise about how to effectively and reasonably compress, store, retrieve, and process information with the help of clever algorithms. And it surely says how one should compute and not only which computations are admissible. However, the reference to this scientific field rather confirms my point of view. Scientists there do not directly state rules about how one should compute. They rather have regulative norms for computations like being fast, cheap, general, reliable, sufficient, etc. There is a continuous trade-off between these norms, and the optimum will vary with the case at hand. The computational rules then try to optimally realize these norms; therein lies their justification. Thus, even this field is no exception to the general pattern of normative dispute sketched above.

11.7 Formal Learning Theory

Among all the philosophical programs developed in formal epistemology one stands out by addressing the very problem of induction with fascinating, yet surprisingly different methods: (somewhat misleadingly) so-called formal learning theory; see Kelly (1996) for an elaboration in book-length, Kelly (2000) for a briefer survey, and Kelly (2004, 2008) for recent progress. This opening sentence already indicates that I am not prepared for a detailed comparison; the formal frameworks are not directly connected. Let me only briefly sketch how different, yet instructive the perspective is that formal learning theory has on inductive problems.

All the epistemological theories I have considered so far, including ranking theory, focused on learning schemes rather than inductive schemes, as I called it in Section 1.1; at least I presented them in this way. That is, I focused on the dynamics of doxastic states, on the revision step transforming a prior into a posterior doxastic state under the influence of experience. The concern then was to find an appropriate format for describing this step and to state and justify rational constraints on it. This was not only the concern of AGM belief revision and ranking theory, but also that of Levi's treatment of expansion and contraction, of Pollock's reason-schemata, and even of Lehrer's undefeated justification. Each of these theories came up with an account of reasons, or arguments, or inductive inferences, directly flowing from the respective mechanism of the doxastic dynamics. One could then exchange arguments concerning the adequacy of these accounts; and so forth.

Formal learning theory does not engage in this kind of discourse and introduces a different perspective instead. It objects that this debate about rational dynamics neglects the crucial criterion for inductive or learning schemes, namely their success or reliability. Does this debate care about whether the dynamics moves us into the right direction, approaches the truth, leads us to true beliefs and hypotheses? Not so far. Was a positive answer tacitly presupposed? Or is it trivial? If so, one would like to know why. The issue was indeed hitherto neglected.

Formal learning theory is a highly sophisticated theory about precisely the relation between inductive methods and truth. An inductive method is, as explained in Section 1.1, a scheme taking finite initial sections of infinite data streams as input and delivering a more or less specific hypothesis about (the future of) the world as output, thus attempting to answer a more or less specific question (i.e., to settle on one of a range of more or less specific possible answers). Inductive methods can then be judged on various dimensions: on the specificity of the questions answered by them; most importantly on their range of success or reliability, i.e., the range of worlds delivering an infinite data stream for which the method eventually gives a true answer; on the speed of their success, i.e., on how fast they settle on a true answer; and on their retraction efficiency, i.e., the number of revisions or retractions required before reaching a true answer. All these issues are amenable to mathematically rigorous theorizing.

It is important to correctly understand here success or reliability. As Kelly (1996) repeatedly emphasizes, success need not be known success; we might hit upon a true hypothesis without knowing it to be true. Indeed, inductive problems typically rule out this kind of guarantee. On the other hand, formal learning theory is not about actual success or actual reliability of belief formation (as it is discussed, e.g., by reliabilists in the theory of knowledge). It is strictly a priori epistemology. Indeed, Kelly (1996) speaks of a transcendental investigation in the Kantian sense, i.e., of the conditions of the possibility of learning the truth. The issue is, for which range of worlds do we know beforehand that the inductive method in question is reliable, i.e., will yield a true hypothesis after finitely many steps. And this issue has an a priori answer; within that range we are guaranteed to eventually reach success, even if unbeknown.

Kelly (1999) looks at specific inductive methods, as proposed in belief revision and ranking theory. Superficially, all looks well; most of the inductive rules, even ones I have criticized, show a satisfying limiting behavior, at least within the setting assumed by Kelly. However, Kelly discovers disturbing facts as well. He shows that the adequate, i.e., eventually successful, projection of hypotheses must be paid for with inductive amnesia, as he says, i.e., with memory loss concerning the past records. Obviously, the matter deserves closer scrutiny.

This, however, is not straightforward, because the formal frameworks of formal learning theory and the kind of theories I have considered so far do not easily square.

This is why I will not further expand on the matter. Ranking theory, e.g., is sophisticated in the short run, but thus far blind in the long run, whereas formal learning theory is unsophisticated in the short run, but most attentive to the long run. This difference just reflects the different interests and issues pursued.

There is no doubt, though, that the issues raised in formal learning theory are most urgent. Kelly (1999) is an attempt to bridge the formal differences from his side. Conversely, one must grant that these issues were indeed neglected in the literature. So ranking theory must face the challenge posed by formal learning theory. And it will. In Section 12.4 I will consider the confirmation of laws from the ranking-theoretic point of view and present relevant convergence theorems. And in the final Chapter 17 I will make a deeper and more general attempt at accounting for the relation between reason and truth, i.e., for the long-run truth conduciveness of our short-run reasons.

11.8 Possibility Theory

Let us finally turn to a few comparisons with theories mainly developed and pursued in Artificial Intelligence. There it was soon recognized that the implementation of something like inductive reasoning beyond deductive reasoning is of overwhelming importance. Moreover, the discipline was not captivated by a probabilistic bias, because probability theory was computationally recalcitrant. Thus, starting in the 70s, a lot of formal models were developed and investigated, as documented by many handbooks most actively edited by Dov Gabbay (cf., e.g., Gabbay et al. 1994 and Gabbay, Smets 1998–2000), the conference and proceedings series *Uncertainty in Artificial Intelligence*, and many other items. Indeed, formal epistemology flourished more in Artificial Intelligence than in philosophy, because the formally oriented researchers are a small minority in philosophy, and because AI researchers have more of a playful, experimental attitude, whereas philosophers want to make sense of what they do before doing anything – a strength and a weakness at the same time. Hence, although ranking theory competes with the developments in this field and has become part of them, it would be hopeless to strive for representative comparisons; a few selective remarks must do.

Let me first take up possibility theory, so-called because it purports to be a neutral account of degrees of possibility (and is thus quite foreign to modal logic as studied by philosophers). It originates from Zadeh (1978), i.e., from fuzzy set theory and hence from a theory of vagueness. Its major elaboration in Dubois, Prade (1988) and many papers thereafter shows its wide applicability, but never denies its origin. I should mention therefore that fuzzy theorizing has never gained a foothold in philosophical accounts of vagueness. The latter form a contested field (cf. e.g., Williamson 1994), but fuzzy logic is not among the serious competitors. The fact that possibility theory originates from something for which it is not suited is not a mere peculiarity, but has also disrupted the formal development. This will be my only complaint.

The formal comparison is straightforward. *Poss* is a *possibility measure* iff it is a function from some algebra of propositions over W into the closed interval $[0, 1]$ such that the following axioms are satisfied.

(11.31) $Poss(\emptyset) = 0, Poss(W) = 1$, and $Poss(A \cup B) = \max \{Poss(A), Poss(B)\}$.

Thus, possibility measures and ranking functions are almost the same. The only difference is one of scale. Full possibility 1 is negative rank 0, (im)possibility 0 is negative rank ∞; indeed, the transformation log *Poss* relative to some positive base < 1 yields a negative ranking function by translating the characteristic axiom of possibility theory into the law of disjunction (5.8b). So far, we just have another instance of the Baconian structure introduced in Section 5.2.

However, Dubois and Prade were well aware that they also need conditional degrees of possibility. They had little intuitive guidance, though. Conditional vagueness is an unknown idea, and degree of possibility is too indeterminate an idea for one to have intuitions about a conditional version. So, their first idea was the same as Shackle's, namely to understand degrees of possibility in a purely ordinal way without arithmetical meaning. Hence, they stipulated.

(11.32) $Poss(A \cap B) = \min \{Poss(A), Poss(B \mid A)\} = \min \{Poss(B), Poss(A \mid B)\}$.

This is nothing but Shackle's assumption (11.5) for his functions of potential surprise. Dubois and Prade went beyond Shackle by turning (11.32) into a definition of conditional possibility. For this purpose they made the additional assumption that, sloppily put, things should be as conditionally possible as possible, i.e., that $Poss(B \mid A)$ is the maximal degree of possibility satisfying (11.32):

(11.33) $Poss(B \mid A) = \begin{cases} P(A \cap B), \text{if } Poss(A \cap B) < Poss(A), \\ 1, \text{if } Poss(A \cap B) = Poss(A). \end{cases}$

Halpern (2003, proposition 3.9.2, Theorem 4.4.5, and Corollary 4.5.8) entails that Bayesian net theory works also in terms of conditional possibility measures thus defined. Many things, though, do not work well. It is plausible that $Poss(B \mid A)$ is between the extremes 1 and $Poss(A \cap B)$. However, (11.33) entails that it can take only those extremes. This is unintelligible. It also entails that, if neither $Poss(B \mid A)$ nor $Poss(A \mid B)$ is 1, they are equal – a strange symmetry. And so on. Such unacceptable consequences spread through the entire architecture.

There is, however, a second way to introduce conditional possibilities (cf. e.g., Dubois, Prade 1988, p. 206), namely by taking numerical degrees of possibility seriously and defining:

(11.34) $Poss(B \parallel A) = Poss(A \cap B) / Poss(A)$.

This looks much better. Indeed, the above logarithmic transformation now carries conditional degrees of possibility into conditional negative ranks. Hence, (11.34) renders possibility and ranking theory formally isomorphic. All theoretical benefits can be gained in either terms.

Still, there remain interpretational differences. Because of its shaky intuitive foundations, the achievements of possibility theory are not easily understood. In particular, the interpretation of degrees of possibility < 1 as disbelief was never envisaged. The unnaturalness of this interpretation is even clearer when we look at the possibilistic analogue of two-sided ranking functions. However, all the philosophical benefits of ranking theory elaborated in the subsequent chapters derive from firmly interpreting ranks in terms of belief or disbelief. They cannot be carried over to possibility theory as long as it wants to remain open for other interpretations, whatever they may be. Hence, possibility theory faces a choice, I think: Either it submits to an interpretation in terms of belief or disbelief, in which case it would do better to adapt the form of ranking theory. Or it will remain in an interpretational grey area severely affecting its significance.

To summarize, I see possibility theory as a technical enrichment, since its formal results are worth studying from a ranking-theoretic point of view; but I do not see it as a philosophical amendment that could produce insights transcending those of ranking theory.

11.9 Dempster–Shafer Belief Functions

The Dempster–Shafer theory of belief functions is certainly one of the major alternatives in the field. I am not aware of any satisfactory thorough-going philosophical discussion of this theory; somehow, it has remained in the domain of theoretical AI. Such a discussion is, I think, much needed. However, it would become quite complex, for the reasons to be mentioned below. Therefore, it is not on my agenda. I have chosen to be brief and to even do without introducing the formalism, well knowing that the briefness of my remarks poorly corresponds to the richness of the theory.

Shafer (1976), the first exposition of the theory, was, at one stroke, a remarkably rich elaboration, an exceptional achievement. The mathematics was unassailable, perspicuous, and computationally manageable. And the interpretational proposals were appealing, to say the least. It contained probability theory as a special case, and thus promised a far-reaching generalization of probability. The bold claim was that the Dempster–Shafer belief functions could form the basic representation of formal epistemology, and this claim had a lot of plausibility. Shafer was not concerned with belief in the intuitive sense that is our focus here. Maybe, though, the theory can embrace even this perspective; the formal apparatus is very flexible. Thus, it might fulfill the prospects of reductionism in the sense of Section 10.1, as I already emphasized at the end of Section 10.2.

Somehow, the past 30 years did not redeem the promises. The theory turned out to be an interpretational chameleon. Many people were captivated by it, tried to make sense of it, and came up with a surprising variety of interpretations. This is why

I cannot strive here for a serious discussion and why I think a *philosophical* discussion, still owing, would be useful.

For instance, the Dempster–Shafer theory grew out of a theory of upper and lower probabilities, as proposed by Good (1962) and used by Dempster (1967, 1968); cf. also Walley (1991). I did not include this theory in my comparisons, since its basic idea is that one's uncertainty is so deep that one is not able to fix one's subjective probabilities completely, but only within some boundaries, and since this idea does not appear to be a genuine extension of the probabilistic stance; or, rather, it appears to be genuine extension in a different direction. Shafer (1976), though, intended to offer a generalization of the theory of upper and lower probabilities. The status of this intention is, however, not so clear. Halpern (2003, p. 279) shows that Dempster-Shafer belief functions are tantamount to so-called inner measures in a specific sense and thus suggests that this generalization does not really transcend the probabilistic point of view.

Haenni (2005, 2009) presents the Dempster–Shafer theory as a mixture of logical and probabilistic reasoning. There are, he says, exogenous variables governed by probability and endogenous variables governed by logic; and if one unites the two kinds of variables in one model, one arrives at Dempster–Shafer belief functions. This model appears to be able to establish reductionist claims – another color of the chameleon. How the distinction between exogenous and endogenous variables in this sense comes about remains, however, unclear in my view.

The characterization and the processing of evidence are the focus of the original account of Shafer (1976). His picture entirely differs, though, from the Bayesian one. Belief change is characterized not by conditionalization, but by Dempster's rule of combination, as he calls it. The prior doxastic state is given by some DS belief function; the evidence is expressed by some simple support function, a special kind of DS belief function; and the posterior doxastic state then is generated by combining the prior state and the evidence according to Dempster's rule. Smets (1998) has generalized the dynamics of DS belief functions still further by his so-called specializations that constitute his so-called transferable belief model. Again, though, the interpretation runs behind the formal generalization.

These remarks point to the tip of the iceberg. The rest of the iceberg would only confirm that the philosophical issue raised here is indeed the nature of evidence, in DS terms, or of reasons, in my terms. The Bayesian point of view on this issue is well known, and the ranking-theoretic point of view, which I will further expand on in Chapter 16, is basically the same. (Also recall Wagner's more generalized conditionalization (9.21), which is intended as a Bayesian bridge to the Dempster–Shafer theory.) However, what precisely is the Dempster–Shafer point of view? As indicated, this is a multiply and ambiguously answerable question that I will leave undiscussed.

Whatever the answer, I have an argument for why ranking theory cannot be subsumed under the Dempster–Shafer theory. That is why we have a dispute that needs to be settled. The argument first presented in Spohn (1988, p. 156) is this:

Negative ranking functions, like Shackle's functions of potential surprise or possibility measures, seem formally to be a special case of DS belief functions; they are *consonant* belief functions as defined and discussed by Shafer (1976, Ch. 10), or rather the degrees of doubt, as Shafer (1976, p. 43) calls them, derived from such belief functions. Intuitive doubt, though, is raised by the characterization in Shafer (1976, p. 219) of consonant belief functions as "distinguished by their failure to betray even a hint of conflict in the evidence"; they "can be described as 'pointing in a single direction'". This might be adequate from the perspective of Shafer's theory of evidence. However, it makes no sense at all in the ranking-theoretic perspective. This indicates that the intended interpretations diverge completely.

The divergence shows also on the formal level. Shafer (1976) was well aware of the relation of his theory to Shackle's. On p. 224 he proves that if *Dou* is the degree of doubt derived from a DS belief function *Bel* by defining $Dou(A) = Bel(\bar{A})$, it derives from a consonant belief function if and only if

(11.35) $Dou(W) = 0, Dou(\varnothing) = 1$, and $Dou(A \cup B) = \min \{Dou(A), Dou(B)\}$

So, here it is again, the Baconian structure. According to (11.35) *Dou* is a function of potential surprise or a negative ranking function, and the consonant counterpart *Bel* is a possibility measure or a positive ranking function. Perfect agreement so far.

The divergence becomes apparent when we look at the dynamics. Shafer (1976, pp. 43 + 66f.) also defines conditional degrees of doubt by:

(11.36) $Dou(B \mid A) = [Dou(A \cap B) - Dou(A)] / [1 - Dou(A)]$.

Equivalently, conditional degrees of doubt $Dou(\cdot \mid A)$ derive from conditional degrees of belief $Bel(\cdot \mid A) = Bel \otimes Bel'$, i.e. the combination of *Bel* and *Bel'*, where *Bel'* is the simple support function given by $Bel'(B) = 1$, if $B \supseteq A$, and $Bel'(B) = 0$ otherwise. According to Shafer, *Bel'* thereby expresses maximal and incorrigible certainty of the evidence A, and $Bel(\cdot \mid A)$ expresses the degrees of belief after thus learning A.

Apart from the normalizing denominator, (11.36) looks like the Definition 5.15 of conditional negative ranks. Congruity still! However, Shafer and I agree that evidence is rarely maximally certain. So, we must rather look at how to treat uncertain evidence. It is here where our theories diverge. Our law (5.25) of conditionalization for negative ranking functions dealt with uncertain evidence by copying Jeffrey's generalized probabilistic conditionalization. Shafer (1976, p. 75), by contrast, captures a single piece A of uncertain evidence by a simple support function S defined by $S(W) = 1, S(B) = s$, if $A \subseteq B \neq W$, and $S(B) = 0$ otherwise. Learning the uncertain piece of evidence carries one from the prior DS belief function *Bel* to the posterior belief function $Bel \otimes S$, the combination of *Bel* and S according to Dempster's rule. The crucial point now is that if *Bel* is a consonant belief function, $Bel \otimes S$ will in general not be consonant; this is easily checked. Thus, whereas negative ranking functions are closed under $A{\rightarrow}n$-conditionalization (5.24), consonant belief functions are not closed under combination with simple support functions according to Dempster's rule. This finally shows that ranking theory cannot even formally be conceived as a special case of the Dempster–Shafer theory, as was already suggested by the incongruous interpretations.

Halpern (2003) sets out a more general consideration. He explains on pp. 92ff. that there are indeed two plausible ways of defining conditional DS belief functions. And on p. 107 he goes on to explain that both kinds of conditional degrees of belief allow for defining Jeffrey's generalized conditionalization. However, p. 107 and p. 114 make clear that both kinds of Jeffrey conditionalization of DS belief functions diverge from Dempster's rule of combination. Thus, there are bewilderingly many ways for defining a dynamics of DS belief functions.

This more general perspective does not help, though, vis-à-vis ranking functions. As is easily seen, consonant belief functions are also not closed under both kinds of Jeffrey conditionalization. So, even these considerations do not open a way for subsuming ranking theory under the Dempster–Shafer theory.

Does this result contradict the fact that ranking functions are equivalent to possibility measures (with their second kind of conditionalization (11.34)), that possibility measures might be conceived as a special case of DS belief functions, and that Jeffrey conditionalization works also for possibility measures as defined by Halpern (2003, p. 107)? No. The reason again is that Jeffrey conditionalization for possibility measures is not a special case of Jeffrey conditionalization for DS belief functions.

To summarize: Whatever the more general epistemological perspective of the Dempster–Shafer theory of belief functions may be, and whether or not it would survive philosophical scrutiny, it neither comprises ranking theory nor offers an account of belief. Thus, it also offers no way of establishing reductionist grounds for belief and probability. The theories are formally and philosophically at cross-purposes.

11.10 A Brief Remark on Non-Monotonic Reasoning

There is no end to possible comparisons; this simply reflects the richness of the relevant theoretical developments in the last 40 years. Let me conclude here with a final remark.

In Section 11.5 I discussed John Pollock's theory of defeasible reasoning because of its anchoring in philosophy. It is, however, only one example out of a large class of non-monotonic logics – "defeasible" and "non-monotonic" stand for the same thing – that have been developed by logicians and in Artificial Intelligence (see again Gabbay et al. 1994 or Gabbay, Smets 1998–2000). Moreover, I explained in that section that the inferential and the dynamic perspective are two sides of the same coin, thus making clear that Pollock's and my theory have the same subject matter and say different things about it. It is obvious what this means: no less than that ranking theory is also closely related to that large field of non-monotonic reasoning and should be compared with it as well. This seems to open another huge chapter.

Thanks to Rott (2001) I can be very brief, though. He offers, among other things, a comprehensive equivalence proof of three theoretical fields: belief revision theory, non-monotonic reasoning, and the theory of rational choice (or revealed preference); see in particular Rott (2001, Ch. 7). I have mentioned the relation between the first and the third field already in Section 4.3. What matters now is the equivalence between the first and the second field. Rott makes clear how far-reaching this equivalence is; there is in fact a one-to-one correspondence between possible revision (or contraction) postulates in belief revision theory and possible axioms and inference rules in non-monotonic logic.

In view of this equivalence we may take my detailed observations comparing ranking and belief revision theory in Chapters 5 and 8 and Section 10.3 as extending to non-monotonic reasoning, and to the extent to which these observations show ranking theory to be superior, this holds as well for the comparison to non-monotonic reasoning. Of course, this short-cut via Rott (2001) can in no way exhaust the present comparison; however, though indirect, it is certainly substantial.

Let me thereby close this chapter. I have developed a varied and unsystematic collection of points, and thus displayed the relations between ranking theory and its rich environment. In the systematic course of this book, however, it was but an interlude. After the general development of ranking theory in Chapters 5–9 it is high time to turn to the most fascinating and rewarding philosophical applications of ranking theory; they will occupy us for the rest of this book.

12

Laws and their Confirmation

12.1 A Plan for the Next Chapters

After the introductory Chapters 1–4, the detailed exposition of ranking theory in Chapters 5–9, and the comparative interlude in Chapters 10 and 11, I now turn to the second half of this book. This exposition has shown that ranking theory, *the* theory of belief, is a rich theory well worth being developed on its own, and we have met many issues and perspectives in need of further inquiry. However, I consider ranking theory first and foremost as a tool for pursuing the goals of philosophical epistemology, goals which are much broader than merely stating the static and dynamics of rational doxastic states. In being such a tool ranking theory unfolds its real power and beauty; there it finds its deeper justification. This is why I will develop the philosophical applications so extensively.

In the first part of the philosophical applications I will be dealing with topics of philosophy of science, laws, dispositions, and causation (and, peripherally, explanation). These chapters are not independent; rather, they present a continuous story that will reach a certain completion only in Chapter 15 on objectification. All the more, some general introductory remarks are in order before going *in medias res*.

The overarching goal of the subsequent chapters is perhaps best described as trying to understand empirical or natural necessity or modality. There are confusingly many modalities. There is logical necessity and relative logical necessity, i.e., relative to some theory or some axioms; they are well understood. There is metaphysical necessity, analyticity, and apriority; only the latter, being of an epistemological kind, will occupy us in the final chapters of this book. There are deontic modalities, clearly not our topic. Some count all the propositional attitudes among the modalities because of their intensionality. And so on. Finally, there is the class I am concerned with: nomic necessity, causal necessity, probability, counterfactuality, and determination, dispositions, powers, forces, and capacities – none of which can be understood as relative logical necessity. They are not the same, but they are closely related and thus subsumable under the common label "natural modality".

Hume's rejection of necessary connexions has firmly anchored natural modalities in our philosophical agenda, forever it seems. Hardly any topic of theoretical philosophy has been so hotly debated over the past centuries as well as in recent decades. The views on it have become ever further-reaching and ever more detailed and refined. But nothing is settled. Here, I want to contribute a further view – or rather, it seems to me (though I will not engage in exegetical discussions) Hume's old view elaborated in detail and at the height of the present debate.

The core idea is that natural modalities are covertly epistemological notions, i.e., that they can only or best be understood through their relation to overtly epistemological notions, the philosophical task being to uncover that relation. For instance, subjective probability is an overtly epistemological notion; and if Lewis (1980, p. 266) is right in claiming that his so-called Principal Principle relating credence and chance is all we know, and need to know, about objective probability, the latter turns out to be a covertly epistemological notion. The father of this idea, is, of course, Hume himself, whose skeptical reasoning culminated in the conclusion that causation is an idea of reflexion (in his sense); cf. Hume (1739, p. 165).

How this idea is to be carried out in detail remains to be seen. However, it is clear that if the elaboration is successful, it is a realization of the program labeled Humean projection by Blackburn (1993) and Grice (1991) and succinctly described by Blackburn (1980, p. 175) in a paper partially about Ramsey:

Ramsey was one of the few philosophers who have fully appreciated the fundamental picture of metaphysics that was originally sketched by Hume. In this picture the world – that which makes proper judgement true or false – impinges on the human mind. This, in turn, has various reactions: we form habits of judgement and attitudes, and modify our theories, and perhaps do other things. But then – and this is the crucial mechanism – the mind can express such a reaction by "spreading itself on the world". That is, we regard the world as richer or fuller through possessing properties and things that are in fact mere projections of the mind's own reactions: there is no reason for the world to contain a fact corresponding to any given projection. So the world, on such a metaphysic, might be much thinner than common sense supposes it. Evidently the picture invites us to frame a debate: how are we going to tell where Hume's mechanism operates? Perhaps everywhere: drawing us to idealism, leaving the world entirely noumenal; or perhaps just somewhere; or nowhere. Hume's most famous applications of his mechanism, to values and causes, are extended by Ramsey to general propositions, which to him represented not judgements but projections of our habits of singular belief, and also to judgements of probability, which are projections of our degrees of confidence in singular beliefs.

I will be less concerned with the general philosophy of this program; that would mean engaging in metaphysical dispute much more thoroughly than is within the scope of this book. Rather, I am concerned with the specific constructive details for which ranking theory will indeed prove to be an ideal tool; and I am convinced that this is, even if tedious, much more helpful to the program than more general philosophizing. In Spohn (2010a) I suggested a projectivistic treatment of objective probability that I will not repeat here. Clearly, the present chapters will provide the full background for that paper, which I could not unfold there. Also, I will not have anything to say about counterfactuality. Though I am convinced that in principle the same strategy applies to it, too, I have repeatedly stated my despair at the linguistic complexities of the counterfactual and conditional idiom. However, I will specifically deal with all the other natural modalities.

Despite my abstention from metaphysical dispute it should be clear that the program of Humean projection attempts to be a full-blown alternative to Lewis' program of Humean supervenience, which has continuously challenged me. One might think that the two programs are at cross-purposes, since the one is purely ontological and the other epistemological. The quote from Blackburn makes clear, however, that this would be a mistake; the two programs compete. In Spohn (2010a) I argued that objective chance indeed remains the "big bad bug" Lewis (1986b, p. xiv) had feared and Lewis (1994) thought he had cured, i.e., that the cure does not work and that he retains more epistemological ingredients than Humean supervenience officially allows. However, Lewis' doctrine is so impressive because it is worked out in such a detailed way. This is a further reason for my emphasis on the constructive details; only in this way are the alternatives at eye-level in the general philosophical arena. Hence, I will have little direct argument with Lewis' doctrine, except in Chapter 14 on causation. It may suffice to have raised the point in these strategic introductory remarks.

12.2 The Problem of Lawlikeness

Let us first consider nomic necessity; it will turn out to be the most simple and basic modality, though it will take a chapter full of formal details to see the simplicity. If unqualified, a law is throughout to be understood as a strict or deterministic law as opposed to a statistical one. So, what is a law? A true lawlike sentence or proposition. Philosophers are accustomed to separate the two dimensions. Kepler's laws have a lawlike status, even though they have turned out to be not exactly true, i.e., literally false. And although there might be general worries as to whether truth pertains to lawlike propositions, the more relevant problem for philosophers of science has been to explain lawlikeness itself. This has turned out to be a surprisingly recalcitrant problem.

However, it is easy to form intuitions. Physicists tend to refer to their proudest examples, conservation laws and symmetry principles, as the proper laws of nature. This tendency is as understandable as it is dangerous. It leaves very few laws and rather conceals the nature of lawlikeness. It defers the status of lawhood to very abstract principles, and the more it is deferred, the more difficult it is to analyze. Many think that the more special laws prove to be lawlike by being derived from the fundamental laws. I think it is the other way around; the fundamental laws inherit that status from all the other laws they systematize – and should hence not form our starting point.

There are all the other laws of physics, well known from its history, even though they have turned out to be derivative and at best approximately true: Kepler's laws, the law of gravitation, the ideal gas law, Hooke's law, etc. Hooke's law about the proportionality of extending force and spring extension is already a complicated case. It is a typical ceteris paribus law obtaining only under narrow conditions, and the deeper explanations in terms of the intermolecular forces within the material a spring is made of have revealed those conditions to a large extent. How one might understand such cases and related phenomena will be discussed only in the next chapter.

Then there are all the other sciences intending to state laws about their part of reality, the laws of chemical reactions, the Mendelian laws, the laws of volcanism, the laws of color vision, the law of supply and demand, and so forth. These are useful examples for sharpening and broadening our intuitions. Lange (2000) takes great care to discuss a large sample of representative laws from the various sciences.

On the other hand, there are all the silly counter-examples that might look like they state laws, but clearly don't: "all of the coins in my wallet today are made of silver", "all mountains in the UK are less than 5000 feet in height", etc. Moreover, there are various doubtful cases, and it is most revealing to study what might be doubtful about them. Lange (2000, p. 13) discusses the fact, e.g., that all native Americans have blood type 0 or A, which looks lawlike, but derives, as presently conjectured, from the fact that all native Americans descend from the first few immigrants crossing the Bering Strait who, by chance, were all of blood type 0 or A. A very nice example, already discussed by Hempel (1948, p. 152), but again suggested to me by Köhler (2004), is the Titius–Bode law about the exponential increase of the distance of the planets from the sun. If it is true for the first planets up to Saturn, must it be true for the next one, Uranus? It was indeed found to be true for Uranus, strongly supporting its lawhood. Neptune, though, turned out to be a problem case. The issue – accident or law? – seems still to be contested. I will return to the case. See also Lange (2000, Ch. 1) for many further examples.

Like many others I will focus here on puzzling pairs of two assertions that look very similar, although it seems intuitively clear that one expresses a law (is *acceptable*) and the other does not (is *bad*). The first pair is:

(12.1) (a) All (past, present, or future) humans on earth are less than 150 years old.
      (b) All persons (presently or ever) in this room are less than 80 years old.

Clearly, (b) is not a law, even if it should be true, whereas (a) may well be a law. What, however, is so different about 80 and 150 or about this room and the earth?

The most famous pair is perhaps:

(12.2) (a) All uranium spheres have a diameter of less than one mile.
      (b) All gold spheres have a diameter of less than one mile.

Again, (b) might well be true of our universe, but everybody denies it to be a law, whereas (a) is at least a derived law; uranium spheres even of much smaller size are (claimed to be) unstable and decay rapidly.

And a final example: I have a bowl at home into which I only put green apples, perhaps for aesthetic reasons, and nobody else happens to put anything into it. On the other hand, biotechnologists have produced a strange kind of bowl. Whenever you put an apple into it, it turns and remains green. It is a mystery how that mechanism works, but it apparently works; at least it has done so many times. So, we have two true assertions:

(12.3) (a) All apples in mystery bowl are green.
        (b) All apples in my bowl are green.

But only the first has a lawlike character. I think all these are acceptable intuitive judgments. How could we account for them? And how could we thus obtain an explication of lawlikeness?

In the 50s when the issue acquired its importance, the focus was on *essential generality*, and the hope was that it could be captured in syntactic or semantic terms provided by extensional logic. This led nowhere. That is the point of (12.1) and (12.2). (12.1a) suggests that there might be laws for very special objects at a very special place, and if one is inclined to reject this suggestion, then (12.2b) makes clear that there might well be universal accidental truths.

One might see a difference between the uranium and the gold spheres lying in the fact that only (12.2a) is part of a more comprehensive systematization. So, the nature of laws might be their *essential systematicity*, as it is almost entailed by Lewis' (1994, Sect. 3) best-system analysis of laws, which he refers back to Ramsey (1928) (we will reach different conclusions in agreement with Ramsey 1929). I am in doubt about this proposal. It reverses the order of analysis; according to it we first need to know what the true laws are before we can then explain lawlikeness (since any nonsense can be systematized by greater nonsense, if one takes the trouble). Moreover, it denies further explanation of the lawlikeness of the axioms of the systematization, the most fundamental laws. Finally, I do not see why there cannot be isolated laws. To be sure, the best-system analysis does not exclude isolated laws (this is why I said that it only almost entails essential systematicity), but they always bring along an unbalance in simplicity and strength. Kepler's laws, for instance, were isolated in the beginning. Did they turn out to be laws only through their derivation from Newton's theory of gravitation? I think not. Were they considered to be laws from the beginning only because of the standing hope that they would be systematized? Again, I think not. The uncontested lawlikeness of Kepler's laws must have a better explanation. Surely, we attempt to systematize laws, and we might say a lot about what it means to systematize. However, this does not exhaust the nature of laws.

Another obvious point is *modality*. The laws in (12.1a + 2a + 3a) speak about all *possible* apples, human beings, and uranium spheres; uranium spheres *must* be smaller than one mile in diameter. By contrast, (12.1b + 2b + 3b) are accidentally true; gold spheres *can* be so large, even if they are not. Alas, the modality used here is exactly the nomic necessity that we are attempting to elucidate; there is no antecedently understood modality we could plug in here. The circle is too short to be instructive.

Since the 70s, then, inquiry has focused on three further features of laws and lawlike propositions, as already suggested in the classic discussion of Hempel (1965, pp. 338–43), but laid out in more detail by van Fraassen (1989, Ch. 2, Sect. 4), Lange (2000, Sect. 1.2), and in many other places. Laws *support counterfactuals*, another aspect of the modal content. If I were to take that red apple and to put it into my bowl, the apple would stay red, and (12.3b) would be false; if I were to put it into mystery bowl, it would turn green, and (12.3a) would remain true. If that huge sphere were to consist of uranium, it would explode; if it were to consist of gold, it would still exist.

Laws have *explanatory force*. The apples in mystery bowl are green because they are in mystery bowl, whereas the apples in my bowl are there perhaps because they are green (if I so strictly attend to my aesthetic standards). This sphere is so small because it is made of uranium, and that sphere is so small, yes, because it is made of gold *and* because it is too expensive or too nonsensical to produce larger ones. The close relation between if-then- and because-constructions is, of course, familiar (cf., e.g., Rott 1986).

Finally, lawlikeness is wedded to *inductive inference*; laws are *confirmed* by, or *projectible* from, their instances. These locutions are usually taken to be synonymous. We will, however, come to distinguish them; the confirmation of laws by observed instances – enumerative induction – and its projection to new instances will turn out to be two different matters. That the first of the ten apples in my bowl is green makes it likelier that all apples in my bowl are green only in the trivial sense that nine apples are less likely to have a different color than ten apples. If, however, we suspect a law or a lawlike mechanism behind mystery bowl, confirmation by the first green apple and projection to the next green apples is boosted tremendously. The same applies to the age of people in the case of (12.1a + b).

No doubt, these examples are a bit shaky, because scientific practice shows us that the confirmation of laws is much more complicated than enumerative induction; it is hard to force examples simply on the basis of the latter. On the other hand, enumerative induction is primitive not only in the sense of being too simple, but also in the sense of being basic. We cannot hope to understand the more complicated forms of scientific inference, if we cannot account for enumerative induction. This is why philosophers tend to focus on such shaky examples and such oversimplified inductive inferences, and I will do so too.

Since the first three ideas – essential generality, essential systematicity, and modality – did not elucidate lawlikeness, can we use these three basic features for further analysis? This is not clear. One might well say that such an analysis would be a case of *obscurum per obscurius*; our intuitions about lawlikeness seem firmer than the intuitions about those features. This view is endorsed by Lewis (1994, pp. 478f.) when he suggests that the best-system analysis of laws entails those features; it is a stand-off, as he says, and I agree. Still, I find it disappointing to take lawlikeness or essential systematicity as primitive; we will see how fruitful it can be to reverse the order of analysis.

If we do so, the next question is which of the three basic features of lawlikeness should be taken as primary, as the starting point of analysis. There is no use in engaging in strategic argument. The only way to resolve the issue is to propose specific analyses and to then compare their degree of success. However, it should be clear what my preferences are. Explanation is closely related to causation, and both seem to be difficult ideas, the analysis of which I will proceed to only in Chapter 14. Moreover, I have expressed my uneasiness about counterfactuals several times. Indeed, my analysis of causation (1983b) led me from Lewis' (1973a) theory of counterfactuals through Gärdenfors' (1978, 1981) epistemic account of counterfactuals (the predecessor of the AGM theory of belief revision) ever deeper into the theory of induction and thus to ranking theory. In any case, the feature relating lawlikeness to inductive inference is clearly the one most directly accessible to ranking theory, which *is* a theory of inductive inference. Hence, this feature will be the starting point for my analysis of lawlikeness. This decision finds strong support, I find, in Lange (2000), for whom the relation between laws and counterfactuals is the heuristic starting point, but who also arrives at induction as the most basic issue.

Before we proceed in this manner, we face, however, a deeper issue. How are we to understand the relation between lawlikeness and its basic features? The prevailing idea has been, it seems, that the inquiry of lawlikeness should lead us to something *entitling* us to use laws in induction, explanation, and counterfactual discourse in the way we do. Thus, for instance, if induction is really the most basic aspect, lawlikeness should be something that *justifies* the role of laws in induction. However, this desire for justificatory insights has resulted in perplexity; there is just no good further justificatory candidate in sight.

The alternative is to say that lawlikeness is exhaustively characterized by its basic features, i.e., if my preference is correct, that it *consists* in the characteristic role in induction. This means the denial of any deeper justification. It should be clear that I endorse this denial; I did so already at the crucial turning point of our discussion in Section 4.2 (cf. also Section 6.4 and the end of Section 11.2). This is, as emphasized, the consequence of inductive skepticism from Hume to Goodman that taught us not to presuppose deeper justification, but to inquire instead about how far rationality carries us in inductive matters. Lange (2000) takes a similar attitude when explaining what he calls the root commitment concerning the inductive strategies associated with laws.

The best-system analysis of laws also denies deeper justification; this is not cause for complaint. However, even then it appears disappointing. It just claims that optimal systematicity entails the three basic features, but does not explain how it does so. The strategy I will pursue will provide more insight into the basic features of lawlikeness and their relation, even though it also rejects further justification from outside the basic features.

We are thus to study the role of laws in induction, i.e., their confirmability and projectibility. Hempel (1945) introduced qualitative confirmation theory, and great efforts were spent on this project in the 50s and 60s. It was, however, abandoned in the 70s, mainly because the efforts were completely unsuccessful. Niiniluoto (1972) gives an excellent survey that displays all the unresolved incoherencies. (A further reason surely was the rise and success of counterfactual logic, that answered many problems in philosophy of science, though not the problem of induction, and thus attracted a lot of the motivation originally directed to an account of confirmation.)

The only remaining viable alternative for confirmation theory was Bayesianism, which thus came to dominate in philosophy of science. It is not clear to me why it still does, why philosophers of science have largely ignored all the options from formal epistemology. Levi's sophisticated epistemology did not widely radiate into philosophy of science; AGM belief revision theory mainly made a career in AI; defeasible and non-monotonic reasoning was something for logicians and not for scientists, and so forth. For instance, as much as I can agree with Lange (2000), he also turns to Bayesianism as a basis of his reflections on induction.

In any case, I believe that philosophy of science went entirely wrong in the last 30 years in this respect; the despair with regard to qualitative confirmation theory was premature. My suggestion is, of course, that ranking theory is suitable for providing such a theory and, if our envisaged strategy works, for thus accounting for lawlikeness. In fact, my suggestion will be that the parallel between probability and ranking theory observed in Chapter 10 fully extends to this field. Bayesianism is related to statistical laws and good for statistical methodology, whereas it is ranking theory that is related to strict or deterministic laws, i.e., the kind of laws that are our primary interest here.

More specifically, my suggestion will be that de Finetti's (1937) story, which offered a perfect account of the relation between subjective and objective probability (although it was intended as an elimination of the latter) carries over to ranking theory and thereby offers an equally perfect account of lawlikeness, of deterministic laws and their confirmation.

This is what I will constructively develop in the rest of the chapter. My discussion will take two dialectic turns, which are possibly not immediately perspicuous: the move away from instantial relevance to what I will call persistence in Section 12.4, and the rise to second-order attitudes in Section 12.5. Some results will appear perfectly natural, others may look artificial, but they are necessitated by the well-founded characteristics of ranking theory. I trust that the picture that emerges in the end will be found to be convincing.

12.3 Laws, Symmetry, and Instantial Relevance

Our study of the inductive behavior of strict laws must start with the fixing of the relevant possibility space W and propositional algebra $\mathcal{A}$ over W. I will assume an infinite sequence $a_1, a_2, \ldots$ of objects that fall under finitely many *mutually exclusive* and *jointly exhaustive* properties F_i ($i \in I$) for some finite index set I (these correspond to the Q-predicates of Carnap 1950/62, §31). We can then state generalizations of the form "all objects are F_1 or ... or F_j" or "none of the objects are F_1 or ... or F_j". This is a quite general framework. The objects might be material objects, F_1 might be "black raven", F_2 "non–black raven", F_3 "black non–raven", and F_4 "non–black non–raven";

and the generalization considered might be "no object is F_2", i.e., "all ravens are black". Or the objects might be actual movements of a point in a finite state space over finitely many points of time. For instance, the moving point might represent the position and the momentum of finitely many bodies measured to the millimeter of a large space, and the points of time might be seconds within a large period. A generalization might then state that each such object, i.e., movement of a point in the state space, has a certain shape, e.g., obeys a certain difference equation.

This framework generalizes upon Spohn (2005a), where I was content with computing the simplest case of two properties F and non-F. Clearly, though, I have not chosen the most general framework; a countably infinite set of objects has to be infinity enough. I will not strive for more mathematical generality, not least because I have not investigated how ranking theory combines with continuum mathematics.

Let us fix the notation we will use. As usual, $\mathbf{N}$ is the set of non-negative integers and $\mathbf{N}^+ = \mathbf{N} \cup \{\infty\}$. Moreover, $\mathbf{N}' = \mathbf{N} - \{0\}$ is to be the set of positive integers. Having exactly one of the properties F_i ($i \in I$) is tantamount to taking a value from I. Hence, we may represent each possibility as a sequence $\mathbf{w} = (w_1, w_2, \ldots)$, where each $w_n \in I$. $\mathbf{w}$ says that the first object a_1 has F_{w_1}, a_2 has F_{w_2}, etc. In short, our possibility space is $W = I^{\mathbf{N}'}$. Equivalently, we may represent the objects $a_1, a_2, \ldots$ by an infinite sequence of variables $X_1, X_2, \ldots$, all taking values in I and generating W in the way defined in Chapter 2 so that $X_n(\mathbf{w}) = w_n$. Since we will be dealing only with natural and hence complete ranking functions, we may allow any subset of W to be a proposition; i.e., $\mathcal{A} = \mathcal{P}(W)$.

We are particularly interested in generalizations or regularities. So, for each $J \subseteq I$, let G_J be the proposition $\{\mathbf{w} \in W \mid \text{for all } n, w_n \in J\} = J^{\mathbf{N}'}$ saying that all objects take values in J, and $G^J = \{\mathbf{w} \in W \mid \text{for all } n, w_n \notin J\} = (I - J)^{\mathbf{N}'}$ saying that no object takes a value in J. The negative formulation is perhaps more useful, since the generalizations $G^i = G^{\{i\}}$ ($i \in I$) can be taken as basic and as defining all other generalizations by $G^J = \bigcap_{i \in J} G^i$. I still avoid speaking of laws, since in the end we will not identify laws with such general propositions or regularities, although they will be closely related.

We will moreover need a way of denoting (sequences of) singular facts. If $i \in I$ and $J \subseteq I$, let us use $\{w_n = i\}$ and $\{w_n \in J\}$, respectively, as short for $\{\mathbf{w} \in W \mid w_n = i\}$ and $\{\mathbf{w} \in W \mid w_n \in J\}$, i.e., for the proposition that the variable X_n takes the value i or some value in J. Similarly for $\{w_{n_1} = i_1, \ldots, w_{n_k} = i_k\}$ and other variations. Finally, let $\mathcal{A}_n$ be the complete algebra of propositions only about the first n objects, i.e., generated by $X_1, \ldots, X_n$ or by propositions of the form $\{w_1 = x_1, \ldots, w_n = x_n\}$.

The conclusion of Section 12.2 set the task of studying the inductive behavior of laws. What might this mean? The first and most plausible guess (which will be disappointed in an instructive way) is that it means studying how such generalizations are confirmed by such sequences of singular facts – where confirmation is now to be taken in a ranking-theoretic sense, i.e., as positive relevance, or the reason relation as defined in (6.1); we will not need any more sophisticated measure of confirmation (as discussed in Merin 2008).

On which ranking function is confirmation to be based? Not on a specific one; our study is supposed to be general. We must, however, make plausible restrictions; otherwise, our investigation will lead nowhere. So, let us first restrict our study to natural and hence complete ranking functions; I will speak about no others. We may thus omit the adjective(s). This restriction is not always necessary; I will explain at the end of Section 12.5 where our results depended on completeness.

Let us secondly assume regularity, since confirmatory relations are undefined for infinite ranks. However, we must loosen our Definition 5.27 when dealing with the uncountable possibility space W. Carnap (1971b, p. 101) also required only that all molecular propositions must have inductive probability > 0. The same effect is produced by

Definition 12.4: A ranking function κ for $\mathcal{A}$ is *regular* iff for all $n \in \mathbf{N}'$ and all non-empty $A \in \mathcal{A}_n$, $\kappa(A) < \infty$.

Regularity is henceforth to be understood in this weaker sense.

Let us thirdly conform to the maxim: "All are equal under the law." Less cryptically, we assume that the relevant ranking function is *purely qualitative* in the sense that it is only the distribution of properties that matters and not the specific objects instantiating the properties. Certainly, the properties themselves are intended to be purely qualitative, too. This intention is, however, not written into the formal framework, and the explication of its meaning would deeply involve us in ontological dispute. That is, grue-like properties are not excluded by the formal framework as such. Of course, our condition on ranking functions is nothing but the classic assumption of *symmetry*, as precisely defined in

Definition 12.5: A ranking function κ for $\mathcal{A}$ is *symmetric* iff for all sequences $x \in W$, all $n \in \mathbf{N}'$, and all permutations φ of $\{1, \ldots, n\}$ $\kappa(\{w_1 = x_1, \ldots, w_n = x_n\}) = \kappa(\{w_1 = x_{\varphi(1)}, \ldots, w_n = x_{\varphi(n)}\})$.

The symmetry assumption has a most venerable history that need not be recapitulated here. Let me only mention that it played a fundamental role in de Finetti's philosophy of probability (under the name "exchangeability"). And it was a basic postulate in all versions of Carnap's inductive logic (in relation to the objects; in relation to the properties it turned out to be more problematic). The most comprehensive treatment is found in van Fraassen (1989, parts III and IV), who indeed went so far as to argue that symmetry should take the key role in scientific reasoning, replacing the allegedly confused idea of lawlikeness. We will attend to this claim. In any case, we restrict our study to regular and symmetric ranking functions.

What counts for symmetric ranking functions is only how many times the properties are instantiated. Let us introduce a bit of notation for this purpose. I will use $\mathbf{n} \in \mathbf{N}^I$ as variables for families or I-tuples $(n_i)_{i \in I}$ of non-negative integers indexed by I. I often use the corresponding italic n for denoting $n = \sum_{i \in I} n_i$. By speaking of I-tuples I sometimes also use I to denote its own cardinality. For $J \subseteq I$ we set $\mathbf{n}_J = (n_i)_{i \in J}$ and $\mathbf{n}^J = (n_i)_{i \in I-J}$, so that $\mathbf{n} = (\mathbf{n}_J, \mathbf{n}^J)$; hence, in particular $\mathbf{n}_i = n_i$. Furthermore, $(\mathbf{n}^i, m_i)$ is to denote the I-tuple $\mathbf{n}$ with n_i replaced by m_i, $(\mathbf{n}^{i,j}, m_i, m_j)$ the I-tuple $\mathbf{n}$ with n_i and n_j replaced by m_i and m_j, etc. $\mathbf{0}$ is to be the I-tuple $(0, \ldots, 0)$. Moreover, $\mathbf{m} \leq \mathbf{n}$ says that $m_i \leq n_i$ for all $i \in I$ and $\mathbf{m} < \mathbf{n}$ says that $\mathbf{m} \leq \mathbf{n}$ and $m_i < n_i$ for at least one $i \in I$. Finally, let us denote by $E_\mathbf{n}$ the proposition that the properties F_i realize with the absolute frequencies n_i ($i \in I$) in the first n objects. With this notation we can first define:

> *Definition 12.6*: A function f from $\mathbf{N}^I$ into $\mathbf{N}$ is *non-decreasing* iff $\mathbf{m} \leq \mathbf{n}$ entails $f(\mathbf{m}) \leq f(\mathbf{n})$ for all $\mathbf{m}, \mathbf{n} \in \mathbf{N}^I$. It is *minimitive* iff for all $\mathbf{n} \in \mathbf{N}^I$ $f(\mathbf{n}) = \min_{i \in I} f(\mathbf{n}^i, n_i + 1)$. It *represents* a regular symmetric ranking function κ or is the *representative function* of κ iff $\kappa(E_\mathbf{n}) = f(\mathbf{n})$ for all $\mathbf{n} \in \mathbf{N}^I$.

Then we have the obvious

> *Theorem 12.7*: Each regular symmetric ranking function κ is represented by some function f from $\mathbf{N}^I$ into $\mathbf{N}$. A function f is a representative function (of some ranking function) if and only if it is non-decreasing, minimitive, and $f(\mathbf{0}) = 0$.

So, our study of regular symmetric ranking functions will reduce to the study of such representative functions.

Symmetry has a surprising consequence. Recall first that one could prove the so-called null confirmation of laws within Carnap's λ-continuum of inductive methods. That is, each confirmation function obeying the so-called λ-principle (that entails regularity and symmetry) gives, a priori and a posteriori, probability 0 to each generalization G_J ($J \subset I$) (see Carnap 1950/62, §110F, or Stegmüller 1973, pp. 501f.). Sloppily put, there is nothing to confirm about laws. This was a shocking discovery. Carnap (1950/62, §110G) thus proposed replacing the issue of the confirmation of laws by the issue of the confirmation of the next instance(s). Who cares for the infinite generalization? What matters are only the instances we meet in our lifetime. In this way, enumerative induction turned into the principle of positive instantial relevance (that was derivable from the λ-principle) and found a Bayesian home (after the failure of qualitative confirmation theory).

Not all were convinced by this step and some instead rejected Carnap's λ-principle. In particular, Hintikka (1966) showed how to do better by proposing a two-dimensional continuum of inductive methods that respects regularity and symmetry and gives positive a priori probability to a generalization that a posteriori converges to 1 or 0, depending on its truth or falsity (cf. Hintikka, Niiniluoto 1976, and Kuipers 1978). I am unsure of present sympathies for that proposal.

Anyway, the issue is radically avoided within the ranking-theoretic context. Given symmetry, there is no difference between belief in the next instance and belief in the generalization about the infinitely many future instances. Suppose that after having observed the first n objects you believe that the next object will not have F_j; that is,

$\kappa(\{w_{n+1} = i\}) = r$ for some $r > 0$. Because of symmetry you then believe that any future object will not have F_i; that is $\kappa\{w_{n+k} = i\} = r > 0$ for any $k \geq 1$. Since κ is complete, the law 5.8(c) of infinite disjunctions holds; and this entails that you believe with the same strength r that *none* of the future objects has F_i; that is $\kappa\left(\bigcup_{k \geq 1}\{w_{n+k} = i\}\right) = r > 0$. I welcome this conclusion, even though it will sound unintuitive for the probabilistically trained. Whether we can return to naïve or unbiased intuitions is not clear to me. In any case, the conclusion follows from the well-founded principles of ranking theory. (One can avoid the conclusion by giving up completeness. However, this is not the point here. Those complaining about the conclusion will also complain about the fact that the next instance not having F_i is as firmly expected as the next billion instances not having F_i.)

In this way, Carnap's problem of the null confirmation of infinite generalizations dissolves in the ranking-theoretic framework, and the transition to instantial relevance that was a doubtful auxiliary move in the Bayesian context turns out to be fully legitimate here. So, the ranking function governing the confirmation of generalizations should obey instantial relevance. This can come in a stronger or weaker form. The *principle of positive instantial relevance* (PIR) says that, given any evidence $E_n \in \mathcal{A}_n$ concerning the first n objects, the $n+1$th object a_{n+1} having the property F_i confirms a_{n+2} having F_i. The weaker principle of *non-negative instantial relevance* (NNIR) requires only that the contrary is not confirmed, i.e., that given any $E_n \in \mathcal{A}_n$ the two-sided rank of a_{n+2} having F_i is not lowered by a_{n+1} having F_i. (For the acronyms and an application within a probabilistic framework to a semantic universal concerning the word "but" and its dedicated translation, see Merin 1996, 1999.)

Confirmation might be given various senses or measures here. Let us consider only two basic senses; there is no point in this chapter in being more sophisticated in this respect. (For more sophistication cf. Merin 2008.) One sense is that the a posteriori two-sided rank (i.e., after evidence E_n) of a_{n+2} having F_i given a_{n+1} has F_i is higher than it is given a_{n+1} lacks F_i. The other sense is that this rank conditional on a_{n+1} having F_i is higher than the unconditional rank of a_{n+2} having F_i:

> *Definition 12.8*: Let κ be a regular symmetric ranking function for $\mathcal{A}$ and τ the corresponding two-sided ranking function. Then κ *satisfies* PIR_c (i.e., PIR in the *conditional* sense) iff for all $E_n \in \mathcal{A}_n$ ($n \geq 0$) and all $i \in I$, $\tau(\{w_{n+2} = i\} \mid E_n \cap \{w_{n+1} = i\}) > \tau(\{w_{n+2} = i\} \mid E_n \cap \{w_{n+1} \neq i\})$. And κ *satisfies* PIR_n (i.e., PIR in the *non-conditional* sense) iff for all $E_n \in \mathcal{A}_n$ and all $i \in I$, $\tau(\{w_{n+2} = i\} \mid E_n \cap \{w_{n+1} = i\}) > \tau(\{w_{n+2} = i\} \mid E_n)$. Finally, κ *satisfies*, respectively, $NNIR_c$ or $NNIR_n$ iff the weak inequality $\geq$ holds in PIR_c or PIR_n instead of the strict $>$.

It is an obvious consequence of the law (5.23a) of disjunctive conditions that PIR_n entails PIR_c, that $NNIR_c$ entails $NNIR_n$, and that none of the reverse entailments holds. Which of the two senses should we prefer? We had already discussed the issue. PIR_c just says that a_{n+1}'s having F_i is a *reason* for a_{n+2}'s having F_i (conditional on E_n) in the sense of Definition 6.1 (or 6.3). There, before (6.2), I already argued that PIR_c is preferable to PIR_n. Likewise, $NNIR_c$ is preferable to $NNIR_n$, as is also clear from the fact $NNIR_n$ is even compatible with the negation of $NNIR_c$, i.e., with *negative* instantial relevance in the conditional sense.

Whichever sense we consider, we are, however, in for unpleasant surprises. A minor observation is this: Whereas symmetry implies NNIR in the probabilistic context (where we need not distinguish $NNIR_c$ and $NNIR_n$) (cf. Humburg 1971, p. 228), this is not so in the ranking-theoretic context; it is easy to construct examples for both, representative functions violating $NNIR_c$ and representative functions satisfying $NNIR_c$. The main observation, though, is: Whereas regularity, symmetry, and the so-called Reichenbach axiom imply PIR in the probabilistic context (cf. Humburg 1971, p. 223), PIR turns out to be unsatisfiable in the ranking-theoretic context even in its weaker sense PIR_c.

Theorem 12.9: There is no regular symmetric ranking function for $\mathcal{A}$ satisfying PIR_c.

A direct proof would be tedious and not particularly revealing. I will defer the proof until (12.20), when we will have collected all the pieces explaining the unsatisfiability of PIR_c.

What does this negative result teach us about lawlikeness? Since the confirmation of the next instances is tantamount to the confirmation of generalizations (concerning all future instances) relative to symmetric ranking functions, and since PIR_c is not generally realizable, we must conclude that generalizations or regularities cannot always be confirmed by their positive instances. As the proof of (12.9) will show, this conclusion cannot be hedged. All generalizations G_j (concerning future instances) can be confirmed by at most finitely many positive instances. Hence, it seems we have reached a dead end; we cannot use enumerative induction, i.e., confirmability by positive instances for characterizing strict laws.

We might settle for the weaker $NNIR_c$, if this is the only ranking-theoretically feasible way. This seems unsatisfactory, though, since it leaves arbitrary room for instantial irrelevance; it seems hard to accept that confirmation by positive instances should fail most of the time. We will eventually discover, however, that we are deceived on this point; we will be able to fully restore positive instantial relevance or enumerative induction, though only after some dialectic labor. For the time being we should accept the negative conclusion.

12.4 Laws and Persistence

We must start anew. What else might it mean to study the inductive behavior of laws? One might conjecture that the basic fault of the considerations so far was that they inquired into the confirmation of generalizations or regularities, while our intention is to apply enumerative induction only to laws, which somehow are something more special. However, the study above gave no hint as to what this specialness might be.

One might also conjecture that the previous considerations were misguided insofar as they looked at PIR_c and $NNIR_c$ conditional on any evidence whatsoever, even evidence that outright falsifies the generalization in question. It is clear, though, that this objection does not literally apply. I carefully stated that what gets confirmed (or not) is always the generalization concerning the future (or unobserved) instances and not the full generalization. However, this sounds like a lame excuse. It still appears odd to worry about the confirmation of a generalization about the future that has been falsified in the past, perhaps many times.

In a way, my point will be that this is not so odd, after all. This has to do with the projectibility of strict laws. I had mentioned in Section 12.2 that the projectibility of a law is usually equated with its confirmability; we project its past success onto the future. This we do with lawlike, not with accidental, generalizations. But we might give projectibility a stronger reading, not as the extension of past observations to future cases, but simply as the continuous application of the law to future cases, whatever the past. I have to explain this dim suggestion.

Let us ask: *What does it mean to believe in a law?* As long as we do not know what a law is, it is hard to answer. It is clear, though, what it is to believe in a regularity or generalization. A generalization is some proposition $G_j \in \mathcal{A}$ ($J \subseteq I$), and to believe in it means $\kappa(\overline{G_j}) > 0$, relative to any negative ranking function κ. However, this belief can be realized in many ways, even if we assume κ to be symmetric; the inductive relations among the various instances can take many different forms. Which form may or should they take? It is clear that as long as one observes only positive instances the belief in G_j is maintained or even strengthened; this is what $NNIR_c$ implies at least for generalizations of the form G^i ($i \in I$). What happens, though, when one observes negative instances? If negative instances of G_j get believed, then G_j gets disbelieved according to any κ; this is a trivial matter of deductive logic. However, with regard only to the future instances, anything might happen according to κ, even if we impose symmetry and $NNIR_c$. Let us more closely look at two paradigmatic (and extreme) responses to negative evidence, in order to better understand the spectrum of possible responses. I call them the persistent and the shaky attitude.

If you have the *persistent* attitude, your belief in further positive instances is unaffected by observing negative instances, i.e., $\tau(\{w_{n+1} \in J\} \mid \{w_1, \ldots, w_n \notin J\}) = \tau(\{w_{n+1} \in J\}) > 0$. If, by contrast, you have the *shaky* attitude, your belief in further positive instances is destroyed by the first negative instance (and due to $NNIR_c$ also by several negative instances), i.e., $\tau(\{w_{n+1} \in J\} \mid \{w_1, \ldots, w_{n-1} \in J\} \cap \{w_n \notin J\}) \leq 0$.

I want to suggest that the different attitudes are distinctive of treating generalizations as lawlike or accidental. Let us look at our puzzling pairs (12.1–12.3). You might believe (12.1b) that all persons ever in this room are less than 80 years old, for whatever reason, perhaps because you have seen only younger ones or perhaps because someone has indicated that admittance to this room is subject to a rule. Your belief might be strengthened by meeting further people younger than 80, but it fades when you find an older one. That is, in this case you do not only believe that not all persons in the room are younger than 80 (this is dictated by logic), but you also lose your confidence that there will be no further exceptions. Your attitude is shaky. By contrast, you also believe (12.1a) that all humans on earth are less than 150 years old. Now you encounter someone claiming to be older. Probably, you would not accept the claim, whatever her credentials are. Perhaps, her birth certificate is a fake. But even if you cannot find fault with her claim and tend to accept it, you would think that she remains an exception and expect the next people you meet to be younger than 150. That is, your attitude tends to be persistent.

Or take (12.2). If you bump into a one mile gold sphere on your intergalactic journey, you would be surprised – and start thinking there might well be further ones. If, however, you stumble upon a uranium sphere that large (and survive it), you would be surprised again; but then you would start investigating this extraordinary case while sticking to your reasons for declaring such a case impossible and expecting no further exception.

If you hear of my bowl you might well believe that all apples in it are green (12.3b). But if I offer you a red apple from it, you lose your trust in the story; the other apples in the bowl might then have any color. By contrast, if you get convinced of mystery bowl (12.3a) and then find a red apple in it, you might think that its mysterious mechanism is not always working properly or that some special apples are apparently immune to the mechanism. But you would continue projecting the rule into the future.

The Titius–Bode law mentioned above is a real-life example where astronomers had split attitudes. Some were shaky and took an apparent counter-example as evidence that it was never a law in the first place and that further counter-examples would be no surprise. Others were persistent or became persistent after the discovery of Uranus, trying to explain or explain away the later apparent counter-example of Neptune and thus to restore the law. As far as I see, the distinction between the persistent and the shaky attitude fits other examples discussed in the literature as well, as listed, e.g., in Lange (2000, pp. 11f.).

I am aware that these illustrations seem only partially convincing. One would be better prepared to say how one would respond to such examples if they were described in more detail, especially concerning the evidence that led one to believe in the relevant generalization in the first place. However, I am painting the picture in black and white in order to elaborate the opposition between the two ideal types of the persistent and the shaky attitude. I admit there is a lot of grey as well. Still, I insist on my ideal types. There are four considerations to elucidate the grey area.

First, our intuitions are likely to be probabilistically biased. A description of the examples in terms of subjective probabilities would certainly diverge from one in terms of beliefs and expectations. This might partially explain a reluctance to accept my presentation above.

Second, my ideal types delimit a broad range of less extreme attitudes. Being shaky means to be very shaky; the belief in further positive instances may instead fade more slowly. And being persistent means to be strictly persistent; the belief in further positive instances may instead take so long to fade that we never come to the point of testing it. Our investigation of symmetry and NNIR in Section 12.5 will give a clearer impression of all the possibilities within this range. Maybe a less persistent or a less shaky attitude fits the examples still better.

Third, it is clear that when confronted with apparent counter-instances to a conjectured generalization, we never either just accept them while rejecting our generalization or persistently write them off. Rather, as already indicated in my illustrations, we carefully study the apparent counter-instances in order to find out whether they really are exceptions and what makes them exceptions. In this way, we hope to be able to state appropriate qualifications that turn the original generalization into an exceptionless law. This process is often accompanied by widening the conceptual field, by considering more properties to be potentially relevant. In short, we take the original law only as a ceteris paribus law. This is an issue I will more thoroughly discuss only in the next chapter, which will certainly contribute to the overall adequacy of my explications.

The fourth point is, I think, the most important. My discussion of the examples was misleading in a way. I simply asked you for your inductive intuitions in all these cases and tried to push them to the persistent or the shaky extreme. But this was an unguarded maneuver, and you may have been reluctant to yield because I was pushing you into an obviously absurd direction. The persistent attitude sticks to its predictions come what may, even given overwhelming counterevidence; it has no capacity to learn. And this is just silly and contrary to any inductive intuitions.

However, this would be a subtle misunderstanding, against which I can guard only after having invited it. I started with the belief in a generalization, observed that it can be realized in many ways in ranking-theoretic terms, and distinguished two extreme ways. Then I suggested that treating the generalization as a deterministic law means having the persistent attitude. That is, *if* you were to firmly believe in the law and nothing else, then you would have the persistent attitude; this was to be my claim. However, you never simply believe in the law and nothing else. You more or less tentatively believe in the law, you always reckon with alternative laws, and you are always in the middle of processing factual information. Your inductive situation is never so unambiguous, and therefore you might have rightfully doubted my description of the examples. Hence, it is important to observe my proviso "if you were to firmly believe only in the law". Then, I submit, the persistent attitude is much more plausible as an expression of lawlikeness.

This observation also preliminarily dissolves the complaint about the silly inability to learn of the persistent attitude. The doxastic attitude that expresses the belief in a law need not display learning capacity by itself. What is rationally required is only that one is not tied to such a doxastic attitude, but is able to change it in response to the evidence. In fact, we have arrived here at a most important point. I already gave the ranking-theoretic account of the change of belief in propositions in Chapter 5, and thus derivatively in this chapter gave the ranking-theoretic account of the confirmation of propositions and of generalizations in particular. If, however, the belief in a law is a certain kind of doxastic attitude, we so far have no account of the change and of the confirmation or disconfirmation of such an attitude, i.e. of the belief in a law. Thus, the first-order account of the confirmation of propositions needs to be complemented by a so far missing second-order account of the confirmation of such first-order attitudes. Thereby, and only thereby, can the complaint be definitely rejected. This will be our task in the next section; I can promise complete success. For the moment, the point was only to distinguish the (so far unexplained, second-order) confirmation of laws from the (ranking-theoretically explained, first-order) confirmation of propositions (and generalizations), thus clarifying possible confusions concerning my claim about what is characteristic of a belief in a law.

My suggestion is idiosyncratic at most insofar as it is firmly couched in ranking-theoretic terms. Apart from this, however, it has ample historic precedent. My chief witness, no doubt, is Ramsey (1929) who states very clearly: "Many sentences express cognitive attitudes without being propositions; and the difference between saying yes or no to them is not the difference between saying yes or no to a proposition" (pp. 135f.). And "... *laws are not either*" (namely propositions – my emphasis) (p. 150). Rather: "The general belief consists in (*a*) A general enunciation, (*b*) A habit of singular belief" (p. 136). This is what has become known also as the view that laws are not general statements, but rather inference rules or licenses, as advanced by Schlick (1931), Ryle (1949, Ch. 5), Toulmin (1953, Ch. 4), and others. And it is exactly what I have suggested above: a belief in a law is not only the belief in a generalization, but is expressed by the persistent attitude that is perfectly characterized as a habit of singular belief.

Of course, one should not read too much of the present context into Ramsey's writings; he was occupied with the problems of his day, for instance, with rejecting Russell's acknowledgment of general facts and with thus finding a different interpretation for general sentences (cf. Sahlin 1991, pp. 138ff.). Sahlin (1997, p. 73) also mentions that Ramsey's view of general propositions was apparently influenced by Hermann Weyl. Still, his attempts to construct a pragmatic theory of belief, which are also reflected in my brief quotation, are surprisingly modern. There does not appear to be a 70 year gap when Lange (2000) says in his much more elaborate book that "the root commitment that we undertake when believing in a law involves the belief that a given inference rule possesses certain objective properties, such as reliability" (p. 189). (Lange then goes on to explain the differences between the older inference license literature and his position which hide in the reference to "certain objective properties such as reliability".)

The inference license view of laws was not generally well received, mainly, I think, because it was hard to see from a purely logical point of view what the difference might be between accepting the axiom "all F's are G's" and accepting the inference rule "for any a, infer Ga from Fa". The only difference is that the rule is logically weaker; the rule is made admissible by the axiom, but the axiom cannot be inferred by the rule. What else besides this unproductive logical point could be meant by the slogan "laws are inference rules" was always difficult to explain.

Still, one might say that the inference license view puts more emphasis on what to do in the single case. This emphasis need not be mere rhetoric. It is reflected, I find, in my notion of persistence and thus receives a precise induction-theoretic basis. We have often been able to observe (e.g., in Section 11.5) that ranking functions can be understood as (possibly quite complex) defeasible inference rules that can be taken to be reliable, even if they are not universally valid.

Hence, the mark of laws is not their universal validity that breaks down with one counter-instance, but rather their operation in each single case that need not be impaired by exceptions. This point is most prominent in Cartwright (1983, 1989) and her continuing forceful efforts to argue that what we have to attend to are the capacities and their (co-)operation, taking effect in the single case. I will elaborate on my concurrence with Cartwright in the next chapter.

Moreover, I should mention that the idea of persistence closely resembles the notion of resilience that is central to Skyrms' (1980, part I) analysis of probabilistic causation. However, since that context is somewhat different and since there is no point now in a comparative discussion, I preferred to choose a different label. (I will briefly return the notion of resiliency in Section 14.6.)

Last, not least, if the relation between counterfactual and ranking-theoretic talk were to be clarified, then I think we would find a close relation between my notion of persistence and the notion of invariance that Woodward (2003, ch. 6) so ably argues to be a replacement for the notion of a law. In any case, when Woodward (2003, pp. 299ff.) explains how Skyrms' notion of resilience and his notion of invariance differ despite their apparent similarity, this reminds me of my above distinction between persistent (or invariant) first-order attitudes and second-order attitudes (that relate to resilience in Woodward's construal – of course, the distinction of levels of attitudes is quite un-Woodwardian).

So much for some important agreements. The most obvious disagreement is with Popper, of course. Given how much philosophy of science owes to Popper, my account is really ironic, since it concludes in a way that it is the mark of laws that they are *not* falsifiable by negative instances (this is the persistent attitude); only accidental generalizations are so falsifiable (this is the shaky attitude). Of course, the idea that the belief in laws is not given up so easily has been familiar at least since Kuhn (1962), and even Popper (1934, ch. IV, §22) insisted from the outset that the falsification of laws proceeds by more specialized counter-laws rather than simply by counter-instances. However, I have not seen the point elsewhere being so radically stripped to its induction-theoretic bones.

Let me sum up this discussion. My core claim was that the belief in a strict law consists in the persistent attitude. The latter in turn is characterized, in formally precise terms, by the ranking-theoretic *independence* of the instantiations of the law, i.e., of the variables $X_1, X_2, \ldots$ In other words, if ξ is any negative ranking function for I, we may define λ_ξ as the independent and identically distributed infinite repetition of ξ, the representative function f_ξ of which is given by $f_\xi(\mathbf{n}) = \sum_{i \in I} n_i \cdot \xi(i)$. If ξ is regular, λ_ξ is regular, too. Of course, λ_ξ is symmetric. To say that λ_ξ satisfies (any version of) NNIR is correct, but misleading, since λ_ξ is distinguished by perfect instantial irrelevance. λ_ξ persists in ξ for the next instance whatever the previous evidence.

Does the core claim help us to say what a law *is*? So far, I was careful only to analyze the *belief* in a law. The tricky point is that the belief in a law turned out, following Ramsey, to be not (merely) the belief in something; the habit λ_ξ of singular belief ξ is just a more complex doxastic attitude. "Belief in a law" is a *not further parsable* phrase. One might object that this shows that the approach is the wrong one to start with, and I would retort by pointing to the unimpressive alternatives. Let us not repeat this. On the contrary, my inclination is to say that such a λ_ξ *is* a law (and not just a belief in a law); but this looks like a rhetorical trick. We are here in a real linguistic predicament.

To resolve the predicament, I propose calling the λ_ξ's *subjective laws*. This is a sufficiently artificial term to indicate the rhetorical move explicitly, not surreptitiously. And there is a twofold justification behind that move. First, we will be able in the next section to fully explain the phrase "belief in a law" in a *parsable* manner, as belief in something, namely exactly in such λ_ξ's; this has to do with the distinction between first-order and second-order attitudes already indicated. Secondly, the adjective "subjective" is not merely to signify that subjective laws are still something doxastic, i.e., entertained by doxastic subjects. The adjective's significance runs deeper. Let me explain.

I emphasized from the beginning that lawlikeness abstracts from the truth dimension of laws. Likewise, the talk of belief in laws is silent about the truth of laws. Thus, it is clear that subjective laws are only possible laws that may or may not obtain. Still, laws in that wider sense must be something that *can* obtain, can be true or false. Subjective laws, however, are generally not of this kind. ξ may so far be *any* ranking function for the index set I of properties or, what comes to the same thing, for the next instance. And we do not yet know what it could mean to say that such a ranking function obtains – is true or false. It can at best be true *of* a subject. Likewise, we do not know what it means to call the subjective law, the infinite independent repetition λ_ξ of ξ, true or false. λ_ξ is as subjective as ξ; this is the real significance of "subjective" here.

In Chapter 15 I will advance an account of the objectification of ranking functions, distinguishing those ranking functions that can be called true or false. Of course, I will call then λ_ξ an objective law if it is based on an objectifiable ξ. This is at present a hardly intelligible statement. It indicates, though, the strategy behind my terminology and the fact that our present topic of lawlikeness will be completed only then. In this chapter, we must be content with having worked up at least to the notion of a subjective law. Let us conclude our efforts with

Definition 12.10: Let Ξ be the set of negative ranking functions for I. For any $\xi \in \Xi$ λ_ξ is to be that ranking function for $\mathcal{A}$ according to which all of the X_n ($n \in \mathbf{N}'$) are distributed according to ξ and are mutually independent, i.e., the representative function f_ξ of which is defined by $f_\xi(\mathbf{n}) = \sum_{i \in I} n_i \cdot \xi(i)$. λ_ξ is called the *subjective law* for $\mathcal{A}$ based on ξ. λ is a *subjective law* for $\mathcal{A}$ iff $\lambda = \lambda_\xi$ for some $\xi \in \Xi$. The set of all subjective laws for $\mathcal{A}$ will be denoted by Λ.

I would like to point out here that $\xi \in \Xi$ and $\mathbf{n} \in \mathbf{N}^I$ are entities of the same formal kind; both are functions from I into $\mathbf{N}$. (There is the difference that ξ must take 0 as a value, whereas $\mathbf{n}$ need not; to get rid of the difference, consider $\mathbf{n} - \min n_i$ instead of $\mathbf{n}$.) This remark appears trifling. However, I think it has a deep significance. It states that ranking functions (something subjective in the mind) and absolute frequencies of exceptions (something objective in the external world) are of the same kind. Subjective probabilities and relative frequencies are also entities of the same kind; and their relation is fundamental to probabilistic epistemology. Thus, the analogy is that subjective ranking functions likewise find their external anchoring in absolute frequencies (of exceptions), a point already indicated in Section 10.3 (under the label "reality aspect") and repeated here as the formal occasion arose. It will be substantially deepened in Chapter 15.

12.5 The Confirmation of Laws

Having studied the ranking-theoretic confirmation of generalizations in Section 12.3 with instructive, even if unexpected results, we concluded that we should instead study the confirmation of laws. Now, a subjective law turned out to be not a proposition, but a special kind of ranking function. What it could mean to confirm such a law is, however, as yet unexplained. Can we make precise sense of it?

Yes, we can. Fortunately, there is clear precedent provided by Bruno de Finetti's philosophy of probability. Given the close similarity between probability and ranking theory, we should attempt to translate de Finetti's account of statistical laws. We will see that this attempt indeed works in a satisfactory way.

What gets the attempt going is the observation that (in the simplest case) a statistical hypothesis consists in a Bernoulli measure for an infinite sequence of random variables according to which these variables are independent and identically distributed. Thus, *what I called a subjective law is nothing but the ranking-theoretic analogue to a Bernoulli measure*! Positivistically minded, de Finetti was suspicious of objective probabilities and of statistical laws hypothesizing them. In his (1937) he thus developed an account which perfectly explains them away. His famous representation theorem showed that each symmetric probability measure for the infinite sequence of random variables is a *unique* mixture of Bernoulli measures. If the symmetric measure expresses your subjective probabilities and the Bernoulli measures represent hypotheses about objective probabilities, your subjective opinion is hence a unique mixture of objective statistical hypotheses, the weights of the mixture representing your credence in these hypotheses. The mixture changes through evidence in such a way that it favors the hypotheses close to the observed frequencies and disfavors the others. In fact, if the evidence converges to a certain limit of relative frequencies, the mixture converges to the Bernoulli measure taking these limiting relative frequencies as objective probabilities (provided this measure is in the so-called carrier of the original mixture). Thus, the learning process satisfies the so-called Reichenbach axiom (under the proviso mentioned). (For all this see, e.g., Loève 1978, Sect. 30.3 and 32.4, Jeffrey 1971, Sect. 10, or the relevant contributions to Jeffrey 1980.)

(Note, by the way, that de Finetti's representation theorem is a special case of the ergodic theorem proved by George D. Birkhoff in 1931. Despite the efforts of Skyrms (1984, pp. 52ff.) it is not so clear, however, whether the more general ergodic theorem can be given the same philosophical significance.)

De Finetti intended his story to be eliminativistic. Since the mixtures are always unique, he could confine himself to talking only of the first-order symmetric subjective probabilities. All the rest, objective statistical hypotheses and mixtures of them expressing second-order subjective probabilities in objective hypotheses, may be taken as mere as-if constructions.

The positivistic motive is not mine. Also, the business about objectivity and subjectivity is not yet ours. I emphasized above that subjective laws are still subjective and that the issue of objectivity will be raised only in Chapter 15. But apart from this, every detail of de Finetti's account is perfectly suited for translation. If my identification of a law with a belief in a law was bewildering, and if my accentuation of this identification by the label "subjective law" was unhelpful, even if honest, my hope is that this translation of de Finetti's philosophy of probability will essentially clear up the confusions. (As we will see, the translation does not work in a perfectly smooth way; there are some niceties that did not show up in the special case of two properties, i.e., $I = \{1, 2\}$, belabored in Spohn (2005a) and hence underestimated there. They add to the list of divergences between probability and ranking theory.)

So, we start from the set Λ of possible subjective laws for $\mathcal{A}$, and we consider a complete negative ranking function ρ for Λ that is to represent our (dis-)belief in laws (or rather in subsets of Λ, i.e., types of laws or law-propositions). This time, this makes proper sense; ρ represents second-order beliefs in independent first-order objects, i.e., subjective laws. We can then mix the possible first-order subjective laws by the second-order attitude ρ, as defined in (5.37), and we thus arrive at some first-order ranking function κ for $\mathcal{A}$ that expresses our (dis-)belief in factual propositions; of course, factual disbelief is never guided by just one law taken as certain.

Referring to such mixtures as weighted, or weighted averages, might raise overly firm probabilistic associations. To avoid such associations, let us call ρ an *impact function*; the subjective law λ has *impact* $\rho(\lambda)$ on the mixture κ. This loosely agrees with our picture in Section 6.3, in which the balancing of ranks is a matter of forces, of pulls and pushes, rather than a matter of weighing. Of course, the impact of λ is larger, the *smaller* $\rho(\lambda)$, since ρ expresses disbeliefs.

Our task is thus to investigate the precise relation between such second-order mixtures of subjective laws by impact functions and first-order attitudes or regular symmetric ranking functions defined for factual propositions (or their representative functions).

Analytically, i.e., within a real-valued framework, the issue would be quite straightforward. There, a representative function would be any non-decreasing function for I-tuples of non-negative reals starting at the origin (minimitivity would have no analogue in the real-valued framework), subjective laws would simply be I-dimensional hyperplanes passing through the origin, the impact function ρ would shift the laws or hyperplanes upwards, and the mixture of all the shifted hyperplanes (as defined in 5.37) is just their lower envelope. This makes clear that, if and only if a non-decreasing function is concave, it can be conceived as the lower envelope of its tangential or supporting hyperplanes. This geometric picture helps intuition enormously, and it makes clear that we are going to move on mathematically well-trodden paths. We only have to translate the picture into our discrete framework. (I am indebted here to Günter M. Ziegler and Friedrich Roesler for crucial hints regarding convex geometry and analysis.)

First, we have to take care of regularity:

Definition 12.11: The impact function ρ for Λ is *proper* iff $\rho(\lambda_\xi) < \infty$ for at least one regular $\lambda_\xi \in \Lambda$ (or regular $\xi \in \Xi$).

Theorem 12.12: The mixture κ of Λ by ρ is regular in the sense of (12.4) if and only if ρ is proper.

Proof: If ρ is proper, κ is obviously regular. Conversely, let ρ not be proper. So, whenever $\rho(\lambda_\xi) < \infty$, there is some $i \in I$ such that $\xi(i) = \infty$. Hence, if $\mathbf{1} = (1, \ldots, 1)$, $\kappa(E_1) = \infty$, since $\lambda_\xi(E_1) = \infty$ for all λ_ξ with $\rho(\lambda_\xi) < \infty$.

Our further inquiry best proceeds by embedding our representative functions mapping $\mathbf{N}^I$ into $\mathbf{N}$ (and thus being subsets of $\mathbf{N}^{I+1}$) into $I+1$-dimensional Euclidean space $\mathbf{R}^{I+1}$. (Recall $\mathbf{R}$ is the set of real numbers and $\mathbf{R}^+$ the set of non-negative reals.) For each set $S \subseteq \mathbf{R}^{I+1}$ the *convex hull* of S is defined as $\mathcal{H}(S) = \{\mathbf{x} \in \mathbf{R}^{I+1} \mid \mathbf{x} = \sum_{k=1}^{n} \alpha_k \mathbf{x}_k$ for some $\mathbf{x}_1, \ldots, \mathbf{x}_n \in S$ and $\alpha_1, \ldots, \alpha_n \geq 0$ such that $\sum_{k=1}^{n} \alpha_k = 1\}$. More specifically, for each non-decreasing function f from $\mathbf{N}^I$ into $\mathbf{N}$ let $\mathcal{H}(f)$ denote the convex hull of $\{(\mathbf{n}, x) \mid \mathbf{n} \in \mathbf{N}^I$ and $f(\mathbf{n}) \geq x \in \mathbf{R}\}$. Thus, $\mathcal{H}(f)$ is not simply the convex hull of $\{(\mathbf{n}, f(\mathbf{n})) \mid \mathbf{n} \in \mathbf{N}^I\}$, but includes all points below ("below" taken relative to the $I+1$th dimension). Clearly, $\mathcal{H}(f)$ is closed and the upper boundary of $\mathcal{H}(f)$ is given by some function g from $(\mathbf{R}^+)^I$ into $\mathbf{R}^+$ such that $g(\mathbf{x}) = \max\{y \mid (\mathbf{x}, y) \in \mathcal{H}(f)\}$. $\mathcal{H}(f)$ is thus what is called the *hypograph* of g. Since $\mathcal{H}(f)$ is convex, g is *concave* (or *convex from below*) in the sense that for all $\mathbf{x}, \mathbf{y} \in (\mathbf{R}^+)^I$ and $\alpha \in [0, 1]$ $g(\alpha \mathbf{x} + (1 - \alpha) \mathbf{y}) \geq \alpha g(\mathbf{x}) + (1 - \alpha) g(\mathbf{y})$. (For all these terms and concepts cf. Stoer, Witzgall 1970 or Rockafellar 1970.)

Let us call the concave function g just constructed the *real spreading* of f and denote it by $f^{\mathbf{R}}$. With it, we can also most naturally explain concavity for the original representative functions:

Definition 12.13: A non-decreasing function f from $\mathbf{N}^I$ into $\mathbf{N}$ is *concave* iff for all $\mathbf{n} \in \mathbf{N}^I$, $f^{\mathbf{R}}(\mathbf{n}) = f(\mathbf{n})$, i.e., iff the real spreading of f is an extension of f. A regular symmetric ranking function κ is *concave* iff its representative function is concave.

In other words, f is concave iff all points $(\mathbf{n}, f(\mathbf{n}))$ are boundary points of $\mathcal{H}(f)$, and f is not concave iff for some $\mathbf{n}$, $f(\mathbf{n}) < f^{\mathbf{R}}(\mathbf{n})$, so that $(\mathbf{n}, f(\mathbf{n}))$ is an inner point of $\mathcal{H}(f)$.

With this, we immediately arrive at a first representation result:

Theorem 12.14: A ranking function κ for $\mathcal{A}$ is regular, symmetric, and concave if and only if it is the mixture of the set Λ of subjective laws by some proper impact function ρ.

Proof: If κ is such a mixture, it is regular because of (12.12) and symmetric because all $\lambda_\xi \in \Lambda$ are symmetric. It is also concave: Let f_ξ be the representative function of the law λ_ξ. Then, $f_\xi^{\mathbf{R}} + \rho(\lambda_\xi)$ is just a non-descending linear function or an I-dimensional hyperplane, and $\mathcal{H}(f_\xi + \rho(\lambda_\xi))$ is just the closed half-space below $f_\xi^{\mathbf{R}} + \rho(\lambda_\xi)$. Since the representative function f of κ is given by $f = \min_\xi (f_\xi + \rho(\lambda_\xi))$, $\mathcal{H}(f)$ is simply the intersection of all the $\mathcal{H}(f_\xi + \rho(\lambda_\xi))$, and each $(\mathbf{n}, f(\mathbf{n}))$ indeed a boundary point of $\mathrm{H}(f)$. Hence, f and κ are concave.

Conversely, suppose κ is regular, symmetric, and concave. We know that any closed convex set C in $\mathbf{R}^{I+1}$ is the intersection of the closed half-spaces containing it (cf. Stoer, Witzgall 1970, p. 98). It is indeed the intersection of the *supporting half-spaces H* of C (defined as minimally containing C; i.e., no closed half-space properly contained in H contains C), since each boundary point of C lies on at least one *supporting hyperplane* (defined as the boundary hyperplane of a supporting half-space) (cf. Stoer, Witzgall 1970, p. 103).

This applies to $\mathcal{H}(f)$, too. Since f and hence $f^{\mathbf{R}}$ are non-descending (i.e., not descending in any direction of a unit vector $(\mathbf{0}^i, 1)$), the hyperplanes supporting $\mathcal{H}(f)$ are non-descending as well. And since all $(\mathbf{n}, f(\mathbf{n}))$ are integer-valued, the gradient or derivative of $f^{\mathbf{R}}$ at any point into any direction $(\mathbf{0}^i, 1)$ is integer-valued as well. That is, the hyperplanes supporting $\mathcal{H}(f)$ can be chosen to be of the form $f_\xi^{\mathbf{R}} + \rho(\lambda_\xi)$ ($\xi \in \Xi$). Hence, $f^{\mathbf{R}}$ is the envelope of the supporting hyperplanes of that form, and since f is concave, i.e., since each $(\mathbf{n}, f(\mathbf{n}))$ is a boundary point of $\mathcal{H}(f)$, $f = \min (f_\xi + \rho(\lambda_\xi))$, the minimum taken over these supporting hyperplanes. So, κ is the mixture of Λ by some ρ (which must be proper according to 12.12).

In a way, this is already our main result. Still, we are not finished at all. We have, later on, to discuss the philosophical content of (12.14). First, however, we should investigate in more detail which kind of mixtures can be used to represent concave ranking functions. For this we need a further piece of notation: For $\xi, \zeta \in \Xi$ let us write $\xi \le \zeta$

iff for all $i \in I, \xi(i) \leq \zeta(i)$ and $\xi < \zeta$ iff $\xi \leq \zeta$ and for some $i \in I, \xi(i) < \zeta(i)$. Then, a note-worthy observation is

> *Theorem 12.15*: If κ is the mixture of Λ by some proper ρ, it is a *finite* mixture of Λ by some ρ in the sense that $\rho(\lambda_\xi) < \infty$ only for finitely many $\xi \in \Xi$.

Proof: If κ is a mixture of Λ by some proper ρ and f its representative functions, then f^R is the envelope of the hyperplanes of the form $f_\xi^R + \rho(\lambda_\xi)$ supporting $\mathcal{H}(f)$. Since f or f^R is concave, these hyperplanes never become steeper. More precisely, if $\rho(\lambda_\zeta) > \rho(\lambda_\xi)$, then not $\zeta \geq \xi$, i.e., $\zeta(i) < \xi(i)$ for some $i \in I$. However, any sequence ξ_1, $\xi_2, \ldots$ in Ξ such that not $\xi_s \geq \xi_r$ for all $r < s$ must be finite, because in each dimension $i \in I$ one can descend from ∞ to 0 only in finitely many steps. Therefore, f^R is the envelope of finitely many hyperplanes of the form $f_\xi^R + \rho(\lambda_\xi)$ supporting $\mathcal{H}(f)$ and κ a finite mixture of Λ.

In other words, Theorem 12.15 states that for any f representing a regular symmetric concave ranking function κ the hypograph $\mathcal{H}(f)$ is a convex polyhedron.

There still remains the question of whether we can say something more specific than (12.15). In particular, it would be nice to know whether the mixture generating a concave κ is unique in some sense, as de Finetti was able to show in the probabilistic case. In that case any tiny weight made a difference. This is different in the ranking-theoretic case. Here the impact of a mixture component λ_ξ may be so small, i.e., $\rho(\lambda_\xi)$ so large, that it does not change the mixture at all. This was clear from (12.15). If the impact of all but finitely many λ_ξ had been chosen to be just large enough, but not infinite, this would not have changed the resulting mixture. Hence, uniqueness of the mixing impact function ρ is never to be expected.

This non-uniqueness of the mixing impact function, i.e., this amount of insensitivity vis-à-vis the precise impacts, is ultimately a consequence of the basic law 5.8(b) of disjunction for negative ranks. One might see here an advantage of the probabilistic approach. The situation is ambiguous, though. One might as well praise this insensitivity, since it allows remote subjective laws (or hypotheses or possibilities in general) to have finite ranks, but no manifest influence on beliefs. We had occasion to observe this blessing in a slightly different guise already in Section 7.2 after (7.15y). (See also Section 14.15 that expands on that point.)

Be this as it may, since we know that a finite mixture in the sense of (12.15) suffices to generate a concave κ, we may hope that there is a suitable sense of a minimal mixture uniquely generating κ. This is indeed the case. The geometric picture is quite simple. Take any concave representative function f and consider the convex polyhedron $\mathcal{H}(f)$ generated by f, which is the hypograph of the real spreading f^R of f. Such a polyhedron has *faces* that are defined as the intersection of the polyhedron with a supporting hyperplane (that intersects the polyhedron only at boundary points – see the definition in the proof of 12.14). These faces can have any dimension from 0-dimensional vertices up to I-dimensional faces spanning the entire supporting hyperplane (cf. Stoer, Witzgall 1970, pp. 38ff.).

Now we look for those supporting hyperplanes (= subjective laws with an impact factor) that intersect with $\mathcal{H}(f)$ at *maximal faces* or faces of maximal dimension in the sense that there is no supporting hyperplane intersecting with $\mathcal{H}(f)$ at a larger or higher-dimensional face. Those *maximally* supporting hyperplanes are clearly those minimally needed for generating $\mathcal{H}(f)$. We do not need more, since all the other supporting hyperplanes meet $\mathcal{H}(f)$ in lower-dimensional faces contained in the maximal faces. And we cannot do with less, since any relative interior point of such a maximal face (which, by definition, is not contained in a lower-dimensional face) is met only by such a maximally supporting hyperplane.

The following characterization of these maximally supporting hyperplanes will prove useful:

Theorem 12.16: $f_\xi^R + \rho(\lambda_\xi)$ ($\xi \in \Xi$) is a hyperplane maximally supporting $\mathcal{H}(f)$ if and only if it is a supporting hyperplane and there is no $\zeta < \xi$ such that $f_\xi^R + \rho(\lambda_\xi)$ is a supporting hyperplane.

Proof: Let $f_\xi^R + \rho(\lambda_\xi)$ be a supporting hyperplane. Thus there is an $\mathbf{n} \in \mathbf{N}^I$ with $f(\mathbf{n}) = f_\xi(\mathbf{n}) + \rho(\lambda_\xi)$. Suppose that for some $\zeta < \xi$, $f_\xi^R + \rho(\lambda_\xi)$ is also a supporting hyperplane. Then ζ can be chosen such that $\zeta(i) < \xi(i)$ and $\zeta(j) = \xi(j)$ for $j \neq i$. Hence, $f(\mathbf{n}^i, n_i + 1) < f_\xi(\mathbf{n}^i, n_i + 1) + \rho(\lambda_\xi)$. Since f is concave, ζ can even be chosen such that $\xi(i) - \zeta(i) = f_\xi(\mathbf{n}^i, n_i + 1) + \rho(\lambda_\xi) - f(\mathbf{n}^i, n_i + 1)$, i.e., that $f(\mathbf{n}^i, n_i + 1) = f_\zeta(\mathbf{n}^i, n_i + 1) + \rho(\lambda_\xi)$. This shows that $f_\xi^R + \rho(\lambda_\xi)$ is not maximally supporting.

Conversely, suppose that $f_\xi^R + \rho(\lambda_\xi)$ is not maximally supporting. Thus there is a supporting hyperplane $f_\xi^R + \rho(\lambda_\zeta)$ meeting $\mathcal{H}(f)$ at a higher-dimensional face. Let $i \in I$ be (one of) the additional dimension(s). Hence, there is in particular some $\mathbf{n} \in \mathbf{N}^I$ such that $f(\mathbf{n}) = f_\xi(\mathbf{n}) + \rho(\lambda_\xi)$, but $f(\mathbf{n}^i, n_i + 1) < f_\xi(\mathbf{n}^i, n_i + 1) + \rho(\lambda_\xi)$. Since f is concave, ζ can again be chosen as before; i.e., the supporting hyperplane meeting $\mathcal{H}(f)$ at a larger face can be chosen to be of the form $f_\xi^R + \rho(\lambda_\xi)$ with $\zeta < \xi$.

Does this description of $\mathcal{H}(f)$ in terms of maximally supporting hyperplanes survive the coarsening to natural numbers and thus apply also to the original representative function f? Yes and no. There is a subtle distinction that does not show up as long as we move within Euclidean space:

Definition 12.17: For any $\xi \in \Xi$, λ_ξ is a *non-redundant component* of the mixture of Λ by ρ iff for some proposition $A \in \mathcal{A}$ min $\{\lambda_\zeta(A) + \rho(\lambda_\zeta) \mid \zeta \in \Xi\}$ < min $\{\lambda_\zeta(A) + \rho(\lambda_\zeta) \mid \zeta \in \Xi - \{\xi\}\}$; otherwise, it is a *redundant component*. Furthermore, λ_ξ is a *latently non-redundant component* of the mixture of Λ by ρ iff for some proposition $A \in \mathcal{A}$ min $\{\lambda_\zeta(A) + \rho(\lambda_\zeta) \mid$ not $\xi < \zeta \in \Xi\}$ < min $\{\lambda_\zeta(A) + \rho(\lambda_\zeta) \mid$ not $\xi \leq \zeta \in \Xi\}$; otherwise, it is a *strongly redundant component*. Finally, the mixture of Λ by ρ is *minimal* iff for each strongly redundant component λ_ξ of that mixture $\rho(\lambda_\xi) = \infty$ and for each latently non-redundant component λ_ξ the mixture of Λ by ρ differs from the mixture of Λ by ρ' whenever $\rho'(\lambda_\xi) < \rho(\lambda_\xi)$ and $\rho'(\lambda_\zeta) = \rho(\lambda_\zeta)$ for $\zeta \neq \xi$.

Let me explain these definitions. λ_ξ is a non-redundant component of the mixture κ of Λ by ρ iff there is a proposition A for which λ_ξ *solely* determines the mixture minimum $\kappa(A)$. Hence, one should think that the required sense of a minimal mixture derives from this notion; all the redundant components should receive impact ∞.

However, this idea does not agree with the geometric picture developed before, for the following reason: For a concave representative function f, all the vertices or extreme points of the hypograph $\mathcal{H}(f)$ are points of the function f, i.e., of the form $(\mathbf{n}, f(\mathbf{n}))$. This is not to say, though, that each $(\mathbf{n}, f(\mathbf{n}))$ is a vertex of $\mathcal{H}(f)$; it can also be a relative interior point of a face of $\mathcal{H}(f)$. Now, at each vertex of $\mathcal{H}(f)$ several hyperplanes supporting $\mathcal{H}(f)$, even several maximally supporting ones, possibly intersect. Hence, it may also be that some maximally supporting hyperplane meets f only at vertices $(\mathbf{n}, f(\mathbf{n}))$ that are contained also in other maximally supporting hyperplanes. In that case, this hyperplane (or the corresponding subjective law) would not be required for generating f – it would be a redundant component – even though it would be required for generating $\mathcal{H}(f)$ or f^R.

We had better stick to the geometric picture, not only because it is mathematically more useful, but rather for the substantial reason that redundant components might become non-redundant through learning; this point will be explained below (see (12.21) and the appertaining explanations). Therefore, Definition 12.17 introduces the notion of latent non-redundancy. λ_ξ is a latently non-redundant component of the mixture κ of Λ by ρ iff at one point A it solely determines the mixture minimum $\kappa(A)$ taken only over those $\zeta \in \Xi$ such that not $\xi < \zeta$. Hence, as (12.17) makes clear, these latently non-redundant components λ_ξ precisely define the hyperplanes $f^R_\xi + \rho(\lambda_\xi)$ maximally supporting $\mathcal{H}(f)$. Since the latter were those minimally needed for generating $\mathcal{H}(f)$, we have actually shown:

> *Theorem 12.18*: For each regular symmetric concave ranking function κ for $\mathcal{A}$ there is a unique impact function ρ for Λ such that κ is the minimal mixture of Λ by ρ in the sense of (12.17).

This is as close as we can get to de Finetti's representation theorem in terms of ranking theory. It should have been clear all along that my account indeed is an ironic commentary to the basic theme of van Fraassen (1989). In no way do I doubt the profound and increasing significance of the notion of symmetry as developed by van Fraassen in grand historic lines as well as in many detailed analyses. I also agree with the direction of van Fraassen's criticism of various current approaches to laws of nature (though I would not subscribe to all of his arguments). However, there is no need to bring the notion of symmetry into a principled opposition to the notion of a law. On the contrary, (12.14) and (12.18) tell us how tightly they are knit together: no laws without symmetry, and no symmetry (and concavity) without laws! The relation is as close in the deterministic case as de Finetti has shown it to be in the probabilistic case.

From the mathematical point of view our results seem perfectly satisfactory; the geometric notion of concavity was the one we needed. However, the philosophical significance is still wanting; we should know what concavity says in terms of inductive behavior or in terms of instantial relevance. Unfortunately, I can offer only a partial answer:

Theorem 12.19: A regular symmetric concave ranking function satisfies $NNIR_c$.

Proof: In terms of the representation function f $NNIR_c$ says for any $\mathbf{n} \in \mathbf{N}^I$ and $l \in I$:

(a) $f(\mathbf{n}^l, n_l + 2) + \min_{i,j \neq l} f(\mathbf{n}^{i,j}, n_i + 1, n_j + 1) \leq 2 \min_{i \neq l} f(\mathbf{n}^{i,l}, n_i + 1, n_l + 1)$.

Suppose that the minimum on the RHS is realized by k. Thus, what we have to show is:

(b) $f(\mathbf{n}^l, n_l + 2) + \min_{i,j \neq l} f(\mathbf{n}^{i,j}, n_i + 1, n_j + 1) \leq 2 f(\mathbf{n}^{k,l}, n_k + 1, n_l + 1)$.

Now, either the minimum on the LHS of (b) is $f(\mathbf{n})$. Then $f(\mathbf{n}^{k,l}, n_k + 1, n_l + 1) = f(\mathbf{n}^l, n_l + 1)$ so that (b) reduces to

(c) $f(\mathbf{n}^l, n_l + 2) + f(\mathbf{n}) \leq 2 f(\mathbf{n}^l, n_l + 1)$.

This, however, is a direct application of concavity. Or the minimum on the LHS of (b) is $> f(\mathbf{n})$. Then $f(\mathbf{n}^l, n_l + 2) = f(\mathbf{n})$ and $f(\mathbf{n}^{k,l}, n_k + 1, n_l + 1) = f(\mathbf{n}^k, n_k + 1)$ so that (b) reduces to

(d) $f(\mathbf{n}) + \min_{i,j \neq l} f(\mathbf{n}^{i,j}, n_i + 1, n_j + 1) \leq 2 f(\mathbf{n}^k, n_k + 1)$.

Since the minimum on the LHS is $\leq f(\mathbf{n}^k, n_k + 2)$, (d) further simplifies to (c) with l replaced by k, which again is entailed by concavity.

(12.19) does not reverse, however; concavity is clearly stronger than $NNIR_c$. After despairing at proving a positive assertion I doubt that there is an equivalent or a sufficient condition expressible in terms of instantial relevance; maybe there is one in terms of arbitrarily long sequences of observations being non-negatively relevant for arbitrarily long sequences of predictions. Thus, I admit that the philosophical justification for the concavity of ranking functions needs to be improved.

In a very special case, though, concavity reduces to $NNIR_c$:

Theorem 12.20: If $I = \{1, 2\}$, i.e., if only two properties F_1 and F_2 are considered, then $NNIR_c$ entails concavity.

Proof: A two-dimensional representative function f is severely restricted in form. For any $\mathbf{n} = (n_1, n_2)$ minimitivity entails that $f(\mathbf{n}) = f(n_1 + 1, n_2) \leq f(n_1, n_2 + 1)$ (or the other way around, admitting the same reasoning). Let $\xi(1) = 0$ and $\xi(2) = f(n_1, n_2 + 1) - f(\mathbf{n})$. Consider then the plane $g = f_\xi^R + f(\mathbf{n}) - f_\xi(\mathbf{n})$: It intersects with $\mathcal{H}(f)$, since $g(\mathbf{m}) = f(\mathbf{m})$ for all $\mathbf{m} = (n_1 + k, n_2)$ $(k \geq 0)$ and $g(n_1, n_2 + 1) = f(n_1, n_2 + 1)$. However, it does not intersect with the interior of $\mathcal{H}(f)$, since $NNIR_c$, or inequality (c) in the proof of (12.19), together with minimitivity entails that $g(\mathbf{m}) \geq f(\mathbf{m})$ for all other $\mathbf{m}$. Hence, g is a plane supporting $\mathcal{H}(f)$. Since $g(\mathbf{n}) = f(\mathbf{n})$, also $f_\xi(\mathbf{n}) = f(\mathbf{n})$. Thus f is concave at $\mathbf{n}$.

In Spohn (2005a), I only considered the case $I = \{1, 2\}$. Hence, things looked deceptively simple there. In particular the notion of concavity did not yet surface; it is required only in the general case.

However, (12.20) finally helps us to see why positive instantial relevance could not be satisfied by symmetric ranking functions:

> *Proof of Theorem 12.9*: For any $i \in I$ PIR$_c$ states a condition on κ referring in effect only to the two properties F_i and non-F_i. Hence, it suffices to consider the thus coarsened setting. Within this setting, PIR$_c$ trivially entails NNIR$_c$, NNIR$_c$ entails concavity (by 12.20), and concavity entails that (the coarsened) κ is a finite mixture of laws (by 12.15). This means that from some $\mathbf{n}$ onwards (the coarsened) f defines a plane, i.e., is identical with some $f_\xi + \rho(\lambda_\xi)$. That is, given $E_\mathbf{n}$, κ is the same as the law λ_ξ and can no longer exhibit positive instantial relevance, but only instantial irrelevance like the law λ_ξ. Thus, PIR$_c$ entails non-PIR$_c$.

Our duplication of de Finetti's philosophy of probability is not yet completed; we have to explore a final important issue. We know now that precisely the regular symmetric concave ranking functions are minimal mixtures of subjective laws. How does the mixture change, though, through evidence? What does evidence teach us about the subjective laws? These questions finally address the topic left obscure in Section 12.4, the confirmation of laws.

The answer can be directly read off from the results reached so far. Suppose that we start with the regular symmetric concave ranking function κ that is the minimal mixture of Λ by some proper impact function ρ and has representation function f, and that we now collect the evidence $E_\mathbf{n}$ about the first n instances. Thereby, we arrive at the a posteriori ranking function $\kappa_\mathbf{n}$. Suppose further, in order to make things simple, that we learn the evidence for sure so that $\kappa_\mathbf{n} = \kappa(\cdot \mid E_\mathbf{n})$. Hence, $\kappa_\mathbf{n}$ is regular and symmetric, too, thus having the representative function $f_\mathbf{n}$, and concave, thus being the unique minimal mixture of Λ by some posterior impact function $\rho_\mathbf{n}$. What can we say about $\rho_\mathbf{n}$ and its relation to ρ? This is a straightforward calculation:

f is generated by the mixture by ρ. Hence, $f(\mathbf{n}) = \min_{\xi \in \Xi} [f_\xi(\mathbf{m}) + \rho(\lambda_\xi)]$. Since $f_\mathbf{n}$ results from f by conditionalization, we simply have $f_\mathbf{n}(\mathbf{m}) = f(\mathbf{n} + \mathbf{m}) - f(\mathbf{n})$, for all $\mathbf{m} \in \mathbf{N}^I$. Therefore

$$f_\mathbf{n}(\mathbf{m}) = \min_{\xi \in \Xi} [f_\xi(\mathbf{n} + \mathbf{m}) + \rho(\lambda_\xi)] - f(\mathbf{n}) = \min_{\xi \in \Xi} [f_\xi(\mathbf{m}) + \rho(\lambda_\xi) + f_\xi(\mathbf{n}) - f(\mathbf{n})].$$

This suggests that we ought to define $\rho_\mathbf{n}$ by $\rho_\mathbf{n}(\lambda_\xi) = \rho(\lambda_\xi) + f_\xi(\mathbf{n}) - f(\mathbf{n})$; this seems to be how the impacts on the mixture get rearranged by evidence.

However, $\rho_\mathbf{n}$ has to be a minimal mixture, and this entails a modification. In the geometric representation, it is clear which one. The hypograph $\mathcal{H}(f_\mathbf{n})$ is obtained from the hypograph $\mathcal{H}(f)$ simply by moving the origin of Euclidean space from $\mathbf{0}$ to $(\mathbf{n}, f(\mathbf{n}))$ and cutting off all parts of $\mathcal{H}(f)$ which fall outside the positive quadrant $\{\mathbf{x} \mid \mathbf{x} \geq \mathbf{0}\}$ by this translation of the origin. Thus, hyperplanes not maximally supporting $\mathcal{H}(f)$

(= strongly redundant components of the mixture of Λ by ρ) remain so with respect to $\mathcal{H}(f_n)$. Hyperplanes maximally supporting $\mathcal{H}(f)$ (= latently non-redundant components) remain so if and only if they intersect with $\mathcal{H}(f)$ at vertices not cut off after the translation of the origin, i.e., at vertices $(\mathbf{m}, f(\mathbf{m}))$ with $\mathbf{m} > \mathbf{n}$. This means that those redundant, but latently non-redundant components λ_ξ of the mixture of Λ by ρ which agree with f only for $\mathbf{m}$ not $> \mathbf{n}$, i.e., for which $f(\mathbf{m}) = f_\xi(\mathbf{m}) + \rho(\lambda_\xi)$ only if not $\mathbf{m} > \mathbf{n}$, become strongly redundant in the a posteriori mixture; these λ_ξ are too steep, as it were, to still be included. This consideration already proves

> *Theorem 12.21*: Define, for all $\xi \in \Xi$, $\rho_n(\lambda_\xi) = \rho(\lambda_\xi) + f_\xi(\mathbf{n}) - f(\mathbf{n})$, if $f(\mathbf{m}) = f_\xi(\mathbf{m}) + \rho(\lambda_\xi)$ for some $\mathbf{m} > \mathbf{n}$, and $\rho_n(\lambda_\xi) = \infty$ otherwise. Then κ_n is the minimal mixture of Λ by ρ_n.

This consideration also explains why we had to distinguish non-redundant and latently non-redundant components of mixtures. According to Definition 12.17 it would have sufficed to mix only non-redundant components in order to produce κ. However, an initially redundant component might become a posteriori non-redundant. This is the case when and only when the component is initially redundant, but latently non-redundant, and some of the components making it initially redundant drop out from the posterior mixture as described in (12.21). Thus, if the initial mixture had contained only the non-redundant components, we might have been unable to define the posterior mixture from it. It is only by including the redundant, but latently non-redundant components from the outset that Theorem 12.21 can work for all posterior κ_n ($\mathbf{n} \in N^I$).

Let me point out a number of consequences of (12.21) and our representation Theorems 12.14 and 12.18. Perhaps the most important is that we can thereby fully restore positive instantial relevance with respect to the subjective laws instead of the next instances. Suppose we have already acquired evidence E_n and now observe the $n+1$th object to have F_i so that the evidence is $E_{n'}$, where $\mathbf{n'} = (\mathbf{n}^i, n_i + 1)$. How do the mixture impacts shift? (12.21) immediately entails

> *Theorem 12.22*: With the notation just introduced and $f_n(i)$ being short for $f_n(0^i, 1)$, we have $\rho_{n'}(\lambda_\xi) - \rho_n(\lambda_\xi) = \xi(i) - f_n(i)$, if $\lambda_\xi(\mathbf{m}) \leq f_n(\mathbf{m})$ for some $\mathbf{m} \geq (0^i, 1)$; otherwise $\rho_{n'}(\lambda_\xi) - \rho_n(\lambda_\xi) = \infty$.

What does this mean? If $\xi(i) > f_n(i)$, i.e., if the $n+1$th object being F_i was more disbelieved according to λ_ξ than in κ_n, then the impact of λ_ξ increases proportionally, i.e., λ_ξ gets proportionally more *dis*believed by the additional observation. If $\xi(i) = f_n(i)$, i.e., if the observation was exactly as unexpected according to λ_ξ as according to κ_n, the impact of λ_ξ does not change. And if $\xi(i) < f_n(i)$, then λ_ξ gets proportionally confirmed, i.e., less disbelieved. This is how we intuitively expect the confirmation of laws to behave. In particular, if $f_n(i) = 0$, i.e., if after evidence E_n the $n+1$th object was not excluded from being F_i, then the laws that also do not exclude this observation keep their impact while all the other laws get disconfirmed according to the strength of their expectation being disappointed. Again, this is as it should be.

The only exceptional case is when not only $\xi(i) > f_n(i)$, but indeed $\lambda_\xi(\mathbf{m}) > f_n(\mathbf{m})$ for all $\mathbf{m} \geq (0^i, 1)$. Then the subjective law λ_ξ is definitely refuted, not because it has met counter-instances, but because the disbeliefs in all possible ways of realizing the properties F_i ($i \in I$) in future objects have become weaker than is representable by λ_ξ. This might seem to be an artificial consequence of my notion of a subjective law; it is, however, theoretically required. It should not distract, though, from our restitution of positive instantial relevance with respect to laws, i.e., on the level of second-order attitudes. This positive relevance is blurred by the mixture and thus weakens to NNIR_c on the level of first-order attitudes. (12.9) has shown us that the weakening is unavoidable, but now we can see that it is only an artifact of the mixture. So, to sum up, we have thus shown within our setting how enumerative induction derives from conditionalization, when it is restricted to laws, as it should.

(12.22) also allows us to study the limiting behavior of ρ_n, as $\mathbf{n}$ increases infinitely. The consideration of a normal case may suffice here. Normally, if n_i goes to infinity with $\mathbf{n}$, i.e., if the property F_i is instantiated infinitely often, $f_n(i)$ will converge to 0, i.e., *be* 0 from some point onwards. (The contrary, i.e., maintaining the disbelief in instances of F_i even in the face of infinitely many instances of F_i, would be weird.) In this normal case, $\rho_n(\lambda_\xi)$ diverges to infinity as well, if and only if $\xi(i) > 0$. In other words, precisely those subjective laws that do not exclude the properties instantiated infinitely often do not drop out of the mixture. All the other ones get disbelieved with ever greater firmness diverging to infinity. This observation rounds off my ranking-theoretic translation of de Finetti's account of the confirmation of statistical hypotheses. It would be most desirable to compare the account presented here with formal learning theory as presented by Kelly (1996, 2008), since it is here where the two theories have the largest topical overlap.

A final consequence of our representation theorems still needs to be considered; it has no probabilistic counterpart and opens interesting perspectives. I discuss it in the last section of this chapter. First, let me conclude this section by briefly recapitulating the effects of restricting this chapter to complete ranking functions.

As far as I can see, completeness was not required for our basic representation Theorem 12.14 and its supplement 12.18 about minimal mixtures. Also, the consequences for the confirmation of subjective laws (12.21 and 12.22) do not depend on it. However, we used completeness in Section 12.3 (after 12.6) when identifying the (dis-)belief in the next instantiation with the (dis-)belief in the corresponding infinite generalization (but we noticed that this identification is no more objectionable than the identity with respect to finite generalizations that does not depend on completeness).

The most important use of completeness, of course, was in Theorem 12.15 about the finiteness of mixtures, where the crux of the argument was that, if ranks are well-ordered (as is required by completeness), descending sequences of ranks must be finite; this was also the crucial point of Theorem 12.9 about the unsatisfiability of positive instantial relevance. If ranks are real-valued, this argument no longer obtains, and in such a setting positive instantial relevance might well make sense. Hence, it might appear as if Theorem 12.9 was just a dramatic device in order to trigger the subsequent considerations in Sections 12.4–12.5. Well, if it served this purpose, it cannot have been so bad a device. All the considerations thus motivated (with the exception of 12.15) also continue to hold within such a more general real-valued setting.

One might therefore think that it would have been better to base this chapter on real-valued ranking functions, all the more as PIR might have been saved by doing so. I am not so sure. In any case, I have thought through all the philosophical applications of ranking theory in terms of natural and hence complete ranking functions (out of habit, I suppose), and I feel that the interpretations developed in these applications are quite sensitive and would need more or less extensive readjustments within a real-valued framework. For instance, the close relation between absolute frequencies and subjective laws, which we noticed after (12.10) and which appears philosophically significant to me, would be lost in the more general setting. Finally, one must never forget the fundamental reason for assuming completeness that I explained after Definition 5.9, namely that an infinite conjunction of truths is still true and an infinite disjunction of falsities still false.

12.6 A Priori Lawfulness?

I just indicated a final, specifically ranking-theoretic, but philosophically significant consequence of the results of Section 12.5. It concerns the special role of the subjective law λ_0 where $0(i) = 0$ for all $i \in I$ so that $\lambda_0(E_\mathbf{n}) = 0$ for all $\mathbf{n} \in N^I$. λ_0 falls under our definition of a subjective law, but it cannot be said to express a strict law; it rather amounts to total agnosticism, expressing belief in lawlessness instead of lawfulness. Is it also contained in the minimal mixture κ of Λ by some impact function ρ as represented by f? Yes, perhaps. It immediately follows from (12.18) that $\rho(\lambda_0) = \sup f$. This can occur in three different ways. Let me explain what they mean.

First, f may increase indefinitely so that $\sup f = \infty$. Thus, $\rho(\lambda_0) = \infty$ as well, and indeed $\rho_\mathbf{n}(\lambda_0) = \infty$ for all $\mathbf{n} \in N^I$. In this case, ρ embodies the maximally firm belief that some genuine subjective law (i.e., different from λ_0) will obtain. This belief is invariable, not refutable by very long sequences of apparently random behavior of the instances distributing over all properties F_i ($i \in I$).

Second, $\sup f$ may be finite, but not vanishing, so that $f(\mathbf{n}) = \sup f > 0$ for some $\mathbf{n}$ and $f(\mathbf{m}) = \sup f$ for all $\mathbf{m} \geq \mathbf{n}$. Then, lawlessness is initially disbelieved with impact $\rho(\lambda_0) = \sup f$. Since $\rho_\mathbf{n}(\lambda_0) = \sup f - f(\mathbf{n})$, it gets less disbelieved with each unexpected observation. After too many disappointments (which may be very many) κ will have eventually lost the belief in lawfulness and any belief about the behavior of the future instances, the belief in lawlessness being the only remaining option.

Third, there is the possibility that $\sup f = 0$. This, however, means that lawlessness is maximally firmly believed from the onset, since all other subjective laws are thereby rendered strongly redundant.

The third alternative is completely unreasonable; it leaves nothing to be believed and nothing to be learned. The first alternative is not so bad, but does not appear very reasonable, either. We might well despair of finding an acceptable law for the empirical realm considered, at least one statable with the given properties F_i $(i \in I)$. So, the second alternative seems to be the most reasonable. However, it may not really be attractive, either. It can always be criticized from both sides, for accepting lawlessness too early or too late.

Before reproaching ranking theory with this unsatisfactory alternative, we must recall, however, what I said in Section 12.4 in commenting on the apparent implausibility of the persistent attitude. One comment, leading to Section 12.5, was that we never fully believe in a single law and thus never actually have the persistent attitude. The other comment can now be repeated. Our normal response is neither that we stubbornly insist in lawfulness with respect to the given properties F_i $(i \in I)$, whatever the evidence (in case sup $f = \infty$), nor that we simply acquiesce in lawlessness with respect to F_i $(i \in I)$, initially or at some later point (in case sup $f < \infty$). Rather, the normal response is to search for suitable enlargements of the given space of properties that allow lawfulness to be (re-)established in a more convincing way. Then the game starts all over again within that larger space of properties, until a further enrichment seems to be required.

Another reasonable response to variegated evidence is to turn statistical and to take a probabilistic attitude. If deterministic laws do not seem maintainable, we might be content with exploring statistical regularities. Then we treat the counter-instances to our conjectured deterministic laws, not as apparent exceptions to be explained away, but as more or less rare events falling under some statistical distribution.

When exactly we switch to that probabilistic response is unclear. There is a grey area dividing ranking and probability theory, the deterministic and the probabilistic attitude, which, as I emphasized at the end of Section 10.3, needs to be cleared up in order to better understand the relationship between probability and ranking theory. However, there is presumably no need to resolve which attitude to take. What we actually do is to promote and try to get the best of both of them; the attitudes pragmatically coexist.

Let me express our observation about λ_0, the belief in lawlessness, in more traditional terms. Kant tried to overcome Hume's objectivity skepticism generally with his transcendental logic and its synthetic principles a priori, and Hume's inductive skepticism particularly with his a priori principle of causality. This principle ascertained only the rule- or law-guidedness of everything going on in nature and was thus also called the principle of the uniformity of nature (cf., e.g., Salmon 1966, pp. 40ff.). As was often observed, this principle does not offer any constructive solution of the problem of induction, since it does not give any direction whatsoever concerning specific rules or laws or specific inductive inferences. Still, it provides, if a priori true, an abstract guarantee that our inductive efforts are not futile in principle. Is it a priori true?

In Section 6.5 I already introduced the notion of apriority into our ranking-theoretic framework, and distinguished between unrevisable and defeasible apriority. This distinction fits perfectly into the present discussion:

Our initial ranking function κ, I have argued, is to be regular, symmetric, and concave (well, the latter was not fully argued). And whenever sup $f > 0$, an extremely reasonable assumption, as we saw, this is tantamount to the disbelief in lawlessness, i.e., $\rho(\lambda_0) > 0$. In other words, the belief in lawfulness is at least defeasibly a priori in the sense of (6.22).

If even sup $f = \infty$, i.e., $\rho(\lambda_0) = \infty$, the belief in lawfulness is taken to be unrevisably a priori in the sense of (6.18). This assumption appeared doubtful, at least with respect to any fixed (finite) set of properties. Still, it might be unrevisably a priori that there is *some* set of properties with respect to which lawfulness – i.e., the principle of the uniformity of nature – holds. I am not prepared to argue about this issue. The foregoing considerations should, however, make the issue more amenable to clear argument.

The principle of the uniformity of nature is not the only explication of the principle of causality. Chapter 14 will allow us to state principles that are more obviously related to causation. Again, the issue of the apriority of such principles will come up, which, however, will be discussed only in Chapter 17. This discussion will finally supersede my present indecision concerning the apriority of lawfulness.

13

Ceteris Paribus Conditions and Dispositions

The core claim of the previous chapter was that belief in a strict or deterministic law is not merely belief in something, but a specific kind of doxastic attitude. Only by introducing the hybrid notion of a subjective law was I able to reestablish the idea that belief in a law *is* belief in something, if only in a subjective law. The objectification theory of Chapter 15 will make this notion more respectable, and this chapter will do so, too.

My contention will be that the theoretical means provided so far allow us to substantiate the obscure notion of a ceteris paribus law, i.e., a law qualified by a ceteris paribus condition. The basic difficulty here is that it is terribly hard to escape the dilemma of either turning the assertion "ceteris paribus, all F's are G's" into something clearly false or indefensible, or collapsing it into the triviality "all F's are G's, unless they aren't" or at least into something true for far too many F and G.

In Section 13.1 I will introduce the topic and outline its importance and difficulties. Section 13.2 will then offer my constructive treatment of ceteris paribus clauses within the ranking-theoretic framework, which I dare say is their natural home. This will lead directly to an account of dispositions in Section 13.3. For the insight of Carnap (1956, p. 69) that the reduction sentences associated with ordinary (i.e., less than pure) disposition predicates hold only ceteris paribus is still valid. At that time, this insight brought discussion to a dead end; but now we will be able to fruitfully utilize it. It means, in effect, that dispositions are characterized by a priori ceteris paribus laws, something apparently odd, but especially suited for a ranking-theoretic analysis. The significance of such an analysis is emphasized by the fact that dispositions are of the same ilk as powers, forces, capacities, etc. The thought that I will thereby be offering a (re-)interpretation of the physical notion of force unsettles me, even though I will not elaborate on this exciting prospect.

13.1 Ceteris Paribus Conditions: Hard to Avoid and Hard to Understand

The history of the topic seems not yet written, but the term "ceteris paribus" as well as the awareness of the phenomenon it designates was clearly around in the 19th century. Modern philosophy of science had an early discussion about laws in history and historic explanations, and thus came in touch with the topic. I came in touch with it through Canfield, Lehrer (1961), who were perhaps the first to point to serious difficulties in that context. However, this paper was taken lightly; the general reaction seems to have been that it should not be too difficult to replace the ceteris paribus condition by some adequate explicit condition, "ceteris paribus" just being a convenient abbreviating device. This was a grave error, though. Somehow, it took 15 years or more till the significance of the topic was properly perceived.

Then, however, the discussion developed all the more vigorously – and confusingly. Everybody seems to agree with everybody on some important points and to disagree on other no less important points. Each attempt to sort things out somehow made them messier. Additionally, the topic was loaded with a dispute about the status of the special sciences as opposed to fundamental science, i.e., physics. Either, each science is indiscriminately and essentially riddled with ceteris paribus laws, in which case there is no difference in status, anyway. Or ceteris paribus laws are somehow special to the special sciences, in which case the task seems to be to show how one can do proper science with such laws, in order to save the status of those sciences.

However, the special sciences are not in need of justification; they have their own unquestioned dignity. The philosopher's aim must be to understand it, not to criticize it. Here, I take this to be merely a side issue, which I will not pursue. My goal will rather be to explain how the notion of a ceteris paribus condition flows directly from the logic of non-probabilistic defeasible reasoning as explicated by ranking theory. If defeasible reasoning really is the basis of the phenomenon, it is no wonder that it is ubiquitous in the sciences, including physics. This is sufficient reason for me to find the phenomenon interesting and important; how essentially it characterizes a given science need not be our additional concern.

Ubiquitous it is indeed. Let us look at some familiar examples: Hooke's law, for instance, about the proportionality of the force applied to a spring and its extension. It needs qualifications in many ways, as was clear from the outset, even though the qualifications could be neither fully nor precisely specified. One must not overstretch the spring, the material the spring is made of must be elastic and homogeneous, its shape regular, the thermal distribution uniform, etc. Physicists have cleared up most of the conditions – all of them? – under which Hooke's law holds and provided deeper explanation in terms of the intermolecular forces within a molecular lattice. After all, our sophisticated technology of spring balances that is (or was) used in each lab and each deli-shop depends on it. Similar stories might be told about many laws of physics: the law of thermal expansion of bodies, the law of falling bodies, the laws of air resistance,

Ohm's law, etc. – typically all those laws we learn in school about the more primitive stages of physics.

The law of falling bodies is perhaps a contested example. Let us, more generally, look at Newton's law of gravitation. As such, it only says which forces, among others, there are in the world, with no condition applied to it. Combined with Newton's second law it also says how bodies move. Now a strong condition must be assumed: the gravitational forces exerted by the other bodies must be the only forces, i.e., add up to the total force acting on a body, or, in other words, the system of bodies considered must be a closed system. Therefore, one is tempted to treat the case analogously to the examples above. In the application, the condition is an open-ended one, since we can never finitely certify its satisfaction. Or one may perhaps classify it as an idealization, since there are no closed systems except the entire universe. Still, the important difference is that the condition is determinately stated in terms of Newtonian mechanics; there is no reference to such-likes or don't-know-what's.

The ideal gas law is another instructive case. It started in the 17th century as the law of Boyle–Mariotte, stating the inverse proportionality of volume and pressure of a given portion of gas and was later supplemented with the proportionality of volume and temperature given constant pressure. So far, it was a law of phenomenological thermodynamics and hence a typical ceteris paribus law hedged by ever better investigated conditions. Only in the 19th century was it derived from, and explained by, the kinetic theory of gases and thus established as an idealization holding for ideal gases, i.e., collections of freely moving extensionless bodies (molecules) interacting only by pushes. Thereby, the indeterminate ceteris paribus condition was replaced by a determinate, but unsatisfiable ideal condition, thus raising the question of how much it deviates from reality.

The example given in Schiffer (1991, p. 2) is: "If a person wants something, then, all other things being equal, she'll take steps to get it." This is a psychological rule of thumb, at best; many kinds of exceptions immediately occur to us; and we have no idea how to complete the list of exceptional conditions. We might take here the same move as with the ideal gas law. We might amend the rule of thumb to the rule of maximizing conditional expected utility and then declare it an idealized law about rational persons free of exceptions. The only (but vast) difference is that in the case of the ideal gas law the idealization is so much better defined, and gives us so much better a conception of what to do in order to undo the idealization.

The same remarks extend to economics. In its microeconomic reduction it is essentially an elaboration of the classic idealization of *homo oeconomicus* and thus subject to the same ceteris paribus conditions as the rule of maximizing conditional expected utility. If not so reduced, the point applies all the more. Indeed, economics apparently was the first discipline to explicitly recognize the need for ceteris paribus clauses.

The favorite example of Lange (2002) is about island biogeography. With the advent of the century of the life sciences it has become fashionable among philosophers of science to consider biochemistry, pharmacology, neurosciences, etc., which are indeed full of examples of ceteris paribus and similar laws. The literature abounds in further, not always congruent examples, thus displaying how pervasive the phenomenon is in our scientific activities.

In fact, it is not a single phenomenon, as the examples display and as most of the literature carefully observes; "ceteris paribus" is just a common label – not the happiest one – for a bunch of related phenomena. As far as I see, the various uses of "ceteris paribus" can be neatly summarized in the formula "other things being equal or normal or absent or ideal". Let me explain:

"Ceteris paribus" literally means "other things being equal". Equal to what? This is a relational phrase, and as long as the relatum is not specified, its meaning is incomplete. The relatum may be some specific case or rather, as is the intention, any given case. Then, however, the ceteris paribus clause just amounts to a general dictum like "equal conditions, equal outcomes" or "equal causes, equal effects", leaving it to empirical inquiry what the relevant aspects of equality might be. Whatever the substantial content of such principles, it seems well taken care of by the symmetry assumption so central to the previous chapter. I suspect, though, that the preferred standards of comparison are not this or that condition, but the normal conditions or the absence of further relevant conditions, i.e., that "ceteris paribus" rather means "other things being normal or absent".

These two formulations fit most of the examples above. Hooke's law holds under normal conditions, i.e., for diligently manufactured springs handled with the usual care. The rules of thumb of folk psychology normally apply; this is why they are such tremendously useful common knowledge. Likewise, the laws of scientific psychology and of economics at best hold under the conditions explicitly specified, along with further unspecified normal conditions.

The law of falling bodies holds in the absence of disturbing influences or interfering variables. The same may be said of the laws of practical rationality. Above, though, I introduced this example as a normally applicable rule of thumb. This shows that the various examples cannot be sharply classified according to their implicit conditions.

The closed system assumption in physics also states that disturbing influences are absent. This assumption is indeed required even by the most fundamental conservation laws. This might explain the temptation to claim that the ceteris paribus clause applies "all the way down". However, there is also an important difference, and this is why Earman et al. (2002) rightly object to this claim: namely the difference that this assumption is fully and determinately stated in terms of theoretical physics and not merely a sweeping exclusion of countless unidentified possible sources of interference.

This remark might also help understanding some confusion about the notion of a proviso as introduced by Hempel (1988). Hempel placed this notion in the empiricist context of distinguishing and relating observational or, as he says more cautiously, antecedently understood and theoretical vocabulary. However, Hempel's paper enormously stimulated the discussion about ceteris paribus conditions. Provisoes were apparently taken to be more or less the same thing. Earman, Roberts (1999, Sect. 3) protest, and again they are right. The point is subtle, though. Hempel's provisoes are often unstated assumptions that can be explicitly stated in theoretical terms. Their point is not that they cannot be made explicit at all, as it is with a genuine ceteris paribus clause, but only that they cannot be stated in observational or antecedently understood terms.

Our concern, in any case, will be these mysterious, apparently unstatable conditions. And here it is clear that we should distinguish the two forms mentioned, "other things being normal" and "other things being absent", even if examples cannot be determinately classified. To grant that there may be a lot of further relevant variables and to vaguely assume them to be realized in some normal way is one thing. To grant that there may be a lot of further relevant variables and to assume them to be absent or to take a non-interfering value 0, although they usually do not, is clearly another thing. Accordingly, I will deal separately with these two forms in the next section.

Idealizations that assume ideal conditions are presumably a further case to be distinguished. At least, the examples I gave look different. The closed system assumption is unlike the assumption of extensionless molecules. And taking rationality as an idealization is presumably not the same as taking rationality (and the beliefs and desires it operates on) as the only causally relevant variable. Accordingly, idealizations usually form a separate chapter in philosophy of science. As such I do not want to contribute to it here. However, when I look, for instance, at the procedures of idealization and concretization as described by Nowak (1992), I do not find them so different from the specification procedures applied to ceteris paribus conditions. Also, I find the distinction intuitively not so clear. What is so different in assuming conditions to be ideal, when I know they are not, from assuming certain factors to take a non-interfering value 0, when I know they do not? So, maybe, our discussion indirectly bears on idealization, too. I should moreover mention that Rott (1991a) already offers a very instructive treatment of idealization in terms of belief revision theory that is particularly suited for dealing with the counterfactual (rather than the approximative) element contained in idealizations (see also Kuokkanen 1994).

So, what is the state of the art in philosophy of science concerning ceteris paribus conditions? Nothing, in a way. As I said, everybody seems to disagree with everybody. At least, though, the attempts may be roughly classified.

The first to be mentioned are Cartwright (1983, 1989, 1999) and many other papers of her not contained therein, because she is the one who first and most persistently elaborated on the significance of the phenomenon. What she offers is less an abstract meaning analysis of ceteris paribus conditions, but more a very rich description of the various ways in which such conditions operate, how they pervade the sciences, and how drastically our conception of laws, of explanation, and of the working of the sciences in general is thereby changed. I concur with many things she says. (Or rather, it is a concordance. I recall our discussions about probabilistic causation, beginning in 1978. I always intended to deal with causation in the single case, whereas she initially talked about probabilistic causal laws, as did all the others participating in the discussion. She came to focus more and more on the single case in the course of her further investigations, and was thereby able to unfold her rich descriptions. Only then did we slowly realize our initial difference. See also her remark in Cartwright 1989, p. 9.) On the other hand, it is obvious that our methods for dealing with the issues diverge considerably; it is hard to see and to describe the similarity of our intentions.

The project of a meaning analysis of ceteris paribus conditions is perhaps most clearly pursued by those attempting to provide truth conditions for statements with ceteris paribus clauses. Somehow, one makes a claim with them; and if so, one should be able to say what the claim really is. The most obvious idea is to say that "ceteris paribus, (all) F's are G's" claims something like "there are conditions of a certain suitable kind C under which (all) F's are G's"; the inability to specify those conditions is thus turned into a positive existence claim. Schiffer (1991, Sect. II) critically discussed this idea, Fodor (1991) and Pietroski, Rey (1995), among others, elaborated on it in various ways, and Earman, Roberts (1999) in turn criticized it extensively, and so forth.

The general trouble with that paraphrase is, as has been observed, e.g., by Mott (1992), that it comes out true for far too many F and G even if one is very restrictive about the suitable kind C. In particular, "ceteris paribus, F's are G's" and "ceteris paribus, F's are not G's" turn out to be true at the same time. One could have been warned. Existential quantifications over predicates, properties, or conditions always fail to do the job, as could perhaps first be observed with the attempts of Kaila (1941) to save reduction sentences for disposition predicates from applying to too many objects (see also the description of the private discussions between Carnap and Kaila in Stegmüller 1970, pp. 221ff.).

Another idea is that "ceteris paribus, F's are G's" says something like "being F is a partial cause of, or plays a causal role for, being G". One can read Cartwright in this way, and it is the variant Woodward (2002) mainly discusses. This idea at least indicates how ceteris paribus law might have an explanatory role, and if one has a story about the confirmation of causal relations, one can extend it to the alleged ceteris paribus laws. Woodward (2002) concludes, though, that, if that is the idea, there is no gain in introducing ceteris paribus clauses. Better leave it aside and focus directly on the causal business. Moreover, the problems mentioned in the previous paragraph do not vanish with this shift. The causal role of a property F may vary with the relevant circumstances, and again one is usually unable to specify the circumstances under which being F is a partial cause of being G. Of course, causal theorists are acutely aware of this problem; but we may leave it open whether it is in good hands with them.

At least in the "other things being normal" reading it seems obvious that "ceteris paribus, F's are G's" should be taken to entail that most F's are G's or that the suitable conditions under which F's are G's mostly obtain. This is what Earman, Roberts (1999, pp. 463ff.) discuss. This move may help with the above-mentioned trivializations, but again at the cost of making the truth conditions of ceteris paribus laws quite indeterminate. The temptation to therefore understand them as vague statistical or probabilistic claims has been severally criticized, e.g., by Hempel (1988, pp. 151ff.). Earman, Roberts (1999) are not happy with this move either and argue that the vague "mostly" is not used in the sciences; it is either avoided or given a precise sense.

At this point, the close similarity of the normality condition with default assumptions may be noticed. It is suggestive to bring the resources of default logic and non-monotonic reasoning to bear on our topic. Given the richness of that literature, it is surprising how little philosophers of science have taken up this connection. Of course, even though these branches of logic mostly present themselves as specifying truth conditions, it remains unclear whether such conditions are thereby actually provided for ceteris paribus laws; one easily slips into giving them an epistemic or pragmatic treatment. This path has been explicitly pursued by Silverberg (1996) and Schurz (2001, 2002); Silverberg applies Lewis-type conditional logic, whereas Schurz prefers a probabilistic semantics along the lines of Adams (1975). In principle, I approve of this move, even though I will use other theoretical means for carrying it out.

The previous point finally suggests that the search for truth conditions might be the wrong goal and that we should rather focus on the epistemic or pragmatic role of ceteris paribus conditions. This suggestion is amply found in the literature. The many case studies display much of this role. This is also the field in which Schiffer (1991) seeks positive insight and where Earman, Roberts (1999) can finally make sense of ceteris paribus laws, if at all. As enlightening as the case studies are, though, the general statements are rough and vague. This discussion reminds me of the earlier stages of philosophy of language where semantics flourished and pragmatics was in effect treated as sort of a waste-basket category, although acknowledged to be in need of theoretical foundations.

The foregoing remarks were not intended as criticism; nor can they take the place of a careful discussion. They should only provide a rough sketch of the main facets of this bewildering issue. If this sketch is as fair as it can be in a few pages, it has displayed how tentative and unconsolidated the situation is. A new approach cannot make it worse, and might have the chance of throwing some light into the thicket. This is what I hope to do in the next section.

That I will not be following the standard ways should be clear, though, from the fact that Chapter 12 was actually an ironic commentary on the discussion sketched. The presupposition of that discussion was that proper laws have determinate truth conditions, and the result was that ceteris paribus laws are somehow improper because of the difficulties of endowing them with determinate truth conditions, and should rather be understood via their epistemic or pragmatic role. If Chapter 12 is right, though, this is not the dividing line; proper laws are not characterized by their truth conditions, either. One might retort that proper laws are at least associated with truth conditions; even if a belief in a law is not the same as a belief in a proposition, it entails such a belief. And, one might go on to ask, which proposition or truth condition is believed when one believes in a ceteris paribus law? This is not as hopeless a question as one might think. Let us see how far we can move towards a positive answer.

13.2 A Ranking-Theoretic Account of Ceteris Paribus Conditions

I concluded above that I would deal only with two forms of ceteris paribus conditions, "other things being normal" and "other things being absent". Let us discuss them separately, beginning with normal conditions.

The basic idea is that subjective laws as defined in (12.10) in Section 12.4 *are* ceteris paribus laws, i.e., that relative to a subjective law we can adequately explain what the qualifying normal conditions are. The execution of this idea is quite straightforward. For this purpose we have to restate the formal framework we used in Chapter 12.

There we had considered a finite family F_i ($i \in I$) of mutually exclusive and jointly exhaustive properties and a sequence $a_1, a_2, \ldots$ of objects having these properties or, equivalently, a sequence $X_1, X_2, \ldots$ of variables taking values in I. As defined in (12.10), a subjective law λ_ξ was then just the independent infinite repetition of some ranking function ξ for I describing the distribution of each X_n ($n \in \mathbf{N}'$). We may neglect, though, the independent repetitions generating the law λ_ξ; all information is already contained in the ranking function ξ describing a single case, X_1 for instance.

Moreover, we should rearrange the properties. We should not consider one large indiscriminate family of properties or one variable X_1; we should rather see it as structured and complex. That is, we should conceive of the complex variable X_1 as being composed of three variables X, Y, Z, each taking finitely many values. Hence, each value $i \in I$ of the variable X_1 is in fact a triple $\langle x, y, z \rangle$ of values of the component variables, I is the cross product of the ranges W_X, W_Y, W_Z of X, Y, Z, and we may assume that each component variable is defined on I so that for $i = \langle x, y, z \rangle$ $X(i) = x$, $Y(i) = y$, and $Z(i) = z$. This might mean assuming a large I, but fortunately we proved our results in Chapter 12 for any finite I whatsoever. Let us call X and Y the *target variables* and Z the *background variable*, for predictable reasons. Of course, the background variable Z might (and usually will) further decompose into many component variables $Z_1, \ldots, Z_m$. There is, however, no need to consider such a decomposition. One unstructured background variable will do for our purposes.

Suppose now that we conjecture a lawful connection between the target variables X and Y. The natural question then is in which way this lawful connection depends on the background conditions described by the background variable Z. One may fill in here any of the examples mentioned in the previous section. X may be the force applied to a spring and Y its extension; or X may be the supply of and the demand for a certain good and Y its price; etc. And Z describes relevant conditions on which the hypothetical connection between X and Y is conjectured to depend. Let us see how we can spell this out in terms of the ranking function ξ for I, the joint ranking distribution of the variables X, Y, and Z.

The conjectured lawful connection between the target variables X and Y is just a certain proposition or hypothesis $H \subseteq I$ that must satisfy some conditions: First, H must actually make a claim, i.e., $H \subset I$. Secondly, H must only be about the variables

X and Y and must not contain any assertion about Z; that is, if $\langle x, y, z \rangle \in H$, then $\langle x, y, z' \rangle \in H$ for all values z' of Z. Thirdly, H must not make any claim about X and Y as such; H does not exclude any value x of X or any value y of Y; that is, for each x there is a y and for each y there is an x, such that $\langle x, y, z \rangle \in H$ for some value z of Z. However, H need not specify a functional relation between X and Y; for a given value x of X H may allow Y to take several or even all possible values. (This will change in Chapter 15; if the hypothesis H is to provide objectifiable reason relations in the sense explained there, it has to be functional; cf. Theorem 15.7.)

That the hypothesis H is conjectured according to ξ means that H is the strongest proposition about X and Y believed according to ξ, i.e., $\xi(\bar{H}) > 0$ and for each x, y such that $\langle x, y, z \rangle \in H$ for some or all values z of Z, $\xi(\{X = x, Y = y\}) = 0$. The belief need not be a strong one; it just means that, if asked for the relation between X and Y, H would be the preferred answer from the point of view of ξ. Given our assumption about H, this entails that $\xi(\{X = x\}) = 0$ and $\xi(\{Y = y\}) = 0$ for all values x of X and all values y of Y; ξ contains no beliefs about X and Y by themselves.

This fully describes the beliefs of ξ, as far as the target variables are concerned. But, of course, ξ also holds beliefs about the background variable and it sees the background and the target variables as connected. This is the issue at which we have to look more carefully.

The basic observation, sketched already in Spohn (2002c, Sect. 4), is this: Even if H is unconditionally believed in ξ, i.e., $\xi(\bar{H}) > 0$, it need not be believed under all conditions. There will be a background condition, i.e., a proposition N only about the background variable Z that is unsurprising or not excluded; that is, $\xi(N) = 0$. In this case, we still have $\xi(\bar{H} \mid N) > 0$; the hypothesis H is still held under such a condition N, which we may therefore call a *normal* condition. However, there might also be a condition E about the background variable Z, given which the hypothesis H about X and Y is no longer held, i.e., such that $\xi(\bar{H} \mid E) = 0$. This entails $\xi(E) = \xi(\bar{H} \cap E) \geq \xi(\bar{H})$, that is, such a background E must be at least as disbelieved as the violation of H. We may therefore call E an *exceptional* condition with which we do not reckon according to ξ.

This is, I find, quite an appropriate schematic description of what actually goes on. We entertain a hypothesis, and this includes the expectation of conditions under which the hypothesis holds, or rather the denial of conditions under which it does not hold. When we nevertheless encounter a violation of the hypothesis, we are surprised, we start inquiring into how this was possible, and we find that some unexpected condition is realized under which we did not assume the law to hold, anyway. Or, still more realistically, it is rather this inquiry that elicits our tacit assumptions and expectations, the ranking function ξ we are operating with. In this way, each ranking function representing the belief in a hypothesis and about the relevant background automatically carries an aura of normal and, conversely, of exceptional conditions that does not surface on the level of beliefs, but becomes visible only by looking at the various levels of disappointed expectations, i.e., negative ranks.

What I am suggesting, then, is that we give the talk of normal conditions an *epistemic* reading: normal conditions are the conditions expected or at least not ruled out. This contrasts with the *trivial* reading according to which the ceteris paribus conditions with regard to a given law or hypothesis H are defined as those conditions under which H is true. It contrasts with the *existential* reading according to which "ceteris paribus" is just an existential quantifier over conditions possibly of a suitable kind. We have seen in the previous section that both ideas do not give us what we want. It also contrasts with the *eliminativistic* reading, as one might call it, according to which "normal conditions" is to be replaced by an explicit list of specific conditions under which H holds. Each such list must remain hypothetical, however; it is only at the end of an inquiry, which we never actually reach, that such a replacement might be carried out with confidence.

It does even not fully agree with the perhaps *natural* reading, according to which the normal conditions are just those that usually or mostly obtain. One might try to get rid of the vagueness of this reading by giving it a statistical twist. However, this leads us astray, as also observed in the previous section. Moreover, we should notice that the difficulty of turning "usually" or "mostly" into an objective statistical assertion does not only lie in its vagueness, but also in its unavoidable egocentricity or indexicality. What obtains usually or mostly is what *we* usually or mostly encounter, given our idiosyncratic location and selection criteria. From this perspective, it is quite natural to take the rare and unusual as the unexpected and the normal and usual as the not unexpected. Thus, we have returned to the epistemic reading.

In order to spell out the epistemic reading – this is the crucial point – we need an adequate account of conditional belief. This is why I think that ranking theory is the natural home for the logic of normal and exceptional conditions. Indeed, my above ranking-theoretic account of these conditions was only a preliminary one illustrating the basic idea. We have to study the matter a bit more closely.

The matter is complicated by the fact that there are several intuitions regarding normality and exceptionality that are not jointly satisfiable. Let us return to our formal framework with the two target variables X and Y, the background variable Z, the hypothesis H about the relation between X and Y, and the ranking distribution ξ over X, Y, and Z entertaining, i.e., believing H; thus $\xi(\bar{H}) > 0$. I suggested that

(13.1)   the condition N for the variable Z is *normal*^{nu} iff $\xi(N) = 0$.

(The superscript "nu", "*not unexpected*", is for distinguishing other senses yet to be discussed.) This entails $\xi(\bar{H} \mid N) > 0$, as desired. I further suggested that

(13.2)   the condition E on Z is *exceptional* iff $\xi(\bar{H} \mid N) = 0$, entailing $\xi(E) \geq \xi(\bar{H})$.

Now one should think that E and $\bar{E}$ cannot both be exceptional. This is indeed implied by (13.2) (the proof is via the law 5.23a of disjunctive conditions). If N is normal, should $\bar{N}$ be not normal? This is implausible; in any case, N and $\bar{N}$ can both be normal^{nu} according to (13.1). Moreover, one should think that E is exceptional if and only if it is not normal. This does not hold, though; non-normalcy^{nu} does not entail exceptionality.

So, should we weaken exceptionality and define E to be exceptional iff $\xi(E) > 0$? No, this only says that E is unexpected. But if still $\xi(\bar{H} \mid E) > 0$, then, it seems, E cannot be reasonably called exceptional with respect to H. So, we had better weaken normalcy and define

(13.3)   the condition N on Z to be *normal** iff $\xi(\bar{H} \mid N) > 0$.

The fact that there can be unexpected normal* conditions is perhaps acceptable. However, this idea is dangerously close to what I called the trivial reading; it would mean that normal conditions for a hypothesis H are just those conditions under which H is believed to be true.

There are more problems. So far, there are many normal conditions. Intuitively, though, we are after *the* normal conditions; the definite article presumably presupposes that there is something like the strongest normal condition. However, in either sense, if N and N' are normal, $N \cap N'$ need not be normal. Similarly, one should expect that, if E and E' are both exceptional, then $E \cap E'$ is exceptional, too. This is not entailed, though, by our definition; we might have $\xi(\bar{H} \mid E \cap E') > 0$. In such a case $E \cap E'$ is intuitively *doubly* exceptional, since the judgment about H is reversed twice.

In order to get a hold, we should perhaps first focus on the maximal conditions, as far as they are captured in our conceptual framework, i.e. on the states of the background variable Z. For each value z of Z the proposition $[z] = \{\langle x, y, z \rangle \mid x \in W_X, y \in W_Y\}$ is such a maximal condition. They fall into two kinds:

(13.4) $[z]$ is a *maximal normal** condition (regarding the hypothesis H) iff $\xi(\bar{H} \mid [z]) > 0$ and a *maximal exceptional* condition iff $\xi(\bar{H} \mid [z]) = 0$.

(In the terminology of Definition 6.3 an exceptional condition is a necessary reason for $\bar{H}$ or against H, and a normal* condition is the contrary.) We may now collect

(13.5)   all maximal normal* conditions $[z]$ in a background condition N^* and all maximal exceptional conditions $[z]$ in a background condition E^*.

The law 5.23a of disjunctive conditions then guarantees that N^* and E^*, respectively, are normal* and exceptional, too. N^* and E^* thus are the weakest such conditions that are pure in the sense of containing only maximal conditions of the same kind. Hence, N^* and E^* might seem to deserve to be called *the* normal condition and *the* exceptional condition.

However, as useful as these observations about maximal conditions may be, N^* again is dangerously close to the trivial reading. Moreover, we still have the problem that there are (maximal) normal* conditions that are intuitively doubly exceptional. However, the fact just exploited, that normalcy* and exceptionality are closed under union or disjunction, helps us to the proposal that I finally endorse (and that I intended in Spohn 2002c, Sect. 4):

So far, we have found no reason to doubt the point that conditions under which the hypothesis H cannot be maintained are exceptional. Now, we have seen that there is a weakest exceptional condition: $E^{\geq 1} = \bigcup \{E \mid E$ is exceptional$\}$ is exceptional, too (the superscript will soon explain itself). This is a distinguished condition we may well call *the exceptional condition*. The distinction carries over to $N^0 = \overline{E^{\geq 1}}$ that may thus be called *the normal condition*; this will be justified, however, only in the course of the subsequent considerations. In any case, N^0 is normal* and normal^{nu}; indeed, N^0 is believed in ξ, since $E^{\geq 1}$ is disbelieved in ξ. More interesting is the observation that $E^* \subseteq E^{\geq 1}$ and $N^0 \subseteq N^*$. The reason is that N^0 will turn out to be that part of N^* not containing any fake normal but actually doubly exceptional maximal conditions, as I am about to explain. This also means that N^0 cannot be denounced as a formalization of the trivial reading (in epistemic disguise), it is a substantially stronger reading.

By the same argument, we are now able to account for the phenomenon of double exceptionality within the range of exceptionality. For $E \subseteq E^{\geq 1}$ we may call E *doubly exceptional* iff $\xi(\bar{H} \mid E) > 0$, i.e., iff the hypothesis H is in turn supported by E. Again, the disjunction of doubly exceptional conditions is doubly exceptional, too, so that we may define $E^{\geq 2} = \bigcup \{E \subseteq E^{\geq 1} \mid E$ is doubly exceptional$\}$ as *the doubly exceptional condition* (always with regard to H). Now, the superscripts start to make sense.

Of course, we must not stop here; if there is double exceptionality, there can be triple exceptionality, and so on. How many degrees of exceptionality there are depends, of course, on the richness of the background variable Z and on the sophistication of our judgment about the hypothesis H (as expressed by ξ). In practice, there will usually not be very many degrees; it is, however, not so difficult to construct contrived examples in which these degrees recursively build up indefinitely. The situation resembles Simpson's paradox, well known in statistics, that actually goes back to Yule (1911) (cf. Cartwright 1983, p. 24) and consists in the bewildering fact that the partial correlation of two random variables can reverse each time a further variable is added to the list of variables relative to which the partial correlation is taken.

In any case, the general definition is straightforward:

(13.6)   We set $E^{\geq 0} = W$, the whole possibility space considered. For odd n we define $E^{\geq n} = \bigcup \{E \subseteq E^{\geq n-1} \mid \xi(\bar{H} \mid E) = 0\}$, for even n we set $E^{\geq n} = \bigcup \{E \subseteq E^{\geq n-1} \mid \xi(\bar{H} \mid E) > 0\}$, and we call $E^{\geq n}$ *the exceptional condition of degree n*.

In our finite setting there must be some m such that $E^{\geq m} \neq \varnothing$ and $E^{\geq m+1} = \varnothing$; we may call m the *depth of exceptionality* of the case at hand.

We may go on to define:

(13.7) $E^n = E^{\geq n} - E^{\geq n+1}$. Thus, N^0 defined above is the same as E^0.

I stick to the letter N in this case in order to emphasize that exceptionality of degree 0 is in fact normalcy. The sequence $E^0 = N^0, E^1, E^2, \ldots$ forms a partition of W, providing a classification of maximal conditions: the maximal condition $[z]$ is *exceptional to degree n* (and *normal* if $n = 0$) iff $[z] \subseteq E^n$. Thereby, we finally arrive at a classification of all background conditions:

(13.8)   A proposition or condition E about Z is *exceptional to degree* n iff n is the minimal degree of exceptionality among the maximal conditions $[z] \subseteq E$. If n is even, we have $\xi(\bar{H} \mid E) > 0$, and if n is odd, $\xi(\bar{H} \mid E) = 0$. A condition N that is exceptional to degree 0 is *normal*.

Thus, *the* normal condition N^0 is believed in any case, while *a* normal condition N may even be disbelieved and not normal[nu]; it must only be not as disbelieved as the falsity of the hypothesis H. Indeed, we have:

(13.9) $[z] \subseteq N^0$ iff $\xi([z]) < \xi(\bar{H})$,

since no $[z]$ with $\xi([z]) < \xi(\bar{H})$ can be exceptional, some $[z]$ with $\xi([z]) = \xi(\bar{H})$ must be exceptional (to degree 1), and then each condition E about Z with $\xi(E) \geq \xi(\bar{H})$ is exceptional, too (at least to degree 1). Thus, in particular, N as well as $\bar{N}$ can be normal, as we had found desirable.

(13.8) also explains the fault in our attempt (13.5) to define the normal and the exceptional condition as N^* and E^*. We have $N^* = \bigcup\{E^n \mid n \text{ is even}\}$ and $E^* = \bigcup\{E^n \mid n \text{ is odd}\}$; hence, this attempt wrongly classified all maximal conditions of an even positive degree of exceptionality.

Finally, the degrees of exceptionality are nicely ordered by their negative ranks. We have $0 = \xi(E^{\geq 0}) \leq \xi(E^{\geq 1}) \leq \xi(E^{\geq 2}) \leq \dots$ so that $\xi(E^n) = \xi(E^{\geq n})$ for all n. Indeed, we can show something stronger (all formal claims in this section are easily proved with the basic theory of Chapter 5):

(13.10)   For even n we have $\xi(\bar{H} \mid E^{\geq n}) > 0$ and $\xi(\bar{H} \mid E^{\geq n+1}) = 0$, hence $\xi(E^{\geq n+1} \mid E^{\geq n}) \geq \xi(\bar{H} \mid E^{\geq n}) > 0$, and thus finally $\xi(E^{n+1}) > \xi(E^n)$.

(13.11)   For odd n we have $\xi(E^{\geq n+1}) \geq \max \{\xi([z]) \mid [z] \subseteq E^n\}$.

Thus, for odd n, we might possibly have $\xi(E^{n+1}) = \xi(E^n)$, provided all maximal conditions in E^n have the same rank. In all other cases, the sequence $\xi(E^0), \xi(E^1), \dots$ is bound even to strictly increase. This, of course, agrees well with our intuition. The higher the degree of exceptionality of a condition, the more unexpected it should be.

No doubt, this investigation of normal and exceptional conditions might be carried on much further. Let us take stock, though. We have spelled out in some detail what I have called the epistemic reading of normal conditions. We have avoided all the false paths we have encountered in our discussion. We have accounted for the obvious phenomenon of double exceptionality (and higher degrees), something which I see only ranking theory as capable of doing. All this does not prove that we are on the right track, but it makes for a strong case.

Given fixed target variables X and Y, a fixed background variable Z, and a fixed belief state ξ about the single case at hand, we were even able to adequately define N^0 and $E^{\geq 1}$ as the normal and exceptional conditions regarding a hypothesis H about the target variables. Thereby, we can also answer the question at the end of Section 13.1. The strongest belief associated with ξ is $\xi^{-1}(0)$, of course, some proposition jointly about the variables X, Y, and Z. However, if we separate the roles of target and background variables, as we did here, then the beliefs most naturally associated with the ranking function ξ first are the belief in the normal conditions N^0, i.e., $\xi(\bar{N}^0) > 0$, and the conditional belief in the hypothesis H given the normal conditions N^0, i.e., $\xi(\bar{H} \mid N^0) > 0$. As we will more carefully investigate in Chapter 15, conditional beliefs themselves are not generally associated with truth conditions; but if they are, they are most naturally associated with the unconditional belief in the material implication "if N^0, then H", i.e., in $\bar{N}^0 > H$, which is something true or false. With the means of Chapter 12, finally, we can expand ξ to a subjective law λ_ξ about all similar cases, which we may hence rightfully call a ceteris paribus law. This is as far as we can get with the association of determinate truth conditions with ceteris paribus laws in the sense of "other things being normal".

The fixity of X, Y, Z, and ξ had to be presupposed, of course; otherwise, our investigation could not even have started. However, this fixity is perhaps the main point raising doubts. The fixation of the target variables is fine; they are dictated by the hypothesis in question. The fixation of the belief state ξ was not a genuine one; the analysis works for all ranking distributions over the variables X, Y, and Z. So, we might consider changing belief states and thus develop a dynamics of normal and exceptional conditions. We might consider many λ_ξ, and mix them in the way discussed in Chapter 12; and we might let the mixture change through experience, thus favor certain λ_ξ, and conceive of normal conditions as they are relative to the favored λ_ξ. And so on. In this way, the relativity of the epistemic reading of normal conditions to a given ranking function is apt for opening the route to such investigations as we have started here.

The main doubt, though, will certainly concern the fixity of the background Z. The background is principally open and not fixable; it contains all the don't-know-what's. Indeed, this openness seemed to raise the issue of ceteris paribus conditions in the first place. The point, though, is that all other accounts of ceteris paribus conditions are painfully wrong when referred to a fixed background. The trivial reading is no better, when the background is fixed. The eliminativistic reading turns each ceteris paribus law into a falsity, since no fixed background will offer a correct substantial condition to replace the ceteris paribus condition (unless the background happens to be complete with respect to the hypothesis at hand). For the same reason, the existential reading turns ceteris paribus laws into falsities within a fixed background. All these accounts derive their prima facie plausibility only by referring to an open background. By contrast, the present account has something positive and substantial to say, given a fixed background. By doing so, it offers the prospect of also saying something substantial about open background conditions. The means for doing so would be to study how normal conditions within smaller and larger backgrounds are related. We are here at a similar point to where we were in Sections 12.4 and 12.6 when we observed the need to widen the conceptual framework for stating generalizations and laws. As there, I will not say more here on the issue of changing conceptual frameworks, of fine- and coarse-graining possibility spaces. At least, however, we have something from which to continue. (See, however, Section 14.9, where I discuss the corresponding problem in relation with causation.)

Let us turn, more briefly, to ceteris paribus conditions understood as conditions about other things being absent. For this purpose, we have to slightly change our formal set-up. Let us consider three target variables X_1, X_2, and Y describing a given single case, and let us neglect any further background into which the present considerations might be embedded; the variables X_1 and X_2 will mutually serve as the only explicit background. As before, let us assume a ranking distribution ξ over the three variables, which may be independently repeated for similar cases and thus generate a subjective law λ_ξ. Finally, let us assume that the range of the variables X_1 and X_2 contains a natural value 0, signifying that they are absent, mute, inactive, or non-interfering. For instance, in order to state a realistic law for falling bodies one might take into account air resistance that may be 0 or larger. How bodies move around also depends on their electric charge, which may be 0 or positive or negative. In a controlled study of the efficacy of a drug some people get a certain amount of the drug, some get a placebo, and some get nothing at all; the last case would be the natural 0. Many variables have a natural 0, but not all. (Temperature, for instance, does not have a natural 0. 0K is a natural 0 of the temperature scale in some sense. However, in the present sense, 0K is a most unnatural condition, in which bodies behave in the most unpredictable ways.)

Suppose we conjecture or (tentatively) believe in a hypothesis H_1 about the relation between X_1 and Y. H_1 has to satisfy the same conditions as before; in particular, it is only about X_1 and Y. However, the implicit assumption now is not that the other variable X_2 is in some normal state, but rather that it is non-interfering — it takes the value 0. Such an implicit assumption is common scientific practice. So far, there is no problem with entertaining H_1.

Problems will emerge when we similarly have a hypothesis H_2 about the relation between X_2 and Y, again implicitly assuming that X_1 is absent and takes the value 0. For, there is then a condition C about the variables X_1 and X_2 – there is indeed provably a largest or weakest such condition – such that $C \cap H_1 \cap H_2 = \varnothing$. If $C = \varnothing$ itself, there is no problem; then H_1 and H_2 are compatible. If, however, $C \neq \varnothing$, H_1 and H_2 are incompatible in the sense of making incompatible assertions about Y for possible joint states of X_1 and X_2. How can we entertain such incompatible hypotheses?

Put differently, this is the problem of the *superposition of laws*, or, if the hypotheses have a causal interpretation, of the *interaction of causes*, as urged in particular by Cartwright (1983, Chs. 2 and 3) and later papers. In mechanics, the problem finds an elegant solution: the total force acting on a body is just the vector sum of the individual forces, each of which is governed by a specific force law. Often, though, the superposition or interaction takes strange and unexpected forms (see Cartwright 1983, pp. 67ff., for an impressive example). Of course, nothing general can be said here about the actual superposition; each case requires hard scientific work we cannot anticipate here. Our problem is only to describe the epistemological situation we are in with the schematic hypotheses H_1 and H_2.

The first thing to say (cf. Spohn 2002c, Sect. 4) is that we can, of course, believe in H_1 as well as H_2; we can have both $\xi(\bar{H}_1), \xi(\bar{H}_2) > 0$. This only entails that we must also believe the range of conflict C to not obtain (in the case at hand), i.e., $\xi(C) \geq \xi(\bar{H}_1)$, $\xi(\bar{H}_2) > 0$. Secondly, ξ contains beliefs about the superposition of H_1 and H_2 in any case. Outside the range of conflict C, it simply sticks to H_1 and H_2. And given C, ξ also makes assumptions about the variable Y, that may conform to H_1 or conform to H_2 or diverge from both or, in the worst case, suspend judgment. In some such way, the ranking function ξ is bound to take a stance towards the superposition; it can do so, though, only by containing conditional beliefs.

So far, however, we have assimilated the case of other things being absent to the case of other things being normal. We have treated the range of conflict C as an exceptional condition for at least one of the hypotheses H_1 and H_2. This was presumably not our intention. The interference of X_2 need not be exceptional with respect to H_1, and vice versa; only, it is so much harder to account for and non-interference so much simpler. Then, however, the range of conflict C will not be unexpected; and we arrive at a contradiction; we cannot believe in H_1 and H_2 and take the incompatible C to be possible. The only way out seems to make the implicit assumption of H_1 and H_2 explicit and to acknowledge that H_1 and H_2 are not the hypotheses we actually hold; what we believe is only "if $X_2 = 0$, then H_1" and "if $X_1 = 0$, then H_2" or, more strongly, "if $\bar{C}$, then H_1 and H_2".

However, the above observation still obtains; ξ will embody some belief or other (or suspense of judgment) about the variable Y even given the critical conditions C. So, this is a second way in which H_1 and H_2 may be superposed in ξ: namely by restricting them to $\bar{C}$ and having some assumptions for the case that C. In fact, the ξ of this second way may be understood as the contraction of the ξ of the first way by $\bar{C}$ (that is believed in the latter ξ).

To summarize: Taking hypotheses ceteris paribus in the sense of other things being absent gives rise to the problem of the superposition of laws or the interaction of causes. My response was trivial, in a way. It was that potentially conflicting hypotheses taken ceteris paribus in this sense are somehow superposed in the doxastic state, in any case. My point was simply that in order to be able to give this trivial response, one has to conceive of doxastic states in a suitable way; and my further point was to schematically describe the superposition in terms of ranking functions. The crucial scientific issue of finding the true superposition remains untouched.

13.3 What We Know A Priori about Dispositions

In the introduction to this chapter I indicated that the topic of dispositions is a natural continuation of the previous topic, for the laws associated with dispositions are typically qualified by ceteris paribus clauses. Let me, however, develop the topic a bit more systematically, so that the bearing of the previous section stands out more clearly.

The dispositional idiom speaking of powers, forces, capacities, etc. has always been part of natural language. It received its first extensive philosophical treatment in Aristotle's metaphysics whose notions of *dynamis* and *energeia* puzzled many interpreters. However, it acquired its modern philosophical bite only in the Enlightenment, where the secondary qualities, powers to act on our minds, unfolded their great epistemological riddles (whereas dispositions or powers in general, called tertiary qualities by Locke, received less attention). Later on, logical positivism might be said to have foundered at adequately dealing with dispositions. In the last 30 years, the idea that the world is fundamentally a world of powers and capacities has gained serious advocates and increasing popularity; maybe we can even dispense with laws in favor of powers. Thus, the issue is as unresolved as ever.

I will be dealing with secondary qualities only in Section 16.3, when trying to come to terms with the special role of perception in the formation of our beliefs; the present section will be of great help there. Here we can start our systematic discussion with the attempts of the logical positivists. (I laid out the basic ideas of this section in Spohn 1997a, with the ranking-theoretic underpinning in mind, but not on paper.)

A dispositional predicate, to speak in the formal mode, like "soluble", "magnetic", "indigestible", "red", "intelligent", or "obedient" denotes a property applying to an object if it would show a certain characteristic response if it were subject to a characteristic test situation. This is as familiar as it is obscure. The program of the logical positivists was to reduce everything non-given, non-observable to the given or observable. (The still unsurpassed short description of this program is given by Quine 1969.) Dispositions clearly belong to the realm of the non-observable; it is not written on the face of an object whether or not it is soluble. Thus the task was set to the logical positivists to define dispositional predicates like "soluble" in terms of observational predicates like "being put into water" and "dissolve" (let us grant that these are observational, though "water" is certainly not). How to do so seems obvious:

(13.12) x is soluble if and only if x dissolves when put into water.

This is supposed to be a definition, an analytic sentence; it is the first basic formulation of what is now called the *conditional analysis* of dispositions.

The problem with (13.12) was the logical positivists' commitment to extensional logic. It is natural to read the definition in (13.12) as a subjunctive conditional, and then (13.12) sounds very plausible (we will have a closer look at the subjunctive reading in the next section). However, the logical positivists could not do this, because in that reading the truth condition of "x is soluble" is not a function of, and thus not reduced to, the truth condition of "x is put into water" and "x dissolves"; rather, the subjunctive imports a further substantial, presumably non-observational, and as yet unanalyzed content. Hence, they read the conditional in the definiens as a material implication, with the disastrous consequence that all objects not or never put into water turn out to be soluble according to (13.12). This consequence proved to be incurable. Thus, Carnap (1936/37) acknowledged that the positivists' – that is, his own

— program of reducing the non-given to the given was doomed and started to develop what eventually became the (final version of the) received view of scientific theories (see, e.g., Suppe 1977, pp. 6–56, for an elaborate description of this development).

Let us look only at the first step of this development. It consisted in weakening (13.12) to the (two-sided) *reduction sentence*:

(13.13)   if x is put into water, then x is soluble if and only if x dissolves.

The obvious disadvantage of (13.13) is that it offers only a conditional definition of solubility, i.e., no definition at all. However, the received view ingeniously turned this into a virtue by formulating an elaborate picture of how non-observational, theoretical language receives at least a partial interpretation in observational language through so-called correspondence rules such as reduction sentences – with the effect that theoretical sentences, if not expressible in observational language, at least receive empirical significance and become confirmable and disconfirmable by observational statements. So loosely stated, this empiricist intention can of course be subscribed to by every scientist. The reason why even the empiricist program, the received view, was finally abandoned was that its precise description of the logical and confirmatory relations between the observational and the theoretical level turned out to be fundamentally flawed, something I will attempt to improve, though only for the paradigmatic case of disposition. However, let us not anticipate.

The obvious advantage of (13.13) is that all its conditionals can be taken as truth-functional material implication and that, thus taken, it apparently makes a true assertion. Moreover it appears to be analytically true, as is fitting for a partial interpretation. It must indeed be so according to the received view; otherwise, one would have to ask how the reduction sentence (13.13) is confirmed in turn, a question presupposing a different, but unavailable partial interpretation of solubility.

All this cannot be quite right, though. A first symptom showed up in multi-track dispositions having more than one characteristic manifestation. Magnetism has often been taken as an example, which induces current in coils and arranges iron chips into characteristic patterns. They pose the problem that two different reduction sentences of the form (13.12) for the same dispositional predicate have synthetic consequences. Hence, not all reduction sentences can be analytic; and since there is no ground for preference, none can be.

More worrisome was the observation that even for single-track dispositions reduction sentences are not literally true. Even soluble stuff need not dissolve in water, for instance when the liquid is already saturated. Even unbreakable glass can break when hitting the ground in a normal way, for instance when an intense high-pitched sound simultaneously rings nearby. There are sundry exceptional and bizarre cases in the actual world, and in other possible worlds all the more. Hence, the reduction sentence needs a bit of slack:

(13.14)   Ceteris paribus, if x is put into water, then x is soluble if and only if x dissolves.

This was also the conclusion of Carnap (1956, pp. 66ff.). He distinguished pure dispositional terms, for which reduction sentences hold strictly, from theoretical terms, for which relevant test procedures work "unless there are disturbing factors" or "provided the environment is in a normal state", and then he observed that many dispositional terms are theoretical rather than purely dispositional. Thus, the discussion lapsed into the mystery of ceteris paribus clauses and into the even greater mystery of how a sentence like (13.14), with such an indeterminate clause, can be an analytic meaning postulate.

These are battles that are long past, one might think. The subjunctive conditional became respectable in the late 60s; it has even been endowed with a standard intensional logic, Lewis' (1973a) system VC (p. 132). This logic automatically makes a conditional dependent on the relevant conditions; recall that a subjunctive conditional is a *variably* strict conditional according to Lewis. Thus, the discursive situation has entirely changed. In principle, we can return to the conditional analysis (13.12) in the subjunctive reading, which is the focus of the more recent discussion (and indeed contested, though for other reasons than foreshadowed by the logical empiricists). This is the field we should enter.

Well, I am not so sure. Let us defer this discussion to the next section. I propose first to directly apply the epistemic account of ceteris paribus conditions in the previous section to (13.14) (along the lines of Spohn 1997a), and to see where this leads us. Only afterwards can our results be fruitfully compared with the more recent discussion.

The application is indeed straightforward. We are again talking about a single case, a specific object. In that case, we are considering three binary target variables, getting into the relevant test situation or not, S or $\bar{S}$, having the disposition in question or not, D or $\bar{D}$, and showing the appropriate response or not, R or $\bar{R}$. The target variables operate on a large background, described by a variable Z. And as before, we have a doxastic attitude ξ over the whole set-up that we may independently repeat to form a subjective law λ_ξ for all similar cases in the usual way. Our target hypothesis is the reduction sentence (13.13), i.e., the proposition $H = S \rightarrow (D \leftrightarrow R)$, in self-explanatory abuse of the logical connectives as set-theoretic operations on propositions. We believe the reduction sentence, i.e., $\xi(\bar{H}) > 0$, and since the ceteris paribus clause in (13.14) is obviously to be understood in the sense of "other things being normal", we believe it given normal conditions, i.e., $\xi(\bar{H} \mid N^0) > 0$, with N^0 as defined in (13.7), and we believe in the normal conditions, i.e., $\xi(\bar{N}^0) > 0$.

So far, this was mere repetition. The crucial difference to the previous section is that the reduction sentence H is not just any empirical hypothesis. Its truth (under normal conditions) and thus our belief in it are somehow of a conceptual nature, and we need to do justice to this nature. The earlier, somewhat suspicious moves consisted in declaring the reduction sentences (13.13) and (13.14) to be analytic. However, these moves were taken before Kripke's (1972) reform of modalities. Afterwards, it was clear that philosophers of an empiricist provenance in particular had often meant apriority when they had spoken of analyticity. So, let us first discuss what is a priori about dispositions; I will return to the more complicated issue of analyticity in the next section.

Still avoiding tackling apriority straight away in general – this must wait until Chapter 17 – let us recall from Section 6.5 that there are two natural notions of apriority in the ranking-theoretic framework; there are defeasibly a priori propositions that are initially believed, though not necessarily maintained, and there are unrevisably a priori propositions that are necessarily believed under all circumstances. The principle of the uniformity of nature was of one or the other kind (relative to the second-order impact function). I emphasized in Section 6.5, though, that these notions of apriority should not be restricted to propositions, but applied to any kind of doxastic features. This point now finds its first application; reason relations can also be defeasibly and unrevisably a priori! The fuller significance of this important observation will be discussed only from Section 16.3 onwards. Presently, let us bring home what I take to be a particularly clear instance of that point.

Let us consider whether our reduction sentence $H = S \rightarrow (D \leftrightarrow R)$ is a priori. We have seen that it might turn out false in a given case; hence it cannot be unrevisably a priori. Initially, though, we do not know anything about a given case, about strange circumstances or exceptional conditions; initially, no reason to mistrust H can have emerged. Hence, it is overwhelmingly plausible to assume that the reduction sentence H is *defeasibly a priori*; this is the correct characterization that had not been available in the old empiricist discussion.

Initially, we do not know even whether or not the object at hand is in the test situation, has the disposition, or shows the response; that is, for each initial ξ, $\xi(S) = \xi(\bar{S}) = \xi(D) = \xi(\bar{D}) = \xi(R) = \xi(\bar{R}) = 0$. This given, we have two equivalent characterizations:

(13.15)   The belief in the reduction sentence $H = S \rightarrow (D \leftrightarrow R)$ is defeasibly a priori, or, equivalently, it is defeasibly a priori that, given S, D is a necessary and sufficient reason for R in the sense of (6.2), and vice versa.

Now, if so much is only defeasibly a priori, what might be unrevisably a priori about dispositions? For an answer, we must study in two steps what happens to the reduction sentence and the appertaining reason relation when we learn more about the given case.

First, suppose we get more information about the background conditions. This possibly initiates the very interplay described in detail in the previous section. We expect to learn that the background is normal, and our experience might confirm this. It might also disappoint this expectation. We might learn that the liquid is already saturated so that soluble stuff does not dissolve in it. We might hear the high-pitched sound and thus no longer exclude even the breaking of unbreakable glass upon hitting the ground. And so forth. We might even learn that the background is doubly exceptional; a strong electric current in the liquid might raise its saturation point. And so on. However, what cannot change through this learning process is the conditional belief in the reduction sentence given normal conditions; this relation remains fixed throughout all information about the background.

We can state this point a bit more precisely with the help of the conditionalization rules, the dynamics of ranking functions stated in Section 5.4. Let ξ be any initial ranking function on the algebra generated by the propositions S, D, and R and the background variable Z; i.e., ξ conforms to (13.15). Let $\mathcal{Z}$ be the subalgebra generated by Z, ζ any regular ranking function for $\mathcal{Z}$, and N^0 the normal conditions for the reduction sentence H relative to ξ as defined in (13.7). Then:

(13.16)   If ξ' is the $\mathcal{Z} \to \zeta$-conditionalization of ξ (see 5.32), then H is believed conditional on N^0 in ξ', and given N^0 and S, D is a necessary and sufficient reason for R relative to ξ', and vice versa.

I have referred to the most general form of conditionalization in order to make (13.16) as strong as possible; note that iterated applications of this form of conditionalization can be summed up in a single application. Still, (13.16) is entirely trivial in view of the fact that $\xi(\bar{H} \mid [z]) > 0$ for any $[z] \subseteq N^0$ according to (13.8) and the further fact that any $\mathcal{Z} \to \zeta$-conditionalization by definition preserves the values $\xi(\bar{H} \mid [z])$ for all $[z]$. In claiming the conditional necessary and sufficient relation between D and R to be preserved as well I have tacitly assumed that nothing is known about S, D, and R, not only initially, but also given each maximal normal condition $[z] \subseteq N^0$. This completes the first step.

(13.16) holds for any kind of background Z explicitly considered in the ranking function ξ. Does this mean that the belief in the reduction sentence and the necessary and sufficient reason relation are unrevisable, given normal conditions? Not quite, as we have to reflect in the second step:

We might receive information about S, D, and R directly and not only via Z. Or we might receive information not explicitly represented in the propositional framework spanned by S, D, R, and Z. We might hope to deal with the latter case by refining the background Z and then applying (13.16) to the more fine-grained background and the appertaining initial ranking function and normal conditions. That might work. However, the information originally not explicitly represented might hold information about S, D, and R that cannot be limited to a refined background and thus violates the presuppositions of (13.16). So, how can information about S, D, and R, be it direct or indirect, but not via the background Z, affect the status of the reduction sentence and the appertaining reason relation?

The reduction sentence need not be maintained, I think. The hitherto unknown object at hand might very much look like other stuff we take to be insoluble; hence we believe it to be insoluble, too. Now we put it into water and it apparently dissolves; at least we can no longer see it after a while; hence we believe it to have dissolved. We are surprised, we investigate the background, and we do not find anything exceptional. As far as we can establish, normal conditions as explicitly represented did obtain. Still, we might stick to our conviction that the object was insoluble, though less firmly than before. So, we might accept the case as an apparent exception to the reduction sentence. There is nothing incoherent about this story; it is very much unlike accepting, say, 2 + 2 = 5.

Moreover, the reason relation need not be maintained as a necessary and sufficient one, for a much simpler reason. Suppose we get the information that the object in question has been put into water and that it has dissolved; perhaps we have seen this or believe we have seen this, perhaps a person we take to be reliable has told us so. Of course, we infer that the object is soluble. Then we presumably believe that the object has dissolved given that it is soluble, as well as given it is in fact insoluble. (This is not the same as believing that the object would dissolve, no matter whether or not it is soluble; I am not talking about belief in (counterfactual) conditionals.) If so, its solubility can no longer be a necessary and sufficient reason for its actually dissolving; it can only be a supererogatory reason in the sense of (6.2).

However, this indicates what is maintained. D is still a reason for R given S, and vice versa (by symmetry 6.7b of the reason relation). The supposition that the object is in fact not soluble at least weakens the belief in its actually dissolving. Many things might be wrong, but perhaps that person was not so reliable, after all, or perhaps our perception was somehow deceived. Likewise, in the former case in which the reduction sentence was given up. We might believe that S and that the normal conditions N^0 obtain, and we might nevertheless believe $\bar{D}$ and R. So, we also believe R given $\bar{D}$. Again, though we would believe R more strongly given D.

The supposition of the normal condition N^0 remains important. Under exceptional conditions not only might the reduction sentence lose its credibility, even the reason relation might reverse. The normal conditions under which elasticity proves itself include normal temperature. There are bodies that lose their elasticity when cooled down and other bodies that become elastic only at high temperatures; thus, for both kinds of bodies the reason relation between disposition and manifestation is reversed under exceptional conditions.

To conclude, the point of my two-step argument is this:

(13.17)   It is unrevisably a priori that, given N^0 and S, D is a reason for R, and vice versa.

Let me be clear, though, about the structure of my argument on the last three pages. I wanted to capture which beliefs or doxastic attitudes are associated with the notion of a disposition for, as is said, conceptual reasons. So, I gave an intuitive argument that (13.15) is included in the notion of a disposition. Then I gave a rigorous argument that (13.15) and the account of normal conditions in the previous section entail (13.16). (13.16) did not quite establish (13.17), but it is, I think, as far as we can get by formal argument. Hence, the last step was, in turn, an intuitive one, arguing that even (13.17) holds on conceptual grounds. Conversely, (13.17) entails (13.15), with the equally plausible assumption that the belief in normal conditions is defeasibly a priori. Indeed, these three claims may be taken as mutually supporting each other.

Note that this is also a kind of explication of what conceptual grounds might mean with respect to dispositional notions. Note, moreover, that the argument makes intelligible how something unrevisably a priori can include a reference to something as vague as normal conditions; of course, this was possible only under the epistemic reading of these conditions. And let me emphasize once more that (13.17) was *not* too lightly achieved by giving normal conditions the trivial reading; it was rather backed up by the substantial and stronger reading elaborated in the previous section. Note, finally, that I have thus given a clear example of how other things than beliefs, in this case reason relations, can be defeasibly and unrevisably a priori.

This is, I think, what philosophers can truly say about dispositions. It is not yet the full truth, as we will see shortly; but it is almost the full truth about the conditional analysis (13.12). Before thus turning to the discussion of the conditional analysis, let me make one further remark.

I mentioned above that the claim of reduction sentences of the form (13.13) is undermined by multi-track dispositions. It seems that the problem persists after our replacement of analyticity by apriority. How can two reduction sentences $H_1 = S_1 \to (D \leftrightarrow R_1)$ and $H_2 = S_2 \to (D \leftrightarrow R_2)$ referring to the same disposition be a priori in any sense? This apparently entails that the proposition $S_1 \cap S_2 \to (R_1 \leftrightarrow R_2)$ is a priori in the same sense. How can it be? After all, that proposition clearly has an empirical synthetic content.

I have no elaborate answer, rather only a hunch. The conceptual a priori is apparently relative to a given conceptual framework and the appertaining propositional framework. Thus, whatever beliefs or doxastic features are defeasibly or unrevisably a priori associated with certain concepts, they cannot be found in all initial or all doxastic states whatsoever, but only in those states mastering those concepts and thus having attitudes towards propositions generated by those concepts. This means in particular that initiality is a relative notion; what counts as an initial doxastic state is relative to the concept considered. Such initial states need not be conceived as *tabula rasa*; they may be otherwise loaded with contents, and such contents may be presupposed in the formation of a new concept.

So it is already with ordinary single-track dispositions. We do not introduce a disposition for arbitrary test situations, say falling to ground or entering a shop, and arbitrary responses, say cheeping or whistling, although we find falling objects that cheep and others that do not cheep, and whistlers and non-whistlers entering a shop. Rather, we have already observed a certain pattern concerning the relevant test situation and response and form the dispositional concept only on the basis of this pattern.

Likewise with multi-track dispositions. There we have observed a more extensive ceteris paribus pattern, including the association $S_1 \cap S_2 \to (R_1 \leftrightarrow R_2)$, between the two characteristic test situations and the two responses, we conjecture one and the same disposition to be responsible for that pattern, and we thus introduce the multi-track dispositional concept D, accompanied by two reduction sentences. This conjecture might turn out to be wrong; the presuppositions for introducing the concept need not hold. But then we lose the dispositional concept and perhaps replace it by two single-track dispositional concepts.

In this way, we might perhaps defend the possibility of two claims of the form (13.15) or (13.17) being, respectively, defeasibly or unrevisably a priori for a given concept, even though the relevant doxastic states have a posteriori propositional content generated by this concept. Perhaps, this response could also have been given in terms of analyticity; but, of course, this would not have solved the other problem with normal conditions, which plagued the empiricist account. (See also my further remarks on the notion of initiality in Section 17.1.)

13.4 The Metaphysics and the Epistemology of Dispositions: A Short Comparative Discussion

Current debate about dispositions rather worries about the inadequacies of the conditional analysis (13.12), even in its subjunctive reading, and draws more or less far-reaching conclusions in particular concerning the metaphysics of dispositions. How does my account (13.15)–(13.17) relate to this discussion? Let us first look at the worries about the conditional analysis.

There is a very simple reason for the inadequacy of (13.12) in its subjunctive reading, which does not require fancy examples; to my surprise I did not find it noted in the literature. The point is that the conditional analysis appeals to the wrong kind of conditions. In the empiricists' reading as material implication it does not refer to any further conditions at all; this is totally insufficient. The subjunctive reading does more; its antecedent implicitly refers to the *actual* conditions of each case at hand. If I define, "for all x: x has D if and only if it would show R, if it were in S", I am in effect saying, for all x, that x has D if and only if it would show R, given x is in S *and* given – but this goes without saying – the conditions x is *actually* in. This is simply wrong, though. If I were to put the soluble x into water that is actually saturated it would not dissolve. Hence, it is not only reduction sentences, it is also the conditional analysis that needs to be hedged by the unloved *normal* conditions. It seems that Lewis (1997), who is more sophisticated otherwise, overlooks the point. Mumford (1998, pp. 87ff.) does not whole-heartedly refer to normal conditions, either. So, one cannot say that the point was simply taken for granted.

It was rather because of a more sophisticated problem presented by Martin (1994) that the conditional analysis – at least in its simple form (13.12) – was agreed to be insufficient. It consists in the fact that dispositions can be finkish, as Lewis (1997) calls them. That is, a disposition can be such that the test situation in which it should manifest makes it go away. A nice example is the chameleon that is blue at night. Being blue is, let us assume, the disposition to look blue in daylight. However, whenever the chameleon is exposed to daylight, it changes its disposition – and turns green. So, contrary to (13.12), it would not look blue in daylight. A more artificial example is the electro-fink, introduced by Martin (1994). A wire can be live or dead, something that manifests when the wire is touched by a conductor. Now, the live wire might be connected to an electro-fink that turns the wire dead as soon as it is touched by a conductor, and conversely a dead wire might be turned live in the very same moment by a reverse electro-fink. Martin concludes that the conditional analysis (13.12) in the subjunctive reading is neither necessary nor sufficient for having the relevant disposition.

Martin (1994, pp. 5f.) therefore considers the natural move to add a ceteris paribus clause to the conditional analysis. However, he rejects this move since he only thinks of the trivial reading of the ceteris paribus clause, and Mumford (1998, p. 83) concurs. Martin and Mumford conclude that dispositions are real properties independently of whether they are given any opportunity to manifest themselves. Thereby, the discussion takes a somewhat unexpected metaphysical turn concerning realism, since the conditional analysis is accused of denying such realism and giving dispositions the status of mere and mysterious potentialities. Lewis (1997) wants to avoid these metaphysical associations, despite defending a more sophisticated conditional analysis; in footnote 11, he expressly says that he has "taken care to bypass . . . the question of what dispositions are". This is an issue I will comment on shortly.

Let us check, though, whether we really have to worry about finkish dispositions. My first response is obvious. It is that we need not fear the trivial reading of normal conditions since we have a better one. And we may well say that an environment that makes a disposition of an object finkish is exceptional. Wires are not normally connected to (reverse) electro-finks. At night when chameleons are blue, daylight is a very unusual condition. I cannot see this response refuted by Martin (1994).

My second response is considered by Lewis (1997) and, I find, not convincingly rejected. It is that it is an open question what it is that has a disposition. Ordinary objects, we think, like glasses, wires, chameleons, or persons. But it might also be something more complex, a set-up, as we might say; recall that the talk of chance set-ups (having probabilistic dispositions or propensities) has become quite familiar. What is a set-up? It is hard to say in general. It might be the wire and everything connected to it. This set-up indeed has the disposition to be dead even if the wire by itself is live, in agreement with the conditional analysis. Lewis (1997, pp. 147f.) calls this a compound disposition, but rejects the construal because it runs counter to his conception of dispositions as being intrinsic properties of their objects. He supports his rejection by funny examples that appeal to funny set-ups, for which it is hard to say what is intrinsic to them. Still, a funny set-up is a set-up that fails to have the disposition the component would have without exceptionally being integrated into the set-up.

So, the reference to normal conditions, that we already require for simpler reasons and that my analysis respects, along with greater clarity about the objects disposed, seem to be able to account for finkish dispositions. How do others deal with that problem?

Lewis (1997) wants to retain and to refine the conditional analysis. His idea is that dispositions are properties intrinsic to their objects that can always be changed from outside by finks, daylight, etc., but would produce the relevant response if not changed. The full analysis looks complicated, Lewis apologizes:

(13.18)   Something x is disposed at time t to give response r to stimulus s iff, for some intrinsic property B that x has at t, for some time t' after t, if x were to undergo stimulus s at time t and retain property B until t', s and x's having of B would jointly be an x-complete cause of x's giving response r. (Lewis 1997, p. 157)

However, if one overlooks the subtleties about timing (as I have done in this chapter) and the subtleties about x-complete causes (see below), (13.18) is quite straightforward.

Mumford (1998) has metaphysical complaints about Lewis' defense of the conditional analysis. Indeed, his book is mainly about the metaphysics of dispositions. But he is clear that before metaphysics can be addressed, the conceptual side must be settled. In response to Martin's problem cases, he proposes to replace the conditional analysis (13.12) by what he calls a functional account of dispositions, namely that "disposition ascriptions are ascriptions of properties that occupy a particular functional role as a matter of conceptual necessity" (p. 77). He further argues that the functionalist should not attempt "to analyze a functional role in terms of conditionals" (p. 87); rather, because of the threat of finkishness, "a functional role is specified in terms of what causal consequences a property or state will produce to what antecedents" (p. 87). Thus he ends up with the following definition of a disposition D to show response G in the situation F:

(13.19) x is $D =_{\mathrm{df}} x$ has some property P (and P is a cause of x G-ing if x is F-ed in conditions C_i). (Mumford 1998, p. 135)

This definition is literally repeated two pages later. Mumford attributes a similar one to Sober (1982).

Basically, I can agree with both (13.18) and (13.19) without betraying my (13.15)–(13.17). Some subtle differences cannot be denied, though.

First, some hairsplitting, not because of disputatiousness, but because it is a nice illustration of an earlier point. Both, (13.18) and (13.19) fall into the traps of the existential reading of ceteris paribus conditions, or something analogous. In Section 13.2 I referred to Kaila (1941) as an early warning example of these traps, who actually tried to save (13.12) in the extensional reading with an 'existentialist' strategy. According to Mumford's (13.19), an object not only has the disposition to G when F-ed, but also the disposition to non-G when F-ed, because there is also another property P' causing non-G-ing upon being F-ed in some conditions C_i' that also satisfy Mumford's specification of these conditions (still to be discussed).

Lewis' (13.18) falls into the same trap. An object disposed to give response r to stimulus s is at the same time also disposed to give response non-r to stimulus s according to (13.18), since it always has a further intrinsic property B' the preservation of which would prevent it from giving response r; Lewis (1997, p. 158) himself points out that one must not keep fixed the whole intrinsic character of an object. (Curiously, on

p. 157 Lewis considers the same point and finds it unobjectionable that an object can be disposed in two opposite ways at the same time. He thinks that this possibility is restricted to finkish dispositions and thus acceptable. If I am right, the possibility is always realized. Then it constitutes an objection.)

In both cases the fault lies in the existential quantification over properties; there always are too many properties, and even too many intrinsic properties, that an object has. Its disposition is rather *the* functional or *the* intrinsic property causing the relevant manifestation under *normal* conditions. The reference to normal conditions, not present in (13.18) and indeterminately present in (13.19), is essential here; deviant manifestations that the object is not disposed to show can occur only under exceptional conditions. This sounds better; we will, however, see at the end of the section that, even with reference to the normal conditions, the use of the definite article is not yet justified.

Is Mumford (1998) really indeterminate about normal conditions? On pp. 87ff. he describes the conditions C_i mentioned in (13.19) as ideal conditions "that can vaguely be understood as 'normal' . . . [and] are not realized only in exceptional circumstances" (p. 89). He is eager to avoid the trivial reading, he emphasizes the context relativity of the ideal conditions with respect to the disposition ascription at hand (hence, there might be also ideal conditions for the counter-disposition to non-G upon being F-ed), and he emphasizes that the relativity entails the possibility of ideal conditions not being normal, but quite extraordinary. I find this indeterminate. The point, I suppose, is that Mumford thinks that nothing much sharper can be said about those conditions, anyway, because of their fundamental openness. There is an exchange between Malzkorn (2000) and Mumford (2001) about this point that mirrors the difficulties of coming to terms with these conditions (Malzkorn explicitly intends normal conditions). As a comment, I can only recommend the employment of my epistemic reading of normal conditions, which was designed to overcome such difficulties.

So much for hair-splitting. Clearly, there are more important differences. The most conspicuous one is that (13.18) and (13.19) speak of causes whereas (13.15)–(13.17) speak of reasons. It would indeed have been very natural to refer to dispositions as causing or – subtle differences might be seen here – as causally explaining their manifestations; most of the discussion centers on the real or alleged causal role of dispositions. Mumford (1998) forcefully argues that this role is real, thus opposing old Rylean views about the conceptual instead of causal nature of dispositional explanations (allegedly tied to the conditional analysis), as well as that of Prior et al. (1982) that dispositions are causally impotent second-order properties (since they allegedly differ from their causal bases, but certainly do not causally overdetermine their manifestations). Lewis (1997) dodges the latter problem in a different way, thus also maintaining the causal role of dispositions. (Note that in this respect there is hardly any difference between (13.18) and (13.19). The "if" in (13.19) is clearly a subjunctive "if". Hence, the conditional and the causal ingredients of (13.18) and (13.19) are quite the same, and give no ground for seeing, as Mumford does, a principled difference between a conditional and a functional or causal analysis.)

I entirely agree (though not with Lewis' particular way out – see below). However, so far I have wanted to completely avoid causal talk about dispositions (perhaps artificially so), for the simple reason that we cannot yet systematically use the causal idiom; how causation can be grasped in ranking-theoretic terms is a task undertaken only in the next chapter. Everything will turn out well, though. We will be able to clear up the bewildering relation between reasons and causes (cf. Sections 14.3–14.4), and thus find that we might as well replace "is a reason for" by "is a cause of" in (13.15)–(13.17) (not as an effect of mere conceptual jugglery, but on the base of a substantial argument in Section 14.14).

When I said in the introduction of this chapter that I am unsettled by the prospect of (re-)interpreting the notion of force, I was referring to this point. In a way, I am claiming with (13.17) that what we a priori know about dispositions or forces – not only about secondary qualities, but about tertiary qualities or powers in general – is that *they move our rational minds*. This sounds very Humean, if not crazily so: an obvious and dangerous confusion. Well, the depths of Hume's claim that causation is an idea of reflexion (in his sense) are still to be fathomed; I hope that the alleged confusion can be successfully clarified. At the moment, I only wanted to signal the direction of my considerations, a direction predictable already from my peculiar notion of a subjective law.

There is a final crucial difference, or rather two differences in one. On the one hand, (13.18) and (13.19) propose definitions and thus make analytic claims about dispositions, whereas (13.15)–(13.17) tell only about apriority. On the other hand, I have mentioned that Mumford's dominant interest is in the metaphysics of dispositions on which I have so far been silent. I am reluctant to enter into discussion of these issues since they open up wide dimensions of philosophical debate, a systematic discussion of which would require another book. However, my treatment of dispositions would be deplorably incomplete if I were to stay silent. So I will compromise as well as I can.

The basic and unoriginal point I want to make is that by dealing with the metaphysics of dispositions we close the gap between (13.15)–(13.17) and (13.18)–(13.19), between a priori and analytic claims. The reason lies in Kripke's (1972) reform of modalities mentioned earlier. It says that the (metaphysically) necessary and the epistemically necessary, or a priori, are two independent modalities, and that the analytic is the a priori necessary. I should emphasize that Kripke's a priori is always unrevisable and never defeasible. Thus, given what is a priori about dispositions, saying what is metaphysically necessary for them amounts to saying what is analytic for them. Here, I simply endorse the Kripkean reform (defending it would be a first chapter of a new book).

However, as far as I see, the current discussion about dispositions is still insufficiently aware of the consequences of the Kripkean reform. The quintessence of this reform is two-dimensional semantics, founded by Kaplan (1977) and Stalnaker (1978), closely related to situation semantics as developed by Barwise, Perry (1983), and most illuminatingly continued by Haas-Spohn (1995), Chalmers (1996, part I), and Jackson (1998) (see, e.g., the many contributions in García-Carpintero, Macia 2006). The basic claim of two-dimensional semantics is that there are epistemic possibilities (or contexts) and metaphysical possibilities (or points of evaluation), that the meaning of each linguistic expression is to be represented as assigning an extension to each pair of an epistemic and a metaphysical possibility (or of a context and an evaluation point), and that there are, hence, two kinds of intensions associated with a linguistic expression: for each epistemic possibility a secondary intension (to use Chalmers' terms) assigning an extension to each metaphysical possibility, and a primary or diagonal intension assigning an extension to each epistemic possibility and its associated metaphysical possibility (where one needs to say what association means here, some think just identity). The secondary intensions (of sentences) tell us what is (metaphysically) necessary in an epistemic possibility, and the primary intensions tell us what is a priori. (To really explain all this would require further chapters, but I will suppress this remark in the sequel.)

The point now is that two-dimensional semantics claims that *all* linguistic expressions and phrases have a two-dimensional meaning; it is the basic format of semantic theorizing. Thus it applies to all predicates or general terms. It is appropriate, I believe, to speak of the primary intension of a general term as the *concept* associated with the term (this is the epistemic component of the two-dimensional meaning) and of its secondary intension (in a given context or epistemic possibility) as the *property* associated with the term (this is the ontological or metaphysical meaning component). Many confuse concepts and properties; many attend to the distinction, but draw it in a different way. I find it thus perfectly explicated in the two-dimensional framework.

This in particular applies to dispositional predicates; any explication must specify their two-dimensional meaning or, equivalently, the concepts associated with them and (for each epistemic possibility) the properties associated with them. Mumford (1998), for instance, does not really think within that framework. Lewis certainly does so all the time, and so does, I suppose, Jackson, one of the authors of Prior et al. (1982). Still, the framework is rarely explicit, to the detriment of the dispute over dispositions. In Spohn (1997a), my first attempt at dispositions, I expressly tried to utilize this framework.

Within this framework, specifying the (two-dimensional) meaning of a dispositional predicate amounts to specifying the concept associated with it (this is the epistemological task) and specifying the property associated with it in each given context or epistemic possibility (this is the metaphysical task). Are we already finished with the epistemological task? No. We have been doubtful about whether (13.18) and (13.19) are true, and hence whether they are a priori statements. Nor does (13.17) offer a complete answer, simply because it does not have the form of a definition. Let us defer the issue to the end of the section and address the metaphysical task pretending that the epistemological task has been finished.

Here, Mumford (1998) is most helpful, even if not couched in the two-dimensional framework. In Chapter 4, he argues convincingly that the distinction between the dispositional and the categorical is not a metaphysical distinction. With reference to Frege, he says it is a distinction in the senses of dispositional predicates (and Frege's senses are best understood as primary intensions or, with respect to predicates, as concepts within the two-dimensional framework). One and the same property may be presented in the dispositional mode as solubility or in the categorical mode as a certain pattern of intermolecular coherence. In Chapter 5, he rejects the idea of metaphysically grounding the distinction in a property dualism. Even then, the question remains of whether it can be said that properties are basically categorical or basically dispositional. No, he concludes in Chapters 7 and 8, opting for what he calls neutral monism – a wild position in the philosophy of mind, but perfectly reasonable in the present context.

So, his result is an identity theory for dispositions, having a disposition is the same property as having the causal base of that disposition. We might thus roughly define, in agreement with (13.18) and (13.19) – but I said we have still to check this – that the disposition D to show R in situation S is the (intrinsic) cause of showing R in situation S. However, we have to read that definite description rigidly. When we assert "D, the cause of R in S, causes R in S", we face the old ambiguity of the referential and attributive reading of definite descriptions (that is perfectly accounted for in two-dimensional semantics). In the attributive reading the assertion is analytic; and it may well have been this reading that misled philosophers in pre-Kripkean times into thinking that dispositional explanations somehow are conceptual explanations and thus of an entirely different nature than causal explanations. In the referential or rigid reading, by contrast, the assertion is an example of Kripke's famous contingent a priori; it makes an ordinary contingent causal claim. Only in this reading does the identity of a disposition with its causal base hold.

So far, I have presented the identity theory as a type–type identity theory. This, however, is not what Mumford wants to claim in the end. Facing the standard objection to type–type identity of multiple realizability, Mumford (1998, pp. 157ff.) prefers the standard move of retreating to the essentially weaker token–token identity theory. I disagree. So, let us look at this objection a bit more carefully.

There is, first, multiple realizability on the level of concepts. We start by presupposing an a priori principle of causality. A response R cannot show up uncaused, and if some objects show R in S and others do not, this must have a differential cause as well. (That is, we neglect the difficulties posed by chance events.) This entitles us to say in a first move that having the disposition D to show R in S is having *some* causal base for showing R in S. Prior et al. (1982) call this a second-order property, though I am not happy with this term. After all, it is a property of objects and thus first-order. Is being spherical second-order because it means consisting of *some* stuff arranged in spherical form? Anyway, Prior et al. identify the disposition with this second-order property that differs from any specific causal base.

Lewis (1997) intends to be neutral on this identity issue. He says on p. 152: "When a glass is fragile, it has two properties. It has some first-order property which is a causal basis for fragility; it also has the second-order property of having some causal basis for fragility or other. We need not say which of these two properties of the glass is its fragility."

Still, I do not find Lewis' position as neutral as he says, since he misleadingly suggests that only the metaphysical side is at issue. Rather, I sense here a confusion of concept and property. Of course, concepts may be multiply realized; water may turn out to be H_2O or turn out to be XYZ. This has never been seen as problematic. So, a disposition may certainly have varying causal bases across varying epistemic possibilities. Therefore, the position of Prior et al. (1982) is prima facie plausible. However, they tacitly turn their plausible conceptual claim into a metaphysical claim (at least I find their discussion on pp. 253f. highly ambiguous between concept and property), and so they derive the metaphysical consequences rightfully criticized as absurd by Mumford (1998, Ch. 5).

Hence, let us explicitly turn to multiple realizability on the level of properties. Our discussion bifurcates again. We may ask whether a disposition, the property associated with a dispositional term in a given epistemic possibility, may be multiply realized, i.e., have multiple causal bases *within* a metaphysical possibility or *across* metaphysical possibilities.

Turning to the first issue, we should first improve our rough description of the concept; we have forgotten the normal conditions. So, it is, more accurately, that having the disposition D to show R in S is having some causal base for showing R in S under normal conditions. I am not quite sure whether this schema always applies. There are difficult cases like intelligence or, better still, obedience, where the search for a causal base looks hopeless and need thus not be implied; perhaps such notions are instead family concepts referring to a vague cluster of capacities and attitudes. It will be better to stick to the simple examples of single-track dispositions.

Can we strengthen our concept of the disposition D in such a way that having the disposition D to show R in S is having *the* causal base for showing R in S under normal conditions? The existential presupposition is met by the a priori principle of causality. The uniqueness presupposition is precisely the issue of multiple realizability within a possible world that we wish to discuss. There is not much to discuss, though. One must simply grant that uniqueness might fail. Nature is kind, or we are well adapted, and we surprisingly often find a single causal base. Surely, though, it might turn out that two different intermolecular mechanisms lie behind the solubility or the fragility of an object. Red-green blindness, for instance, has exactly two different causal bases; it consists in having either the more red-sensitive pigments or the more green-sensitive pigments in both the 'red' and the 'green' cones (cf. Hardin 1988, pp. 76ff.; see also Section 16.3, where I elaborate on this example for other purposes).

However, in such a case the disposition simply has a disjunctive causal base, a disjunctive essence. I cannot see a relevant difference from the discussion about natural kind terms or substance names. Here, we also assume that there is a single substance underlying the phenomenal bundle associated, say, with the concept of water. We were right, water is H_2O. But the assumption might be disappointed; there might be two kinds of water in our world, H_2O and XYZ, and they are both water. The example of the two kinds of jade mentioned by Putnam (1975, p. 241) is a telling real-life example. There might even be hopelessly many substances underlying the concept of water. In that case we would give up on a hidden nature of water, and the concept of water would also be the property of being water. Perhaps this is what we have to say about obedience in the case of dispositions.

Thus, the crucial case deciding about the type–type identity theory concerning dispositions is the second issue of multiple realizability across metaphysical possibilities. Can we project the causal base(s) we have found a disposition to have in a given epistemic possibility to other counterfactual possibilities, or may a disposition have other causal bases there? In other words, do we have to read the definite description "the causal base for showing R in S under normal conditions" in a referential or in an attributive way?

Sugar is soluble and has the causal base of solubility. Now suppose that in a counterfactual possibility there is some stuff that is clearly sugar, i.e., has the same chemical composition as our sugar and thus has the causal base of solubility. However, when it is put into water there – our water! – it (almost) never dissolves. Should we call sugar insoluble there? Well, we have forgotten about the normal conditions. If sugar hardly ever dissolves in that world, then, it seems, the normal conditions in that world are such that sugar does not dissolve when put into water. However, are the normal conditions in that world those that mostly obtain within that world? This would mean reading "the normal conditions" attributively. This does not make much sense, though. I had emphasized that normality is an indexical or egocentric notion that refers to what is normal to *us* in *our* environment. Detached from such a context, normality is not meaningful. Thus detached, we could only say that everything in our environment is extremely exceptional, since the earth is such an extraordinary place in our universe.

So, we have to apply the referential reading of "the normal conditions". Then the strange behavior of sugar in that counterfactual possibility may have a simple explanation. Sugar is certainly soluble there; it is only that there normal conditions hardly ever obtain, and this is why sugar hardly ever dissolves there. If this were an adequate description of the case, it would support a referential reading of "the causal base".

However, we may go on to suppose that it is our normal conditions that are usually present in that counterfactual possibility and that sugar still hardly ever dissolves when put into water. Now the example is becoming fanciful. Should we say then that the causal base of solubility is different from the actual causal base and hence that sugar is not soluble there? Or should we rather say that sugar there has the actual base of solubility and is therefore soluble, though this causal base is somehow causally inoperative? Clearly, such a counterfactual possibility is governed by different causal laws than the actual possibility. Should we thus make our causal laws part of our normal conditions? Then we could handle the case in the same way as before. However, we do not usually consider the laws to be part of the conditions. I am unsure what to think.

On the one hand, I have made clear that the type–type identity theory, the issue of whether a dispositional property is identical with its causal base, depends exactly on how we would talk about such a counterfactual case of alleged multiple realizability across metaphysical possibilities. On the other hand, I confess that I have been unsure since my (1997, Sect. 6). The case is just too far-fetched to allow firm intuitions. So, I plead for semantic indecision. This means, though, that there is a space devoid of argument, a space which may be filled by a decision. If this is so, then I opt for the type–type identity theory for dispositions. If this theory holds in all more or less familiar cases and if it would founder only with that undecided far-fetched case, then we may decide to stick to it also in that case. Spoils to the victor, as David Lewis used to say.

From this position on the metaphysical side, we should finally look back at the epistemological side. Did we clear up the concept associated with a dispositional predicate? No, not really. (13.17) specified the a priori inferential role of such a concept, but not the concept itself. I have criticized (13.18) and (13.19) for their "existentialist" twist and their insufficient respect for normal conditions. Then I tentatively suggested several times that we should instead characterize the dispositional concept by a definite description (the referential reading of which then refers to the associated property). Which definite description has crystallized from our discussion so far? Something like this:

(13.20) x has D, the disposition to show R in S, iff x has that (possibly disjunctive) intrinsic property that causes x to show R in S and normal conditions, or x's having of which is a reason for expecting x to show R given x is in S and in normal conditions.

This will not do, however. The point is that there always are many such reasons and many such causes, even given the further restrictions, and that the definite article is therefore unjustified. Let me explain this in the more familiar terms of causes. Lewis (1997, p. 156) already considers the problem that there are partial causes and complete causes. Partial causes are also causes, but they might be accompanied by the wrong supplementary partial causes, so that (13.20) gets mixed up. Lewis tries to solve the problem by alluding to what he calls an x-complete cause in (13.19).

However, there is not only the problem of partial causes; there is also the problem of causal chains. Suppose the definite description in (13.20) were to refer to a well-defined property D. Now there will be some causal chain mediating between x's having D and x's showing R in S and normal conditions. Thus, x will be in some intermediate intrinsic state D^+ that also causes R. Similarly, the fact that x has D has causes, too, and indeed intrinsic causes, say D^-, that hence cause R as well. So, why does the definite description in (13.20) refer to D rather than D^- or D^+? I do not see any good answer. One should think that there is some a priori guidance as to which stage of the causal process leading to R we identify as (the causal base of) the disposition. In Spohn (1997b, p. 374) I came across the same problem with respect to the physical property of being 'appeared to redly'. I have seen neither awareness of, nor progress on the issue.

Without this progress we are not justified in maintaining (13.20). I conclude that, besides the intrinsicality of dispositions to the set-ups having them, (13.17) (or its causal variant) is so far our only safe assertion of what we know, unrevisably a priori, about dispositions.

14

Causation

Inductive inference is mainly causal inference, laws typically are causal laws, and dispositions certainly are causes of their manifestations. It is high time for ranking theory to face this topic and to account for the causal idiom.

This topic is indeed what ranking theory was originally made for. I was deeply impressed by probabilistic theories of causation as presented by Reichenbach (1956), Good (1961–1963), and in particular Suppes (1970), and their clearly superior means for dealing in a most sophisticated way with relevance considerations, which, as so convincingly shown by Salmon (1970) and others, were badly missing in Hempel's theory of explanation. I contributed to this discussion in my (1976/78), with what may now be seen as a quite explicit predecessor to the theory of Bayesian nets (see also Spohn 1980). The deterministic approach could not compete. Regularity theories clearly failed. Mackie (1974, Ch. 8), a paradigm of circumspection, was quite indeterminate about what might replace them. Counterfactual theories of causation did much better. Still, I did not find them to be equally sophisticated. It was a handicap, as we will see below, that the reference to obtaining conditions is always implicit in subjunctives. Moreover, I found Lewis' (1973a) objectivistic interpretation of counterfactuals in terms of similarity spheres unilluminating, and felt the same way with regard to his account of causation; I found Gärdenfors' (1978) epistemic interpretation much more appealing. Perhaps the most disturbing fact, though, was that I could not see the unity in all this work on causation. This seemed intolerable to me. Maybe there are variants of causation, but they must be variants of a single core idea! Struggling with the principle of categorical matching (5.1), which appeared to me to be the central failing in Gärdenfors' account, ranking theory somehow came to my mind, and so I was able to present in my (1983b) an account of deterministic causation that perfectly paralleled my account of probabilistic causation. I became dissatisfied with details of the latter and improved it in my (1990); it is the account I still maintain. There, I mentioned on p. 125 that my probabilistic account could be routinely turned deterministic, and since nobody seemed to take notice of this remark I presented an updated version of the deterministic account in my (2006b).

In any case, the feasibility of reestablishing the unity of the theory of causation is evident from Chapters 5, 7, and 10; whatever one thinks is right for the one side can be directly carried over to the other side. Surely, the crucial question is *what* is right for either side. I will develop here the deterministic side and treat the probabilistic side mainly by implication. Of course, one might debate whether such unity is really needed, and whether it should be reestablished in this particular way. I will not engage in abstract discussions of this point. However, I submit that the mere feasibility of re-establishing unity is a very strong prima facie argument for my account, the force of which is still insufficiently realized. And even if this or that particular claim or move I will make is disputable or wrong, it should become clear, at least, that ranking theory is a new and excellent tool for developing the theory of deterministic causation.

This chapter somehow became the most unbalanced of the entire book, even in length; I could not confine myself to presenting just an outlook on some further application of ranking theory. Taking notice of all the literature on causation, which has grown tremendously in the last 40 years and even at an accelerating rate in the last decade, I realized that I can essentially stick to my old views, that they contain insights I still find worthy of attention, that they can be better expressed than I have done, and that I can develop them in some important respects. So I decided to do my best to explain these points, even at some length, and to defend them on the level of the present discussion. My only excuse is the paramount importance of the topic.

Here is, very briefly, the program for this chapter. First, in Section 14.1, we have to get clear about time and about the objects of the causal relation, which I will take to be atomic facts, and thus to fix the conceptual framework of our inquiry. Section 14.2 is a preliminary discussion, sketching why I will not thoroughly consider various approaches to causation that are quite prominent in the current debate; sufficiently many approaches will remain for comparative purposes. Section 14.3 will argue for my basic explication of causation that is, I find, firmly grounded in the history of the topic. We will find reason to split up the topic into direct and indirect causation. Section 14.4 thus defines and defends my explication of direct causation in ranking-theoretic terms, the cornerstone of my account; Section 14.5 illustrates it with some more or less problematic examples. Section 14.6 will more carefully investigate the circumstances of direct causal relationships, a topic amenable to theoretical treatment, but badly neglected in the literature. So far, the account will have looked only at atomic facts and their negations, i.e., at binary variables. Section 14.7, still occupied with direct causation, will undo this restriction and discuss some special problems arising with multi-valued variables and their realizations.

At this point, there are two ways to go. I decided to first stick to the direct case, but to extend the investigation from direct causation between facts to direct causal dependence between variables. In this way, I will connect up, in Section 14.8, with the theory of Bayesian and ranking nets (presented in Chapter 7) and its causal ambitions. I have a dissenting opinion on the status of the basic axioms of this theory; for me they are analytic consequences of my basic explication. This claim needs further defense, provided in Section 14.9 through a discussion of what frame-relative causal depend-ence as explicated so far might teach us about absolute or "real" causal dependence, a topic which has not received proper attention in the literature. My hope is that this settles the essential point of dispute I have with Spirtes et al. (1993). Section 14.10 extends the previous two sections by considering simultaneous causation and thereby presenting my recent thoughts about how to not reject, but to accommodate inter-active causal forks within my account, thus hoping to reach agreement also with Cartwright (2007) and to advance the dispute about conjunctive and interactive forks.

The second way to proceed after Section 14.7 is obvious, namely, to return to causa-tion between facts and consider indirect causation. If we could assume that causation is transitive, we could be very brief. However, precisely this assumption is at issue; Section 14.11 explains all the conflicting demands on indirect causation. In Section 14.12, I will defend my view that it is best to stick to transitivity, i.e., to define causa-tion in general as the transitive closure of direct causation. This defense will be quite a theoretical one, as are most parts of this chapter. For, one important side purpose of mine is to reclaim the weight of theoretical considerations over the discussion of examples and intuitions. Still, we will return to some further examples in Section 14.13 in order to see how my account can cope with them in comparison with others. (How-ever, completeness will not be my aim; there are just too many puzzling examples in the literature, even if they are often similar.)

Causation and explanation are two closely intertwined topics. I did not plan a twin chapter on explanation. However, I could not suppress some remarks on the notion of (causal) explanation and on inference to the best explanation, which many claim to be the basic inductive rule in the sciences. Since this book is mainly about induction, I indeed should not suppress them. So, these remarks are found in the two appendices 14.14 and 14.15.

So much for preview. Still, let me point right away to the philosophically most pressing issue. As is to be expected, I will explain causal relations relative to a ranking function, i.e., to something that can so far be understood only as a doxastic state. Thus, like Hume, I take causation to be an idea of reflexion; I am bound to claim that causa-tion is in the eye of the beholder. Many found Hume confused, read him as a plain regularity theorist, or outright rejected his account. Kant was thereby moved to his Copernican revolution in metaphysics and provoked even greater bewilderment about his sense of reality. I am fully aware that this point is, perhaps with justification, taken as a knockdown argument. This point was addressed in my (1983b, Sect. 6.5), even set on the right track, but the account remained sketchy and unsatisfactory. This was one of two reasons why my (1983b) remained unpublished.

Still, I must ask the reader for patience. In this chapter, I will develop only my subjectivistic account and simply neglect this deeply disturbing point. My implicit argument will be, of course, that the subjectivistic account works so well and is super-ior to objectivistic accounts in so many points of detail that we should try to preserve these advantages and somehow accommodate the subjectivistic approach with our objectivistic intuitions. This accommodation will be the task of Chapter 15. I am aware that I am burdening that chapter with high expectations; I did so already in Chapter 12. You will have to judge whether I have built a castle on sand that finally collapses or whether the whole construction turns out sound and stable in the end.

14.1 The Conceptual Framework

We must start our inquiry by fixing the conceptual framework within which we want to talk about causation. Let me first make clear that we will be dealing with singular causation, with particular causal processes. I am not asking how parental income affects kids' educational and professional opportunities, I am considering how my income affects my first son's educational opportunities, or, rather, how the development of my income gradually affects the development of my first son's education. I am interested how energy consumption, say, in 2001–2005 influenced global temperature in 2006, not how energy consumption generally influences global temperature. I focus on the causal processes in this mechanical clock, not on how mechanical clocks work in general. Generalization is a most complicated business. Basically, though, we can assume here that our beliefs about a given single case are governed by some ranking function ξ, and that we can then generalize to the corresponding subjective law λ_ξ, as explained in Definition 12.10.

In speaking about the single case, I prefer the language of variables, as carefully introduced in Chapter 2. The given single case is described by a (*conceptual*) *frame*, as I often say, a set U of specific, not generic variables, e.g., my possible annual income in 2001, in 2002, etc., and my first son's possible education in 2001, in 2002, etc. Each variable $X \in U$ has a range W_X of possible values, and the Cartesian product $\times \{W_X \mid X \in U\}$ is the possibility space characterizing the single case at hand, about which we can make all sorts of claims in the form of propositions $A \subseteq W$. As before, we will assume U and W to be finite, so that the algebra of propositions considered is the power set $\mathcal{P}(W)$ of W. It is certainly desirable to remove this restriction; however, I have simply nothing to say about how ranking theory coheres with continuum mathematics as commonly used not only in physics. My only excuse is that this generalization seems more of mathematical than of conceptual interest (though questions of infinity are always conceptually intriguing, too).

When speaking about causal matters, we must assume determinate temporal relations, without this I would not know how to even begin. That is, I will assume a finite set T of (extended) points of time that is linearly ordered by the relation $<$ of precedence; $\leq$, correspondingly, stands for "not later than" and $\approx$ for "simultaneous with". I will further assume that each specific variable $X \in U$ is realized at a time $\sigma(X) \in T$. I will usually write $X < Y$ and $X \leq Y$ instead of $\sigma(X) < \sigma(Y)$ and $\sigma(X) \leq \sigma(Y)$ in order to express that X precedes or is realized before (or at the same time as) Y. Thus, conceived as a relation between variables, $\leq$ is a weak order on U.

T need not consist of specific points of time; in particular, I do not presuppose metric time. It suffices to consider T as a set of linearly ordered temporal positions. For instance, a specific game of chess certainly is a causal process, and the natural variables to consider are all the possible moves of the game. The exact points of time at which the moves occur may be taken to be irrelevant; what matters is only the temporal order of the moves. It would certainly be most interesting to generalize T from a linearly ordered set of temporal positions to a partially ordered set of space-time points, say, a Minkowski space; but this is beyond my ambitions.

No doubt, my assumptions are quite natural. They are also simplifying, however. The realization of a variable is usually not instantaneous, but takes some time, and the temporal interval is not sharply defined. Thus, the variables need not be temporally well-ordered, they may overlap, with complicated causal interactions in the overlap. I neither deny nor deal with such complications. Many social science applications seem to claim to be able to get by without temporal assumptions at all, either because of such complications or because they deal with generic variables that have no temporal position, anyway. It is an utter mystery to me how they can possibly arrive at any causal conclusion. In any case, my assumptions hold for the basic case to be investigated first. Any hope of dealing with more complicated cases can only build on that.

My assumptions also entail that we are not after a causal theory of time that attempts to infer the temporal from the causal relations. This is a fascinating philosophical project, no doubt, but a demanding – too demanding – one, on which I will be silent. Rather, I simply presuppose the temporal relations.

I mentioned in Chapter 2 that I will use X, Y, Z, etc., to denote both members and subsets of U. There I also introduced the notion of an X-measurable or, in short, an X-proposition, for $X \in U$; informally A is such a proposition if it is only about the variables in X and says nothing about variables outside X. For any $X \in U$, let us call X-propositions about a *single* variable X and different from W and $\varnothing$ *atomic propositions*. I propose to take such atomic propositions as the *relata of the causal relation*; they are to be possible causes and effects.

This proposal needs a bit of defense. Let me first finish, though, the introductory formal business. We will need short ways of referring to various sets of variables. So, for $X \in U$, let $\{< X\} = \{Z \in U \mid Z < X\} = \{Z \in U \mid \sigma(Z) < \sigma(X)\}$ be the past of X. $\{\leq X, \neq X\}$ denotes the set $\{Z \in U \mid Z \leq X, Z \neq X\}$, the past and present of X without X itself. Similarly, if $\sigma(X) = t \in T$ or if A is an X-proposition, then $\{< t\} = \{< A\} = \{< X\}$. Further uses of this abbreviating device will be self-explanatory.

Finally, for any possibility or possible world $w \in W$ and any set of variables $X \subseteq U$, $^w[X]$ is to be the proposition $\{w' \in W \mid w'$ agrees with w on $X\}$. I will call $^w[X]$ an *X-state*; thus X-states are maximally specific X-propositions or the atoms of the algebra of X-propositions.

In combination, we may say that $^w[< t]$ denotes the past of t in w; $^w[\leq X, \neq X]$ is the proposition stating what has happened in w before and besides X; $^w[> A]$ is the future of A in w – where we omit the curly brackets, if unambiguously possible. And so forth. This notation will prove to be most useful.

Now, why should we consider atomic propositions to be possible causes and effects? With this question we enter the confusing, indeed agonizing discussion about the nature of causal relata. I will try to be brief. (In Spohn 1983b, Sect. 4.3 and 4.4, I dealt more extensively with these issues – as far as I could at that time.) First, in statistical, econometric, and social literature (for a prominent example, see Granger 1969, 1980), but not only there, it is common usage to speak of variables themselves causing other variables. This is literally meaningless and an abuse of language. What is obviously meant is that one variable causally depends on another. This is, as we will see in Section 14.8, directly explicable in terms of the basic causal relation between atomic propositions (which, to be sure, is not considered in that literature). I do not think that the abuse is harmless. Here, we must strictly abstain from it, since we want to talk about causal relations between both propositions and variables.

Commonly, it is taken as axiomatic that it is events that are causes and effects. This opens several opportunities for confusion. One is that probability theory and the related literature speaks of events as having probabilities, and if causal relations are (partially) explained as probabilistic relations, then causes and effects are also events. However, the probabilistic usage is just another abuse of ordinary language. Probabilistic events are simply sets of possibilities, what we call propositions, whereas ordinary-language events are certainly something else – and something for philosophers to struggle with.

This opens the main opportunity for confusion. Roughly, one may say that events are objects, with all the indefinitely many properties attached to them they actually have. In order to render this precise, though in an artificial way, Quine (1985) iden-tified events with space-time regions. Perhaps we should take the event to include everything that happens within its space-time region. That would mean to take all the intrinsic properties of events as essential to them. This looks too strict. Therefore, Lewis (1986b, Ch. 23) introduces so-called standards of fragility that tells how much an event may change without losing its identity, i.e., how fragile an event is, and he argues that the standards of fragility actually used are vague as well as context-dependent. Perhaps the best and most comprehensive discussion of the issues concerning our ordinary notion of an event may be found in Bennett (1988).

In any case, these short remarks make clear that events are categorically different from facts, states of affairs, propositions and the like. Events are concrete particulars with ordinary properties, whereas facts or propositions state that (concrete) objects have ordinary properties or relations, and thus are objects in turn, but abstract ones with different kinds of properties. The clearest formal crystallization of this categorical difference is that algebraic or logical operations such as negation, disjunction, or con-junction apply to propositions, but not to events; there are no negative or disjunctive events, and conjunctive events are at least doubtful.

In view of this intriguing ordinary-language analysis of events, one must not forget, though, that it is an open question whether the notion thus analyzed is the one best suited for causal theorizing; one cannot simply stick to the axiom that causes and effects are events and at the same time insist on the ordinary notion of an event. Thus, the issue becomes more complicated – or rather eases, I find, for the further complications open room for stipulation.

For instance, Vendler (1967), one of the first careful considerations of the issue, concludes that in ordinary talk causes are events, but effects may be events or facts. Lewis (1973b), the founding paper of the counterfactual analysis of causation, takes events e to be causal relata, but all the theoretical work is done by the propositions of the form $O(e)$ saying that e occurs. The insistence that events e rather than propositions $O(e)$ are the causal relata thus appears to be a negligible supplement, and the categorical difference emphasized above becomes marginal. Kim (1973) proposes an analysis of events for causal purposes and ends up identifying events with triples of a property, an object, and a time. Such events are exactly the same as my atomic propositions. Thus, the front lines blur further.

Let me jump to a very recent formulation of an event-based theory of causation, Lewis (2000/04). For the first time, Lewis there emphasizes the so-called alterations of an event. Something may happen in various ways; the most fine-grained ways are alterations, maximally fragile events. An event need not be so fragile, though; it may be realized in a slightly different way. Then the same event occurs, though a different alteration; Lewis calls this a version of that event. If the divergence is larger, the fragile event "breaks"; a different event occurs instead, and the alteration occurring is called an alternative to the original event. Hence, an event is representable as the set of its versions that is disjoint from the set of its alternatives. Lewis thus maintains that causes and effects are events. It is obvious, though, how indistinguishable from our variable talk Lewis' event talk has become; the alterations of an event constitute a variable, and events are atomic propositions.

So, even without digging more deeply into the ontological issues about events, I can affirm that our assumption of atomic propositions as causal relata is perfectly conventional, agreeing with many, if not most other conceptions. Formally, it will be essential for us that we can perform algebraic operations on the causal relata, and this condition is obviously satisfied.

Still, if I should have given the impression that we only face a choice between essentially equivalent ways of talking, this would not be quite correct. In several ways, our choice makes a difference. One difference not to be underestimated is a sociological one, as it were. Throughout the literature of the natural and social sciences dealing with causation, the language of variables and their states prevails. The authors in these fields have little understanding for all the ontological subtleties philosophers worry about, and they have difficulties recognizing that philosophers are talking about the same subject matter. Woodward (2003, pp. 3f.) also complains about an increasing distance between scientific and philosophical discourse about causation. I want to minimize the distance by siding with the scientific terminology. (The distance between foundational and applied studies will remain, and treating causation with ranking-theoretic means is most unfamiliar, anyway.)

The point has another important facet. I have assumed a fixed (finite) conceptual frame U of variables, and our study of causation will be carried out within this fixed frame. This is what scientific studies do as well. By contrast, the treatment of events does not seem to fit into such a fixed frame. If causes are events and events have indefinitely many features essentially, as suggested above, then indefinitely many features possibly contribute to the event's causal relations. Of course, the event's having a certain feature is a proposition, and if it is that proposition that stands in causal relations, our framework might do. If, however, the event itself stands in causal relations, as event theorists claim, indefinitely many of its features are possibly responsible for them, and no fixed frame is likely to capture this.

One might indeed wonder how much truth about causation can be found within a fixed conceptual frame. Conversely, one might wonder how far causal theorizing can get without referring to a fixed frame. The latter worry is by far the dominant one for me. So we will have to see what we can say about the first worry. The topic will be addressed in detail in Section 14.9.

The choice of language is not neutral in a final respect; there are specific causal problems that can be stated in the language of events, but not in the language of variables. These problems are related to the fact that there are no negative events. Event theorists thus ponder about causation by absences or omissions, where the cause does not exist, in a way, and about preventions, where the effect does not exist, in a way. The propositional framework is, prima facie, not concerned with these problems, since it contains unrestricted negation. So, does the event theorist worry about illusory problems? Or is the propositional theorist blind by nature for real problems? I will leave these questions untreated in this chapter.

14.2 A Preliminary Discussion

Having thus sufficiently settled issues concerning the conceptual framework, at least for our purposes, let us move on to the substantial issue. What is the causal relation? This is the issue we wish to carefully unpack.

The history of the topic is as old as philosophy, extremely complex, and not something to which I could do justice here. The ancient times were dominated by Aristotle's fourfold doctrine of causation, the long-lasting influence of which faded in the Renaissance. Its history was intertwined with that of occasionalism, which also lasted for more than a millennium, which particularly flourished in Arabic philosophy, which was deeply entangled in the mysteries of God's almightiness, but which shifted the emphasis to the notion of an effective cause, the only one surviving in the modern times from Aristotle's distinctions under the label "cause". I will not attempt to connect up with these old doctrines. (For an excellent account of occasionalism see Perler, Rudolph 2000.)

Approaches to the topic changed dramatically during the Enlightenment. It is fair to say, though, that the final overthrow was effected by David Hume's philosophy, in particular his criticism of causal necessity, which has dominated the history of the topic until the present day. Hume was instructively ambiguous (and many have struggled with these ambiguities; see, e.g., Mackie 1974, Ch. 1, Beauchamp, Rosenberg 1981, Ch. 1, and Beebee 2006). Hume's associationistic theory of causation, which I think is his primary one, simply seemed crazy; causation cannot be an idea of reflexion and thus essentially relativized to a subject's grasp. And Hume's regularity theory simply seemed too weak; causal necessity cannot be reduced to constant conjunction. However, the opponents of Hume were surprisingly feeble, and his victory nearly total. For several decades at the beginning of the 20th century, the attitude even prevailed that it was best not to talk of causation at all, lest one wished to engage in disreputable metaphysics.

Of course, this could only be an overreaction. Hempel (1948) taught us how to again talk sensibly of explanation, and via this detour, talk of causation returned en masse soon afterwards. Since then, we have seen a wealth of ideas and conceptions, analyses, and hypotheses concerning causation put forward; and everybody is entitled to feel confused by the evolving vigorous debate between the many conceptions. I cannot do justice to this wealth in this chapter. As a poor substitute, let me briefly explain why I dismiss various important strands of the discussion. This will also help to explain the core issues on which I will then focus.

The first conception I would like to put to one side is the so-called transference theory of causation, the idea that causation consists in the transfer of energy, momentum, or other conserved quantities, as proposed and elaborated by Aronson (1971), Fair (1979), Dowe (1992, 2000), Salmon (1994), Kistler (1998), and others. For me this idea has too physicalistic a ring. It might well be, indeed I believe it is so, that all causation is ultimately physical, in particular that mental causation is nothing but physical causation. This might even be an a priori truth, as Jackson (1998, 2005) claims, given an appropriate notion of "physical". However, even if these reductionistic claims are true in principle, they do not help at all in understanding causation in most areas where it is at work, in particular in human affairs and social domains.

The last remark is unfair, though; it raises expectations that this idea was never meant to fulfill. At this point it is useful to distinguish the nature and the concept of causation. Putnam (1975) explained us that the nature of water is not a kind of meaning of "water" that we grasp or have in the head (he was still reluctant to speak of concepts). In his wake, two-dimensional semantics generally distinguished between the nature and the concept of something, between the horizontal or C-intension and the diagonal or A-intension of a name denoting it. We had already seen the usefulness of this distinction in Section 13.4 on dispositions. Against this background it is a natural, but still intriguing idea that this distinction also applies to causation and that we should think about both the concept and the nature of causation.

The literature on causation is rarely explicit on this point. (For an explicit embedding of causation within two-dimensional semantics see Grenda 2008.) Many investigations clearly keep to the conceptual side, but many investigations also appear ambiguous to me; they proceed as if this distinction did not exist. Indeed, it is an open question whether it does exist. Often concept and nature coincide; they come apart only when the nature is hidden. (Beware, though! Being hidden is thereby defined, and natures are much more often hidden than we think. In Spohn (1997b) I argued, for instance, that not only the nature of colors, but even that of color appearances, is hidden to us.)

The conception that causation consists in something like energy transfer is, in any case, clearly only about the metaphysical side, about the *nature* of causation. As such it makes good sense and is a valuable contribution. At the same time, it confirms the suggestion that the conceptual and the metaphysical issue indeed come apart. However, it is precisely for this reason that I do not further consider it. For in this chapter I am unequivocally dealing only with the *concept* of causation. Therefore, there seems to be no point of contact with the energy transfer view. Still, I am at variance with it. In Section 15.7 I will instead arrive at the unoriginal conclusion that, roughly, the nature of causation consists in the obtaining of regularities or laws. In our world the basic laws might be physical conservation laws, and the nature of causation would then consist in them. However, this claim is at best true a posteriori.

There are two prominent conceptions that I find to be ambiguous between metaphysics and conceptual analysis and thus not clearly focused on the basic conceptual issues. One is the so-called process theory of causation, as championed by Salmon (1984); it is, of course, closely related to the transference theory. According to it, causal influence essentially propagates within processes; when we know what processes are, we have a sufficient grip on causation. Salmon has many illuminating things to say about processes and how they are distinguished from pseudo-processes. One crucial feature, though, is that processes are able to transmit so-called marks, an ability Salmon (1984, pp. 142ff.) initially characterized in counterfactual terms. He was criticized for this (e.g., by Kitcher 1985), and he tried to circumvent the problem (e.g. in Salmon 1994). I remained unconvinced (see also the critical discussion in Dowe 2000, pp. 72ff.). At best, one might say that he appealed to particularly controlled counterfactuals, which, however, are still counterfactuals from a logical point of view. This makes his attempts doubtful as a conceptual analysis. Why not immediately surrender to the counterfactual analysis, if one cannot avoid it, anyway? I have not seen any good analytical attempts at this question. Therefore, I will seek argument directly with the variants of the counterfactual analysis.

The other conception I have in mind is the so-called mechanistic conception, which is one important key to the Enlightenment's revolution in the 16th century and its abandonment of the medieval tradition. No doubt, it has been deeply entrenched in our scientific practice and intellectual background ever since. Within the analytic discussion, however, it was only recently reintroduced as a philosophical account, by such authors as Glennan (1996), Machamer et al. (2000), and others. According to it, causal relations unfold in the underlying mechanisms. Laws or regularities are always in danger of overlooking the mechanisms sustaining them. Of course, mechanisms must be understood here not in a narrow, mechanical, but in a wide sense; there are also, biological, economic, political mechanisms, etc.; and even gravitation is not really what one would take to be prototypical of a mechanism.

The shift of emphasis brought about by the reintroduction of the mechanistic conception is certainly healthy. Nothing enhances scientific understanding better than clearing up the working mechanisms behind the manifold of phenomena in meticulous detail; this *is* causal analysis in the best sense. For instance, I am convinced that the most severe handicap in philosophy of mind is that we still have, and are perhaps bound to have, only the coarsest understanding of the working mechanisms of the mind. As long as this is so, we can only shiver with the abyss, the explanatory gap, and all claims to the supervenience of the mind on matter, though certainly true, must remain philosophical speculations. The only way to make progress here is to try to uncover as many of those working mechanisms in as fine detail as possible.

However, from a logical point of view, this only means that most causal relations are indirect and best understood by splitting them up into many small links. Conceptual analysis is thereby only shifted to these small links, to the direct causal relations in the mechanism, or to the "bottom-out level", as Machamer et al. (2000, pp. 13f.) call it (which is, as they emphasize, relative to the field and the investigation at hand). As such, the mechanistic conception is silent on the issue of what constitutes the small causal links – or proposes to apply its strategy to these relatively direct links at a deeper level until it has reached rock-bottom and thus finally hit upon the nature of causation. However, this does not improve its standing as a conceptual analysis. Perhaps its late reintroduction into the analytic discussion is due to this long-felt analytical weakness. In any case, in Section 14.4 we will find a specific analytic reason for focusing first on direct causation. Later on, we will extensively consider causal nets, which are, within the approach I am working in, the counterparts of mechanisms (although I will not return to the certainly important question of how adequate a counterpart they really are).

The other accounts of causation in the field I am aware of clearly aim at a conceptual analysis. There are variants of the regularity theory, there are a bunch of counterfactual analyses, there are the ramifications of the probabilistic approach, there is the interventionist account accompanied by the so-called structural equations or structural modeling approach: they are my neighbors, and we will be in continuous conversation with them. No anticipatory remarks are required. However, the most recent of all these accounts, interventionism, gives rise to two further and quite different remarks.

First, interventionism appears to be a variant of the agency account of causation. The point here is not the distinction of agent from event causality or, in medieval terms, of immanent from transeunt causality. Agent causality, if opposed to event causality, is, I believe, a philosophical chimera, a forlorn attempt to solve philosophical problems in action theory that can be better solved otherwise. No, the agency view of causation, as introduced by Collingwood (1940, pp. 296ff.) and Gasking (1955), elaborated by von Wright (1971, Sect. II.8–II.9, and 1974), and recently defended, e.g., by Price (1991) and Menzies, Price (1993), says that causation is an anthropocentric notion insofar as it builds on the notion of human agency (and not merely on the notion of activity that might have a much wider application). Very crudely, I first grasp how to do things, i.e., how I cause things by certain movements, and I thereby grasp causal relations in general, what it is for causes to bring about effects. As an ontogenetic account of concept formation this may even be true, and as an evolutionary story it is quite plausible that only living organisms that move around in order to satisfy their needs develop the concept of causation. As a conceptual analysis, however, this move always seemed misguided to me. I do not want to engage in argument here. However, it always seemed obvious to me that the concept of causation is entirely part of our cognitive or epistemic furniture, of our representations or theories of the external world. It is indispensable to our conative and practical furniture, but in no way based on or derived from it.

The following story provides the crucial thought experiment (envisaged by Dummett 1964, pp. 338ff., though with a somewhat different intention): Imagine a being that has rich cognitive and epistemic capacities, but no practical interests, no ability and no desire to move and act, say, sort of a tree endowed with a big brain and far reaching sense organs (and perhaps even standing in communicative exchange with other gifted trees in order to expand its sensory input), but unable to change anything about its situation. (Poor tree; but we need not endow it with pains as well.) Could such a tree have the concept of causation (as well as a grasp of counterfactuality)? I say yes (and the theory I am going to develop will explain why). Not that this would be a good argument; intuitions about this fancy tree are weak and induced by theory, as is my "yes". Saying yes is, however, quite compatible with granting that we are so much better able to investigate causal relations, to get below the surface of phenomena, precisely because of our agency, through our ability to cognitively devise and practically perform experiments, to intervene in the course of events. Still, the thought experiment shows that agency is not analytically involved in the concept of causation. (Dummett, by the way, also says yes; he only thinks that the tree could not conceive of a direction of causation. I do not see why; in any case, the tree's dynamics of belief has a temporal direction, and it could have a notion of time and of time's arrow.)

The interventionist account appears to be a variant of the agency account of causation. But it is not; Woodward (2003, pp. 123ff.) clearly distances himself from any such allegation. Interventions may be actual interventions, they may be hypothetical or counterfactual interventions, and the notion of intervention thus generalized becomes a theoretical notion within the interventionist theory. For this reason, I instead classify the interventionist account as a variant of the counterfactual approach, in agreement with the intentions of Woodward, who sees his account as devising clear rules for the potentially sweeping counterfactual talk. I am inclined to think that interventionism precisely captures the conceptual truth inherent in the agency account of causation.

The advance made in causal theorizing by the notion of intervention vis à vis the notion of agency brings me to my second remark, which is of utmost methodological importance. On the one hand Woodward (2003, pp. 45ff.) explains causal relations in terms of interventions, on the other hand he is very clear that even the merely hypo-thetical feasibility of interventions rests on causal presuppositions that he carefully describes on pp. 98ff. So, his declared aim is to describe the conceptual entanglement of causation and intervention, not to give a reductive analysis of causation. He is thus a prominent example of an attitude that has become quite fashionable in the last 20 years; maybe it has even won the upper hand in causal theorizing. Glymour (2004, p. 779) proposed two grand labels for the opposition that is at issue here:

Philosophical theories come chiefly in two flavors, Socratic and Euclidean. Socratic philo-sophical theories, whose paradigm is *The Meno*, advance an analysis (sometimes called an "explication"), a set of purportedly necessary and sufficient conditions for some concept, giving its meaning; in justification they consider examples, putative counterexamples, alternative ana-lyses and relations to other concepts. Euclidean philosophical theories, whose paradigm is *The Elements*, advance assumptions, considerations to warrant them, and investigate the consequences of the assumptions. Socratic theories have the form of definitions. . . . Euclidean theories have the form of formal or informal axiomatic systems.

It is clear that Glymour's sympathies are with the Euclidean method. Like him I have limited sympathies for the manner in which the Socratic method is mainly executed in contemporary analytic philosophy, where the analysis is taken to be the end point of the efforts, not as the starting point of theory construction in which the analysis must ultimately prove itself. However, I cannot find anything wrong with the Socratic goal in itself; at least concerning causation, I definitely want to maintain the aim of analysis. One concern will be that the Euclidean method tends to leave the status of the basic assumptions or axioms unclear. Since they would be circular as definitions, Woodward (2003) apparently takes his basic characterizations at least as meaning postu-lates. By contrast, Spirtes et al. (1993) rather take their basic principles to be very general and widely applicable empirical hypotheses. As a philosopher, one would like to determine the status, even if it does not make a difference to theory construction. Another point will be that I believe that the initial, and perhaps crucial, argument for the Euclidean method can be specifically dissolved.

However, in the end abstract argument will not decide the methodological issue. What I intend to provide in this chapter is a kind of existence proof. If we actually find an acceptable analysis of causation, I am sure no Euclidean will complain. So, let us set to work and straightforwardly pursue our analytic or explicatory enterprise. Then we see what the alleged existence proof amounts to.

14.3 Causes and Reasons

Let us start by fixing the presuppositions of the causal relation. As to *causation and reality*, it is clear that A can be a cause of B only if both A and B actually obtain. So, in view of the previous discussion, causation is a relation not between arbitrary atomic propositions, but between *atomic facts*.

Let us be equally short about the relation between *causation and time* and simply assume that A can be a cause of B only if A is not later than B. General relativity theory has inspired fantasies about backwards causation, and so do certain obscure quantum effects. All this is far beyond my ken. Let me simply state that none of the subsequent theorizing would work without this assumption. This is at the same time my main response to the problem of the asymmetry of the causal relation. The causal asymmetry is basically grounded in the asymmetry of temporal precedence. This response is conventional, contested, and of no issue for us in the sequel.

This leaves us with simultaneous causation, an equally difficult topic; recall that Newton's theory of gravitation eventually foundered on its unintelligible assumptions about remote action. Hence, we might exclude simultaneous causation as well, thus completing the purely temporal account of causal asymmetry. However, I prefer to take a more pragmatic attitude and to allow simultaneous causation. Note that I only required that the members of T are linearly ordered; they may, however, be extended intervals of time. So, if two variables are 'simultaneous' within the coarse-grained representation T, one may 'really' precede the other, and so we may have 'simultaneous' causation. This is why I have no principled reason against simultaneous causation (within restricted or coarse-grained frameworks like ours), even though it will be representable here only as symmetric interaction. Starting in this liberal way, we can still exclude simultaneous causation or even simultaneous variables altogether, if suitable (for instance, in order to get nicer theorems); I will always be explicit about my assumptions. In Section 14.10 I will address the issue more systematically.

Let us move on to the substance. The traditional idea is that A is a cause of B if A somehow necessitates B. This hides various problems. One problem is what *making* necessary might mean here. If B is necessary, anyway, it cannot be said to have been *made* necessary by A. Necessitation rather means that, given A, B is necessary, whereas given non-A or not given A – this is a permanent ambiguity – B is not necessary.

The crucial problem, of course, is what making *necessary* might mean. Hume despaired of this problem, and his skeptical solution was that necessitation is merely habit or determination of thought; A urges us to believe B, and non-A does not. Dissatisfied, many sought for a, roughly put, more realistic understanding of necessitation. We will look at their attempts.

Another point is that it seems equally adequate to say that A is a cause of B if A somehow makes B possible in the sense that B is impossible without A, but possible with A. If nothing is contingent given its past, the two explanations of causation coincide. Thus there is nothing to decide for determinists. If contingency given the past is possible, however, we have to take a stance. The usual stance, that I endorse, is to leave the issue open, i.e., to accept both explanations.

A final crucial point, presumably thought of all along, but forcefully made explicit by Mill (1843, Book III, Ch. V, §§1–5, in particular §3), is that if A is to necessitate B, A must be a complete cause of B. Complete causes, however, are usually complex sets of conditions; many things must come together to produce a certain outcome. This is not our ordinary notion of a cause. We rather speak of partial or contributory causes that are complete only in the limiting case. Now, a partial cause by itself does not necessitate its effect. The formula usually accepted in response is that A is a cause of B if A necessitates B under or given the obtaining circumstances. This will do as a first move. It is, however, quite unclear what the obtaining circumstances of a given case are, and whether they are really as well defined as the use of the definite article suggests. This issue will occupy us throughout Section 14.6.

Let us draw a preliminary conclusion. It is:

(14.1) A is cause of B iff A and B obtain, A precedes B, and A raises the metaphysical or epistemic status of B given the obtaining circumstances.

This leaves out the contested issue of simultaneous causation. The status raisings we have discussed so far are from the impossible or contingent to the necessary and from the impossible to the contingent or necessary. It always seemed clear that such raisings are to be explained in terms of conditional modal status. Hume's skepticism, however, promotes the epistemic reading of that modal status, and his challenge is to come up with an acceptable metaphysical reading.

Except for the clumsy phrasing, (14.1) has been endorsed by almost all writers on causation at least in the last four centuries. Often, causes are referred to as necessary and/or sufficient conditions, where the "and/or" reflects the ambiguity already mentioned and conditions have been at least tacitly understood as relevant conditions. This sounds like well-defined mathematical talk, but of course, as Mackie (1974, Ch. 2) made amply clear, conditionality is just as problematic as necessitation in this context.

Regularity theorists explicate the metaphysical status in (14.1) as nomological status: A is a cause of B iff A or $\bar{A}$ and the obtaining circumstances (but not the latter alone) nomologically imply, respectively, B or $\bar{B}$. One of their problems is whether nomological entailment refers to the true regularities, in which case their account is clear, but clearly false, or to the true laws, in which case they sink into the problems discussed in Chapter 12. Even then, however, they seem unable to adequately deal with various problem cases like causal preemption or causal overdetermination. (See, e.g., Mackie (1974, pp. 43ff. and pp. 81ff.). However, see also Baumgartner (2008b, 2009) and Baumgartner, Graßhoff (2004), the most considered recent defense of the regularity theory.)

Counterfactual theorists explicate the metaphysical status in (14.1) as counterfactual status: A is a (necessary) cause of B iff B would not have obtained given that A had not obtained. Sufficient causes are awkward to describe in counterfactual language. The reference to the obtaining circumstances is implicit in the counterfactual. The main point, of course, is that theories of the subjunctive conditional such as Lewis (1973a) seem to provide precisely the account of conditional necessity required for (14.1). This was the basic idea of Lewis (1973b). At that time he probably did not dream that the approach would get so complicated as displayed in Collins et al. (2004) (under his prominent influence).

Interventionist or manipulationist accounts can be taken to adhere to (14.1) as well. At least, Woodward (2003), perhaps their most articulate elaboration, is very explicit in subsuming interventionism under the counterfactual approach; it is only that counterfactuals there receive a special interpretation that excludes backtracking and allows miracles in an unproblematic way (cf. Woodward 2003, pp. 138ff.).

One very important reason for stating (14.1) in the way I did is that probabilistic theories of causation can be subsumed under it, too, simply by interpreting the modal statuses in (14.1) as probabilities. Thus, A is a cause of B iff A raises the probability of B under the obtaining circumstances. This can certainly be seen as the basic idea of *all* probabilistic theories of causation, however variegated they become in the end. Of course, the ambiguity between metaphysical and epistemic statuses reappears here in the interpretation of probabilities as in some way objective or subjective.

This shows that (14.1) is indeed suited for preserving the unity of the theory of causation. It is the kernel we should seek to adequately develop. As hinted in Section 14.2, the debate about causation has recently become most multifarious, no doubt in response to problems with the prevailing theories and to discussions in the empirical sciences. Thus we saw that a number of accounts have emerged that do not clearly fit or clearly do not fit with (14.1). However, I find problematic the failure of large parts of the present discussion to attempt to connect up to something like (14.1) and thus to the mainstream of causal thinking in recent centuries. This is a connection definitely worth maintaining, and this is what I will try to do by developing (14.1).

We should also be aware that causal talk is governed by a most complex pragmatics. Usually, we point out only the salient, surprising, or informative causes, hoping that the rest of the causal story can be easily inferred; but this is not to say that the tacitly understood conditions, contributors, or mediators are not causes. Since Anderson (1938) we have become used to distinguishing the causal field that forms the constant background of the causal processes we intend to study from these processes themselves, whereby we can focus on the latter; but this is not to say that the causal field does not contain causally relevant or active circumstances, i.e., causes. The familiar distinction between triggering and structural causes is similar; here, the structures will be acknowledged to be causes as well, even if the triggering causes tend to be considered the genuine ones. In human affairs, causal talk is inextricably entangled with responsibility and guilt; but it is clear that responsibility is much narrower than causation. Thus, causal talk is highly context-dependent and interest-relative in various ways. I will flatly ignore all these complexities and aim, in the apt phrase of Hall (2004, pp. 227f.), at an egalitarian notion of cause, at the basic meaning of causation as preliminarily captured in (14.1); and it should also be clear that a study of these complexities presupposes an adequate analysis of this basic meaning.

My next move is predictable. Of course, I will propose the interpretation of the modal statuses in (14.1) as ranks. Rank raising is exactly what reasons do according to Definition 6.1. So, we get:

> (14.2) *A* is a cause of *B* iff *A* and *B* obtain, *A* precedes *B*, and *A* is a reason for *B* given the obtaining circumstances.

(14.2) determinately shifts to the epistemic reading; being a reason is a doxastic relation relative to some ranking function ξ. The talk of reasons here replaces Hume's talk of a habit or determination of thought, for which he has been so devastatingly denounced as psychologistic, in particular by German philosophy after Kant. My talk of reasons brings the perspective of rationality back into the picture, which, I am convinced, Hume never intended to leave. Otherwise, (14.2) is as skeptically imprinted as Hume's thinking. You will learn to love it.

Only with (14.2) will the deterministic-probabilistic unity of the theory of causation be completed. (14.1) provides the basic idea; however, when the behavior of modal statuses given by regularity, counterfactual, and probability theories diverge, the divergences spread through the theories generated by them. No such divergence will arise when we use (14.2) to build the theory of deterministic causation on ranking theory.

According to (14.2) causes are a specific kind of conditional reasons. On the one hand, this simply reflects the epistemic turn. On the other hand, it revives deep analogies suggested by traditional terminology. The Latin *causa* was a generic term that divided into various kinds; the term *ratio* was often used exchangeably. There were first the four Aristotelian causes, *causa formalis*, *causa materialis*, *causa efficiens*, and *causa finalis*. After their decline, another distinction became more important, that between *causa cognoscendi* and *causa fiendi*. German provides a similar generic term, *Grund*, and the distinction is then between *Vernunft-* or *Erkenntnisgrund* and *Grund des Werdens* (cf. Schopenhauer 1847). This way of talking is outmoded, and fairly so; the ordinary terms *reasons* and *causes*, *Gründe* and *Ursachen*, are good enough. Still, the terminological connection has been lost, while the substantial relation between reasons and causes has become all the more confusing, not only in the specific context of action theory (where the issue is a different one), but also in the present general context. (14.2) promises to uncover that substantial relation.

A final point before getting to the real business: Ranking theory naturally distinguishes four different kinds of rank raisings; in (6.2) we defined supererogatory, sufficient, necessary, and insufficient reasons. Therefore, we will be able to equally naturally distinguish supererogatory, sufficient, necessary, and insufficient causes. Necessary and sufficient causes are the focus of the traditional accounts that instantiate (14.1). However, I have never seen envisaged something like supererogatory or insufficient causes; they belong to the peculiar potential of ranking theory. A peculiarity indeed, one might think; they will, however, prove to play an important positive role.

14.4 Direct Causes

Having treated the relation between time and causation stipulatively and having interpreted necessitation in a Humean way, the only place left to render (14.2) precise is with the obtaining circumstances. So, what should count as the obtaining circumstances for the relation between the cause A and its effect B?

A natural answer, suggested already by Mill (1843, Book III, Ch. 5, §3), would be to say that those circumstances consist of all the other causes of the effect B. All of them? This appears too much. The idea is rather that the circumstances consist of all the other causes of B coordinate with A (that together form a total cause of B). The meaning of being coordinate is perhaps not fully clear, but it certainly excludes two kinds of causes of B: those that are also causes of A and those that are effects of A and thus causally mediate between A and B. The former may nevertheless be admitted as circumstances; they seem neither to be required nor to do any harm (except within a regularity account). The latter, though, must be definitely excluded; if the causal steps mediating between A and B were to count among the circumstances, the circumstances would determine B by themselves and the causal or status-raising contribution of A could not show up. So, we end up saying that the circumstances in question consist of all the other causes of B not caused by A.

This is, roughly, the thesis for which Cartwright (1979) forcefully argues. And she insists that the circumstances must not contain more. In Cartwright (1979, p. 432) she says that "partitioning on an irrelevancy can make a genuine cause look irrelevant, or make an irrelevant factor look like a cause" and goes on to illustrate this alleged possibility. The crushing conclusion is clear: any explanation of the circumstances must refer to the notion of causation. The circle is vicious, and the explication of causation along the lines of (14.1) is doomed.

I feel that this argument had a devastating influence on the discussion of causation. In the 70s great efforts went into trying to realize the Socratic ideal and to give an analysis of causation. This changed dramatically in the 80s, less because the former efforts were not completely convincing and grew weary, but rather because they now appeared bound to be unsuccessful. Cartwright's paper certainly was a major cause for this change. The work that realized the Euclidean ideal by theoretically unfolding the relation between causation and other notions (like probability, space, and time) has prevailed since. Scheines (2005, p. 285) shares the diagnosis, unsurprisingly, since he is one of the main promoters of the Euclidean method: "Although there will always be those unwilling to give up on a reductive analysis of causation, by the mid-1980s it was reasonably clear that such an account was not forthcoming. What has emerged as an alternative, however, is a rich axiomatic theory."

We are standing at the crossroads here. Should we continue to strive for an explication of the notion of causation that results in an explicit definition? Or should we rather settle for the axiomatic method that might be said to yield an implicit definition, an analysis only in the weaker sense of unfolding the conceptual net in which this notion is embedded? As I said, I am heading for the more ambitious project, and I have indicated my reasons. The specific reason is that I have believed since my (1976/78, Sect. 3.3) that the difficulty presented by Cartwright's argument is surmountable. The general reason is simply that such an explication seems feasible to me; of course, we can judge this only at the end of our work.

How can we reject Cartwright's argument? As far as the allegedly misleading conditioning on an irrelevancy is concerned, Cartwright's example has been sufficiently criticized, e.g., by Eells, Sober (1983, p. 42) who also allow conditionalization on irrelevant factors. I do not pursue this example here, since I think I can give a proof of the innocence of irrelevant conditions in Section 14.6.

The other point that the circumstances of the causal relationship between A and B must not contain facts mediating between A and B is serious. However, it clearly does not apply in cases where A is a *direct* cause of B; then the causal path from A to B contains no mediating steps. Hence, it appears that we have a non-circular explanation of the obtaining circumstances at least for direct causation. This is what I will pursue in this and the next sections. The extension of this explanation to indirect causation is a most contested matter, which I will take up only in Sections 14.11–14.12.

What exactly is the non-circular explanation? Well, the circumstances must consist of facts, of course. The only facts that can be a priori excluded from the circumstances are those that are realized after the effect B. All facts that are realized before the cause A or between A and B, however, are potentially relevant; and if they are irrelevant, they can be included anyway, as we just said. As to simultaneous conditions, if we start pragmatically by allowing simultaneous causation, we also have to include the conditions simultaneous with the effect B as potentially relevant. So we arrive at:

> *Definition 14.3*: Let $X, Y \in U$, A be an atomic X-proposition, B an atomic Y-proposition, $w \in W$, and ξ a negative ranking function over W. Then A is a *direct cause* of B in w relative to ξ – in symbols: $A \xrightarrow[w]{+} B$ – iff $w \in A \cap B, A \leq B$, (i.e., $X \leq Y$), and A is a reason for B given $^w[\leq Y, \neq X, Y]$ w.r.t. ξ. A is, respectively, a *supererogatory, sufficient, necessary, necessary and sufficient,* or *insufficient direct cause* of B in w relative to ξ – in symbols: $A \xrightarrow[w]{e+} B, A \xrightarrow[w]{s+} B, A \xrightarrow[w]{n+} B, A \xrightarrow[w]{ns+} B$, and $A \xrightarrow[w]{i+} B$ – iff $w \in A \cap B, A \leq B$, and A is a supererogatory, sufficient, necessary, necessary and sufficient, or insufficient reason for B given $^w[\leq Y, \neq X, Y]$ w.r.t. ξ.
>
> For the sake of completeness let me add: A is a *direct counter-cause* of B in w relative to ξ, i.e., $A \xrightarrow[w]{-} B$, iff $w \in A \cap B, A \leq B$, (i.e., $X \leq Y$), and A is reason against B or for $\bar{B}$ given $^w[\leq Y, \neq X, Y]$ w.r.t. ξ. A is *directly causally relevant* to B in w, i.e., $A \xrightarrow[w]{\pm} B$, iff $A \xrightarrow[w]{+} B$ or $A \xrightarrow[w]{-} B$. Finally, B is *directly causally independent* from A, i.e., $A \xrightarrow[w]{0} B$, iff not $A \xrightarrow[w]{\pm} B$.

The definition is implicitly relative to the given conceptual frame U. It obviously must be; otherwise, there would be nothing to define. Still, this relativity is disturbing, and we will carefully address it in Section 14.9. The definition is explicitly relative to a possibility or world w; what the facts and thus the actual circumstances are depends, of course, on the given possibility. And, conforming to (14.2), it is relative to a ranking function ξ. I will use the letter "ξ", as in Chapter 12, in order to emphasize that the frame U is to capture a particular causal situation that may be quite complex and involve many variables, but still constitutes a single case. I do not speak about repetitions, generalizations, or causal laws, though I do suggest that this is a simple step, once we have successfully dealt with the single case. Or to be explicit: If ξ describes the causal relations in the given single case, then the law λ_ξ as defined in (12.10) is the causal law that generalizes to all like cases. Of course, causal laws may only be ceteris paribus laws. We may embed all of our considerations about the single case into a background of normal conditions, as explained in Section 13.2. The corresponding generalization will then produce only a ceteris paribus causal law. This idea, though it will remain implicit in the rest of the chapter, should always be kept in mind.

One should not confuse counter-causation with prevention. Prevention is usually taken as a success term, whereas counter-causation is not by definition; if $A \xrightarrow[w]{} B$, B still obtains in w. We might then say that B happens *despite* A. This way of expressing the matter shows that counter-causation is a most familiar phenomenon (see, e.g., Rott 1986 for more elaborate remarks on the semantics of "although", "despite", and similar locutions). In the neuron diagrams so popular in the counterfactual approach, there are causing and preventing arrows, and for no good reason, just per fiat, the preventing arrow always beats the causing arrow. In other words, counter-causation is simply not considered there.

Note that a direct causal relationship between simultaneous facts is allowed by (14.3) and, as already mentioned, amounts then to direct interaction; if $A \approx B$ and A is a direct cause of B, then B is a direct cause of A. Perhaps there are real interactions; if not, this consequence merely acknowledges that our framework cannot tell cause and effect apart in such a case. As announced, Section 14.10 will deal more carefully with this issue.

The core of (14.3) is that it counts the entire past and present of a direct effect except the direct cause itself as obtaining circumstances. This seems to be much too much. The point, though, is that (14.3) is free of explicit or implicit circularity. The circumstances are so rich because we have applied such a weak notion of potential relevance. The idea is to introduce stronger notions of potential relevance on this basis, which will lead to more narrowly and thus more adequately confined circumstances provably equivalent to the rich circumstances. This strategy will be carried out in Section 14.6.

One might think that a direct cause must temporally immediately precede its direct effect, so that there is no time in between for circumstances to be realized. This would dispel Cartwright's circularity worries in the case of direct causes for simpler, purely temporal reasons. However, this move would be premature. We have so far made no special assumptions about U and ξ; no assumptions of naturalness are written into U and ξ. For instance, U and ξ might represent two temporally intertwined causal processes at once. Imagine a chess player simultaneously playing two games against two opponents. Every (temporally) odd move belongs to the first game, every even move to the second. Causal influences, even direct ones, run from odd to odd and from even to even moves (let us exclude interactions of the two games in the mind of the simultaneous player). This example violates the assumption of the immediate precedence of direct causes.

So we should avoid this assumption. One might complain that my example of the two chess games was somehow causally disorderly and that the assumption should hold at least in causally well-ordered situations. Yes, indeed. We must, however, not start with the assumption of causal well-orderedness. Rather, (14.3) will eventually enable us to say what causal well-ordering might mean, and the assumption of causal well-orderedness will then turn out to have a surprising role in the objectification theory in Sections 15.4–15.5, and in my discussion of the principle of causality in Section 17.4.

Doubts about the role of time in (14.3) might come from another direction. Realistically, one might say, time is continuous, and in continuous time there are no direct causes. Hence, it is futile to seek entry into causal terminology via (14.3). However, this puts the cart before the horse. Continuous time is a very complicated case, which we should try to approximate by ever more fine-grained discrete time, so that we can see what evolves in the limit. This is precisely how the theory of continuous stochastic processes is standardly developed (cf., e.g., Breiman 1968). I do not know where this strategy would lead us in the case of ranking theory. In any case, we should start with discrete time, and then the notion of direct causation makes perfect sense.

So far, my argument for (14.3) has been that it avoids circularity in the notion of obtaining circumstances. There is a more positive argument, though, that was given by Suppes (1970, Ch. 2) and, when completed, directly yields (14.3).

Suppes (1970, p. 12) starts with the notion of a prima facie cause: A is a *prima facie cause* of B iff, in our terminology, A and B are atomic facts, A precedes B, and A raises the probability of B or is a (probabilistic) reason for B. Indeed, when the presuppositions for a causal relation are satisfied, we tend to interpret correlations in a causal way. If 1 percent of those inoculated had a bout of cholera, but 12 percent of the non-inoculated did, then inoculation apparently impedes cholera. If the lights turn green when a car approaches the crossing and there are no other cars at the crossing, then apparently the light turns green *because* the car is approaching (they presumably are smart lights). We know that *post hoc ergo propter hoc* is a fallacy, but it is a most tempting one.

A fallacy it is indeed. Statistical literature abounds in funny examples where a correlation is spurious and no causation is involved. Suppes utilizes this for his explicatory strategy. He says that the prima facie cause A of B is *spurious* iff there is a further atomic fact preceding A, conditional on which A and B are independent (Suppes 1970, p. 23); otherwise, A is a genuine cause of B. And the prima facie cause A of B is *indirect* iff there is a further atomic fact temporally between A and B, conditional on which A and B are independent; otherwise, A is a direct cause of B (Suppes 1970, p. 28). Thus, the learning of facts might change or qualify the prima facie judgment about the causal relation between A and B.

As is clear from the statistical literature, and as I argued in my (1976/78, Sect. 3.3) and repeated several times, Suppes neglects a further cause of change in one's causal assessment: There can also be spurious independence; A may prima facie not be a cause of B, but given some circumstance, some further atomic fact preceding B, A may raise the probability of B and thus turn out to be a *hidden* cause of B. To take a somewhat worn-out example (poor Freud!): Vis-à-vis a certain kind of neurosis, psychoanalytic treatment might appear entirely inefficient; on average, remission time after treatment is the same as spontaneous remission time. However, this is not the full truth; an analysis of the data shows that the treatment in fact halves the remission time of the poor and doubles that of the wealthy patients. So, it has a massive causal influence, which cancels out on average, and is hence hidden. Of course, the calculation of regression and partial correlation coefficients in statistics is intended to detect just such hidden influences.

Thus, we have found three ways in which our causal assessment can change or modify. Are there more? In my (1983b, Ch. 3) I attempted to give a complete overview of all possible changes of our causal assessment. I found five, two that concern only indirect causes, and the three ways above, with the only difference being that there are cases in which a prima facie cause is rendered spurious, not indirect, by facts between the cause and the effect (see also the overview in my 1983b, p. 217). Hence, even under that more comprehensive investigation of the issue, the following argument remains correct:

The point is that the above three ways of changing a causal assessment develop a characteristic interplay. *Prima facie*, A may be a cause of B (we could just as well start with the reverse assumption). Some fact preceding A may show, *secunda facie*, A to be a spurious cause of B. Then a further fact preceding A may reveal, *tertia facie*, that A is a hidden cause of B. And so forth. This dialectics comes to a halt only after we have taken into account *all* facts preceding A.

Suppose we have concluded that A is a cause of B, after all. Now, a second dialectics begins. Some fact between A and B may suggest that A is only an indirect cause of B (or even spurious, if I was right in my 1983b, pp. 78ff.). A further fact may show the direct causal relation from A to B to be hidden. And so forth again. (That this dialectics always threatens is, of course, the basic moral of Simpson's paradox, about which statisticians have worried at least since Yule 1911.) Only after considering not only all facts preceding A, but also all facts between A and B, i.e., the whole past of the potential effect with the exception of the potential cause itself, is our assessment of the causal relation between A and B guaranteed to stabilize and to change no further. If A is a reason for B given all this, then A is a direct cause of B. This is exactly what (14.3) says.

Perhaps our relevance and irrelevance judgments change further when we also consider facts in the future of B; however, our causal judgment is not thereby changed. To be explicit, the final step of my argument is this: If there is no possible way whatsoever that our causal judgment about A and B can change, then we have arrived at the true judgment. Thus, (14.3) is true. Note, though, the implicit frame-relativity in this argument; the talk of *all* facts preceding A and between A and B is well-defined only if the domain of quantification, i.e., the frame U of variables is fixed. (Section 14.9 will precisely try to come to grips with ever expanding frames.)

Of course, it is impossible for us to know and to take into account the whole past of the effect (unless we consider only very small frames). Therefore, our causal judgment is always bound to be preliminary. This is not just general fallibilism concerning empirical matters; it has, as explained, a specific reason in the logic of causation. We may certainly hope that such reversals of judgment occur at most three or four times, and mostly this will be so; but there is no guarantee whatsoever. More complicated situations are easy to realize. Lewis attempts to capture them on paper by his neuron diagrams. Electrical circuits realize them in many ways. For me, the most impressive visualization of causal relations is presented in those huge hangars where millions of dominoes are most artfully arranged, take hours to all fall down and realize all of the causal situations causal theorists have ever dreamed of (an example which apparently inspired the cover design of Woodward 2003).

At the same time, my argument for (14.3) makes Hume's and my epistemic reading of (14.1) more intelligible. Our prima facie assessment and all further preliminary judgments are of an epistemic nature, indeed trivially so; they refer to reasons, to positive relevance, in the sense explicated. Then it is only natural that the final judgment does so, too, that causes indeed are a kind of conditional reason and thus a kind of objectified reason (in the sense of Jeffrey (1965, Sect. 12.7) who proposed that objective probabilities are subjective ones conditional on the true member of some partition of the possibility space). I am only suggesting by this that the epistemic reading is not so far off the mark; in Chapter 15 I will strengthen the point.

Apart from this epistemic move, (14.3) agrees well with the literature. For instance, Good (1961–1963), the first elaborate attempt at a probabilistic definition of causation, took the whole past of the cause and the true laws of nature as the circumstances on which to conditionalize the probability of the effect. He did so because he still thought that his definition need not proceed from a distinction between direct and indirect causation. However, if we neglect his reference to the true laws of nature (which would indeed spoil our explicatory enterprise) and if we make the assumption that direct causes immediately precede their direct effects (which I avoid for the reasons explained above), then Good's proposal obviously agrees with (14.3).

(14.3) also agrees with the counterfactual approach. There we say, roughly, that A is a cause of B iff B would not have occurred, had A not occurred. Thus, the circumstances of A's causing B may be taken to be the facts tacitly assumed to hold, together with the counterfactual premise non-A, i.e., the facts cotenable with non-A. In the causal case, facts after the effect B are not required to be cotenable; and all facts before the cause A should be cotenable. Moreover, in the case of direct causation the facts between A and B should be either cotenable with non-A or irrelevant to B or both, because none of the facts revoked can, together with A, have an influence on B.

The matter is clearer in the interventionist interpretation of the counterfactual approach. Woodward (2003) relates his explication of causal concepts to a fixed set of variables arranged in a given Bayesian net. On this basis he states (on p. 55):

(DC) A necessary and sufficient condition for X to be a direct cause of Y with respect to some variable set $\mathbf{V}$ is that there be a possible intervention on X that will change Y (or the probability distribution of Y) when all other variables in $\mathbf{V}$ besides X and Y are held fixed at some value by interventions.

The similarity of (DC) to (14.3) is obvious. Woodward is primarily interested in type-level or generic causation, and is thus reluctant to presuppose temporal relations between variables. If he took temporal relations into account, I see no reason why he could not give up the requirement that the variables in $\mathbf{V}$ posterior to Y are also held fixed, in which case no difference would remain concerning the obtaining circumstances.

The only – albeit decisive – difference consists in my epistemic turn. The true interventionist refers to actual wiggling: "Wiggle X, and see whether Y wiggles!" Woodward, well aware that actual intervention is often impossible, retreats to counterfactual wiggling. (14.3), by contrast, is about doxastic wiggling, as it were; and it turns the actual or counterfactual fixing of the circumstances by intervention mentioned in (DC) into conditionalization, i.e., into suppositional fixing, which, however, has the same force.

Does it really have the same force? There are two concerns. First, as explained, my interest in proposing (14.3) is to thereby get a non-circular entry to causal terminology. By contrast, Woodward is very explicit about the circularity of his analysis. Causal relations are explained by interventions, in (DC) and elsewhere in his book. However, when it comes to explaining what interventions are, Woodward (2003, p. 98) uses causal terminology heavily. He carefully argues why he thinks this is a virtue, why causation and intervention can only be jointly understood through each other. He insists, though, on pp. 104ff. that his analysis is not viciously circular; in order to explain what an intervention on X with respect to Y is, he need not refer to the presence or absence of a causal relationship between X and Y. So, it seems that (DC) and (14.3) cannot agree.

However, Glymour (2004) doubts that the circularity is ineliminable. His doubt seems well justified at least in the case of (DC). Woodward's (2003, pp. 98f.) causal conditions I1–5 on interventions, which are there to ensure that all the changes brought about by the intervention on X with respect to Y are mediated through X alone, and not accompanied or disturbed by concomitant changes, seem to run empty in the case of (DC). This is so because (DC) does not only assume an intervention on X with respect to Y, but also interventions on all the other variables in $\mathbf{V}$, fixing their values; there is then simply no space left for concomitant changes to blur the direct causal relation between X and Y. (Well, (DC) leaves no such space within $\mathbf{V}$; it does not yet exclude concomitant changes outside $\mathbf{V}$. This remark shows that (DC) is as frame-relative as (14.3) and that we should presumably try to get rid of this frame-relativity. However, this will be our topic only in Section 14.9.)

Hence, I cannot see any difference in this case between specifying values for all variables in **V** except Y by intervention, as (DC) does, and simply supposing, or conditioning on, these variables having those values, as (14.3) does.

Still – this is the second concern – didn't we learn that intervening and conditioning are two entirely different things? This is one cogent way of putting the decision-theoretic lesson taught by Newcomb's problem, as paradigmatically set out by Meek, Glymour (1994) (see also my considerations in Spohn (1978, Sect. 5.1–5.2) to the same effect). The lesson is correct, no doubt. In probabilistic terms, conditioning on some variable X taking the value x leads to the distribution conditional on $\{X = x\}$, whereas intervening on X to take the value x has a different result; it leads to the corresponding truncated distribution (cf. Pearl 2000, p. 72) or manipulated distribution (Spirtes et al. 1993, p. 79). The same holds in ranking-theoretic terms. However, this is not to say that the two distributions would be unrelated or incomparable. Conditioning is still the basic notion, as is clear from the fact that the distribution resulting from one or several interventions is *defined* in terms of conditional probabilities of the original distribution (see the truncated factorization of Pearl 2000, p. 72, or the manipulation theorem of Spirtes et al. 1993, p. 79).

This definition indeed entails that the difference between intervening and conditioning collapses in the case of (DC) and (14.3). This is so because the truncated or manipulated distribution preserves precisely those conditional probabilities (or ranks) for the variables not manipulated or intervened upon that were considered in (14.3) for the effect variable Y, i.e., the probabilities for Y conditional on its entire past. This is indeed the point of the truncated or the manipulated distribution. Hence, stating (DC) in terms of intervention increases its vividness; its substance, though, does not go beyond that of (14.3) stated in terms of conditioning.

I will continue my comparison with the interventionist approach when it comes to indirect causation in Section 14.11. The summary so far, in any case, is that we have found essential agreement between (14.3) and many of the prominent alternative approaches.

14.5 Some Examples

Let me illustrate (14.3) with some more or less problematic examples. Most puzzle cases concern indirect causation; some of them will be discussed in Section 14.13. But there are also puzzles for direct causation. Let us see how (14.3) deals with them.

Example 14.4, Causes and Symptoms: The basic feat of each theory of causation is to distinguish between causes and symptoms, between causal chains and causal forks. The regularity theory already has surprising difficulties with this feat (see the discussion of Mill's example of the London and Manchester hooters in Mackie 1974, pp. 83ff.; for a defense, see Baumgartner, Graßhoff 2004, pp. 99ff.). The counterfactual approach has no problems: If the mediating cause had not occurred, the effect would have not occurred, either; but if the symptom had not occurred, the effect would still have occurred. For probability theory the feat is also easy; it is ideally suited for stating screening-off relations. This treatment can be immediately copied by ranking theory:

Let us neglect the circumstances; let us just consider three atomic propositions, A, B, and C, and the very small world w with $\{w\} = A \cap B \cap C$ in, which these propositions are facts. A, B, and C can be causally related in many ways. (In my 1983b, Sect. 3.4, I was so crazy as to count 4^{16} possible causal relations, most of them of a highly dubious combinatorial and indirect nature.) The two ways under discussion can be graphically represented thus:

(14.4a) *causal chain*   (14.4b) *conjunctive fork*

(Salmon (1978) claimed that there also are so-called interactive forks. This is indeed a crucial issue, which I will separately discuss only in Section 14.10.) The two situations are easily distinguished in ranking-theoretic terms:

Suppose that $\xi(A) = \xi(\bar{A}) = 0$, leaving us to specify only the ranks conditional on A and $\bar{A}$. One specification is this:

(14.4c)

$\xi(\cdot \mid A)$	C	$\bar{C}$
B	0	1
$\bar{B}$	2	1

$\xi(\cdot \mid \bar{A})$	C	$\bar{C}$
B	1	2
$\bar{B}$	1	0

where the entries in the tables give the values of $\xi(A \cap B \mid C)$, etc. A is a direct cause of B in w, in fact a necessary and sufficient one, since $\xi(B \mid A) = \xi(\bar{B} \mid \bar{A}) = 0$ and $\xi(\bar{B} \mid A) = \xi(B \mid \bar{A}) = 1$. B is a direct cause of C in w, again a necessary and sufficient one, since $\xi(C \mid A \cap B) = \xi(\bar{C} \mid A \cap \bar{B}) = 0$ and $\xi(\bar{C} \mid A \cap B) = \xi(C \mid A \cap \bar{B}) = 1$. Moreover, C and A are conditionally independent, given B and given $\bar{B}$ w.r.t. ξ; thus, according to (14.3), no direct causal influence runs from A to C. This is indeed the simplest ranking-theoretic representation of a causal chain; note that ξ here simply counts how many times the obtaining causal relations are violated (two times, for instance, in the sequence A, $\bar{B}$, and C).

The other specification to be looked at is this:

(14.4d)

$\xi(\cdot \mid A)$	C	$\bar{C}$
B	0	1
$\bar{B}$	1	2

$\xi(\cdot \mid \bar{A})$	C	$\bar{C}$
B	2	1
$\bar{B}$	1	0

As before, A is here the only direct cause of B in w, in fact a necessary and sufficient one, since $\kappa(B \mid A)$ etc. are the same as in (14.4c). Now, by contrast, A is also a necessary and sufficient cause of C, since $\xi(C \mid A \cap B) = \xi(\bar{C} \mid \bar{A} \cap B) = 0$ and $\xi(\bar{C} \mid A \cap B) = \xi(C \mid \bar{A} \cap B) = 1$. Moreover, B and C are conditionally independent given A and given $\bar{A}$ w.r.t. ξ; thus, according to (14.3), no direct causal influence runs from B to C. We might also say in Reichenbach's language that A screens off B from C. So we now have the simplest example of a conjunctive fork in ranking-theoretic terms in which the ranks again count the violations of causal relations (two violations, for instance, when A, $\bar{B}$, and $\bar{C}$ occur); the more violations, the more disbelieved. Of course, (14.4a + b) can be realized by many different distributions of ranks.

When it comes to more recalcitrant examples, there are essentially two strategies that one finds applied again and again. Salmon (1980, p. 64, and 1984, pp. 194ff.), has already clearly identified them as "the method of more precise specification of events" and "the method of interpolated causal links". The methods consist in redescribing the example either, as I will say, by *fine-graining events* (or *variables*) or by *fine-graining causal chains* (or *graphs*). The first strategy takes recourse to multi-valued variables, in our terminology, the complications of which will be discussed in Section 14.7, the second gets involved in issues of indirect causation that are taken up systematically only in Section 14.11. In my view, which I more carefully argued in Spohn (2008, Sect. 16.5), these methods of fine-graining the description of examples are usually inferior to trying to cope with examples on the given level of description; the former is theoretically conservative, whereas the latter, if successful, leads to fruitful theoretical insights. In any case, ranking theory adds a third strategy which one might call *fine-graining of theory*; we will see its beneficial operation after the next example.

A good heading for a larger cluster of problem cases is *redundant causation*. A and B redundantly cause C iff the following holds: if neither A nor B had been realized, C would not have been realized; but if only one of A and B had not been realized, in the presence of the other, C would still have been the case. The following paradigmatic ranking tables, in terms of the two-sided ranking function τ derived from the basic negative ranking function ξ, are instructive:

(14.5)

$\tau(C \mid \cdot)$	B	$\bar{B}$
A	1	-1
$\bar{A}$	-1	-1

$\tau(C \mid \cdot)$	B	$\bar{B}$
A	1	0
$\bar{A}$	0	-1

$\tau(C \mid \cdot)$	B	$\bar{B}$
A	1	1
$\bar{A}$	1	-1

(a) *joint necessary and sufficient causes*   (b) *joint sufficient, but not necessary causes*   (c) *redundant causes*

Cases (a) and (b) are ordinary ones, but case (c), redundant causation, is a big problem for most theories of causation. Pure regularity theory indiscriminately declares both A and B to be a cause of C, since both regularities hold: whenever A, then C, and whenever B, then C. Amended regularity theory, taking the circumstances into account, indiscriminately denies both *A* and *B* as causes of *C*, since, given *B*, *A* is not relevant for nomically implying *C*, and vice versa for *B*. A naïve counterfactual analysis would also have to indiscriminately deny both *A* and *B* as causes of *C*; this is how redundant causation was explained above. However, no indiscriminate answer will do, since, as we will see, various cases of redundant causation are to be distinguished and varying answers to be given. Mostly, these differences are brought to the fore through fine-graining of causal chains. However, some cases of redundant causation, symmetric as well as asymmetric, do not seem to involve such fine-graining. Let us look at three cases in the rest of this section.

Example 14.6, (Symmetric) Overdetermination: One such case is where two (or more) independent causal processes go to completion and produce the effect in question. For instance, to avoid the notorious cruel firing squad, the prince sings a love song (*A*), and accompanies it by playing the mandolin (*B*), in order to wake up the beloved princess (*C*). Any performance would have been sufficient for doing so, and hence we have a case of overdetermination. We would like to say that both performances are causes of the awakening, but we have seen above that neither the amended regularity nor the counterfactual analysis can say so.

A standard solution invokes fine-graining of events. Of course, the princess wakes up either way; but the manner in which she wakes by hearing both performances differs slightly from the manner when she hears only one performance. Thus, *A* and *B* turn into ordinary joint causes of the manner in which *C* is realized. In a way, though, we do not thereby take the example at face value; we asked why the princess woke up, not why she woke up in a specific manner. Lewis (1986b, pp. 197ff.) explains why he does not want to fully rely on this strategy, and I agree.

Indeed, Lewis (1973b, footnote 12) already declares such cases of overdetermination useless as test cases because of a lack of firm intuitions, something he almost literally repeats in his (2000/04, p. 80). In his (1986b, pp. 207ff.), he is more optimistic, and agrees with Bunzl (1979) that fine-graining of causal chains shows most alleged cases of overdetermination to reduce either to ordinary joint causation or to preemption via an intermediate Bunzl event, as he calls it. (The remaining cases, if there are any, might then be resolved by his doctrine of quasi-dependence, which he withdraws, however, in Lewis 2000/04, pp. 82ff.) All in all, overdetermination puts a lot of strain on counterfactual analyses.

Halpern, Pearl (2005a, pp. 856ff., example 3.2) are able to account for the intuition that both overdetermining causes *A* and *B* are indeed actual causes of *C*. Roughly, they say that the causal influence of *A* on *C* is masked by what they call a structural contingency, i.e., the presence of *B*, and comes to the fore only under different circumstances, i.e., the absence of *B*. Their account verifies this informal description. At a cost, though; their account then has difficulty in distinguishing this case from a case where *A* is actually not a cause of *C* at all, but would have been under different circumstances, difficulties they can resolve only by invoking fine-graining of causal chains again.

These theoretical difficulties are in strange disharmony with the ease with which at least prima facie cases of overdetermination can be produced; they are quite common in everyday live. Overdetermination seems so plain a case that we should be able to take it at face value without any theoretical gymnastics. Ranking theory enables us to do so, by fine-graining neither events nor causal chains, but theory, as I called it. In Definition 6.2 we distinguished supererogatory reasons and could thus define supererogatory causes in (14.3). This is exactly what overdetermining causes are, as the following table displays:

(14.6)

$\tau(C \mid \cdot)$	B	$\bar{B}$
A	2	1
$\bar{A}$	1	−1

overdetermining causes

According to this table, each of A and B would have been a necessary and sufficient cause of C in the absence of the other. In the presence of the other each is still positively relevant to, i.e., a cause of C; but then each can only be a supererogatory cause. I find this account perfectly natural and simple. If a sufficient cause occurs, it is unbelievable that the effect does not occur, as was the case in (14.5a–14.5b). If in a case of overdetermination the effect does not occur, at least two things appear to have gone wrong at once; and this is at least doubly unbelievable, as represented by (14.6).

Ranking theory helps us to the notion of a supererogatory cause. This notion, and hence this solution, is not available to any other account of causation. However, there is no gain without costs. It can be true or false that the princess wakes up, but it cannot be, as it were, doubly true or doubly false. This is the reason why the notion of a supererogatory cause is foreign to all realistic accounts of causation; it makes sense only in an epistemic account, as proposed here. Thus, the case of overdetermination throws us back to the fundamental issue of the epistemic relativization of causation, which we have deferred to Chapter 15. However, the simplicity of my account of overdetermination might favorably incline us towards that relativization, and we will see in Section 15.5 that to some extent we can even make realistic sense of overdetermination.

Example 14.7, Preemption by Cutting: Preemption is a generic label for cases of asymmetric redundant causation; it is a venerable stumbling block for regularity theories as well as counterfactual analyses. A occurs, and in the absence of B it would bring about C. However, B interferes, breaks A's influence, and brings about C by itself. The classic example introduced by Hart, Honoré (1959, p. 219f.) is the story of the desert traveler, which starts with the first assassin pouring poison into the traveler's water keg, continues with the second assassin drilling a hole in the keg, and sadly ends with the traveler's death in the desert. Clearly, the second assassin killed him, the first only tried. For reasons to be explained in a moment, Lewis (2000/04, p. 80) coined the term "preemption by cutting" for such cases.

Regularity theories have great difficulty in treating the two assaults differently. Prima facie, counterfactual theories do, too; at least, it is not true that the traveler would not have died, if the hole had not been drilled in the keg. Fine-graining of events helps; dying from thirst and dying by poison are two different kinds of death, after all. It is doubtful, though, whether this always works. However, fine-graining of causal chains helps as well; this was Lewis' (1973b, p. 567) solution. Insert some link between the second assault and the traveler's death, say, the traveler's having or not having something to drink in the desert. Then it is true that, if the hole had not been drilled, the traveler would have had something to drink, but it is also true that, if the traveler had had something to drink, he would not have died. If we moreover assume transitivity of causation – a topic left to Sections 14.11–14.12 – the second assault is a cause of the death. I am fully satisfied by this solution, and it can, of course, be duplicated in the present account.

However, this is only the beginning of a long story of ever more cases of preemption. Lewis (1986b, pp. 200ff.) calls such easy cases early preemption and distinguishes them from cases of late preemption, where the effect C brought about by the cause B is itself the one that prevents the causal chain from the preempted cause A from going to completion in C. Those troubled by this possibility have always referred to temporal vagueness as to when exactly an event happens; the very same event could also have occurred a little bit earlier or later. I have only been able to see such cases as entailing backwards causation, an interpretation supported by the familiar neuron diagrams representing such cases, and as forcing us to be more precise temporally: there are two possible effects, C_1 and C_2, and the earlier C_1 forestalls the latter C_2. In any case, it is clear that there cannot be any literal translation of late preemption into my framework. (For a fuller rejection of late preemption see Spohn 2006b, pp. 110f.)

Example, 14.8, Preemption by Trumping: Schaffer (2000) suggests that there are not only cases of preemption by cutting, in which the causal chain from A heading to C is cut before reaching C, but also cases of so-called preemption by trumping. Bas van Fraassen invented an earthly example, described by Lewis (2000/04, p. 81) as follows:

The Sergeant and the Major are shouting orders at the soldiers. The soldiers know that in the case of conflict, they must obey the superior officer. But as it happens, there is no conflict. Sergeant and Major simultaneously shout "Advance!"; the soldiers hear them both; the soldiers advance. Their advancing is redundantly caused: If the Sergeant had shouted "Advance!" and the Major had been silent, or if the Major had shouted "Advance!" and the Sergeant had been silent, the soldiers would still have advanced. But the redundancy is asymmetrical: Since the soldiers obey the superior officer, they advance because the Major orders them to, not because the Sergeant does. The Major preempts the Sergeant in causing them to advance. The Major *trumps* the Sergeant.

Again, one is tempted to apply fine-graining of causal chains. Somehow, there must be something like a switch, possibly in the brains of the soldiers, that responds discriminately to the various shoutings. Halpern, Pearl (2005a, p. 875) explicitly specify a model of this sort that successfully accounts for the asymmetry between the major and the sergeant within their structural-model approach, that is, according to which the major's – but not the sergeant's – shouting is an actual cause of the soldiers' advancing. This success can certainly be duplicated within other accounts of causation.

However, it again seems that fine-graining of causal chains should not be necessary to do justice to this case. Surprisingly, this is also the attitude of Lewis (2000/04, p. 81), perhaps because he is confident that his new account of causal influence needs no help from such fine-graining. Without that help, Halpern, Pearl (2005a, p. 874) intuit that the case in fact resembles symmetric overdetermination, perhaps because their theory delivers the conclusion that both shoutings are causes of the soldiers' advancing, the only difference being that the major's shouting is what they call a strong cause (a technical term in their account), while the sergeant's is not. It seems we have here two further cases of theory-laden intuitions.

The ranking-theoretic account of trumping is straightforward, even in the binary case, where the major's and the sergeant's only alternatives on which the soldiers' advancing (= A) depends are either to shout "Advance!" (= M_1, S_1) or to be silent (= $\bar{M}_1$, $\bar{S}_1$). Look again at the following two-sided ranks:

(14.8a)

$\tau(A \mid \cdot)$	S_1	$\bar{S}_1$
M_1	2	2
$\bar{M}_1$	1	−1

Trumping, binary case

According to this table, the sergeant's shouting is a necessary and sufficient cause of the soldiers' advancing, if the major is silent, and vice versa. In view of this, the major's shouting can only be a supererogatory cause, given that the sergeant shouts. But it is the only cause, since the sergeant's shouting is ineffective in the presence of the major's shouting. Here, trumping gets expressed by the two-sided rank 2 that trumps, as it were, rank 1. Of course, we thereby get entangled again in the epistemic relativization; at least, though, (14.8a) can account for the asymmetry between the major and the sergeant in the intended way, even within the most coarsest setting, in contrast to all other accounts.

Lewis instead appeals to fine-graining of events. The major and the sergeant can give many different commands, and the major always overrules the sergeant. Considering the many possible commands, Lewis (2000/04, pp. 92f.) is able to account for the asymmetric causal influence of the major and the sergeant. This move is certainly reasonable. However, it is not a move to not-too-distant alterations of the causes, as Lewis describes them with intended vagueness (which need not, and in this case cannot, be understood as versions of the same causes); it is really a move to multi-valued variables. And then Lewis' notion of causal influence resembles the notion of actual causal dependence between variables to be introduced in Section 14.8; he is no longer talking of causation between events or facts, as Collins (2000, Sect. 4) also notes critically. Halpern, Pearl (2005a, pp. 874f.) also apply the multi-valued variable setting and thereby arrive at the weak asymmetry mentioned above.

What does ranking theory say about that setting? Let us decompose $\bar{M}_1$ into M_0 = "the major is silent" and M_{-1} = "the major shouts: 'Retreat!' ", and similarly for $\bar{S}_1 = S_0 \cup S_{-1}$. Let us assume that $\tau(A \mid M_i \cap S_j)$ is the truth value of A (1 = true, -1 = false) given $M_i \cap S_j$; so at least we do not invoke different strengths of belief. Still, it is not sufficient to observe that different conditions produce different outcomes, as Lewis as well as Halpern and Pearl essentially do. (14.3) requires us to check for positive relevance, i.e., to compare, e.g., $\tau(A \mid M_1 \cap S_1)$ with $\tau(A \mid \bar{M}_1 \cap S_1)$. The latter rank is not yet determined by the assumption above; we need to know $\tau(M_i \mid S_1)$ as well. Assume these ranks to be 0; what the major does is open, even given S_1. This is perhaps the most natural specification. An easy calculation then yields:

(14.8b)

$\tau(A \mid \cdot)$	S_1	$\bar{S}_1$
M_1	1	1
$\bar{M}_1$	0	-1

Trumping, multi-valued case

Thus, as desired, the sergeant's shouting "Advance!" is not a cause of the soldiers' advance. By contrast, the major's shouting "Advance!" is a sufficient, but not a necessary cause of the soldiers' advance. This is as it should be. If the major had not shouted "Advance!", then the soldiers might or might not have advanced, depending on whether the major had been silent or shouted something else.

These examples aptly show how, already in the case of direct causation, the ranking-theoretic account (14.3) provides us with greater expressive means than all rivals. These means allow us to take our intuitions at face value without further ado. Of course, the modeling of examples is hardly ever unique; as Halpern, Pearl (2005a) emphasize again and again, there often are several plausible alternatives, and several manners of causal talk are thus representable. Still, I submit that ranking theory enriches our modeling options in plausible and unprecedented ways.

14.6 The Circumstances of Direct Causes

(14.3) was based on the largest possible conception of obtaining circumstances, in order not to leave out anything potentially relevant and in order to render (14.3) non-circular. Intuitively, however, these circumstances are far too large. How could the entire past of the effect, starting with the big bang, so to speak, potentially intrude in the local relation between the cause and the effect? Still, (14.3) was a wise move. On its basis, and *only* on its basis, can we unfold the ambiguities in the notion of obtaining circumstances; we will be able to distinguish six significant notions of circumstances and investigate their relations. This inquiry is worthwhile in itself, for it will dispel our worries about (14.3), and it will advance my argument against Cartwright's claim regarding the circularity of all explications of causation (the argument can be finished only in Section 15.5). In a way, it parallels our inquiry in Section 13.2. There we studied other kinds of conditions, normal and exceptional ones. In both cases, though, it is the close theoretical look at conditions and circumstances that is revealing.

In Spohn (1990, Sect. 4) I conducted this inquiry for the probabilistic case; in the ranking-theoretic case it is essentially the same, although it shows some new facets. One new facet is that it is advisable in the ranking-theoretic case to distinguish binary or yes–no variables taking only two values and multi-valued variables taking more than two values. The causal behavior of the latter has its intricacies, which arise already in relation to the present inquiry. Therefore let me restrict the present inquiry to the case where all variables in U are binary. The next section will be devoted to multi-valued variables; at the end of that section I will briefly explain how most of our present results can be preserved in the case of multi-valued variables.

We based (14.3) on the following explication of the obtaining circumstances:

Definition 14.9: Let $X \leq Y, A \leq B$, and $w \in A \cap B$ be as in (14.3). Then the *temporally possibly relevant circumstances* of (the direct causal relation between) A and B in w are defined as $C_w^{++}(A, B) = {}^w[\leq Y, \neq X, Y]$.

This is the weakest sense of potential relevance, referring only to the temporal relations, and thus yielding the largest kind of circumstances. On the basis of (14.3) we can, however, introduce stricter senses of relevance leading to smaller kinds of circumstances. (Terminology is bound to be ambiguous here. Let me explicitly state that I call circumstances large when they are long conjunctions of atomic facts and hence small sets of possibilities, and I call them smaller when they are shorter conjunctions or larger sets of possibilities; I hope this is the more intuitive choice.)

The relevance of a fact or a variable to the relation between A and B may not only depend on the temporal relations, but also on the doxastic attitude, i.e., on the ranks (or probabilities) involved. With the help of the next definition we can thus specify a stricter sense of potential relevance:

Definition 14.10: Let B be a Y-proposition, $w \in B$. A variable $Z \in U$ is *actually directly causally relevant* to B in w relative to ξ iff ${}^w[Z] \xrightarrow{\pm}_w B$, or rather, iff $Z \leq Y, Z \neq Y$, and not $B \perp_\xi Z \mid {}^w[\leq Y, \neq Y, Z]$ (cf. the notation introduced in Definitions 7.1 and 7.5; the present mixing of propositions and sets of variables as arguments of the ternary relation $\perp_\xi$ is self-explanatory). $R_w(B)$ is then to denote the set of all variables actually directly causally relevant to B in w; and $R(B)$ is to denote the set of all variables directly causally relevant to B in *some* world. Thus, $R(B) = \bigcup_{w \in W} R_w(B) = \{Z \in U \mid Z \leq Y, Z \neq Y,$ and not $B \perp_\xi Z \mid \{\leq Y, \neq Y, Z\}\}$.

Note that the assumption of binary variables is important for (14.10). Since Z is binary, $^w[Z]$ is the only Z-fact. If Z were multi-valued, however, it might happen that some Z-fact, though not $^w[Z]$ itself, is a direct cause or counter-cause of B. Then only the second statement of the definiens of actual direct causal relevance applies, and so $R_w(B)$ and $R(B)$ also make sense without this assumption.

An immediate crucial consequence of (14.10) is this:

(14.11) $R(B)$ is the smallest subset R of $\{\leq Y, \neq Y\}$ such that $B \perp_\xi \{\leq Y, \neq Y\}$ $- R \mid R$.

This is so because (14.10) says that $Z \in \{\leq Y, \neq Y\} - R(B)$ iff $B \perp_\xi Z \mid \{\leq Y, \neq Y, Z\}$; this entails (14.11) via intersection, the graphoid property (7.8f) of conditional independence. We might also say that $R(B)$ is the minimal set of variables not later than B screening off from B all the other variables not later than B; and if we were to speak in graph-theoretic terms, we would say that $R(B)$ is the set of parents of Y. This suggests another sense of relevant circumstances, namely those possibly relevant only on the basis of temporal relations *and* ranks:

Definition 14.12: Let X, Y, A, B, and w be as in (14.3). Then the *ranking-theoretically possibly relevant circumstances* of (the direct causal relation between) A and B in w are defined as $C_w^+(A, B) = {}^w[R(B) - \{X\}]$.

We will soon see that this kind of circumstances is equivalent to the first kind in a specific sense. The point then will be that $C_w^+(A, B)$ usually contains only a small part of the facts contained in $C_w^{++}(A, B)$ and is thus apt to soothe our worries about overly largely conceived circumstances.

However, we cannot stop with (14.12). What one has in mind is actual, and not merely possible, relevance; it should suffice to consider the actually relevant circumstances. Here is a first attempt at explication: (14.3) can be interpreted as giving the truth condition of the sentence "A is a direct cause of B", i.e., as specifying when this sentence is true in a world w. Viewed in this way, it seems plausible to say that the actually relevant circumstances of A's being a direct cause of B just consist in the fact that A is a direct cause of B, i.e., in the set of worlds in which A and B are so related; likewise for "direct counter-cause" and "direct causal irrelevance". Thus:

Definition 14.13: Let X, Y, A, B, and w be as in (14.3). Then the *smallest actually relevant circumstances* of (the direct causal relation between) A and B in w are defined as $C_w^-(A, B) = \{w' \in W \mid A \xrightarrow[w']{+} B$ iff $A \xrightarrow[w]{+} B$ and $A \xrightarrow[w']{-} B$ iff $A \xrightarrow[w]{-} B\}$. If A is a sufficient direct cause of B in w, we may more specifically define those smallest circumstances as $C_w^-(A, B; s) = \{w' \in W \mid A \xrightarrow[w']{s+} B\}$. And similarly where A is a supererogatory, necessary, or insufficient direct cause (or counter-cause) of B in w.

However, these smallest circumstances are at best partially useful, as the following observation shows:

(14.14)   Each of positive relevance, negative relevance, and irrelevance of A to B given $^w[\leq Y, \neq X, Y]$ is compatible with each of positive relevance, negative relevance, and irrelevance of A to B given $C_w^{-}(A, B)$. However, if A is, respectively, a sufficient or a necessary direct cause of B in w according to (14.3), A is also a sufficient or a necessary reason of B given $C_w^{-}(A, B; s)$ or $C_w^{-}(A, B; n)$.

Both claims follow from the law (5.23a) of disjunctive conditions and the fact that it cannot be strengthened. Thus, the causal relations as defined in (14.3) are not necessarily preserved under the smallest actually relevant circumstances; they are only preserved in the special cases of sufficient and necessary direct causation. This fact will prove useful also in the remarks on explanation in Section 14.14.

For general purposes, however, we have to modify (14.13). (14.13) said that the actually relevant circumstances of A's being a direct cause of B just consist in the fact that A is a direct cause of B. Now, in view of the difficulties stated in (14.14), it seems rather that they consist in the fact that A is a direct cause of B in the way it actually is – where this clause refers to the specific numerical change of the ranks of B and $\bar{B}$ that is actually due to A. This idea is captured in

> *Definition 14.15*: Let X, Y, A, B, and w be as in (14.3). Then the *smaller actually relevant circumstances* of (the direct causal relation between) A and B in w are defined as $C_w^{-}(A, B) = \{w' \in W \mid \xi(B' \mid A' \cap {}^w[\leq Y, \neq X, Y]) = \xi(B' \mid A' \cap {}^w[\leq Y, \neq X, Y])$ for each $A' \in \{A, \bar{A}\}$ and $B' \in \{B, \bar{B}\}\}$.

Again, the law (5.23a) of disjunctive conditions directly entails:

(14.16)   For each non-empty $\{\leq Y, \neq X, Y\}$-proposition $D \subseteq C_w^{-}(A, B)$, $A' \in \{A, \bar{A}\}$, and $B' \in \{B, \bar{B}\}$, $\xi(B' \mid A' \cap D) = \xi(B' \mid A' \cap {}^w[\leq Y, \neq X, Y])$ and hence $B \perp_\xi {}^w[\leq Y, \neq X, Y] \mid A' \cap D$ for $A' \in \{A, \bar{A}\}$. $C_w^{-}(A, B)$ is indeed the largest $\{\leq Y, \neq X, Y\}$-proposition for which this is true.

Thus, $C_w^{-}(A, B)$ represents the smallest circumstances such that conditionalization on them necessarily agrees with conditionalization on any more largely conceived circumstances. I take this as the proof indicated above that, contrary to Cartwright (1979), conditioning or partitioning on irrelevancies does no harm. We might also say with Skyrms (1980, part IA) that $C_w^{-}(A, B)$ makes the ranks of B and $\bar{B}$ given A and $\bar{A}$ maximally resilient over the rest of the actual past of B. It thus seems that we have hit upon a reasonable explication.

Let us therefore study $C_w^{-}(A, B)$ a bit more closely. One valuable piece of information concerns which states of which sets of variables are contained in $C_w^{-}(A, B)$:

(14.17)   Let X, Y, A, B, and w be as in (14.3). Then we have for each $w' \in C_w^{-}(A, B)$ and $Z \subseteq \{\leq Y, \neq X, Y\}$ that $^{w'}[Z] \subseteq C_w^{-}(A, B)$ iff $B \perp_\xi \{\leq Y, \neq X, Y\} - Z \mid A' \cap {}^{w'}[Z]$ for each $A' \in \{A, \bar{A}\}$.

Again, this is a direct consequence of (14.15). The observation points to a useful distinction within $C_w^-(A, B)$. Each $w' \in C_w^-(A, B)$ differs from w on some variables. The only interesting differences are those in $\{\leq Y, \neq X, Y\}$, because outside that set, i.e., on the future of Y, the possibilities in $C_w^-(A, B)$ may arbitrarily vary, anyway. Thus, let $Z = \{V \in \{\leq Y, \neq X, Y\} \mid w'(V) = w(V)\}$ and $Z' = \{V \in \{\leq Y, \neq X, Y\} \mid w'(V) \neq w(V)\}$. Now, the distinction is this: One case is that w' is in $C_w^-(A, B)$ because all variations of w on Z' are in $C_w^-(A, B)$, i.e., because ${}^w[Z] \subseteq C_w^-(A, B)$ or, equivalently, $B \perp_\xi Z' \mid A' \cap {}^w[Z]$ for each $A' \in \{A, \bar{A}\}$. The other case is that these conditional independencies do not hold. In this case, w' is, in a sense, only accidentally in $C_w^-(A, B)$, i.e., not because the variables in Z' do not matter to B given ${}^w[Z]$ and A or $\bar{A}$. Rather, the variables in Z' do matter; it is only that in some particular realizations of Z' the relevant conditional ranks come out the same as for w, and that w' represents one such realization of Z'.

This suggests that the actually relevant circumstances of A and B in w should be conceived in a somewhat larger way, namely as comprising only all the arbitrary variations of w in $C_w^-(A, B)$:

Definition 14.18: Let X, Y, A, B, and w be as in (14.3). Then the *larger actually relevant circumstances* of (the direct causal relation between) A and B in w are defined as $C_w(A, B) = \bigcup \{{}^w[Z] \mid Z \subseteq \{\leq Y, \neq X, Y\}$ and $B \perp_\xi \{\leq Y, \neq X, Y\} - Z \mid A' \cap {}^w[Z]$ for each $A' \in \{A, \bar{A}\}\}$.

It will soon become clear why this is my preferred sense of the actually obtaining circumstances of a direct causal relation.

As an immediate consequence of (14.16) and (14.14), the five notions of obtaining circumstances introduced so far are related in the following way:

(14.19) $C_w^{++}(A, B) \subseteq C_w^+(A, B) \subseteq C_w(A, B) \subseteq C_w^-(A, B) \subseteq C_w^{--}(A, B)$; and if D and D' are any of these circumstances except $C_w^{--}(A, B)$, then $\xi(B' \mid A' \cap D) = \xi(B' \mid A' \cap D') =$ for each $A' \in \{A, \bar{A}\}$ and $B' \in \{B, \bar{B}\}$.

In other words, instead of the largest possible circumstances $C_w^{--}(A, B)$ we may as well base our explication (14.3) of direct causation on the much smaller circumstances $C_w^+(A, B)$, $C_w(A, B)$, or $C_w^-(A, B)$ (and in the case of sufficient or necessary direct causation even on the still smaller $C_w^-(A, B; s)$ or, respectively, $C_w^-(A, B; n)$) and thereby arrive at exactly the same direct causal relations. This dispels the concern about $C_w^{++}(A, B)$ being too large.

However, we have still not returned to the most natural, albeit circular suggestion with which we opened Section 14.4. Was the conjunction of all the other direct causes and counter-causes of B not another feasible sense of the obtaining circumstances of A and B? Yes, certainly. Let us fix it in:

Definition 14.20: Let X, Y, A, B, and w be as in (14.3). Then the *ideal actually relevant circumstances* of (the direct causal relation between) A and B in w are defined as

$C_w^*(A, B) = \bigcap\{D \mid D$ is a Z-proposition for some $Z \in \{\leq Y, \neq X, Y\}$ and $D \xrightarrow[w]{\pm} B\}$
$= {}^w[R_w(B) - \{X\}]$; i.e., the conjunction of all direct causes and counter-causes of B in w besides A.

Note that the equation $C_w^*(A, B) = {}^w[R_w(B) - \{X\}]$ holds only because of the assumption of binary variables; see my remark after (14.10).

For the moment, "ideal" means something bad. The trouble is that we cannot prove that $C_w^*(A, B) \subseteq C_w^-(A, B)$. Thus, the relevant ranks, given the ideal circumstances, might well differ from those, given the circumstances in the senses accepted so far. How can this happen? This is clarified by a more positive result:

(14.21) $C_w(A, B) \subseteq C_w^*(A, B)$, and the identity holds if and only if $B \perp_\xi \{\leq Y, \neq X, Y\} - R_w(B) \mid A' \cap {}^w[R_w(B) - \{X\}]$ for each $A' \in \{A, \bar{A}\}$.

The proof essentially requires writing out the relevant definitions. (14.21) says that the identity holds only if the variables individually independent of B, given the actual states of all the other variables in the past of B, are also collectively independent of B, given A and the rest of the actual past of B, as well as given $\bar{A}$ and the rest of the actual past of B. Both aspects of this condition are easily violated.

Such a violation occurs precisely in the strange case where some fact D is directly causally irrelevant to the effect B, but nevertheless relevant to $A \xrightarrow[w]{\pm} B$. Suppose I am running a business. The business might run well (B) or badly ($\bar{B}$). Actually, I have no competitor (A), but I might have one ($\bar{A}$). And I might set a high price for my product (D), selling less of it, or a low one ($\bar{D}$), selling more of it. Then it is plausible that D is causally irrelevant to B, since in the absence A of competitors, my business will be equally profitable in either case. Still, D is a relevant circumstance. If I set a high price (D), the absence A of competitors is crucial for my economic success B, whereas the importance of A is lowered or even vanishes, if I set a low price ($\bar{D}$).

The possibility that these two causal roles of D come apart must be simply admitted, I think. One may consider this as anomalous, but we should not try to exclude the anomaly by definition. Rather, we should study how causal structures as defined by (14.3) behave in general and how much more nicely causal structures behave in particular when they do not show this anomaly.

The assumption under which this anomaly is excluded is already stated in (14.21); it is the identity of $C_w(A, B)$ and $C_w^*(A, B)$. This explains why I have called $C_w^*(A, B)$ the ideal circumstances of A and B in w; $C_w^*(A, B)$ specifies how the circumstances ideally are, but need not be. Herein also lies the deeper reason why the larger actually relevant circumstances $C_w(A, B)$ are my preferred explication of the obtaining circumstances; among all the otherwise equally acceptable explications this is the only one that allows for the statement of the assumption that circumstances are ideal. This assumption will acquire some theoretical importance in Section 14.12 (and it will turn out to be a consequence of the objectification theory offered in Section 15.5).

14.7 Multi-Valued Variables

In the previous section, the discussion of circumstances was restricted to binary variables in order to keep it fairly perspicuous. Let us take at least a brief look at the problems involved with the undoing of this restriction. So, let us allow the variables in U to be multi-valued and assume in particular that $X, Y \in U$ with $X \leq Y$ are multi-valued. The issue then is what our basic explication (14.3) tells us about the direct causal relations between X-facts and Y-facts.

Of course, multi-valued variables are considered throughout the literature. However, the issue has rarely been posed in the way I just did. The main reason is that causal theorizing that has emerged from, or been related to, statistics has always focused on causal dependence between variables, contrary to their misleading language. Then the size of variables does not matter. (Or rather, it is the binary variables that show peculiarities in the probabilistic context, due to the independence law (7.11).) Even Suppes (1970, Ch. 5), who had an early awareness of the topic, was ambiguous; his now forgotten notion of a quadrant cause, which will prove helpful below, really describes a special form of causal dependence between variables. The complementary reason is, I think, that the more philosophical literature that attended to causal relations between facts or events was rarely explicit about the size of variables. For instance, I observed in Section 14.1 that events, if they have versions and alternatives as conceived by Lewis (2000/04), correspond to propositions about multi-valued variables; but this point is not specifically addressed in counterfactual theorizing about causation.

What is so problematic about causal relations between facts about multi-valued variables? This is easily explained. Both the algebra $\mathcal{A}(X)$ of X-propositions and the algebra $\mathcal{A}(Y)$ of Y-propositions, may be very large in this setting, and the ranking-theoretic and probabilistic dependence and independence relations between them may thus display the complicated structures stated in Theorems 6.7, 7.2, and 7.4. Of course, these also hold, given the obtaining circumstances, and hence, following (14.3), the structure of direct causes and effects is equally complicated. For instance, it may happen that the X-facts A and B are both direct causes of some Y-fact, but the X-fact $A \cap B$ is not or is even a direct counter-cause; similarly for effects. Or some X-fact may be a direct cause of some Y-fact and a direct counter-cause of some other Y-fact; and so forth. All the prima facie disturbing, but unavoidable observations from Section 6.2 about which laws do *not* hold for the reason relation apply here as well, and perhaps appear even more worrisome.

I see four responses. One might feel justified in leaving this awkward topic and focusing on the more tractable topic of causal dependence between variables. This response is understandable, bur understandably not ours. Another response is to blame (14.3) for such consequences, and to outright restrict it to binary variables. However, there is no reason why the explicatory strategy leading to (14.3) should have stood under this restriction. Moreover, this negative response gives us no positive clue as to what to do instead.

A better idea is to draw attention to so-called contrastive emphasis in causal claims, as observed by Dretske (1977), and in explanations, as observed by van Fraassen (1980a, Ch. 5, §§ 2.8 and 4.3). The point is that even in ordinary discourse we often deal with what may be conceived as multi-valued variables, and that ordinary language then provides means for expressing which facts about such variables we are interested in. The point was elaborated in an exemplary way by Hitchcock (1996). I will only sketch it here.

Let us look at an example of van Fraassen (1980a, p. 127) and consider the question "Why did Adam eat the apple?" Put so neutrally, it is not clear which variable is alluded to. Other persons might have eaten the apple, Adam might have eaten other fruits, or he might have done other things with the apple. Taking all three variations together we get a three-dimensional variable, with many propositions pertaining to it. Only a few of them are important to us, and it is convenient to refer to them by some contrastive emphasis. Usually, we do not ask the neutral question; we rather ask "Why did *Adam* eat the apple?" or "Why did Adam *eat* the apple?" or "Why did Adam eat the *apple*?", or we might even topicalize the phrase in question. The effect of the contrastive emphasis is to reduce what might at first appear to be a multi-valued variable to a less multi-valued or even binary variable taking, e.g., the values "Adam" and "somebody else" (or "Eve", if she is the only other person around).

For van Fraassen (1980a), this point belongs to the pragmatics of explanation, and even the pertinent relevance relation (cf. van Fraassen 1980a, p. 143) is mere pragmatics – perhaps because he is skeptical of causation, anyway. However, by reading the relevance relation in a probabilistic way and providing a probabilistic account of causation, Hitchcock (1996) brings out the systematic core of the phenomenon. The same strategy works when the contrastive emphasis lies not in the explanandum or the description of the effect, but in the description of the cause, as in Dretske's examples. Of course, we may replace probability by ranking theory and thereby simply take over Hitchcock's treatment of contrastive emphasis. (Woodward (2003, pp. 145f.) makes a similar move with respect to his interventionist theory of causation.)

I prefer, however, to sketch a final, quite different response to the problem of the causal relations between multi-valued variables. In general, these relations may be as messy as indicated above. In common cases, though, the variables and their causal relations may be much more well-behaved, at least concerning natural facts about them, and it is our task to describe such cases.

Let me discuss only one such common case. It is characterized by the assumption that the ranges W_X and W_Y of the variables X and Y in question are both linearly ordered by some relation $\leq$. This is the case when X and Y are real-valued magnitudes, but the assumption applies much more widely. As such, this assumption is of no help, but let us further assume that the values the variables take are correlated in our beliefs. One common form of that correlation is this – where τ is to denote the two-sided ranking function corresponding to the negative ranking function ξ underlying our inquiry:

Definition 14.22: Y is *monotonically related* to X w.r.t. ξ iff for each $y \in W_Y$ the function f_y defined by $f_y(x) = \tau(Y \geq y \mid X = x)$ is non-decreasing. Y is *strictly monotonically related* to X w.r.t. ξ iff for each $y \in W_Y$ except the smallest y (if there is one) that function f_y is increasing.

As an example, take the variable X to measure the hours Jim spent preparing for the examination and the grade Jim got as the variable Y. And then think about such questions as whether Jim passed the examination (got at least a "D" grade) because he prepared for (at least?) 100 hours. (14.22) is the ranking-theoretic translation of the notion of a prima facie regress cause proposed by Suppes (1970, p. 63). His accompanying Theorem 17 (p. 63) translates into this:

Theorem 14.23: If Y is monotonically related to X w.r.t. ξ, then for all $x \in W_X$ and $y \in W_Y$ $\{X \geq x\}$ is not a reason against, or non-negatively relevant for, $\{Y \geq y\}$ w.r.t. ξ. If Y is strictly monotonically related to X w.r.t. ξ, then for all $x \in W_X$ and $y \in W_Y$ $\{X \geq x\}$ is a reason for $\{Y \geq y\}$.

Proof: For the first assertion we have to show for all x, y (cf. Definition 6.1) that $\tau(Y \geq y \mid X \geq x) \geq \tau(Y \geq y \mid X < x)$, i.e., that

(a) $\xi(X \geq x, Y < y) - \xi(X < x, Y < y) \geq \xi(X \geq x, Y \geq y) - \xi(X < x, Y \geq y)$.

For the given y we can decompose f_y into $f_y = f^+ - f^-$, where $f^+(x) = \xi(Y < y \mid X = x)$ is non-decreasing and $f^-(x) = \xi(Y \geq y \mid X = x)$ is non-increasing. Hence, $\xi(X \geq x, Y < y) = \min_{z \geq x} \xi(X = z, Y < y) = \min_{z \geq x} [\xi(X = z) + f^+(x)]$; and similarly for the other terms of (a). Thus, (a) translates into:

(b) $\min_{z \geq x} [\xi(X = z) + f^+(z)] - \min_{z < x} [\xi(X = z) + f^+(z)] \geq$
       $\min_{z \geq x} [\xi(X = z) + f^-(z)] - \min_{z < x} [\xi(X = z) + f^-(z)]$.

Suppose that the four minima in (b) are, respectively, reached at x_1, x_2, x_3, and x_4. Then (b) says:

(c) $\xi(X = x_1) + f^+(x_1) - \xi(X = x_2) - f^+(x_2) \geq \xi(X = x_3) + f^-(x_3) - \xi(X = x_4) - f^-(x_4)$.

Now replace in (c) the minimum taken at x_2 by the larger value taken at x_4; thereby, the LHS of (c) gets smaller. And replace the minimum taken at x_3 by the larger value taken at x_1; thereby the RHS of (c) gets larger. The two substitutions yield:

(d) $\xi(X = x_1) + f^+(x_1) - \xi(X = x_4) - f^+(x_4) \geq \xi(X = x_1) + f^-(x_1) - \xi(X = x_4) - f^-(x_4)$,

which reduces to

(e) $f^+(x_1) - f^+(x_4) \geq f^-(x_1) - f^-(x_4)$.

Since $x_4 < x \leq x_1$ and f^+ is non-decreasing, the LHS of (e) is ≥ 0; and since f^- is non-increasing, the RHS of (e) is ≤ 0. So, (e) holds, this proves (c), and thus (a). If Y is strictly monotonically related to X, the analogous proof applies.

We might look for various improvements of Theorem 14.23. For instance, even if Y is only monotonically related to X, not strictly, $\{X \geq x\}$ will usually be not only non-negatively, but even positively relevant to $\{Y \geq y\}$; it would be tedious, though, to describe the precise conditions. Moreover, for each interval $[x_1, x_2]$ there should be an interval $[y_1, y_2]$ such that for all $[x, x'] \subseteq [x_1, x_2]$ and $[y, y'] \supseteq [y_1, y_2]$ $\{x \leq X \leq x'\}$ is non-negatively or even positively relevant to $\{y \leq Y \leq y'\}$. One might also inquire into the threshold values of x and y that distinguish the X-propositions in question as insufficient, necessary, sufficient, or supererogatory reasons for the relevant Y-propositions. To exemplify my strategy, however, (14.23) is good enough.

The last step is to proceed from reasons (Suppes' prima facie causes) to causes, by embedding the considerations into the obtaining circumstances (in anyone of the appropriate senses of the previous section). If $X \leq Y$ and Y is (strictly) monotonically related to X, given the circumstances, we might conclude with (14.23) that any fact of the form $\{X \geq x\}$ is not a direct counter-cause or even a direct cause of any fact of the form $\{Y \geq y\}$. (Of course, necessary and/or sufficient causal relations between X and Y are more restricted.)

In this way, we have a choice. We can either blame our basic explication (14.3) for allowing for messy or unintuitive causal relations between facts about multi-valued variables. Or we can blame the violation of tacit assumptions for messy or unintuitive consequences and try to state, as I just did, plausible assumptions that avoid such consequences. The latter strategy must not be taken to immunize (14.3), but it is, I take it, by far the more informative and fruitful alternative. I will apply it again and again.

This strategy finally helps to close the gap left in the previous section. We had restricted our investigation of the obtaining circumstances to binary variables, since not all of our results carry over to multi-valued variables. But if the multi-valued variables are monotonically related, the results do carry over. Let me briefly make the case:

The alternative formulation of Definition 14.10 of the actual direct causal relevance of a variable $Z \leq Y$ to some Y-fact B applies to multi-valued variables. Therefore, the definitions and observations (14.11)–(14.19) also hold in the multi-valued variable setting. However, Definition 14.20 contained a claim, namely, that the conjunction of all direct causes and counter-causes of the Y-fact B in w besides the X-fact A, i.e., the ideal actually relevant circumstances $C_w^*(A, B)$ amount to $^w[R_w(B) - \{X\}]$. This is not generally the case for multi-valued variables. However, if the Y-fact B is of the form $\{Y \geq y\}$ (or $\{Y < y\}$) and if all the variables $Z \in R_w(B) - \{X\}$ are strictly monotonically related to Y, given the circumstances, then we have: if z is the value $Z \in R_w(B) - \{X\}$ actually takes in w, then $\{Z \geq z\}$ is a direct cause and $\{Z \leq z\}$ a direct counter-cause of B (or vice versa), and their intersection is $\{Z = z\} = ^w[Z]$. Thus, under these assumptions $C_w^*(A, B) = ^w[R_w(B) - \{X\}]$ holds also in the multi-valued variable case, and so does observation (14.21). (From Section 4 of Spohn (1990) it is clear that these complications do not arise in the probabilistic setting.)

14.8 Causal Dependence between Variables, Markov Conditions, and the Principle of the Common Cause

After this brief look at the possibly complicated causal relations between facts about multi-valued variables and at ways of simplification, we may return to the first response mentioned in the previous section and turn to investigating only the causal relations between the variables themselves and not the finer details of the causal relations between facts. We can define the former immediately from the latter, which are basic. A first notion would be that the variable *Y directly causally depends* on the variable *X actually* (*in w*) iff some *X*-fact is actually (in *w*) a direct cause of some *Y*-fact. This hybrid notion of actual causal dependence might well deserve scrutiny. Indeed, a critical view on Lewis (2000/04) might conclude that his recent notion of causal influence amounts to no more than that hybrid notion. However, here I will not further dwell on that notion. We are rather interested in whether *Y* causally depends on *X* at all, i.e., in some worlds or under some obtaining circumstances, thus avoiding all reference to actual facts. This leads us, we might say, to the pure underlying causal structure independent of particular facts. This notion is captured in

> *Definition 14.24*: Let $X, Y \in U$ and $X \leq Y$. Then *Y directly causally depends on X relative to* ξ iff some *X*-proposition is a direct cause or counter-cause of some *Y*-proposition in some world $w \in W$ relative to ξ, i.e., iff not $Y \perp_\xi X \mid \{\leq Y, \neq X, Y\}$.

(I defined the same notion in Spohn 1983b, Definition 6.22, and its probabilistic counterpart in Spohn 1976/78, p. 118, Definition 19.)

This, or rather the probabilistic counterpart, is the basic notion behind all statistical methodology when it tries to get at causation. Look at path analysis from Wright (1921, 1934) onwards (cf., e.g., van de Geer 1971), at econometrics from Haavelmo (1943) onwards (cf., e.g., Granger 1969, 1980), at causal reasoning with Bayesian nets (cf. Glymour et al. 1987 and Pearl 1988) and the huge recent literature inspired by it: it is always this notion that is in the center. Within the probabilistic context, the abstraction from actual facts is perhaps excusable, since the realization of facts might there be taken to be ultimately random. Still, the notion is a derived one, derived from (14.3), as it is required by conceptual order.

Beware, though. It is not quite the notion usually investigated. Although (14.24) is about variables and not about facts, it is still about specific variables characterizing a given single case and endowed with a fixed temporal order. By contrast, all the statistical literature I have been referring to, and even large parts of the philosophical literature, are at least ambiguous at this point and mainly deal with generic variables. This shift might be explained by statistical needs. The sampling of statistical data looks at many cases, and statistical inference then aims at hypotheses about larger populations. Genuine single-case propensities do not seem to be objects of study in the various applications from biology to economics. The shift entails further differences we will have to attend to. From the point of view of causal theorizing, however, the shift appears problematic; I will firmly stick to my present explicatory strategy, to the single case, ranking-theoretically – and thus subjectively – represented.

The fundamental observation about direct causal dependence is almost obvious from Definition 14.24:

> *Theorem 14.25*: Suppose that there are no two simultaneous variables in U, and let $\rightarrow$ denote direct causal dependence relative to the ranking function ξ. Then $\langle U, \rightarrow, \xi \rangle$ is a ranking net according to Definition 7.29.

> *Proof:* If there are no simultaneous variables in U, the temporal order $<$ is a strict linear order. This given, Definition 14.24 specifies a list of direct causes in the sense of Definition 7.24(b) and thus entails via intersection, i.e. the graphoid property (7.8f), that ξ satisfies the Markov condition (7.26) and the minimality condition (7.28) w.r.t. $\langle U, \rightarrow \rangle$ and the temporal order $<$.

So far, this theorem is restricted to the absence of simultaneous variables. When unfolding its consequences in this and the next section I will stick to this restriction. Whether and how the picture changes when simultaneous variables are allowed will be discussed only in Section 14.10.

Theorem 14.25 is of paramount importance. In Chapter 7 we had developed the basics of the theory of Bayesian and ranking nets. There, I again and again slipped into causal talk, partly because it was natural, partly because I was following established terminology, but I emphasized that this talk was so far unjustified. Now, if our present explicatory strategy is sound, (14.25) tells us that this talk is well founded. I presented that theory as a model of inductive inference. Now we see that it indeed provides a theory of causal inference; all the causal implications suggested there do apply.

This indicates the relevance of (14.25). Following Pearl (1988) and Spirtes et al. (1993), the literature on Bayesian nets as a method of causal inference boomed, and it continues to do so. The probabilistic counterpart of (14.25) provides a conceptual basis for these activities, and (14.25) itself suggests that all this theorizing might be transferred to ranking theory. To do so would constitute a huge working program that is beyond my powers and the scope of this chapter. I will continue to focus on the foundations.

Let me first mention an important difference: If variables are taken in the generic way, it is often hard to give them a univocal temporal order. For this reason, Spirtes et al. (1993) aim for a theory of causal inference without temporal presuppositions. Statistical data deliver relative frequencies, statistical inference then provides statistical probabilities and in particular conditional probabilistic dependencies and independencies among variables (though huge samples might be required to provide them in a reliable way), and causal inference finally tries to establish causal relations among variables on this sparse basis. This is, of course, a much more ambitious and difficult task than I am pursuing here; I do not forgo the benefits of temporal structure.

For instance, Spirtes et al. (1993, Ch. 4) struggle with what they call *statistical indistinguishability* or what is later called *Markov equivalence*, the fact that two different DAGs can agree on the same probability distribution (cf. 7.29), a relation that was completely characterized by Verma, Pearl (1991), which Andersson et al. (1997) proposed capturing with the notion of an *essential graph*, a graph structure that captures everything common to all Markov equivalent DAGs that agree with a given distribution. Thus, causal inference from statistical probabilities does not get beyond the causal structure as represented in those essential graphs. If, by contrast, one feels free to assume temporal order, as I do, one need not engage in these intricate considerations; that order and the probability distribution determine a unique DAG as causal structure and not a larger set of Markov equivalent graphs. This is a good example of how differences in basic assumptions ramify.

Keeping such differences in mind, we should have a closer look at least at the basic import of Theorem 14.25 on causal theorizing, which will also facilitate our subsequent discussion. I could do so by recommending a rereading of Chapter 7, and in particular of Section 7.3, in light of the causal interpretation we have now justified. It is better, though, to summarize the essential points, rearranged for the present purposes.

To begin with, since the directed edges or arrows in a graph now express direct causal dependence, we may read DAGs as causal graphs, as suggested after (7.19). Their basic property of acyclicity hence amounts to the exclusion of causal loops. This is not to deny feedback processes, though; feedback processes unfold in time and appear to form causal loops only on the level of generic variables and not in terms of specific variables.

This entails that we may give a causal interpretation to all the graph-theoretic notions introduced in Definition 7.20. The set $pa(X)$ of parents of a variable $X \in U$ is the set of variables on which X directly causally depends or, as I will say, the set of *direct causal predecessors* of X. The natural extension that is either stated or implied in all of the Bayesian net literature is this:

> *Provisional Definition 14.26*: *Causal dependence* is the transitive closure of direct causal dependence. That is, $X \in U$ *causally depends* on $Y \in U$, or Y is a *causal predecessor* of X, relative to ξ iff Y is an ancestor of X, i.e., $Y \in an(X)$ in the associated ranking net $\langle U, \to \xi \rangle$.

I call this definition provisional because it is not yet justified; for that purpose we would first have to take a stance towards indirect causation. This we will do in Sections 14.11–14.12. There we will indeed arrive at a different Definition 14.75 of (direct or indirect) causal dependence between variables, which we will prove to agree with the provisional (14.26) only under the conditions of Theorem 14.76. Hence, insofar as we rely on (14.26) in this section, it also stands under these conditions – or weaker conditions, if Theorem 14.76 can be improved. However, these conditions are not so restrictive as to essentially diminish the value of the present considerations. So, let us accept (14.26) for the next three sections.

To continue on terminology, X is a descendant of Y iff X causally depends on Y, and the set $nd(X)$ of non-descendants of X is the set of variables not causally dependent on X. Moreover, we may say that X and Y are causally connected iff they are connected, i.e., iff there is a path between X and Y in the causal graph $\langle U, \rightarrow \rangle$.

In the causal interpretation (7.26) turns into what Spirtes et al. (1993, pp. 54f.) call the causal Markov and minimality conditions:

(14.27)   *Causal Markov condition* (1st version): In the causal graph $\langle U, \rightarrow \rangle$ generated by the ranking function ξ according to (14.24) we have for all $X \in U$: $X \perp_{\xi} \{<X\} - pa(X) \mid pa(X)$.

That is, each variable is rendered independent from the rest of its past by the set of its direct causal predecessors. And the more impressive version (7.27), that is the one stated by Spirtes et al. and that we have seen to follow from (7.26), turns into:

(14.28)   *Causal Markov condition* (2nd version): For all $X \in U$, $X \perp_{\xi} nd(X) - pa(X) \mid pa(X)$.

That is, each variable is rendered independent by its direct causal predecessors even from all variables not causally dependent on it. Finally, (7.28) becomes:

(14.29)   *Causal minimality condition*: For all $X \in U$ the set $pa(X)$ of direct causal predecessors of X is the smallest set $Y \subseteq \{< X\}$ such that $X \perp_{\xi} \{<X\} - Y \mid Y$, i.e., such that (14.27) and (14.28) hold.

(14.27–14.29) simply restate the content of Theorem 14.25 in a more detailed way; they are thus an immediate consequence of (14.24) and ultimately of our basic explication (14.3). That's the crux. The status of these conditions is quite contested in the literature, as we will see, whereas I hold the particularly strong claim that they are analytically true. This discrepancy demands discussion.

In order to facilitate this discussion, we should analyze these very rich conditions into more perspicuous parts; then we might better see what is critical about them and hence about my explication that entails them. The traditional and most enlightening decomposition is the following one.

First, given (14.27), the causal minimality condition (14.29) is clearly equivalent to:

(14.30)   *Ranking dependence*: For each $X \in U$ and $Y \in pa(X)$ it is not the case that $X \perp_{\xi} Y \mid pa(X) - Y$.

That is, a variable ranking-theoretically depends on each of its direct causal predecessors given all the other direct causal predecessors. This is how direct causal dependence should show up in ranking-theoretic dependence. Note that this precisely corresponds to Cartwright's thesis introduced at the beginning of Section 14.3, when transferred to the level of direct causal dependence between variables. Now, however, it is an immediate consequence of our explication, whereas on the level of direct causation between facts it could be shown to hold only under the additional assumptions stated in (14.21).

Note, moreover, that direct causal dependence of X on Y does not entail unconditional dependence between X and Y, or dependence conditional on any set not comprising $pa(X) - Y$. The conditional dependence in (14.26) might vary for various realizations of $pa(X) - Y$, and thus average out for smaller or other conditioning sets. It is true that for $Y \in pa(X)$ X and Y cannot be d-separated by any set of variables. However, this reminds us only of the fact familiar from Section 7.3 that graph-theoretic and ranking-theoretic dependence need not coincide. Such other ranking-theoretic dependencies would hold only if ξ were faithful to $\langle U, \rightarrow \rangle$ in the sense of (7.30), something for which there is no causal guarantee.

Similarly, ranking-theoretic dependence need not spread out further than direct causal dependence. That is, if Y is a parent of X, and Z a descendant or causal successor of X, we might nevertheless have $Z \perp_\xi Y \mid pa(X) - Y$. As before, the indirect causal influence Y has on Z might average out and thus not show up in a ranking-theoretic dependence. This is one of the anomalies we will have to scrutinize when discussing indirect causation. Again, though, Y and Z can only be d-separated by causal intermediaries of Y and Z such as X. Hence if ξ were faithful to $\langle U, \rightarrow \rangle$, the case just envisaged could not occur.

So much for simplifying the causal minimality condition. The causal Markov condition is best decomposed into two assumptions:

(14.31)   *Narrow Markov condition*: For all $X \in U$, $X \perp_\xi an(X) - pa(X) \mid pa(X)$.

That is, X is screened off by all of its direct causal predecessors from all its indirect causal predecessors. (14.31) is an obvious weakening of (14.27). The term "Markov property" and its variants were originally established in the theory of stochastic processes that dealt with very simple causal situations, i.e., with linear causal chains only. There it denoted the absence of after-effects, as it was said; its content is given by the simple slogan that the (immediate) future is independent of the past given the present. It is not so clear how to generalize this established usage to more complex situations or arbitrary causal graphs. (14.28) offers the strongest generalization. (14.31), by contrast, is the most conservative adaptation; it states only the separation of the (immediate) causal future (X) from the causal past ($an(X) - pa(X)$) by the causal presence ($pa(X)$). Hence the label "narrow". Since (14.31) is the most conservative version, it is also the least contested. In fact, I have not seen doubts raised against the narrow Markov condition; we should just accept it and welcome it as a consequence of our basic explication.

The other component into which the causal Markov condition splits is the principle of the common cause, or rather a generalized version that is slightly unperspicuous. Let us work up to it. First, we need a further bit of graph-theoretic terminology and a stronger notion of causal connection:

Definition 14.32: Two variables $X, Y \in U$ are *causally coordinate* iff neither $X \in an(Y)$ nor $Y \in an(X)$. A set $V \subseteq U$ of variables is *causally coordinate* iff any two variables in V are causally coordinate. For any two causally coordinate variables $X, Y \in U$, $cc(X, Y) = an(X) \cap an(Y)$ is the set of *common causal predecessors* of X and Y, and $pcc(X, Y) = \{Z \in cc(X, Y) \mid$ there is a child Z' of Z such that $Z' \in \{X, Y\} \cup an(X) \cup an(Y) - cc(X, Y)\}$ is the set of *proximate common causal predecessors* of X and Y. X and Y are *causally joined* iff $cc(X, Y) \neq \varnothing$ (and thus $pcc(X, Y) \neq \varnothing$). If $V \subseteq U, X \notin V$, and $V \cup \{X\}$ is causally coordinate, then $cc(X, V) = \bigcup_{Y \in V} cc(X, Y)$ is the set of *common causal predecessors* of X and V, and $pcc(X, V) = \{Z \in cc(X, V) \mid$ there is a child Z' of Z such that $Z' \in \{X\} \cup V \cup an(X) \cup \bigcup_{Y \in V} an(Y) - cc(X, V)\}$ is the set of *proximate common causal predecessors* of X and V.

With this, we are able to state our variants of the common cause principle. Recall its original statement by Reichenbach (1956, Ch. 19). Roughly, he claimed that every correlation has a causal explanation. This seems to be a modern and perhaps more tenable version of the venerable principle of causality; we will come back to this comparison in Section 17.4. Reichenbach connected the principle with issues about the direction of time; however, this is not my ambition. More precisely, Reichenbach stated that if two events (or propositions) are positively correlated, then either one event is a cause of the other or there is a third event, a common cause, that is positively relevant to both and renders the two conditionally independent.

This sounds problematic in various ways. The required conditional independence suggests that the common cause is a complete one; however, a complete common cause will rather consist of a set of events. We have also seen that a cause need not be unconditionally positively relevant to its effect. Presumably, it is the wrong idea, anyway, to try to state the common cause principle on the level of event or fact causation; the level of causal dependence between variables is more suitable. In fact, one may read Reichenbach as making claims about (two-valued) variables. Then, the claim is that if two variables are probabilistically dependent, either one causally depends on the other or there is a (set of) variable(s) on which both causally depend and which renders them probabilistically independent. (In Spohn (1993) I more carefully discussed this interpretation and derived it from my explication of causal dependence.) The ranking-theoretic translation is this:

(14.33)   *Reichenbach's common cause principle*: For any two variables $X, Y \in U$, if not $X \perp_\xi Y$, then either one causally depends on the other relative to ξ or they are causally joined, i.e., $cc(X, Y) \neq \varnothing$, and

    (a) $X \perp_\xi Y \mid cc(X, Y)$,

    (b) $X \perp_\xi Y \mid pcc(X, Y)$, and indeed

    (c) $\{X\} \cup an(X) - an(Y) \perp_\xi \{Y\} \cup an(Y) - an(X) \mid cc(X, Y)$, and

    (d) $\{X\} \cup an(X) - an(Y) \perp_\xi \{Y\} \cup an(Y) - an(X) \mid pcc(X, Y)$.

(c) and (d) are powerful, but straightforward strengthenings of (a) and (b). If the proximate common causal predecessors were direct ones, (a) and (b) would do. In general, though, they are not, and then the full strength of (c) and (d) will be required. There is an ambiguity between (a) and (b) and between (c) and (d). If the "common cause" is to be complete, (a) seems to be the more literal translation. However, (b) seems to be more economic and thus to conform to our intuitions even better. We will resolve the ambiguity soon.

In any case, all versions of (14.33) are consequences of the causal Markov condition (14.27) and thus ranking-theoretic theorems holding for all ranking functions ξ:

> *Proof of (14.33)*: We have to realize that if X and Y are causally coordinate, the relevant sets are d-separated by $cc(X, Y)$ and by $pcc(X, Y)$: Take any path from $\{X\} \cup an(X) - an(Y)$ to $\{Y\} \cup an(Y) - an(X)$. It can either avoid $pcc(X, Y)$ and thus $cc(X, Y)$. It can do so, however, only by containing a collider, and neither the collider nor any of its descendants are in $pcc(X, Y)$ or $cc(X, Y)$. Hence, the path is blocked by $pcc(X, Y)$ and by $cc(X, Y)$ (cf. 7.21). Or the path crosses $pcc(X, Y)$ and thus enters $cc(X, Y)$. At least one of the nodes where it crosses must be a chain node or a fork node. Hence, it is blocked, too. This means that these sets are d-separated by $cc(X, Y)$ and by $pcc(X, Y)$, and since ξ agrees with $\langle U, \to \rangle$ due to (14.23), d-separation entails their conditional independence (c) and (d) w.r.t. ξ. That X and Y are causally joined, i.e., $cc(X, Y) \neq \varnothing$, finally follows from the premise that not $X \perp_\xi Y \mid \varnothing$.

In the next section we will dwell further on the philosophical significance of Reichenbach's principle. At the moment we are still occupied with decomposing the causal Markov condition. (14.33) is not yet strong enough for this purpose. (14.33) deals only with two causally joined variables having a common causal predecessor. However, it may happen that many variables have common predecessors, and all of the former are rendered independent by the latter. This is the content of

(14.34)   *Strong common cause principle*: Let $V \subseteq U$, $X \notin V$, and $V \cup \{X\}$ be causally coordinate; moreover, let $X^* = \{X\} \cup an(X) - cc(X, V)$ and $V^* = V \cup \bigcup_{Y \in V} an(Y) - cc(X, V)$. Then

     (a) $X^* \perp_\xi V^* \mid cc(X, V)$, and

     (b) $X^* \perp_\xi V^* \mid pcc(X, V)$.

In other words, for each variable X and a causally coordinate set V, this variable and its "private" causal predecessors (= X^*) are independent from all the variables in the set and their "private" causal predecessors (= V^*) given their (proximate) common causal predecessors; the ambiguity between (a) and (b) is the same as before. If (14.33) is plausible, (14.34) should be so, too. In any case, the derivation of (14.34) from (14.27) is exactly the same as that of (14.33).

Having thus decomposed the causal Markov condition (14.27), we can now begin reversing the entailments. A picture helps to illustrate the strength of three principles we are discussing; the picture simplifies the situation by looking only at two variables X and Y and their ancestors:

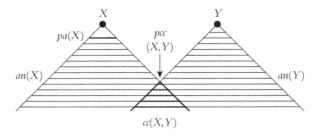

Figure 14.35

Here, the peak X is screened off by the top layer $pa(X)$ from the rest of the X-cone; this is the narrow Markov condition. The lightly hatched areas are screened off from one another by the intersecting cone or its boundary; this is the common cause principle. And both entail that the peak X is screened off by the top layer $pa(X)$ from the entire hatched area, i.e., from everything else, including the Y-cone; this is the causal Markov condition. This entailment is formally stated in

> *Theorem 14.36*: The causal Markov condition (14.27) is equivalent to the narrow Markov condition (14.31) and the strong common cause principle (14.34a).

> *Proof:* We have already seen one direction. For the reverse direction let $X \in U$, let V_1 be the latest variable $\{< X\}$ causally coordinate with $X, \ldots, V_k$ the latest variable in $\{< X\}$ causally coordinate with $X, V_1, \ldots, V_{k-1}$; let this process end with V_n, and set $V = \{V_1, \ldots, V_n\}$. Thus $V \cup \{X\}$ is causally coordinate and, with the notation of (14.34), $X^* \cup V^* = \{\leq X\}$. (14.34a) says $X^* \perp_\xi V^* \mid cc(X, V)$. This entails $X \perp_\xi V^* \mid an(X)$ via (7.8d) (weak union). And this, together with (14.31) $X \perp_\xi an(X) - pa(X) \mid pa(X)$, entails $X \perp_\xi V^* \cup an(X) - pa(X) \mid pa(X)$ via (7.8e) (contraction).

Let me be clear here about the various entailments. All principles discussed follow from the basic Definition 14.24. Hence, one might think they are trivially equivalent, as all theorems are. However, this is not how (14.36) is to be read. (14.36) does not refer to the causal graph generated by ξ. Rather, it takes its graph-theoretic notions relative to any given causal graph and then makes the substantial assertion that if ξ satisfies (14.31) and (14.34a) relative to that graph, it satisfies (14.27) as well, i.e., it agrees with that causal graph. This also explains why we could not yet benefit in the proof of (14.36) from the correspondence between graph-theoretic and ranking-theoretic independence.

Moreover, this clarifies the relation between the two versions of the strong common cause principle (14.34). Given the narrow Markov condition, we now see that version (a) entails version (b) via the causal Markov condition. However, the reverse entailment, even given (14.31), does not hold, as a simple counter-example shows: Take the simple graph $X \to Y \to Z, W$. (14.31) asserts $Z \perp_\xi X \mid Y$ and $W \perp_\xi X \mid Y$; and (14.34b) asserts $Z \perp_\xi W \mid Y$. However, this cannot entail (14.34a) $Z \perp_\xi W \mid \{X, Y\}$, since no independencies among three variables entail anything about four variables. Thus, version (b) turns out to be weaker than version (a), perhaps contrary to appearances, and it is version (a) that we need.

All of this nicely illustrates the consequences of our basic explication (14.24). More importantly, it clarifies the burden of argument. Of course, I welcome these consequences and take them as confirming my explication. Other philosophers defend these principles, perhaps under certain assumptions, without accepting the explication. Still others reject the causal Markov condition, whereby the causal minimality condition loses its foundation. Since the narrow Markov condition is unshaken, the focus of attack must be and has been the common cause principle.

This is indeed the punchline. I take these principles to be analytic; hardly anyone else does. How on earth could they be analytic? Hence, I have to defend my view against those who accept the principles, but for other than semantic reasons; this is the task of the next Section 14.9. The other issue is whether the common cause principle is at all true; this will occupy us in Section 14.10. To defend those principles first as analytic and then as true appears to be the wrong order. At the beginning of Section 14.10 I will be able to explain my seemingly strange choice.

14.9 Frame-Relative and Absolute Causal Dependence

The point that separates me from those taking the causal Markov condition to be only synthetically or hypothetically true is quite subtle; after all, we essentially agree. Spohn (2001b) was my first attempt to explain the point; I hope to make it clearer here.

What is the attitude I oppose? Let us look at its main articulation, the SGS theory, as I call it here (due to the major book Spirtes, Glymour, Scheines (1993) that was followed by many papers, dissertations, etc. mainly produced at the CMU philosophy department), which keeps entirely within the probabilistic context.

The main ground SGS have for accepting the causal Markov and the causal minimality condition (14.28 + 29) and, less firmly, the faithfulness condition is, as it were, an empirical one; they say that "their importance – if not their truth – is evidenced by the fact that nearly every statistical model with a causal significance we have come upon in the social scientific literature satisfies all three" (Spirtes et al. 1993, p. 53). Reason enough to work out their general mathematical theory! Of course, this is no accident; SGS have an explanation for their observation that even allows the elimination of most putative counter-examples.

They acknowledge, though, that the conditions are not universally true. The clearest exception is presented by the quantum domain where we find correlations between, say, the spin of two elementary particles jointly produced, but now in space-like separation: correlations felt to be paradoxical by Einstein, Podolsky, and Rosen, confirmed in the famous experiments of Alain Aspect, and proved to not be reducible to a common cause by Bell's inequality. However, they are not really worried by this apparent exception; this does not prevent their conditions from holding in most or all other domains.

By contrast, I should indeed worry deeply about these quantum correlations when I claim the causal Markov condition to be analytic, and thus true without exception. However, I will be silent on them as well. The turmoil of traditional conceptions caused by quantum theory is great, and not to be dealt with briefly; it is simply beyond my scope and that of this book. If my discussion succeeds within the classic domain, I am happy enough.

Having cleared this point away, what is SGS' explanation for the substantial success of their conditions? They hold, they say, for causally sufficient systems or set of variables. This is a crucial notion for them:

> *Explanation 14.37*: A frame U of variables is *causally sufficient* iff each proximate common causal predecessor of two or more variables in U is in U as well.

Why I call this an explanation and not a definition will become clear later on. To require inclusion of the proximate common predecessors is enough; otherwise, causally sufficient systems would be forced to extend indefinitely into the past. Spirtes et al. (1993, p. 45) qualify this explanation; a common causal predecessor X need not be included in a causally sufficient system iff the distribution over the joint causal successors is the same for all values that X actually takes in the population considered (as mentioned, they always deal with populations in the statistical context). We may ignore here this qualification and stick to (14.37).

Spirtes et al. (1993, p. 44) grant that almost any two variables might be, or might be made, causally dependent. "A dictator could . . . arrange circumstances so that the number of childbirths in Chicago is a function of the price of tea in China." Likewise, any two apparently causally dependent variables might have a common causal predecessor. Indeed, this was the doctrine of early or pure occasionalism, before it tried to make limited room for human agency and responsibility. Since all power is divine and earthly causation a mystery, each causal relation could be brought about only with God's help; He was the direct cause of everything and the proximate common cause of every apparent cause and effect (cf. Perler, Rudolph 2000, in particular Ch. 1).

Thus, it appears that causal sufficiency is too hard to achieve. What, then, makes SGS so confident that it applies widely? The general answer, I suppose, is that although it is easy to gerrymander non-sensical, causally insufficient systems, this is not what science is interested in. In their respective fields, scientists have acquired great skills at selecting causally sufficient systems and recognizing failures of sufficiency.

However, SGS do not merely rely on scientists' skills. They give two kinds of arguments. One is about deterministic and pseudo-indeterministic systems. A *deterministic* system is one where all endogenous variables, i.e., those having some parents, are functions of their parents. Such a system is obviously causally sufficient with respect to its endogenous variables, since the deterministic parents form a complete set of direct causal predecessors. If we assume, as Spirtes et al. (1993, p. 51) do, that all exogenous variables with no parents are jointly probabilistically independent, an assumption that is amenable to statistical confirmation, we may infer that they have no unrepresented common causal predecessors, either. Hence, under this assumption deterministic systems are causally sufficient. And, SGS prove, they satisfy the causal Markov condition!

The result extends to what Spirtes et al. (1993, p. 52) call *pseudo-indeterministic* systems. These are systems that are not deterministic, but can be extended to a deterministic one by adding new (error) variables and new arrows starting only at new variables such that no new (error) variable is a common causal predecessor of two old variables. Hence, when the extended system is causally sufficient, the original pseudo-indeterministic one is so, too, and again it provably satisfies the causal Markov condition. Steel (2005) attempts to further generalize this result.

The other kind of argument consists in an inspection of alleged violations of the causal Markov condition. In all cases (except the quantum theoretic ones) SGS argue that the counter-examples draw on causally insufficient systems and collapse once the systems are extended to sufficient ones (cf. Spirtes et al. 1993, Sect. 3.5.1). Thus, they also apply the methods of fine-graining variables and causal chains in order to get rid of unwanted examples.

At the same time, these arguments are what SGS offer in support of the claim that causally sufficient systems satisfy the causal Markov condition (14.28). Note that this claim cannot be a theorem of the SGS theory; (14.28) is an axiom of this theory and can at best be informally justified by that claim. However, this does not hamper its validity. On the contrary, I will arrive at it from my side, too.

My sketch would be incomplete without mentioning that SGS theorizing has lately become much more cautious about causal sufficiency. Latent or hidden or unmeasured variables (this is all the same) are everywhere, and they can always blur the causal pattern that shows up in a given set of measured variables. Causal theorizing needs to get a grip on such disturbances.

There are, in effect, two kinds of disturbances, and if Bayesian net theory gets causal relations right, there cannot be more. The first disturbance is produced by genuine latent variables that might entail causal insufficiency. One must always reckon with latent variables that are outside the realm of measured variables, but create dependencies within that realm that appear to be causal, since one observes only the distribution marginalized with respect to the latent variables. This illusion can be corrected only by explicitly conditioning on those variables. The other disturbance is created by so-called selection variables that also lie outside the realm of measured variables, that, unlike the latent variables, causally depend on the measured variables, and that have to take on a certain value in order for the measured variables to be observed at all. This selection bias can only be neutralized by explicitly modeling it. (I mention the selection bias only in order to give a fair impression of the development of SGS theorizing. Later on, I will neglect it again, since it appears to arise only in the statistical context, and not in ours.)

Example 14.38: Richardson (1998) gives a nice example that makes both kinds of problems immediately clear. Look at the following causal graph:

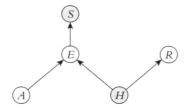

Richardson (1998, p. 234) tells this story about the graph:

> The graph represents a randomized trial of an ineffective drug with unpleasant side-effects. Patients are randomly assigned to the treatment or control group (*A*). Those in the treatment group suffer unpleasant side-effects (*E*), the severity of which is influenced by the patient's general level of health (*H*), with sicker patients suffering worse side-effects. Those patients who suffer sufficiently severe side-effects are likely to drop out of the study. The selection variable (*S*) records whether or not a patient remains in the study. Thus for all those remaining in the study *S* = *StayIn*. Since unhealthy patients who are taking the drug are more likely to drop out, those patients in the treatment group who remain in the study tend to be healthier than those in the control group. Finally health status (*H*) influences how rapidly the patient recovers (*R*). This example is of interest because, as should be intuitively clear, a simple comparison of the recovery time of the patients still in the treatment and control groups at the end of the study will indicate faster recovery among those in the treatment group. This comparison falsely indicates that the drug has a beneficial effect, whereas in fact, this difference is due entirely to the side-effects causing the sicker patients in the treatment group to drop out of the study.

The true causal pattern is supposed to be captured by the full causal graph. The quotation makes clear, though, that as long as you look only at the observed variables *A*, *E*, and *R*, a different, and misleading, picture emerges. If you neglect the latent variable *H*, there appears to be a causal dependence between *E* and *R*, though the direction might be unclear. And if you neglect the selection variable *S* and the implicit conditioning on values of *S* biased by the collection of data, *A* appears to have an influence not only on *E*, but also on *R*.

The potential presence of latent and selection variables hence poses a tremendous problem to the causal interpretation of the observed distribution over measured variables. A considerable part of recent SGS theorizing has been devoted to this problem. And it is completely solved. The answer lies in what are called *maximal ancestral graphs* (MAGs), first conceived in Richardson, Spirtes (2002). We need not engage now with these more complicated graphs. Still, their point should be clear. Whenever you have a causal graph or DAG ⟨*O* ∪ *L* ∪ *S*, →⟩, *O* being the observed, *L* the latent, and *S* the selection variables, and a distribution *P* that agrees with that graph, then the conditional independence structure of the marginalization of *P* to *O* is represented by a MAG over *O* (where representation now means something more complicated than agreement, i.e., the Markov and minimality condition in the case of DAGs). Conversely, each MAG and each distribution represented by it can be extended by latent and selection variables to form a Bayesian net. See Richardson, Spirtes (2003) for precise results. Finally, we must not forget that SGS theorizing does without temporal structure. So, if the full P over the set $O \cup L \cup S$ could be observed, the inferred causal structure would be given only by the essential graph of the class of Markov equivalent DAGs. This point deepens on the level of MAGs; again, one needs to suitably define and fully characterize an equivalence notion for MAGs. Even this problem has been solved (cf. Zhang 2006). These are remarkable results, elucidating causal inference even in the absence of causal sufficiency. As this sketch makes clear, though, the three SGS conditions are still at the basis of this sophistication; all claims about MAGs are derived from the assumption that the underlying DAG satisfies those conditions.

SGS theorizing is a fascinating cosmos that I fully accept. What is my complaint then? It is still the Socratic worry about the Euclidean method. It is clear how causally insufficient systems might violate the SGS conditions. What, though, grounds the assumption that causally sufficient systems satisfy them? What *are* the grounds of the SGS conditions? There are seven overlapping ways to go.

A first way is simply not to care about further grounding. This is partially the attitude Spirtes et al. (1993) take. If a substantial and widely applicable theory can be built on the SGS conditions, it is good strategy to develop this theory, and to not wait for further clarifications and justifications. This is perfectly reasonable; but it does not, and SGS never meant to, deny the legitimacy of further justificatory attempts and of further explanations of the surprisingly wide applicability.

A second idea is to somehow derive the SGS axioms from still more basic conditions. Hausman, Woodward (1999) may serve here as a paradigm. Their argument is that the interventionist conception (cf. Section 14.4) entails the condition of modularity, as they call it, which in turn entails the causal Markov condition. Their idea is roughly this: Suppose you have a real system represented by a Bayesian net. The causal relations in the system are supposed to show up in (counterfactual) interventions, where proper interventions have to satisfy certain causal presuppositions in turn (whence the alleged conceptual circularity). By an intervention on a variable you outright force that variable to have either a certain value or a certain distribution independently of the (other) causal predecessors of the variable. Which overall distribution correctly describes the thus manipulated system? For the original system, the overall distribution could be given by the factorization (7.35). For calculating the manipulated system – this is agreed by all sides – you have to replace the term for the manipulated variable in that factorization by the forced value or distribution. If this really works, if the real manipulated system is correctly described by simply replacing the relevant term in the factorization and leaving the other terms unchanged, then the Bayesian net correctly captures the causal relations in the system. Hausman, Woodward (1999) call such a system *modular*. In their Sections 7 and 10, then, they proceed to give a kind of derivation of the causal Markov condition from this modularity condition, which might, as hoped, deliver a deeper grounding. (Whether they really succeed is another issue.

In their derivation they also appeal to causal sufficiency (14.37) and to Reichenbach's common cause principle (14.33), and so one wonders whether the modularity condition is really needed or whether the uncontested narrow Markov condition (14.31) might do.)

However, even if this second idea could be realized in some way, it leaves open the status of the more basic premises and hence does nothing to clarify the status of the SGS conditions themselves. The further ways try to be more explicit on this score.

So, a third attitude is to claim that there is no further explanation either for the SGS conditions or for more basic conditions; it is simply an empirical fact that those conditions apply widely. There are basic laws of causal dependence, just as there are, say, basic laws of gravitation that have no deeper grounds. In support, one might allude to the distinction between the nature and the concept of causation introduced in Section 14.2 and say that the SGS conditions, or some more basic conditions, belong to the nature of causation, the discovery of which is a matter not of analysis, but of empirical inquiry, which might also elucidate why the conditions are widely, but not universally applicable. However, such a claim about the nature of causation would be very different from those we met in Section 14.2; and in fact I have not found anyone explicitly making such a claim. The SGS conditions certainly have a conceptual feel for them (and more basic conditions, if such there are, must have it, too). In any case, this chapter is about the possibly indeterminate or tacit, but certainly not hidden concept of causation; and the task of this section is to find out about possible conceptual grounds of those conditions.

A fourth position would be to deny just those conceptual grounds, on the basis of the alleged explicatory circles. However, for my part I have rejected two such allegations, Cartwright's claim that the circumstances of a causal relation create an explicatory circle (cf. Sections 14.4 + 6) and the interventionist claim that (counterfactual) intervention and causation are conceptually entangled (cf. the end of Section 14.4). A Euclidean need not share this denial, because even though such circles prevent a reductive analysis, they need not discredit the causal axioms as a means of conceptual analysis.

Indeed, this is a fifth way to go. We might look for conceptual grounds within those circles, i.e., for a conceptual web connecting causal dependence with other notions and containing the SGS conditions as integral part. While Hausman, Woodward (1999) put forward their modularity condition more as an empirical hypothesis that might be violated, Woodward (2003) is clearly engaged in conceptual investigations in this sense. In this case, the basic assumptions and all their consequences are claimed to be meaning postulates. This is fine with me. The basis of such claims ultimately consists in conceptual intuitions. I have no better basis, even though I am proposing explicit definitions. The only point is that I see no need to grant conceptual circles. Therefore I claim to be able to say what causation *is* – certainly something to be preferred, *if* feasible.

Though Glymour (2004) happily welcomes Woodward (2003) into the Euclidean camp, he approves less of the clarification of conceptual circles. On p. 790, he suggests that direct causal dependence is the only primitive notion and that interventions should be definable by it, along the route of Woodward (2003). The notion of intervention is not part of my formal theory. However, if Glymour's suggestion holds and I could accordingly introduce that notion, I am sure that the postulates Woodward (2003) endorses would turn out to be theorems of my account as well.

The sixth strategy for further grounding the SGS axioms is perhaps the most natural one. These conditions seem to appeal to an independently or antecedently understood notion of causal dependence. So, try to state that notion! If you do it well, you should be able to say whether or under which conditions it generates those axioms. The main idea – I am not aware of any alternative having been pursued – is, of course, to appeal to a notion of deterministic causation and to work out its probabilistic consequences. Papineau (1985) was perhaps the first to explicitly realize this strategy, although Railton's (1978) attempt at probabilistic explanation and Lewis' attempt to house chancy causation within the counterfactual approach in his (1986b, pp. 175ff.) might be subsumed here as well. Nothing like the causal Markov condition, though, was forthcoming from these attempts. However, the SGS argument about deterministic and pseudo-indeterministic systems sketched above might and should be understood precisely in this vein. It appeals to an antecedently given notion of deterministic causation and guarantees the SGS conditions for deterministic systems, or systems thus extendable.

My objection is obvious: this appeal is illusory. This chapter *is* about deterministic causation, and its point is that we run into exactly the same issues on the deterministic side. Of course, my comparative discussion of other accounts of deterministic causation was so far insufficient. However, if there is any truth in my ranking-theoretic perspective, then we have to take a stance towards the status of the SGS axioms and similar conditions in the deterministic context as well, and this time we cannot respond by appealing to yet another notion of causation. So, I think there is simply no convincing realization of the sixth strategy.

The seventh and final way is the one I want to propose. It is almost the same as the previous way; the only difference is that I do not claim to proceed from an independent notion of causal dependence; rather, my slogan is the title of my paper (2001b): Bayesian nets are all there is to causal dependence – or essentially so. What does this mean? My position seems unclear so far, or even inconsistent. After putting quantum-theoretic puzzles to one side, I said I would argue that the causal Markov condition is true by definition, yet I have agreed to SGS' claim that it can be violated in case of causal insufficiency. How can I do so? Let me more precisely explain what my analysis really is, at least with respect to causal dependence.

The crucial point is that Definitions 14.24 and 14.26 offer only a frame-relative notion of causal dependence. This seems alright insofar as *direct* causal dependence is concerned. Refinements of the conceptual frame might turn direct into indirect relations, and in a continuous refinement we might no longer find any direct causal dependence. However, according to (14.24) direct causal dependence is frame-relative not only due to possible omissions of mediating variables, but also because of incomplete representations of the circumstances, and this point carries over to causal dependence (14.26) as well. This seems to be an unacceptable consequence; there is no such frame-relativity in our ordinary notion of causal dependence, not even in a hidden way. In fact, none of the authors discussed above accepted such a far-reaching frame-relativity. Somehow, my analysis seems to import unwanted elements.

I agree that we have an absolute notion of causal dependence. What I want to argue, though, is that we can gain an understanding of that absolute notion only through the frame-relative one. Let me be brief on the critical "only"-part of this claim. Here, my general argument is implicit in my criticism of other conceptions of causation, as already contained in Sections 14.2–14.6. And as to such conceptions that are more specifically able to ground the SGS conditions, I expressed my doubts in the preceding discussion. So, let me turn to the constructive part and explain what understanding of the absolute notion I can offer.

Apparently, the only way to get from the one to the other notion is to say that absolute causal dependence is causal dependence relativized to the right conceptual frame, to the one that allows for the correct representation of the causal dependencies taken absolutely; this is trivial. Of course, absolute causal dependence must also be taken relative to something like a, or the, true ranking function; what this might mean is, however, an issue for the next chapter. Presently, we are dealing only with frame-relativity. So, what is the right frame? This must be the universal conceptual frame, it seems, comprising all variables whatsoever; no smaller frame is guaranteed to get all things right. Let us fix this in

Explanation 14.39: *Y causally depends* on *X* (*simpliciter*) iff *Y* causally depends on *X* within the universal frame (and relative to a true ranking function on the universal frame). (All the relative definitions and conditions (14.23–14.30) thereby acquire an absolute sense.)

I call this an explanation rather than a definition, since the universal frame is not a formal entity; we are transcending here the limits of formal theory (as we did already in 14.37). Human imagination is unbounded, and so is the realm of empirical concepts or variables; what the universal frame might be is not well defined. In friendlier terms, we might say that the universal frame is an ideal limit we are approaching by improving and completing our necessarily partial world picture. This implies that absolute causal dependence is an ideal limit notion, too (it is so for the further reason of referring to something like a true universal ranking function).

True as all this might be – I think it really is – it is more evasive than illuminating. Can we say something more informative than that truth comes to the fore in the ideal limit? Does frame-relative causal dependence teach us anything about the absolute notion? Yes. Instead of appealing to the unmanageable universal frame, we might look at well-defined smaller and larger frames. Starting from a small frame, though, we cannot say much about an enlarged frame, because the probabilities or the ranks over the enlarged frame may be anything, as long as their relevant marginalization agrees with those over the small frame. However, the other way around the issue becomes meaningful. Given a large frame and the causal dependencies within it, relative to a given ranking function, we are able to say how these dependencies appear within a smaller frame.

The issue I am going to explore, hence, is the same as the one I sketched at the beginning of this section regarding the SGS theory. There, the issue was that of what surfaces from a larger causal graph and Bayesian net within a smaller frame of observed variables, and it was aggravated by the fact that probabilities alone determine DAGs and MAGs only up to Markov equivalence. Our task is fortunately much simpler, since time and probabilities or ranks determine a unique DAG on a large frame, and time and marginal probabilities or ranks again determine a unique DAG on a smaller frame. Hence, we do not have to deal with anything like the theory of maximal ancestral graphs. (This is why I did not explain, but only mentioned it.)

In the following, I will consider only reductions of a larger frame by a single node or variable; larger reductions can be generated by iterating single reductions. The question we have to answer therefore is: How does a causal graph $\langle U, \to \rangle$ change if a single node Z is deleted from U? The answer arises from

Definition 14.40: The causal graph $\langle U^*, \to^* \rangle$ is called the *single reduction* of a causal graph $\langle U, \to \rangle$ by the node Z iff:

(a) $U^* = U - \{Z\}$,

(b)   for all $X, Y \in U^*$, $X \to^* Y$ iff either $X \to Y$, or not $X \to Y$ and one of the following three conditions holds:

  (i) $X \to Z \to Y$ (call this the *IC-case*), or
  (ii) $X < Y$ and $X \leftarrow Z \to Y$ (call this the *CC-case*), or
  (iii) $X < Y$ and there is a variable $V < Y$ such that $X \to V \leftarrow Z \to Y$ (call this the *N-case*).

In other words, the reduced graph contains all the arrows of the unreduced graph not involving the deleted variable Z. And it contains an arrow $X \to^* Y$ where the unreduced graph contains none exactly when Y is rendered *indirectly causally dependent* on X by the deleted Z:

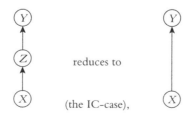

reduces to

(the IC-case),

or when the deleted Z is a common causal predecessor of X and Y:

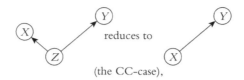

(the CC-case),

or when X is the *neighbor* of such a CC-case involving Y (so that an N-shaped subgraph results:

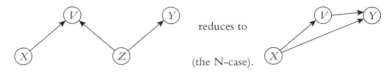

(the N-case).

The N-case is always accompanied by a CC-case. Note the importance of the temporal relation $V < Y$; if Y were to precede V, we would only have a CC-case involving Y and V, but not X.

The adequacy of this definition is shown by

Theorem 14.41: Let $\langle U, \rightarrow, \xi \rangle$ be a ranking net, let $\langle U^*, \rightarrow^* \rangle$ be the single reduction of $\langle U, \rightarrow \rangle$ by $Z \in U$, and let ξ^* be the marginalization or restriction of ξ to U^*. Then the causal graph agreeing with ξ^* is a (proper or improper) subgraph of $\langle U^*, \rightarrow^* \rangle$. If ξ is faithful to $\langle U, \rightarrow \rangle$, then it is $\langle U^*, \rightarrow^* \rangle$ itself that agrees with ξ^*.

Proof: Suppose that Y directly causally depends on X relative to ξ^*, i.e., that not $X \perp_\xi Y \mid \{< Y, \neq X, Z\}$. (Since ξ and ξ^* are identical within U^*, the same conditional independence statements within U^* hold relative to them.) Since ξ agrees with $\langle U, \rightarrow \rangle$, this means that X and Y are not d-separated by $\{< Y, \neq X, Z\}$ in $\langle U, \rightarrow \rangle$. This is so when $X \rightarrow Y$ or when the IC-, the CC-, or the N-case obtains. It is easily checked that, when not $X \rightarrow Y$, i.e., when each path between X and Y in $\langle U, \rightarrow \rangle$ is blocked by $\{< Y, \neq X\}$, and when Z has any other position relative to X and Y, then each path between X and Y in $\langle U, \rightarrow \rangle$ is blocked also by $\{< Y, \neq X, Z\}$. Thus, not $X \perp_\xi Y \mid \{< Y, \neq X, Z\}$ entails $X \rightarrow^* Y$. This proves the first assertion.

Now, suppose conversely that Y does not directly causally depend on X relative to ξ^*, i.e., that $X \perp_\xi Y \mid \{< Y, \neq X, Z\}$, and suppose further that ξ is faithful to $\langle U, \rightarrow \rangle$. Hence, X and Y must be d-separated by $\{< Y, \neq X, Z\}$ in $\langle U, \rightarrow \rangle$. Therefore, as we have just seen, neither $X \rightarrow Y$ nor one of the IC-, CC-, or N-cases can obtain; so $X \rightarrow^* Y$ does not hold. This proves the second assertion of the theorem.

In the case that ξ is not faithful to $\langle U, \rightarrow \rangle$ the theorem cannot be strengthened, because a lot of conditional independencies might then hold that are not foreseen by d-separation. Thus, d-separation might tell us $X \rightarrow^* Y$, even though $X \perp_\xi Y \mid \{< Y, \neq X, Z\}$ that excludes a direct causal dependence of Y on X with respect to ξ^*. If, however,

ξ is faithful to $\langle U, \to \rangle$, this situation cannot arise, and (14.41) provides a complete answer about the behavior of single reductions. Note, though, that even if ξ is faithful to $\langle U, \to \rangle$, ξ^* need not be faithful to $\langle U^*, \to^* \rangle$. Indeed, ξ^* cannot be faithful to $\langle U^*, \to^* \rangle$ if the N-case applies, since in that case (when only the four variables X, Y, Z, and V are involved, as depicted) we have $X \perp_\xi Y$, though X and Y are not d-separated by $\varnothing$ in the reduced graph.

Having thus seen how causal dependencies known to hold for a larger frame appear within a smaller frame, we should again reverse the perspective; we can now tell how our necessarily restricted causal view may develop through enlargement of the frame. According to (14.41), a direct causal dependence $X \to^* Y$ in the restricted frame may remain unchanged by a (single) refinement of the frame, or it may turn into an indirect causal dependence, or X and Y may turn out to be causally joined (in the sense of 14.32). A triangle $X \to^* V \to^* Y$ and $X \to^* Y$ may evolve into an N-shaped structure, but only if $X \perp_\xi Y \mid \{ < Y, \ne X, Z \}$. And even if there is no arrow between X and Y in the restricted causal graph, we might have one in the refined causal graph, simply because (14.41) only guarantees that the causal graph given by ξ^* is a subgraph of the single reduction of the graph apparently true of the large frame.

This is not very perspicuous. The matter becomes clear and simple, though, when we allow ourselves to assume that in the process of in(de)finitely refining our frame we are dealing only with ranking functions that are faithful to the causal graphs they induce on their respective frames; let us call this the assumption of *universal faithfulness*. I fear this is a strong assumption. It might be acceptable on the basis of the observation (7.32) that unfaithful probability distributions are rare, a point which presumably also applies to ranking functions. If the refinement also concerns the value ranges of the variables, unfaithfulness gets even more exceptional. However, we are now assuming the faithfulness of in(de)finitely many ranking functions; this is a very bold step.

Still, let us presuppose universal faithfulness for the sake of argument. Then we can first neglect the somewhat confusing N-case that we have seen to be realizable only by unfaithful ranking functions. And secondly, we need not reckon with a relative direct causal dependence between two variables within a smaller frame showing up only in a larger frame; the relative causal dependence must be seen from the beginning, i.e., even in the smallest frame consisting only of those two variables. Thus, through all frame extensions, we only have to deal with the IC- and the CC-case. Therefore, we are entitled to state

> *Assertion 14.42*: Assuming universal faithfulness, $X \to Y$ within any given frame means, in absolute terms, that Y causally depends on X *or* that X precedes Y and is causally joined with Y; and if $X \to Y$ does not obtain within the given frame, this means, in absolute terms, that Y does not causally depend on X.

Again, this is not a formal theorem, since it transcends formal limits; but I have made clear why we may assert it nevertheless. (14.42) resembles an absolute version of Reichenbach's common cause principle (14.33); it is, of course, of the same kin. One might wonder, therefore, why (14.33) did not presuppose (universal) faithfulness. This is so, first because (14.33) only told about how $X \to Y$, holding relative to the minimal frame $\{X, Y\}$, unfolds in extensions, and not about arrows in less than minimal frames, and secondly because the second, converse statement in (14.42) depends crucially on universal faithfulness.

So, when I defined relative direct causal dependence in (14.24) and theoretically developed it afterwards, this might have been misleading. What I was really talking about, in the almost automatically understood absolute sense, was causal dependence simpliciter (this was clear, anyway) *or* causal join simpliciter (this might have been the misleading part). My justification – not just my excuse – is that we could approach the absolute sense only by studying the frame-relative case in the previous section; the present clarification could only be given afterwards.

As mentioned above, the occasionalist might be pleased by this observation; what he assumes to be true is thus acknowledged to be at least a possibility, namely that every apparent causal dependence is merely a causal join created by God, that God indeed is the universal, not only the first mover. A possibility it is, indeed, not to be ruled out on the basis of causal theorizing. Still, there is clearly no reason to assume any such causal structure.

I close the circle of my considerations by returning to the SGS theory. In retrospect it should be clear why I have called (14.37) only an explanation of causal sufficiency. In the relative sense, each frame is trivially causally sufficient. The notion is substantial only in the absolute sense that transcends formal theory. However, on the basis of (14.42) we can immediately proceed to

Assertion 14.43: Let us still assume universal faithfulness, and let U be a causally sufficient frame (in the absolute sense of 14.37). Then, if $X \to Y$ within U, Y causally depends on X (in the absolute sense), and if not $X \to Y$ within U, Y does not causally depend on X.

This is so, simply because causal sufficiency rules out the alternative of a causal join that is still allowed in (14.42).

(14.43) hence justifies the basic assumptions of the SGS theory. In causally sufficient systems the arrows and the non-arrows do mean what they are supposed to mean, and hence the causal Markov and the causal minimality condition do not only hold in the relative sense in which they are derivable from my explications, but also in the absolute sense in which they are intended. In other words, I entirely endorse the SGS theory. My suggestion, though, is that it can be successfully supplemented by what they call the Socratic method, but which they shy away from. Analysis does add to axiomatic theory.

I am not entirely satisfied by my preceding argument, on two scores. First, I do not like to so heavily depend on universal faithfulness. It is not clear that it is ineliminable from the argument. It is very clear, though, that the argument would have to take a much more complicated form without this assumption; maybe weaker assumptions are required in any case. Second, I have given the argument only on the level of causal dependence between variables. Obviously, however, it should be carried out on the level of causation between facts. A minor reason for not having done so is that we do not yet know what (direct *or* indirect) causation is; recall that our discussion so far was based on the provisional acceptance of Definition 14.26 of causal dependence. The major reason is that I would not know how to transfer the argument to the more basic level of causation between facts (even assuming the explication I am going to put forward in Section 14.12); the formal details simply get very complicated on that level.

Still, I think that my argument is illuminating, as far as it goes. It redeems my promise above that I would provide a grasp of the intended absolute notion of causal dependence via the frame-relative versions. And it confirms my slogan: Bayesian nets, or, for that matter, ranking nets, are all there is to causal dependence (besides space and time, of course).

14.10 Interactive Forks, Simultaneous Causal Dependence, and Chain Graphs

So far, I have argued that all the principles of Section 14.8 are analytically true in the relative sense and, assuming universal faithfulness, even in the absolute sense. Now I want to turn to the concerns about whether they are true at all; as our discussion in Section 14.8 made clear, these concerns are mainly directed against the common cause principle (14.33) and (14.34). This seems to be the wrong order. How can I start discussing the truth of those principles after allegedly having established their analyticity? Things are even worse; when, in the end, I have saved the common cause principle, one might say that I have done so only by changing its meaning. This sounds confusing. Recall, though, that Sections 14.8–14.9 were governed by the assumption that there are no simultaneous variables. Under this assumption I stand by all of my previous claims. Without this assumption things change, but not too much, and in a way more easily explainable subsequent to the previous discussion of the simultaneity-free case. This is why I chose the present order of discussion.

Let us not jump ahead, though; let us first try to understand the concerns about the common cause principle and their not so obvious relation to the issue of simultaneous causation. These concerns were first voiced and always maintained by Wesley Salmon (cf. his 1978, 1980, pp. 65f., or 1984, Ch. 6), and they were vigorously taken up by Nancy Cartwright in various papers, many of which are collected in her (2007). Salmon, Cartwright, and their supporters have remained a minority, though; they mainly faced incredulity and attempts to refute their examples and arguments. Still, they were unimpressed. As always in such a case, the hunch is that the debate is plagued by hidden misunderstandings, and that, properly sorted out, both sides are right in some way. This is what I want to argue in the sequel. (The insight is new for me, too; until recently, I fully sided with the majority.)

What was Salmon's concern? He claimed, in his terminology, that there are two kinds of causal forks in which an event, a common cause, has two distinguishable effects: *conjunctive forks*, in which the two effects are rendered independent by their common cause, and *interactive forks*, in which the two effects remain connected or correlated even given the common cause. Clearly, the former conform to the common cause principle and the latter contradict it. An important motive for Salmon was, of course, the quantum theoretic EPR paradox; there we are confronted with real interactive forks. However, the attitude I took at the beginning of Section 14.9 is widely held: let us wait and see how the quantum-theoretic puzzles resolve in their entirety! Therefore, Salmon did not merely rely on the EPR case. On the contrary, he was convinced that we face interactive forks everywhere. In particular, whenever joint effects are governed by a conservation law – this is a common situation in classical physics – they and their common cause form an interactive fork; the effects remain correlated through the conservation law. Similarly, Cartwright urges us to realize that interactive forks are not exotic, but perfectly ordinary phenomena, which causal theorizing must account for.

Salmon's (1980, p. 65) main example is very graphic; it is taken from deterministic particle mechanics, i.e., billiards. Suppose the cue ball hits two other balls at once, thus (partially) transferring its momentum to them. The transfer is governed by the conservation of total momentum, and therefore the movements of the two balls are correlated; even given the push by the cue ball, if the one ball were to move differently, the other would have to move differently, too. Cartwright (1993, pp. 115ff.) presents an everyday story about two factories producing a chemical and, as a by-product, a pollutant, and explains that the example might very well constitute an interactive fork.

These examples were not found to be very convincing; rather, they have been taken to be critical instances of the two methods we have encountered several times, fine-graining of events and fine-graining of causal chains. Spirtes et al. (1993, pp. 62f.) apply the first method to Salmon's cue ball example and find the common cause to be incompletely specified. As soon as the position and the momentum of the cue ball when it hits the other balls are precisely specified, the further paths of the two balls are independently determined, and the correlation vanishes. Hausman, Woodward (1999, pp. 560ff.) apply both methods to Cartwright's example; if the mechanisms in those factories were described in more detail, the persistence of the interactive fork would be an utter mystery.

I agree with the critics; the specific examples indeed do not survive fine-graining (with the exception of the EPR paradox, where Bell's inequality proves that there can be no adequate fine-graining). So, Hausman, Woodward (1999, p. 569) seem right when they conclude that their considerations at least place the onus on Cartwright to explain why one should take the possibility of interactive forks seriously. On the other hand, they seem to sense the futility of their considerations when admitting in the same paragraph that "they are hardly more than a re-assertion of the naturalness of the Markov condition and do not constitute an independent argument".

I share this feeling as well. In Section 14.5 I already expressed my doubts about simply relying on the method of fine-graining the description of examples. The moral applies here, too. I think Cartwright would flatly deny that the onus is on her; her message, which is rather obscured by the diversion to examples, is simply that there is *absolutely no reason* to assume that causal forks are always conjunctive and that the joint effects are independently distributed. The transition probabilities from parents to children in a causal graph may take *any form whatsoever*, either in our representing minds or as set by nature itself; the onus is on those who want to prescribe a certain form. This is her challenge. And once one grasps it, it seems there is no answer to it.

So, let us see how the matter presents itself from our frame-relative point of view. To simplify, let us look at three variables X, Y, and Z such that $X < Y < Z$, still excluding simultaneities for a while. In our account we had no restrictions at all on ranks or probabilities, since they do not have to conform to a given causal graph; they rather determine the graph. Now, if X, Y, and Z are to form an interactive fork, we must have not $Y \perp_\xi X$ so that Y causally depends on X, not $Z \perp_\xi X \mid Y$ so that Z causally depends on X, and not $Y \perp_\xi Z \mid X$; otherwise, the fork would be conjunctive. So, we end up with the following causal graph:

(14.44a)

Interactive Fork?

(14.44a) says that Z causally depends on Y, too, in our frame-relative sense. This does not sound entirely wrong; after all, according to (14.42) frame-relative causal dependence means causal dependence *or* causal join, and Y and Z *are* causally joined. However, if Z does not causally depend on Y, (14.42) says that there is a further common causal predecessor of Y and Z, and this is clearly inadequate; all common causal predecessors, namely X, are already in the picture. Within the restricted frame $\{X, Y, Z\}$ there is nothing more to say, and the interactive fork is, so far, indistinguishable from a situation where the three arrows really stand for causal dependence.

However, the distinction comes to the fore within a refinement of the causal graph (14.44a). If the process is to develop from X to Y on the one prong of the fork and from X to the later Z on the other prong, it should pass on the latter through a mediating stage Z'. Z' helps us to the desired distinction. If Z were to causally depend on Y, taking account of Z' would not eliminate the ranking dependence between Y and Z. However, if X, Y, and Z form an interactive fork, we should expect that $Z \perp_\xi \{X, Y\} \mid Z'$; the causal influence only runs from X to Y and from X through Z' to Z, and, given Z', the move to Z is independent from the past of $Z' (= X)$ and from what happens besides $(= Y)$. Thus, even the causal Markov condition is reestablished w.r.t. Z; it is screened off by its parent Z' from all its non-descendants X and Y. This is the argument against interactive forks from fine-graining causal chains.

However, it is clear, and it was clear to Hausman, Woodward (1999, p. 563), that this argument is so far of no avail. The interactive structure must reappear in $\{X, Y, Z, Z'\}$ in the form:

(14.44b)

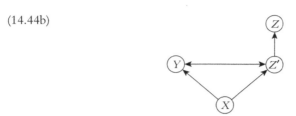

Interactive Fork Explained Away and Reemerging

Here, the direction of the arrow(s) between Y and Z' depends on the temporal relation of Y and Z'. The point gives rise to a certain dialectics between eliminating and re-establishing the interactive fork. The dialectics continues indefinitely, and it appears to be undecided which side wins. The conclusion, also drawn by Martel (2003), seems to be that the issue needs to be shifted into the context of continuous time where we can properly study the results of the infinite dialectics.

I have persistently refused this shift in this book, thinking that the finite case must first be entirely cleared up; the flight into infinity does not help much. So it is here. If Y is earlier than Z, it is natural to assume, even within finite time, that the process from X and Z has intermediate stages and indeed a stage simultaneous with Y. This starts the dialectics as indicated, but it is clear where it comes to a halt: at the first time after X. There, the process from X to Y passes, say, Y_1 and that from X to Z passes, say, Z_1, so that we have the following minimal constellation:

(14.44c)

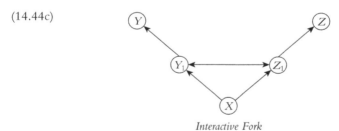

Interactive Fork

(14.44c) cannot be further reduced or refined simply because there is no time represented between X and Y_1, Z_1. In that minimal constellation, we again find that X, Y_1, and Z_1 form an interactive fork, and we have no further way of explaining it away. Hence, any finite dialectics is won by the proponent of interactive forks.

Note that we now have a genuine double arrow between Y_1 and Z_1, something not yet explained and still lacking a graph-theoretic account. So far, it is merely an expression of a causal dependence between Y_1 and Z_1 that can be represented only as simultaneous causal interaction and cannot be given a direction because temporal asymmetry is our only clue to causal asymmetry. This situation is the consequence of a symmetric treatment of the two prongs of an interactive fork, which, of course, is more natural than an asymmetric treatment as presented in (14.44a).

One might complain that the minimal constellation still does not adequately represent what an interactive fork is intended to be; intuitively, such a fork does not contain a piece of simultaneous causal interaction. However, we can apparently not improve upon this representation within any given finite frame. This is how the two topics, interactive forks and simultaneous interaction, get mixed up.

In order to make progress we should first study what simultaneous interaction might mean within the account presented here. That is, we should finally allow simultaneous variables and pursue the consequences for our causal theory. They are neither too trivial nor too grave, and take a while to explain, but they can be understood as an illuminating variation and extension of the theory developed in Sections 14.7–14.8. Only after this development will I return to the issue of how our understanding of interactive forks is thereby advanced.

The first and most important point to note is that our basic explications, (14.3) and (14.24), already allowed for simultaneous variables and thus for simultaneous causation and causal dependence. The observations in Sections 14.6–14.7 applied to that case, too; it was only the developments in Sections 14.8–14.9 that were restricted to the simultaneity-free case. Thus, given that we trust in our basic explications, we just have to develop the general theory and to check which modifications are required in order for the results obtained so far carry over.

Under a certain assumption, they fully carry over. Let us define:

Definition 14.45: Let U be a finite frame that is temporally weakly ordered by $\leq$, and ξ a ranking function for U. Then U is *phase-separated* by ξ iff for all $X \in U$, $X \perp_\xi$ $\{\approx X\} - \{X\} \mid \{\leq X\}$.

In other words, in a phase-separated frame, all simultaneous variables are rendered jointly independent by their past. In view of weak union and intersection, the graphoid properties (7.8d + f), phase-separation is equivalent to causal independence of simultaneous variables, i.e.:

(14.46) U is phase-separated by ξ if and only if for all $X \in U$ $\{\approx X\}$ is causally coordinate.

Hence, with phase-separation the graph induced by ξ does not contain any arrows between simultaneous variables and is still a DAG; so, all the results of Sections 14.8–14.9, in particular the common cause principle, continue to hold.

One might think that simultaneous variables must be causally coordinate, at least if one assumes that there is no genuine simultaneous causation. However, causal coordination as well as simultaneous causation is to be understood in a frame-relative sense. This point will enable us to make sense of the idea of giving up phase-separability, and to maintain the common cause principle even then.

In my (1994a, pp. 224ff.), I derived Reichenbach's common cause principle from phase-separability (that I called condition *L*, for "locality") for the general case (the simultaneity-free case is trivially phase-separable). I there claimed that phase-separability by itself is not a causal condition. Now I think this was euphemistic; the conditions are too close, as (14.46) shows. Still, it is a good question whether phase-separability is somehow more easily accepted than the common cause principle; at least it looks very perspicuous and natural.

In my view, its acceptance is connected with the issue of what is to be treated as *one* variable rather than two. Here, I essentially left the issue to the model-builder and made use of the fact that, formally, there is not much of a difference between a single variable and a set of (simultaneous) variables. Indeed, a general answer is presumably hopeless. In my (1994a, p. 224) I suggested that phase-separability might be taken as a defining characteristic of single variables; this would be a good reason to accept it (and hence the common cause principle in the general, simultaneity-allowing case as well). From their interventionist point of view Hausman, Woodward (1999, pp. 565f.) take a similar attitude; this is their modularity condition. A single variable should be manipulable all by itself; a process that leads through a realization of a single variable should be "independently disruptable" there.

On the other hand, what constitutes a variable in a given scientific field depends on the entire conceptualization of that field. Multifarious considerations play a role there, and naturalness of conceptualization hardly follows a single criterion. In particular, if applicable, we have strong spatial intuitions concerning variables. Their realization should take place not only at a relatively specific time, but also at a relatively well-defined location; a spatially scattered or widely extended variable would be odd. However, this intuition might be at variance with the intuition of independent disruptability or of phase-separability; this is what quantum theory teaches.

Hence, we will do better to not enforce phase-separability by definition; we should study the issue of simultaneous variables without this assumption. As indicated, we will still be able to save the common cause principle and even to bring it into harmony with the existence of interactive forks.

So, let us allow now for (frame-relative) causal dependence between simultaneous variables. As already emphasized, this must always be symmetric causal dependence or causal interaction, since temporal asymmetry is our only means of establishing causal asymmetry in our framework. Thus, if the arrows in a causal graph are to preserve their interpretation as frame-relative direct causal dependence, we now may have double arrows between simultaneous variables, i.e., edges with two arrow-heads. This results in a kind of graph we have not yet studied. What might the effects on causal theorizing be?

Fortunately, all is well. The resulting graphs have a name, *chain graphs*, and formally they are now as well understood as the simpler DAGs dealt with so far; they are quite easy to grasp on the basis of DAGs. Let me explain their theory as far as is needed here.

We are now considering graphs with two types of edges, *hybrid graphs*, for short. We can best describe them by a finite set U of nodes and two irreflexive relations $\rightarrow$ and $\leftrightarrow$ in U, *directed* and *undirected* edges, such that $X \rightarrow Y$ precludes $Y \rightarrow X$ as well as $X \leftrightarrow Y$. A hybrid graph with $\rightarrow = \varnothing$, i.e., no directed edges, is an *undirected graph*; a hybrid graph with $\leftrightarrow = \varnothing$, i.e., no undirected edges, is a *directed graph* (this is the kind we had considered so far).

Moreover, we say that a hybrid graph $\langle U, \rightarrow, \leftrightarrow \rangle$ is *consonant* with a given weak temporal order $\leq$ on U iff, for all $X, Y \in U$, $X \rightarrow Y$ implies $X < Y$ and $X \leftrightarrow Y$ implies $X \approx Y$.

A *route* in a hybrid graph $\langle U, \rightarrow, \leftrightarrow \rangle$ is a sequence $X_1, \ldots X_n$ of nodes in U such that for all $i = 1, \ldots, n - 1$ either $X_i \rightarrow X_{i+1}$ or $X_i \leftarrow X_{i+1}$ or $X_i \leftrightarrow X_{i+1}$. Note that a node may recur in a route. If this does not happen, i.e., if $X_1, \ldots, X_n$ are distinct nodes, then we call the route $\langle X_1, \ldots, X_n \rangle$ a *path* (a notion we already encountered in 7.20e). A route or a path is *undirected* iff all its edges are undirected. A set $V \subseteq U$ of nodes is *symmetrically connected* in $\langle U, \rightarrow, \leftrightarrow \rangle$ iff there is an undirected path from each $X \in V$ to each $Y \in V$. The maximal symmetrically connected sets (having no symmetrically connected proper supersets) are called the *connectivity components* of $\langle U, \rightarrow, \leftrightarrow \rangle$. Clearly, the connectivity components form a partition of the set U of nodes. Finally, a path $\langle X_1, \ldots, X_n \rangle$ ($n \geq 4$) in $\langle U, \rightarrow, \leftrightarrow \rangle$ is a *directed cycle* iff $X_1 = X_n$, for all $i = 1, \ldots, n - 1, X_i \leftrightarrow X_{i+1}$ or $X_i \rightarrow X_{i+1}$, and for at least one $i = 1, \ldots, n - 1, X_i \rightarrow X_{i+1}$. Now the central concept is this:

Definition 14.47: A hybrid graph $\langle U, \rightarrow, \leftrightarrow \rangle$ is a *chain graph* iff it contains no directed cycles.

It is quite obvious that a hybrid graph is a chain graph if and only if its connectivity components can be ordered as a *chain*, i.e., such that $X \leftrightarrow Y$ only if X and Y belong to the same component and $X \rightarrow Y$ only if X is a member of a component preceding the component containing Y according to that order. This explains the terminology.

It should be clear why this concept is central, for we have

Theorem 14.48: Let U be a finite set of variables weakly temporally ordered by $\leq$ and ξ a ranking function on U. If we have $X \rightarrow Y$ iff Y directly causally depends on X and $X \leftrightarrow Y$ iff X and Y directly causally depend on each other relative to ξ according to (14.24), then $\langle U, \rightarrow, \leftrightarrow \rangle$ is a chain graph.

This is so simply because $\langle U, \rightarrow, \leftrightarrow \rangle$ is consonant with $\leq$ and the required order of the connectivity components of $\langle U, \rightarrow, \leftrightarrow \rangle$ is given by their temporal order and an arbitrary ordering of simultaneous components. Hence, for our general account of causal dependence, chain graphs are the relevant objects of study.

This point is left obscure in the pertinent literature, the reason again being the preoccupation with generic variables and the accompanying neglect of temporal order. The original point of introducing chain graphs was relatively clear. The history of representing conditional independence structures by abstract graph-theoretic models began, it appears, with Hammersley, Clifford (1971), Besag (1974), Speed (1979), and

Darroch et al. (1980). They studied so-called Markov fields represented by undirected graphs (which I have ignored in this book); and the interest and interpretation was first in spatial structures, i.e., in such things as distributions over lattices (in the physical sense), and was generalized later on. Directed acyclic graphs came under scrutiny soon after, in research resulting in eminent work like Pearl (1988) and Lauritzen, Spiegelhalter (1988); they strongly attracted interest, presumably because of their causal interpretability.

Soon, Lauritzen, Wermuth (1989), and Frydenberg (1990) had the idea of combining the two kinds of structures, and thus invented chain graphs. However, their interpretation, and in particular that of the undirected edges, remained contested. Typically, the undirected edges were drawn as lines, not as double arrows as I have, indicating non-causal (?) "associations", the nature of which is unclear; if it were known, arrowheads could perhaps properly be added. Formally, of course, their interpretation is fixed by the independence relations or Markov properties they are supposed to represent. Even these can take at least two forms, but as Richardson (1998) argues, none of them fits to the ignorance interpretation of undirected edges. More precisely, "we may not wish to put a directed edge between two directly associated variables X and Y", but only an undirected one, because "the association may have arisen due to presence of an unmeasured confounding variable, some artefact of the way that the sample was selected, or a feed-back relationship, or we may believe that the association is causal but not know whether X causes Y or vice versa" (Lauritzen, Richardson 2002, p. 326). Lauritzen, Richardson (2002) go on to argue that the first two reasons for using undirected edges are not captured by chain graphs, but, as indicated in the previous section, by maximal ancestral graphs, that the last reason does not fit, either, and that only the feed-back interpretation is appropriate in certain cases. Their summary is that "the apparent simplicity of chain graphs belies the subtlety of the conditional independence hypothesis that they represent" (p. 321).

These discussions of chain graphs are highly illuminating, but they are not directly relevant to us, since they are couched in a framework of generic variables. As soon as we move into a framework of temporally ordered singular variables, the interpretational difficulties evaporate, and (14.48) provides a perfectly clear interpretation of chain graphs.

With this interpretation in mind, let us proceed with the theory of chain graphs as far as necessary. We find all the required information in Studeny, Bouckaert (1998), who meticulously explain how to carry over all the results of Section 7.3 from DAGs to chain graphs.

First, (14.25) and (14.48) immediately suggest that chain graphs can be generated from basic causal statements, from a list of total or direct causes, just as DAGs were generated according to (7.24) and (7.25a).

Definition 14.49: Let U be a set of variables weakly temporally ordered by $\leq$. Then L is a *list of direct causes* (for U and $\leq$) iff $L \subseteq U \times U$ such that if $\langle X, Y \rangle \in L$, then

$Y \in \{\leq X, \neq X\}$, and if moreover $Y \in \{\approx X\}$, then $\langle Y, X \rangle \in L$; $\perp_L$ is to be the corresponding set of independence statements $\{X \perp Y \mid \{\leq X, \neq X, Y\} \mid Y \in \{\leq X, \neq X\}$ and *not* $\langle X, Y \rangle \in L\}$. Similarly, L' is a *list of total causes* (for U and $\leq$) iff L' is a set of pairs containing for each variable $X \in U$ exactly one pair $\langle X, Z \rangle$ with $Z \subseteq \{\leq X, \neq X\}$ such that, if $X \approx Y, \langle X, Z \rangle \in L', \langle Y, Z' \rangle \in L'$, and $Y \in Z$, then $X \in Z'$; for such a list $\perp_{L'}$ is to be the corresponding set of independence statements $\{X \perp \{\leq X, \neq X\} - Z \mid Z \mid \langle X, Z \rangle \in L'\}$.

Each list of total or, respectively, direct causes for U and $\leq$ determines exactly one chain graph, and vice versa; this is as trivial as it was in the case of (7.25a). Still, this abstract reformulation will prove useful after the next step.

This step is prepared by defining the slightly more complicated neighborhood relations and route features in a chain graph.

Definition 14.50: Let $\langle U, \rightarrow, \leftrightarrow \rangle$ be a chain graph. Then we say for $X, Y \in U$:

(a) X is a *parent* of Y or Y a *child* of X iff $X \rightarrow Y$; $pa(Y)$ is the set of parents of Y (as in 7.20a–b).

(b) X is a *neighbor* of Y iff $X \leftrightarrow Y$; $ne(Y)$ is the set of neighbors of Y.

(c) $bd(Y) = pa(Y) \cup ne(Y)$ is the *boundary* of Y.

(d)   A route $\langle X_1, \ldots, X_n \rangle$ in $\langle U, \rightarrow, \leftrightarrow \rangle$ is *weakly descending* iff for all $i = 1, \ldots, n - 1$ either $X_1 \rightarrow X_{i+1}$ or $X_1 \leftrightarrow X_{i+1}$.

(e) X is an *ancestor* of Y and Y a *descendant* of X iff there is a weakly descending route from X to Y; $an(Y)$ is the set of ancestors of Y and $nd(Y)$ the set of non-descendants of Y (similar to 7.20c–d).

(f) $\langle X_i, \ldots, X_j \rangle$ $(i \leq j)$ is a *section* of the route $\langle X_1, \ldots, X_n \rangle$ iff it is a maximal undirected subroute, i.e., for all $k = i, \ldots, j - 1, X_k \leftrightarrow X_{k+1}$, but neither $X_{i-1} \leftrightarrow X_i$ nor $X_j \leftrightarrow X_{j+1}$. It is *collider section* iff it is a section and $X_{i-1} \rightarrow X_i$ and $X_j \leftarrow X_{j+1}$; otherwise, it is a *non-collider section*. (Note that, e.g., $X \rightarrow Y \leftarrow X$ is a route, too, and that $\langle Y \rangle$ is a collider section of that route.)

With the help of these notions, Studeny, Bouckaert (1998, Sect. 4) define a graphical independence or separation criterion in analogy to d-separation (7.21); they call it c-separation. Note that a finite chain graph contains infinitely many routes due to arbitrary recurrences of nodes. A separation criterion referring to all routes thus appears difficult to compute. Of course, the infinity is spurious, and in order to prove this, Studeny and Bouckaert give a more complicated definition of c-separation. However, they prove on p. 1453 that it is equivalent to the following, conceptually more perspicuous

Definition 14.51: Let $\langle U, \rightarrow, \leftrightarrow \rangle$ be a chain graph and X, Y, and Z three subsets of U. A route from X to Y is *blocked* by Z iff it contains a collider section not intersecting with Z or a non-collider section intersecting with Z. X and Y are *c-separated* by Z, in symbols: $X \perp_c Y \mid Z$, iff all routes from X to Y are blocked by Z.

This is very similar to d-separation; the only difference is that instead of nodes of paths we now refer to sections of routes. Therefore, we need not consider the descendants of the colliders on a path, as (7.21) did, since for any descendant of a collider section of a route there is a larger route containing this descendant as a collider section (as indicated in 14.50f).

The notion of c-separation is exactly what we want. This is first established by

Theorem 14.52: Let L be a list of direct causes for U and $\leq$, L' be the corresponding list of total causes for U and $\leq$, and $\langle U, \rightarrow, \leftrightarrow \rangle$ be the unique chain graph generated by either list. Then, for all $X, Y, Z \in U, X \perp_c Y \mid Z$ if and only if the statement $X \perp Y \mid Z$ is in the graphoid closure of the set $\perp_L$ or the set $\perp_{L'}$ of independence statements.

This is just consequence 3.1 of Studeny, Bouckaert (1998, p. 1447) (which refers to a different separation criterion that is shown to be equivalent to c-separation in consequence 4.1 on p. 1453). This generalizes Theorem 7.25b–c to chain graphs. The effect is the same; it says that we can axiomatize c-separation, i.e., the relation $\perp_c$, by a list of direct or total causes and the graphoid properties as derivation rules. The latter are again at the heart of the matter. Note that, even for a list of total causes, we have to refer to graphoid closure and not only to semi-graphoid closure, as in (7.25b).

More importantly, c-separation is exactly what we want when it comes to relating a chain graph to probabilities and ranks. Again, we consider the triplet of conditions of increasing strength:

Definition 14.53: Let $\langle U, \rightarrow, \leftrightarrow \rangle$ be a chain graph consonant with $\leq$, P a strictly positive probability distribution P over U and ξ a negative ranking function for U.

(a) P and ξ satisfy the *causal Markov condition* w.r.t. $\langle U, \rightarrow, \leftrightarrow \rangle$ iff, respectively, for each $X \in U$, $X \perp_{P,\xi} \{\leq X, \neq X\} - bd(X) \mid bd(X)$.

(b) P and ξ satisfy the *causal minimality condition* w.r.t. $\langle U, \rightarrow, \leftrightarrow \rangle$, or they *agree with* $\langle U, \rightarrow, \leftrightarrow \rangle$, iff, respectively, for each $X \in U, bd(X)$ is the smallest set $Z \subseteq U$ such that $X \perp_{P,\xi} \{\leq X, \neq X\} - Z \mid Z$.

(c) P and ξ satisfy the *faithfulness condition* w.r.t. $\langle U, \rightarrow, \leftrightarrow \rangle$, or P and ξ are *faithful to* $\langle U, \rightarrow, \leftrightarrow \rangle$, iff, respectively, $\perp_P = \perp_c$ and $\perp_\xi = \perp_c$.

Hence, by definition, if $\langle U, \rightarrow, \leftrightarrow \rangle$ reflects direct causal dependence w.r.t. ξ (or P) according (14.48), ξ (and P) agrees with $\langle U, \rightarrow, \leftrightarrow \rangle$ in the sense of (14.53b). Because of this interpretation, I immediately qualified these conditions as causal.

Clearly, we may strengthen the causal Markov condition in the same way that (7.27) strengthened (7.26):

(14.54)   If P and ξ satisfy the causal Markov condition w.r.t. $\langle U, \rightarrow, \leftrightarrow \rangle$, then we have, respectively, for each $X \in U, X \perp_{P,\xi} nd(X) - bd(X) \mid bd(X)$.

The proof is the same as that of (7.27).

The main point, though, is that, once again, c-separation, or in view of (14.52), conditional independence as implied by a list of direct or total causes, is all that is probabilistically or ranking-theoretically implied:

Theorem 14.55(a): For each chain graph $\langle U, \rightarrow, \leftrightarrow \rangle$ there is a strictly positive probability measure P over U that is faithful to $\langle U, \rightarrow, \leftrightarrow \rangle$.

Conjecture 14.55(b): For each chain graph $\langle U, \rightarrow, \leftrightarrow \rangle$ the set of probability measures over U *not* faithful to $\langle U, \rightarrow, \leftrightarrow \rangle$ has Lebesgue measure 0 within the set of strictly positive measures agreeing with $\langle U, \rightarrow, \leftrightarrow \rangle$.

Conjecture 14.55(c): For each chain graph $\langle U, \rightarrow, \leftrightarrow \rangle$ there is a negative ranking function ξ for U that is faithful to $\langle U, \rightarrow, \leftrightarrow \rangle$.

(14.55a) is proved by an ingenious 17 page construction in Studeny, Bouckaert (1998, Sect. 5). I am not aware that (14.55b) has been proved in the literature; it would, however, be very surprising, if Theorem 7.32 were not to generalize to chain graphs. Likewise, I have cautiously called (14.55c) a conjecture, although the proof of Hunter (1991a) for the ranking-theoretic part of Theorem 7.31 should carry over to chain graphs along the lines of the construction of Studeny and Bouckaert.

These results (if they obtain) allow a preliminary conclusion: The dialectical situation concerning my account (14.24) of direct causal dependence is exactly the same for the general case including simultaneous variables as it has been for the restricted case excluding simultaneous variables; we only need to consider the boundary of nodes in a chain graph instead the set of parents of nodes in a DAG. In particular, both versions of the causal Markov conditions, w.r.t. chain graphs (14.53a) and (14.54), are entailed by Definition 14.24.

So, again, the question is whether this Markov condition is defensible. In a way, our argument remains unchanged. In Section 14.8 we approached the issue by breaking the complex Markov condition into two parts, the narrow Markov condition (14.31) and the strong common cause principle (14.34). This analysis continues to hold:

(14.56)   *Narrow Markov condition* w.r.t. a chain graph $\langle U, \rightarrow, \leftrightarrow \rangle$ and a negative ranking function ξ for U: for all $X \in U$, $X \perp_{\xi} an(x) - bd(x) \mid bd(x)$.

Again, (14.56) modifies (14.31) only by referring to the boundary of X instead of its parents.

As to the common cause principle, we need not even change the wording. All the graph-theoretic terminology introduced in (14.32) simply carries over to chain graphs. However, we must observe that ancestorship now has a wider (and indeed unexpectedly wide) meaning, since it runs along neighborhood, too, and not only along parenthood. For instance, in the graph (14.44c) Y_1 and Z_1 are both common causal predecessors of the causally coordinate Y and Z. Moreover, Y_1 and Z_1 themselves are not causally coordinate in the required frame-relative sense (even if one might think that they should be so in an absolute sense); hence, they do not constitute a counter-instance to the common cause principle. Therefore, the strong common cause principle (14.34), thus understood, remains as a consequence of the causal Markov condition (14.53a); the proof goes through with c-separation just as well as with d-separation.

Finally the decomposition is complete, as before:

Theorem 14.57: The causal Markov condition (14.53a) is equivalent to the narrow Markov condition (14.56) and the strong common cause principle (14.34a).

The proof is again identical to the one of (14.36).

Still, something has changed. The common cause principle is now fully compatible with the existence of interactive forks, at least in our frame-relative sense. This is best seen with Reichenbach's version (14.33), which stated that any two correlated variables are either causally joined or one is a causal predecessor of the other. The change is that the latter clause must now be understood as including the case where the two variables are mutual causal predecessors of each other. Moreover, the existence of interactive forks does not affect the independence of the "private" causal predecessors of two causally joined variables, given their common causal predecessors. The point is only that, if the causal join originates in an interactive fork instead of a conjunctive one, the common causal predecessors do not merely consist of the origin of the fork; if they did, indeed, the common cause principle would not hold. Rather, the common causal predecessors are enriched to form a triangle consisting of the origin and the initial parts of the prongs still connected by a double arrow, as depicted in (14.44c); from this triangle the "private" parts of the prongs diverge. With the common causal predecessors thus enriched, I think, even Nancy Cartwright would approve of the common cause principle.

All these observations apply to the frame-relative sense (14.24) of causal dependence. This is why they might not appear to be fully adequate. Above, we already noticed that there is no real interaction involved in interactive forks; indeed, one might well deny the reality of simultaneous mutual causal dependence. Similarly, one might say that the enriched common ancestorships are an artifact of our representation, introduced in order to save the common cause principle. In reality, one prong of an interactive fork does not have causal predecessors on the other prong.

In order to address these worries we must make the same move as in the previous section; we must find out what the frame-relative sense of simultaneous mutual causal dependence teaches us about interaction and interactive forks in an absolute sense. The procedure is the same; we have to study coarsenings or reductions in order to learn how our reduced perspective might be refined or extended.

We can even keep the classification of cases; it is only that they further multiply, depending on whether or not the variables considered are simultaneous:

Definition 14.58: The chain graph $\langle U^*, \rightarrow^*, \leftrightarrow^* \rangle$ is the *single reduction* of the chain graph $\langle U, \rightarrow, \leftrightarrow \rangle$ by the node $Z \in U$ iff the following holds:

(a) $U^* = U - \{Z\}$,

(b)   if $X, Y \in U^*$ and $X \rightarrow Y$ or $X \leftrightarrow Y$, then, respectively, $X \rightarrow^* Y$ or $X \leftrightarrow^* Y$,

(c)   if $X, Y \in U^*$ and neither $X \rightarrow Y$ nor $X \leftrightarrow Y$, then

    (i)   if either $X \rightarrow Z \rightarrow Y$ or $X \rightarrow Z \leftrightarrow Y$, or $X \leftrightarrow Z \rightarrow Y$, then $X \rightarrow^* Y$, and if $X \leftrightarrow Z \leftrightarrow Y$, then $X \leftrightarrow^* Y$ (*the IC-cases*),

    (ii)   if $X \leftarrow Z \rightarrow Y$, then $X \rightarrow^* Y$ for $X \prec Y$ and $X \leftrightarrow^* Y$ for $X \approx Y$ (*the CC-cases*),

    (iii)   if there is a variable $V \preceq Y$ such that $X \rightarrow V \leftarrow Z \rightarrow Y$, then $X \rightarrow^* Y$ (*the N-case*),

(d) $X \rightarrow^* Y$ and $X \leftrightarrow^* Y$ hold only in the cases described in (b) and (c).

Thereby, Theorem 14.41 carries over from DAGs to chain graphs:

Theorem 14.59: Let $\langle U, \rightarrow, \leftrightarrow \rangle$ be a chain graph agreeing with the ranking function ξ, let $\langle U^*, \rightarrow^*, \leftrightarrow^* \rangle$ be the single reduction of $\langle U, \rightarrow, \leftrightarrow \rangle$ by $Z \in U$, and let ξ^* be the marginalization or restriction of ξ to U^*. Then the chain graph agreeing with ξ^* is a subgraph of $\langle U^*, \rightarrow^*, \leftrightarrow^* \rangle$, and if ξ is faithful to $\langle U, \rightarrow, \leftrightarrow \rangle$, then $\langle U^*, \rightarrow^*, \leftrightarrow^* \rangle$ itself agrees with ξ^*.

Once more, the proof is that of (14.41), with d-separation replaced by c-separation.

As far as the interpretation of arrows in a chain graph is concerned, we can therefore simply stick to our conclusions of the previous section. Causal dependence and causal sufficiency in the absolute sense can be explained as in (14.39) and (14.37). And under the assumption of universal faithfulness, we can reaffirm (14.42), that $X \rightarrow Y$ within a given chain graph means that Y is later than X and either causally depends on X or is causally joined with X, where the latter possibility is excluded (14.43) if the chain graph is causally sufficient.

The only point left, and the crucial one within this section, is the interpretation of double arrows in a chain graph. Their frame-relative sense was clear. But what might they signify in an absolute sense? There are three possibilities.

A first possibility is that a given double arrow $X \leftrightarrow^* Y$ either persists in all refinements of the frame or unfolds into a symmetric connection according to the last of the IC-cases in (14.58c). In this case it signifies genuine interaction, i.e., mutual causal dependence in the absolute sense. This might be a possibility to reckon with. Or one might try to exclude it on a priori grounds. We might accept the irreflexivity and the transitivity of causal dependence and thus derive its asymmetry. Or we might assume that distinct variables are somehow spatially separated and that causal influence needs time to cross space, and conclude on this ground that genuine interactions in the absolute sense do not exist. I will leave this issue open.

A second possibility is that a double arrow $X \leftrightarrow^* Y$ simply vanishes in refinements of the frame. Then it indicates that X and Y are causally coordinate and joined through a conjunctive fork; this was the second CC-case in (14.58c). Thereby, the double arrow would be explained away as a mere correlation via the common cause principle, just as could happen to a single arrow.

There is, however, a third possibility that has been in the center of this section; a double arrow $X \leftrightarrow^* Y$ might also disappear in refinements of the frame, but reappear on a more proximate level, just as illustrated in the process (14.44a–c). If this process continues through all refinements, the double arrow $X \leftrightarrow^* Y$ can be understood as a trace of an interactive fork in the absolute sense.

Thus, the difference between a conjunctive and an interactive fork can be captured in the following way: The origin of a conjunctive fork is already a single node in the discrete case, and is preserved as such through all refinements. By contrast, the base of an interactive fork in the discrete case is a thick one, as it were, an interactive triangle that, as described, gets smaller and smaller through refinements and finally contracts to a single node, distinguishable from the origin of a conjunctive fork not in a graphic way, but only by the interactive distribution of ranks or probabilities.

In this way, we are forced neither to explain away the double arrow by a conjunctive fork nor to admit genuine interaction. Rather, the extension of our account to simultaneous variables and chain graphs in the manner indicated allows us to accommodate interactive forks without betraying the basic principles of the SGS theory, the causal Markov condition and its parts, the narrow Markov condition and the strong common cause principle. These principles or conditions may still be claimed to hold for all finite and presumably all discrete frames. However, if the foregoing considerations are correct, the common cause principle(s) need not be preserved under continuous refinements, which may result in conjunctive *or* in interactive forks (or even genuine interaction).

14.11 Indirect Causes: Conflicting Demands

Let us turn to the second task in the development of our basic explication (14.3) of direct causation. I have done my best to make it plausible and to corroborate it by its implications for direct causal dependence. However, it only dealt with a single causal step; now we urgently need to consider how the single connections expand to a universal causal nexus constituting, in Hume's famous phrase, the cement of the universe. We should at least have a notation for what we are searching for. So, in a continuation of the notation introduced in (14.3), $A \xrightarrow{\ ...+\ }_{w} B$ is to denote that A is a (direct or indirect) cause of B in the (small) world w. More notation is to follow; this, however, is the basic relation we seek to explicate.

If we assume that we know what direct causation is, as we do, the issue stands or falls with the issue of the transitivity of causation.

(14.60)   *Transitivity*: If $A \xrightarrow{\ ...+\ }_{w} B$ and $B \xrightarrow{\ ...+\ }_{w} C$, then $A \xrightarrow{\ ...+\ }_{w} C$.

Certainly, being a cause is being either a direct or an indirect cause or possibly both. What else? Let us further assume that indirect causation cannot extend farther than the transitive closure of direct causation; this is eminently reasonable as long as we keep within discrete time and avoid continuous time where direct causation makes no sense.

Given these assumptions, (14.60) immediately completes our task; causation must be the transitive closure of direct causation. The question, of course, is whether we should accept this very simple and elegant conclusion.

Many reasons speak against this conclusion. Indeed, the issue of the transitivity of causation is extremely bewildering. Historically, I am not aware of any explicit discussion of the issue; transitivity seems to have been taken for granted. Indeed, if we accept some kind of regularity account of causation and roughly reduce causation to nomic implication (given the circumstances), transitivity seems an obvious consequence. In the same vein, Lewis (1973b) outright assumed transitivity of causation as an axiom, after discovering that counterfactual dependence, which seemed sufficient, but not necessary for causation, is not transitive. However, at that time confusion was already beginning to grow. Suppes (1970, p. 58) stated as a theorem that, contrary to naïve intuition, probabilistic causation is not transitive. Many examples turned up in the probabilistic field (see, e.g., Hesslow 1976) that underscored Suppes' claim. However, the rejection of transitivity was not unanimous. In Spohn (1990) I argued that probabilistic causation *is* transitive, a conclusion I will endorse here as well. And the Bayesian net literature on causal dependence apparently took the transitivity of causal dependence for granted (explicitly, e.g., in Spirtes et al. 1993, pp. 44f.).

The theory of deterministic causation has belatedly taken a similar turn. Although initially deterministic theorists at least appeared to stand firm on transitivity, doubts were sown, e.g., by McDermott (1995) and Hall (2000), that led to the acceptance of non-transitivity, e.g., by Hitchcock (2001) and Halpern, Pearl (2005), or even to the conclusion drawn by Hall (2004) that there appear to be two notions of causation, one transitive (production) and one non-transitive (dependence). The ironic consequence was that at each time deterministic and probabilistic theorists were mainly in opposite camps on this issue. As I said, the issue is extremely bewildering. I believe, though, that a careful study of the probabilistic and the deterministic discussion would reveal a broad and thorough-going parallel between the types of arguments used.

I have reservations about the predominant methodology for approaching the topic, which is driven mainly by examples and intuitions. Hall (2004, pp. 227ff.) makes extensive critical remarks about that methodology, which I fully endorse. Even in his papers, though, I find the weight of theory underestimated. There is not only the clash of intuitions and principles within examples; there is even provable contradiction between principles. In Section 11.3 I argued for the indirect and multi-faceted relationship between intuitive examples and general principles of belief revision and contraction, with the result that the former are hardly ever conclusive for the latter; there is always a huge pragmatic and explanatory leeway. Within the theory of causation the situation is quite similar. For this reason this and the next section will focus on theoretical issues raised by transitivity (14.60). Only in Section 14.13 will I confront theory with some representative examples.

One may arrange the difficulties and incompatibilities one finds in such problem cases in various patterns. For instance, Suppes (1984, pp. 55–70) devotes a whole section to "conflicting intuitions" concerning causation, Salmon (1988) proposes a somewhat different schema of conflicts, and Hall (2004) proposes yet another. Such patterns are helpful – one finally sees a structure – and at the same time dangerous – one's perspective is thereby narrowed. The balance is delicate. The pattern I developed in Spohn (1990, Sect. 5) was a response to the situation I found then, but I think it is highly pertinent even to the more recent debate within the deterministic domain. So, it is still the most instructive pattern I can propose; it is also the one closest to our basic conceptualization (14.2–14.3).

The first strand of the pattern I am going to lay out consists in a purely structural view of causation. What would be a nice and plausible structure for causation to have? Clearly, transitivity provides a perfect answer. If we add irreflexivity – nothing is a cause of itself – we can infer asymmetry – there is no genuine interaction, as considered in the previous section. All this is very pleasing. Note that on this picture, causation is extremely far-reaching; the causes, say, of my existence and my doings spread through my past light cone back to the big bang. This is simply the truth, though. Of course, when it comes to the important, informative, or salient causes, we lose that structure; but we decided to put the pragmatics of causal talk to one side. We might also say that causal influence comes in degrees, an idea that can be made precise very well in a probabilistic context. Then, although the causal connection between adjacent members in a causal chain might be very strong, causal influence might fade through very long chains. However, it does not fade entirely. It may sink below any limit of measurability by many orders of magnitude; but it is not nil. Of course, we can set an arbitrary threshold below which we deny causal influence; but this again belongs to the pragmatics of causal talk.

So, the question is rather: Is there any alternative to transitivity from a purely structural point of view? In my (1990, pp. 133ff.) I turned this question into a tedious exercise, trying out various stronger notions of causal chains that need not be transitive. The result was disappointing: I could not find any convincing or stable position that denied transitivity.

This does not settle the issue, though; there is more to consider than structure. Let us start with the idea that indirect cause and effect must be connected by a causal chain. In fact, there might be a whole net spanning between cause and effect; but the minimum is a chain. So, we only need to say what a causal chain is. Here are three structural characterizations for further use:

Definition 14.61:

(a) $\langle A_1, \ldots, A_n \rangle$ is a *weak causal chain* in the (small) world w iff $A_1 \xrightarrow[w]{+} A_2 \xrightarrow[w]{+} \ldots \xrightarrow[w]{+} A_n$.

(b) $\langle A_1, \ldots, A_n \rangle$ is a *connected causal chain* in w iff it is a weak causal chain in w and if for all $i < j$, $A_i \xrightarrow[w]{+} A_j$.

(c) $\langle A_1, \ldots, A_n \rangle$ is a *strict causal chain* in w iff it is a connected causal chain in w and if for no $j \geq i + 2$, $A_i \xrightarrow[w]{+} A_j$.

Clearly, each connected causal chain can be shortened to a strict causal chain; thus, being connected by some connected causal chain is equivalent to being connected by some strict causal chain. Moreover, if we assume transitivity (14.60), each weak causal chain is a connected causal chain, so that (14.61) contains no distinctions. Without transitivity this implication does not hold, and we are free, and indeed required, to define $\xrightarrow[w]{\cdots+}$ in some other way.

So, what else might hold a causal chain together? Besides the structural consider-ations, an important intuition is that a causal chain is a Markov chain or, if this should sound too determinate, *Markovian* in some sense. When discussing causal dependence between variables this intuition was already crucial; now we want to get it down to the level of causation between facts. Thereby we transcend mere structure; this intuition refers essentially to an independence notion offered by some probability measure or, in our case, some ranking function. A first attempt to account for this idea is this:

Definition 14.62: Let $X_1, \ldots, X_n$ be variables in U and ξ a ranking function for U. Then $\langle X_1, \ldots, X_n \rangle$ is a *Markov chain* iff for all $i = 2, \ldots, n-1$, $X_{i+1} \perp_\xi \{X_1, \ldots, X_{i-1}\} \mid X_i$. And $\langle X_1, \ldots, X_n \rangle$ is a *causal Markov chain* in w iff it is a Markov chain and a weak causal chain in w – where $\langle X_1, \ldots, X_n \rangle$ is a *weak causal chain* in w iff $\langle {}^w[X_1], \ldots, {}^w[X_n] \rangle$ is.

The first part is established usage, and the second part attempts to voice a Markovian notion of a causal chain. It leads us to an alternative to transitivity (14.60).

(14.63)   *Causal Markov chain condition*: Let $A = {}^w[X]$ and $B = {}^w[Y]$. Then $A \xrightarrow[w]{\cdots+} B$ iff there exists $Z_1, \ldots, Z_n \in U$ ($n \geq 0$) such that $\langle X, Z_1, \ldots, Z_n, Y \rangle$ is a causal Markov chain in w.

A causal Markov chain $\langle X_1, \ldots, X_n \rangle$ is a connected causal chain, since any part $\langle X_i, \ldots, X_j \rangle$ is a causal Markov chain as well, and indeed a strict causal chain, since ${}^w[X_i]$ cannot be a direct cause of ${}^w[X_j]$ for $j \geq i + 2$. However, causal Markov chains fail to satisfy transitivity (14.60), since connecting two shorter Markov chains does not necessarily result in a larger Markov chain.

(14.63) tries to capture the Markovian idea that causal processes proceed stepwise; each stage determines the next stage independently of the previous stages. This idea seems to be able to amend and correct the purely structural point of view. However, does (14.63) adequately express this idea? I do not think so. If there is only one causal path from the indirect cause A to the indirect effect B, i.e., if there is only one weak causal chain from A to B, (14.63) might be adequate (it is not even then, as we will see immediately after 14.65). If, however, there is more than one such path, a situation we can and should not exclude at all, (14.63) cannot be satisfied. If two facts are connected by two different weak causal chains, we can expect neither of them to be a causal Markov chain, simply because the conditional independence required by the one is violated through the other. This point is familiar from the previous sections where we considered causal nets instead of chains on the level of variables for precisely this reason.

Should we therefore say that indirect cause and effect are connected by an entire causal net and not only a chain, and should we then characterize that net as somehow Markovian? We might try to do this. However, this strategy looks complicated; we should not yet give up our much simpler strategy of trying to find an adequate notion of a causal chain. Indeed, it seems clear how we can do better. Direct cause and effect were connected by positive relevance not absolutely, but within the given context, i.e., given the obtaining circumstances. Likewise, we cannot expect the Markovian conditional independence characteristic of indirect causation to hold absolutely; it must be embedded in the context of the past course of events, too. In my (1990, p. 136) this idea led me to the following notion:

> *Definition 14.64*: Let $X_1, \ldots, X_n \in U$, $w \in W$, and ξ a ranking function for U. Then $\langle X_1, \ldots, X_n \rangle$ is a *w-Markov chain* iff $X_1 \leq \ldots \leq X_n$ and for all $i = 2, \ldots, n - 1$, $X_{i+1} \perp_{\xi} \{X_1, \ldots, X_{i-1}\} \mid X_i, {}^w[\leq X_{i+1}, \neq X_1, \ldots, X_{i+1}]$, i.e., iff the conditional independence characteristic of a Markov chain holds only if the rest of the past (and presence) of X_{i+1} in w is additionally given. Moreover, $\langle X_1, \ldots, X_n \rangle$ is a *causal w-Markov chain* iff it is a *w*-Markov chain and a weak causal chain.

The idea is then to require indirect cause and effect to be connected by a causal *w*-Markov chain:

(14.65)   *Causal w-Markov chain condition*: Let $A = {}^w[X]$ and $B = {}^w[Y]$. Then $A \xrightarrow[w]{\cdots+} B$ iff there exists $Z_1, \ldots, Z_n \in U$ ($n \geq 0$) such that $\langle X, Z_1, \ldots, Z_n, Y \rangle$ is a causal *w*-Markov chain.

This condition helps with the problem noted for (14.63). If two (or more) weak causal chains lead from A to B, none of them could be expected to form a Markov chain in the sense of (14.62). However, each can be expected to form a *w*-Markov chain, i.e., to display the relevant conditional independencies when the other chains are kept fixed as they are in (14.64). The amendment is required even if there is only one weak causal chain from A to B and even if we were to consider, as envisaged, the whole causal net spanning between A and B; in both cases the embedding context needs to be taken into account as well.

(14.65) is the best way I have found to do justice to our Markovian intuition concerning causal chains. Again, though, it is clear that causal *w*-Markov chains are strict causal chains in the sense of (14.61c) and violate transitivity (14.60); two shorter causal *w*-Markov chains need not combine to form a larger one. Thus, structural and Markovian intuition still clash. How is the conflict to be resolved?

It is too early to say. We have not yet introduced the most important intuition in the field, the idea that a cause is somehow *positively relevant* to its effect. Indeed, this was the fundamental idea (14.1) from which we started, and which we then restricted to direct causation in order to solve the circularity problems presented by the obtaining circumstances. Clearly, though, the idea is intuitively not so restricted; we rely on it generally. That we should do so is also suggested by the observation that the direct/indirect distinction is frame-relative; what is a direct cause and thus positively relevant in a coarser conceptual frame is an indirect cause in a refined frame and should thus preserve its positive relevance.

In order to spell out this idea, we have to tackle the issue that we had avoided by introducing the direct/indirect distinction, namely of saying what the circumstances of an indirect causal relationship are. Note that it is precisely because of the Markovian intuition that the simple answer we found for direct causes no longer works; given the mediating links, which are also part of the history of the effect, we do not expect the indirect cause to be positively relevant. When now tackling this issue, we need not yet be concerned by the circularity problem raised by Cartwright (1979). We only seek to render precise the positive relevance idea; whether and how it might help in defining indirect causation is to be seen later.

The most plausible strategy for tackling the issue is already contained in my remarks above: an indirect cause should turn into a direct one and thus reveal its positive relevance to the effect after omitting the mediating causal chain(s). Since we are still undecided about what a causal chain really is, this strategy results in a host of possible specifications. Let me list all reasonable possibilities that we can state so far (cf. also Spohn 1990, p. 137):

(14.66)   *Positive relevance conditions*: Let $A = {}^w[X]$ and $B = {}^w[Y]$. Then $A \xrightarrow{\cdots+}{}_w B$ holds relative to the frame U iff $A \xrightarrow{+}{}_w B$ holds relative to the reduced frame $U - V$

   (a)   for some $V \subseteq \{\geq X, \leq Y\}$,
   (b)   for some $V = \{Z_1, \ldots, Z_n\}$ such that $\langle X, Z_1, \ldots, Z_n, Y \rangle$ is a weak or a connected or a strict causal chain or a causal w-Markov chain,
   (c)   for all $V = \{Z_1, \ldots, Z_n\}$ such that $\langle X, Z_1, \ldots, Z_n, Y \rangle$ is a weak or a connected or a strict causal or a causal w-Markov chain,
   (d)   for $V = \{Z \mid Z$ is a member of a weak or a connected or a strict causal chain or a causal w-Markov chain from X to $Y\}$,
   (e)   for $V = \{\geq X, \neq X, \leq Y, \neq Y\}$.

Even if we neglect the complications in (b), (c), and (d) due to our indecision concerning causal chains, these are five different specifications, all of which we find proposed in the literature. (14.66a) is the definition of an indirect cause given by Suppes (1970, p. 28). As a sufficient condition for $A \xrightarrow{\cdots+}{}_w B$, it is certainly too weak. However, as a necessary condition it seems to be the inalienable minimum of the positive relevance idea. Nevertheless, we will see below that it should be rejected.

(14.66e) is the proposal of Good (1961–1963) (if we neglect simultaneous variables), by which he hoped to explicate direct and indirect causation all at once. As I mentioned in Section 14.4, this may pass muster for direct causation, if direct causes immediately precede their effects. For indirect causation, however, it does not seem acceptable. Look at the following numerical example, where $A < B < C$ are the only three facts obtaining in the small world w and where τ is the two-sided ranking function corresponding to ξ:

(14.67a)

$\tau(B\mid\cdot)$		&	$\tau(C\mid\cdot)$	B	$\bar{B}$	$\Rightarrow$	$\tau(C\mid\cdot)$
A	1		A	1	-1	A	1
$\bar{A}$	-2		$\bar{A}$	-1	2	$\bar{A}$	2

The first table says that $A \xrightarrow[w]{ns+} B$, i.e., that A is a necessary and sufficient direct cause of B in w. The first row of the second table says that $B \xrightarrow[w]{ns+} C$, too. The first column of the second table is chosen such that even $A \xrightarrow[w]{ns+} C$. In other words, A is a necessary and sufficient direct cause of C, on the one hand, and linked to C by a chain of necessary and sufficient direct causes, on the other hand. Still, A might fail to be an indirect cause of C according to (14.66e). If we choose the fourth entry of the second table as shown, a little calculation yields the third table, according to which A is even negatively relevant to C. The same result might easily be produced in more familiar probabilistic terms.

(14.66d) is perhaps the best approximation to the position of Cartwright (1979); see there, in particular, clause (iv) of her condition *CC*. Maybe, though, she considered only the case where there is only one causal chain from the cause A to the effect B. In that case, (b), (c), and (d) of (14.66) obviously coincide. If there is more than one causal chain, (b), (c), and (d) might arbitrarily diverge (only (c) is guaranteed to be stronger than (b)), and it is not obvious what the choice should be.

This may suffice as a selection of instantiations of (14.66) from the probabilistic camp, where, to be sure, causal dependence between variables has attracted much more attention than our present concern, causation between facts. The efforts within the deterministic camp can also be understood as attempts to realize (14.66). Of course, none of the approaches found there possesses the notion of positive relevance, as provided by ranking theory and as required by (14.66). However, as remarked immediately after (14.1), they have notions that attempt to fill a corresponding role. (To be sure, my basic criticism is that all these substitutes are too poor, but this is not the present point.)

Within the counterfactual approach, for instance, the corresponding notion is counterfactual dependence, of course. It cuts across our distinction of direct and indirect causation. As I explained at the end of Section 14.4, the counterfactual approach agrees with Definition 14.3 of direct causation. Usually, however, indirect effects counterfactually depend on the indirect causes as well. And this relevance or dependence is best captured by (14.66d); at least it seems that precisely the states of the variables in V as specified in (14.66d) are not cotenable with the counterfactual assumption that the indirect cause does not occur. (Still, Lewis (1973b) thought that counterfactual dependence does not exhaust indirect causation because of the missing transitivity, whence he axiomatically added the latter.)

The assumption that counterfactual dependence is at least a sufficient condition for causation is taken by Hall (2004, p. 225) to be the undoubted hallmark of the counterfactual analysis. However, a variation of the above numerical example (14.67a) against (14.66e) works against this assumption as well:

(14.67b)

	$\tau(B \mid \cdot)$			$\tau(C \mid \cdot)$	B	$\bar{B}$			$\tau(C \mid \cdot)$
A	1	&	A	0	1	$\Rightarrow$	A	0
$\bar{A}$	−1		$\bar{A}$	1	−1		$\bar{A}$	−1

Here, the first column and the first row of the second table say that both A and B are necessary direct counter-causes of C. Moreover, the first table says that A is a necessary and sufficient direct cause of B. In other words, A is a direct as well as an indirect counter-cause of C (the latter insofar as A is a direct cause of a direct counter-cause of C; this is the most reasonable thing we could mean by "indirect counter-cause"). Hence, there is no way to call A a cause of C. Still, with the fourth entry of the second table as shown, the first two tables entail the third, and anyone having this τ would have to assert that if A had not occurred, C would not have occurred, either – in contradiction to the alleged hallmark of the counterfactual analysis.

The recent so-called structural equations or structural modeling approach to causation, as developed in Hausman, Woodward (1999), Pearl (2000), Hitchcock (2001), Woodward (2003), and Halpern, Pearl (2005a,b), is a more elaborate attempt at realizing (14.66) on the deterministic side. Structural equations are an invention of economics and social sciences and widely applied elsewhere, e.g., in evolutionary biology and population genetics. Usually, they are amended by so-called error terms that are conceived as random variables, and precisely for this reason a sophisticated statistical methodology towers over them. Since causal assumptions are built into structural equations, this methodology amounts to a testing of causal hypotheses, something deeply refined by Spirtes et al. (1993) and subsequent work. All this is fundamentally statistics. Thus, it was in a way a surprising, in another way a surprisingly late, discovery that the approach has a lot to teach about deterministic causation. In a way, the discovery is an irony; it seems that theories of deterministic causation gain more inspiration from probabilistic sources than from their own. (Recall that ranking theory, too, was born from the idea of copying theories of probabilistic causation.)

What, then, is the basic idea of the structural equations approach? Well, if the statistical error terms are omitted, what is left is a set of deterministic laws that is so arranged as to establish a causal order among the variables related by the laws. Indeed, structural equations generate causally interpretable DAGs, just as Bayesian nets do. Thus, they are meant to be causal laws describing generic causation. By starting from structural equations so interpreted, the approach does not aspire to a fundamental analysis of causation and is hence a child of the more recent Euclidean approach to causation (cf. Section 14.2). Still, given this starting point, the approach can specify a clear model of what we mean by an intervention; this leads to an interventionist account of causation, the project of Woodward (2003). At the same time, it delivers a clear model of non-backtracking counterfactuals as they are required for causal theorizing; this is elaborated by Pearl (2000, Ch. 7) and continued by Halpern, Pearl (2005a,b). Thus, it pulls the counterfactual approach from its logical abstractions down to scientific practice. Finally, presupposing generic causation, it tries to give an account of singular causation under the label "actual causation", something that is quite remote from its statistical origins. This is the topic we are interested in.

Take, e.g., Hitchcock (2001). He considers the DAG generated by a set of structural equations and a specific evolution of the system where each variable in the DAG takes a certain value. Now he asks whether X's taking the value x is a cause of some descendant Z's taking the value z. This, he suggests on p. 287, depends on whether there is what he calls an active causal route from X to Z. The latter (cf. p. 286) is a descending path, a causal chain, from X to Z such that the value of Z counterfactually depends on the value of X if all the variables on other descending paths from X to Z are fixed at their actual values. Thus, Hitchcock clearly proposes a variant of (14.66b).

This idea is also endorsed by Woodward (2003, p. 77). Earlier in the book, when he is not yet concerned with actual causation, Woodward distinguishes two notions. On p. 50 and p. 59, he defines a variable X to be a *contributing cause* of the variable Y if there is some chain of direct causal relationships from X to Y such that the value of X makes some difference to the value of Y, provided that the values of all variables not on this path are fixed (by intervention). Hence, we get from contributory causation to actual causation simply by replacing the existentially quantified reference to some values of the variables by a reference to their actual values. And on p. 51, he defines a variable X to be a *total cause* of a variable Y iff there is a possible intervention on X that makes some difference to Y, when none of the intermediate variables is fixed to a given value. Thus, contributory causes exemplify (14.66b) and total causes (14.66d).

We again find the same basic idea in Halpern, Pearl (2005a, p. 853), though in a considerably generalized form. And so I might go on. (For a more extensive discussion of the structural equations approach see Spohn (2010b). One criticism given there is that this approach provides too poor a notion of relevance. For instance, it provides no way for a fact to be negatively relevant to another fact and thus captures relevance *simpliciter* rather than positive and negative relevance.)

For the moment, though, we have sufficient evidence of how prominent positive relevance conditions (14.66) are in probabilistic as well as deterministic theorizing on causation. What is their relation to the other important ideas concerning indirect causation introduced in this section?

The fundamental observation is that all versions of (14.66) violate transitivity (14.60), at least if we proceed from Definition 14.3 of direct causation. A further variation of the above numerical examples suffices to show this, where $A < B < C$ are still the only three facts obtaining in the small world w:

(14.67c)

	$\tau(B \mid \cdot)$			$\tau(C \mid \cdot)$	B	$\bar{B}$			$\tau(C \mid \cdot)$
A	1	&	A	0	−1	$\Rightarrow$	A	0
$\bar{A}$	−1		$\bar{A}$	1	1		$\bar{A}$	1

Now, the first table says that $A \xrightarrow{\ ns+\ }_{w} B$, i.e., that A is a necessary and sufficient cause of B in w. The first row of the second table says that $B \xrightarrow{\ n+\ }_{w} C$. Hence, with transitivity, A should be an indirect cause of C in w. Still, the first column of the second table shows that A is a direct counter-cause of C; i.e., $\bar{A}$, if it had happened, would have been a sufficient direct cause of C. Moreover, we might choose the fourth entry of the second table such as to imply the third table, which shows that $\bar{A}$ is sufficient reason for C. Thus A is negatively relevant to C in any case, whether given B or not given B, and according to all versions of (14.66) we cannot have $A \xrightarrow{\ \cdots+\ }_{w} C$. Again, the same negative result may be obtained in more familiar probabilistic terms.

Of course, this conclusion has been clear all along; the conflict has been deeply disturbing for the theory of causation. The structural modeling approach is precisely an attempt at a constructive non-transitive theory of deterministic causation, in order to account for the apparent intuitive violations of transitivity. In view of the conflict, the hope is that the positive relevance conditions at least harmonize with the Markovian intuition. This hope is nourished by the following probabilistic and ranking-theoretic

Theorem 14.68: Let $X_1, \ldots, X_n$ be binary variables, $A_i = \{X_i = 1\}$, $\bar{A}_i = \{X_i = 0\}$ ($i = 1, \ldots, n$), let P be a probability distribution and ξ a ranking function for $\{X_1, \ldots, X_n\}$ and let $\langle X_1, \ldots, X_n \rangle$ be a Markov chain relative to P and ξ. Then:

(a) If $P(A_{i+1} \mid A_i) - P(A_{i+1} \mid \bar{A}_i) = p_i$ for $i = 1, \ldots, n - 1$, then $P(A_n \mid A_1) - P(A_n \mid \bar{A}_1)$ $= p_1 \cdot \ldots \cdot p_n$. Thus, in particular, if each A_i is positively relevant to A_{i+1}, A_1 is positively relevant to A_n, too.

(b) If $\xi(\bar{A}_{i+1} \mid A_i) = r_i > 0$ and $\xi(A_{i+1} \mid \bar{A}_i) = s_i > 0$ so that A_i is a necessary and sufficient reason for A_{i+1} (for $i = 1, \ldots, n - 1$), then $\xi(\bar{A}_n \mid A_1) = \min_{i<n} r_i$ and $\xi(A_n \mid \bar{A}_1) = \min_{i<n} s_i$, so that A_1 is a necessary and sufficient reason for A_n as well.

(a) is very familiar; for a proof see, e.g., Good (1980), and for more general probabilistic results see Eells, Sober (1983, pp. 49ff.). The ranking-theoretic counterpart is easy to prove; however, it is only so simple because of its restriction to necessary and sufficient reasons (we came across a similar point in (11.30)). (14.68) confirms our intuition that probabilistic causal influence gets weaker, but never entirely fades in larger causal chains and that deterministic influence through a causal chain is as weak as its weakest link.

However, (14.68) does not achieve the desired harmony. If additional variables are dispersed between $X_1, \ldots, X_n$, then the positive relevancies assumed in (14.68) need not indicate direct causal relationships according to (14.3). Moreover, we had concluded that we should rather focus on w-Markov chains and not on Markov chains. To my knowledge no corresponding theorem is available for w-Markov chains; I will try to provide one in the next section.

So, as I suggested earlier, we find deep theoretical conflicts. Transitivity (14.60) is incompatible with both the causal w-Markov condition (14.65) and the positive relevance condition (14.66), and the relation between the latter two is less clear than we might have hoped for. This was my conclusion in my (1990, p. 137), and even now

I find that these theoretical conflicts are at the root of most of the many confusing examples (some of which I will discuss in Section 14.13).

14.12 Causation and its Transitivity

What to do? We might look at all the problematic examples and might try to find an optimal balance of judgment. I am skeptical, though. I would very much prefer a theoretically principled resolution of the conflicts. What could this be? The point which I presented in my (1990, p. 138) (and to which I was helped by a long discussion with Karel Lambert) is very simple:

At first, one might think that transitivity (14.60) is a particularly strong structural assumption, which one is well-advised not to put at the start of the explicatory enterprise. However, it is in fact the weakest possible assumption about indirect causation. For what counts is conceptual strength, and the conceptually weakest notion of causation is that with the widest extension. And if we assume transitivity, we indeed arrive at the causal relation with the widest extension (as long as we stick to the fundamental claim that causation cannot extend further than single direct steps carry us). The crucial theoretical point then is that theory should start with the weakest notion of causation, in order to allow for the study of stronger notions on its foundation. If, by contrast, we were to start with any of the stronger notions, the weaker notions would be suppressed. This is a telling consideration. Therefore I proceed to my official

> *Definition 14.69*: Let X, $Y \in U$, A be an atomic X-proposition, B an atomic Y-proposition, $w \in W$, and ξ a negative ranking function over W. Then A is *cause* of B in w w.r.t. ξ – in symbols: $A \xrightarrow[w]{\cdots+} B$ – iff A stands to B in the transitive closure of $\xrightarrow[w]{+}$. A is a *counter-cause* of B in w w.r.t. ξ, i.e., $A \xrightarrow[w]{\cdots-} B$, iff $A \xrightarrow[w]{-} B$ or for some proposition C, $A \xrightarrow[w]{\cdots+} C \xrightarrow[w]{-} B$. A is *causally relevant* to B in w w.r.t. ξ, i.e. $A \xrightarrow[w]{\cdots\pm} B$, iff A stands to B in the transitive closure of $\xrightarrow[w]{\pm}$. Finally, A is *causally irrelevant* to B in w w.r.t. ξ, i.e., $A \xrightarrow[w]{\cdots 0} B$, iff not $A \xrightarrow[w]{\cdots\pm} B$.

The definition of counter-causation (already suggested after (14.67b)) is, I think, the most plausible one. If counter-causation is to be allowed at all, then the counter-causal influence seems to end with the realization of the counter-effect, and to not extend beyond. The issue is not so important, though. The other notions do not require further comments. I have not attempted to generalize the notions of an additional, sufficient, necessary, and insufficient direct cause to the indirect case, since these notions refer to specific forms of positive relevance and the spreading of positive relevance through (weak) causal chains was seen to be quite unclear.

We will see in the next section how well (14.69) stands the test of examples. In this section, I want at least to begin to show what the theoretical advantage of starting with the weakest possible notion might be. The point is that it now becomes a theoretical matter how to build stronger conceptions on the weak foundation, i.e., how to specify conditions under which (14.69) does justice to the Markovian and the positive relevance conditions. Eells, Sober (1983) took the opposite route. They started with a positive relevance notion of causation and then studied assumptions along the line of Theorem 14.68 under which that notion satisfies transitivity. In a way, this procedure leads to the same conclusions. The difference, however, is that the required assumptions then merely receive a formal statement, whereas my procedure allows them to be stated in causal terms; they can thus be understood as specific conditions of causal well-behavedness.

Let us see what this means, and let us address the easier task first: Under which conditions are causal chains Markov or rather w-Markov chains? Note that we no longer need to distinguish weak and connected causal chains, since they coincide according to (14.69); only strict causal chains are still stricter. A first answer is straightforward. We need

Definition 14.70: $\langle X_1, \ldots, X_n \rangle$ is a *chain of causal relevance* in w relative to ξ iff $^w[X_1]$ $\xrightarrow[w]{\pm}$ $^w[X_2]$ $\xrightarrow[w]{\pm}$ $\ldots$ $\xrightarrow[w]{\pm}$ $^w[X_n]$, and it is a *strict chain of causal relevance* in w iff moreover for no $i < j$, $^w[X_i]$ $\xrightarrow[w]{\pm}$ $^w[X_{j+1}]$.

Then a first result is provided by

Theorem 14.71: The following two assertions are equivalent:

(a)   For each $w \in W$ all chains $\langle X_1, \ldots, X_n \rangle$ of causal relevance in w are strict.
(b)   For each $w \in W$ all chains $\langle X_1, \ldots, X_n \rangle$ of causal relevance in w are w-Markov chains.

For a proof of the probabilistic counterpart see Spohn (1990, Theorem 11).

This is a perspicuous theorem. It says that, if causal relevance spreads strictly in stages in all worlds so that no worlds contain direct as well as indirect causal relations between any two atomic facts, then all these chains are w-Markovian and thus have the preferred Markov property. The only flaw in (14.71) is that the order of the quantifiers is not quite right; there is no reason to expect that all possible worlds are causally well-ordered in this way. Hence, it would be nicer to have a universal equivalence instead of an equivalence of universal statements. This requires an additional assumption; a suitable one that I have found is that what I have called the larger actually relevant circumstances $C_w(A, B)$ (cf. 14.18) are ideal, in the sense of (14.20) – i.e., equal to $C_w^*(A, B)$:

Theorem 14.72: Let $X_1, \ldots, X_n$ be binary variables and $\langle X_1, \ldots, X_n \rangle$ a chain of causal relevance in w. Assume that for all $i < n$, $C_w^*(^w[X_i], \, ^w[X_{i+1}]) = C_w(^w[X_i], \, ^w[X_{i+1}])$. Then $\langle X_1, \ldots, X_n \rangle$ is a w-Markov chain iff it is a strict chain of causal relevance.

For a proof of the probabilistic counterpart see again Spohn (1990, Theorem 12).

These theorems are not deep, but nice. And they are certainly plausible; intuitively, it is precisely the existence of direct bypasses to causal chains that violates the Markovian intuition.

The issue of the preservation of positive relevance within causal chains is much more complicated. In my (1990, pp. 139ff.), I carried out a somewhat tedious investigation with some success. Here I will only try to explain one main result.

Let us consider a sequence $\langle X_1, \ldots, X_n \rangle$ of binary variables and a world w in which all $A_i = \{X_i = 1\}$ $(i = 1, \ldots, n)$ are realized. If it forms a Markov chain, Theorem 14.68 applies, and all is fine. However, we have found that causal chains are instead w-Markov chains, and we have just described conditions under which they are. So let us assume that $\langle X_1, \ldots, X_n \rangle$ forms a (causal) w-Markov chain. Then, however, we get the problem that the characteristic conditional independencies refer to a different condition for each X_i. Thereby the applicability of (14.68) is compromised. This suggests regaining (14.68) by, as it were, equalizing the different conditions. This means searching for some set $V \subseteq U$ of variables such that for all $i = 1, \ldots, n$ the positive relevance of A_i to A_{i+1} as well as the characteristic Markov independence of X_{i+1} from $\{X_1, \ldots, X_{i-1}\}$ given X_i holds, not only under the condition $w[\leq X_{i+1}, \neq X_1, \ldots, X_{i+1}]$ that varies with i, but also under the constant condition $w[V]$. If we find such a set V, then we can apply (14.68) conditional on $w[V]$.

Which properties should the equalizing set V be expected to have? The first requirement is that $X_1, \ldots, X_{n-1} \notin V$; $w[V]$ is to be conceived as providing the circumstances of the indirect causal relation of A_1 to A_n via $A_2, \ldots, A_{n-1}$, and the mediating links must not belong to those circumstances. Secondly, it seems that $w[V]$ has to preserve the circumstances of each link $A_i \xrightarrow[w]{+} A_{i+1}$ $(i = 1, \ldots, n-1)$ in the causal chain (in some suitable sense of "circumstances", as amply provided by Section 14.6); if we do not preserve those circumstances, it is unpredictable what will happen to the conditional positive relevance of A_i to A_{i+1}. Thirdly, however, we have to omit from $w[V]$ not only direct information about A_{i+1} $(i = 1, \ldots, n-1)$ – this is achieved by the first requirement – but also any indirect information about A_{i+1} that exceeds the information provided by its circumstances; such indirect information may again confound the positive relevance of A_i to A_{i+1}.

The problem now is that the second and the third requirement work against each other. This is clear in abstract; the second requirement is about what is to be included in $w[V]$, the third about what is to be excluded. More specifically, by including the circumstances of each link $A_i \xrightarrow[w]{+} A_{i+1}$ we make sure that there is no further confounding information about the past of A_{i+1}. However, information about the future of A_{i+1} might be confounding as well, and the circumstances of some later link $A_j \xrightarrow[w]{+} A_{j+1}$ $(j > i)$, which we have to include in $w[V]$, too, might be such future information. This conflict might be unsolvable. So, what we have to do is to find conditions under which the circumstances of later links do not provide inadmissible information about earlier links. If we succeed, we know that under those conditions a suitable equalizing set V exists.

The exclusion of confounding future information can be stated with the help of the following notion:

Definition 14.73: The variables X and Y are *causally connected in* the world w *within* the set $Z \subseteq U$ of variables iff there are $Z_1, \ldots, Z_n \in Z$ such that $Z_1 = X$, $Z_n = Y$, and for $i = 1, \ldots, n - 1$ ${}^w[Z_i] \xrightarrow[w]{\pm} {}^w[Z_{i+1}]$ or ${}^w[Z_{i+1}] \xrightarrow[w]{\pm} {}^w[Z_i]$.

This is just the notion (7.20f) of connectedness, relativized to worlds and connecting sets of variables. (It is not the notion of causal join in 14.32.) Causal unconnectedness in a lot of worlds implies a lot of conditional independence, as required by the above suggestions. A first attempt to realize these suggestions is stated in

Theorem 14.74: Suppose that U does not contain simultaneous variables, that $X_1 < \ldots < X_n$ are binary variables in U, $A_i = \{X_i = 1\}$, $\bar{A}_i = \{X_i = 0\}$, $(i = 1, \ldots, n)$, and $w \in A_1 \cap \ldots \cap A_n$, that P is a probability measure and ξ a ranking function for U, and that $\langle X_1, \ldots, X_n \rangle$ is a w-Markov chain w.r.t. P and ξ. Suppose moreover that $V \subseteq U$ is a set of variables such that for all $i = 1, \ldots, n$:

(a) $X_i \notin V$,
(b) $R(X_i) - \{X_{i-1}\} \subseteq V$ (where $R(X_i)$ is the set of parents or direct causal predecessors of X_i; see (14.10 + 11)),
(c)   each $Y \in V$ with $Y > X_i$ is not causally connected with X_i within $\{> X_i, < Y\}$ in all $w' \in {}^w[V \cap \{< X_i\}]$.

Then $\langle X_1, \ldots, X_n \rangle$ is a Markov chain conditional on ${}^w[V]$ w.r.t. P and ξ. Thus we can transfer (14.68) and get in particular:

(d)   If $P(A_{i+1} \mid A_i \cap {}^w[< X_{i+1}, \neq X_1, \ldots, X_i]) - P(A_{i+1} \mid \bar{A}_i \cap {}^w[< X_{i+1}, \neq X_1, \ldots, X_i]) = p_i$ $(i = 1, \ldots, n - 1)$, then $P(A_n \mid A_1 \cap {}^w[V]) - P(A_n \mid \bar{A}_1 \cap {}^w[V]) = p_1 \cdot \ldots \cdot p_n$.
(e)   If $\xi(\bar{A}_{i+1} \mid A_i \cap {}^w[< X_{i+1}, \neq X_1, \ldots, X_i]) = r_i > 0$ and $\xi(A_{i+1} \mid \bar{A}_i \cap {}^w[< X_{i+1}, \neq X_1, \ldots, X_i]) = s_i > 0$ $(i = 1, \ldots, n - 1)$, so that A_i is a necessary and sufficient direct cause of A_{i+1} in w, then $\xi(\bar{A}_n \mid A_1 \cap {}^w[V]) = \min_{i<n} r_i$ and $\xi(A_n \mid \bar{A}_1 \cap {}^w[V]) = \min_{i<n} s_i$, so that A_1 is a necessary and sufficient reason for A_n given ${}^w[V]$.

However, such a set V exists only if in all $w' \in {}^w[V]$ $\langle X_1, \ldots, X_n \rangle$ is the only chain of causal relevance from X_1 to X_n.

So, this is my proposal to transfer the simple Theorem 14.68 for Markov chains to the conceptually required, but much more involved case of w-Markov chains. For a proof of the probabilistic part see Spohn (1990, Theorems 14 and 17); it is easily extended to the ranking-theoretic counterpart. We might well call ${}^w[V]$ the circumstances of the indirect causal relationship between A_1 and A_n. Thus, we can maintain our basic characterization (14.2) even for indirect causes, namely that a cause is positively relevant to its effect in the obtaining circumstances. Note, however, the many provisos under which this result stands. One should check the proof to find out the extent to which one can undo the exclusion of simultaneous variables. If the variables forming the chain are multi-valued, things get still more complicated. So too, when there is more than one chain of causal relevance between indirect cause and effect; in this case, no such set V defining the circumstances exists. I do not know whether there are ways of generalizing this kind of result and/or rendering it more perspicuous.

Note, moreover, that clause (b) of (14.74) says that the ranking-theoretically (or probabilistically) possibly relevant circumstances of each direct causal link in the sense of (14.12) are contained in $^w[V]$. These might be quite large circumstances, thus making it harder to satisfy clause (c) as well. In my (1990, Theorem 16) I managed to improve the situation by referring to the larger actually relevant circumstances in the sense of (14.18) and assuming them to be ideal in the sense of (14.20). The basic pattern, however, remains the same.

I am well aware that these are insufficient glimpses of a theory of indirect causation. In the case of causal dependence between variables DAGs and d-separation, or chain graphs and c-separation, provide a lot of information about conditional dependence and independence, indeed complete information in case of faithfulness. A firm theoretical grip has been established on these matters. By contrast, when it comes to causation between facts the theoretical basis is given by conditional dependence and independence of propositions, and this is simply a much messier field. And it is an unduly neglected field. Probabilistic theorists neglect it because they are mainly or only interested in variables. And deterministic theorists neglect it because they mainly work within the counterfactual approach, which is essentially unfit for making circumstances and conditions explicit. But this is what one must do in order to really dig into the matter. So, even if my glimpses are insufficient, they point, I hope, in the right direction.

These remarks about theorizing about variables and about facts remind me of a point that we raised with the provisional Definition 14.26 and that we should finally settle. Sections 14.8–14.10 crucially depended on the assumption that causal dependence between variables is also transitive. Is this assumption really justified? We might think that we can decide this issue by conceptual stipulation, as we did in the case of causation between facts. This would be a mistake, though. Once we have made up our mind about causation between facts, we are no longer at liberty concerning variables. Rather, we get from indirect causation to indirect causal dependence in the very same way we got from direct causation to direct causal dependence. That is, the following definition, which finally replaces (14.26), appears fully adequate:

> *Definition 14.75*: Let ξ be a ranking function on the frame U and $X, Y \in U$. Then Y (*directly* or *indirectly*) *causally depends* on X w.r.t. ξ iff there is a X-proposition A, a Y-proposition B, and a world w such that $A \xrightarrow{\;\pm\;}_{w} B$ w.r.t. ξ.

Given (14.75), it becomes a theoretical question whether causal dependence between variables is transitive. The answer is clearly no! In addition to the X-proposition A and the Y-proposition B as in (14.75) we might have a Z-proposition C and a world w' such that $B \xrightarrow{\;\pm\;}_{w'} C$, so that Z causally depends on Y. However, this does not entail that there is a world w'' such that $A \xrightarrow{\;\pm\;}_{w''} C$; neither w nor w' might be suited to this. (Thanks to Christian Bender for pointing this out to me.)

Could we simply skip the transitivity of causal dependence and plug (14.75) into Sections 14.8–14.10? Certainly not. The narrow Markov condition (14.31) and (14.56) would not be affected. However, the common cause principles (14.33 + 34) can no longer be expected to hold under Definition 14.75. For an obvious consequence of (14.75) is that causal dependence is a possibly proper subrelation of the transitive closure of direct causal dependence. Thus, on the basis of (14.75), there might be fewer causal predecessors of a variable than we assumed in Sections 14.8–14.10, and the common causal predecessors of two or more variables might be much less. This is critical, since most of Sections 14.8–14.10 depended on the common cause principles.

What to do? We might replace "causal dependence" by "transitive closure of direct causal dependence" throughout Sections 14.8–14.10, even if the latter exceeds the former. Or we might apply our previous strategy and finally investigate the conditions under which the two notions coincide:

There are two reasons for the failure of the transitivity of causal dependence between the three variables X, Y, and Z. One reason is that the variable Y is multi-valued. Then the variable X might influence a certain subset B of the range W_Y of Y, whereas values of Y outside B are independent of X. On the other hand, it might be just those values of Y outside B that influence Z, whereas any value within B does not make a difference to Z. Then we do not find a single world in which X has an indirect influence on Z. (Lewis (2000, pp. 93ff.) discusses a similar situation.) However, this situation cannot occur if all the variables involved are binary.

The other reason for non-transitivity lies in the circumstances. Whether X influences Y depends on the circumstances, i.e., at most on how the other parents of Y are realized. Likewise, whether Y influences Z depends on how the other parents of Z are realized. Now, if Y and Z have parents in common, we might get an incongruence; the sets of values of those parents under which the two influences, that of X on Y and that of Y on Z, show up might be disjoint, and transitivity fails again. However, this situation cannot occur whenever the parents of any variable Y and the parents of any descendant Z of Y are disjoint (in our times a much better ground for calling certain graphs moral). In that case we can, as it were, compose the worlds in which the transitivity of causal dependence shows up from separate parts, i.e., from the separate circumstances of the single steps. In fact, we have thereby proved

Theorem 14.76: If all variables in the frame U are binary and if direct causal dependence between variables is intransitive, then causal dependence in the sense of (14.75) is the transitive closure of direct causal dependence in the sense of (14.20).

The intransitivity of direct causal dependence is similar to the condition that all causal chains or all chains of causal relevance are strict, which proved to be important in (14.71). If direct causal dependence proceeds in temporal stages, i.e., always from one point of time to the next, it is guaranteed to be intransitive. So, this condition should be satisfied at least in sufficient refinements of the conceptual frame. And the condition that all variables are binary can certainly be relaxed. For instance, in the probabilistic context the assumption of a normal distribution of the variables should do, because then uncorrelatedness implies independence. In the ranking-theoretic context, features like monotonic relatedness (as discussed in Section 14.7) or single-peakedness (as used in Section 11.3 in the discussion of the recovery postulate for belief contraction) might do as well. I leave these issues for further investigation.

Are our considerations in Sections 14.8–14.10 thereby invalidated? I do not think so; as just indicated, the prospects that they hold under sufficiently wide conditions are good. Still, (14.76) points to an important caveat, which becomes smaller, the more we improve on (14.76).

14.13 Some Further Examples

In Section 14.5 we discussed some problematic cases related to direct causation. So, let me now discuss some examples related to indirect causation. The problem is: there are just too many, often similar and often different. In my (1983b, Ch. 3) I tried to give a systematization of all possible examples; in retrospect, that collection was relatively complete (the main neglect was examples of causation by omissions). However, since the theoretical defense of transitivity (14.60) has been my main intent, I will confine myself to some examples that have been raised specifically against transitivity.

In relation to indirect causation, I see two basic problems that I do not believe have been sufficiently emphasized in the literature. One problem is that A can actually be a cause of B, while non-A, if it had happened, would also have been a cause of B. In my (1983b, Sect. 3.2) I called such causes A relays that simply activate one or another causal route to B. Nowadays they are called *switches*, and their importance is generally recognized. They are almost ubiquitous; for instance, they operate in each case of ordinary early preemption by cutting. Recall, e.g., the case of the desert traveler (example 14.7), where the second assault acts as such a switch. If the hole had not been drilled into the keg, this would have caused the poison to stay in the keg and thus the traveler to die anyway.

The other problem is that A can be a cause as well as a counter-cause of B; i.e., if non-B had happened, A would have been also a cause of non-B. We might call such a cause A *(causally) ambiguous*. Within the probabilistic realm Hesslow (1976) introduced this kind of problem with his example of the contraceptive pills, which may be read as a counter-example to transitivity, but is better understood as causally ambiguous. On the one hand, contraceptive pills probabilistically cause, or raise the danger of, or contribute to causing, thrombosis. On the other hand, they prevent thrombosis by preventing pregnancies that also raise the danger of thrombosis. This example caused quite some turmoil, which has somehow calmed, but not really settled, to my knowledge. Of course, such examples can be reproduced within the deterministic realm.

These possibilities border on paradox, and indeed for direct causation they are utter impossibilities, intuitively as well as for all theories of causation I know of. (This is why preemption by cutting essentially involves indirect causation.) The impossibilities extend to indirect causation as long as one adheres to some version of the positive relevance condition (14.66). If one renounces that condition, the only plausible reason for this being the adherence to transitivity (14.60), then one must acknowledge these paradoxical cases as real possibilities. Thus, these cases throw us back to the issue of transitivity.

I have stated my theoretical reasons for accepting transitivity in the previous two sections. Lewis (2000, pp. 98f.) puts forward three further reasons (or rather four; see his point about causation in history). I share all of them. He was aware of the counter-examples, but Hitchcock (2001) and Hall (2000) have strongly reinforced them. Let us first inspect Hitchcock's cases.

Example 14.77, Failure of Composition? A first example is taken from McDermott (1995): A terrorist, who is right-handed, must push a detonator button at noon to set off a bomb (P). Shortly before noon, a dog bites his right hand (D). Unable to use his right hand, he pushes the detonator with his left hand. The bomb explodes as planned (E). Hence, if the dog had not interfered, the terrorist would not have used his left hand, and if he had not pushed the button with his left hand, the bomb would not have exploded. These are causal relations. Transitivity forces us to conclude that the dog bite is a cause of the explosion. This seems absurd; the bomb would have exploded whatever the dog does.

Hitchcock (2001) carefully discusses various modelings of this example, among them the one just suggested, namely that the dog acts as a switch, as it were, activating either the terrorist's right or his left hand. I agree with his conclusion that the simplest DAG $D \rightarrow P \rightarrow E$ is also the most adequate, where D and E take two possible values "yes" and "no" and P three possible values "right push", "left push", and "no push"; let us denote the corresponding propositions by $D_1, D_0, E_1, E_0, P_r, P_l, P_0$, and $P_1 = P_r \cup P_l$. His diagnosis then is that the two causal steps from D_1 to P_l and from P_l to E_1 "fail to compose", and his causal analysis in terms of active routes within structural models (cf. the very brief description between (14.67b) and 14.67c)) is able to construe this as a case of non–transitivity.

I think he is right, but ranking theory reveals that this is not even a case of alleged transitivity. The following distribution ξ of negative ranks appears simple and natural:

$\xi(P_j \mid D_i)$	P_r	P_l	P_0		$\xi(E_k \mid P_j)$	E_1	E_0
D_1	1	0	1		P_r	0	1
D_0	0	1	1		P_l	0	1
					P_0	1	0

Since the DAG represents a Markov chain, the second table does not depend on D. A little calculation yields:

$\xi(P_j \cap E_k \mid D_1)$	E_1	E_0	$\xi(P_j \cap E_k \mid D_0)$	E_1	E_0	$\xi(E_k \mid D_i)$	E_1	E_0
P_r	1	2	P_r	0	1	D_1	0	1
P_l	0	1	P_l	1	2	D_0	0	1
P_0	2	1	P_0	2	1			

The last of these five tables shows that the bomb is expected to explode no matter what the dog does. The first table shows that in the assumed course of events D_1 is indeed a necessary and sufficient cause of P_l, as D_0 would have been of P_r. However, it also shows that D_1 is *not* a cause of P_1. We need the third and the fourth table in order to calculate $\xi(E_k \mid D_i \cap \bar{P}_j)$. When we compare these values with the second table, we find only that P_1 is a cause of E_1, but neither P_r nor P_l is. In other words, the issue of transitivity does not even come up in this example.

Example 14.77 thus teaches three things: First, it confirms my earlier warning that the distribution of positive relevance can become tricky when we deal with multi-valued variables. It secondly shows that we must look at positive relevance and not only at relevance, as the structural modeling approach essentially does (cf. my remark before 14.67c). Thirdly, it indeed displays a failure of composition and thus of transitivity, however, not of causation between facts, but rather of causal dependence between variables; we saw in (14.76) that there is a distinction to be made. Hence, the dog bite exemplifies a violation of (14.76) by multi-valued variables.

Example 14.78, Cancellation Along Different Routes? There is another type of example that Hitchcock (2001) thinks undermines the transitivity of causation. It comes in two forms, a good and a bad one (for transitivity). The good one is a normal case of a back-up system or early preemption. A trainee who is learning to parachute is standing at the hatch of the airplane, one mile up in the air. He jumps, falls, and safely lands 10 minutes later. The jump is a cause of the landing. But he might have been anxious and refused to jump. In that case, the supervisor would have pushed him out of the plane and followed him. Again, the trainee would have safely landed.

In this case, the landing does not counterfactually depend on the jumping. Still, as Hitchcock (2001, pp. 287f.) explains, there is an active causal route from the jumping to the landing in the sense that if the alternative causal routes were fixed at their actual values or states, we would find a counterfactual dependence (or what corresponds to it in the structural modeling approach). This is so, because if the trainee had not jumped and if the supervisor had done the same as what he actually did, namely not pushing him, he would not have fallen and landed. Hence, the jump is a cause of the landing, indeed a direct one, as long as we consider only the three variables jumping, pushing, and landing. I entirely agree. In my account, the jumping is positively relevant to the landing given the remaining actual course of events, i.e., the non-pushing. So far, so good.

Now look at a prima facie similar case taken from a preliminary version of Hall (2004). A boulder breaks off and begins rolling towards a hiker. Before it reaches him, the hiker sees the boulder and ducks. Thus, the boulder sails over his head without doing any harm. The hiker survives and continues hiking. Clearly, the boulder's falling is a cause of the hiker's ducking, and the latter is a cause of his survival. Transitivity dictates that the boulder's falling is a cause of the hiker's survival, too; but intuition strongly revolts against this conclusion.

Too bad for intuition, or for theory? Well, not for Hitchcock's theory. He finds no active causal route from the boulder's rolling to the hiker's survival, i.e., no counterfactual dependence if the alternative causal routes were fixed at their actual values or states. The point is that there is nothing to fix; the boulder's not falling would have been a direct cause of the hiker's staying alive. Hence, the falsity of "if the boulder had not dislodged, the hiker would not have survived" is already decisive against the causal claim of transitivity.

I am not convinced. Hitchcock notices that, on the level of variables, the structure of the two examples is the same; they are both characterized by a DAG of the form

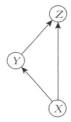

– where $\langle X, Y, Z \rangle$ is $\langle$jumping, pushing, landing$\rangle$ in the one case and $\langle$rolling, ducking, surviving$\rangle$ in the other (the gerunds here stand for binary variables taking values "yes" or "no"). Still, the examples are different. However, their only difference is that the variables are realized in different ways. Had the boulder not fallen, this would have been a cause of the hiker's survival, just as the jumping caused the landing. If the trainee had not jumped, that would have caused the supervisor to push him and that in turn the trainee's landing, though again the refusal to jump is not a cause of the landing according to Hitchcock and, presumably, according to intuition as well.

However, Hitchcock is perfectly aware that his account crucially depends on there being a direct causal dependence of Z on X. Intuitively, the two causal routes from X to Z cancel each other, resulting in $\{Z = \text{yes}\}$ however X realizes. And intuitively, the efficacy of the one causal route should show up when the other causal route is discounted, i.e., held fixed at its actual state. However, the direct route cannot be discounted; one cannot focus on the indirect route $\langle X, Y, Z \rangle$ alone, since the direct route $\langle X, Z \rangle$ cannot be eliminated from the picture.

The situation changes immediately when one grants symmetry to the two routes by turning both of them indirect:

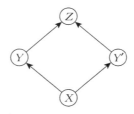

In this representation, Hitchcock grants, we find no failure of transitivity; $\{X = \text{yes}\}$ is a cause of $\{Z = \text{yes}\}$ along the one route, and $\{X = \text{no}\}$ would have been a cause of $\{Z = \text{yes}\}$ along the other route. Therefore, he closely looks at the refined representation. What might Y' stand for? The only suggestions he finds appear bizarre and result, he says, in counterfactual reasoning only trained philosophers would engage in. He is right. Let us look only at one attempt in the hiker's story:

One problem is that in a more realistic setting there are two almost continuous processes that are intertwined. The boulder has a continuous path, and there are many times at which the hiker can perceive the boulder and duck in time, if necessary. Let us make this complexity artificially discrete and consider the kinematic state of the boulder only at two times. At the first time, the boulder is either at rest or approaching dangerously; this is X. This, let us suppose, is the only time when the hiker can perceive the boulder. Subsequently, he does or does not duck; this is Y. The second time is when it is much too late for the hiker to duck, if necessary. Then, again, the boulder is either at rest or about to strike the hiker; this is Y'. So, the hiker does or does not survive; this is Z. In the actual course of events Y' is such that the boulder is about to strike the hiker. And given this we see the causal relevance of X to Z; given this state of Y', if the boulder had been at rest at the first time, it would have killed the hiker (because he would not have ducked). This is unassailable – and bizarre. It requires us to suppose that the boulder is first at rest and then about to strike the hiker, contrary to the laws of physics. Or – this would be much fairer – we should appeal to the Markov property and say that, independently of what the boulder is doing at the first time, it would kill the hiker if it were about to strike him, unless he ducked.

At this point I am willing to trade intuition against theory. I think intuitions about these cases are nourished by the pragmatics of causal talk, whereas, as emphasized in section 14.3, we intend an egalitarian notion of cause that does not look at salience or informativity; and in the egalitarian perspective, the normal Y (ducking) has the same causal standing as the bizarre Y' (being about to strike the hiker or not, whatever the prior kinematic state). In this perspective, it seems perfectly adequate to say that, however X is realized, it would have been a cause of $\{Z = \text{yes}\}$, and thereby to acknowledge that X has a special causal role, namely that X acts as a switch both positions of which lead indirectly to the same result.

The alternatives are no better. One alternative, Hitchcock's, would be to grant a causal role only to one position of the switch and deny it to the other. However, this simply hides the special role of the switch. Moreover, we saw that this asymmetry is not stable under refinements of the frame. I prefer to give a pragmatic explanation of the apparent asymmetry of the positions of the switch. One explanation, which I did not pursue here, would be to try to characterize something like the overall effect of X on $\{Z = \text{yes}\}$ (what Woodward (2003, p. 51) calls being a total cause; cf. above before (14.67c)); and then one might find that one position of the switch X (the boulder's not being dislodged) is much better or safer in bringing about the result (the hiker's survival).

Another alternative, which Hitchcock would oppose as well, is to say, in order to preserve symmetry, that no position of the switch X is a cause of $\{Z = \text{yes}\}$, since $\{Z = \text{yes}\}$ comes about anyway. This would be no less odd, though. For instance, looking also at the stages of the causal process before the switch, we could then have a weak causal chain (in the sense of 14.61a) from those early stages to the result. However, this would be a chain with gaps, in the sense that some members of the chain are causes of the result and others are not (namely those that mark switches); and these gaps may lie in the middle of the chain. This would be queer, too.

No, the least distorting way of describing such cases is the one proposed above. We should allow switches, and if "cause" connotes "non-switching cause", we should make this connotation explicit. In any case, we should stick to transitivity.

Example 14.79, Double Prevention? In a number of papers Ned Hall has put a lot of argumentative weight on cases of double prevention. Let us see how they fare in our framework. First of all, there seems to be no mystery about prevention: A (directly) *prevents* B iff A is a (direct) cause of $\bar{B}$. The only caveat is that there may be many ways to realize $\bar{B}$; they, as well as B itself, may all be values of a multi-valued variable Y. In that case one must beware of the pitfalls of multi-valued variables and check whether one is interested in the causation of $\bar{B}$ or some more specific Y-fact.

It is clear, though, that this first move is already a distortion of Hall's intentions. He prefers event talk in causal matters, and though he wants to make room for causation of and by the omission of events, this is something that goes beyond the primary causal relation between events. In this way of talking a prevention is the causation of an omission of an event. It is also clear, though, that we cannot reproduce this distinction within our fact-oriented talk of causation; the distinction between positive and negative facts has persistently proved to be elusive. Under the egalitarian view of causation that Hall recommends, positive and negative facts, that are hardly distinguishable anyway, should be considered equally eligible as causes and effects. So, the above definition of prevention is the best we can give within our framework.

So, why is double prevention particularly problematic? Let us look at the preferred example of Hall (2004, p. 241):

Suzy and Billy ... get involved in World War III. Suzy is piloting a bomber on a mission to blow up an enemy target, and Billy is piloting a fighter as her lone escort. Along comes an enemy fighter plane, piloted by Enemy. . . . Billy spots Enemy, zooms in, pulls the trigger, and Enemy's plane goes down in flames. Suzy's mission is undisturbed, and the bombing takes place as planned. If Billy hadn't pulled the trigger, Enemy would have eluded him and shot down Suzy, and the bombing would not have happened.

This is a case of double prevention; Billy's action prevents Enemy from preventing Suzy's bombing. So, is Billy's action a cause of Suzy's bombing? One might hesitate to say yes. One might also consider how Hall continues on p. 242:

Here she [Suzy] is, in one region, flying her plane on the way to her bombing mission. Here Billy and Enemy are, in an entirely separate region, acting out their fateful drama. Intuitively, it seems entirely unexceptionable to claim that the events in the second region have no causal connection to the events in the first — for isn't it plain that no *physical* connection unites them?

In other words, it seems particularly difficult to understand how causal influence can spread beyond omissions; it is as if sound were able to bridge a vacuum. Hall tries to back up this intuition by discussing how the thesis that counterfactual dependence implies a causal relation — to which one appeals when one accepts that Billy's action is a cause of Suzy's bombing — gets into conflict with locality and the intrinsicness and transitivity of causation.

However, I fail to see the basic difficulty. Just as the presence of oxygen is a cause of the fire, on equal footing with the lightning in an egalitarian perspective, so the stability of flying conditions is a cause of the success of Suzy's mission. In a way, this stability means that nothing happens (to Suzy); however, it is possibly hard work to guarantee that nothing happens, and this is Billy's contribution. I find the allusion to the separate regions with their separate going-ons misleading. The regions do not spatially overlap, but they are causally and indeed physically connected, by quiet air that would not have stayed quiet, had Billy not intervened. In sum, what we ordinarily describe as a case of double prevention is, from my theoretical point of view, an ordinary causal chain (which we may refine by filling in the physical details).

In his further discussion, Hall considers even more complicated scenarios. For instance, on pp. 246f. he turns the prevented prevention into a switch:

Early in the morning on the day of the bombing, Enemy's alarm clock goes off. . . . If it hadn't, he never would have woken up in time to go on his patrolling mission. . . . It follows that if Enemy's alarm clock hadn't gone off, then Billy would not have pulled the trigger. But it is also true that if Billy hadn't pulled the trigger, then the bombing would never have taken place. By transitivity, this ringing is one of the causes of the bombing.

I grant that this sounds odd. However, it does so because the alarm clock acts as a switch. If it had not gone off, it — it alone, as it were — would have prevented Enemy's intervention and thus caused Suzy's not being attacked (we might even say that the overall effect of its not going off is overwhelmingly preventive). Since it went off, it provoked Billy to prevent Enemy's intervention. So, either way it leads to Suzy's not being attacked. Maybe the oddity is boosted by embedding the switch in a double prevention. However, I argued above how best to deal with switches. And so it is best here, too, to explicitly specify how the alarm clock enters the causal history of Suzy's bombing. This can be done by tracing all of the actual and potential direct causal relationships in that history. Then we know all there is to know about that history. Whether or not some ordinary usage of "cause" then fits to the thus specified role of the alarm clock is, I find, an idle question that is of little help in advancing causal theorizing.

Hall (2004) feels urged to draw theoretical consequences from his examples and arguments. If a notion of causation based on counterfactual dependence is incompatible with his three conditions of locality, intrinsicness, and transitivity, which seem as well-founded as that notion, then there should be another notion of causation that does satisfy these conditions. He calls that *production* and tries to characterize it. The term "production" and the surrounding metaphors reminds of the mechanistic conception of which I gave a pessimistic assessment in Section 14.2; it throws us back into accounting for the dependence inherent in each single causal step. Indeed, this is not what Hall has in mind. Rather, he characterizes his sense of production in terms of minimal sufficient conditions and thus broadly in terms of the current investigation. Hence, I think the moral stated at the beginning of Section 14.12 applies here as well. We should begin with the broadest sense of causation as defined in (14.69), thus accept all the contrived stories as describing causal relations, and then turn the description of the conditions under which further conditions that may be imposed on causal relations are realized into a theoretical issue.

14.14 Appendix on Explanation

In view of the relevance of ranking theory to the theory of causation displayed in this chapter, it is plausible that this relevance extends to the theory of explanation. It might be unfolded in another long chapter, presumably with many repetitions. I will here restrict myself to some brief remarks.

How straightforward the extension is depends, of course, on how close the relation between causation and explanation is. Some say, very close; for instance, Lewis (1986b, Ch. 22) argues that explanation is essentially causal explanation. The point was opened to debate by Hempel (1965) himself. In Section 2.2 he argues that the deductive-nomological model of explanation comprises more than causal explanation and supports his case, among other things, with a puzzling example (p. 353): namely that Fermat's principle of least time entails that a beam from point A in one medium to point C in another medium must cross the border at a specific point B and that the latter fact is thereby explained. Elsewhere he discusses dispositional and theoretical explanations and other forms, all not obviously or obviously not of causal form. The most careful pleading on behalf on non-causal explanation I know of is Bartelborth (1996, pp. 330–345).

I remain unconvinced. I still think that getting an explanation of some phenomenon is learning a cause of it, and having an explanation of it is knowing, or believing one knows, a cause of it; this is the basic connection. Let me simply accept it here. Thereby, our account of causation immediately entails an account of (causal) explanation. Explaining B by A hence means believing that $A \xrightarrow{-+}{w} B$, i.e., in the proposition $\{w \mid A \xrightarrow{-+}{w} B\}$. Since we saw that (direct or indirect) causation is formally not easily tractable, it might be a good idea to confine ourselves to direct explanation, i.e. to the belief in direct causes. Then we arrive at the explication I proposed in Spohn (1991, p. 177):

Definition 14.80: Let $X, Y \in U, A$ be an X-proposition, B a Y-proposition, ξ a negative ranking function for U, and τ the corresponding two-sided ranking function. Then A *directly explains* B relative to ξ iff $\tau(\{w \mid A \xrightarrow{+}{w} B\}) > 0$; and A *directly explains B as necessary* or *as possible* relative to ξ iff, respectively, $\tau(\{w \mid A \xrightarrow{-+}{w} B\}) > 0$ or $\tau(\{w \mid A \xrightarrow{n+}{w} B\}) > 0$.

Is belief in a cause good enough for explanation? Or should we rather require knowledge of a cause (the critical point being that I have not said what knowledge is)? I do not really worry about this issue. Hempel insisted that the explanans (and hence the explanandum) be true, whereas others were content with the explanans being accepted or acceptable. There is not really an issue here; "explain" is simply ambiguous between a factive and a non-factive reading. And so I may go for the non-factive reading, which is within my reach.

(14.80) opens a way for studying the important and peculiar role explanations play in our every-day and scientific epistemic efforts. Let me only make two small observations:

First, do explanations always provide reasons? This is the issue, raised by Hempel (1965, p. 368), of whether "an adequate answer to an explanation-seeking why-question is always also a potential answer to the corresponding epistemic why-question" – transferred to our framework. Hempel affirmed this in his famous thesis of the structural identity of explanation and prediction (that also contained the problematic converse claim, which was not endorsed by Hempel). Having formal explications of the notions involved we can prove the plausible half of the identity thesis, with restrictions:

(14.81)   If A directly explains B as necessary relative to ξ and $\tau(\{w \mid A \xrightarrow{s+}{w} B\} \mid \bar{A})$ ≥ 0, then A is a sufficient reason for B relative to ξ; and if A directly explains B as possible relative to ξ and $\tau(B - \{w \mid A \xrightarrow{n+}{w} B\} \mid \bar{A}) < 0$, then A is a necessary reason for B relative to ξ.

Proof: (Cf. also Spohn 1991, p. 188 and footnotes 54 and 55.) As to the first claim, let $C = C_w^{--}(A, B; s)$, i.e., the smallest actually obtaining circumstance of A and B in w in the sense of (14.13) for some w such that $A \xrightarrow{s+}{w} B$ (they are the same for all such w). Thus $\{w \mid A \xrightarrow{s+}{w} B\} = A \cap B \cap C$. Hence, (14.14) tells us that

(a) $\xi(\bar{B} \mid A \cap C) > 0$ and

(b) $\xi(\bar{B} \mid \bar{A} \cap C) = 0$.

Since A directly explains B, let A, B, and C are believed in ξ, implying $\xi(A) = 0$ and $\xi(\bar{C}) > 0$; hence $\xi(A \cap \bar{C}) > 0$ and so

(c) $\xi(\bar{C} \mid A) = \xi(A \cap \bar{C}) - \xi(A) > 0$.

The additional premise of (14.81) says that

(d) $\xi(C \mid \bar{A}) = 0$.

Now we can see that (a) implies $\xi(\bar{B} \cap C \mid A) > 0$ and (c) implies $\xi(\bar{B} \cap \bar{C} \mid A) > 0$. Hence $\xi(\bar{B} \mid A) > 0$. Moreover, (b) and (d) imply $\xi(\bar{B} \cap C \mid \bar{A}) = 0$ and hence $\xi(\bar{B} \mid \bar{A}) = 0$. This says that A is a sufficient reason of B.

As to the second claim, let $D = C_w^{--}(A, B; n)$ for some w such that $A \xrightarrow[w]{n+} B$ (it again does not matter which such w we take). Thus $\{w \mid A \xrightarrow[w]{n+} B\} = A \cap B \cap D$. Again, (14.14) tells us that

(e) $\xi(B \mid A \cap D) = 0$ and

(f) $\xi(B \mid \bar{A} \cap D) > 0$.

Since, as before, A, B, and D are believed in ξ, we have $\xi(A \cap D) = 0$ and so

(g) $\xi(D \mid A) = 0$.

The relevant additional premise says

(h) $\xi(B \cap \bar{D} \mid \bar{A}) > 0$.

Now, (e) and (g) imply $\xi(B \cap D \mid A) = 0$ and hence $\xi(B \mid A) = 0$. Moreover, (f) entails $\xi(B \cap D \mid \bar{A}) > 0$, and this, together with (h), entails $\xi(B \mid \bar{A}) > 0$. This means that A is a necessary reason for B.

The appeal to (14.14) shows that a similar theorem for direct explanation in general is not available. And since the positive relevance of indirect causes is not guaranteed, it is difficult to generalize to indirect explanation.

As has been the case so many times in this book, the point of the exercise (14.81) is to turn philosophically interesting claims into theorems, thus formally strengthening the checks and balances, and at the same time to control for hidden premises. We could see that (14.81) does not hold without additional premises, and I have chosen the most obvious ones.

A still simpler, but perhaps more significant observation is this: Our proof of (14.81) proceeded from the fact that, if A directly explains B as necessary or possible, then A is, respectively, a sufficient or necessary reason for B, given the smallest circumstances believed to be actually obtaining. From there (14.81) moved to A being an unconditional reason for B. However, we might also move in the other direction. (14.16) then tells us that A remains a sufficient or necessary reason for B, given any stronger information about the past of B, i.e., about $\{\leq Y, \neq X, Y\}$. (Or we might look at modifications of 14.16; formal possibilities again multiply at this point.) In Section 17.3 I will describe this point by saying that A provides a *stable* reason for B, where stability means that A's positive relevance for B cannot disappear, in a specific sense. As noted after (14.16), this notion resembles the notion of resiliency introduced by Skyrms (1980, part IA). In Spohn (1991), I developed the idea that such stable reasons are epistemologically particularly valuable, and so are explanations that provide such stable reasons; this will be further discussed in Section 17.3.

A final remark: There is a strange kind of reflexivity in (14.80). Whether $A \xrightarrow[w]{+} B$ obtains is judged relative to the ranking function ξ; so the proposition $\{w \mid A \xrightarrow[w]{+} B\}$ is relative to ξ; and then this ξ-relative proposition must be believed according to ξ. This looks odd, and indeed it is not what I intended. The idea rather is this: First, such a ξ represents a causal hypothesis about a given single case, perhaps an objectifiable causal hypothesis in the sense of the next chapter. Such a ξ is extended to a causal law λ_ξ for all similar cases as defined in (12.10). Our beliefs about each single case are then given by a mixture κ of all such causal laws as envisaged in Section 12.5. So, finally, the explanatory relations are to be judged relative to the mixture κ and not relative to each individual ξ. (Note that Section 12.5 fully carries over to this chapter. If the frame U consists of finitely many finitely-valued variables, each ξ as considered here spreads over a finite possibility space, and our copy (12.18) of de Finetti's representation theorem applies.)

Gärdenfors (1980) has proposed such a second-order approach to probabilistic explanation. It seems highly plausible that the basic idea that explaining is the knowing or giving of causes is to be developed along this line. However, the actual development is another matter, and not one for an appendix.

14.15 A Further Appendix on Inference to the Best Explanation and Innocence of the Worst Explanation

I feel obliged to add finally some remarks on inference to the best explanation. This is obviously a topic for another chapter I cannot afford. However, I cannot simply be silent, either. Many think that Bayesianism is only a partial story of scientific inference; and even more think that enumerative induction and the hypothetico–deductive method are too simplistic as models of scientific inference. The poorer the alternatives appear, the stronger the view that abduction or inference to the best explanation (IBE, for short) is the more advanced or suitable methodology, the way sciences actually proceed in their inductive inferences. It is rich and obscure at the same time, rich because it refers to all our explanatory practices and our methods for assessing them, and obscure because philosophers of science have a hard time saying what explanations are and an even harder time saying how their goodness might be assessed. However, precisely because of its richness and obscurity it seems most suitable for capturing scientific practice. If this is so, and if I claim that ranking theory is a, or the, basic theory of inductive inference (besides Bayesianism), then there is a tension I must not leave without comment.

So, is IBE somehow implicit in ranking theory? Yes, the basic story seems quite simple to me; the complexity of the inferential method of IBE lies in the complexity of the scientific applications and not in the labyrinths of inductive theory.

What is the basic story? If we were to move within a probabilistic context, every Bayesian statistician, I guess, would agree that, at bottom, IBE is Bayes' theorem (3.5), which says that the posterior probability of some hypothesis is proportional to its prior probability and to its likelihood. Lipton (1991, pp. 61ff.) has convincingly argued that the goodness of the explanation a hypothesis provides for the data essentially depends on two factors, its likeliness and its loveliness, and fills these labels with rich content. Okasha (2000) proposes that, basically, the likeliness of a hypothesis may be identified with its prior probability and its loveliness with its likelihood. And so the two kinds of inference come to agree.

This goes too quick, one might object. Well, not really. My impression is that the majority thinks that the similarity between IBE and Bayes' Theorem is sound and substantial, and even the opponents seem to argue against the strong appeal they themselves feel. Look, for instance, at the opposition between IBE and conditionalization (= Bayes' Theorem) that van Fraassen (1989, Ch. 6) attempts to construct. I find his construction to be utterly forced and thus implicitly confirming that appeal. The main criticisms the similarity meets derive from the fact that many applications of IBE do not clearly, or clearly do not, move within a probabilistic context. Salmon (2001) finds IBE too ill-specified a story about confirmation and much prefers Bayesianism for providing that story. Conversely, Psillos (2004) takes IBE to be a sound rule of acceptance, something not deliverable from probability theory.

Still, most believe in the similarity, if not congruence (see, e.g., Niniiluoto 2004 and McGrew 2003). Also Lipton (2004, Ch. 7) takes what he calls an irenic view. He does not quite agree with Okasha's identifications; for instance, the loveliness of an explanation may also guide the prior probability of a hypothesis. And if we move outside a strictly statistical context, where a statistical hypothesis defines its likelihood by itself, we need to more fully explain what is explanatory about the likelihood. And so, he argues, Bayesianism and explanationism do not contradict, but rather supplement each other.

Now, of course, I propose to move the discussion from the probabilistic into the ranking-theoretic context. (In a similar spirit, Schurz 2011 moves abduction into the belief revision context.) There we find Bayes' Theorem for negative ranks:

(14.82) $\kappa(A_k \mid B) = \kappa(B \mid A_k) + \kappa(A_k) - \kappa(B)$ (= 5.23d),

where $A_1, \ldots, A_n$ are n pairwise disjoint and jointly exhaustive propositions (the alternative hypotheses), where B is another proposition (representing the data), and where we may or may not spell out the normalizing $\kappa(B)$ by the formula of the total negative rank:

(14.83) $\kappa(B) = \min_{i \leq n} [\kappa(B \mid A_i) + \kappa(A_i)]$ (= 5.23c).

According to (14.82), the prior (im)plausibility of the hypothesis A_k and the im(plausibility) of the data B given the hypothesis A_k (minus normalization) add up to its posterior (im)plausibility; the better the first two values, the better the last. As a first primitive account of IBE, this does not sound so bad. And, as called for by Psillos (2004), (14.82) indeed states IBE as a rule of acceptance.

Look, for instance, at the notorious Uranus–Neptune case. Let B be the measurement of the irregularities in the orbit of Uranus. Their discovery was quite surprising; so, initially $\kappa(B) > 0$. Now we consider various hypotheses about how B is possible. First, there are various additions of auxiliary hypotheses to Newton's gravitational theory A, which we initially believe, so that $\kappa(A) = 0 < \kappa(\bar{A})$. One such amendment is A_1, that our list of planets (and other heavy bodies) in the solar system is already complete. However, $\kappa(B \mid A_1)$ is quite large; there might be unknown measurement errors, or some other remote ways of making A_1 compatible with B; but they are very implausible. Another amendment is A_2 that there is a further planet, Neptune, at a specific place. This is as surprising as B itself; we thought we knew of all of the planets of our sun; hence $\kappa(A_2) = \kappa(B)$. However, given A_2, B is to be expected; hence, $\kappa(B \mid A_2) = 0$. We might be more historically adequate and also consider less accurate calculations leading to vaguer determinations of Neptune's position, i.e., to vaguer versions of A_2. We might think of other celestial bodies perturbing Uranus and thus try amendment A_3. We might even give up Newton's theory A, play around with the gravitational constant or the law of gravitation, and thus look at various alternatives A'; but they would all be very surprising, indeed more surprising than the data, so that $\kappa(A') > \kappa(B)$, and they all fit more or less badly, so that $\kappa(B \mid A') > 0$. We might close our considerations with the catch-all hypothesis A^*; however, it is most implausible that none of the many hypotheses we have thought of applies; and so, even more clearly, $\kappa(A^*) \gg \kappa(B)$. Hence, as I have told the story, A_2 is the only hypothesis for which $\kappa(A_2 \mid B) = 0$ according to (14.82); that is, we even have $\kappa(\bar{A}_2 \mid B) > 0$, and so A_2 is accepted after observing B.

Of course, (14.82) is only the first primitive step in accounting for IBE. It needs to be spelled out what is explanatory about the term $\kappa(B \mid A_k)$, just as it needs to be spelled out in the probabilistic case what is explanatory about the likelihood. Even if my elaborations are far from complete, I have said many things in this chapter that might deepen this point. Certainly, one should go second-order, as indicated in the previous appendix. The hypotheses of which the explanatory value is to be assessed are not just propositions, they are first-order laws, as explained in Chapter 12. Perhaps the compatibilist considerations of Lipton (2004, Ch. 7) also make good sense within the ranking-theoretic framework. And so forth.

However, I am far from attempting to actually give an account of IBE. This would require a lot of additional constructive and comparative work. I only wished to dispel the worry that there might be a sophisticated form of inductive inference, prevalent in science and highly praised by philosophers of science, that is out of reach of the epistemology elaborated here. I believe that Bayes' Theorem transferred to ranking theory sets us on the right track, a track we should use, if we want to avoid shrouding IBE in mystery.

Above, I mentioned in passing the catch-all hypothesis A^* or, if it should take a specific form, the worst explanation that we might have to take recourse to after all other and better explanations have failed. For Bayesian methodology, the catch-all hypothesis poses an obvious problem that is often discussed, but not satisfactorily solved (the problem originally appears in Shimony 1970, pp. 95ff., 118ff., and plays a crucial critical role, e.g., in Earman 1992, Sect. 7.2 and pp. 228ff.). On the one hand, the catch-all hypothesis must have some non-vanishing prior probability; otherwise, it could never become eligible after all the other hypotheses have failed. On the other hand, it is hard, if not impossible, to assess its likelihood. So, it seems quite indeterminate how the catch-all hypothesis enters the formula of total probability (3.4e) and hence Bayes' Theorem, even though it must enter it in some non-negligible way.

I am not interested in solving all of the problems of the Bayesian. What I want to point out is that this is no problem for ranking theory. The catch-all hypothesis A^* has a higher negative rank than all of the other more respectable hypotheses (and this is even more so for each of the specific (initially) worst explanations into which the catch-all hypothesis might decompose if we should ever come to the verge of considering them in despair). According to (14.83) it *in no way* affects the degree of (dis)belief in the possible data B; this is the crucial difference to the probabilistic case. I already illustrated this point with (7.15x) in my story about Sherlock Holmes and his alarm in Section 7.2.

That is, within ranking theory, the catch-all hypothesis and its possible specifications are genuine, theoretically justified, not merely pragmatically tolerated don't-cares. We might well wait on detailing their epistemological role until we come to the point where we feel forced to seriously take them into consideration. Or we might say, and nobody could detect a difference, that the ranks of the catch-all hypothesis and its possible specifications are firmly buried in the unfathomable depths of our ranking functions and thus are dispositions that, in all likelihood, will never get an occasion to show up even in the entire history of humanity. The point I want to make is that by choosing the latter description we open up the possibility of subsuming all kinds of cognitive progress, even the revolutionary kinds, under the dynamic epistemological framework provided by ranking theory. This is a fascinating or a silly idea; I am not going to probe it any further.

15

Objectification

15.1 The Fundamental Problem

Whatever the merits in detail of the previous three chapters, they are threatened by a dark shadow of doubt. I have shifted to subjective relativization to an extent that seems utterly absurd. The upshot of Chapter 12 was my notion of a subjective law introduced in (12.10). Despite a lot of tentative subjective inclinations in the literature, I am not aware of anybody who has so radically envisaged such an appalling notion. Still worse, as long as I explain causation only relative to a ranking function, I seem to endorse the view that causation is only in the eye of the beholder.

This view is, of course, David Hume's, or rather part of his ambiguity:

The idea of necessity arises from some impression. There is no impression convey'd by our senses, which can give rise to that idea. It must, therefore, be deriv'd from some internal impression, or impression of reflexion. There is no internal impression, which has any relation to the present business, but that propensity, which custom produces, to pass from an object to the idea of its usual attendant. This therefore is the essence of necessity. Upon the whole, necessity is something, that exists in the mind, not in objects. (Hume 1739, p. 165).

He is perfectly aware of the harshness of his position. Two paragraphs later he writes:

I am sensible that of all paradoxes, which I have had, or shall hereafter have occasion to advance in the course of this treatise, the present one is the most violent ...

Another two paragraphs later he lets an objector speak:

What! the efficiency of causes lie in the determination of the mind! As if causes did not operate entirely independent of the mind, and wou'd not continue their operation even tho' there was no mind existent to contemplate them, or reason concerning them. Thought may well depend on causes for its operation, but not causes on thought. This is to reverse the order of nature, and make that secondary, which is really the primary.

In order to escape this attack, he reverts to his regularity theory two more paragraphs later:

As to what may be said, that the operations of nature are independent of our thought and reasoning, I allow it; and accordingly have observ'd, that objects bear to each other the relations of contiguity and succession; that like objects may be observ'd in several instances to have like relations; and that all this is independent of, and antecedent to the operations of the understanding.

Ever since we have vacillated between Hume's two definitions of causation as a natural and as a philosophical relation, as he calls it, "which are only different by their presenting a different view of the same object" (Hume 1739, p. 170). Only? No! They open a deep schism between an objective and a subjective, a mind-independent and a mind-relative, notion of causation. And somewhere in this schism one must take a position. (I am emphasizing the tension here. For a more considerate assessment of Hume's puzzling remarks on natural and philosophical relations see Beebee (2006, Ch. 4).)

Of course, one may ignore the issue. This might even be wise policy. For instance, one can develop the theory of probabilistic causation and treat many important issues there without taking any stance about which notion of probability the theory is based on. One may be ecumenical and say that the theory works for different notions and thereby receives different interpretations, as Suppes (1970, pp. 13ff.) has suggested. Similarly, the theory of Bayesian nets works for any notion of probability: a subjective, statistical, or a propensity notion. Still, interpretational indecision means avoiding Hume's problem.

Simply taking a one-sided position is, in my view, another way of ignoring the issue. The prevailing attitude, which of course is only reasonable, is to take an outright objectivistic stance; we find it in the regularity theory and in the counterfactual theory (though the objectivity of the truth conditions of counterfactuals is a delicate issue), and it is also at the base of process and mechanistic theories, or the structural equations approach (though the latter's acknowledged model-relativity might diminish objectivity claims).

My fundamental worry about all these objectivistic positions is twofold. One worry is that they do not get the logic of causation right, at bottom because none of them has a workable notion of positive relevance. In a way, the entire Chapter 14 was an extended argument for this point.

The other worry is a more general feeling that purely objectivistic approaches do not do full justice to the peculiar all-important epistemological role of the notion of causation. The objectivistic explanation of that role can only be that the world is full of energy or processes or mechanisms or laws and hence, in some way or other, full of causal relations, and that we therefore take such a deep interest in them as cognizing subjects. This explanation is at the same level as the point that our world is full of fellow humans and that we therefore take a deep (practical and theoretical) interest in human relations. The explanation is insufficient, I feel; the epistemological involvement of causation runs still deeper. It is foreign to me to call it a pure category of thought; but as an honorary title this would not be too exalted.

The other one-sided position is to simply acquiesce in an epistemologically relativized notion of causation. This is prominently argued, e.g., by Hilary Putnam who says that "salience and relevance", which lie at the base of explanation and thus of causation, "are attitudes of thought and reasoning, not of nature. To project them into the realist's 'real world', into what Kant called the *noumenal* world, is to mix objective idealism . . . and materialism in a totally incoherent way" (Putnam 1983b, p. 215). Similarly, van Fraassen (1980a, pp. 112ff.) sees the theory of causation almost entirely absorbed by a theory of explanation that can be understood only in subjectively or pragmatically relativized terms; only an empty objective characterization of "the causal net = whatever structure of relations science describes" (p. 124) remains. This subjectivistic strategy might do more justice to the epistemological role of causation. However, it can only succeed, if it does not only try to make us believe that the realistic intuition about causation is a confusion or illusion, but offers us a plausible account or a convincing substitute of it; that intuition is hardly negotiable.

So, it seems, the task is to reconcile and integrate the two one-sided positions. If, indeed, the relation between the objectivistic or realistic and the subjectivistic or epistemological side does not simply consist in the subject's grasp of real causation, as the objectivists suggest, then the direction of analysis should presumably be reversed, i.e., an objectivistic account of causation should be reached by some kind of objectification of the subjectivistic one. This is the path I wish to take.

There are not so many models for doing so. One might indulge into Kant's complicated doctrine of transcendental idealism, in which the present objectification task is meshed with other and, in the Kantian context, more salient ones concerning space, time, the self, and other objects. I will not attempt to reconstruct the Kantian ways; my (timid?) preference was always to sidestep these paths and the baroque edifices through which they pass, and to see whether other ways lead to the same or close points, from which one has perhaps a better view of the Kantian paths. In the recent discussion the need for mediation between the objectivistic and the subjectivistic side is still felt to be lively, no doubt. Salmon (1984) is certainly one of the most forceful attempts to meet this need by defending an ontic conception of causation without losing the virtues of an epistemic conception. The objective Bayesianism of Williamson (2005) is quite a different attempt. I will not engage in a comparative discussion; my proposal, though straightforward, seems to be unlike all the attempts I know of.

As indicated in Section 12.1, my proposal will follow projectivistic lines. As I said there, the challenge seems less to defend projectivism as a position in the philosophical spectrum, but rather to constructively spell out what the position might be, so that one knows what exactly is being defended. This will be my focus.

In principle, I will simply pursue the traditional ways of David Hume. For the ambiguity between his two definitions of causation already offers a solution to our problem. One might take his subjective associationist account of causation as a natural relation as basic, and one might then objectify it to the extent to which our associations can be explained or justified by existing regularities, as required by causation as a philosophical relation. Insofar as our associations do not have such an objective basis, causation is not objectifiable. However, Hume always conceived of our associations as explainable or justifiable in this way, and thus the two definitions converge.

Of course, this won't do as such. Still, what I intend to do is to simply take that idea, and raise it to the much more sophisticated and precise account of belief and association (= reason) offered by ranking theory. This turns out to be feasible, though it raises problems that Hume could not have seen and that require some ingenuity. I will not immediately focus on causation, although this will be my main ambition. Rather, I will attempt in Section 15.2 to generally explain the extent to which ranking functions are objectifiable, or which features of them are so. Then we will see in Section 15.3 that reason or relevance is not objectifiable in this sense; this confirms Putnam's position. This negative result will not carry over to causation, though; Sections 15.4–15.5 find causation to be objectifiable to a surprisingly large extent; to that extent one can defend a regularity theory of causation. The constructive details will be quite intriguing. Section 15.6 closes my constructive efforts with the result that subjective laws, as explained in (12.10), are objectifiable to the very same extent as that to which the underlying ranking function for the single case is objectifiable. Section 15.8 will take philosophical stock. This chapter essentially relies on Spohn (1993a), but I will indicate where I go beyond it.

15.2 What Could Objectification Mean?

The word "objectification" occurred frequently in the previous section, but so far it is entirely unclear what it could mean. Let me therefore start with two simple observations:

First, the having of a certain belief is a subjective matter; it applies to some doxastic subjects and not to others. However, the belief itself can be true or false, and its truth is an objective matter, as objective as truth is, and in no way subjectively relativized. Still more explicitly: Let $@ \in W$ be the actual (small) world or possibility. Then the belief in A is true iff the proposition A itself is true, i.e., iff $@ \in A$; the objective truth condition of the belief in A is just the proposition A itself; propositions are nothing but truth conditions. This is a trivial observation, deriving from the obvious one–one correspondence between beliefs and the propositions they are about. (Recall, however, all my cautious remarks about propositions as objects of belief in Chapter 2.)

The problem of objectification now is whether this simple observation carries over from beliefs to other doxastic states, and in particular to ranking functions. If we know the extent to which other such states can be objectively true or false, we know the extent to which they are objectifiable. Subjective probabilities, for instance, are not objectifiable in this sense, as I noted in Section 10.3. (It is, indeed, terribly difficult to say what else objective probabilities could mean; see Spohn 2010a for my view on that matter.) Do ranking functions fare better? This is not a trivial issue, either, as the second observation teaches us:

Prima facie, one might say that ranking functions have truth conditions as well. A negative ranking function κ has the core $\kappa^{-1}(0)$ (cf. Section 5.2), and A is believed in κ iff $\kappa(\bar{A}) > 0$, i.e., iff $\kappa^{-1}(0) \subseteq A$. Hence, we might say that κ is true iff all its beliefs are true, i.e., iff $@ \in \kappa^{-1}(0)$. That is, its truth condition is just its core. It is clear, though, that this offers only very partial objectification. The correspondence between ranking functions and their truth conditions in this sense is not one–one, but grossly many–one, since the grading of disbelief given by a ranking function can arbitrarily vary below its core without affecting its truth condition.

Thus, objective truth can so far not distinguish between ranking functions having the same truth condition, but differing wildly in their dynamic and inductive behavior, in their reason and relevance structure, and so forth. In other words, all the aspects that make ranking functions epistemologically interesting fall short of the objectification provided by the first simple observation. Hume's two definitions of causation and their apparently simple relation conceal this problem, and hence do not offer any further guiding line. So, can we do any better and somehow objectify – attach some truth to – our inductive strategies? This is the deep issue, and it has a positive answer, I believe. The deep answer will be, roughly, that our inductive strategies are objectifiable just insofar as they can be cast into a suitable causal structure.

Let us not run ahead, though. There are so far two lessons provided by our initial observations. The first lesson is that the attempt to objectify doxastic states has to refer to a certain feature or aspect of these states, and it is this feature on which objectification focuses. In the first observation, the simple feature considered was the set of beliefs contained in a ranking function. In the second observation, though, we saw that we might as well consider other features, e.g., the relation of direct causation relative to a ranking function. This is what we will do.

The second lesson is that objectification may be only partial. It would be complete if one had a one–one correspondence between all doxastic states considered and something objective like a truth condition. However, such a correspondence may be impossible to reach, and indeed is. Rather, only doxastic states of a particular kind will yield to such a one–one correspondence, namely only those states that behave uniformly with respect to the feature considered. For example, if only ranking functions κ are considered that behave uniformly with respect to belief (or disbelief), say, for which $\kappa(A) = m > 0$ whenever $\kappa(A) > 0$, then these ranking functions are easily put into a one–one correspondence with propositions or truth conditions. But then only such ranking functions, and not the other ones distinguishing different grades of disbelief, may be called objectifiable with respect to belief (or disbelief); and in this sense ranking functions are only partially objectifiable with respect to belief.

With these lessons in mind, we can proceed to the proposed explication of what it means to objectify ranking functions. It consists of two steps. The first step is to specify a natural association of propositions, which have objective truth conditions, with features of ranking functions. Here, an *n-place ranking feature* is just any *n*-place relation ($n \geq 1$) relative to a ranking function or, alternatively, any *n*+1-place relation the *n*+1th place of which is taken by a ranking function. The most natural association I can think of is given by

Definition 15.1: Let R be an n-place ranking feature, and $d_1, \ldots, d_n$ be any objects in the domain of R. Then the proposition $O(R; d_1, \ldots, d_n)$ *associated with* $R(d_1, \ldots, d_n)$ is to be the strongest (or smallest) proposition A for which the obtaining of $R(d_1, \ldots, d_n)$ w.r.t. a negative ranking function κ implies A being believed in κ, i.e., which is such that for each negative ranking function κ: if $R(d_1, \ldots, d_n)$ w.r.t. κ, then $\kappa(\bar{A}) > 0$.

The significance of this definition is best shown by examples (that are too obvious to require proofs):

Theorem 15.2:

(a) If $R_b(A)$ obtains w.r.t. κ iff $\mathbf{B}(A)$, i.e., A is believed in κ, then $O(R_b; A) = A$.

(b) If $R_{cb}(A, B)$ obtains w.r.t. κ iff B is believed in κ conditional on A, i.e., iff $\kappa(\bar{B} \mid A) > 0$, then $O(R_{cb}; A, B) = A \to B$ (where $A \to B = \bar{A} \cup B$ is again defined as the set-theoretical operation representing material implication for propositions).

(c) If $R_{er}(A, B)$ obtains w.r.t. κ iff A is a supererogatory reason for B w.r.t. κ (cf. (6.2)), then $O(R_{er}; A, B) = B$.

(d) If $R_{sr}(A, B)$ obtains w.r.t. κ iff A is a sufficient reason for B w.r.t. κ, then $O(R_{sr}; A, B) = A \to B$.

(e) If $R_{nr}(A, B)$ obtains w.r.t. κ iff A is a necessary reason for B w.r.t. κ, then $O(R_{nr}; A, B) = \bar{A} \to \bar{B}$.

(f) If $R_{ir}(A, B)$ obtains w.r.t. κ iff A is an insufficient reason for B w.r.t. κ, then $O(R_{ir}; A, B) = \bar{B}$.

(g) If $R_r(A, B)$ obtains w.r.t. κ iff A is a reason for B w.r.t. κ (cf. (6.1)), then $O(R_r; A, B) = W$.

(h) If $R_{ec}(A, B, w)$ obtains w.r.t. κ iff A is a supererogatory direct cause of B in w relative to κ (cf. 14.3), then $O(R_{ec}; A, B, w) = C_w^{++}(A, B) \to B$ (where $C_w^{++}(A, B)$ are the temporally possibly relevant circumstances of A and B in w defined in (14.9)).

(i) If $R_{sc}(A, B, w)$ obtains w.r.t. κ iff A is a sufficient direct cause of B in w relative to κ, then $O(R_{sc}; A, B, w) = C_w^{++}(A, B) \cap A \to B$.

(j) If $R_{nc}(A, B, w)$ obtains w.r.t. κ iff A is a necessary direct cause of B in w relative to κ, then $O(R_{nc}; A, B, w) = C_w^{++}(A, B) \cap \bar{A} \to \bar{B}$.

(k) If $R_{ic}(A, B, w)$ obtains w.r.t. κ iff A is an insufficient direct cause of B in w relative to κ, then $O(R_{ic}; A, B, w) = C_w^{++}(A, B) \to \bar{B}$.

(l) If $R_c(A, B, w)$ obtains w.r.t. κ iff A is a direct cause of B in w relative to κ, then $O(R_c; A, B, w) = W$.

Of course, one might think of further ranking features; in the context of this book the above are certainly the most natural and interesting. We will return to all of these examples.

So, far, we have associated a proposition with a ranking feature R only as it applies to given items $d_1, \ldots, d_n$. However, this is immediately extended to an association of a proposition with the ranking feature itself.

> *Definition 15.3*: Let R be an n-place ranking feature. Then the proposition $O(R; \kappa)$ *associated with R in relation to* the negative ranking function κ is defined as the proposition $O(R; \kappa) = \cap\{ O(R; d_1, \ldots, d_n) \mid R(d_1, \ldots, d_n)$ obtains w.r.t. $\kappa\}$.

(15.1 + 3) imply that $O(R; \kappa)$ is believed in κ. Indeed, $O(R; \kappa)$ is the strongest proposition believed in κ in virtue of how the feature R is realized in κ.

The second step is to reverse the procedure and to inquire whether a ranking function can be uniquely reconstructed from the proposition $O(R; \kappa)$. This allows the transference of the objective truth and falsity of $O(R; \kappa)$ to κ itself. Given this unique reconstructibility, but only with this, we may call κ objectifiable w.r.t. the feature R. This uniqueness requires that the feature R is realized in κ in some uniform way; I would not know how to encode different ways of realizing R into the proposition $O(R; \kappa)$. Therefore we need to refer to such a uniform specification of R.

Let us call the ranking feature S a *specification* of the ranking feature R iff S is a subrelation of R. Usually, S will have to be a *numeric* specification in order to achieve the required uniformity, i.e., it will have to fix specific ranks for all the ranking terms occurring in the definition of R, such that the relation R is satisfied. For instance, $\{A \mid \kappa(\bar{A}) = 5\}$ is a (numeric) specification of the 1-place feature $\{A \mid \kappa(\bar{A}) > 0\}$, i.e., of belief; and $\{\langle A, B\rangle \mid \kappa(\bar{B} \mid A) = 2$ and $\kappa(\bar{B} \mid \bar{A}) = 1\}$ is a (numeric) specification of the 2-place feature $\{\langle A, B\rangle \mid \kappa(\bar{B} \mid A) > \kappa(\bar{B} \mid \bar{A}) > 0\}$, i.e., of supererogatory reasons.

Moreover, we will find that the reconstruction possibly works only under certain conditions. Therefore we need to refer to such conditions as well.

Now I am finally prepared for the statement of my abstract explication of objectifiability. Examples will follow soon:

> *Definition 15.4*: Let S be any specification of the n-place ranking feature R; and let F be any condition on the items in the field of R. Then a negative ranking function κ is *objectifiable with respect to R*, or *an objectification of R, under the specification S given condition F* iff, given $O = O(R; \kappa)$, κ is the unique ranking function such that the following holds:
>
> (a) for all $d_1, \ldots, d_n$ $R(d_1, \ldots, d_n)$ obtains w.r.t. κ iff $S(d_1, \ldots, d_n)$ obtains w.r.t. κ,
> (b) for all $d_1, \ldots, d_n$ $R(d_1, \ldots, d_n)$ obtains w.r.t. κ iff $d_1, \ldots, d_n$ satisfy condition F and $O \subseteq O(R; d_1, \ldots, d_n)$.

I omit reference to condition F iff F is empty and reference to the specification S by existential qualification.

Thus, I slip into two ways of talking. Sometimes I say that a ranking function is objectifiable w.r.t. some ranking feature, and sometimes I say that the ranking feature itself is objectifiable, i.e., objectified by some ranking function. Both ways of talking seem appropriate.

Clause (a) of (15.4) expresses the realization of R in κ in a specific way. Clause (b) says that the realization of R in κ is determined by the condition F and the proposition $O(R; \kappa)$. The uniqueness clause finally guarantees that κ can be uniquely reconstructed from the information specified in (a) and (b). This entails that not just any specification will do, but only a numeric or similar one that enforces the uniformity required for uniqueness.

Unconditional objectifiability means that one can infer from $O(R; \kappa)$ alone which $d_1, \ldots, d_n$ the feature R holds for w.r.t. κ. Then one may indeed call κ as objectively true or false as $O(R; \kappa)$ is. However, it may be that condition F is needed for this inference. One might certainly find conditions that trivialize this inference, and then conditional objectifiability is trivialized as well. Hence, how much objectivity is reached by conditional objectification depends on how objective condition F itself is. We will see some instances of the operation of condition F in the sequel.

All this is much too abstract, though, to assess its significance. Let us return to all the examples introduced in Theorem 15.2 and investigate the extent to which they allow objectification. This will occupy us in the next four sections.

15.3 The Non-Objectifiability of Reasons

Belief was my introductory example, and the suggestion was that it is objectifiable. This is confirmed by our explication. Let $R_b(A)$ obtain w.r.t. κ iff $A \neq W$ and $\kappa(\bar{A}) > 0$ (this slightly modifies (15.2a) in an evidently required way), and for some $m > 0$, let $S_b(A)$ obtain w.r.t. κ iff $A \neq W$ and $\kappa(\bar{A}) = m$. Then κ is an objectification of R_b under S_b iff for all $w \in W$ with $\kappa(w) > 0$, $\kappa(w) = m$. In short, a posteriori belief is unconditionally objectifiable according to (15.4).

This is intuitively expected. However, a ranking function is uniquely reconstructible from its beliefs only if it does not differentiate among the disbelieved possibilities and propositions, and such a ranking function is an inductively barren doxastic state. If one is in such a state and receives and accepts some information that was initially disbelieved, then one believes only the information accepted and nothing more. This is devastatingly cautious inductive behavior.

For *conditional belief*, the situation is essentially the same. Let us again slightly modify (15.2b), and let $R_{cb}(A, B)$ obtain w.r.t. κ iff not $A \subseteq B$ and $\kappa(\bar{B} \mid A) > 0$, i.e., B is believed in κ given A. For some $m > 0$ let $S_{cb}(A, B)$ obtain w.r.t. κ iff not $A \subseteq B$ and $\kappa(\bar{B} \mid A) = m$. Then it is easily verified that R_{cb} is objectified under S_{cb} by the very same ranking functions that before objectified R_b under S_b. Hence, there are only poor unconditional objectifications of conditional belief, because the objectifying ranking functions have only one level of genuine conditionality; that is, conditional on something disbelieved only that condition and its logical consequences are believed. This is tantamount to the inductive barrenness of those objectifying ranking functions.

Of course, these remarks immediately carry over to conditionals, i.e., conditional sentences. They reflect within the present framework the results of Lewis (1976) and

Gärdenfors (1988, Sect. 7.4), that conditional beliefs or probabilities can be conceived as beliefs in, or probabilities for, conditional propositions only on pain of trivialization. Indicative conditionals are often assumed to just express conditional belief; they suffer directly from these negative results (cf., e.g., Edgington 1995, Sect. 5–6). If other, i.e., counterfactual conditionals are to be more apt for objective truth conditions, they must either ground outside factual truth, or the grounding in factual truth must take a more sophisticated form than presently envisaged. Lewis (1979b) attempted to give such a more sophisticated grounding for richer, multi-leveled similarity orderings – this grounding, though, referred to the actual laws. (This would be an illegitimate move within the story I am unfolding, and one might well wonder how Lewis' best-system analysis of laws confers on them their peculiar counterfactual force – see my worries in Section 12.1.) We will find such a sophisticated grounding when we get to causation; counterfactual conditionals, when understood causally, will turn out to be truth evaluable.

When we turn to reasons, things get even worse. Look first at *supererogatory reasons*, i.e., at R_{er} as defined in (15.2c). No unconditional objectification of R_{er} can exist, for the simple reason that $O(R_{er}; A, B) = B$ and there is hence no way of telling from $O(R_{er}; \kappa)$ whether A or $\bar{A}$ is to be a supererogatory reason for B relative to κ. The same applies to *insufficient reasons*, i.e., to R_{ir} as given in (15.2f), since A is an insufficient reason for B iff $\bar{A}$ is a supererogatory reason for $\bar{B}$. So, a fortiori, there is no way of objectifying the relation R_r (cf. 15.2g) of being a *reason* simpliciter.

The cheap reason for the non-objectifiability of supererogatory and insufficient reasons does not apply to sufficient and necessary reasons. Still, there is no hope. Let us look at R_{sr} (cf. 15.2d), the relation of being a *sufficient reason*. Here we have $O(R_{sr}; A, B) = A \rightarrow B$. So, if κ is to be unconditionally objectifiable w.r.t. R_{sr}, A would have to be a sufficient reason for B according to κ iff $O(R_{sr}; \kappa) \subseteq A \rightarrow B$. This entails that the set of sufficient reasons for B according to κ would have to be a (complete) ideal; i.e., that set would have to be non-empty, it would have to be closed under finite (arbitrary) union, and it would have to contain all the subsets of its members. This looks quite different from the properties of the reason relation we stated in (6.7). Indeed, we find that for almost all ranking functions there is some proposition for which the set of sufficient reasons is not an ideal:

(15.5) If W contains at least three possibilities and there is some $w \in W$ with $\kappa(w) > 0$, then there are propositions A, A', and B such that A is a sufficient reason for B, $\varnothing \neq A' \subset A$, and A' is a not a sufficient reason for B.

Proof: Let w_0, w_1 be two other possibilities in W and $\kappa(w_0) = 0$. Choose B such that $w \in \bar{B}$ and $\kappa(\bar{B}) = \kappa(w)$, i.e., $\{w' \mid \kappa(w') < \kappa(w)\} \subseteq B$. Moreover, choose A such that also $w \in \bar{A}$ and $\kappa(\bar{A}) = \kappa(w)$. Hence, $\kappa(A \cap \bar{B}) \geq \kappa(w)$ and $\kappa(A) = 0$, so that $\kappa(\bar{B} \mid A) > 0$; and $\kappa(\bar{A} \cap \bar{B}) = \kappa(\bar{A})$, so that $\kappa(\bar{B} \mid \bar{A}) = 0$. In other words, A is a sufficient reason for B. Now set $A' = A - \{w_0\}$. Thus $\varnothing \neq A' \subset A$, and $\kappa(\bar{A}' \cap \bar{B}) = \kappa(w)$ and $\kappa(\bar{A}')$, so that $\kappa(\bar{B} \mid \bar{A}') > 0$. So, A' cannot be a sufficient reason for B and is perhaps no reason at all.

This shows that the sufficient reason relation R_{sr} is not unconditionally objectifiable.

This conclusion immediately carries over to *necessary reasons*, since A is a necessary reason for B iff $\bar{A}$ is a sufficient reason for $\bar{B}$.

Compare this with the case of conditional belief. There, objectification also requires that the set of conditions under which some proposition is believed is a (complete) ideal. This is indeed the case for the objectifying ranking functions we found there. However, this tiny success does not carry over to sufficient reasons, because being a sufficient reason consists in conditional belief plus no belief under the contrary condition.

These negative results concerned unconditional objectification. Does the device of conditional objectification help? The general problem is that of reconstructing the sufficient reason relation R_{sr} from all material implications entailed by $O(R_{sr}; \kappa)$. Perhaps the condition F can make a suitable selection from all these material implications. But how should it do so? I have no idea. I see only a trivial proposal: we might take F to be R_{sr} itself according to κ. It was this proposal that I had in mind when I emphasized that the objectivity reached by conditional objectification is as good as the objectivity of the condition itself. That trivial condition is as subjective as the ranking function κ referred to, and I see no way to do better.

So far, these results were perhaps not so surprising, but they are in any case unimpressive. There is no substantial objectification in sight. This will dramatically change when we turn to causation.

15.4 The Objectification of Causation: First Method

Let us return, for the purpose of this section, to the formal set-up used in Chapter 14 (cf. Section 14.1). That is, U is to be a finite set of finite variables (with X, Y, Z, etc., as members or subsets); W is the possibility space thereby generated; $\leq$ is the weak temporal order on U (extendible to the set of atomic propositions); and the negative ranking function for U, the objectifiability of which we are about to explore, will again be denoted by ξ, in order to signal that the formal set-up is to represent a given single case and nothing else. Many causal relationships will obtain relative to ξ; our question is whether and when ξ is objectifiable with respect to them, that is – since it suffices to consider direct causation – whether the ranking features R_{ec}, R_{sc}, R_{nc}, R_{ic}, and R_c (15.2h–l) are objectifiable.

Since a direct causal relationship *simpliciter* is not associated with any specific belief – recall (15.21): $O(R_c; A, B, w) = W$ – *direct causation* is as little objectifiable as reasons *simpliciter*, namely not at all. This is neither disappointing nor discouraging. Obviously, we must specify the kind of direct causal relationship. However, it seems that *supererogatory* and *insufficient direct causation* (15.2h+k) are as unobjectifiable as supererogatory and insufficient reasons, and for the same reason. So, as might have been expected, sufficient and/or necessary direct causation (15.2i+j) are our only hope. This is how I have presented the matter in Spohn (1993a).

On reconsideration, I think I can do better. I will present two objectification methods, the one given in my (1993a), which focuses on sufficient direct causation and, in the next section, a new one that also covers supererogatory direct causation. Each method has its benefits and its costs, as I will make explicit. The fact that there are several methods shows that there is no unique correct answer to the request for objectification. The task requires constructions, and I have worked out two of them. I welcome all improvements, which might well exist.

The objective counterparts of the various types of causal relations as they obtain relative to ξ are $O(R_{ec}; \xi)$, $O(R_{sc}; \xi)$, $O(R_{nc}; \xi)$, and $O(R_{ic}; \xi)$. First note that $O(R_{sc}; \xi) = O(R_{nc}; \xi)$. For, whenever A is a sufficient direct cause of B in w, $\bar{A}$ is a necessary direct cause of $\bar{B}$ in w' that differs from w only in realizing $\bar{A}$ and $\bar{B}$ instead of A and B, and then (15.2i+j) shows that $O(R_{sc}; A, B, w) = O(R_{nc}; \bar{A}, \bar{B}, w')$. By the same argument we have $O(R_{ec}; \xi) = O(R_{ic}; \xi)$. In short, by attending only to sufficient and supererogatory direct causation, we automatically deal with necessary and insufficient direct causation as well.

So, one idea, the one pursued in Spohn (1993a), is to focus on $O(R_{sc}; \xi)$. The task is then to find methods and conditions for rediscovering all the sufficient direct causal relationships, as they obtain relative to ξ, from $O(R_{sc}; \xi)$. Let us call $O(R_{sc}; \xi)$ *the strict causal pattern of* ξ, and let us denote it by $CP(\xi)$. Keep in mind that $CP(\xi)$ is a proposition that is objectively true or false and not a set of causal relationships, as the name might suggest. In Spohn (1993a, p. 241), I called $CP(\xi)$ the causal law of ξ. This now appears misleading to me in two ways. First, according to (12.10) a law is so far something subjective, a ranking function, and not an objective proposition. Second, a law should possibly apply to many cases, whereas the pattern considered now is bound to the single case at hand. Only in Section 15.6 will we consider the generalization of patterns to laws.

Another idea, which we will scrutinize in the next section, is to focus on $O(R_{sc}; \xi)$ $\cap$ $O(R_{ec}; \xi)$; let us call this the *full causal pattern of* ξ and abbreviate it by $CP^+(\xi)$. Then the task would be to find methods and conditions for rediscovering from $CP^+(\xi)$ all of the sufficient *and* the supererogatory direct causal relationships as they obtain relative to ξ. If I am right, this is indeed possible. Note that, in view of the above observation, this would mean rediscovering *all* of the direct causal relationships of ξ. Working out the first idea, which will occupy us in this section, will prepare us well for the second.

Before starting, though, we have to clear away some trouble caused again by multi-valued variables. The trouble becomes evident when we put Sections 14.7 and 15.3 together. We started Section 14.7 with the observation that, for two variables X and Y, the causal relations between X-propositions and Y-propositions may have as complex a structure as the reason relation has it according to (6.7); we then considered conditions that might reduce this complexity. In Section 15.3 we observed that, due to that complexity, none of the four kinds of the reason relation is objectifiable. Therefore, the causal relations between two variables cannot generally be objectifiable, either. In my (1993a), I overlooked this problem.

There is an exhaustive answer, which I will formally present before commenting on it. Let us focus on two variables $X, Y \in U$ with the ranges W_X and W_Y. And let us forget about the circumstances; i.e., we will deal only with reason relations between X- und Y-propositions. Everything we find out about them also holds conditionally, given fixed circumstances, and thus for direct causal relations between X and Y (we applied this strategy already in Section 14.7).

Let R_{XY} be the sufficient reason relation R_{sr} restricted to X-propositions A and Y-propositions B, and let S_{XY} be a specification of R_{XY} such that for some $m > 0$, and for all $A \subseteq W_X$ and $B \subseteq W_Y$, $S_{XY}(A, B)$ iff $\xi(Y \in \bar{B} \mid X \in A) = m$ and $\xi(Y \in \bar{B} \mid X \in \bar{A}) = 0$. Our question is: can R_{XY} be objectified under S_{XY}? Yes, it can:

Definition 15.6: Y is *one–one correlated* to X w.r.t. ξ iff the relation $\{\langle x, y \rangle \mid x \in W_X, y \in W_Y$, and $\xi(X = x, Y = y) = 0\}$ is a one–one function or bijection from some subset D_X of W_X onto some subset D_Y of W_Y. And Y is *uniformly one-one correlated* to X w.r.t. ξ iff Y is one–one correlated to X w.r.t. ξ and, for some $m > 0$, $\xi(X = x, Y = y) = m$ when it is > 0.

In other words, Y is one–one correlated to X w.r.t. ξ iff ξ believes Y to be a specific one–one function of X (possibly defined only for some subset of W_X). Then our question is answered by

Theorem 15.7: ξ is an objectification of R_{XY} under the specification S_{XY} given the condition on the first relatum A of R_{XY} (the putative reason) that $\varnothing \neq A \subseteq D_X$ iff Y is uniformly one–one correlated to X w.r.t. ξ (with the relevant $m > 0$).

Proof: Let Y be so correlated to X, and let f be the one–one function from D_X onto D_Y thus given. If $B \subseteq W_Y - D_Y$, there cannot be a sufficient reason for $\{Y \in B\}$, since $\xi(Y \in B) = m$ and thus for all non-empty $A \subseteq D_X$, $\xi(Y \in \bar{B} \mid X \in A) = 0$. Hence, suppose that $B \cap D_Y \neq \varnothing$. Let $A^* = f^{-1}[B]$. Now, we intend to show that for any A satisfying the condition of (15.7), i.e., for any A with $\varnothing \neq A \subseteq D_X$, $\{X \in A\}$ is a sufficient reason for $\{Y \in B\}$ iff $A \subseteq A^*$:

If $A \subseteq A^*$, thus satisfying that condition, then $\xi(X \in A) = 0$ and $\xi(X \in A, Y \in \bar{B}) = m$, since for all $x \in A, f(x) \in B$; so, $\xi(Y \in \bar{B} \mid X \in A) = m$. Moreover, if $(X \in \bar{A}) = 0$, then for some $x \in \bar{A}, f(x) \in \bar{B}$, so that $\xi(X \in \bar{A}, Y \in \bar{B}) = 0$, and thus, in any case, $\xi(Y \in \bar{B} \mid X \in \bar{A}) = 0$. Hence, $\{X \in A\}$ is a sufficient reason for $\{Y \in B\}$. Suppose, conversely, that $\varnothing \neq A \cap \overline{A^*} \subseteq D_X$. Then there is an $x \in A \cap \overline{A^*}$ for which, then, $f(x) \in \bar{B}$, so that $\xi(Y \in \bar{B} \mid X \in A) = 0$. Thus, $\{X \in A\}$ is not a sufficient reason for $\{Y \in B\}$.

Therefore, if Y is uniformly one–one correlated to X w.r.t. ξ, then for any $B \subseteq W_Y$ the set of X-propositions that are a sufficient reason for $\{Y \in B\}$ and satisfying the condition of (15.7) (plus $\varnothing$) is an ideal and thus amenable to objectification, as shown before (15.5).

Now, assume that Y is not uniformly one–one correlated to X w.r.t. ξ. If Y is one–one correlated to X, but not uniformly, then ξ cannot conform to the given specification S_{XY}. If Y is not one–one correlated to X w.r.t. ξ, though ξ is uniform otherwise, then for some $B \subseteq W_Y$ the set of X-propositions that are a sufficient reason for $\{Y \in B\}$ (plus $\varnothing$) is not an ideal and hence not objectifiable:

Let $B \subseteq D_Y$ such that for some $y \in B$, $\xi(X = x, Y = y) = 0$ for more than one $x \in W_X$. Then, according to the reasoning above, for $A^* = \{x \in W_X \mid \xi(X = x, Y = y) = 0$ for some $y \in B\}$, $\{X \in A^*\}$ is a sufficient reason for $\{Y \in B\}$. However, there also exists some $A \subset A^*$ such that $B = \{y \in W_Y \mid \xi(X = x, Y \in y) = 0$ for some $x \in A\}$ and $\xi(X = x, Y = y) = 0$ for some $x \in \bar{A}$ and $y \in B$. Thus, $\xi(X \in \bar{A}) = 0$ and $\xi(X \in \bar{A}, Y \in \bar{B}) = \xi(Y \in \bar{B} \mid X \in \bar{A}) = m$; that is, $\{X \in A\}$ is not a sufficient reason for $\{Y \in B\}$, as it would have to be, if the set of those reasons were an ideal.

Theorem 15.7 requires a few comments: First, being uniformly one–one correlated is a much stronger condition than being monotonically related as defined in (14.22). This is not surprising; after all, objectification imposes much stricter demands than we aimed for in Section 14.7.

Still, (15.7) might be surprising in view of (15.5), which proved sufficient reasons to be almost never objectifiable. The point why (15.7) could be more successful is that (15.7), or R_{XY}, does not look at any sufficient reasons whatsoever for Y-propositions, but only at those reasons that are X-propositions. This restriction on the form of reasons made (15.7) possible.

Next, observe that (15.7) is a first proper instance of conditional objectification, a proper one, because the relevant condition (that $\varnothing \neq A \subseteq D_X$ holds for the putative reason A) is entirely objective. That is, it depends on the beliefs in ξ determining the domain D_X of f, and thus on ξ; however, we know already that these beliefs are objectifiable and indeed objectified by the ξ that uniformly correlates X and Y. Therefore, Theorem 15.7 is fully satisfactory.

Is (15.7) a philosophically deep theorem? I am not sure. It seems to be so, but it also looks too frugal to be of more than formal significance. It at least promises a kind of justification of our incessant search for functional relationships between multi-valued variables: they are the only ones according to (15.7) that objectify our inductive reasoning from one variable to the other. This is an answer to a question I never had before. Once I pose the question, however, I notice that it has not found many answers. Of course, the easy answer is that functional relations are the strictest and hence maximally informative relations. However, what distinguishes the maximum besides being the maximum? And why should we be required to reach the maximum? Here, (15.7) might add some weight. Certainly, the issue deserves to be pondered further.

This concludes my interlude on multi-valued variables. We now have a choice. We may carry on our attempts to objectify direct causation generally, i.e., for all kinds of variables. This would entail the employment and adaptation of (15.7) in all the required places, which will be numerous. This would render our further enterprise very unperspicuous. Therefore I prefer the other alternative, namely to restrict all further considerations in this and the next section to *binary variables*, for which (15.7) is automatically satisfied. It is certainly plausible that the considerations thus restricted can be adequately generalized with the help of (15.7); but the cost of my choice consists in leaving, the precise way in which it helps.

Having said all this, let us return to our first, thus restricted, task of rediscovering all sufficient direct causal relationships relative to ξ from the strict causal pattern $CP(\xi)$ of ξ. That is, we now set out to objectify R_{sc} (15.2i). The construction we are to undertake must be based on some suitable specification S_{sc} of R_{sc}. There seems to be only one obvious proposal, namely that S_{sc} is such that for some $m > 0$, $S_{sc}(A, B, w)$ obtains relative to ξ iff $A \preceq B$, $w \in A \cap B$, $\xi(\bar{B} \mid A \cap C_w^{++}(A, B)) = m$, and $\xi(\bar{B} \mid \bar{A} \cap C_w^{++}(A, B)) = 0$ (see (14.9) for the temporally possibly relevant circumstances $C_w^{++}(A, B)$). This is the specification we will use.

Now, our task has three steps. First, we must find a characterization of those propositions that are suitable as causal patterns. Second, we must find a method for reconstructing suitable ranking functions respecting the above specification from such causal patterns. And third we must prove that the construction really works, i.e., falls under our explication (15.4) of objectification. Let us take one step after the other.

First step: Let $P \subseteq W$ be any proposition, let Z be some set of simultaneous variables, $Y \in Z$, and B some Z-proposition. Then C is called a *P-sufficient condition* of B iff C is a $\{\preceq Y\} - Z$-proposition (about the past and residual present of B — we will use the notation introduced in Section 14.1 throughout) and $P \cap C \subseteq B$. *The P-sufficient condition* $SC_P(B)$ of B is defined as the union of all the P-sufficient conditions of B, i.e., as the weakest $\{\preceq Y\} - Z$-proposition C such that P entails $C \to B$. This leads to

Definition 15.8: The proposition P is a *pattern of succession* iff for each variable $Y \in U$ and Y-proposition $B \neq W$ the following holds:

(a)   if Y is a temporally first variable, then $SC_P(B) = \varnothing$,
(b)   if Y is not a temporally first variable and X some variable immediately preceding Y, then $SC_P(B)$ is a $\{< Y\}$-proposition (only about the past of B) and there is a $\{\approx X\}$-proposition $A \neq W$ such that $SC_P(B) = A \cup SC_P(A)$.

If so, there is exactly one such $\{\approx X\}$-proposition that will be called *the immediate P-sufficient condition* $ISC_P(B)$ of B.

Usually we have $ISC_P(B) \neq SC_P(B)$, simply because whatever is sufficient for $ISC_P(B)$ is also sufficient for B. Intuitively, a pattern of succession is a conjunction of material implications of the form "if now A, then next B" that does not entail any categorical proposition about a single variable. The latter assertion is somewhat hidden in (15.8). For P alone to entail such a proposition B, we need to have $SC_P(B) = W$; but this is excluded in clause (a) for the temporally first variables and in clause (b) for all the other ones (by requiring $ISC_P(B) \neq W$).

Another useful assertion about the form of patterns of succession is:

(15.9)   For each pattern P of succession we have: For all $w \in W$ and $Y \in U$, if $E_{w,Y}$ is the strongest proposition such that $P \cap {}^w[< Y] \subseteq E_{w,Y}$, then $E_{w,Y}$ is a *conjunction term* in the sense that, if $\{\approx Y\} = \{Y_1, \ldots, Y_n\}$, then there are Y_i-propositions B_i $(i = 1, \ldots, n)$ such that $E_{w,Y} = B_1 \cap \ldots \cap B_n$. Therefore, $P = \cap\{SC_P(B) \to B \mid B$ is a Y-proposition for some $Y \in U\} = \cap\{ISC_P(B) \to B \mid B$ is a Y-proposition for some $Y \in U\}$.

Proof: According to a pattern P of succession $SC_P(B_i)$ must be a $\{< Y\}$-proposition for each B_i $(i = 1, \ldots, n)$. Hence, conversely, we have $E_{w,Y} = \cap\{B_i \mid {}^w[< Y] \subseteq SC_P(B_i)\}$; that is, $E_{w,Y}$ is a conjunction term as claimed. (If, for instance, $E_{w,Y} = B_1 \cup B_2$, then ${}^w[< Y] \cap \bar{B}_1$ would be a P-sufficient condition of B_2, and P hence not a pattern of succession.)

Now, for any proposition P we have $P = \cap\{{}^w[< Y] \to E_{w,Y} \mid w \in W, y \in Y\}$. If, however, all $E_{w,y}$ are conjunction terms, then the material implication can be rearranged to yield $P = \cap\{SC_P(B) \to B \mid B$ is a Y-proposition for some $Y \in U\}$. The last equation then follows from Definition 15.8.

Such patterns of succession are plausible candidates for fulfilling the role of causal patterns. This needs to be confirmed in our second step, that starts from such patterns and then attempts to reconstruct ranking functions that have them as their causal patterns. In Spohn (1993a), my only idea was that, for each world w, the ranking function reconstructed from a pattern P of succession should count how many atomic facts implied by P and their respective pasts failed to obtain. In the next section, we will pursue another idea richer in presuppositions and results, namely the idea that the ranking function counts how many causal arrows implied by P are violated. However, let us start with the first idea:

Definition 15.10: Let P be a pattern of succession. Then P is *violated* by the variable $Y \in U$ in the world $w \in W$ iff $w \in ISC({}^w\overline{[Y]})$, i.e., iff the immediate P-sufficient condition of ${}^w\overline{[Y]}$ obtains in w. And the ranking function ξ_P *strictly associated with* P is defined by $\xi_P(w) = m \cdot v_w$, where v_w is the number of variables violating P in w.

The ranking functions thus associated have a much richer inductive life than those objectifying (conditional) plain belief, since there may occur as many violations of P as there are variables, and ξ_P may take as many different values. Hence, the predictive and retrodictive behavior of these ranking functions under various conditionalizations may be quite complex. It is, however, characterized by two simple properties: First, the frame U is phase-separated by ξ_P (cf. 14.45); i.e., simultaneous variables are independent, conditional on their past. Second, any Y-proposition is expected by ξ_P to obtain whenever its immediate P-sufficient condition is satisfied. Thus, ξ_P displays a peculiar inductive persistence already familiar to us from subjective laws: No matter how often P is violated in the past, ξ_P invariably expects that no further violation(s) will occur.

The third and final step is to check whether these ranking functions are really suitable for the objectification of sufficient direct causation. Are they? Not generally; there are two problems entailing instructive conditions.

The first problem is that P need not be the strict causal pattern of ξ_P. We are guaranteed that $P \subseteq CP(\xi_P)$, since whenever A is a sufficient direct cause of B in w w.r.t. ξ_P, $A \cap C_w^{++}(A, B)$ is also a P-sufficient condition of B. However, the converse need not hold. For an illustration, look at the case (14.5c), where A and B redundantly cause C (in $\{w\} = A \cap B \cap C$). Here, the pattern P of succession is $A \cup B \rightarrow C$, and ξ_P is given by the table (14.5). However, A is a sufficient cause of C only in the absence of B, and B only in the absence of A. Thus, $CP(\xi_P) = (A \cap \bar{B}) \cup (\bar{A} \cap B) \rightarrow C$. This was precisely the problem of redundant causation, where neither of the two facts A and B is a cause of C in the presence of the other.

Surely, though, objectification requires the identity of P and $CP(\xi_P)$. When does it hold? This is answered by:

(15.11) Let P be a pattern of succession. Then $P = CP(\xi_P)$ iff for each $Y \in U$ each Y-proposition $B \neq W$ has at least one sufficient direct cause in each $w \in ISC_P(B)$ w.r.t. ξ_P. This condition is equivalent to the condition that for each $Y \in U$, each Y-proposition $B \neq W$, and each $w \in ISC_P(B)$, X being a variable immediately preceding Y, the intersection of all subsets Z of $\{\approx X\}$ such that $^w[Z] \subseteq ISC_P(B)$ is non-empty. Hence, let us call a pattern satisfying the equivalent conditions an *intersecting pattern of succession*.

Proof: Concerning the equivalence of the two conditions, recall that we are dealing only with binary variables so that each cause of B has the form $^w[X]$. Then it is easily seen that $^w[X]$ is a sufficient direct cause of B in $w \in ISC_P(B)$ w.r.t. ξ_P iff X is a member of each subset Z of $\{\approx X\}$ such that $^w[Z] \subseteq ISC_P(B)$.

Now, let these conditions be satisfied. Then $CP(\xi_P) = \bigcap \{A \cap C_w^{++}(A, B) \rightarrow B \mid R_{sc}(A, B, w)\}$ holds relative to $\xi_P\} = \bigcap \{ISC_P(B) \rightarrow B \mid B$ is a Y-proposition for some $Y \in U\} = P$ according to (15.9). Conversely, suppose that these conditions are violated for some Y-proposition B and some $w \in ISC_P(B)$. Then $CP(\xi_P) \subseteq \ ^w[< Y] \rightarrow B$ does not hold in virtue of w, and, according to ξ_P, there is no other causal relationship that brings about this inclusion. However, $P \subseteq \ ^w[< Y] \rightarrow B$, and hence $P \neq CP(\xi_P)$.

Again, (15.11) seems to be a philosophically deep assertion insofar as it says that some version of the principle of causality is presupposed by objectification, or at least by the present method. The fact that our second method will get along without this assumption (see 15.24) rather suggests, however, that the apparent philosophical depth is only an artifact of the present method. In any case, we will leave the question of which attitude to really take towards the principle of causality until Section 17.4.

The point of giving two equivalent conditions in (15.11) is this: The first intuitively intelligible and philosophically interesting condition appears to be of doubtful objectivity, since it seems to be about causation and thus a subjective condition about ξ_P. This impression is proved wrong by the second equivalent condition, since it is exclusively about the logical form of the pattern P of succession and thus entirely objective.

(15.11) solves the first problem of objectification. There is a second problem, though, namely that it is not generally possible to rediscover a ranking function from its causal pattern. The problem is that the causal pattern is unable to distinguish whether three facts A, B, and C form a causal chain or a causal fork (see (14.4) for illustration), at least in the case where A is a necessary and sufficient direct cause of B so that the causal pattern leaves undecided whether B is a sufficient direct cause of C (causal chain) or A is (causal fork). We saw in (14.4) that this distinction presented great difficulties for regularity theories, but for neither the counterfactual analysis – which, however, is of doubtful objectivity – nor our ranking-theoretic account, as displayed in (14.4c+d).

The only natural way I see to get rid of this problem is to assume that direct causes *immediately* precede their direct effects. (See, however, Baumgartner (2008a) for a different response to this problem.) This assumption allows us to make the required distinction, since it excludes the possibility of A, B, and C forming a causal fork when B precedes C, or of them forming a causal chain when B and C are simultaneous. And it allows the distinction to be made in an objective way, since it is only about temporal relations. So, let us give it a name:

(15.12) *Condition TIDC* (temporal immediacy of direct causation): $R_c(A, B, w)$, and thus in particular $R_{sc}(A, B, w)$, only if A immediately precedes B.

Of course, TIDC entails the exclusion of simultaneous causation, which, however, we already denied objectification through the assumption that patterns of succession were strict causal patterns. There are more familiar and perhaps better reasons for the exclusion of simultaneous causation, pertaining in particular to spatial relations and the limited speed of causal influence; however, the present consideration offers a new and perhaps attractive reason.

So, can we proceed to our first objectification result? Yes, almost. There is still a small technical problem with the temporally first variables. Since propositions about them do not have (frame-relative) causes, causal relations cannot fix any ranks for them. However, we have to fix them somehow in the objectifying ranking functions, and the most natural fixing is to say that the objectifying ranking functions abstain from any beliefs whatsoever about the temporally first variables, as the ξ_P already do. So, the point is that objectification must enforce this fixing, and this can only be done by slightly extending the relation R_{sc} to be objectified. Hence, we define $R'_{sc}(A, B, w)$ to obtain relative to ξ iff either $R_{sc}(A, B, w)$ obtains relative to ξ or B is about a temporally first variable and $\xi(B) = 0$. The specification S'_{sc} of R'_{sc} is to be defined accordingly.

Now we can finally state our first main result about the objectification of sufficient causation:

Theorem 15.13: The ranking function ξ is an objectification of R'_{sc} under the specification S'_{sc} given condition TIDC if and only if $\xi = \xi_P$ for some intersecting pattern P of succession.

Proof: Let P be an intersecting pattern of succession. Obviously, ξ_P satisfies clause (a) of Definition 15.4. It is also clear from (15.8) and (15.10) that ξ_P satisfies clause (b) of (15.4), since P is intersecting. Note that we require condition TIDC here; we might have $P \subseteq A \to B$ and $P \subseteq B \to C$, so that $P \subseteq A \to C$. However, the latter material implication is not to represent a direct causal relationship, which is precisely excluded by that condition.

In order to see that ξ_P is the only ranking function satisfying clause (a) and (b) of (15.4), observe first that, due to (15.9), simultaneous variables are bound to be independent conditional on their past w.r.t. to any ranking function satisfying clause (b) of (15.4). Observe further that for any ranking function ξ satisfying (a) and (b) of (15.4) the value $\xi({}^w[Y] \mid {}^w[< Y])$ is thereby determined for each $w \in W$ and $Y \in U$ to be equal to $\xi_P({}^w[Y] \mid {}^w[< Y])$ (namely, $= m$ or $= 0$, depending on whether or not ${}^w[< Y] \subseteq ISC_P(\overline{{}^w[Y]})$). Observe finally that the extension of R_{sc} to R'_{sc} requires $\xi(B) = 0$ for any proposition B about the temporally first variables and any ξ satisfying clause (b) of (15.4). However, all three observations together uniquely determine ξ to be equal to ξ_P.

This proves the "if"-direction of (15.13). For the "only if"-direction suppose that ξ is the only ranking function satisfying (a) and (b) of (15.4) with respect to R'_{sc} and S'_{sc}. (b) clearly entails that $CP(\xi)$ is an intersecting pattern of succession. It is clear then that (a) forces ξ to be of the special form ξ_P, where $P = CP(\xi_P)$.

Note, by the way, that it was essential for (15.13) to proceed from our basic explication (14.3) of sufficient direct causation, which allowed for simultaneous causation. If we had ruled out simultaneous causation by definition, no condition on causal relations could say anything about simultaneous variables and force on them the conditional independence needed for the unique reconstruction of ξ_P from the sufficient direct causal relationship obtaining relative to it.

So much, then, for my findings in Spohn (1993a), slightly corrected (in particular by (15.9), the validity of which for all patterns of succession I had not realized). Reconsidering the entire issue, I now think that there is another interesting way of objectifying causation, which is the subject of the next section.

15.5 The Objectification of Causation: Second Method

The crucial difference between the second and the first method of objectifying causation becomes clear when we look at cases of causal overdetermination (example 14.6). Suppose the prince sings a love song (A), accompanies it with the mandolin (B), and still the princess does not wake up ($\bar{C}$). Since C is overdetermined by A and B, $\bar{C}$ is doubly unexpected; this is how, within my epistemic relativization, I could simply account for overdetermination with table (14.6). However, this account cannot be duplicated with the objectified ranking functions proposed above. If $\bar{C}$ happens, just one variable violates the relevant pattern of succession, just one fact goes wrong.

Hence, the objectifying ranking function that simply counts factual exceptions cannot count more, and reduces overdetermining causes (14.6) to redundant causes (14.5c). The point that $\bar{C}$ would be doubly wrong, as it were, does not make objective sense in this perspective.

All of this is quite plausible, and I offered it as an explanation of the fact that objectivistic theories of causation have so much trouble with overdetermination. Meanwhile I think that the argument is not inescapable. Maybe we are not only able to count the number of violations of singular predictions based on a pattern of succession; maybe we are also able to count the number of violations of causal arrows contained in a pattern of succession. And in the case of the prince and the princess we should count $\bar{C}$ as violating two causal arrows. If we could count in this way, then we should also be able to objectify overdetermination, i.e., supererogatory causes. Here is my proposal for doing so:

We have to go through the same three steps as with the first method. The first step is the same as before; patterns of succession as defined in (15.8) are also the appropriate objective base for our second method. The difference will be that the patterns of succession are now to serve not as strict, but as full causal patterns representing sufficient *and* supererogatory and hence *all* direct causes.

It seems that we are about to objectify two ranking features at once, a case not provided for by (15.4). A simple trick will do: Consider the 4-place ranking feature $R_{esc}(A, B, w, k)$ with $k \in \{e, s\}$, which is defined as $k = e$ and $R_{ec}(A, B, w)$ or $k = s$ and $R_{sc}(A, b, w)$; i.e., k operates here as a parameter for the kind of direct causal relationship. Then $O(R_{esc}; \xi) = O(R_{sc}; \xi) \cap O(R_{ec}; \xi) = CP^+(\xi)$. So, we are about to objectify R_{esc} while talking about sufficient and supererogatory causes as before.

The difference must show up in the second step when we seek an alternative association of ranking functions with patterns of succession. So far, when such a pattern implies $ISC(^w[Y]) \to {}^w[Y]$, the case in which $\overline{{}^w[Y]}$ occurs while $ISC(^w[Y])$ obtains counted as *one* violation of predicted facts. Now, we want to say that many causal arrows may hide in this material implication, so that many arrows are violated when the material implication turns out to be false. How can we do so? Only by stipulating a logical form of causal arrows to be objectified. The only form that I find suggested in the literature is that it can only be *a minimal conjunctive combination of singular causes, i.e., atomic propositions, that brings about an effect* and that only material implications of that logical form can be interpreted as causal.

This is best explained with our above example of overdetermination: There, the full causal pattern was $A \cup B \to C$. However, it does not have the required logical form because of the disjunctive antecedent. The only way to give it the logical form required for causally interpreting material implication is to write it as $(A \to C) \cap (B \to C)$. Thus, two causal arrows hide in this causal pattern, both of which are violated when $A \cap B \cap \bar{C}$ occurs. In other words, redundant causes (14.5c) are automatically interpreted as overdetermining causes (14.6).

The objectification method thus indicated can be stated in full generality with the help of the familiar notion of an INUS condition, invented by Mackie (1965). That is, for any possible effect B, the immediate P-sufficient condition $ISC_p(B)$ of B, which was used by the first method in an unanalyzed way, is now to be represented as an INUS condition of B. In order to do so properly and to uncover the assumptions needed, we have to recall some basic facts about irredundant and minimal disjunctive normal forms, a field which has mainly been studied by computer scientists, as a way of saving computing resources (for the following, cf., e.g., Schneeweiss 1989, Ch. 4–5):

We are still dealing only with binary variables. Take any set $V \subseteq U$ of variables and any V-proposition D.

> *Definition 15.14*: A *disjunctive normal form* (DNF) of D is any representation $D = (D_{11} \cap \ldots \cap D_{1l_1}) \cup \ldots \cup (D_{k1} \cap \ldots \cap D_{kl_k})$, where for each $i = 1, \ldots, k$ all $D_{i1}, \ldots, D_{il_i}$ are atomic propositions about different variables. Each disjunct, i.e., each conjunction of atomic propositions about different variables in V is called a *conjunction term*.

So, the V-states, the logically strongest V-propositions, are conjunction terms of maximal length. Usually, a V-proposition D will have many DNFs. However, when we require all conjunction terms of a DNF of D to be of maximal length, i.e., V-states, then exactly one DNF remains that is called *the canonical disjunctive normal form* of D.

Now we wish to go in the other direction and make the conjunction terms of a DNF of D and the DNF itself as small or short as possible. The first step is clear:

> *Definition 15.15*: Call a conjunction term $D_i = D_{i1} \cap \ldots \cap D_{il_i}$ a *prime implicant* (PI) of D iff $D_i \subseteq D$, but no shorter conjunction is a subset of D.

So, the PIs of D are the shortest conjunction terms we can use in a DNF of D. Moreover, it is clear that the disjunction of all PIs of D is a DNF of D, the largest, though, we can build from all the PIs. The question is how to build shorter or shortest ones.

There is no unique answer, and this will also hamper our objectification task. Look, for instance, at $D = (A \cap B) \cup (\bar{A} \cap C)$. Clearly, $A \cap B$ and $\bar{A} \cap C$ are two PIs of D, but there is a third one, $B \cap C$. So, there are two DNFs of D consisting only of PIs, one built from the first two PIs, and one built from all PIs. Let us see what this means in causal terms by assuming that D is the immediate P-sufficient condition of some possible effect E. The first of the two DNFs suggests that A acts as a switch regarding E: in the position A, A and B are the direct causes of E, and in the position $\bar{A}$, $\bar{A}$ and C are. The other DNF suggests that some overdetermination is involved as well: if both B and C obtain, they, too, are causes of E. Thus, if, e.g., A obtains, there are two causal paths to D, and B contributes to both of them. How should we resolve the ambiguity between these two causal situations?

Well, the example is highly problematic, anyway. In Section 14.13 we learned to accept switches. However, we did not, and could not, accept direct switches. If A acts as a switch, its two positions ("yes" and "no") should turn on distinguishable causal paths; if, however, E is the direct effect both times, there are not two causal paths to distinguish, and hence there is no switch to discern. So, the D of our example is a dubious immediate P-sufficient condition. Still, if the ambiguity is to be resolved, it seems clear how to do so: $B \cap C$ is a redundant conjunction term of the longer DNF of D, as is clear from the fact that deleting it produces the shorter DNF. So, if the aim is to find shortest DNFs we must not allow such redundant parts.

This motivates the following

Definition 15.16: A DNF of D is an *irredundant disjunctive normal form* (*IDNF*) of D iff all its conjunction terms are prime implicants of D and necessary parts of the DNF in the sense that no shorter disjunction of fewer conjunction terms is a DNF of D, i.e. equivalent to D.

So, the idea is to go for such IDNFs. The idea is not helpful, though, since IDNFs are not unique:

Look at the example given by Quine (1952), where

$$(15.17a) \quad D = (A \cap \bar{B}) \cup (\bar{A} \cap B) \cup (B \cap \bar{C}) \cup (\bar{B} \cap C).$$

Obviously, (15.17a) shows an IDNF of D. It requires only a little calculation, though, to see that there are two further IDNFs of D, namely

$$(15.17b) \quad D = (A \cap \bar{B}) \cup (B \cap \bar{C}) \cup (\bar{A} \cap C), \text{ and}$$
$$(15.17c) \quad D = (\bar{A} \cap B) \cup (\bar{B} \cap C) \cup (A \cap \bar{C}).$$

This example leaves us helpless; there is no hint whatsoever which of these three IDNFs to prefer, if we want to read them causally (i.e., as the IDNF of the immediate P-sufficient condition D of some possible effect E). Of course, we may simply stipulate that, say, (15.17b) is the one representing the intended causal relations. However, in our objectification task we must not stipulate causal structure, we must infer it from logical structure; and this seems entirely impossible in the example given.

I conclude that the required inference works only if we do not allow any such ambiguities. They cannot arise only if the only DNF consisting of PIs is the one consisting of all PIs, i.e., if the latter is already an IDNF. This is captured in

Definition 15.18: A V-proposition D is called *PI-maximal* iff the disjunction of all prime implicants of D is the only IDNF of D.

In this case, the maximal IDNF of D is at the same time the unique minimal or shortest DNF of D. There is a useful characterization of PI-maximal propositions:

Definition 15.19: A PI D_i of D is a *core prime implicant* of D iff there is a V-state $D^* \subseteq D$ such that D_i is the only PI of D for which $D^* \subseteq D_i$.

Clearly, each IDNF of D must contain all core PIs of D, and hence we have:

$$(15.20) \quad \text{A } V\text{-proposition } D \text{ is PI-maximal iff all its PIs are core PIs.}$$

This observation suggests a stronger notion, which will indeed be needed later on:

Definition 15.21: A *V*-proposition *D* is called *PI-independent* iff all its PIs are logically independent, i.e., if $D_1, \ldots, D_k$ are all PIs of *D*, then $D'_1 \cap \ldots \cap D'_k \neq \emptyset$ for all $D'_i \in \{D_i, \bar{D}_i\}$ $(i = 1, \ldots, k)$.

Clearly, PI-independence is stronger than PI-maximality, since it requires more logical combinations of PIs to be non-empty.

After this excursion to disjunctive normal forms, let us return to our objectification task. First, we extend (15.18) and (15.21) to

Definition 15.22: Let *P* be a pattern of succession. Then we call *P* itself *PI-maximal* iff for each *Y*-proposition *B* $(Y \in U)$, $ISC_P(B)$ is PI-maximal. And *P* is *PI-independent* iff for each *Y*-proposition *B* $(Y \in U)$, $ISC_P(B)$ is PI-independent.

Thus, we have reached the first goal we sought:

Definition 15.23: Let *P* be a PI-maximal pattern of succession. Then for each (atomic) *Y*-proposition *B* $(Y \in U)$ *the P-INUS condition of B*, denoted by $INUS_P(B)$, is the unique irreducible disjunctive normal form of $ISC_P(B)$.

Hence, the point of our excursion was simply this: on the one hand, the INUS condition for some possible effect *B* had to be an IDNF of $ISC_P(B)$; on the other hand, this is an unambiguous instruction only under the presupposition of PI-maximality.

If the *P*-INUS condition of *B* has at least two conjunction terms each of which consists of at least two conjuncts, then indeed each atomic proposition occurring in $INUS_P(B)$ is an insufficient, but necessary part of an unnecessary, but *P*-sufficient condition for *B*, as Mackie (1965) conceived it in what was then the most sophisticated version of the regularity theory of causation. (Of course, the limiting cases with fewer disjuncts or conjuncts are also covered by Mackie's definition.) So, a merely terminological difference is that what I call *the* INUS condition of *B* is already the complex of all of the INUS conditions for *B* in Mackie's sense. A substantial difference is that Mackie required that the INUS condition for *B* is indeed a sufficient *and* necessary condition for *B*. There seems to be no need for this requirement. *B* has the immediate sufficient condition $ISC_P(B)$, and $\bar{B}$ has $ISC_P(\bar{B})$. Of course, $ISC_P(B) \cap ISC_P(\bar{B}) = \emptyset$ must hold. However, $ISC_P(B) \cup ISC_P(\bar{B}) = W$ need not hold; *P* may allow cases in which it makes no prediction about *B*. A second substantial difference is that all of the atomic propositions to be found in $INUS_P(B)$ have to be simultaneous, something not presupposed by Mackie.

Now, we can finally implement our second objectification strategy and say what the counting of violations of causal arrows amounts to:

Definition 15.24: Let *P* be a PI-maximal pattern of succession. Then the ranking function ξ_P^+ *fully associated with P* is defined by $\xi_P^+(w) = m \cdot \sum_{Y \in U} v_w(Y)$, where $v_w(Y)$ is the number of violations of causal arrows by $^w[Y]$ in *w*, i.e., the number of conjunction terms or prime implicants in $INUS_P(^w[Y])$ obtaining in *w*.

Let us make clear what the assumption of PI-maximality really means in causal terms. According to (15.20) it says that for each possible effect B, all prime implicants of $ISC_P(B)$ are core PIs. This means that for each PI of $ISC_P(B)$ there is at least one possibility in which that PI is the only true PI and thus lists the only direct causes B has within that possibility (which then are sufficient direct causes). In other words, if we allow for several direct causal paths overdetermining a direct effect, the PI-maximality of patterns of succession guarantees that there is always a situation in which a given direct causal path is the only actual or active causal path leading to this effect; it is always possible to isolate that causal path and to exclude the other paths also leading to the effect. This seems to be an extremely reasonable causal requirement on the logical form of patterns of succession.

This explains at the same time what the stronger assumption of PI-independence needed below amounts to. It says that for each subset of a set of causal paths that possibly overdetermine a direct effect, i.e., that are possibly jointly realized, there must always be a situation in which precisely the paths of the subset and none of the others are actual or active. Assume, for instance, that $ISC_P(B) = INUS_P(B) = (A_1 \cap A_2) \cup (A_1 \cap A_3) \cup (A_2 \cap A_3)$. Here, PI-maximality is obviously satisfied, but not PI-independence; there is no way, e.g., to make the first two PIs true, but the third false. In other words, there are ways in which B is sufficiently caused, but there are no ways in which B is overdetermined by two paths only; overdetermination, if it occurs, must be threefold. Here, I am less sure whether it is reasonable to exclude such situations and hence require PI-independence. However, I do not see a way of avoiding this assumption below.

(15.24) completes step two for our second objectification method. Step three is to prove that the method really works. We proceed as before. First we have:

(15.25)   If P is a PI-maximal pattern of succession, then $P = CP^+(\xi_P^+)$.

Proof: First we have: Whenever A is a sufficient direct cause of B in w w.r.t. ξ_P^+, then $A \cap C_w^{++}(A, B)$ is a P-sufficient condition of B; and conversely, if $A \cap C_w^{++}(A, B)$, but not $C_w^{++}(A, B)$ by itself, is a P-sufficient condition of B, then A is a sufficient direct cause of B in w w.r.t. ξ_P^+. Next, it is also clear that, if A is a supererogatory direct cause of B in w w.r.t. ξ_P^+, then $C_w^{++}(A, B)$ by itself is a P-sufficient condition of B.

The tricky case is the last one: Suppose $C_w^{++}(A, B)$ (i.e., the past of B in w without A) by itself is a P-sufficient condition of B. How are we to tell whether A or $\bar{A}$ or neither is a supererogatory direct cause of B in w w.r.t. ξ_P^+? It is $INUS_P(B)$ that decides about the three possibilities: Let $D_1, \ldots, D_j$ be the PIs of $ISC_P(B)$ compatible with $C_w^{++}(A, B)$. Suppose further that A is a conjunct of D_1, say, and $\bar{A}$ a conjunct of D_2, i.e., $D_1 = A \cap D'$ and $D_2 = \bar{A} \cap D''$ for some conjunction terms D' and D''. This entails that $D' \cap D''$ must also be among the PIs of $ISC_P(B)$ compatible with $C_w^{++}(A, B)$. However, it cannot be a core PI, since whenever $D' \cap D''$ is true in some possibility, either D_1 or D_2 is true as well. So, the supposition contradicts the PI-maximality of P. In other words, $INUS_P(B)$ has at most one conjunction term compatible with $C_w^{++}(A, B)$ and containing either A or $\bar{A}$ as a conjunct. If there is no such term, neither A nor $\bar{A}$ is a direct cause of B in w w.r.t. ξ_P^+. If there is one and it contains A, A is a supererogatory direct cause of B in w w.r.t. ξ_P^+; otherwise, $\bar{A}$ is.

For the concluding theorem we face the same problems as we did for (15.13). First, there was the problem of objectively distinguishing between causal chains and causal forks, and I adopt here the same solution as before by imposing condition TIDC (15.12). Second was the technical problem about temporally first variables. Hence, define again $R'_{esc}(A, B, w, k)$ to obtain relative to ξ if either $R_{esc}(A, B, w, k)$ obtains relative to ξ or B is about a temporally first variable and $\xi(B) = 0$.

Finally, we must fix the required specification S'_{esc} of R'_{esc}. S'_{esc} cannot be a full numeric specification, since it must be open to multiple overdetermination. The following specification will do: for some $m > 0$ let $S'_{esc}(A, B, w, k)$ obtain relative to ξ iff either B is about a temporally first variable and $\xi(B) = 0$, or $R'_{esc}(A, B, w, k)$ obtains relative to ξ and $\xi(\bar{B} \mid A \cap C_w^{++}(A, B)) = m + \xi(\bar{B} \mid \bar{A} \cap C_w^{++}(A, B))$. However, as we will see in the proof below, it will do only in combination with the assumption of PI-independence. Only then does the way in which ranks build up in multiple overdetermination seem to be uniquely determined. I suspect that there are cleverer constructions at this point; however, I am happy to have found one that works at all. Now we can state our final

> *Theorem 15.26*: The ranking function ξ is an objectification of R'_{esc} under the specification S'_{esc} given condition TIDC (15.12) if $\xi = \xi_P^+$ for some PI-independent pattern P of succession.

> *Proof*: Let P be a PI-maximal pattern of succession. Then ξ_P^+ obviously satisfies clause (a) of (15.4). It is also clear from (15.8), (15.24, and (15.25) that ξ_P^+ satisfies clause (b) of (15.4).
>
>   The fact that ξ_P^+ is the only ranking function satisfying clauses (a) and (b) of (15.4) is entailed by the same argument as in the proof of (15.13). First, (15.9) again entails that simultaneous variables are independent conditional on their past w.r.t to any ranking function satisfying clauses (a) and (b) of (15.4). Next, it is clear that for any ranking function ξ satisfying (a) and (b) of (15.4) the value $\xi({}^w[Y] \mid {}^w[< Y])$ is thereby determined for each $w \in W$ and $Y \in U$ to be equal to $\xi_P^+({}^w[Y] \mid {}^w[< Y]) = m \cdot v_w(Y)$, since PI-independence allows the counting of how many PIs of $INUS_P({}^w[Y])$ are violated by ${}^w[< Y]$. Finally, the extension of R_{esc} to R'_{esc} requires $\xi(B) = 0$ for any proposition B about the temporally first variables and any ξ satisfying clause (b) of (15.4). These three points together suffice to uniquely determine ξ to be equal to ξ_P^+.

Note that (15.26) is weaker than (15.13), insofar as it states PI-independence only as a sufficient condition of objectifiability. This expresses my uncertainty about whether PI-independence is really required for objectification. Let me thereby conclude our admittedly – but unavoidably – tedious constructive work. We should take stock of what we have achieved.

Let me first once more summarize our general strategy. $CP(\xi)$ and $CP^+(\xi)$ served as objective counterparts of the causal relations as they subjectively hold relative to ξ.

A lot of material implications, even only temporally directed ones, hide in $CP(\xi)$ and $CP^+(\xi)$. Our problem was then to sort out, from all of these material implications, those that could and should be interpreted causally. Mainly, our solution was to rely heavily on temporal relations. On the subjective side, we required a pattern of succession to be decomposable into stages, i.e., we required the sufficient condition of any atomic proposition to be decomposable into an immediate and a mediate part. And on the objective side, we imposed the corresponding condition TIDC in (15.13) and (15.26), that direct causes immediately precede their direct effects. Without this massive temporal restriction, the required sorting out would have been impossible.

Even this did not suffice. Not all patterns of succession could be used for the strict method, but only those satisfying some principle of causality. And then the strict method objectified only sufficient (and necessary) direct causal relationships. The full method was more powerful in objectifying all direct causal relations. However, it did so only by introducing a strong assumption of a new kind, i.e., by restricting the logical form of causally interpreted material implications. This restriction looked natural, but I have avoided a justificatory discussion of it.

I believe that the objectification methods have a great potential for clarification. It is no surprise that the history of the philosophy of causality is full of confusions over issues of objectivity and subjectivity. When the terrain is so uncertain, it is hard to correctly position all of the notions and ideas one might have. In combination with my subjective analysis my account of objectification makes a proposal here:

(i)   For instance, consider the idea that a direct cause immediately precedes its direct effect. I have argued after Definition 14.3 that this idea must not be part of an analysis of causation. Still, it seemed natural. And now we can understand its naturalness by its being required for objectification; objectified causal relations must conform to it, at least according to our two objectification methods. (When discussing principles of causality in Section 17.4, condition TIDC will again play an important role.)

(ii) Or look at simultaneous causation. For various reasons – cf. Sections 14.3 and 14.10 – it was reasonable not to exclude it on conceptual grounds. Still, there are great reservations about it. The standard argument against it is directed against action at a distance and is thus of a spatial nature. Maybe, though, it is simply not objectifiable; our two methods, in any case, excluded simultaneous causation. There are perhaps cleverer methods that allow it; I doubt it, though.

(iii) The point extends to interactive forks. In Section 14.10 I have proposed a way of accounting for them. This account made heavy use, however, of frame-relative simultaneous causation. So, perhaps, the objectifiability of interactive forks is at least problematic. This might be a further reason for the reluctance they meet.

(iv) I have already mentioned causal overdetermination as a further case in point. The strict method of objectification was unable to account for it. One might say, too bad for the method, fortunately we have a better one. Or one might say, as I already suggested, that this is part of the explanation of why objectivistic accounts have so many difficulties with overdetermination.

(v) Quite a different point was suggested by (15.7). The search for functional relationships between two multi-valued variables has no subjective foundation (besides the desire for maximal information); but it seems required for objectifying our inductive inference from one to the other variable and might thus have objective grounds.

(vi) Here is another example: It is not difficult to confirm that at least according to the second objectification method the identity (14.21) between the larger actually relevant circumstances (14.18) and the ideal circumstances (14.20) of direct causal relationships always holds; that is, objectively, circumstances seem to have to be ideal. So, the attractiveness of assuming ideal circumstances might well have objectivistic grounds.

(vii) A final point: Our objectification methods only dealt with direct causation. We saw in Sections 14.11–14.13 that indirect causation is a messy topic and in particular that it is not so clear how to generalize the different kinds of direct causation to indirect causation. I gave my reasons why I accept the transitivity of (direct or indirect) causation. I wonder, though, how the transitivity issue presents itself in the objectification perspective. It seems straightforward to conclude that objectified causation must be transitive, too, whereas it might be very unclear how to find objective criteria for deciding when causation is transitive and when not. The case should be thought through from this perspective as well.

I will not try to develop these points, though they would deserve it. I only wanted to demonstrate that my rigorous account of objectifying a subject-relative notion of causation opens a new perspective on various live issues. The fact that I could list seven such issues shows that this perspective is indeed a rich one.

To make my point in a still more general way: The way in which I proposed to conceive of objectification made clear from the outset that the objective counterpart of my subjectivistic account could only be some version of the regularity theory of causation. This is nothing but Hume's precious heritage! However, regularity theories have fallen into disgrace as an analysis, and they did so for strong reasons. What I claim, and have outlined in detail, is that the regularity theory can and should be resurrected, not as an analysis, but within the perspective of objectification, and that this perspective provides grounds for deciding about the precise form of the regularity theory. (It would therefore be important to compare my present approach to the regularity theory with Baumgartner, Grasshoff (2004) and Baumgartner (2008b, 2009) – presently the most subtle and vigorous defense of the regularity theory – but this is a comparison I have no space for here.)

Having thus reached considerable, though perhaps still improvable success with the objectification of causation, we should try to understand what this success really amounts to. Before proceeding with this appraisal, however, let me present the last part of my objectification story about (subjective) laws.

15.6 The Conditional Objectifiability of Laws

The last part is very short and simple. Let us return to the formal set-up introduced at the beginning of Section 12.3. There, we had characterized a given single case by the possibility space I. The possibility space for its infinite repetition was given by $W = I^{N'}$, and the variable X_n defined by $X_n(\mathbf{w}) = w_n$ ($\mathbf{w} \in W$) was the projection of W onto its n-th component that described the n-th repetition. Then we considered a negative ranking function ξ on I for the given single case and defined λ_ξ in (12.10) to be the independent and identically distributed infinite repetition of ξ. I argued in detail that these λ_ξ's may be appropriately labeled subjective laws, but I also granted that this almost sounds like a *contradictio in adjecto*. This impression can now be dissolved. For objectification can easily be extended to such subjective laws:

Let R be any n-place feature of ranking functions on I, S be any specification of R, and F any condition on the items in the field of R. It is obvious how to explain the corresponding ranking feature R^∞, specification S^∞, and condition F^∞ for the possibility space $W = I^{N'}$ of repetitions. Then we have

Theorem 15.27: The subjective law λ_ξ on W is an objectification of R^∞ under S^∞ given condition F^∞ if and only if ξ is an objectification of R under S given F.

I take this to be obvious. If ξ is not objectifiable, the objectifiability of λ_ξ founders at each single repetition. And if ξ is objectifiable, then λ_ξ extends this success to all repetitions simply by adding up the objectifications of the single instances. In other words, subjective laws are objectifiable precisely to the extent in which the treatment of the single case is objectifiable; this is what conditional objectifiability is to mean in the section heading. The previous sections provide two applications of (15.27):

If the single case I has no further structure, then, according to Section 15.3, only beliefs and conditional beliefs about I are objectifiable (by very simple, i.e., two-valued ranking functions ξ). The laws λ_ξ then objectify the corresponding beliefs in *generalizations* "it is always so", for instance in "all ravens are black" (if I consists of the four possibilities spanned by "raven" and "black"). Much more interesting, though, is the case where the single case I has some internal structure in terms, say, of temporally ordered variables. If we can objectify the causal relations obtaining in that single case as some causal pattern having the form of a pattern of succession, then and only then we can generalize the causal pattern to a causal law objectifying the causal relations in all like cases. And this objectification is just as substantial as that of the single case. So, this is how I would finally explicate the notion of a *causal law*: subjectively as a ranking function λ_ξ where the single case has the causal structure defined by ξ (that may or may not be objectifiable), and objectively simply as the generalization of the corresponding causal pattern or pattern of succession, provided that ξ is objectifiable.

15.7 Concluding Reflections

Let us summarize the philosophical strategy behind this chapter. In Section 12.1, where I sketched my plan for Chapters 12–15, I said that I would pursue a projectivistic account of natural modalities, and the two modalities I have dealt with in detail were nomic and causal necessity (and thus by implication determination). I have extensively discussed (perhaps forbiddingly) subjectivistic accounts of these modalities in Chapters 12 and 14. And now the objectification methods proposed are intended to substantiate the projectivistic metaphor. How do they do so?

On the objective side, we just have generalizations (in the case of laws), causal patterns (in the single case), and causal laws (in the repeated case) that are ordinary propositions, true or false like all other ordinary propositions. However, these propositions acquire an objective modal force through being objective counterparts to subjective modalities (which are just features of ranking functions or, more plainly, of the dynamics of our beliefs). How they acquire their modal force – how they are projections of the subjective modalities – is precisely described by the methods of objectification, and in this way this modal force becomes as intelligible as the subjective modalities.

To be explicit, an objective (possible) law is a true or false generalization backed up by an objectifiable persistent ranking function, and an objective causal pattern (or law, if generalized) is a true or false pattern of succession backed up by a ranking function that is objectifiable w.r.t. its (subjective) causal relations. Or as I titled my (1993a): causal laws are objectifications of inductive schemes. I consider it a strength of this account that the objective counterparts are as plain as possible, just generalizations in the case of laws or just (some version of) the regularity theory in the case of causation. Of course, we could now introduce objective modal operators, one for nomic necessity and one for (sufficient) direct causation; but this appears to be not much more than an appendix to the account given so far.

We can give another twist to our objectification story; it perhaps provides a sort of explanation for why we have the notion of causation at all: We saw in Chapter 6 how closely the notion of a reason is tied to the dynamics of belief; as soon as we reflect on this dynamics we have that notion. Section 15.2 showed that the objective counterpart of reasons simply consists in material implications. However, not all material implications we believe constitute reasons, and Section 15.3 made clear that it is impossible to sort out the relevant material implications. Now in our subjectivistic approach (direct) causes simply were a particular kind of conditional reasons, and Sections 15.4–15.5 proved that if we assume a specific temporal and logical form for these conditional reasons, we can place them in a one–one correspondence with objective material implications. So, it seems the causal relation is just the well-formed objectifiable part of our much richer and more disorderly reason relation. In other words, if we want to objectify our inductive strategies, if we want to align our dynamics of belief to the real world, we have to attend to causation, to the objectifiable part of our reasons. This is what the notion of causation is for.

Well, I did not prove this; I did not show that to conceive of causal relations is the *only* way to find objectifiable order in our rich and messy realm of reasons. Apparently, though, it is our human way, and this chapter has explained in detail that and how it works.

This is also why I doubt that we will find a deeper, possibly hidden nature of causation, as considered in Section 14.2. What this chapter suggests is that the objective nature of causation derives from objectification. Objectively, we have nothing but the true regularities on which we try to base our objectifiable inductive strategies. We may find such regularities on any level and in every science. We may believe such regularities to be reducible to those of fundamental physics, and we may set out to find appropriate reductions. However, whether or not the regularities are causal does not depend on that reducibility. If so, there need not be a common "substance", as it were, to causal laws. This point would certainly need a more thorough-going discussion. However, since I stayed away in this book from systematically treating metaphysical issues, I am content with this hint at what seems to be a consequence of this chapter.

Only now can I clear up quite a different obscure remark of mine. In Section 10.3 I discussed the philosophical convergences and divergences between probability and ranking theory. There I mentioned the truth aspect as a point of superiority of ranks over probabilities. This superiority has deepened in this chapter, since our objectification methods showed that much more about ranking functions can be true or false than merely the beliefs contained in them. At the same time, this chapter has clarified what I there called the objectivity aspect.

What is still obscure are my remarks about the reality aspect. I worried in Section 10.3 about the point that probabilities seem so much better grounded in reality than ranks; probabilities are somehow firmly anchored in relative frequencies, an anchor which ranks do not have. Well, only apparently; I dimly suggested that ranks are correspondingly grounded in exceptions. This suggestion becomes clearer when we look at the objectifying ranking functions that I defined in (15.10) and (15.24) and that I often used in my examples in Sections 14.5 and 14.13. These functions simply counted the number of violations or errors or exceptions that occurred in the world w in comparison to the given pattern of succession concerning a given single case (multiplied by the parameter m). And then the corresponding objectifiable subjective laws simply counted the number of exceptions occurring in the potentially infinite replications of that single case. This point also relates to my remark after (12.10) that ranking functions and (differences of) absolute frequencies are of the same formal kind. So, in this way objectified ranking functions may indeed be grounded in reality. However, whether this anchor in reality is as good and useful as the anchor of probabilities in relative frequencies remains to be discussed.

So, as I said at the end of the introduction to Chapter 14, after all these laborious presentations I finally leave it to the reader to judge whether I have built a huge castle on sand or whether my constructions seem sound and stable.

In Sections 15.2–15.6 I have been exclusively occupied with getting my account of objectification straight without looking at the many possible implications that we have passed by on our way. Hence, many things still await further consideration and elaboration:

(i)   One should scrutinize the precise version of the regularity theory that results from the objectification.

(ii)   One should explore the objectifiability of further ranking features; I mentioned simultaneous causation and interactive forks (as presumably negative examples) and the transitivity of causation (as a presumably positive example).

(iii) We have lost sight of ceteris paribus laws in this chapter. Their objectification would be another most interesting topic. It might succeed by relating multiple exceptionality as defined in (13.6) with exceptions as used in this chapter.

(iv)   We should carefully study the inductive behavior of the objectifying ranking functions of Section 15.4 (beyond the remarks I made after 15.10).

(v)   What are the repercussions on their updating algorithms for ranking nets as discussed in Section 7.4?

(vi)   Of course, my de Finettian story about mixtures of subjective laws in Section 12.5 should be thought through on the basis of objectifiable laws. It seems that we should want our doxastic state to be not just any mixture of subjective laws, but only a mixture of objectifiable laws. Is there any interesting characterization of such mixtures?

(vii) Finally, if ranks and probabilities bear so many similarities, does my account of objectifying ranks provide any lessons for the still insufficiently understood notion of objective probability? (This is an issue I left open in Spohn (2010a) as well.)

And so forth. I find the multitude and variety of open issues daunting. On the other hand, it is certainly welcome that my chapters open, rather than close, a rich and specific research agenda.

16

Justification, Perception, and Consciousness

The previous four chapters were devoted to applications of ranking theory in philosophy of science. The last two chapters of this book will be occupied with epistemology. However, this distinction should be downplayed, just as this book attempts to bridge the difference between the theory of knowledge and the theory of belief, probabilists and non-probabilists, objectivistic and subjectivistic attitudes, or foundationalists and coherentists (as we will see).

Looking at contents rather than labels, we might say that the previous four chapters were devoted to the inferential aspect of ranks (and probabilities), to the role and workings of (inductive) inferences in our cognitive system. Basically, they were about conditional ranks, about conditional and unconditional dependence and independence, the formal structure of which we had studied already in Chapters 6 and 7 and the pervasive and overwhelming significance of which came to the fore in the applications in Chapters 12–15. This will continue in the last two chapters.

Still, something is missing. Our scheme of inferences is not merely a big switching yard in which we can move to and fro and up and down. Somehow, our inferences must start somewhere in order to arrive somewhere; somehow we form unconditional beliefs in the end; having only conditional beliefs would be useless. Of course, I started to tell the ranking-theoretic story of belief formation in Section 5.4 when I explained the dynamics of ranking functions in terms of various conditionalization rules. No doubt, this offered an explanation or justification of the posterior beliefs through the prior beliefs and the information received in between; it thus accounted for just one explanatory step. In the introductory Section 1.1 I already emphasized the advantages of this apparent modesty and thereby motivated the transition from inductive to learning schemes that is so central for the dynamic perspective of this book. And in Chapter 5 the iterability of this step was the crucial argument for introducing ranking functions. With this, we can indefinitely expand this kind of explanation or justification. However, the iterability in no way reduces the twofold superficiality of that explanation: I was indeterminate about what induces belief change; I always spoke somewhat vaguely of information received with some firmness. And the other part of the explanans, the prior belief state, is of the same kind and equally in need of explanation as the explanandum, the posterior state; iteration only iterates, but does not settle the point. We must deepen our account on both scores.

This means we must address traditional philosophical topics, of course. The information received is ultimately perceptual information; this somehow is the ultimate foundation of all belief change. And repeated referral to the prior belief state must ultimately refer us back to initial or a priori belief states; there is no way further back. So, these are the two topics rounding out the epistemological picture developed here: perception, which will occupy us in this chapter, and apriority, the topic of the concluding chapter.

These *are* traditional philosophical topics overburdened with views, arguments, and problems; I cannot treat them in fairness without doubling the length of this book. So, let me be clear what I will be doing in these final chapters. My primary goal is to unfold my epistemological picture, and my introductory remarks have made clear that by treating these two topics this picture will receive a certain completion. I think that this completion will be a fairly strict consequence of the account given so far. Of course, I will need to assume some further principles for which I will argue, insofar as they are not perfectly plain. After having made explicit these principles my account will, however, be almost as rigorous as in the previous chapters, and my goal is to develop this almost rigorous story. Thereby, I am bound to take a stance towards the ample views, arguments, and problems. Yet, a full defense of that stance is beyond the scope of this book. The relating and comparing I will undertake is, rather, in order to place my account in the rich spectrum of possible positions and to show that many views might need only small corrections in order to make them compatible or even agree with mine. In short, once more I can only hope to strike the right balance between brevity and circumspection. The views I am going to present are essentially contained in miscellaneous papers of mine; I will refer to them in due course. Here I have tried to arrange their relevant parts into a more coherent, perspicuous, and sometimes more explicit line of reasoning, which, however, does not exhaust those papers.

Turning now to the topic of this chapter, it seems clear that in order to understand the origins of belief formation, we first need to better understand its goal. The goal presumably is knowledge or, given my reluctance to attempt an explicit analysis of that term, at least true belief. Whether belief tends or is, in some sense, even guaranteed to be true, will be a recurring topic throughout this and the next chapter. And in the normative, rationalistic perspective taken here, belief always is justified belief; the accidents of actual belief and mere epistemic luck are not our concern. Now I have said a lot about the relation of one belief (or rather proposition) being a reason for another; I have, however, not said anything explicit about the property of a belief being justified. So, justification will be the topic to be addressed first, in Section 16.1 (which will thus contain at least implicit comments on knowledge). This section will also begin my sketch of the field of epistemological positions within which the subsequent considerations will be placed.

Only then can we address the origins of belief. Perception will be the obvious starting point of the discussion in Section 16.2. We will see that frame-relativity is an important issue here, as it has been so many times before. In order to escape such relativizations we will continue the discussion of seemings, lookings, or appearances in general in Section 16.3. This will be most uncertain terrain full of tempting a priori claims, and our task will be to find the right claims and avoid the wrong ones. This will not be easy, as so many seem to have been almost, but not completely right. However, our discussion of dispositions in Chapter 13 will prove to be a perfect preparation for this task (recall that secondary qualities are specific dispositions). Still, we will not yet have reached rock-bottom. In order to do so, we will have to take one step further and to dig into the mysteries of consciousness. By all common-sense standards, perception is mainly or even essentially conscious perception; appearances fill phenomenal consciousness. So, any discussion of the origins of belief must ultimately face this topic. I will face it in Section 16.4.

16.1 Justification

A good starting point of our discussion is the justification trilemma, usually called the Agrippan trilemma in honor of its ancient origins in Sextus Empiricus or even Aristotle. Albert (1991, p. 15) coined the label "Münchhausen trilemma" which I find most graphic (for those familiar with the stories of Baron von Münchhausen); I will stick, though, to the historic reference. Roughly half of epistemology textbooks start with it; BonJour (1985, p. 18) calls it "perhaps the most crucial [problem] in the entire theory of knowledge". It is very simple:

If I am to justify some belief or claim of mine, I have to provide some reasons or grounds in support of it; the belief is justified only if it has such grounds. However, if those grounds were not justified in turn, they could not confer any justification to that belief. This seems to be an obvious or even analytic truth about justification. So, let us accept:

(16.17)   *The Trilemma Generator.* A belief is justified only if one has reasons supporting it that are justified in turn.

This seems to force us into the first horn of the trilemma: Having to give reasons for the reasons already specified we are caught in an *infinite regress* that we can never exhaust and that never produces any grounds for us to eventually stand on. Nobody (except Klein 1999) has ever acquiesced in that regress. Can it be escaped? Yes, we might try the second horn and stop the regress at some basic grounds that are self-justifying. Or, if this sounds odd, we might say that there are some basic grounds that, exceptionally, are not in need of justification. Some disrespectfully call this *dogmatism*. The more familiar and positive label is *foundationalism*. In any case, the task is then to more specifically characterize those alleged basic grounds and to clear up the mystery of how they can be exempt from justificatory demands. If one despairs of resolving this mystery, one might finally reach for the third horn and accept that in the process of specifying reasons one must sometimes (always?) return to reasons already adduced at some earlier stage. This might be denounced as a justificatory *circularity*, or one can acknowledge it as the basic logic of *coherentism*. However, these odd circles are in urgent need of a positive account, apparently a difficult task. So, these are the unpleasant options; one may specify them, one may mix them, and there are dozens of variant positions. (See also the more careful treatments in BonJour 1985, Sect. 2.1, Haack 1993, Plantinga 1993a, Ch. 4, Audi 2003, Ch. 7, or Ernst 2007, Ch. 6.)

All, or almost all, of those participating in this discussion – I am no exception – are in the grip of a certain picture, sometimes made explicit. This picture, which motivates the trilemma generator (16.1), may be called the *hydrodynamic conception of justification*. The picture first imagines a justificatory network. The network consists of propositions, possible objects of belief and justification, which are represented by bulbs. Moreover, the propositions or bulbs entertain justificatory relations. Hence, the network also contains channels or pipes between the bulbs, the width of which corresponds to the strength of the justificatory relations (which may obviously vary). As such the picture does not say anything about the structure of the network, whether it is like a tree with thick and thin branches or like an electric circuit with resistors, capacitors, and transistors, or whatever. The important point is that the network is entirely empty at first; nothing is believed so far. Now, however, we pour some viscous fluid into the network, which is called *warrant*. The fluid diffuses through the pipes; most of the bulbs remain empty, i.e., unjustified and hence unbelieved; but many bulbs get filled by warrant and thus believed, and the amount of fluid they contain represents the strength of their justification.

As I said, most are under the spell of this picture, and I will find no fault with it. Disagreement starts about the precise structure of the network; of course, this is a disagreement about the logic of justification. The more important disagreement, though, is about where that fluid comes from. Foundationalists say that it is produced in special basic bulbs – call them *origins* (in order to terminologically distinguish them from the sources of belief change as defined in (3.13) and (5.36); we will discuss the relation of those sources to the present origins in the next sections). And then the issue is what the origins are: a priori propositions, perceptual beliefs, sense-data, perceptions, or whatever. Some of these proposals are clearly internalist in the sense that the origins are also part of the belief network. If the origins are sense-data, they are usually not taken to be beliefs and hence are internalist only in the weaker sense of being *in* the cognitive subject, though the belief network as such would be fed from outside. There is also the clearly externalist view that the origins indeed consist of external facts, presumably facts perceived, though this might mean stretching justifications too far. If perceptions are taken as origins, it is not so clear where to locate them, presumably somewhere between facts perceived on the one end and perceptual beliefs on the other; it is not so clear what perceptions *are* (see, however, the next section). And so on.

Coherentists instead say that the fluid is somehow produced by the network itself, in the pipes, by a special kind of clustering, or whatever. The idea is mysterious, and the production looks like a miraculous *creatio ex nihilo*. The deeper question, though, is what the metaphor of the production of the warrant fluid could mean here in the first place. We will soon see that there are more internalist/externalist distinctions than the one above, and we will even find a place for so-called contextualism in the picture.

However, we should not try to discuss the picture and the positions we can locate in it as such; I find informal discussions – such as, e.g., those referred to above – hard to assess, because they all proceed on partially shaky grounds. Rather, we should see how the picture looks in the light of ranking theory. This will occupy us for the rest of this section.

The first and, in a way, most important observation will not come unexpectedly: For all I know, ranking theory is the *only* good proposal for turning the metaphor of pipes relating bulbs containing the warrant fluid into theory. And it is so precisely through the notion of one proposition being a reason for another as explicated in Section 6.1. "The only" is perhaps too strong a claim. Probability theory, for instance, might have served the same purpose. It has, however, not been well received among theorists of knowledge (see, e.g., Plantinga 1993, Ch. 6), precisely because the doxastic states it represents do not even contain beliefs that can be true or false; this was the consequence of the lottery paradox (cf. Section 3.3). As I have made clear, ranking theory was designed to escape this objection. There are other proposals. Pollock (1995) is certainly a major effort to render the network metaphor precise. The inhibition nets studied in Leitgeb (2004) might be understood as another proposal; and so on. I will not reopen now the comparative discussion. Rather, my claim is directed to the dominant literature in the theory of knowledge, which makes little effort to provide any comparable proposal; my criticism of Lehrer's account of justification in Section 11.4 was only a symptom of this. Though this literature contains rich discussions of the nature of reasons and justification, I find it shockingly poor concerning the precise working of reasons.

When one surveys this literature (I am following now my discussion in Spohn 2001a), one essentially finds four conceptions of the reason relation; I know of no variant falling outside of this classification (if complemented by my remarks on the subjective/objective dimension below). One is the *positive relevance* conception, which I have emphatically endorsed in this book. Usually, it presents itself as a reflection of the probabilistic paradigm, i.e., Bayesian confirmation theory, and thus has acquired little persuasiveness among theorists of knowledge, since they are not aware of any variant account of positive relevance they could use, such as the one given by ranking theory. This explains their unfortunate preference for other conceptions of the reason relation.

The main conception is certainly the *deductive* (or demonstrative) conception, according to which reasons deductively entail what they are a reason for. This is, no doubt, the traditionally dominant conception, and since the emergence of modern –

now classical – logic it enjoys a precise formal treatment that has put it far ahead of all alternatives. I find the present attitude towards it to be ambiguous. Officially, most would grant that it is too narrow. There are weak and strong reasons, but deductive logic only explicates the strongest, i.e., cogent reasons. There is clearly more to inference than deduction; and to treat each inference as a hidden deductive one proceeding from premises left tacit does not appear to be a plausible strategy (an appearance confirmed in Section 15.3, since the general non-objectifiability of reasons stated there can also be interpreted as their general non-representability through modus ponens). However, most authors, when illustrating reasons, fall back on examples with deductive reasons. (See, e.g., BonJour (1985), who routinely illustrates justifications with deductive arguments, even in his Section 6.3 on coherentist observation.) This is why it still seems to be the main conception, at least subconsciously. It is apparently taken to be the only useful substantial account, although known to be insufficient.

I can discern a somewhat different third conception of reasons, which may be called the *computational* one. It says that reasons imply that which they are a reason for according to some specific rules of inference. Extensionally, it is almost congruent with the deductive conception, since the inference rules of deductive logic are the dominant model. The basic idea is, however, a different one; it is that reasoning or inference is a rule-governed activity that is explicated by stating the rules. The idea need not be confined to classical logic; there are now many well-elaborated variants of non-classical and non-monotonic logics (cf., e.g., Gabbay et al. 1994). The reference to such logics is problematic, however, since they form a quite experimental field, which is not at all settled (see my very brief remarks in Sections 11.5 and 11.10). And I explained my philosophical reservations about such computational conceptions, i.e., concerning their normative defectiveness, in Section 11.6. So, at present, this conception is not really fruitful, and I doubt that it can be. (Of course, if the rules of inference were derived from some more basic conception, they would be most valuable.)

The last main conception of reasons we find in the literature is the *causal* one; it finds much sympathy within naturalized epistemology. According to it, a belief is a reason for another belief if, basically, the former is a (partial) cause of the latter. Of course, one may and should add various sophistications to the basic idea; for instance, one would have to address the problem of so-called deviant causal chains, which pertains not only to agency or perception, but also to belief formation. There is an important truth in this conception. We take reasons to be efficient; they should not be mere rationalizations of a mental reality that works according to entirely different principles. The idea has some affinities with the computational conception. In some way or other, actual belief formation works according to some mechanisms or rules that will be computational ones in some broader (or perhaps only forbiddingly broad) sense. Conversely, if some variant of the computational conception had a basis in mental reality, we could simply say that the earlier computational stages are causes and thus also reasons of the later stages.

Again, though, this conception is not helpful at all, for two reasons. First, it is, I think, a conception without real content. We know only the roughest outlines of the actual causation of beliefs. We have no good account of actual computational thought processes; a little introspection reveals how rambling our mind is. And when belief is essentially dispositional, it is even more difficult to say what causes what. Second, as such the causal conception is silent on the normative dimension of the notion of a reason; actual doxastic causes have to pass a normative evaluation in order to count as reasons. Everybody is aware of this fact; only the most radical naturalists would deny it. Therefore, most take the causal aspect only as an ingredient of the reason relation and attempt to supplement the normative aspect. Then, however, we leave the confines of the causal conception. We do not arrive at the required supplement by looking only at persons we take to be rational and by studying how the causation of belief formation works in them. We can only do it the other way around. We first try to state the normative principles; they wear the trousers. Then we can add a causal story that would apply to persons conforming precisely to these principles. In so doing we respect the causal conception of reasons, as we should do. But we do not take it as basic.

So, to summarize, we had better return to the positive relevance conception. In ranking theory it is fully worked out in a way that is utilizable in epistemology. And it embraces all the other conceptions. Theorem 6.6 proved that the notion of a deductive reason is a special case of reasons in our positive relevance sense. Moreover, Section 7.4 showed that it conforms to computational ideas; that section provided a perfect algorithm for the propagation of reasons, even though it sharply differs from rules of deductive inference. Finally, it provides a causal story about ranking-theoretic minds. For such minds, the conditionalization rules (5.25) and (5.33) can be understood as causal laws according to which the source of a belief change (in the sense of 5.36) causes the entire change at one swoop. And if we wished to have more detail, we might interpret the propagation algorithm of Section 7.4 as a causal story about the propagation of belief change in rational belief networks. These stories can be improved, no doubt; but they already offer more of a causal account than is usually found.

There is another highly important dimension with respect to which the literature on reasons is deeply divided and indeterminate: the *subjective/objective* dimension. We all yearn for objective criteria telling us what is a reason for what, we all talk of *good* reasons as opposed to *bad* reasons that are no reasons at all. However, we mainly have intuitive convictions, but little in the way of theoretical ideas about what good reasons might be. Certainly, good reasons have to satisfy the formal rules of rationality, whatever they are. Presumably, however, those formal rules do not exhaust what we take to be reasonable or not. So good reasons should also satisfy more substantial standards of rationality, if we could only say what they are. Moreover, good reasons are rich or even complete. That is, the richer they are, the better; a single reason, except perhaps a deductive one, is rarely a good one. Finally, we may take good reasons to be true; usually we at most grant that a false reason might have been a good reason, if it had been true. This is why the view that only facts can be reasons has become so popular.

Ranking theory has a clear position on the subjective-objective dimension. On the one hand, I claim as much objectivity for the normative constraints built into ranking functions as there can be for such constraints. If ranking functions, roughly, embody nothing but conditional consistency (as explained after 5.19) or dynamic consistency (cf. Section 8.5), then the objective force of these constraints seems undeniable to me. On the other hand, ranking theory leaves maximal subjective leeway within these constraints, and therefore what is a reason for what depends entirely on the specific subjective state (with the exception of deductive or, more generally, of a priori reasons, some of which we encountered already in Section 13.3).

As for the two moves towards objectivity, requiring good reasons to be rich and true, they can be acknowledged in ranking theory straightaway. Certainly, whenever we have an argument about some matter, we rarely disagree about what is a reason for what. Rather, we dispute which reasons and counter-reasons are to be accepted, i.e., true. We could easily turn this into a defining characteristic of a narrower notion of being a reason. However, this move would still have to build on our basic positive relevance notion of being a reason.

What about good reasons being rich or even complete? Well, any ranking function lays out a complete scheme of reasons (within the algebra of propositions considered); there is nothing to complete. However, in Section 6.1 I carefully distinguished between (one proposition) being a reason (for another) and (a subject's) having a reason; and the richness or completeness requirement refers to the latter: To judge a matter we should have as complete reasons as possible. Of course, this requirement of total evidence, as it has also been called (by Carnap 1950, pp. 211f.), is hard to satisfy; it is difficult or impossible to investigate the truth of *all* reasons and counter-reasons there might be for a certain assumption. Therefore we should rather say: the more complete the reasons we have, the better they are, the better the judgment based on them.

Beyond these points, the prospects for objectively good reasons are dim. "Good" certainly has the emotivist meaning of approval; when I attest that someone has good reasons, I thereby endorse these reasons as well. Virtue epistemologists have tried to lay down criteria for such approval, but I find their conclusions too unspecific to be helpful. Certainly, there is strong social pressure for intersubjective agreement on what is a good reason for what. However, the objective base for this agreement is less clear. It is quite open how to decide disputes about what is a reason for what, which we fictitiously have with the grue/bleen-reversed people and sometimes have actually; that is, it is quite open whether the formal rules of doxastic rationality can be supplemented by more substantial standards. When discussing the first principles of belief revision leading to (4.11) I decided to only subjectively fill the space left by the minimal rationality requirements and not to search for a more objective base – a fundamental decision, as I emphasized, and in retrospect a rewarding decision, I believe. On the whole, the literature on good reasons and related matters has not revived my hope for this forlorn case. Instead, I can refer to the objectification theory of Chapter 15, which showed the extent to which reasons can be objectified, in particular in the special conditional form of causes. I will not discuss, though, whether that objectification theory can be related to the present issues in a satisfying way. The next chapter on a priori reasons will offer another partial answer.

So far, we have discussed reasons, and the reasons why I recommended their explication in Chapter 6 (in addition to those already given there). However, the trilemma generator (16.1) was about justification; we somehow have to move on from the relation of being a reason to the property of being justified. We must face the topic of justification that I have eagerly avoided up to this chapter. Does ranking theory help us with this move?

Yes, in a weaker and a stronger way. First, Bayesians often suggest that their notion of coherence might stand in as a weak substitute for justifiedness, where subjective probabilities are coherent if and only if they obey the axioms of probability. This does not have much to do with what coherentists call coherence. What they mean is strongly intuitive, but in fact quite unclear, and I do not see that the recent vigorous discussion (cf., e.g., Bovens, Hartmann 2004, Olsson 2005, and the exchange of papers quoted there) has led to an accepted or acceptable conclusion. Still, it is not really an accident that the same term appears twice here. If subjective probabilities are coherent, each proposition is assigned its proper position and receives precisely the probability it must receive given the positions, i.e., the probabilities of the other propositions. To that extent, it coheres with the other probabilities, and one may call its probability justified. Of course, this notion of coherence applies to ranking theory as well. According to the ranking-theoretic laws, each proposition is assigned the degree of (dis)belief it must receive, given the degrees of (dis)belief of other propositions; the balance of reasons as described in Section 6.3, for instance, settles precisely on that degree of (dis)belief. In this sense, that degree of (dis)belief coheres with the other degrees and is justified.

The stronger way adds only a little to this consideration. It consists in the following observation: Look at a positive, not a negative ranking function β. $\beta(A) = 0$ means that A is not believed, $\beta(A) > 0$ means that A is believed, and the larger $\beta(A)$ is, the more firmly A is believed. Now compare this with the above hydrodynamic picture of justification. I find that the positive ranking function fits this picture perfectly; its degrees of belief correspond well to the amount of the warrant fluid in the bulbs. If there is no fluid in a bulb, it is not justified at all and hence rationally not believed; and the more fluid is in the bulb, the more strongly it is justified and hence believed.

Here, the strength of the pipes corresponds to conditional positive ranks; the thicker the pipe, the larger that conditional rank. Or, one should allow for inhibitory pipes as well, and then the strength of the pipes corresponds to the degree of relevance, while the direction of the relevance decides which kind of pipe there is. So, if $\beta(B \mid A)$ = $\beta(B \mid \bar{A})$, there is no pipe from A to B, and no positive or negative amount of warrant can flow from A to B. This suggests that we should carry our explication a step further and to equate the two notions:

(16.2)   *Justification explained*: Degrees of justification or justifiedness are degrees of (justified, rational) belief and hence positive ranks.

Of course, the hydrodynamic picture is only a picture, which is open to many interpretations and thus cannot establish the present proposal. However, I complained above that there are hardly any precise explications of the hydrodynamic picture at all; this point was also at the bottom of my criticism of Lehrer in Section 11.4. The fact that ranking theory makes a precise proposal at this point is good news, given this background. And if we accept it, (16.2) agrees well with Theorem 11.23 regarding Lehrer's notion of justified acceptance. (By the way, subjective probabilities seem less suited for the explication of this picture, since there is no determinate probability that could be said to represent an empty bulb.)

Although we can thus reconstruct the network of bulbs and pipes and their filling with warrant, the crucial (metaphorical) question of where the warrant fluid comes from remains as yet unanswered. This will be the topic of the remaining sections of this chapter. The conditionalization rules in Section 5.4 might suggest a foundationalist answer, whereas the formal properties of the reason relation stated in Section 6.2, in particular its symmetry and hence undirectedness, have a coherentist ring. We will see that the truth is a bit more complicated.

Let me guard at this point against two possible misunderstandings of this section. First, I have just apparently identified justified belief with belief. However, I did not do so generally; I do not want to deny the existence of unjustified beliefs. The point is that I am only talking about rational beliefs that are backed up by a justification structure in form of a ranking function. Then the identification is in a way trivial.

Secondly, since this section is about justified belief and also about justified true belief, I have often slipped (and will slip) into talking of knowledge as well. Officially, however, this section is only about justification and not about knowledge. I certainly accept justified true belief as a necessary condition of knowledge; however, I am silent on its prospects as an analysis of knowledge. So far, I do not consider reasons and the justifications they provide to be a guarantee, not even a weak guarantee or a sign of the actual truth of the beliefs thus justified; their truth might still be due to mere epistemic luck. Hence, I am simply bypassing Gettier-type problems; they play no role in this or the subsequent sections. We will find some guarantee in the strongest sense in Section 16.4, and I will try to say more on the relation between reasons and truth, though not on knowledge, in Section 17.3.

In the rest of this section let me explain how we can find a home for other types of positions concerning justification (and thus knowledge), and let me make two preparatory observations. The first is that the above account is firmly internalistic with respect to justification. To be precise: So far I was undecided whether, within a foundationalist conception, the origins or basic bulbs are external or internal. Given their justificatory input, however, all further justification is internal, since it is entirely a matter of reasons relative to the subject's ranking function. Here I assume that each subject is aware of his own ranking function; we are not talking of subconscious belief. This might seem to be a strong assumption. However, in the more familiar case of mere belief, I always took for granted the static reflection principle of doxastic logic, i.e., the logical truth of "*a* believes that she believes that *A* iff *a* believes that *A*". There are prima facie plausible counter-examples, but every example I have seen uses two different notions of belief in the iteration and is thereby invalidated. Moreover, awareness of something need not mean having it before one's inner eye. If belief is dispositional, awareness of it is dispositional as well. The same holds for one's ranking function, and then awareness of it no longer seems demanding. (Recall also our discussion of reflection in Section 9.2 and in particular the principle of auto-epistemic transparency (9.9).) My fuller reason for my assumption will unfold in Section 16.4. In any case, so far my justification internalism fully agrees with the majority intuition, as forcefully argued by BonJour (1985, sec. 3.2–3.3), Lehrer (1990, Ch. 8), and others.

The other observation is that there are two degrees of freedom built into the above account of justification. One parameter is that we might count a bulb as justified not as soon as it is not empty, but only when it contains a sufficient amount of the warrant fluid, where we may set the threshold of sufficiency as we like. Of course, this is nothing but the parameter z introduced in (5.14), where I said that belief might be more strictly represented by $\beta(A) > z$ for some $z \geq 0$. Now, when we said that, rationally, belief is ipso facto (internally) justified belief, this amounts to conceiving of z as a threshold of justifiedness.

The other degree of freedom lies in the underlying conceptual frame or possibility space. One might think that one should refer to the relevant subject's present universal frame and to its present universal doxastic state, i.e., its ranking function for that universal frame. The subject's universal frame is perhaps not so ill-defined as the universal frame simpliciter considered in Section 14.9. Still, it is very rich and pretty indeterminate. What is the set of *all* concepts a subject possesses at a certain time? What is the set of *all* propositions composable from these concepts? What then is the space of *all* possibilities conceivable for the subject having those concepts? In view of such hardly answerable questions it is legitimate and advisable to refer to a specific, more or less restricted frame, or the associated space of possibilities or the associated algebra of propositions, and to the subject's ranking function for that frame or algebra. Which restriction we consider is up to our choice; however, we should proceed according to the invariance principle (5.3).

Let me illustrate the operation of the two free parameters, the propositional algebra and the threshold of justification, with an example that might arise in debate with the skeptic. Let us not consider Moore's notorious example that I know I have two hands, since this knowledge is grounded in memory and proprioception rather than in my present visual perception of my hands. Let us instead take a simpler case where only the present and one sense modality are involved: I am sitting at a desk, there is a laptop on the desk, I look at it, it looks to me like a laptop, and so I believe, very firmly, that there is a laptop on the desk in front of me. Do I know or, rather, am I justified in believing that there is a laptop on the desk? Yes, of course, we would normally say. However, the skeptic now enters the scene and asks silly questions: "Are you sure? Aren't you dreaming? How do you know there isn't an evil demon seducing you into an illusion?" And so forth. Do we then retreat from attributing to me knowledge about the laptop? Even though there was nothing wrong with the original attribution?

Let us represent my doxastic state after the perception in some plausible ranking-theoretic way. Let A be the proposition that there is now a laptop on the desk in front of me; let $\Phi_{I,now}(A)$, or $\Phi(A)$ for short, be the proposition that it now appears to me as if there were a laptop on the desk in front of me; let us accept for the time being that such propositions like $\Phi(A)$ are also in the domain of belief or ranking functions (this might well look doubtful, but I will more carefully argue for a positive conclusion in Section 16.3); and let β be my positive ranking function after my perception of the laptop. So, we have $\beta(A) \gg 0$; I firmly believe A. In the usual discussions of such stories, it is also assumed that $\beta(\Phi(A)) > \beta(A)$; my appearances are more certain to me than external reality (though, again, this sometimes appears doubtful to me). Should we not say that they are maximally certain so that $\beta(\Phi(A)) = \infty$? This is a subtle issue that I will extensively discuss in Sections 16.3–16.4. For the moment let us reckon with the possibility that I mistook my impressions as the appearance of a laptop; what I took as looking like a laptop might really be looking like something else. Then, my certainty of $\Phi(A)$ is less than maximal.

This does not yet complete the specification of β. Let us also ask what I would believe under the disturbing supposition that A is in fact false. Then I would think there is something seriously wrong with the case, presumably with the apparent laptop, and I might develop various hypotheses as to the cause, but again the usual assumption is that I would still maintain my belief, though it is weakened, that it at least appeared to me as if A. Conversely, let me make the even more disturbing supposition that I indeed somehow mistook my impressions and that it did not, in fact, appear to me as if A. Then there would be something critically wrong with me. The usual assumption is that under this supposition I would also cease to believe A. Summarizing all this in a table for the corresponding negative ranking function κ (that makes the conditional ranks more perspicuous), we get, for instance:

κ	A	$\bar{A}$
$\Phi(A)$	0	50
$\overline{\Phi(A)}$	80	70

It is easily checked that all my qualitative assumptions are satisfied by these figures; in particular, we have $\beta(\Phi(A) \mid \bar{A}) = 20$ and $\beta(\bar{A} \mid \overline{\Phi(A)}) = 10$.

Now, the above-mentioned relativizations of ranking-theoretic justification help us in our argument with the skeptic in obvious ways well known from the literature. When the skeptic asks: "Are you sure?", he might be taken as simply raising the standards of justification. In everyday discourse, one counts as being justified in believing A when $\beta(A) > 0$ or, say, $\beta(A) > 10$; of course, the levels are never clearly fixed. However, this is not good enough in the skeptical context, where the threshold is raised to, say, 60 or more, and I no longer count as being justified in believing A. So, I would know or justifiedly believe that there is a laptop on the desk according to the one standard (since $\beta(A) = 50 > 10$, say), but would not know or justifiedly believe that skeptical possibilities do *not* obtain according to the stricter standards. In our short story the skeptical possibility is $\Phi(A) \cap \bar{A}$, but $\kappa(\Phi(A) \cap \bar{A}) = 50 \leq 60$. I do not see why we should not grant so much to the skeptic.

Note, by the way, that the closure principle, i.e., the assumption that knowledge or at least justified belief is deductively closed, holds on each level of justification according to this account; this is precisely how ranking theory was designed. The majority of epistemologists will certainly be pleased by this observation; we need not give up that principle, as famously proposed by Dretske (1970) in order to escape skeptical conclusions.

The other questions of the skeptic – "aren't you dreaming?", etc. – point in a prima facie different direction. They suggest that one knows or justifiedly believes something only if one is in a position to exclude *relevant* alternatives (in which that something is false), where it is up to us what counts as relevant alternatives (a suggestion elaborated, e.g., by Lewis 1996). This choice is precisely the choice of the underlying frame or possibility space mentioned above. We may neglect skeptical possibilities; for instance, our frame may not attend to variables about seemings or appearances that allow us to state skeptical possibilities such as $\Phi(A) \cap \bar{A}$, and within such restrictions skeptical questions do not arise.

However, we should ask what it means to *exclude* an alternative. This sounds like a yes–no affair; exclusion seems to come with a guarantee. Of course, this is not so; exclusion is a matter of degree. By all normal standards I can exclude the possibility that there is an evil demon deceiving me, even without having searched for it in every corner of the universe, just as I know that there is no cat in my house without having looked in every room. In our example, the skeptical possibility $\Phi(A) \cap \bar{A}$ can thus be excluded, and this is even more so for the possibilities where I am seriously confused, i.e., where $\overline{\Phi(A)}$ although I believe $\Phi(A)$. In this case, only the alternative $\Phi(A) \cap A$ remains and hence counts as known or justifiedly believed. When the standards are raised, though, such possibilities can no longer be excluded. (In richer conceptual frames, richer stories about the exclusion of alternatives could be told, of course.) Thus it seems that the exclusion of relevant alternatives operates in the same way as the setting of a justification threshold.

After these remarks about how internalism and the two degrees of freedom are built into the ranking-theoretic account of justification, I can finally explain how I think this account can accommodate two further major positions within the theory of knowledge, namely externalism and contextualism, both of which come in many forms and often together.

Turning first to *externalism*, I explained my justification internalism and the potential little externalist loophole at its base, to be discussed in Section 16.4. However, we must grant that every part of the internalist picture can be externalized; so there are many forms of externalism. To recall, an *externalist* is someone who claims that at least some conditions that turn true belief into knowledge are external conditions, i.e., about something external to the mind of the doxastic subject. Since justifiedness is usually taken to be one of those conditions (before Gettier the only one), a *justification externalist* is one who assumes externalist elements in justification.

My picture provides much space for such ideas. I have emphasized my entirely subject-relative conception of reasons and mentioned, under the label of "good reasons", our multifarious attempts to reach intersubjective agreement on and objective grounds for our subjective reasons. This fully carries over to justification. When judging the justifiedness of someone's belief, we might do so on the basis of his subjective structure of reasons, or on the basis of what *we* take to be reasons, or by alluding to some allegedly objective ground. Often the three perspectives agree and need not be distinguished, and often they are mixed up, though they should be distinguished. Our intuitions are not attuned to these distinctions and might be easily seduced by appropriate examples into assuming the one or the other perspective. Hence, the dispute over justification externalism does not appear fruitful to me; both sides are correct, in a way. We should acknowledge, though, that the basic notion is the internalist one, which needs to be developed, for instance along the proposed lines of ranking theory. Then, the various ways of intersubjectivization and objectification, as far as they are available, can and should be supplemented.

This remark also applies to other forms of externalism. Perhaps the notion of justification is so tightly wedded to its accessibility by the subject herself that whatever external conditions of knowledge there are (besides truth) one should not claim them to be about justification. Therefore, externalists, from Goldman (1979) onwards, have propounded the alternative key notion of *reliability*. Whether a subject's belief formation works reliably is clearly something the subject might misperceive; this reliability is an objective matter, or at least something we judge. However, when *we* start judging the reliability of the subject's belief formation, we just replace the subject's ways and standards of justification by our own ways and standards. And striving for a more objective notion of reliability is hampered by a reference class problem akin to that encountered in statistics (cf., e.g., Feldman 1985). Hence, what motivates reliabilism is, as far as I can see, nothing but the three perspectives above; and again I think that the subjective perspective is theoretically basic, although it is legitimate to explore the extent to which the other perspectives are feasible.

The picture becomes even more complicated with *contextualism*. A *contextualist* holds that knowledge, and perhaps also justification, is context-dependent, and then everything depends on what the context, the relevant contextual parameters, and the precise context-dependence are supposed to be. Sometimes the doxastic subject itself is assumed to be moving in a context, so that the relevant context parameters are subject-relative ones. Above I explained how the justifiedness of a belief depends on two internal parameters, the underlying possibility space and the threshold of justification. Surveying the examples that have been provided for the dependence of knowledge on the subject's context, I have regularly found one or both of these parameters at work. Hence, it seems to me that this kind of dependence is well represented in the ranking-theoretic account of justification.

However, it has been rightly remarked that these relativizations should not be called context-dependence. The latter primarily applies to linguistic meanings, where the dependence refers to the context of the speaker, which in our case is the context of the ascriber of knowledge or justifiedness. Again, the question is what the relevant contextual parameters should be; they are not those usually found for linguistic meanings. However, we can once more externalize our internal picture. We could say that it is the ascriber who sets the threshold of justification and he who chooses the possibility space to be considered (and both may somehow be implicit in his ascription). Again, the examples in the literature suggest that the context of ascription may be characterized by these two parameters. We can also contextualize reliabilism and say that whether a subject is to be called reliable depends on some contextual parameters that provide, as it were, the relevant reference class (this is how, for instance, the well-known barn example is usually treated). However, I still see no other factors at work.

So, my overall impression is that the confusing manifold of positions in the theory of knowledge essentially results from partial (attempted) externalizations and objectifications of the various aspects of the internalistic picture that I have outlined and that I take to be basic. Of course, this conclusion should be much more carefully defended, with a detailed comparison of the positions in question and a detailed inspection of the examples adduced in their favor. However, I will not engage in this business, since my interest was only to clear away the confusing manifold in order to be able to focus in the following sections on what seems more important and basic to me: the Agrippan trilemma, the structure of justification and the debate between foundationalism, the traditional leader, and coherentism, the challenger. This interest may have been satisfied by this section, if only minimally.

16.2 Perception

Let us return, then, to the hydrodynamic picture of justification and to the question of where the warrant fluid comes from. The answer appears obvious: somehow, it comes from perception. All mental activity depends on perceptions; without perceptions we could not acquire any concepts whatsoever; perceptions are the elixir of our minds. This is not quite the right sort of dependence, though. Sure, we could not even acquire a priori beliefs without perceptions, since even a priori beliefs presuppose concepts, and concept acquisition rests on perception; however, the justification of a priori beliefs, if there is any, does not depend on perceptions or perceptual evidence (more on this in Section 17.1). Still, all our a posteriori beliefs are justificatorily dependent in this way, and we need to more specifically describe *this* dependence. So our task, finally, is to understand the role of perception in belief formation.

It might seem that we have already extensively dealt with this topic, at least implicitly. This is not true, though. In Section 5.4, I introduced various conditionalization rules, the laws of the dynamics of ranking functions. However, I was vague there concerning the causes of doxastic change. I did not speak of perception; I rather used such deliberately broad phrases like "the subject receives (and accepts) the information that . . ." and called this the source of belief change. In Section 7.4, I detailed the processing of information by studying its propagation in ranking nets; but, again, the impact of evidence at a certain node of the net was simply taken as given and not further elucidated. Section 9.3 was my last move so far in constructively addressing the topic, this time in an auto-epistemic perspective. I emphasized there the importance of the notion of a subjective protocol and of assuming a set TE of possible pieces of total evidence that were not ordinary propositions, but things we may come to learn, and hence were somehow auto-epistemic propositions. However, even though the formal role of TE was clear, I did not say more about what these "things we may come to learn" are. So I have, in fact, avoided the topic so far, and we must see now how we can integrate perception into our epistemological account.

Of course, this is not to say that we should focus now on any scientific, physiological, and psychological details of perception. As rich and fascinating a topic as this would be, it is neither scientifically nor even conceptually prepared to tell us anything about the relation of perception to justified belief. No, we may − and can presently do no better than to − stick to ordinary-language based philosophy of perception.

In this setting, it is a common theme that perception verbs like "see", "smell", "feel", etc., occur in various grammatical forms: there is object perception, "I hear a dog", event perception, "I hear a dog bark" or "I hear how a dog barks", aspect perception, "I hear a dog as barking", and fact perception, "I hear that a dog barks". The doxastic implications of object, event, and aspect perception are not so clear and in any case very weak or even non-existent; therefore they are not of further interest to us. Fact perception is the idiom relevant to us; if I perceive that A, it is analytically implied that A is actually the case and that I believe that A. Indeed, fact perception is the paradigm of knowledge; if I perceive that A, it is analytically implied that I know that A, and, insofar knowledge is at least justified true belief, that A is the case and I justifiedly believe that A. This is what everybody using our terms must say.

Folk theory, as embedded in the meaning of our terms, makes even stronger claims. The following is, I think, the core of our folk theory of perception. It is also called the causal theory of perception (cf. Grice 1961), at least with respect to object perception (for a possible caveat in the fact perception version stated here, see below):

(16.3)   *The Causal Theory of Perception*: If *a perceives* that A, then the fact A is a cause of a's believing A.

(16.3) cannot be strengthened to a biconditional. There are lots of ways in which the fact A could cause the belief in A that we would not call perception of the fact A; such cases again surface under the heading "deviant causal chains" (cf., e.g., Coates 2000). The rules telling us which ways of A's causing the belief in A are deviant are quite unclear; this is the vague part of our folk theory of perception.

Still, (16.3) most plausibly explains why perception is the paradigm of knowledge. (16.3) says that perception entails true belief. And in a quite intuitive sense, it entails that the belief is well justified. If it is the fact A itself that causes my belief in A, what better foundation could there be for my belief? None, it seems. However, this point transcends the boundaries of internalist justification; the observer can tell then whether I am justified, whereas I might have a hard time distinguishing whether I have actually perceived that A or whether I only believe that I perceived that A. This is, of course, the basic motive of all externalist conceptions of knowledge and justification.

Indeed, (16.3) is essentially externalist in the sense that (16.3) can be assumed or believed *only* by external observers, but not by the subject a herself. In Section 16.4 we will find a clear sense in which a subject cannot think about the causes of her own present beliefs (i.e., of having those beliefs), but only about their reasons (i.e., the reasons for their contents, not the reasons for having them); see also Benkewitz (2011, sect. 5.3), who discusses this point in detail. At present, we should not close this loophole of an externalist element at the origins of belief, even though I pleaded for justification internalism in general in the previous section.

Although perception is a paradigm of knowledge, (16.3) does not necessarily serve as a general model of knowledge. Even externalists concerning knowledge do not claim that knowledge of A is always caused by A itself. The clearest case in point is whether we can have knowledge of the future, which can obviously not be caused by the future. Many think that we know that the sun will rise tomorrow; if so, knowledge must be essentially wider than perception. I am not so sure; I think there is a good sense of knowledge according to which we can have knowledge about the past, but not about the future. However, since knowledge is not our topic, we do not have to take a stance on these issues.

Our interest is rather in accurately grasping how the dynamics of belief as described by the conditionalization rules is driven by perception. The obvious suggestion is, of course, that whenever I learn that A and conditionalize accordingly, i.e., whenever the source of my doxastic change in the sense of (3.13) and (5.36) is the proposition A, I have either perceived that A or perceived something else (perhaps not explicitly represented in the conceptual frame considered) from which I inferred that A. Saying that, however, points to its problems. There is indeed a difference between information reception and perception with respect to factivity; information can turn out to be misinformation without ceasing to have been information, while fact perception cannot turn out to have been misperception. Otherwise, though, the talk of perception is no less infected by relativities than the talk of information reception.

I am referring here to the obstinate, though failed, attempts of the logical empiricists to identify something like an observation language that describes what is perceived, and that serves as a base of all theoretical language. Hempel's (1971) replacement of talk of observation vocabulary by talk of the antecedently understood vocabulary is the paradigmatic acknowledgment of this failure. His example was that the man on the street sees that there are dark clouds in the sky with a strange yellowish tinge (apparently a sign of the upcoming thunderstorm), whereas the physicist sees that a huge electrical voltage has built up in the clouds. We may ask, indeed, whether I see that a bear has crossed the snowfield, or whether I only see that there are footprints of a bear in the snowfield? The truth is I see both; there is no sense in trying to tell what I really see. The so-called accordion effect, an important discovery about the description of actions (see Davidson 1971), equally applies to perception, though it is a good question how far the accordion can be extended.

Indeed, Armstrong (1968, p. 230) rejects the application of the causal theory of perception to fact perception on the grounds that it would be quite common to say, for instance, that I can see that it will rain. In such cases, (16.3) would be obviously wrong; if we accept them, the accordion could expand not only backwards to causes, but also forward to effects, just as Goldman (1967) argued regarding knowledge in general. Maybe. However, such cases are at least dubious. For all central cases – and for our purposes – we may leave it at (16.3).

The accordion effect entails: If object, event, and aspect perception are not epistemologically basic and if we had hoped that at least fact perception would give us access to the basis of belief formation, then that hope is thwarted. Perhaps not all facts can be perceived, but many can be perceived, and many without a claim to being epistemologically basic in anything stronger than a relative sense.

Maybe our fault was to refer to fact perception in general; we would have been better to refer to the facts that are *directly* perceived. However, this moves us from the frying pan into the fire. First, one should note that talk of perception is perfectly ordinary language, whereas direct perception rather is a philosophical term of art. And there has hardly been a term that has caused so much philosophical confusion as this one. Berkeley drove us (most admirably in the first dialogue of his 1713) to the absurd, though apparently inescapable conclusion, that the only things we directly perceive are our own ideas and sense impressions in particular. Huge philosophies were built on that conclusion, not only by Berkeley. And this conclusion is definitely queer (though this does not automatically denounce all the further implications drawn from this conclusion).

(16.3) is a good place to start in explaining the pitfalls of so-called direct perception. If perception is characterized by (16.3), then *direct perception* is presumably characterized thus:

(16.4)   If *a directly perceives* that A, then the fact A is a direct cause of a's believing that A.

By now, though, we know full well that direct causation is frame-relative; this is the crux of the matter with (16.4). Indeed, the relativity doubles in the case of (16.4):

First, the external fact A may be a more or less direct cause. In a way, I directly perceive only the footprints of the bear in the snow, not the bear itself. Saying that leads us onto a slippery slope, though. One might then be tempted to place the external cause ever more proximately to the doxastic subject. Maybe, it is only a distribution of the photons close to my eyes that I see directly. The ultimate step would be to locate the proximate cause within the subject itself. Then one arrives at the claim that I directly perceive only my own sense impressions or sense data. However, one must attend to whether one is still talking about fact perception. The photons may be a more direct cause of my beliefs than the objects reflecting them; still, if I do not believe that the photons before my eyes behave in such and such a way, I do not perceive any facts about the photons. Again, in philosophers' speak I have sense impressions, but this need not mean that I believe my sense impressions to be such and such (although I will claim precisely this in Section 16.4). In any case, one must beware of the skeptical move that one really perceives only what one directly perceives (in some more or less absolute sense) and thus beware of all the disastrous consequences of that move for knowledge or even ontology (cf., e.g., the still refreshing criticism of Austin 1962).

The last remarks point to the other frame-relativity hidden in (16.4). It may also be the doxastic effect that is more or less direct. In fact, it has often been observed that the directness of perception shows in the proximity not of the cause, but of the effect. That is, if I directly perceive that A, then my belief in A is direct, as it were, i.e., a non-inferential belief not inferred from or caused by other beliefs. Berkeley (1713) was already aware of this point when he said that "the senses make no inferences". This sounds more reasonable and less dangerous than the search for proximate causes. Indeed, when we look for the source of a doxastic change according to the conditionalization rules, we apparently find it in direct perception in precisely this sense. According to (3.13) and (5.36) the source is a doxastic attitude itself, which is somehow externally caused, but not inferred from other beliefs or attitudes; only the further doxastic changes are inferred from that source.

However, this sense of directness or non-inferentiality is again frame-relative. One might see here the same relativity as before; the directness of effects is as relative as that of causes. The relativity thus refers to the conceptual frame of the observer who judges the causal relations. Or the relativity might be seen as referring to the conceptual frame imputed to the subject. Which frame or space of possibilities is considered as the domain of the subject's doxastic attitudes? What is a non-inferentially acquired belief, relative to a coarse frame, might well be an inferred belief, relative to a more detailed frame. This relativity is well expressed in Theorems 3.12 and 5.35. There, the source of a doxastic change is unique only relative to the propositional algebra $\mathcal{A}$ considered. Take, e.g., a change on $\mathcal{A}$ and another change on the more fine-grained algebra $\mathcal{A}' \supseteq \mathcal{A}$ which, however, agrees with the change on $\mathcal{A}$ in the sense that the doxastic attitudes on $\mathcal{A}'$ restricted to $\mathcal{A}$ are the same as those on $\mathcal{A}$ before and after the change. Hence, we are considering not two changes, but only one change described in two different, more or less detailed ways, according to the invariance principle (5.3). Still, the source of the change on the richer $\mathcal{A}'$ will usually differ from the source of the change on $\mathcal{A}$, and indeed will ranking-theoretically determine the latter.

So, even this view of direct perception entangles us in relativizations. The basic bulbs of our hydrodynamic picture seem to be only relatively basic. Absolutely basic bulbs can be discovered, it seems, only by appealing to the subject's universal frame. However, in the previous section I was distrustful of this move and pleaded instead for the relativization of reasons and justifications. The situation has not improved.

All in all, talking of fact perception instead of information reception, as we did in previous chapters, has not really helped us in detecting whatever truth there is to foundationalism. Only the causal theory of perception (16.3) – the point that, in fact perception, a fact causes the belief in it – is a valuable observation that will prove to be important in Sections 16.4 and 17.2.

16.3 The Schein–Sein Principle

There is a way, it seems, of avoiding reference to the ill-defined universal frame of a subject. I occasionally slipped into mentioning a subject's sense impressions or sense-data, sometimes also called percepts, and it seems that, whatever a subject's universal frame may be, these are the things that are at the basis of all belief formation. They do not have, however, the best philosophical reputation. Maybe they have been reified in problematic ways, maybe they come with problematic assumptions about consciousness (recall, e.g., the devastating criticism in Quine 1969). In any case, we cannot simply refer to them in a naive way.

There is, however, a perfectly plain ordinary-language practice of talking about sense-data and the like (though not using these terms). Ordinary language provides two large classes of perception vocabulary. In our attempt to connect up with our conditionalization rules, the laws of belief change, we have, it seems, focused on the wrong class, namely on verbs like "perceive", "see", etc. that relate the perceiving subject to some external object, event, or fact (that, accordingly, is implied to exist). However, there is another large class, the appearance vocabulary of verbs like "appear", "seem", "look", "sound", etc., which, again, can be used in various grammatical constructions. For some sense modalities, we even have the same verbs in both classes, e.g., "smell" and "feel". "I smell a rotten egg" and "I smell that there is a rotten egg in the bowl" are perception talk, whereas "it smells to me like a rotten egg" is appearance talk. In any case, let us focus now on this kind of vocabulary. After our failure in the previous section, this is a more promising attempt to get at the foundations of a posteriori belief in a more absolute sense.

Again, though, this means moving on extremely slippery ground. Appearance terms come in various grammatical constructions that often have subtly different meanings. For instance, "something looks red to me" and "something looks to me as if it were red" usually mean the same thing; but under blue light, the first may be false in some sense and "something looks violet to me" true in some sense, while the second may still be unambiguously true and "something looks to me as if it were violet" unambiguously false. With "something looks square to me" we find less ambiguity; it always means the same as "something looks to me like a square", I believe, since we never say "something looks trapezoid to me", even if this is, in some sense, the way the square lying in front of me looks. Of course, it is not only simple qualities that appear to us; the constructions may be complex and the appearances rich and specific, as in "it looks like a (rusty, worn-out) car (Mercedes) to me", "it looks like the aftermath of an earthquake", "it looks as if vandals have taken over the city", etc. Poets are very imaginative at describing our rich manifold of impressions.

We must reduce complexity in a fair way. Hence, I would like to settle for the locution "it appears to me as if A" or, since the generality will be important, "it appears to the subject a at time t as if A"; $\Phi_{a,t}(A)$ is to denote the proposition thereby expressed. This is dangerous in two ways. First, as I will discuss in a moment, "appear" has more unintended connotations than verbs like "look" and "sound". And second it means focusing on a locution that is not entirely representative of the wide range of appearance terms.

This is outweighed by four advantages. First, "appear" is applicable to all sense modalities, even though my examples will usually refer to looks. Second, the "as if"-construction allows for the widest range of complements; it has the best prospects of being nearly representative. More important is the third point, that the complement is propositional. By focusing on $\Phi_{a,t}(A)$, we can right away proceed to the issue we are interested in, the issue of how the propositions A and $\Phi_{a,t}(A)$ are epistemically related; for other locutions, the propositional counterpart would have been less clear and straightforward. Finally, $\Phi_{a,t}(A)$ is in no way relational, unlike the other locutions. Phrases like "it looks red to me" or "it looks like a car" usually presuppose that "it" refers to some real object; like perception talk, they still express a relation between an external object and the subject. By contrast, the truth condition of $\Phi_{a,t}(A)$ is internal to the subject a; it lies entirely in the character of her sense impressions and can thus be satisfied even if a is hallucinating. The "it" in "it appears to me as if A" is merely expletive as in "it rains".

So, let us accept this restriction. For ease of expression, I may slip back to other locutions as well, even though I am officially speaking only about $\Phi_{a,t}(A)$. As far as I can see, the subsequent discussion does not suffer from this restriction; *mutatis mutandis*, the results would be the same if we were more attentive to the full range of appearance terms.

There is another well-known classification of our descriptions of appearances, not along grammatical, but along semantic lines (though the overlap is large); I am referring to the three readings of appearance terms, the epistemic, the comparative, and the phenomenal reading, as introduced by Chisholm (1957, Ch. 4).

According to the epistemic reading, it appears to me as if A if and only if what I sense suggests to me, or makes me believe, or tends to make me believe in the absence of counter-evidence, that A. The verb "seem" most clearly tends to this reading; which is why I will avoid it. "Appear" is more open to the other readings; still, the epistemic connotation is strong, in particular in connection with the "as if"-construction. However, it is clear that I want to, and indeed have to, exclude this reading; $\Phi_{a,t}(A)$ is to be an entirely non-doxastic state of affairs, about which we, and a himself, can have ordinary first-order beliefs. Otherwise, we could not discuss the epistemic place of $\Phi_{a,t}(A)$ and its epistemic relation to other propositions such as A. Of course, its epistemic entanglement may be so close that one might be tempted to make it part of its content. Clearing up this entanglement, as I hope to do, can thus at the same time secure non-epistemic readings of $\Phi_{a,t}(A)$. In any case, the subsequent discussion stands or falls with the possibility of such non-epistemic readings, and my use of "appear" should always be understood in such a way.

Indeed, the comparative and the phenomenal reading are both intended to be non-epistemic. They are subtly different. If something looks like a car, it looks similar to the way other cars usually look; this is the comparative reading. If, in this reading, it appears to me as if A, then it appears to me in the way realizations of A usually appear to me. There are countless ways in which cars may look; there is no specific quality that is common to all car appearances. Rather, what unites all car appearances is that they are all appearances of cars, among them mostly typical ones. The phenomenal reading is different. If something looks red to me, there is a specific phenomenal way it looks to me, the characterization of which does not require a comparison with (the appearance of) other red things. If it phenomenally appears to me as if A, then there are some more or less complex phenomenal characteristics of my impressions that are captured by "as if A".

Even if these two readings are clearly distinct, we need not decide between them. As far as I see, the subsequent considerations hold under both readings, since both are non-epistemic. I should also add that I do not think that we are really dealing here with three different readings of appearance terms; they do not have three different meanings. In Spohn (1997b, pp. 361ff.), I argued that appearance terms are hidden indexicals; they have a context-dependent meaning, and there are three different types of contexts in which those terms take on the three readings. Here, this point is of no further relevance, so I need not go into the details. The only remark worth making is perhaps that, according to this argument, the epistemic reading is required only in the case of absent qualia, i.e., in the strange context where most of us are zombies lacking qualia. We may grant that no one can exclude this case for sure; so it is at least a possible context. However, each of us knows for sure that he or she is not a zombie, but has qualia, so that this strange context is irrelevant when a, who may be any of us, thinks about what to believe about $\Phi_{a,t}(A)$, i.e., about herself. Hence, even in that perspective we may exclude the epistemic reading of $\Phi_{a,t}(A)$.

I hope to have thus sufficiently and not artificially delineated the intended meaning or content of $\Phi_{a,t}(A)$. What, then, is the domain of $\Phi_{a,t}$, the range of propositions A for which $\Phi_{a,t}(A)$ makes sense? If we allow all three readings, anything or almost anything can appear to me as if it were so; for instance, the continuum hypothesis might seem true to me. However, the continuum hypothesis is not accompanied by characteristic impressions. So, the non-epistemic readings enforce some restrictions on the domain of $\Phi_{a,t}$. They are quite vague, though. Certainly, if A is expressed by some observation sentence, as intended by the logical empiricists, then it will be in the domain of $\Phi_{a,t}$. Clear examples would be "there is a laptop on the desk" or "the pointer points to 3.7". However, as emphasized in the previous section, there does not seem to be a well-defined observation language. This is no concern for us. In the absence of clear criteria, we may simply rely on our common sense understanding of what can be perceived by us or appear to us, and we can tolerate its vagueness.

In any case, we will henceforth assume that, for each admissible proposition A, $\Phi_{a,t}(A)$ is also in the propositional algebra to be considered. So, we are dealing with a rich conceptual frame, but it need not be a subjectively or objectively universal frame. After all of these preparatory discussions we can finally turn to our proper topic, the epistemological place of propositions of the form $\Phi_{a,t}(A)$. (The subsequent considerations essentially follow Spohn 1997/98.)

Our general analysis of dispositions in Section 13.3 has prepared us well for this topic. Recall our findings there. We concluded in (13.15) that it is defeasibly a priori that, given the relevant test situation or stimulus S, the assumption or proposition that an object has the disposition D is a reason for the assumption or proposition that it shows the relevant manifestation or response R – and vice versa, since the reason relation is symmetric. On the basis of our account of normal conditions in Section 13.2 we further concluded in (13.17) that the same is unrevisably a priori ceteris paribus, i.e., given also that normal conditions obtain. If this holds for dispositions – Locke's tertiary qualities – in general, then it applies to secondary qualities as well, to dispositions to appear to us in a certain way – for instance, to redness. For this case, we would start with the reduction sentence

(16.5)   an object is red iff it looks red to those who look at it,

where looking at it is the test situation (= S) for the red object (= D) to show the behavior of looking red (= R).

It is obvious that we have to hedge (16.5) with normal conditions (consisting of bright daylight, normal observers, etc.) and should rather claim:

(16.6)   an object is red iff it looks red, when looked at under normal conditions.

We have seen in Section 13.3 that, while (16.5) is only defeasibly a priori, (16.6) is unrevisably a priori. However, Section 13.4 showed that (16.6) need not be analytic; that depends on the resolution of hidden ambiguities in (16.6), as I more carefully argued in Spohn (1997b, pp. 367ff.).

Now, by the same reasoning as in Section 13.3 we arrive at:

(16.7)   Given that an object is looked at, the proposition that it looks red is a defeasibly a priori reason for the proposition that it is red, and vice versa.

This corresponds to (13.15). I spare myself the unrevisably a priori version that refers to normal conditions.

The next step is as obvious as it is crucial. Not only do some objects look colored to us, the world incessantly appears to us in this and that way, at least as long as we are awake and our awareness is directed outwardly; the entire universe is disposed to appear to us. If so, we may generalize (16.7) and claim: On condition that some external situation is attended to, the proposition that it appears as if A is a defeasibly a priori reason for (the proposition that) A, and vice versa. The condition looks clumsy; I will comment on it in a moment. First, however, we need to be more explicit about the subjects and the times involved and thus arrive at what I call

(16.8)   *The Schein–Sein Principle* (*defeasible version*): Given that the person a attends at t to some external situation, $\Phi_{a,t}(A)$ is, for the person b, a defeasibly a priori reason for A, and vice versa.

As in the case of dispositions, this is accompanied by an unrevisable version referring to normal conditions:

(16.9)   *The Schein–Sein Principle* (*unrevisable version*): Given that a attends at t to some external situation under normal conditions, $\Phi_{a,t}(A)$ is, for b, an unrevisably a priori reason for A, and vice versa.

This sounds much stronger than (16.8), but in fact it is not; the unrevisability is compensated by the volatility of the normal conditions, which we were able to grasp in Section 13.2 only in an epistemically relativized way. Hence, I will focus on (16.8) and refer to it even without the distinguishing adjective. Let us also stick to the German label that is much nicer than the English "appearance-being principle".

Person b may be different from a, or the same as a; the latter case will be the more interesting one, of course. The appearance is specified to occur at time t, whereas the reason relation is not temporally indexed, since it is claimed to hold for all possible initial doxastic states (recall Definition 6.22 and the preliminary elucidation of "initial"). However, it should also hold for b's doxastic state at time t' if b's information up to t' allows the initial reason relation to be maintained, as will normally be the case. Again, t' might be $\neq t$ or $= t$. Let us refer to the case where $b \neq a$ or $t' \neq t$ as the *non-reflexive case* of (16.8), and the case where $b = a$ and $t' = t$ as its *reflexive case*.

It is presupposed in (16.8) that A is in the domain of $\Phi_{a,t}(A)$. Is A simultaneous with the appearance, i.e., realized at t? Or is A rather an indexical proposition referring to "now"? Both sound plausible. However, in view of the potential flexibility of the domain of $\Phi_{a,t}(A)$ we should not assume this; perhaps the past (and if the above caveat about (16.3) is correct, even the future) can now phenomenally appear to me in some way. Nothing will depend on it; hence we need not bother about this issue, interesting as it is.

A word about the condition stated in (16.8): It seems clear that under the contrary condition that the person a is not attending to any external situation at all, nothing can be assumed a priori about the relation between external events and a's appearances. If so, we also cannot derive anything about an unconditional reason relation. However, often and, particularly in the reflexive case, the condition will be given to the believer b in a way that does not defeat the initial reason relation, which will then also hold unconditionally.

I think that the Schein–Sein Principle is universally true, i.e., in all the cases just unfolded. If the generalizing steps up to (16.8) were correct, this indeed follows from my argument in Section 13.3. However, the principle is philosophically too important to be merely based on that argument. We should directly inspect the various cases and check for their plausibility. Having thus ascertained the veracity of the principle, we can afterwards ponder its philosophical significance.

The most innocent case is the one where a and b are different persons. In that case (16.8) seems perfectly reasonable. In the one direction, it says that we initially trust the senses of others. If they make us believe, by credible assertions or whatever, that certain things looked to them to be a particular way, we also believe that these things actually were that way. This conclusion can only be obviated by specific counter-reasons that were not initially given. Conversely, if A is an observable state of affairs and if a is attending to the situation in which A realizes, then it should normally appear to a as if A. Again, special reasons are required for assuming otherwise.

The interpersonal case where $a \neq b$ is the epistemologically less exciting one, of course. What is relevant to our search for the foundations of belief formation is the intrapersonal case where $a = b$. However, it largely resembles the interpersonal case. It does so when we focus on the doxastic state of a at some time $t' \neq t$; this also covers a's relevant initial doxastic state, which a is in before t. The resemblance results from the fact that my doxastic situation vis-à-vis my past or future self is not essentially different from the situation vis-à-vis other persons. In thinking about the doxastic relation between past or future facts and past or future ways things appear, I cannot see an important difference between the case where those things are supposed to appear to you and the case where they appear to me. (Of course, it makes a great difference for me who is actually having the experiences.)

So, as expected, the difficult case of (16.8) is the reflexive case where $a = b$ and $t = t'$. One might argue that the case cannot really arise, since as soon as we start reasoning about or from how things appear to us, the appearance is already past, and we can reason about it only via recollection. However, this would require the assumption of an unattainable precision in the times and temporal extension of experiences and mental states; moreover, it would import a procedural or computational perspective on reasoning that we rejected in Section 16.1. We should not simply block the reflexive case with such a problematic argument. So, let us face it squarely.

The reflexive case seems odd on two scores. The first score is that it is difficult to entirely ban the epistemic reading of $\Phi_{a,t}(A)$. Replace, for the moment, $\Phi_{a,t}(A)$ by $\mathbf{B}_{a,t}(A)$, the proposition that a believes at t that A, and let a at t be me presently, in order to emphasize the reflexive case. Under this replacement, (16.8) requires something strange of me, namely that I think about my reasons for $\mathbf{B}_{I,now}(A)$. Here, reasons are always reasons to believe (more strongly). So, what could my reasons be for believing that I believe A? The question sounds confused (as has been observed several times; for a particularly careful analysis see Benkewitz 2011, sect. 5.3). None, would be one reasonable answer. I know what I believe; we are not talking about subconscious beliefs that are as inaccessible to me as to others, and about the presence and content of which I might start speculating. A more sympathetic answer would be: these reasons are the same reasons as my reasons for A. However, this is actually to reject the original question and to construe it as the sensible question asking for my reasons for believing A. So, it seems that facts about my own present beliefs fall outside the scope of my own present reason relation. In the next section I will attempt to give a more thorough argument for this conclusion.

Now, the reflexive case of (16.8) is not as outright bad as the case just discussed. However, as soon as we take $\Phi_{a,t}(A)$ to (weakly) imply $\mathbf{B}_{a,t}(A)$ (or something like that), we slip into the bad case. This is why I have so strongly emphasized that we have to read $\Phi_{a,t}(A)$ in a non-epistemic way. It must be an open question which beliefs accompany our appearances, even though we move so effortlessly from one to the other that it is difficult to keep them apart.

However, even if we firmly exclude any epistemic ingredient in $\Phi_{a,t}(A)$, the reflexive case of the Schein–Sein Principle remains odd. My present beliefs about my present appearances of the form $\Phi_{I,now}(A)$ are then ordinary first-order beliefs. They are a posteriori, not a priori; they arise in that very moment and are not grounded in some initial belief state. However, once I have the appearance and thus acquire the belief in it, this belief seems maximally certain and strictly unrevisable (cf. 6.13–6.14); at most, it may, and usually will and must, be forgotten later on (given our cognitive economy). However, when there are reasons for or against a belief of mine, I may possibly receive them and thus revise (the strength of) my belief. In other words, there are no reasons for or against strictly unrevisable beliefs (cf. 6.17), and the Schein–Sein Principle again seems false in the reflexive case.

This argument is mistaken, though. As soon as the content of my belief is given by $\Phi_{I,now}(A)$, by "it appears to me now as if A", where A is in turn, as I have made clear, some linguistically expressible proposition, there is room for error and revisability. The classic passage making this point is found in Austin (1962, pp. 112f.):

Reflections of this kind apparently give rise to the idea that there is or could be a kind of sentence in the utterance of which I take no chances *at all*; my commitment is absolutely minimal; so that in principle *nothing* could show that I had made a mistake, and my remark would be 'incorrigible'.

But in fact this ideal is completely unattainable. There isn't, there couldn't be, any kind of sentence which as such is incapable, once uttered, of being subsequently amended or retracted. . . . I may say 'Magenta' wrongly either by a mere slip, having meant to say 'Vermilion'; or because I don't know quite what 'magenta' means, what shade of color is called *magenta*; or again, because I was unable to, or perhaps just didn't, really notice or attend to or properly size up the color before me. Thus, there is always the possibility, not only that I may be brought to admit that 'magenta' wasn't the right word to pick on for the color before me, but *also* that I may be brought to see or perhaps remember, that the color before me just wasn't *magenta*. And this holds for the case in which I say, 'It seems, to me personally, here and now, as if I were seeing something magenta', just as much as for the case in which I say, 'That is magenta.' The first formula may be more cautious, but it isn't *incorrigible*.

And it is clear that Austin's point does not hinge on the utterance. If I now believe that it now appears magenta to me, I might be wrong and be corrected for the same reasons (with the exception of the slip); actually, the color impression I now have might be different and not magenta.

The general point is that by believing $\Phi_{I,now}(A)$, I am classifying my impression as being as if A, and I might thereby misclassify it, not by a slip of a tongue that cannot occur in belief, but by some other failure. This can happen with any proposition A (for which $\Phi_{I,now}(A)$ is well-defined), but that it can happen even with color terms is, of course, particularly impressive.

It seems that the failure must lie in the conceptual or linguistic processing of the impression; this is what Austin indicates. Clearly, this is one possible source of error. However, it can also occur in the generation of the impression; this is a possibility I find even more telling. I have in mind the case of so-called pseudo-normal vision, which is a scientifically realistic case of inverted qualia. It goes as follows:

As is well known, the human retina is trichromate; it contains a lot of cones, each of which is equipped with one of three kinds of pigments. All three pigments are sensitive to large parts of the visible spectrum, but in varying degrees. The maximal sensitivity of the pigments lies, respectively, in the red, the green, and the blue segment of the spectrum. Hence, the pigments are called R-, G-, and B-pigments and the cones containing them R-, G-, and B-cones. A crucial link between the activation of the cones by incoming light and the ensuing color sensation is provided by the so-called opponent process theory. It says that the activity of the R- and the G-cones is compared closely behind the retina and that the color impression is reddish or greenish depending on how much the activity of the one sort of cones outweighs that of the other sort of cones. Moreover, the activity of the R- and the G-cones is summed up and compared with the activity of the B-cones. Again, the resulting impression is yellowish or bluish depending on whether or not the sum of the R- and the G-activity outweighs the B-activity. It is important here not to get confused about the classification underlying the labels R, G, and B. The pigments so labeled are classified according to their chemistry, while the opponent process theory gives a functional criterion for classifying cones as R, G, and B; they are so classified according to their subsequent wiring. (For details see, e.g., Boynton 1979, chs. 7 and 8; Ch. 10 explains that there is actually considerable variation in human color vision; matters are far more complex than I need to represent here.)

The interesting point for us now is that the opponent process theory has a nice explanation of dichromatism or red–green blindness. It is simply that for some reason both the R- and the G-cones contain the same pigments, so that their activity is always the same and no impression tends to be reddish or greenish. Note that this explanation presupposes the independence of the classifications of pigments and cones just stated. Now, Piantanida (1974) advanced a most interesting hypothesis: The foregoing makes clear that dichromatism may come in two forms; either the R-pigments are contained also in the G-cones, or the G-pigments are contained in the R-cones. Very roughly (for details see again Boynton 1979, pp. 351–358), Piantanida conjectured that both forms are due to genetic defects, that these defects are located in different genes and therefore statistically independent, and that there is hence a slight chance of both defects occurring at once (which Piantanida estimated to be 1.4 per thousand among males). Persons with both defects are called pseudo-normal.

Are pseudo-normals handicapped in any way? Not in the least; their discriminatory and recognitional powers are the very same as those of normal persons, and they have learned to talk as everybody does. Perhaps you are one of them, and you never realized! It would not be easy to find out; no behavioral test could tell, and investigating your genes would deliver quite indirect evidence that relied on sophisticated scientific hypotheses.

The truth of the story is not important for us; what matters is only that it is scientifically plausible and testable, and thus substantiates airy fantasies about inverted qualia. Hilbert (1987, p. 92) seems to be the first to have mentioned pseudo-normality in the philosophical literature. However, only Nida-Rümelin (1993, Ch. 4, and 1996) has fully realized its philosophical significance; cf. also our dispute in Nida-Rümelin (1997) and Spohn (1997b). For the interesting point is how we should talk about pseudo-normals:

It seems obvious that their reddish and greenish sensations are reversed (turquoise becomes violet, etc.). One might speculate that there is a compensatory mechanism deeper in the brain, similar to the one coming into operation when we wear glasses that turn everything upside down. However, in contrast to the latter case, there is no need for such a mechanism in the color case. Let us stick to the simple story, even though it is full of unconfirmed assumptions. Hence, red peppers look green to pseudo-normals and green peppers red; this is the appropriate way to talk about them, I think. One might sense an ambiguity here, and refer to Chisholm's three readings. Indeed, I would grant that red peppers look red to pseudo-normals in the epistemic as well as in the comparative reading. However, the phenomenal reading is clearly applicable to pseudo-normals, and in that reading red peppers clearly look green to them. And this is how my predicate Φ is to be understood in this context (even if my official "as if" locution is misleading in this case).

Still, pseudo-normals, having no insight into their abnormality, sincerely assert that red peppers are red and that they appear red to them; that is what they have learned. Hence, we should not hesitate to say that they believe that red peppers are red and even that they believe that red peppers appear red to them. One might sense de re/de dicto ambiguities here. However, in a clear de dicto sense this is what they believe, and more importantly, also in the sense of Explanation 2.6 (that tried to get rid of the semantic-epistemological entanglement).

This completes my point. We have here another clear case in which actual appearances and beliefs about them fall apart, another clear story according to which my belief that it now appears to me as if A turns out false, and I might learn that this is so. In a way, I like it more than Austin's example, which turns on somehow making a mistake. Pseudo-normals, however, cannot be said to make any mistake; they have correctly learned everything, and they are full members of our linguistic community; it is only that they are in justifiable error about themselves. (The case might still sound confusing, and one might suspect that we should first give a much more careful analysis of the meaning of color terms in order to uncover all potential ambiguities and thus to gain more clarity. This is what I tried to do in Spohn (1997b). Here, the foregoing might do, I hope, without going into the much longer story.)

To summarize, if all these arguments are acceptable, I have established the Schein–Sein Principle (16.8) and (16.9), even in its reflexive case. How does the principle contribute to our understanding of the basis of belief formation, of the origin of the warrant fluid, to use that metaphor again? Piles of literature could be marshaled here for comparative discussion. Let me just pick four illustrative points.

The first point is that the Schein–Sein Principle may be suited for soothing *skeptical worries*. All traditional epistemologists were foundationalists and took impressions to be somehow basic. So did Hume, and unwilling to turn to idealism – i.e., to mentalize or idealize realistic ontology – he found himself entangled in skepticism about the external world, as displayed in his (1739, part IV, Sect. 2) and reflected in his (1748, Sect. XII). No step of reason seemed to carry us from our impressions to beliefs about reality. Hume particularly worried about real objects that essentially have properties like continuity and independence, which can never be displayed in our impressions and which hence cannot, it seems, be justifiedly postulated. Kant tried to overcome this scandal by his distinction between things in themselves and things as appearances, which produced more bewilderment than acceptance. Now, the notion of an object is a large topic of its own that is not on our present agenda. Generally, however, we may say that the skeptical gap is bridged precisely by the Schein–Sein Principle.

The crucial reason why this bridge is able to carry us lies, I think, in its flexibility, i.e., in its merely defeasible apriority. Usually, the bridge has been conceived too rigidly, and then it was too short or it broke. For instance, when we try to conceive of reality as a logical construction out of impressions (as Berkeley and his idealistic followers did and as Carnap 1928 ventured in the spirit of methodological solipsism), we may rightly wonder whether we have ever left the realm of ideas. Or one might try to argue that there are analytic connections between appearance and reality; for instance, von Kutschera (1994) thought that, if there are no logical relations, there might at least be probabilistic connections of an analytic kind. Thus put, this sounds unintelligible, even though there is some truth to it, as we will see shortly. Only Pollock (1995, pp. 52ff.) got it right in my view, where he treats the perceptual input of his defeasible reasoning machinery. Or to quote Pollock, Cruz (1999, p. 201): "Having a percept at time t with the content P is a defeasible reason for the cognizer to believe P-at-t." The only difference is that he takes this to be an instance of direct realism, since for him a percept (a sense impression) is not a belief or the content of a belief, although it has a perceptual content that may also be a doxastic content. (More on this difference below.)

Hume thought that reality is causally inferred from our impressions and trenchantly argued that causal inference cannot be (logically) necessary inference: so, there is no rigid bridge. The only conclusion he saw was that causal inference is habitual inference, a conclusion resulting in skepticism and justificatory void, as measured against traditional standards. In a way, my conclusion is the same: it is only – this is what the entire book is about – that the inferences that are belittled as habitual form a rich and rational edifice, as Hume certainly intended.

In the preceding paragraph, I presented the bridge as if it had only one direction, from appearance to reality. I did so only to please the skeptic, who only feels safe with his impressions. However, this was inadmissible; the bridge is, of course, a two-way bridge; the reason relation in (16.8) is symmetric, as always. This is as it should be. It is not clear to me that the belief in $\Phi_{I,now}(A)$ is always primary and the belief in A is inferred. It may be the other way round. I see, for instance, a car passing fast; I hardly glimpse it. Somehow, I know or firmly believe it is a Porsche. When asked how it did look to me, I say I cannot really tell; it went so fast; apparently, it looked to me as if it was a Porsche. This is not an artificial example, I think. Usually, beliefs in appearances and beliefs in external facts go hand in hand; there is little evidence to establish a temporal, causal, or inferential order between them. This is why I have routinely avoided referring to such order.

What is indeed reversed is the order of learning. We first learn realistic language, and then we learn to attend to our appearances as such and to describe and classify them with the help of the realistic language already mastered; to differentiate appearance talk from realistic talk is a late conceptual achievement. (A *locus classicus* for this point is Quine 1960, §1, "Beginning with Ordinary Things".) This is not to say, though, that the order of reasons would be reversed; it only entails that the symmetric reason relation can be extended to satisfy the Schein–Sein Principle only when these conceptual means are acquired and thus an initial doxastic state with respect to those means is reached.

It seems that I am suggesting that we may overcome skepticism with respect to the external world simply by postulating, a priori, a bridge principle denying the skeptical challenge. This looks cheap. Recall, however, the credentials of the principle. They spring from Section 13.3 about what we know a priori about dispositions. This was not necessarily analytic knowledge, as we saw in Section 13.4, but it was conceptual knowledge in the sense that the relevant defeasibly a priori reason relation belongs to our concept of a disposition. This appeared unassailable, and all I proposed was to generalize it to reality's universal disposition to appear to us. Hence, the Schein–Sein Principle is of a conceptual nature as well, just as my rendering of the reduction sentences. (This is why I said that there is some truth to von Kutschera's (1994) analyticity claims.) The only difference is that in the case of ordinary dispositions the manifestations are likely to be conceptually mastered before the dispositions, whereas in the present case, as just observed, the disposition is likely to be conceptually mastered before the manifestation. However, this learning order, if such it is, is without relevance to the conceptual relation between disposition and manifestation.

Moreover, this conceptual relation is not analytic; fortunately so, since this seems too strong and just wrong. It is only defeasibly a priori, and this leaves us plenty of room for learning when and under which circumstances our appearances can presumably be trusted and when they presumably deceive us, i.e., what the normal conditions are for the defeasibly a priori relation between appearance and reality. This seems to be the right way of understanding the relation. As mentioned, Pollock sees the matter in the same way; he even proposes a specific defeater. However, he simply postulates his version of (16.8) quoted above: ". . . nothing justifies this. This is one of the basic principles of rational cognition that make up the human cognitive architecture" (Pollock, Cruz 1999, p. 201). What I have supplemented, then, are the above credentials that add conceptual apriority to defeasibility.

A second context for the Schein–Sein Principle (16.8) is provided by the paradigmatic discussion among logical empiricists about so-called *protocol sentences* in the 1930s; this was in fact a discussion about the foundations of empirical knowledge. That discussion displayed a characteristic indecision between a physicalistic and a phenomenalistic basis, as it was called. The physicalists emphasized the intersubjective character of science. For them, only observation sentences like "there is a laptop on the desk" or "the pointer points to 3.7", which every observer can verify, were acceptable as basic. Indeed, they provide a common-sense basis that usually needs no defense; doubts about such observations are usually answered not by further argument, but by the request to look more carefully. Still, such doubts may be legitimate and need not be decidable just by watching (because, as was granted later on, the pure observation language is a kind of fiction); indeed, my example of the laptop is already a borderline case (perhaps it is a perfect fake). Hence, the phenomenalists, emphasizing the certainty or indubitability of the basis, were driven to take sentences about sense data as basic, ordinarily expressed by "it looks to me as if there were a laptop on the desk" or "the pointer appears to be pointing to 3.7". However, any scientist starting his paper with such confessions would be ridiculed. And so the debate was torn between intersubjective verifiability and absolute certainty.

If the Schein–Sein Principle (16.8) is correct, that debate seems to be a Schein debate. An observational proposition A is not indubitable or unrevisable, but neither, as we have seen, is the phenomenal proposition "it appears to me as if A" (even though the reasons for doubt may be very far-fetched). Instead of a single indubitable base, we have a two-fold fallible base, the parts of which mutually support each other; and as emphasized above, the mutuality is really to be taken serious. Hence, one might say that the Schein–Sein Principle realizes a kind of coherentism. It provides a pervasive coherentist link as a crucial building block of our empirical worldview, from which further coherentist links spread to other propositions about the external world more remote from observation; Chapter 7 on ranking nets helped to explicate the metaphor of spreading. Experience may refine this building block or even replace it in special cases, but it is the guaranteed point of departure of all our inquiries.

For a third point of comparison, let us look at BonJour (1985), who is critical of foundationalism in its various forms and hence promotes a *coherentist account of observation*. In the critical part (his Section 2.3) he ponders the appropriate characterization of basic beliefs and arrives at this (on p. 31):

...a particular empirical belief B could qualify as basic only if the premises of the following justificatory argument were adequately justified:

(1) B has feature φ.
(2)   Beliefs having feature φ are highly likely to be true.

Therefore, B is highly likely to be true.

He goes on to argue that: first, (1) and (2) cannot both be a priori (since the conclusion is empirical or a posteriori); second, that the argument from (1) and (2) must be accessible to the doxastic subject himself; and hence, third, that "B is not basic after all, since its justification depends on that of at least one other empirical belief" (p. 31).

This translates into our scheme in the following not perfect, but quite smooth way:

(1*) B is of the form "it appears to me now as if A" and is accompanied by the belief that I am now attending to an external situation.
(2*) = the Schein–Sein Principle.
Therefore*, A is more likely to be true, or the appearance as stated in B is more likely to correspond to reality.

This translation specifies and confirms BonJour's characterization. The second part of (1*) is a posteriori, and since (2*) is only defeasibly a priori, the "therefore*" is defeasible, too; it relies on the further implicit a posteriori premise that the relevant normal conditions may be assumed to obtain. Thus construed, the translated argument works. Moreover, the translation has replaced BonJour's "high likeliness" notion of justification by our positive relevance notion of reasons. Finally, the translation diverges from the original by not speaking of the same belief in (1*) and after "therefore*". This precisely reflects the point that the Schein–Sein Principle yields a two-fold base, which, by thus incorporating an element of coherentism, does not exactly satisfy BonJour's criteria for foundationalism.

Therefore, I also largely agree with BonJour's positive coherentist account of observation (in his Chapter 6), which he summarizes in the following general sort of justificatory argument for what he calls cognitively spontaneous beliefs (p. 123):

(1)   I have a cognitively spontaneous belief that P which is of kind K.
(2)   Conditions C obtain.

(3)   Cognitively spontaneous beliefs of kind K in conditions C are very likely to be true.

Therefore, my belief that P is very likely to be true.
Therefore, (probably) P.

I am not happy with the notion of a cognitively spontaneous belief, and to that extent, am not happy with the argument, either. As if I would suddenly become aware – to my surprise, as it were – of some belief I have, keep it in suspension while I sought a justification for it, then find an argument for its truth, and thus accept it. I find this picture distorted and think that the alleged spontaneous belief was not really a belief in the first place, at least according to my rationalistic picture that does not distinguish between belief and acceptance.

Otherwise, though, the same translation holds as before. Now, (3) translates into the Schein–Sein Principle. The conditions C in (2) are either the normal condition or some specific normal conditions, and thus make explicit what I said was implicitly presupposed in the previous translation. If we then take cognitively spontaneous beliefs to be beliefs about appearances, something BonJour does not admit, we arrive at our present account of the basis of belief formation.

To summarize, so far I share BonJour's (1985) criticism of foundationalism, and I highly appreciate his insistence on the role of all the tacitly acquired and refined assumptions about the observational conditions under which the observations may be (more or less) trusted; we continuously apply a most sophisticated and mostly implicit theory of perception – i.e., about the relation of appearance and reality – that denounces all straightforward forms of foundationalism. My account in this section has done justice to this insight, I think. It goes beyond BonJour's account in possessing the notion of defeasible apriority and hence in specifying the defeasibly a priori nucleus of that theory of perception in form of the Schein–Sein Principle.

The fourth point, finally, is a comment on *direct and representative realism*. The Schein–Sein Principle (16.8) certainly agrees with what is usually called representative realism. This is explained by BonJour (2007, Sect. 2.2) as "the view ... that our immediately experienced sense-data, together with further beliefs that we arrive at on the basis of them, constitute a *representation* ... of an independent realm of objects – one that we are, according to the representationalist, justified in believing to be true". And this justification derives from the fact "that the *best* explanation ... is that those experiences are caused by ... the character of a world of genuinely independent material objects". As should be clear by now, I do not like the representationalists' suggestion of a unidirectionality of justification that can also be sensed in the quote just given. Moreover, the appeal to the inference to the best explanation is usually left quite vague by the representationalists; I suggest using the account given here in Section 14.15. Indeed, (16.8) is content with referring only to reasons and not to explanations and thus is not yet occupied with questions of the *best* explanation. However, the reason relation between appearance and reality is guaranteed to hold a priori under normal conditions. If the representationalist were to also have a priori grounds for taking reality to be the best explanation of appearances, and thus for excluding skeptical explanatory hypotheses (cf. BonJour 2007, Sect. 2.2.2), I would be happy to transfer them to the present account. These are minor points, though; otherwise, my agreement with the representationalists is quite obvious.

Does this imply that I am opposed to direct realism? No. True, the two kinds of epistemological realism are usually seen as alternatives. "According to direct realism, in veridical cases we directly experience external material objects, without the mediation of . . . sense-data", and "the justification or reasons for beliefs about material objects that result from sense experience do not depend on the sort of inference from the subjective content of such experience that the representative realist appeals to" (BonJour 2007, Sect. 2.3). Rather than an opposition, I see frame-relativity at work again, as I did in Section 16.2. Of course, if propositions about her own sense impressions are not in the subject's domain of beliefs, then neither the reflexive case of the Schein–Sein Principle nor the ensuing justifications are available to her; only the observer can tell that the material objects cause or induce her to have the appropriate beliefs via her impressions. This is the picture of the direct realist. However, as soon as those propositions are considered to be in the grasp of the subject and thus in the domain of her beliefs, we can return to the representationalist's picture. So, direct and representative realism seem perfectly compatible.

This is a point where I finally disagree with Pollock. He confesses to be a direct realist: "The fundamental principle underlying direct realism is that perception provides reasons for judgments about the world, and the inference is made directly from the percept rather than being mediated by a (basic) belief about the percept" (Pollock, Cruz 1999, p. 201). One thing the quote shows is that the substances behind the labels are a bit unsteady. The main point, though, is that by conferring the justificatory or inferential role from the belief about the percept to the percept itself, Pollock admits to being a justification externalist; for him the reason relation must extend beyond the realm of beliefs. This is something I have denied; it cannot be integrated into my ranking-theoretic account of justification.

So, the Schein–Sein Principle, (16.8) and (16.9), and the ensuing explanations present my account of the foundations of empirical or a posteriori beliefs; in a nutshell, I have opted for a peculiar two-fold coherentist base. However, I have not yet completed my account; a final step, not so harmonious with BonJour's (1985) coherentist or Pollock's defeasibilist intentions, is still missing. Before taking this step, let me conclude this section with a variant of the Schein–Sein Principle extended to propositions about belief and at the same time restricted to the non-reflexive case. We will need it only in Chapter 17, but it is best introduced here.

In this section, I carefully avoided any epistemic reading of $\Phi_{a,t}(A)$ so that A as well as $\Phi_{a,t}(A)$ were ordinary first-order propositions or objects of reasoning without overt or hidden epistemic content. The reason was, as explained, that I wanted the Schein–Sein Principle to also hold in the reflexive case; this was crucial for my discussion of foundationalism and coherentism. Of course, though, these propositions are closely connected with beliefs about them, and in the non-reflexive case we may clearly reason about the having of those beliefs as well. This reasoning again has a defeasibly a priori core that is captured in:

(16.10)   *The Schein–Sein-Belief Principle*: Let $\mathbf{B}_{a,t}(A)$ be the proposition that a believes at t that A, and let A be a proposition such that $\Phi_{a,t}(A)$ is well-defined. Moreover, assume the non-reflexive case in which either $b \neq a$ or the reason relations are considered at a time $t' \neq t$. Then, given that a attends at t to some external situation, each of the four propositions A, $\Phi_{a,t}(A)$, $\mathbf{B}_{a,t}(\Phi_{a,t}(A))$, and $\mathbf{B}_{a,t}(A)$ is a defeasibly a priori reason for b for each other of the four propositions.

This is the defeasible version. Again, the unrevisable version also holds, with the stipulation of normal conditions (which might be different for each of the defeasible relations). We may moreover strengthen (16.10) by arranging the four propositions (or the four variables they realize) in a little ranking net, indeed a Markov chain; the assumptions that $\mathbf{B}_{a,t}(\Phi_{a,t}(A))$ and $\mathbf{B}_{a,t}(A)$ are conditionally independent from A given $\Phi_{a,t}(A)$, and that $\mathbf{B}_{a,t}(A)$ is conditionally independent from A and $\Phi_{a,t}(A)$ given $\mathbf{B}_{a,t}(\Phi_{a,t}(A))$, are certainly plausible. The assumption will play a role in a moment.

This principle fits our folk theory (16.3) of perception well. If all four propositions obtain, as well as normal conditions so that any deviant causation is excluded, and if the four propositions indeed form a ranking net as suggested, then A is not only a cause of $\mathbf{B}_{a,t}(A)$, but a indeed perceives at t that A. I will not spell out the tedious details, but it looks (an epistemic use of "look") as if the reason relation of (16.10) can be extended to a causal relation according to (14.3) and as if the folk theory (16.3) of perception can be reversed in the presence of normal conditions.

Why should we accept the Schein–Sein-Belief Principle (16.10)? Insofar as (16.10) is just the Schein–Sein Principle (16.8), I have already answered the question; this was the point about dispositions and their manifestations. However, this point does not extend to the doxastic propositions covered by (16.10); beliefs can be understood as dispositions to act, but it seems weird to conceive of them as manifestations of sensations or the external facts. Hence, there must be other patterns of defeasible apriority to back up the rest of (16.10).

Concerning the relation between $\Phi_{a,t}(A)$ and $\mathbf{B}_{a,t}(\Phi_{a,t}(A))$, I argued above that the truth of the one is no guarantee of the truth of the other. However, the circumstances under which the two propositions come apart were quite exceptional, as illustrated. In particular, for person a to believe that it appears as if A presupposes that a has acquired the pertinent conceptual means, and the defeasibly a priori assumption of person b is that persons have mastered their concepts correctly.

Concerning the relation between $\mathbf{B}_{a,t}(\Phi_{a,t}(A))$ and $\mathbf{B}_{a,t}(A)$ in b's reasoning, b need only apply the reflexive case of the Schein–Sein Principle to a at t; if the one belief is a reason for the other for a, then a's having the one belief is more likely for b, given a's having the other. Here, it is presupposed that what are normal conditions for a are also normal conditions for b. Moreover, b is taken to assume that a's actual belief is rational; again, I think that the rationality of other persons belongs to the common defeasible a priori.

Finally, the remaining three defeasibly a priori reason relations among the four propositions follow from the three relations just argued and the assumption that these propositions form a Markov chain. So, in particular, A is for b a defeasibly a priori reason for $\mathbf{B}_{a,t}(A)$ and vice versa (provided A is such that it can appear to a as if A), as is stated in (16.10).

The preceding discussion was presented informally. I think I have provided all the means needed to tell a more precise version of the same story. To do so would be useful; it would be clearer then which assumptions I have made along the way and whether I have succeeded in at least mentioning all of them. Here, I will not try to satisfy such ambitions. The centrality of the Schein–Sein Principle should also have been clear on the informal level, as well as my frequently repeated point that the defeasible a priori is the appropriate kind of apriority, with which many claims appear correct that would otherwise sound wrong.

16.4 Consciousness

The previous section has given the impression that the Schein–Sein Principle and the two-fold defeasible base it provides is the largest concession we can make to the foundationalist. The traditional or strong foundationalist, however, will not be satisfied. He dreamt of infallible foundations and thought that impressions or sense data or conscious experience would somehow provide them. Under the pressure of the kinds of argument which I presented in Austin's version as well as my own in the previous section, strong foundationalists are almost extinct, it seems; they have been replaced by a sophisticated and bewildering spectrum of positions, with a variety of labels either from epistemology or from philosophy of mind, which all try to do without infallibility (or indubitability or maximal certainty or unrevisability) and still try to do justice to such hardly deniable phenomena as consciousness or awareness, introspection, transparency of the mind, privileged access, or first-person authority (which are certainly similar, even though they can and must be distinguished). In the previous section I told half of the story positioning myself within that spectrum.

The other half needs to be told because we still have not reached rock-bottom; we still have to dig one level deeper. I sense the force of the complaints of the strong foundationalist about all these ersatz infallibilisms; his dream, he will insist, is not a delusion. Confronted with such examples as in the previous section, demonstrating the corrigibility of $\Phi_{I,now}(A)$ even in the reflexive case, he might well grant these examples and say that by describing the content of my impressions by such phrases as "it appears to me as if A" one already appeals to conventions of public language that govern the usage of "as if A". Then it is not surprising at all that all kinds of things may go wrong. The subject may not fully master the convention, as in Austin's case; the pseudo-normal, though fully mastering the conventions, is unaware, like everybody else, of his exceptional condition, making red peppers appear green to him, just as the few actual, though unnoticed samples of XYZ are not water according to our conventions; or wherever else the error may lie. In any case, this only proves to the strong foundationalist that appearances thus described are not what he calls indubitably and immediately given.

The immediately given, though it meets with a lot of skepticism nowadays, was philosophically well respected for a long time. Impressions were taken as immediately given in the 17th and 18th century. Russell's philosophy essentially built on the distinction between knowledge by acquaintance and knowledge by description, the potential epistemological soundness of which might have been blurred by the semantic weirdness of his claim that logically proper names refer only to things we are acquainted with, which are none of the ordinary things we have names for. The logical positivists had extensive discussions about the immediately given (though not under this label). I have already mentioned the proponents of a phenomenalistic basis; the role that sense data played in Ayer (1940), which was in the focus of Austin's (1962) criticism, would be a relevant example. I have always been sympathetic to the so-called *Konstatierungen*, as introduced and described by Schlick (1934). And so forth.

It seems to me that what all these philosophers thought cannot simply be wrong; such authority must have grasped a grain of truth, at least. I am well aware that this is no argument; still, it is the point that moves this section. Though the immediately given, as we just saw, has received many labels, a central characterization that all its defenders could agree on is that it consists of contents of consciousness; it is precisely what is presented in consciousness. Hence, our task will be to determine what sense we can make of such contents of consciousness. I will put forward a bold hypothesis, perhaps overly bold by present lights. Still, I will defend it in three parts. I will first try to secure its correct understanding. Then I will explain what those contents are according to the hypothesis, implying that they indeed exist and agree with what we would ordinarily call conscious. These parts essentially follow Spohn (2005b). Finally, I will unfold their epistemological consequences, which will be quite limited, in a way, and will thus please only the current mainstream, but not the strong foundationalist. Here, I also follow Spohn (1999a, Sect. 6). My suggestion, in the end, will be that the hostility with which contents of consciousness currently meet is rooted in a common false view of those epistemological consequences. By contrast, the epistemological picture developed in this chapter will be able to assign a reasonable place to those contents, to the immediately given.

Narrowing down our topic to contents of consciousness is not immediately helpful, though. For, what are they? The first answer is that they are those things of which we say we are aware or conscious of. However, ordinary usage is variegated and indeterminate; moreover, the adjectives "conscious", "aware", and the German "bewusst", for instance, are not fully interchangeable. In particular, "I am aware of A" often says no more than "I know A"; certainly, though, not each item of knowledge should qualify as a possible content of consciousness.

Philosophical usage diverges from, or rather makes more specific, ordinary usage – often a dangerous move. However, the specification philosophers have in mind seems clear enough. For most possible contents of knowledge, A, we do not know whether or not A holds, simply because we know so little. By contrast, possible contents of consciousness are not taken as doubtful or unknown in this way. If one is given to me, I know I have it; and if one is not given to me (though it possibly could) I know I do not have it. The usual paradigm is pain: If I am in pain, I feel – that is, I am aware of – my pain; if I feel no pain, I have no pain, however painful my bodily condition may be. Conversely, if I feel pain, I have pain, however phantom-like or inexplicable it may be. This relation is so close that it seems odd (or even ungrammatical, as Wittgenstein (1953, §246) suggested) to say: "I know I am in pain". Shoemaker (1990) speaks here of those contents as weakly self-intimating (or even strongly, if the implication about the absence of those contents holds as well). Other cautious labels speak of privileged access or first-person authority. Though it is clear that there are many states of mind that do not even qualify for these weak distinctions, I do not see why one should be so defensive about contents of consciousness, why we should not apply the traditional stronger labels of indubitability or infallibility. Indeed, I would like to sharpen the issue and propose the following explication or definition:

(16.11)   *The Conscious Essence Principle*: C is a (*possible*) *content of consciousness* of subject a at time t if and only if, necessarily, C iff $\mathbf{B}_{a,t}(C)$ (that is, a believes C at t).

This explication, as simple as it is, requires a number of explanatory remarks, seven in fact; this is the first of the three parts of my argument about consciousness:

First, the preliminary discussion referred to knowledge, whereas in (16.11) I returned to speaking of belief. I had to return; "belief" is the basic term of this book, and I have abstained from a deeper analysis of knowledge. However, later on I will briefly argue that the belief mentioned in (16.11) is indeed knowledge.

Second, contents of consciousness are relativized to *subjects* and *times*. The latter is required by (16.11) since not only my beliefs, but also the contents of consciousness possible for me, can change in time (if I become deaf, my consciousness is impoverished, and by taking drugs I can, so they say, expand my consciousness), and since it is only the possible contents of consciousness at t that can be essentially related to my beliefs at t.

As to the subject a, it may be anything. If a is a stone, it cannot have any beliefs and hence no contents of consciousness. If I cannot know what it is like to be a bat, this is so because there are contents of consciousness a bat can have, but I cannot. In the sequel, though, I will speak only of contents of consciousness of persons like us. An interesting side issue is whether the subject a may be a group or some other social entity. Colloquially, we speak of something like collective consciousness. Often, this may mean no more than common knowledge. Perhaps, though, the Conscious Essence Principle fits even then. For instance, if Lewis (1969, part II) is right to argue that the conventions of a community include common knowledge of those conventions, this entails that they are contents of collective consciousness in the sense of (16.11). And if language is constituted by the linguistic conventions in this sense, language is part of the collective consciousness (of the Hegelian spirit?) as well. However, I will not further pursue this idea.

Third, the most crucial question about the Conscious Essence Principle is, of course: what does the variable C stand for? For (16.11) to make sense, C must be a proposition; only propositions can be possible contents of consciousness. But which kind of propositions? I will more thoroughly discuss below what our contents of consciousness are; this will also show how (16.11) fits into the present ramified discussion about consciousness. However, one possible confusion needs to be cleared up right away:

Certainly, phenomenal consciousness is to be among the central applications of (16.11). How, though, can it have propositions as contents? Usually, one would have said that it is sensations, feelings, and the like that are presented in phenomenal consciousness, or perhaps sounds or smells or colors (though the latter already have a bias towards objectivity, leaving room for misperception). And these are objects, a strange kind of object perhaps, tokens in some sense, but certainly different from propositions, another strange kind of object. At this point, it is important to distinguish what has been called *transitive* and *intransitive* consciousness (cf. Rosenthal 1986, 1997). When we say that I feel a certain pain or sense a certain desire, and hence am conscious of it, we are speaking of consciousness in the transitive mode. However, we might as well take the content of consciousness in these cases to be the fact *that* I have that pain or that desire; then we speak of consciousness in the intransitive mode. Understood in this way, it is no mystery how those contents can be propositions; of course, I may believe (*pace* Wittgenstein) that I have that pain. Hence, in proposing the Conscious Essence Principle I am only referring to intransitive consciousness, and will continue to do so, even if I may slip into the transitive mode. This understanding, however, entails no restriction of our discussion, since I take the two modes to be interchangeable.

At this point, I should, fourthly, emphasize the importance of referring to *possible* contents of consciousness. Our usage of "conscious" is factive. If I am conscious of a feeling (transitively speaking), I have that feeling; that feeling actually exists. Yes, of course; in one sense it is trivially true that only actual contents of consciousness exist. However, this does not preclude the existence of possible contents, in the way all possibilities exist. And it is important that (16.11) refers to them as well. (16.11) also claims that, if a counterfactual content of consciousness were given to me, I would ipso facto believe in it. Conversely, if I do not believe in it (intransitively speaking), I ipso facto do not have it (transitively speaking), even though I might have had it. Only in this way can we hope that the Conscious Essence Principle provides a general characterization of contents of consciousness.

Note, by the way, that the principle entails that the contents of consciousness possible for a at t are algebraically closed. I just explained that if C is such a content, $\bar{C}$ is so as well. And if C and C' are two such contents, $C \cap C'$ is so as well, since belief is closed under conjunction and hence the necessary equivalence in (16.11) is also closed under conjunction. Keeping firmly in mind that we are speaking about possible contents of consciousness in the intransitive mode, this point loses its prima facie oddity.

Fifth, the preceding discussion has already suggested that contents of consciousness (in the intransitive sense) are propositions of the form "I am in pain (now)" or "it (now) appears red to me" (though we still have to discuss the linguistic expressibility of those contents), i.e., that they are, in any case, *indexical* or *egocentric* propositions. Indeed, it seems there is no way to necessitate beliefs about who I am or when now is, and hence, if C were an eternal, non-egocentric proposition, (16.11) could not apply to it. This is no problem, though. Up to Chapter 15 indexical propositions were nowhere relevant, and so I ignored them. However, Chapter 2 was explicitly liberal enough to also allow indexical propositions as objects of belief; thus (16.11) does not transcend the framework we adopted.

Still, the point might provoke bewilderment about the Conscious Essence Principle. Even if C must be an indexical proposition, the proposition that a believes C at t is a non-indexical proposition (about an indexical belief of a). How, though, can an indexical and a non-indexical proposition be necessarily equivalent, as (16.11) seems to claim? This is indeed impossible. However, one must be careful to read (16.11) correctly. If C is indexical, so is "C iff $\mathbf{B}_{a,t}(C)$" and hence also "necessarily, C iff $\mathbf{B}_{a,t}(C)$". In the latter assertion, however, the indexicals "I" and "now" figuring in the first occurrence of C take wide scope over "necessarily"; and so, after assigning the values a and t to these indexicals, there is no logical obstacle to the possible truth of the necessity claim of (16.11).

Sixth, how is belief to be understood in the context of the Conscious Essence Principle? Should I not grant that there is something like unconscious or subconscious belief? Yes, certainly. So, should I not restrict (16.11) to conscious belief? Yes, again. However, we are not thereby caught in a circle, as I will explain below. Rather, we will see that it follows from (16.11) that it refers only to conscious belief. Moreover, one should note that the distinction between conscious and unconscious belief is not the same as the one between occurrent and dispositional belief. Conscious belief may well be dispositional, and unconscious belief is a more awkward disposition (in which one behaves as having that belief while granting neither the belief nor the behavior). Indeed, my highly dispositional conception of belief given in Explanation 2.6 will shortly be seen to perfectly cohere with this chapter and with (16.11).

Seventh, finally, I must disclose how "necessarily" is to be understood in the Conscious Essence Principle. Logical necessity, being a matter of logical form, does not apply. Since we are talking about epistemic matters, some weaker or stronger epistemic necessity, defeasible or unrevisable apriority, say, might be intended. Nothing of this sort. The epistemic relation between contents of consciousness and their being known is already stated in (16.11) in the clause modally qualified; to read the modal qualification itself in an epistemic way would involve an ascent to second-order beliefs. This is not intended, though I will argue that the unrevisable apriority of the equivalence in question is entailed by (16.11). It might be tempting to read "necessarily" as causal necessity. However, we have seen this to be a defeasible kind of necessity, as Hume already knew. The effect might fail to occur, even given the cause; the circumstances might not have been right. So, might it be that one is in a state of consciousness, be in pain, say, without knowing it? This is not what we intended to claim; the knowledge is not merely an effect of the conscious state. Similarly, the necessity in question is not merely one of psychological law.

Our discussion has already made clear: we are talking here of metaphysical necessity, as forcefully introduced by Kripke (1972). Pain *is* felt pain, *including* awareness of it; without belief in present pain there cannot be any pain. And generally having a content of consciousness is the very same as believing one has it. There is no metaphysical possibility in which the content is given, but not the belief in it, and vice versa. Just as being water and being or consisting of H_2O is the *same* physical state for some piece of matter, so is a state of consciousness and knowing or believing it the *same* mental state of a conscious subject. Benkewitz (2011, sect. 3.8) speaks here of a belief token being modally indistinguishable from its content (and develops this observation for phenomenal and intentional consciousness, as I will do below); see also Benkewitz (1999, pp. 206f.), where he conceives of contents of consciousness in the same way and calls them internal contents. (I acknowledge that his views on the present matters basically convinced me, and thus led to the considerations I am presenting here.)

This reading of "necessarily" and thus of the Conscious Essence Principle is strong, perhaps surprisingly strong. However, I have indicated why it is the intended one. We will see that it provides a number of insights, yet does not lead to unacceptable consequences.

The point also explains the label I have chosen. The label might be grammatical nonsense. Still it ambiguously refers to the two necessities contained in (16.11). On the one hand, (16.11) is a definition and thus intends to grasp the essence of possible contents of consciousness as a kind. On the other hand, it says that the essence of each individual content consists in being identical, i.e., essentially identical with the belief in it. The label "Conscious Essence Principle" is to signify both points.

Having thus explained every critical item of the Conscious Essence Principle, the next question is whether there are in fact any contents of consciousness, as explicated there. This is the second part of my discussion of (16.11). We find a confusing variety of forms of consciousness in the literature, and each of them exists. Of course, my claim is that (16.11) fits at least to the central forms. Among those central forms are *phenomenal consciousness* of impressions, sensations, feelings, and the like, and *intentional consciousness* of thoughts, beliefs, hopes, fears, desires, and the like. (16.11) applies at least to these forms, as I intend to show.

Let us first turn to contents of *phenomenal consciousness* that strike us most: I see a speck of a particular shade of red, I have a throbbing headache, I feel my indignation rise, etc. In the very moment I have such sensations or feelings, I realize this and thus believe I have them, in an intuitive sense, and indeed in the sense introduced in

Section 2.3. For, whenever I am in such a phenomenal state, I have the ability to notice changes of this state, i.e., to tell whether or not a possible phenomenal state is like the one I am presently in, and thus whether or not such a possible state passes the belief test specified in Section 2.3.

This ability lasts for a while. Some sensations may burn into my memory so that I know for years what they were like. Most sensations, however, are so transient that a few moments later one no longer remembers what exactly they were, and thus they fail the belief test of Explanation 2.6. In any case, while having the sensation I have that ability.

Of course, matters are often quite indefinite. Our attention to our sensations varies tremendously, and so does our ability to notice changes. However, the degree of attention is part of the sensational state. When I am so focused on working on my manuscript that I do not realize that I am hungry, I *am* not hungry. Psychologists love to inquire into the grey areas; they present, for instance, stimuli so briefly that attention cannot work properly, if at all. Thereby, they uncover a lot of half- or unconscious processes. In a way, this is highly interesting. On the other hand, it does not say much about our common sense notion of consciousness, which is not built for the grey areas. We should rather work on describing the black and the white cases in order to better delineate those grey areas.

Also, I do not mean to imply that all phenomenal experiences are conscious. Pains and sense impressions are so; if one does not realize one has them, one does not have them. By contrast, one may adore or despise somebody – possibly strong feelings – without being aware of it; and our self-assessment of moods and other darker or subtler feelings is often poor. Only conscious sensations and feelings are conscious.

Hence, it seems that the Conscious Essence Principle fits smoothly with our understanding of phenomenal consciousness. Claims about the self-intimation of, the privileged access to, or the first-person authority concerning phenomenal states assert the same point in a more cautious way. I prefer (16.11), which is bolder, clearer, and more traditionally minded. How could one object to it? Well, I have not found many objections in the literature:

Pollock (1986, p. 32) explicitly considers (16.11) in the strong way intended here, namely as the claim that the state "of being appeared to in a certain way" is the very same as the state "of thinking of being appeared to in that way". He calls this the identity thesis and takes it to be utterly implausible, since the thesis would entail a negative identity saying, e.g., "that not having a certain red impression would include thinking of not having it". He goes on to discuss plausible weakenings of the identity thesis that he confronts with similar objections. The crux of his arguments, though, is always the same; it takes the thinking of being in a certain state as an accompanying occurrent mental process, as the word "thinking" suggests. Then he is right, of course; in this sense, I cannot think of the countless negative characterizations of my sensations. However, this only entails that he has misinterpreted the reflexivity of consciousness. If one replaces Pollock's notion of thinking by my dispositional notion of believing, the identity thesis avoids such problems. *Not* experiencing such and such is a content of consciousness, too, since the relevant subject, when performing the belief test proposed in Section 2.3, would confirm *not* sensing such and such. (In the second edition, Pollock, Cruz (1999), the corresponding passage is indeed deleted.)

The main worries about (16.11) spring precisely from such considerations as presented in the previous section. If $\Phi_{a,t}(A)$ is only a defeasibly a priori reason for $\mathbf{B}_{a,t}(\Phi_{a,t}(A))$ (and vice versa), as the Schein–Sein–Belief Principle (16.10) asserts, then the two propositions may come apart, and cannot be identified. This shows that $\Phi_{I,now}(A)$ cannot be a content of consciousness (for a at t).

This is a conclusion we should indeed accept, I think. In the proposition that it appears to me as if A, my appearances are brought under concepts that are not under my exclusive control, but are at least partially deferred to my linguistic community. As we saw, various things might go wrong then, even in the seemingly innocent case of color appearance terms. Hence, genuine contents of phenomenal consciousness cannot be linguistically described; they are ineffable. This sounds queer and even wrong. In fact, I do not want to deny that we describe our impressions all the time, with our ordinary words, mostly truly and sometimes falsely. Still, none of those descriptions exactly captures the contents of phenomenal consciousness, the raw feels or pure sensations. The best I can linguistically do is to use a demonstrative; I can say: "it appears thus to me", where "thus" is accompanied by a kind of inner pointing to something in my mind and inaccessible to others. However, no description exactly grasps the reference of "thus", already for the simple reason that it is hardly possible to find words for the specific impression I have, say, for the specific purple iridescent shade of sepia I am just perceiving, but also for the deeper reasons introduced in the previous section. Still, that it appears *thus* to me right now represents a proposition, a set of possibilities, even if it cannot be perfectly delineated by linguistic means.

It has been often observed that $\Phi_{a,t}(A)$ and $\Phi_{a,t}(thus)$ need to be distinguished, and many have concluded that the latter, the pure qualia or the raw feels, are indeed ineffable (cf., e.g., Block 1990 and Dennett 1988); I also understand Schlick's (1934) *Konstatierungen* to be of the second elusive kind. I emphasize this distinction here because the epistemic behavior of the two kinds of proposition is very different. The first kind is engaged in defeasible reasoning as displayed in the previous section, whereas only the second are contents of consciousness. In any case, when pleading that (16.11) applies to phenomenal consciousness as ordinarily understood, I was referring to the second kind, and then my pleading seems correct.

What about *intentional consciousness*? Here, the first point to emphasize is that my conscious mental activities are not limited to my sensations or feelings. I think or try to grasp a thought, I decompose a number into its prime factors, I seek for words and try to recollect something, etc., and all of this is conscious to me. Such activities do not feel a certain way; that is why they are usually not subsumed under phenomenal consciousness, but rather are said to have an intentional character. The label "monitoring consciousness" has been introduced here, though it has been used in varied ways (cf.

Block 1994, p. 213). In any case, that form is quite similar to phenomenal consciousness, and everything I just said about the latter seems to apply to such conscious occurrent mental activities as well.

The main form of intentional consciousness, which refers to propositional or intentional attitudes and thus in particular to conscious beliefs and desires, is quite different. At any time I have many conscious beliefs and desires that I do not and could not actually have in mind. Conversely, unconscious beliefs and desires, which definitely exist in various forms, are not simply those the contents of which are not presently before one's mind. Even conscious beliefs and desires are mostly or most of the time only dispositional, their usual manifestation being to sincerely say yes when asked whether one has them. For years, for example, I continuously had the conscious wish to finish this book; however, this is not to say that I thought of it every second.

Thus, there is a sharp difference between phenomenal and intentional consciousness. The need to distinguish various notions of consciousness has very often been put forward (cf. e.g., Rosenthal 1986, pp. 334ff., Block 1994, pp. 216f., Burge 1997, or Kemmerling 1998). I am not so sure. The simplest explanation for the various usages of a word is still that it always carries the same meaning. This also applies to the case at hand. At any rate, the Conscious Essence Principle adequately captures intentional consciousness as well.

This follows from the belief test I introduced in Explanation 2.6, according to which belief is indeed reflexive. Whenever a centered possibility or possible world conforms to all of my beliefs, I must be able to locate myself in the center, i.e., the center must be occupied by someone who could be me, for all I know, who hence shares all my beliefs, and therefore thinks himself that that possible world might be the actual one. In this sense of "believe", then, "I believe that p" and "I believe that I believe that p" represent the very same proposition, the very same set of possibilities, as was embodied in the static reflection principle of doxastic logic by its initiator Hintikka (1962, pp. 23ff.) and as is now confirmed by (16.11). The having of a belief in this sense is a possible content of consciousness.

This consideration entails, moreover, that the belief referred to in the Conscious Essence Principle is a conscious belief. For (16.11) generates an iteration. If A is an actual content of consciousness for me, I believe A, and I also believe that I believe A, etc. Thus, intentional consciousness agrees with what has been called "higher-order-thought consciousness" (cf. Rosenthal 1986, p. 336). However, higher-order belief is again to be conceived as dispositional and not as an occurrent thought process. Moreover, the iteration is spurious in a way, i.e., only syntactic. According to (16.11), the iteration does not refer to an infinite number of belief states that are mysteriously bound to coexist, but always and only to one and the same belief state; this is the deep point of referring to metaphysical necessity in (16.11).

One might conversely wonder how one can ever have an unconscious belief according to my characterization of belief. Let us look at an example: John regularly and sincerely affirms that all is well with his marriage. However, his friends feel that he is mentally estranged from his wife; his peculiar behavior indicates that he has already given up on his marriage. This is perhaps a case in which one can plausibly say that John unconsciously believes that his marriage has failed, though he has not yet realized this.

As it should, this unconscious belief fails our belief test from Section 2.3. Let John's counterpart in a centered possible world behave in the same peculiar way. John would ascribe the same unconscious belief to that counterpart. Perhaps, John would even not exclude the possibility that he is that counterpart; he need not recollect every single piece of behavior he has shown and might thus grant the possibility of having unconscious beliefs. However, he would not accept that he is this counterpart. Another counterpart in another possible world that behaves normally according to his conscious beliefs would also be a doxastic possibility, indeed a more plausible one. If John were to realize how strangely he is behaving, he could exclude such normal counterparts, and his belief would no longer be unconscious.

I take it that *mutatis mutandis* the same is true of conscious and unconscious desires and other propositional attitudes. (If we are right in Kusser, Spohn (1992), the case of desires is still more complicated, though.) Here, however, only the observation that conscious beliefs fall under (16.11) will be of further relevance.

All of these considerations seem to show that the Conscious Essence Principle captures our notion of consciousness (or at least essential aspects of it) well. And I cannot forbear the remark that this principle closely resembles the original synthetic unity of pure apperception that Kant (1781/87, B136) declares to be the supreme principle of understanding. It says that the "I think" must be able to accompany all my representations, intuitions as well as judgments. Intuitions are part of phenomenal consciousness; judgments are part of intentional consciousness; "be able" indicates that the "I think" need not go through one's mind all the time; and "must" may be interpreted as making the modal claim of the Conscious Essence Principle. It is not my intention to use the principle in the way Kant uses his synthetic unity of apperception. Still, the historic reference shows that the Conscious Essence Principle connects up to our philosophical heritage. Let us accept it for the rest of this section and work out its consequences.

A first remark is that the identification of contents of consciousness with beliefs in them gives them a very peculiar status that has been observed in the literature for a long time. Wittgenstein (1953, §§288ff.) pleaded that the application of epistemic vocabulary like "I know", "I doubt", etc., to "I am in pain" is excluded from the "pain" language game. However, that may only be a pragmatic exclusion; it may mean that the epistemic vocabulary cannot add to, or detract from, the assertion of "I am in pain". One might also refer to Hume's principle of the constancy of content (Pears 1990, p. 53) that governs belief in matter of fact. It says that the belief contents (the ideas for Hume) must be able to remain the same while the attitudes (belief or mere imagination) change or, in other words, that state of belief and content of belief are ontologically independent. This principle has widely been accepted (without thereby accepting Hume's problematic account of belief). (See again Benkewitz (2011, sect. 3.8), who also argues for the modal distinguishability, in his terms, of state and content of belief.) Rightly so; however, if (16.11) is correct, the principle is violated precisely by contents of consciousness. Should one conclude, then, that contents of consciousness are not genuine contents of belief? No, I prefer the converse conclusion that contents of consciousness are an exception to Hume's criterion, and by definition the only one. Otherwise, we could not even state the Conscious Essence Principle in the way we did.

A second remark is that the Conscious Essence Principle obviously makes a claim about the essence of contents of consciousness, namely that they are ipso facto believed. Does this contradict materialist prejudices that mind reduces to or supervenes on matter, and thus that mental properties have a material essence? Not at all. The essence of something is presumably the collection of all its essential properties, which may be many. So, contents of consciousness can have essential properties of a physical kind and still be characterized by (16.11). Materialism is not denied by (16.11).

A welcome consequence of the Conscious Essence Principle is that contents of consciousness are ipso facto not only believed, but also known. Having been non-committal about knowledge, I cannot really prove this. Still, the point seems obvious. First, (16.11) entails that beliefs in contents of consciousness are always true. Are they justified in the sense of having reasons? No, as we will see in a moment. Still, they are perfectly warranted. Contents of consciousness guarantee the truth of the beliefs in them. Indeed, the truth guarantee is absolutely reliable and cannot be compromised; if contents of consciousness and beliefs in them are identical, no wedge whatsoever can be driven between them. Hence, it seems appropriate to call those contents knowledge.

The main and last issue we have to discuss, though, is how the Conscious Essence Principle fits into the ranking-theoretic account of belief and justification and how it enriches our picture of the basis of belief formation.

First, it is clear that beliefs in the contents of consciousness are a posteriori; one may or may not have them, depending on which of the possible contents of consciousness are actual. Second, it follows from (16.11) that one is never unopinionated about any of one's possible contents of consciousness; at any time, one knows of each whether or not one has it. In view of my dispositional account of belief, this consequence is not as bold as it might seem.

If those contents are either believed or disbelieved, they must receive some ranks, and the crucial question is which ones. There is only one answer:

(16.12)   If $\tau_{a,t}$ is the two-sided ranking function of a at t, then for any possible content C of consciousness of a at t, $\tau_{a,t}(C) = \pm\infty$.

Assigning C two different ranks would result in two different belief states that cannot both be identical to C itself. So, we must select a specific rank that is characteristic of the belief in contents of consciousness, and that rank must be fixed and unrevisable. Hence, the extreme two-sided ranks $+\infty$ and $-\infty$ are the only non-arbitrary choice.

Also, if (16.11) is to hold, the belief in C must fall outside any possible range of vagueness of belief. In view of the vagueness observed in (5.14) the extreme ranks are again the only choice. This comes close to a strict proof of (16.12).

Again, I take this to be a welcome consequence; (16.12) expresses the infallibility or indubitability or unrevisability of beliefs in contents of consciousness that was already observed at the beginning of this section. However, this entails at the same time that these beliefs are excluded from the space of reasons; they cannot have reasons and they cannot be reasons for other beliefs. I will consider in a moment what this means for phenomenal consciousness; but let us first look at intentional consciousness.

Of course, I can and do have reasons for all of my ordinary first-order beliefs (and desires, which were not our topic in this book), but my belief that I have all of those ordinary beliefs (and desires) cannot in turn have any reasons. So, when asked why I believe A, I can only interpret this as a request for my reasons for A, but I cannot present reasons and hence causes or explanations for the fact that I believe A (as already stated when discussing the meaningfulness of the reflexive case of (16.9) in the previous section). I certainly can do so regarding the beliefs of others, or regarding my own past beliefs, and I can reason about my likely future beliefs. My own present beliefs, however, are a peculiar blind spot for me in this respect. This peculiarity has often been noticed (according to Rosefeldt (2000, pp. 169ff., it was already a constant issue for Kant; see also Benkewitz (2011, sect. 5.3) for a most careful discussion), and it seems to be well explained by (16.11–16.12). (In practical reasoning, by the way, we find the same blind spot concerning our own possible actions. When deliberating on our possible actions, we only evaluate them and do not take an epistemic or even explanatory attitude towards them. Or so I argued in a decision-theoretic context in Spohn (1977, pp. 114ff.), a thesis later labeled "deliberation crowds out prediction"; cf. Levi (1986, pp. 65f.). In my view, this thesis is crucially related to freedom of the will; cf. Spohn (2007).)

Similar observations apply to contents of phenomenal consciousness. At each time I am awake, something somehow appears to me; $\Phi_{I,now}(thus)$ is true, and so is the identical infallible belief in $\Phi_{I,now}(thus)$. So, there *is* an infallible basis of our belief formation, as the strong foundationalists have always insisted. This basis has a very special position regarding the dimension of internalism vs. externalism. An externalist basis seemed attractive since it delivers a genuine *fundamentum in re*; on the other hand, only an internalist basis seemed to be able to be part of the realm of belief and thus to unfold the expected justificatory power. Now we can see that this is a false opposition, falsified precisely by the strange ontological double status of $\Phi_{I,now}(thus)$ as a belief and a non-belief at the same time. $\Phi_{I,now}(thus)$ is an external *fundamentum in re*, at least external to my beliefs, although not in the external world. On the other hand, it is part of the realm of my beliefs. Thus, in my description contents of phenomenal consciousness provide the crucial ontological and epistemic link connecting beliefs with the world or at least the realm outside beliefs.

All this sounds too beautiful to be true. Do I thereby fully reestablish foundationalism? Certainly not; we already saw the price there is to pay. Though contents of phenomenal consciousness reach into the realm of belief, they do not and cannot enter the space of reason, precisely because of their infallibility; the basis I have provided is justificatorily barren. Reasons do not behave as the strong foundationalist has thought, and therefore he cannot proceed as desired.

How, then, can we doxastically proceed from those contents of phenomenal consciousness? In two ways. First, as time goes by, the proposition $\Phi_{I,now}(thus)$ transforms into the proposition $\Phi_{I,past}(thus)$, which is no longer a content of consciousness and thus very much an object of reasoning. Though the belief in $\Phi_{I,now}(thus)$ is infallible and unrevisable, it can be forgotten and, with a few exceptions, will be; no memory is able to store such huge amounts of information as our senses provide. What remains is usually no more than a vague recollection of the kind of impressions we had, helped also by the second point, to be mentioned in a moment. Usually, at best only the external facts related to our past impressions are of any interest; but if we want to, we can argue about our past impressions just as about any other fact in the world. In any case, the change of $\tau_{a,t}(\Phi_{I,now}(thus)) = \pm\infty$ into a non-extreme $\tau_{a,t+1}(\Phi_{I,now-1}(thus))$ through forgetting is the first way of entering the circle of reasons. It should be clear that there cannot be any ranking-theoretic law for this kind of change.

The second way, of course, is that the extreme $\tau_{a,t}(\Phi_{I,now}(thus))$ induces a non-extreme $\tau_{a,t}(\Phi_{I,now}(A))$ for some proposition A. Unless it is very novel or very strange, the pure impression is instantaneously brought under concepts and classified as being as if A, for some A. Usually, this is an immediate and effortless process. Or rather, it may be a process in our brain; in our minds the two propositions $\Phi_{I,now}(thus)$ and $\Phi_{I,now}(A)$ are so tightly connected (for an appropriate A) that it takes efforts to separate them.

The crucial point about this induction is that it cannot, in my picture, be a rational process; $\Phi_{I,now}(thus)$ cannot be a reason for $\Phi_{I,now}(A)$. Bringing impressions under concepts, Kant's synthesis, is something we somehow do and have learnt to do; it is something I have to accept as given within my epistemological account and as not subsumable under ranking-theoretic laws. It is not part of that account to say how we conceptualize our impressions and how much we trust in our conceptualization; in $\tau_{a,t}(\Phi_{I,now}(A)) = n$ both A and n are determined somehow, but not by the doxastic state of a prior to t as far as ranking theory can represent it. This is the deeper truth about the parameter n characterizing simple conditionalization (5.24). (And it might show, by the way, that the device proposed after (7.45) for treating multiple pieces of evidence was not just an algorithmic trick, but has perhaps a substantial interpretation.)

Indeed, in view of the considerations in the previous section it need not be the belief in $\Phi_{I,now}(A)$ that is induced by $\Phi_{I,now}(thus)$; it may as well be the belief in A itself; as I argued there, neither of these two beliefs is primary *per se*. Besides again respecting the issue of frame-relativity, the point is rather that once we look at those induced beliefs (and all the further beliefs they are a reason for), the coherentist picture developed in the previous sections takes over, which provided only fallibilistic double foundations via the Schein–Sein Principle.

This is how my account combines externalism, foundationalism of a strong, infallible form and of a weaker, fallible form, and coherentism, all of which have been extensively argued for in the literature. My combination grants that every side is right in some way, and it may help sorting out in which way. Thus, there certainly are many similarities of my account with others. Still, the combination I propose is peculiar and, as far as I know, not to be found elsewhere in the literature. I will not expand my comparative discussion beyond the many remarks I have already made in this chapter. Any more careful or critical reflection of the issues pondered here has, however, to take a stance towards the grounds of my specific combination of familiar positions: the identification (16.2) of degrees of justification with degrees of belief in Section 16.1, the Schein-Sein Principle in Section 16.3 (in its defeasible and in its unrevisable versions 16.8 and 16.9), and the Conscious Essence Principle (16.11) of this section.

17

The A Priori Structure of Reasons

In the previous chapter we have explored one kind of input to our cognitive system, the continuous and contingent input from outside via our perceptions, and we have explored it to its deepest limits, as far as philosophers can tell. (For other disciplines mysteries start there.) In this final chapter I would like to explore the other kind of cognitive input. The first kind of input must meet with some kind of cognitive structure on which it operates; and that structure must be there beforehand. It is this structure that provides the other kind of input to our cognitive evolution. The structure that our cognitive system contributes by itself is traditionally called the a priori. So, this is the concluding topic of this long book.

The previous remarks sounded as if I were after a genetic story, which could only be a fictitious one from the mouth of a philosopher. Therefore, I should emphasize once more that my concern is exclusively with a normative picture of our cognitive evolution, as far as it is normatively accessible. As emphasized several times, the normative story is thoroughly intertwined with the empirical story; it is, however, not our present task to clear up this entanglement.

I already introduced two notions of apriority in Section 6.5, which we have used throughout the book. As a starting point for this chapter I will recapitulate in Section 17.1 what we have determined so far about apriority. A recurring point was certainly that the conceptual is one main source of the a priori — this is why it is so hard to distinguish the a priori from the analytic. However, this topic has been well worked-over and is not the one on which I wish to focus. My interest in this chapter is rather to explore another source of the a priori — of which I find very little awareness in the contemporary discussion — which clearly separates the a priori from the analytic, and which may hence help the Kantian project, though in quite un-Kantian ways: namely the a priori structure of reasons.

The first exploration, in Section 17.2, investigates the structure of reasons as such. It starts from a basic empiricist principle that derives much of its compelling force from being almost true by definition. We will see that it entails, first, what I will call the special coherence principle, which seems to me to embody the weak, but tenable core of verificationism and, second, what I will call the general coherence principle, which is one way of expressing the unity of science.

The second exploration, in Section 17.3, investigates the relation of reason to truth. This is one of the deepest mysteries of theoretical philosophy. I can see no other way than to simply postulate that relation to hold a priori. Recall, though, that I indicated already in Section 2.3 that there are two notions of truth, a correspondence notion and, in default of a more appropriate term, a pragmatic or internal notion. Whatever I will say about the relation of reason and truth will pertain only to the latter notion. And then the postulational move will turn out not to be as cheap as it may at first seem. The section will discuss a basic reason-truth connection and some interesting ramifications.

The foregoing a priori principles will finally entail some principles of causality, not quite the traditional principles, but more or less weaker versions. This is, as a kind of appendix, the topic of Section 17.4. These principles are, of course, also a consequence of my account of causation in Chapter 14. The point of my derivations will obviously be to catch up with the Kantian synthetic a priori in a rigorous, though un-Kantian way. Thereby, my account of belief, of its dynamics, its input, and its starting points, which has piled up to a very long and strenuous argument, will finally have reached a certain – though still preliminary – state of completion.

17.1 Once More: Unrevisable and Defeasible Apriority

The two notions of unrevisable and defeasible apriority were introduced in Section 6.5 and repeatedly used since. To recall, (6.18) and (6.22) defined a feature of a doxastic state to be unrevisably a priori iff all possible rational doxastic states have it, and defeasibly a priori iff all possible initial doxastic states have it; derivatively, then, these two notions could be carried over to propositions (or judgments or sentences). Let me briefly recapitulate: Were these the notions appropriate to our needs?

Emphatically, yes. We were motivated to distinguish the two notions by the ambiguity of the traditional slogan of the a priori as being independent of all experience. This move is also historically adequate insofar as we historically find both notions; the probabilists' use of the a priori, in any case, was always the defeasible one. The distinction stood out particularly clearly in our discussion of dispositions in Section 13.3 (which radiated to Section 16.3), where reduction sentences turned out to be defeasibly a priori, since normal conditions hold defeasibly a priori, and could be turned into unrevisably a priori sentences by explicitly conditioning on normal conditions.

Also, the generalizing move of speaking of a priori features was well taken. It certainly conformed to Kantian intentions, even though for Kant human understanding essentially operates with concepts and judgments, so that the only a priori contributions of the understanding were the pure categories and the a priori principles associated with them (of course, for him intuition and its relation to understanding provide further sources of the a priori). However, how a doxastic state is to be conceived is an open issue; it need not compartmentalize into intuition, understanding, and reason, and in this book the conception was indeed different. If so, notions of apriority should be flexible enough to fit any description of the form of our mind. This is the point of generally referring to doxastic features as a priori. Indeed, we exploited this generality in many clear examples of defeasibly and unrevisably a priori conditional reason relations in Section 13.3 and 16.3. It was also exploited in Sections 12.5–12.6 where we could interpret certain features of symmetric first-order ranking functions, via their mixture representation, as expressing a defeasibly or unrevisably a priori second-order belief in lawfulness.

Finally, the dynamic approach to apriority was certainly appropriate. Perhaps it is a virtue of this approach that it brings into sharper focus a problem that I have avoided so far: the problem of where to locate the origin of the doxastic evolution to which "a priori" apparently refers, when it is temporally or genetically loaded. The problem hides in the obscure talk of initial doxastic states in my definition of defeasible apriority. One may think that unrevisable apriority is not affected by this problem; this would be an error, though, as we will see in a moment. The only thing I have said so far about the problem was at the end of Section 6.5, namely that initiality should be taken to refer to some relative starting point, e.g., to the start of some specific inquiry.

Let me try to be a bit more determinate on this point. It seems obvious that, typically, belief contents are conceptually structured in some way. That all bachelors are unmarried should turn out to be unrevisably a priori, indeed analytic, but it cannot be believed by someone who has not acquired the concept of a bachelor and does not know what a bachelor is. This, it seems, refutes our original explication: that all bachelors are unmarried is not believed in all possible doxastic states; at best it is believed in all possible doxastic states mastering the concept of a bachelor. What can and cannot be believed in a given doxastic state is relative to the conceptual means acquired by or in this state.

This point also helps us towards the relativized notion of initiality we were looking for. Initiality is relative to concept acquisition. Thereby a certain amendment of our original explications in Section 6.5 is suggested:

> *Explanation 17.1*: A feature of a doxastic state is *unrevisably a priori* iff each possible rational doxastic state capable of having this feature has it.

For instance, a doxastic state is capable of having beliefs involving the concept of a bachelor if and only if it has acquired that concept. Each such state must believe, then, that bachelors are unmarried; otherwise one could not say that the state has acquired the concept. And so that proposition is unrevisably a priori. Likewise, one can entertain the reason relation between a disposition and its manifestation conditional on normal conditions only when one has acquired the dispositional concept, and one could not be said to have acquired that concept unless one has grasped this reason relation. Hence, this relation is an unrevisably a priori feature.

Similarly for the defeasible a priori:

Explanation 17.2: A feature of a doxastic state is *defeasibly a priori* iff each possible initial doxastic state capable of having this feature has it — where a doxastic state is *initial* with respect to this capacity iff the state has (acquired) the capacity, but not further exercised or applied it.

My primary examples are again dispositional concepts. By acquiring such a concept one not only learns that the relevant reason relation or the appertaining reduction sentence is unrevisably a priori given normal conditions, one also learns that it obtains defeasibly a priori, if taken unconditionally (or only conditionally on the test situation). This is part of having the concept. This defeasibly a priori feature may, however, vanish through the exercise of the conceptual ability, e.g., through applying the concept to a given case, observing the environment of the case, possibly noticing that it is not normal, etc. Everything one learns through that exercise is no longer part of a doxastic state that is initial with respect to the dispositional concept. I will mention other examples below.

Before complaining about how little progress these formulations achieve, let me note that these explanations, with their reference to conceptual capacities, sit relatively well with the noncommittal view of propositions laid out in Section 2.2 and mostly used later on. For one could say that each possibility space is associated with its own conceptual means, so that each conceptual refinement generates refined doxastic possibilities that replace the former possibilities.

However, (17.1–17.2) do not seem to sit well with the more determinate view of doxastic possibilities I tentatively explained in Section 2.3, since those possibilities are fixed as maximal from the outset. However, even if those possibilities are maximally fine-grained, the propositional algebra over them can be more or less coarse-grained, and it is the algebra mastered by a doxastic state that displays its conceptual and hence its discriminatory capacities. For instance, if Putnam, as he confesses in his (1975, pp. 226f.), could not tell elms from beeches, if his elm concept is the same as his beech concept, he could not distinguish worlds with elms and beeches interchanged. That is, in the counterfactual test described in Section 2.3, he could, of course, easily distinguish such interchanged worlds, and he could easily acquire discriminatory concepts of elms and beeches; after all, they in fact look very different. However, with his present concepts he could never say of any pair of worlds thus interchanged that one of them might be actual, but not the other; any proposition he could believe in that conceptual state would contain either both worlds or none. (This remark, however, neglects semantic deference, which could help even if perception and private investigation fail.)

The point, then, is: that even according to the more determinate view of Section 2.3, we can distinguish different conceptual capacities underlying different doxastic states by looking at the propositional algebras mastered by those states. So, even on this view, initiality of doxastic states with respect to certain capacities is well defined — insofar as it is well defined at all.

The latter, though, may well be doubted. To begin with, there is a gray area in my characterization of initiality in (17.2) which seems unavoidable. How can a subject have or acquire a certain capacity without at least having attempted to exercise or apply the capacity? And how good or successful do those repeated attempts have to be in order to finally make the capacity attributable to the subject? These questions apply equally when the capacity consists in the possession of a concept. There is no good answer; there is no invisible boundary that the gradual learning process crosses at some point in time. So I see no further way of sharpening my notion of relative initiality. Still, the intention of my qualification in (17.2) − of not having further exercised or applied the capacity to have a certain doxastic feature in an initial doxastic state − should be clear. For instance, in the case of concept possession, it at least means that one does not have relevant information about new cases to which the concept is to be applied.

However, the obscurities of the picture I have started to develop reach much deeper, and in order to get clearer about it, we would have to thoroughly explore such issues as: What exactly are concepts? And precisely how are concepts acquired and eventually possessed? What is conceptual change? And so forth. These issues have received tremendous philosophical attention in the last twenty or even forty years, and a lot of philosophical skepticism as well. For instance, Kuhn (1962) and Feyerabend (1962) made strong claims about the incommensurability of scientific theories before and after scientific revolutions, while Quine's (1951) suggestion of a confirmation holism had the in-built tendency to transform into a meaning holism, with which, e.g., Fodor (1987, Ch. 3) paradigmatically struggles. As a consequence, an extensive literature trying to separate conceptual or meaning change from theoretical or belief change in general has emerged, but one may well say that the prospects are dim. Still, I have not lost hope concerning these issues, and we explained in Haas-Spohn, Spohn (2001) and Spohn (2008a, Ch. 15), though still quite insufficiently, how we would like to approach them in conformity with the more determinate view of Section 2.3.

However, this is not a business we should attend to now. My considerations should only serve two modest preliminary goals:

One goal was to at least begin rescuing the notion of initiality from its obscurity by relativizing it to stages of concept acquisition and to give (relatively) initial doxastic states an intelligible place in our cognitive evolution. Thereby, I have also given more determinate content to my notion of defeasible apriority. And I have shown that, contrary to appearances, both kinds of apriority are *dynamic* notions that may and do evolve and change with our conceptual development.

For instance, I am impressed by Putnam's (1962, pp. 372f.) notion of relative or contextual apriority and his point that Euclidean geometry is relatively a priori in his sense. This observation fits well into my scheme. We may indeed say that Euclidean geometry *was* unrevisably a priori in the sense explained here, as long as this geometry was the only conceptualization of space at our disposal. Kant *was* right; no experience whatsoever could disprove Euclidean geometry. This changed with the extended conceptual means provided by the invention of non-Euclidean geometries; only then could we even begin to ask whether experience conforms to this or that geometry. In this way, even the unrevisable a priori may be subject to change, i.e., be relative in Putnam's sense. (See Friedman (2001) for an elaboration of this point; he also rejects the Kantian a priori, if taken in an absolutely unrevisable sense, and construes it in a relative sense which he calls constitutive apriority.)

The other goal was to preliminarily clarify the relation between the a priori and the conceptual. It was always clear and accepted that the conceptual is one source of the a priori. However, it is not clear right away how this fact may fit into the dynamic explication of apriority given here. This relation is now explicit in (17.1–17.2). Indeed, it thereby comes to the fore that there are no less than three different kinds of conceptual truths:

There are, first, analytic truths. Kripke (1972, p. 264) defined analytic truths to be a priori necessary truths (the "a priori" here qualifies the "necessary"; the adjectives are *not* to be read conjunctively, as it is often, perhaps carelessly, said). Thus he taught us that there are, secondly, a priori, but contingent truths, famously exemplified by "I am here now" or by "the standard meter is one meter long". My favorite example is "the first to climb Mt. Everest climbed Mt. Everest" that is analytic in the attributive reading and a priori, but contingent in the more suggestive referential reading of the definite description; Sir Hillary need not have climbed Mt. Everest. It thus illustrates that we are dealing here with a systematic phenomenon that is treated well in two-dimensional semantics. Hence, it is clear that these a priori, but contingently true sentences are still conceptual truths; one can be unhappy with Kripke's narrow explication of analyticity only if one prematurely identifies the analytic and the conceptual. And it is also clear that in the Kripkean context apriority is always unrevisable apriority.

Therefore I wish to emphasize that there is also a third kind of defeasibly a priori conceptual truths; that is, I should speak of propositions that are defeasibly a priori on conceptual grounds (since they may and sometimes do turn out false). My primary examples have always been reduction sentences for dispositions (unqualified by normal conditions); however, there are more, I think. For instance, for ostensively acquired concepts like "dog", the existence claim "there are dogs" is defeasibly a priori. In this case, it is true, but it may turn out to be false. Take, e.g., the ostensively learned term "witch". Of course, one can ostensively acquire that term only by also coming to believe that there are witches. Later on one learns that witches are essentially possessed by the devil, and since there is no devil, as we know today, there cannot be any witches. In the fifties, there was a vigorous discussion of the so-called paradigm case argument (cf. Watkins 1957/58 and the other papers collected in that *Analysis* volume), precisely about the conceptual or analytic nature of such statements, which is perhaps adequately captured by the notion of defeasible apriority – an unfamiliar notion at that time.

As interesting, indeed urgent as it would be to further ponder all these remarks, I have made them vivid (in the required brevity) only in order to contrast them with the main purport of this chapter. So far I have addressed the conceptual origins of the a priori. It is Kant's merit to have claimed other sources of the a priori manifesting themselves in his famous synthetic principles a priori. His arguments were vulnerable, and his distinctions appeared unintelligible to many. Long prevailed the logical empiricists' denial of these distinctions and their claim that there is only empirical truth and truth by convention. This dominance was finally overcome by Kripke's reconceptualization of metaphysical and epistemological modalities. Still, all this had nothing to do with Kant's intentions. As I just tried to indicate, Kripke's apriority still was of an exclusively conceptual origin, whereas Kant's notorious necessary conditions of the possibility of experience intended to more generally refer to the constitution of our mind, our intuition and our understanding, and not merely to what comes along with the concepts we happen to acquire.

Following Kant, I contend that there indeed are more sources of the a priori than the conceptual, and I will try to constructively unfold some of them in the rest of this chapter. It will be obvious that I am not pursuing specifically Kantian paths, but I may reach similar goals.

17.2 Empiricism and Coherence

What might the further source of the a priori be? Well, rationality itself. I already observed in (6.19), almost in passing, that all principles of theoretical rationality are unrevisably a priori. This is pleasing, but quite empty. If we only knew what those principles were! The basic principles (4.1) and (4.2) I started with were consistency and deductive closure. We learned in Sections 5.3 and 8.5 that ranking theory embodies not much more, namely conditional or dynamic consistency. We have learned, though, that those principles, or the structure they entail, were surprisingly consequential. Or we may take probability theory and its basic axioms to represent our form of thought, and again we know about the rich consequences of these basic principles. As we extensively discussed in Chapters 3, 5, and 9, both structures may be enriched by powerful dynamic rules for processing new information. All this is most substantial, on the one hand.

On the other hand, the problem is that these basic forms still admit of arbitrary amounts of craziness. This is well known for the probabilistic case. Carnap's persistent attempts at an inductive logic, from (1950) to (1971b, 1980), are the paradigm in point. He started out with strong convictions, leaving only his (1952) λ-continuum of inductive methods, and he ended up with a bunch of weaker principles: regularity, symmetry, positive relevance assumptions, the Reichenbach axiom, and a quite tentative theory of so-called analogy influence. After this, the research program tottered: there was Hintikka's two-dimensional continuum (see Hintikka 1966), solving the so-called problem of null confirmation of laws (recall, however, my remarks after Theorem 12.7), there were extensions of symmetry principles (cf., e.g., Skyrms 1991, Festa 1997), but as far as I know, not much more. Certainly, the objective Bayesianism of

Williamson (2005) is a very strong doctrine in the same spirit; it would require further scrutiny. Perhaps one should still speak of a substantial on-going research program. Even then, however, some dissatisfaction remains concerning the fact that all of this proceeds strictly within probabilistic constrictions; the wider relevance of these considerations for other fields of epistemology is somehow walled off.

A noteworthy exception is the work of Pollock (1990, 1995). It determinately leaves the probabilistic confines and is full of substantial a priori principles. I have sufficiently explained (in Sections 11.5–11.6) why I am working in a different framework, so I will not engage in further comparisons in this chapter. However, it should be clear that we are basically acting in concert.

By contrast, I find nothing much to attend to outside of formal epistemology. Beyond the principles we have already accepted, the proposals we find there at best amount to rules of thumb, not well suited for theorizing. Believe as many (relevant) truths and as few falsehoods as possible! Yes, of course; but the slogan is hardly helpful. Be coherent! I do my best, but that slogan is even worse. And so forth. Believe the truth! This is much better and entirely clear – if we only knew what the truth is. We will see in Section 17.3 how close we may come to this maxim.

However, ranking theory is no better off so far. We have the basic theory with its many consequences. I have discussed regularity. In Sections 12.3–12.5 we considered symmetry within the same context as in probability theory. There, we derived positive relevance from the additional assumption, normatively not yet clearly justified, of convexity. I have not considered a ranking-theoretic analogue to the Reichenbach axiom. Can we state substantial, normatively compelling postulates of rationality that go beyond the basic structure? I think so, as I am going to argue. My argument will follow my somewhat scattered attempts in Spohn (1991, 1999a, 2000a) at coming to grips with that issue. I assume that the subsequent considerations could also be carried out within a probabilistic framework. However, I find that their import and their persuasiveness stand out much more clearly in terms of belief and hence in the ranking-theoretic framework. Therefore I will confine myself to the latter.

In order to work up to the first principle I wish to propose, we should recall that (6.14) and (6.16) distinguished between strict unrevisability (which applies under all conditionalizations whatsoever) and unrevisability (which appears under all conditionalizations with respect to evidential sources – where I left indeterminate what the latter are). In (6.21) we observed that unrevisably a priori propositions are strictly unrevisable. And recently, in (16.12), we concluded that the contents of consciousness are strictly unrevisable, too. These were the two exceptional cases. What about the rest? Let us give it a name:

> *Definition 17.3*: A proposition is *empirical* iff it is neither unrevisably a priori true or false nor a content of consciousness.

What is the epistemic status of empirical propositions? The above observations don't say. Since empirical propositions are not unrevisably a priori, different doxastic states may take different attitudes towards them; this is true by definition, and trivial. More interesting is whether one and the same subject may, or should be able to, change his doxastic attitude towards them, i.e., whether his ranking function is, or should be, such that it is able to change its degrees of belief in empirical propositions. The latter may be taken to require that empirical propositions should not be strictly unrevisable w.r.t. rational ranking functions. Note that this postulate already goes beyond our basic definitions. It is still not very interesting, though; as I defined strict unrevisability in (6.14), it simply says that ranking functions should be regular with respect to empirical propositions. However, the ability to change degrees of belief may be given a stronger reading; we may even require that empirical propositions should not be unrevisable, as defined in (6.16); evidential sources should be able to change the degrees of belief in them. The gain in strength is still vague, since the restriction to evidential sources is still indeterminate. Nevertheless, it is more consequential. So, let us state:

(17.4)   *The Basic Empiricist Principle*: All empirical propositions in the domain of a rational ranking function κ (i.e., grasped by it) are revisable, i.e., not unrevisable w.r.t. κ.

Why should we accept this principle as a rationality postulate, and hence as unrevisably a priori? My way of introducing it should show that it is only a slight strengthening of what we have accepted anyway, partially by definition. So, it certainly looks convincing. However, I have no deeper justification; basic principles must start somewhere.

The grand label I have chosen suggests, though, that most philosophers and most scientists have taken it for granted for centuries. If we put to one side the two exceptional cases, unrevisably a priori propositions and contents of consciousness, it says that everything else should be under the control of evidence; whatever we believe about empirical matters, it should allow of being affected, confirmed, or defeated by evidence. Empiricists of various brands have also held stronger and perhaps doubtful views. For instance, they initially required verifiability instead of possibility of confirmation, and later on they tried to use unhelpful accounts of confirmation. (17.4), however, seems to be an unassailable core, sharable by everyone. In Spohn (2000a), I called (17.4) the principle of non-dogmatism. Indeed, it says, in a way, that we must not be dogmatic about any empirical claim. We may also call (17.4) a basic principle of learnability. Our mind must be open to learning about all empirical matters.

The Basic Empiricist Principle still looks weak. However, in the rest of this section I would like to draw some consequences which appear significant to me, and which I will call the Special and the General Coherence Principle. Roughly, they say that all of our empirical beliefs must cohere in the sense of being tightly connected by reason relations. This is very vague, and we first need to work up to the intended precise formulation of those principles before considering how we may prove them on the basis of (17.4).

As a first step, recall our observation in (6.17) that strictly unrevisable propositions are and have no reasons, as we explicated them in (6.1); the same was true of contents of consciousness. In analogy to our approach to (17.4), we may hence postulate the converse, i.e., that all empirical propositions, which we have assumed to be not strictly unrevisable, do have reasons and, by symmetry, are reasons for other empirical propositions. However, this is entirely trivial; propositions that are not strictly unrevisable always have deductive reasons as defined in (6.4).

So the idea should rather be that there is at least one inductive, i.e., non-deductive reason for each empirical proposition. Again, though, this is almost trivial:

(17.5)   If the negative ranking function κ on the complete algebra $\mathcal{A}$ over W takes at least three values $\neq \infty$, i.e., at least two values $\neq \infty$ besides 0, then each proposition $B \in \mathcal{A} - \{\emptyset, W\}$ not of the form A or $\bar{A}$ for some atom A of $\mathcal{A}$ has an inductive reason.

Proof: First note that the restriction on B is required, since, for any atom A of $\mathcal{A}$, A and $\bar{A}$ can only both be deductive reasons or counter-reasons for other propositions. This entails that (17.5) is non-trivial only if $\mathcal{A}$ has at least four atoms; for three or less atoms, there is no such B. Now assume B to be as required. Recall that C is a reason for B iff $\kappa(B \cap C) + \kappa(\bar{B} \cap \bar{C}) < \kappa(B \cap \bar{C}) + \kappa(\bar{B} \cap C)$. Suppose, without loss of generality, that $\kappa(B) = 0$. Now, either, for all atoms $A \subseteq B$, $\kappa(A) = 0$ so that for any C $\kappa(B \cap C) = \kappa(B \cap \bar{C}) = 0$. In this case, the atoms in $\bar{B}$ take at least two values so that is possible to choose C such that $\kappa(\bar{B} \cap \bar{C}) < \kappa(\bar{B} \cap C)$. Or we have $\kappa(A) > 0$ for some atom $A \subseteq B$. Then it is possible to choose C such that $\kappa(B \cap C) < \kappa(B \cap \bar{C})$ and $\kappa(\bar{B} \cap \bar{C}) \leq \kappa(\bar{B} \cap C)$, where again all four intersections are non-empty. In either case C is an inductive reason for B, and vice versa.

Hence, simply asking for inductive reasons will not do; the coherence produced by reasons must be given a stronger reading. In order to arrive at the intended principles I propose returning to the generation of the possibility space W through a set U of variables, as introduced in Section 2.1 and used in so many chapters. Perhaps there are a priori variables that generate only unrevisably a priori propositions; and perhaps variables relating to mathematical matters are such. Certainly, there are 'conscious' variables that generate only contents of consciousness. Let us call any other variable *empirical*. And let U consist only of empirical variables; we may neglect a priori and conscious variables in the present context. Finally, let us recall our usage that, for any $V \subseteq U$, a V-proposition is a proposition in the algebra $\mathcal{A}(V)$ generated by V. Then a stronger notion of coherence is provided by the idea that no variable $X \in U$ is independent of all the other variables, i.e., that X-propositions are always accompanied by reasons extrinsic to X. This is indeed a substantial idea, not trivializable in the way (17.5) was. Let us state it explicitly:

(17.6)   *The Special Coherence Principle*: For any empirical variable $X \in U$ and any empirical X-proposition A there is a $U - \{X\}$-proposition that is a reason for A w.r.t. any rational ranking function κ for U.

(This is somewhat stronger than mere dependence between X and $U - \{X\}$ which would require only some X-proposition to have an extrinsic reason.)

The special principle (17.6) certainly has the same empiricist credentials as the Basic Empiricist Principle (17.4). In Spohn (1983b, Sect. 6.3), when I first contemplated such principles, and also in Spohn (1991, Sect. 4), I called the Special Coherence Principle a weak, but tenable, version of the positivists' verifiability principle, according to which a statement's meaning consists in its method of verification and, hence, a statement is meaningful if and only if it is verifiable. Now, verification has turned out to be too strong a notion in this context; confirmation was instead acknowledged to be an appropriate substitute (see, e.g., Carnap's (1956) attempt to base empirical significance on confirmability or predictive relevance). If we accept this replacement and employ our notion of confirmation, we arrive at (17.6), guaranteeing that each atomic statement is meaningful by having at least one inductive reason (not involving that statement). (Haas (2011) elaborates a similar weaker explication of empirical significance with the means of belief revision theory.)

This is certainly suggestive. It indeed suggests a semantic line of justification of the Special Coherence Principle. Following the verification theory of meaning and its successors, we may construe the meaning of a sentence or statement not as its truth condition, but rather as its verifiability or acceptability or justifiability condition, and we may explicate the latter in terms of our notion of reasons or confirmation. That is, if we define propositions to be sentence meanings, propositions would be individuated not as sets of possibilities, as understood in this book, but by their reason relations to other propositions. Then there could be at most one proposition standing in no reason relation at all (presumably the a priori true proposition) or two, if we postulate negations. All others would have to satisfy the Special Coherence Principle.

I am reluctant, though, to endorse this line of reasoning. For one, I am bewildered by the thorough-going semantic holism thereby entailed. How can something be individuated by its reason relations to other things that are in turn individuated only by such relations? Davidson (1969) conceived of events in a similarly holistic way, but did not explain it any further. Perhaps we can make peace with semantic holism in the way proposed by Esfeld (2001). I am very unsure. (Cf. also Fodor's criticism of semantic holism in his 1987, Ch. 3.)

Mainly, though, I think that no actually available attempt at a semantics of acceptability conditions or the like presents a constructive alternative to a semantics of truth conditions; these attempts hardly get beyond metaphorical descriptions. For instance, if the relevant reason relations include deductive relations, then, as we saw above, such attempts get trivialized. And as long as we have no good account of other justificatory relations, the project is stuck, anyway. So these attempts may be helped by the ranking-theoretic account of justification. Maybe; how exactly is very open. The formal literature provides two nice examples for generating at least the Boolean structure of propositions from epistemological structures, namely Popper (1934/89, Neuer Anhang *IV), proceeding from conditional probabilities, and Gärdenfors (1988, Ch. 6), proceeding from the AGM theory of belief revision. Still, these are at best first steps. (Decision-theoretic semantics, as developed by Merin (1997, 1999, 2006b), is highly relevant in this context, too, though it is a somewhat different enterprise.)

In view of these problems, I prefer to stay away from semantic arguments for (17.6), and to present purely epistemological arguments below. This is not to entirely repudiate such semantic ideas. On the contrary, as with many others, I feel the temptation to do semantics in terms of something like justifiability conditions, and am supportive of the arguments that suggest it. Indeed, I take my account of dispositions in Sections 13.3–13.4 to be a paradigm for how certain reason relations can, by being a priori, belong to a concept. And I see my rich epistemological considerations in this book as hopefully preparing better grounding for such semantic ideas. If I were to pursue those ideas, I would, as indicated in Section 2.3, 13.4, and 17.1, do so within the framework of two-dimensional semantics. However, at present none of this can be relied on.

Stating the special principle in the above way immediately suggests a generalization. What is so special about the partition $\{X, U - \{X\}\}$? Nothing. (17.6) does not lose a bit of its plausibility when it refers to any partition of U:

(17.7)   *The General Coherence Principle*: For any non-empty proper subset V of U, V and $U - V$ are dependent w.r.t. any rational ranking function κ for U.

My epistemological argument below in favor of (17.6) will generalize to (17.7), whereas I would not know how to argue for (17.7) on semantic grounds; unlike (17.6), it does not even look like a semantic principle. This is my final reason for staying away from semantic arguments for the coherence principles.

(17.7) is a bit weaker than (17.6), insofar as it does not claim each V-proposition to have an extrinsic reason. Perhaps some variable X is so deeply embedded in V, as it were, that X-propositions only find reasons within V; therefore the weaker existential statement in (17.7) is appropriate. Otherwise, however, (17.7) is much stronger than (17.6). If we think of U arranged in a ranking net, then (17.7) requires the net to be connected in the sense of (7.20g), whereas (17.6) still allows the net to separate into many components (cf. 7.20f), as long as each of these components contains at least two variables. This graphically explains why I call (17.6–17.7) coherence principles and why only the general principle fully represents coherence; for this reason, I am keen to establish it. In Spohn (1999a), where I had an idea about how to do this, I somewhat pompously said on p. 159 that the General Coherence Principle affirms the unity of science or the unity of our picture of the world, and it does so, I contend, in an unrevisably a priori manner. Pompous or not, this is what it does.

What is my argument leading from the Basic Empiricist Principle (17.4) to the coherence principles? Let us start with the special principle (17.6), and assume A to be an empirical X-proposition ($X \in U$). According to (17.4), $\kappa(A) < \infty$ for any rational ranking function κ. How can $\kappa(A)$ be revisable, that is, change through experience by conditioning on some evidential source, as required by (17.4)? Well, by obtaining reasons for or against A, by coming to believe or, more generally, by changing the degree of belief in other propositions that are positively or negatively relevant to A. End of proof?

Not quite; the quick argument captures the gist of the matter, but it leaves a gap. A may itself be or belong to the evidential source, and thus change its rank directly under the impact of evidence, unmediated by reasons. And it may lack consequences by not being a reason for other propositions. This strange case would agree with (17.4), but violate (17.6).

The exclusion of this case is somewhat complicated by the unclear reference of (17.4) to evidential sources. What qualifies as such a source? In Chapter 16, we reflected amply on this issue. In Section 16.4, we concluded that the rock-bottom level of the contents of consciousness does not provide evidential sources; they are inductively barren. The next level discussed in Section 16.3 is much better; we may fix evidential sources to be of the form $\Phi_{I,now}(A)$ or A itself, provided it can appear (to me now) as if A. Then, however, the Schein-Sein Principle closes the gap, since A and $\Phi_{I,now}(A)$ are propositions pertaining to different variables and thus, either way, the evidential source entertains reason relations as required by the Special Coherence Principle.

There is still a tiny loophole. The defeasible version (16.8) of the Schein-Sein Principle only claims defeasible apriority for the relevant reason relation and hence holds only for initial ranking functions and not for all, whereas the unrevisable version (16.9) only claims there to be a conditional reason relation for all ranking functions. However, even this final loophole may be closed with the help of the following observation:

(17.8)   Let C be a reason for B given D w.r.t. κ. Then either $C \cap D$ or $C \cup \bar{D}$ (or both) is an unconditional reason for B w.r.t. κ.

Proof: Let $\kappa(B \cap C \cap D) = a$, $\kappa(B \cap \bar{C} \cap D) = b$, $\kappa(\bar{B} \cap C \cap D) = c$, $\kappa(\bar{B} \cap \bar{C} \cap D) = d$, $\kappa(B \cap \bar{D}) = x$, and $\kappa(\bar{B} \cap \bar{D}) = y$. C being a reason for B given D then translates into

(a) $a + d < b + c$,

$C \cap D$ being a reason for B into

(b) $a + \min(d, y) < \min(b, x) + c$,

And $C \cup \bar{D}$ being a reason for B into

(c) $\min(a, x) + d < b + \min(c, y)$.

The sizes of a, b, and x may now be arranged in six different ways; likewise, the sizes of c, d, and y. In all 36 cases thus generated it may be easily checked that (a) implies (b) or (c).

To return, the unrevisable version (16.9) said that $\Phi_{I,now}(A)$ is a reason for A given the normal condition N^0 (cf. 13.7) for all ranking functions. Hence, setting $B = A$, $C = \Phi_{I,now}(A)$, and $D = N^0$, (17.8) tells us that A must also have an unconditional reason.

We may summarize the argument by stating

Assertion 17.9: If possible evidential sources only contain propositions of the form $\Phi_{I,now}(A)$, or of the form A for which $\Phi_{I,now}(A)$ is defined, then the Basic Empiricist Principle (17.4) and the Schein–Sein Principle (16.9) entail the Special Coherence Principle (17.6) (for all those ranking functions having those possible evidential sources in their domain).

In Spohn (1991, Sect. 4) I simply postulated the Special Coherence Principle, having no strict argument; I also thought it could be claimed to hold only within the ill-defined universal frame containing all variables whatsoever. (17.9) thus brings double progress. Still, in Spohn (1999a) my argument was slightly different, further shifting the demands on the domain of the relevant ranking functions. Since it will be the only argument I am able to carry over to the General Coherence Principle (17.7), we have to look at it as well; it is only a slight and slightly less obvious variant of the argument just given:

The first step is the same as before, leaving the same gap about evidential sources. How else can we close the gap? Well, in Section 16.2 we developed a more liberal, frame-relative understanding of evidential sources (that was the first and weakest attempt at analyzing the foundations of belief). According to it, any proposition may belong to an evidential source as long as it turns out to be so under an appropriate coarsening of the most fine-grained frame that generates all relevant phenomenal propositions of the form $\Phi_{I,now}(A)$. This coarsening need no longer contain these phenomenal propositions. Hence, we can no longer use the reflexive case of the Schein–Sein Principle (16.9) for our argument, as we did just before. Instead, however, we can use the (unstated unrevisable version of the) Schein–Sein-Belief Principle (16.10), which still claims there to be an a priori reason relation between $\mathbf{B}_{a,t}(A)$ and A, between the external fact A and a's belief at t in A, even if the phenomenal details have fallen victim to the coarsening. The price to pay for this more liberal understanding of evidential sources is, however, that the use of (16.10) works only in the non-reflexive case, where either a is not the subject to which the Special Coherence Principle (17.6) is to be applied or t is not the time at which the Special Coherence Principle is to be applied. That is, the implicit subject of (17.6) has to think about what other subjects or she herself at other times may believe; i.e., such propositions have to be in the domain of her (present) ranking function. Given this, the same reasoning goes through as before, including the step from conditional to unconditional reasons with the help of (17.8). The conclusion is

Assertion 17.10: If possible evidential sources are understood in a frame-relative way, then the Basic Empiricist Principle (17.4) and the (unrevisable version of the) Schein-Sein-Belief Principle (16.10) entail the Special Coherence Principle (17.6) (for all ranking functions having the propositions A and $\mathbf{B}_{a,t}(A)$ in their domain for all members A of those frame-relative evidential sources).

A more succinct description may run as follows: Frame-relative evidential sources contain what is directly perceived in the frame-relative sense. Then, as soon as the implicit subject has grasped the basic theory of perception (16.3–16.4), which pertains to the non-reflexive case, evidential sources must stand in reason relations, namely to propositions describing others' beliefs about those sources.

The reference to others' beliefs or to one's own beliefs at other times looks like a detour, and hence the first argument may seem preferable. However, as far as I can see, only this detour helps us to an argument for the General Coherence Principle. How does it run?

Let us consider any partition $\{V, U - V\}$ of the given frame U (of empirical variables), and let us see how we find reason relations between V- and $U - V$-propositions, i.e., between $\mathcal{A}(V)$ and $\mathcal{A}(U - V)$. One case is that all possible evidential sources (in the frame-relative sense) are in $\mathcal{A}(V)$. Suppose no $U - V$-proposition bears a reason relation to $\mathcal{A}(V)$. Then all $U - V$-propositions would be unrevisable, in contradiction to the Basic Empiricist Principle (17.4). Likewise for the case where all possible evidential sources are in $\mathcal{A}(U - V)$.

So, consider the final case where we find evidential sources in $\mathcal{A}(V)$ and in $\mathcal{A}(U - V)$, and let A belong to some evidential source in $\mathcal{A}(V)$ and B belong to some evidential source in $\mathcal{A}(U - V)$. Finally, let the implicit subject have propositions of the form $\mathbf{B}_{a,t}(C)$ in the domain of his ranking function. Then, as just observed, the Schein–Sein-Belief Principle (16.10) deems A to be a reason for $\mathbf{B}_{a,t}(A)$ and B to be a reason for $\mathbf{B}_{a,t}(B)$. Suppose that these belief propositions are, respectively, in the same part of the partition as the propositions the beliefs are about; otherwise, we would already be done.

Now, the trick I want to propose is to consider the proposition $\mathbf{B}_{a,t}(A \cap B)$. It seems clear that, if A is a reason for $\mathbf{B}_{a,t}(A)$ and B a reason for $\mathbf{B}_{a,t}(B)$, then A and B are both reasons for $\mathbf{B}_{a,t}(A \cap B)$ for the implicit subject; both A and B make it more plausible that a believes the conjunction $A \cap B$ at t.

Where, however, is the proposition $\mathbf{B}_{a,t}(A \cap B)$? It is perhaps not clear which variables describe the doxastic state of a at t. Let us try the two most suggestive ideas. The most coarse-grained proposal would be to assume a single variable with a rich range consisting of all possible doxastic states of a at t. Then, however, both A and B are reasons to assume that a is in a certain state at t. And since this single variable must be either in V or in $U - V$, at least one reason relation goes across the partition. (In this case, we need not even refer to the belief in $A \cap B$.) The most fine-grained proposal would be to assume a separate variable for each $\mathbf{B}_{a,t}(C)$ or for a's degrees of belief in C at t. Thus, $\mathbf{B}_{a,t}(A \cap B)$ is an X-proposition for some variable X. Again, however, either $X \in V$ or $X \in U - V$, and at least one reason relation crosses the partition. Thus we have shown

Assertion 17.11: If possible evidential sources are understood in a frame-relative way, then the Basic Empiricist Principle (17.4) and the Schein–Sein-Belief Principle

(16.10) entail the General Coherence Principle (17.7) (for all ranking functions having the propositions A and $\mathbf{B}_{a,t}(A)$ in their domain for all members A of those frame-relative evidential sources).

What should we think of the trick, as I called it, that I applied in the above argument? I am not sure. It may have a deep significance; but I am not prepared to develop it. In any case, it makes clear why we could not generalize the first argument for the Special Coherence Principle. In contrast to objects of belief, objects of phenomenal appearance are not closed under conjunction. That is, it may appear to me as if A, and it may appear to me as if B. This, however, does not entail that it can appear to me as if $A \cap B$. Sometimes, this may make sense; usually, it will not. (In Section 16.4, I pointed out that possible contents of consciousness are algebraically closed. However, this only means that, if $\Phi_{I,now}(A)$ and $\Phi_{I,now}(B)$ are such contents, $\Phi_{I,now}(A) \cap \Phi_{I,now}(B)$ is so as well. It is silent about the meaningfulness of $\Phi_{I,now}(A \cap B)$.)

Still, one may be surprised by the externalization of the unification via the beliefs of some third person or of the subject himself at some other time. One may have expected a more direct argument: that I myself, or rather the many egocentric variables I have for all of the various states of mine that I am aware of, including my doxastic state, must be located in the partition $\{ V, U - V \}$, and that no dividing line violating the General Coherence Principle can exist that would either cut through myself, i.e., the set of my egocentric variables, or forever separate a set of non-egocentric variables from all of my egocentric variables.

Such an argument would certainly have a Kantian ring. The egocentric variables should only refer to the states of my consciousness, not to arbitrary empirical states of mine, and then the point that the dividing line cannot cut through that set of egocentric variables should derive from something like the unity of consciousness. This unity in turn seems to be provided by Kant's idea that the "I think" must be able to accompany all my thoughts and ideas (1781/87, B132), i.e., by what Kant calls the original synthetic unity of pure apperception and declares to be the supreme principle of understanding lying at the base of all our judgments.

However, as explained in Section 16.4, this supreme principle rather resembles the Conscious Essence Principle as applied to conscious beliefs (or judgments). In any case, if contents of consciousness are bound to be inductively barren, as also explained in Section 16.4, the unity of consciousness, whatever it is, does not entail the unity of science as explicated in the General Coherence Principle. This is, it seems to me, the reason why we had to externalize the unifying force to the beliefs of someone other than my present self.

The coherence principles thus established are only a start. For instance, so far the principles entirely operate on a global level. It would certainly be worthwhile to study the many details of the local a priori structure of reasons. This would presumably require going into all of the semantic considerations that I have put to one side, into all of the specific reason relations that are a priori associated with specific concepts. As mentioned, my treatment of dispositions would be my paradigm for such an inquiry, but I do not know how far it would carry us.

For the time being, though, let us be content with the global observations we have made. There are some directions for further development on the global level, two of which I will pursue in the two final sections.

17.3 Reason, Truth, and Stability

One direction is to look for the connection between reasons and truth. What has our almost obsessive search for reasons to do with truth? Why should we suppose reasons to be truth-conducive? This is about the most fundamental question we can ask; without a positive answer all of the workings of our reasoning machinery, as we have described in ample detail, just seem idle.

Well, we did find a truth guarantee for our most basic beliefs, i.e., actual contents of consciousness; indeed, we found the best guarantee there could be, namely identity. That is, those beliefs have to be true because they are identical with their contents. However, the step from (belief in) $\Phi_{I,now}(thus)$ to (belief in) $\Phi_{I,now}(A)$, the non-inferential entry into the realm of reasons, was already burdened with uncertainty – this is what it means to enter the realm of reasons! All further moves within that realm are burdened all the more. Why should these moves tend towards the truth?

The urgency of the question seems clear, but not its content. It is not easy to see what at all is at issue here. From the first-person perspective the answer seems obvious: Reasons induce belief; and to believe something is to believe it to be true – so much is tautological. Therefore, reasons bring me closer to the truth; this is what I have to think and say. What more could or should I say from my perspective?

Apparently, the force of the question only appears in the third-person perspective. You, or God, or the scientist, may respond to me: "Sure, this is what you have to say; you always believe that you believe the truth. However, we would like to question whether this is really so, whether, and to what extent, your impressive rational powers dispose you to find out about the actual truth." And so the scientist may start her inquiry, just as she may investigate whether, and to what extent, migrant birds find their way home. Of course, the point of that inquiry is that it opens the skeptical gap; perhaps the scientist finds, contrary to my expectations, that, my powers and efforts notwithstanding, I am very bad at truth tracking.

However, I said already in Section 1.1 that I do not wish to systematically engage with the skeptical challenge here, even though my constructively oriented strategy has produced various implicit comments. In this spirit I will abstain from reviewing how that challenge might be treated within the third-person perspective. Thus, I will not look at the many illuminating things that have been contributed from an evolutionary or a cognitive perspective, and which impressively display the pre-established harmony between our environment and our cognitive machinery. I also neglect these scientific perspectives because I have not found them to specifically comment on our search for reasons.

Indeed, I do not see how the third-person perspective could instruct our normative issues, which essentially belong to the first-person perspective. I expressed this doubt already in Section 1.1 when criticizing Quine's (1969) claim that setting the ultimate aim, truth, would only leave the task of (cognitive) engineering (which would be a task within the third-person perspective). This appeared to be wrong; rather, we have to explore our rich normative discourse, which cannot be assumed from the outset to reduce to a single aim. That is to say: I cannot employ the scientist as my trainer and ask her: "Have I done well?", and insist after her disappointing answer: "How can I do better?" The child can ask his teacher. However, the first-person perspective is not my private one, it is that of humanity, not only of actual humanity, but of the entirety of potential participants in normative discourse. And that entirety cannot ask someone on the outside. Surely we can try to figure out in that discourse what the external scientist, or God, would say. However, this would just be another move within the internal normative discourse.

To illustrate the point: BonJour (1985) is acutely aware of the very problem and attempts to answer it by his metajustificatory argument in Section 8.3. However, his considerations do not concern the scientist who already knows the truth and wonders how we manage to come so close to it. Rather, he wants to justify our epistemological practice within the first-person perspective. He starts from the observation that we can apparently keep our belief system remarkably coherent and stable over the long run, despite continuous, potentially irritating input through evidence. And then he argues that this observation shows that our beliefs are very likely to be true, the argument being an inference to the best explanation (IBE): the coherence and stability of our beliefs over the long run, given continuous input, is best explained by their truth.

The argument is obviously circular. BonJour uses the very same (justified?) inductive practices in order to argue that those practices are truth-conducive. Now I do not wish to assess BonJour's argument, which certainly goes far beyond the first-person perspective trivialities stated above. Neither do I wish to criticize it for its circularity, which indeed seems unavoidable to me. Rather, the case should only illustrate the inescapability of normative considerations. We have to assess the normative status of our inductive practices, including IBE, and an argument like BonJour's can only serve as an internal argument that positively increases our normative trust in those practices, including IBE. (Of course, I would like to recall at this point that I indicated how IBE may be reconstructed within ranking theory in Section 14.15.)

Therefore, I think that the issue of the truth-conduciveness of reasons must also be approached within the normative first-person perspective. If we do so, the next question we should pose is which notion of truth is relevant for the task. The answer seems clear: there is only one well-articulated theory of truth, namely the correspondence theory of truth and its modern semantic and deflationary variations. However, I think that the correspondence theory of truth is simply not suited for the enterprise. All of our reasoning pertains to propositions made out of doxastic possibilities, and in Section 2.3 I took a stance on what doxastic possibilities are, namely Lewisian possible worlds (plus a center and possibly further placeholders for objects). If so, then our reasoning activities have to relate to a notion of truth that applies to Lewisian possible worlds. And as I suggested there, this notion is not the correspondence notion. That notion applies to Wittgensteinian possible worlds which contain all the facts to which beliefs, sentences, or whatever may correspond or not. By contrast, it makes no sense to speak of such things as beliefs or sentences as simply corresponding to a huge object such as a Lewisian world. Rather, such a Lewisian world is excluded or not excluded by our beliefs, and this exclusion is the result of a maximally experienced and considered judgment, as sketched in Section 2.3. Moreover, as also suggested there, we cannot transfer the correspondence notion from Wittgensteinian to Lewisian worlds by establishing a mapping between Lewisian and Wittgensteinian worlds, since such a mapping presupposes a notion of truth appropriate for Lewisian worlds (cf. also Spohn 2008a, pp. 12ff.).

In a nutshell, we might say that the correspondence theory of truth is part and parcel of the third-person perspective. God or the scientist can apply it in judging our cognitive relation to the external world. We, however, move within our normative first-person perspective, and if we want and have to get at the truth-conduciveness of reasons from within that perspective, we need a notion of truth suited for that perspective. I am not denying the correspondence theory; this would be absurd. Rather, as expressed in Section 2.3, I think that there are two notions of truth, each for its own purpose, which can peacefully coexist, and which indeed complement each other within two-dimensional semantics.

In Section 2.3, I also indicated that this other notion of truth appropriate for Lewisian worlds is something like the pragmatic notion, or perhaps a variation for which another label fits better. In any case, it was ill-specified. However, whatever precisely it is, we face an objection from BonJour (1985), which he couches in coherentist terms (because this is his context). The objection is that to accept a coherence theory of truth in order to show the truth-conduciveness of, in his case, coherentist justification is "both commonsensically absurd and also dialectically unsatisfactory" (p. 109.). It would be commonsensically absurd because it is infected by too much idealism, and dialectically unsatisfactory because the metajustification, the argument for the truth-conduciveness of coherentist justification, would be too easy:

obviously if truth is long-run ideal coherence, it is plausible to suppose that it will be truth-conducive to seek a system of beliefs which is as coherent as one can manage to make it at the moment (p. 109);

and moreover circular:

If . . . the only rationale for the chosen concept of truth is an appeal to the related standard of justification, then the proposed metajustification loses its force entirely. It is clearly circular to argue both (1) that a certain standard of epistemic justification is correct because it is conducive to finding truth, conceived in a certain way, and (2) that the conception of truth in question is correct because only such a conception can connect up in this way with the original standard of justification. (pp. 109f.)

Because of this objection he turns in his Chapter 8 to providing a metajustification in terms of the correspondence theory of truth. However, if I am right, this in effect results in a doubtful mixing of perspectives. The justificatory demands are raised within the first-person perspective and cannot then be answered in the third-person perspective.

In fact, I am not really moved by BonJour's objection. I am not worried by commonsensical absurdity, since I acknowledge the correspondence theory as well, and certainly do not lean towards odd ontological consequences of an idealistic brand; the pragmatic theory of truth need not have such consequences. Neither am I worried by circularity. As I said, it seems unavoidable to me to some extent, and is indeed not avoided by BonJour's recourse to the correspondence theory of truth.

What I do worry about is the allegedly too easy argument – not about its being too easy, but about there being an argument at all. For we must grant that the quotations above from BonJour (1985) are full of unredeemed promises. We do not really know what a coherence theory of justification is, nor what a coherence theory of truth is. So, how can we connect them by argument? Perhaps the best elaboration of the coherence theory of truth is that of Rescher (1973, 1985); however, it rests on a theory of plausibility indexing which goes back to Rescher (1964). In Section 11.1, I acknowledged the latter as a predecessor of ranking theory, but no more; so we cannot rely on it, either. In Section 2.3, I referred instead to the pragmatic notion of truth as the notion applicable to Lewisian worlds. However, it is in no better condition; as I granted there, we are dealing here with suggestive and important, but unsatisfactorily elaborated ideas. So, I have maneuvered myself into a predicament. On the one hand, I have argued that the truth-conduciveness of reasons must be assessed relative to a notion of truth different from the correspondence notion, and on the other hand we find there no workable base at all for that assessment.

To escape this predicament, my proposal will be, roughly, to not argue for, but simply to postulate the truth-conduciveness of reasons. The idea is that we thereby get a better understanding, even if only a partial one, of the notion of truth involved, and gain further insight into the structure of reasons, if they are to be truth-conducive in this way. Let me explain what this hazy proclamation is supposed to mean.

I started this section by suggesting that our topic is trivialized from the first-person perspective: the reasons I receive direct my beliefs and thus, as I have to believe, direct me towards truth. However, this triviality does not exhaust the first-person perspective. Of course, at each moment I think that my present beliefs are true; this is what it means to have them. Still, I know well enough that they might turn out to be false; if they are really true, they must survive all further stages of learning. So, within the first-person perspective, what the truth really is only emerges in a dynamic setting. As so often in this book, this is the setting we have to attend to.

From this point of view we must first claim that each truth is believable, not in the static sense that there is some doxastic state in which it is believed – that is trivial – but in the dynamic sense that each doxastic state may come to believe it. If truth, in the intended internal sense, is determined by the maximally experienced and considered judgment, it must become to be believed on the way to that judgment. Therefore we must secondly claim that for each truth there is a true reason. Otherwise, the first claim could not be maintained. Let me state these principles a bit more precisely before further discussing their content.

To begin with, we again must move within the ill-defined universal frame, as we already did in Sections 2.3 and 14.9. We must do so because, even if the existence of true reasons for a given truth should be guaranteed a priori, this guarantee cannot hold for any limited fixed frame. As far as we can presently tell, the reasons may be arbitrarily remote. Hence, let U^* be the ill-defined set of all variables whatsoever and W^* the set of all complete doxastic possibilities whatsoever, i.e., of all centered Lewisian possible worlds (neglecting the placeholders for objects).

Moreover, one of those doxastic possibilities must be the actual one; let it be denoted by $@^*$. Thus, a proposition $A \subseteq W^*$ is true iff $@^* \in A$. To be sure, this is only a formal representation of the internal truth notion we are after, not an explication. Whether A is believed, and hence whether this belief is true, still appeals to the vaguely conceived limit of inquiry, sketched in explanation 2.6, which alone decides about internal or pragmatic truth. Also, it is important to note that $@^*$ is not a rigid designator for the actual universe we are living in; it must rather be conceived as a variable over W^*. After all, we do not know in which possible universe we live in, and the feasibility of reasons must not hold accidentally, as it were, only for our actual universe. Perhaps, the range of the variable $@^*$ should be restricted, say, to all conceptualizable possibilities; recall that there may be insensible and unconceptualizable Lewisian worlds, as noted in Section 2.3. We will find further reasons below for possibly restricting the range of $@^*$.

Now, what should it mean that each truth is believable? As in the previous section, let me restrict attention to empirical truths about single variables, in order not to trivialize the possibility of reason finding. Since such a truth is empirical, it need not be believed. Of course, it can be believed; but this is not the intended sense of believability. The intent rather is that each doxastic state should be able to come to believe that truth, i.e., that there are experiences and, in consequence, revisions of that state that result in the belief in this truth. This is still not specific enough, though. There should not only be some possible experiences and revisions with that effect. It must be possible to have the required experiences in the actual world $@^*$. That is, the doxastic dynamics through which one comes to believe that truth should not merely be possible in the wide sense used in the previous section, but *actually possible* within the actual world, whichever it is. This is the intended sense of "-able" in taking truths to be believable. Thus, we have arrived at

(17.12)   *The Basic Belief-Truth Connection*: Let $X \in U^*$ be an empirical variable and A be an empirical X-proposition with $@^* \in A$. Then for any rational ranking function κ for U^* there exists a sequence of experiences available in $@^*$ such that κ changes through those experiences into some ranking function κ' (according to our rules of conditionalization) in which A is believed.

Note that this is much stronger than the Basic Empiricist Principle (17.4), which requires only the revisability of (atomic) empirical propositions through some possible experiences. Rather, (17.12) requires that true empirical propositions must be revisable so as to be believed through actually possible experiences.

(As an aside, let me remark that I do not believe that (17.12) is affected by the knowability paradox of Fitch (1963), precisely because of the dynamic conception of believability embodied in (17.12), which is missing in the knowability paradox. It may well be that A and that I (presently) do not know that A. The alleged paradox results from the fact that I cannot (presently) know this. Hence, not everything can be (presently) known. However, I may well learn both that A and that I did not know A. I could argue in the same way if we started from the more dramatic fact that A and nobody ever actually will know that A (which is certainly the case for some A). This fact is still counterfactually believable in my dynamic sense.)

A direct consequence of (17.12) is

(17.13)   *The Basic Reason-Truth Connection*: Let X and A be, as before, such that $@^* \in A$. Then, for each rational ranking function κ there is a $U^* - \{X\}$-proposition B such that $@^* \in B$ and B is a reason for A w.r.t. κ.

In short, for each atomic empirical truth there is a true reason. In fact, (17.12) even entails that for each atomic fact there is a true sufficient reason. However, I shall not make use of this strengthening of (17.13) in the following.

The Basic Reason-Truth Connection (17.13) parallels and strengthens the Special Coherence Principle (17.6) in the same way as did (17.12) vis-à-vis (17.4). (17.6) follows from (17.13) simply because any atomic empirical proposition is true in some $@^*$; and if it has a reason that is true in $@^*$, it has a reason. Why is (17.13) a consequence of (17.12)? The derivation of (17.6) from (17.4) was circumstantial because both were stated for variable frames and because we had to secure the existence of suitable evidential sources (or other suitably related propositions) within those frames. Now, however, we move within the universal frame U^* where this problem does not arise. So, if there is no true reason at all for A, there cannot exist actual experiences that move us to believe A. Trivial as these observations may be, let me state them explicitly for book-keeping:

Assertion 17.14: Within the universal frame U^*, the Basic Belief-Truth Connection (17.12) entails both the Basic Empiricist Principle (17.4) and the Basic Reason-Truth Connection (17.13), each of which in turn entails the Special Coherence Principle (17.6).

Why should we accept (17.12) and (17.13) as unrevisably a priori principles that hold for all rational doxastic states? Well, I find them obvious; they fit to the informal characterization of internal truth as being reached in the maximally experienced and considered judgment. However, this is just to grant that I have no deeper justification, just as in the case of (17.4). As I said there, basic principles have to start somewhere. There are two ways of looking at these principles, both of which are apt.

The first way is to take them as conceptual truths about truth, truth in the intended internal or pragmatic sense. Of course, they do not define this notion, but they provide at least a minimal characterization that seems unassailable as far as it goes. Truth in that sense must be accessible to experience and reason, and the principles specify some minimal way in which it is so accessible.

The Basic Reason-Truth Connection (17.13) may well be seen as an ingredient of a coherence theory of truth, since it says that truths must cohere, i.e., be related by the reason relation, though so far only to a minimal extent. We will see whether (17.13) may be strengthened in acceptable ways.

Without doubt my strongest ally is internal realism, as amply described and argued for by Putnam (1978, 1980, 1981). Indeed, these texts inspired me to extract from them some formally more precise principles. Putnam concludes that the ideal theory must be true, and if one looks at how he describes the ideal theory in Putnam (1978, p. 125, and 1980, pp. 473f.), it is clear that the ideal theory must satisfy (17.12) and (17.13). If, contrary to (17.12), a proposition cannot come to be believed after ever so many actually possible experiences that are all part of the ideal theory, then this proposition cannot belong to the ideal theory and thus be true. And if, contrary to (17.13), a proposition meets with no true reason, no support in any part of the ideal theory, it can again not belong to the ideal theory.

Because of the strong inspiration I should add more careful comparative remarks. However, they are difficult and would lead us too far astray. This is so because Putnam's considerations are couched in semantic terms. His theories are linguistic entities, and he is profoundly concerned with the interpretation of theories: how a theory's terms acquire reference, how its sentences acquire truth, and how all this is embedded in an account of the understanding of language. A careful comparison would have to engage with these semantic issues much more deeply than I can afford here.

Let me make just two remarks: First, I am not speaking of theories as linguistic entities (or as internal mental representations). Rather, the theories I am speaking of are beliefs or, more precisely, belief types as individuated by their contents, i.e., propositions or sets of possibilities. And, as I construed these matters in Section 2.3, there is no gulf between theories in this sense, i.e., beliefs, and the external world, and there is no desperate need to close this gulf. This is so because I take beliefs to be vastly counterfactual dispositions to respond to possible worlds, one of which is the actual world. It thus seems that I am not concerned by the problems raised by Putnam (1980). Theories in his sense, even ideal ones, may have unintended interpretations that make them true in almost any given world. Theories in my sense can have no unintended interpretations. The disposition *is* the intention, and it manifests in the application.

This remark seems to agree with Putnam's (1980), pp. 481f.) solution of those problems (if there can be agreement across such different frameworks). For him, these problems arise when the interpretation of theories – and thus reference and truth – are separated from our use of language and our understanding of the theories. Internal realism is intended to prevent this separation, and to let understanding fix the interpretation. Similarly, I would run into these problems if I were to reduce beliefs to their linguistic representation. But beliefs are here conceived as beliefs-in-use, as it were, as dispositional and sometimes actual responses to real and possible situations or worlds.

Secondly, however, I do not agree with Putnam in putting his internal realism into opposition with metaphysical realism. Both kinds of realism peacefully coexist, just as pragmatic and correspondence truth coexist in two-dimensional semantics. The ideal theory always refers to a given epistemic possibility or centered Lewisian world $@^*$. Then we can indeed say that this theory, the fully experienced and considered judgment about $@^*$, can no longer turn out to be false; there is simply nothing left that could change the judgment. So, for all $@^*$ the ideal theory about $@^*$ just picks out $@^*$, i.e., it is the singleton proposition $\{@^*\}$. In short, it is unrevisably a priori that the ideal theory is true.

Still, I see no difficulty in counterfactually supposing, even from the point of view of $@^*$, that the ideal theory about $@^*$ is false. It might always be that there is a reality behind the reality of $@^*$, although there is none in $@^*$ itself. There always is a possible world w which, in some way or other, contains $@^*$ and in which we can have exactly the same experiences as those that would be complete in $@^*$. Thus we would have the same theory about w that would be the ideal one about $@^*$.

This theory would have to be true of w in many ways. Many propositions will stand unrefuted by a reality behind reality. However, there will also be many propositions in that theory that are falsified by w; after all $w \notin \{@^*\}$. In particular, many propositions in that theory will implicitly or explicitly refer to the limit of inquiry that this theory would have reached in $@^*$, but does not reach in w, and at least those propositions may be false in w. General propositions about which kinds of things there are or are not in a world explicitly refer to such a limit, and may be false in w, though true in $@^*$. And claims about essences and causal relations implicitly refer to such a limit. For instance, this theory will make claims about what water really is. These claims are true in $@^*$, but might be false in w, where the nature of water is more deeply hidden. Or w might be the occasionalist's world in which God is a common cause of each pair that this theory would correctly claim to be cause and effect in $@^*$.

So, in a nutshell, the ideal theory exemplifies Kripke's contingent a priori. It is a priori, but not necessarily true. Hence, we may share Putnam's claim that there is no room for the metaphysical realist's skepticism. However, from our actual world view which is finite and restricted doubt cannot be eliminated. There might always turn out to be a reality behind the reality that presents itself to that finite view. We do not know what @* is, and it might also be the w only counterfactually supposed above. (For careful reflections about Putnam's argument that we cannot be brains in a vat, indeed the most careful ones I know of, see Müller (2003).)

This must suffice as digressive remarks about Putnam's internal realism, which at the same time provided some background on the first way of understanding principles (17.12) and (17.13) as conceptual truths about internal or pragmatic truth. It would be one-sided, however, to focus on this understanding of the principles. There is a second way of understanding them. They should also be taken as substantial principles constraining rational ranking functions, i.e., postulating that such functions must relate truths by reason, and in particular must relate truths to actual experience or evidential sources, and they must do so however the actual truths and evidential sources turn out to be. That this is a genuine constraint is shown by the fact that the Basic Reason-Truth Connection is actually stronger than the Special Coherence Principle; there is no way to infer the former from the latter. Viewed in this way, they are strengthened learnability principles. Empirical propositions have to be learnable not only in the sense that (degrees of) belief in them can be changed, but in the stronger sense that the changes are biased in the right way towards truth. To this extent, they have the same credentials as their weaker counterparts in the previous section.

Indeed, I am not sure whether the principles constrain not only rational ranking functions, but also the range of the variable @* over W^*. Of course, for each @* $\in W^*$ there is ranking function satisfying (17.12) and (17.13). It is, however, an open question whether there is a ranking function that satisfies (17.12) and (17.13) for all @* $\in W^*$. With respect to (17.12) and (17.13) this may seem likely; the answer will be negative, though, for some of the stronger principles below. I am not sure how to respond to such a situation. It may indicate that a principle that is not universally satisfiable by any ranking function is too strong. Or it may present the challenge of finding ranking functions that maximize the set of worlds @* for which the principle in question holds. In any case, this issue is most important, but also technically delicate, and I shall not further pursue it here.

Be this as it may, these principles and the subsequent variations should be seen as what they are, both, conceptual principles about internal truth and substantial principles of (theoretical) rationality. It seems that the more we advance our account of rationality, the better we understand internal truth; and conversely, grasping internal truth helps us to further our account of rationality. I do not claim that this entanglement is inescapable; at present I simply do not see whether it might be escaped. The entanglement is not objectionable, though, and perhaps even natural or welcome. In lieu of alternatives, the aim can only be to search for better or stronger principles, to work out their consequences, and to see whether they withstand our critical normative examination.

Well, having started in this way, improvements immediately suggest themselves. The Basic Reason-Truth Connection looks extremely weak insofar as it requires only one true reason for another truth. The slogan "truth must be believable" suggests, rather, that the totality of true reasons in favor of some true atomic proposition A outweighs the totality of true reasons against A. Otherwise, one could not ultimately believe A. This is again what Putnam's ideal theory requires; if, in total effect, the rest of the ideal theory does not speak in favor of A, A cannot belong to the ideal theory. Indeed, it seems this is exactly how far we can go, no more and no less. The question is how we might render precise the indeterminate talk of outweighing. Perhaps we can capture it by saying that the entire rest of the truth must be a reason for A. Borrowing the notation introduced in Section 14.1 and letting $^{@*}[U^* - \{X\}]$ stand for the rest of the world $@*$, i.e., the way in which all variables in $U^* - \{X\}$ are realized in $@*$, we can formally express this in the following way:

> (17.15)   *The Total Reason-Truth Connection*: Let X and A be as in (17.12) such that $@* \in A$. Then according to any rational ranking function κ $^{@*}[U^* - \{X\}]$ is a reason for A.

Again, we might require that $^{@*}[U^* - \{X\}]$ is even a sufficient reason for A. In the case of (17.15) it is clear that there can be no ranking function that satisfies it for all $@* \in W^*$. Another reason why I am unsure about (17.15) is that what (17.15) really does is to weigh the conjunction of all truths (besides A) against its negation. Perhaps this is not what we originally meant by weighing the positive reasons against the negative reasons. These points do not convincingly refute the Total Reason-Truth Connection. We should keep it as a candidate. But we should also look for other ways of strengthening the basic principles.

I propose to advance in the following direction. As already mentioned at the end of Section 6.1, there is not only a dynamics of belief, but also a dynamics of the reason relation; if B is a reason for A, it can cease to be so after further information, and vice versa. This observation points to another weakness of the Basic Reason-Truth Connection (17.13). For (17.13) requires there to be a reason B for the proposition A in question that is true in $@*$, but B may cease to be a reason for A after further evidence becomes available in $@*$. This may happen sometimes, but it should not happen all the time, and there is nothing in (17.13) to prevent that critical situation.

Let us look at the situation a bit more carefully. Consider a true X-proposition A, a true $U^* - \{X\}$-proposition B, a ranking function κ_0, and an infinite sequence of experiences exhausting all the items of evidence available in $@*$ and generating from κ_0 an infinite sequence $\kappa_1, \kappa_2, \ldots$ of ranking functions. Now three things may happen to the reason relation between B and A: (i) it may vacillate for some time (or no time at all) and then stay on the positive side forever; or (ii) it may vacillate for some time (or no time at all) and then stay on the non-positive side forever; or (iii) it may vacillate forever.

If B is of kind (i), this is fine. If B is of kind (ii) or (iii), this is no tragedy; maybe there are other true reasons for A of kind (i). However, the situation becomes odd, if all true $U^* - \{X\}$-propositions relate to A in the way (ii) or (iii) according to κ_0. How can there still be a true reason for A according to each κ_n, as postulated by (17.13)? If all true $U^* - \{X\}$-propositions are of kind (ii), then this can happen only if there is an infinity of ever-new reasons for A, while all old reasons are burnt. If at least two true $U^* - \{X\}$-propositions are of kind (iii), there need not be a continuous rejection of old reasons for A in the sequence $\kappa_1, \kappa_2, \ldots$; there may instead be a continuous alternation of a finite number of reasons, each of which is, however, denied as a reason for A infinitely many times. Both situations seem utterly odd. Each time, when asked why one should believe A, one rejects the previous reasons and either produces new reasons or returns to old ones that were given up at an earlier stage. These do not seem to be acceptable processes for tracking internal truth. Truth is detectable or believable only under more stable conditions; i.e., only when there are true reasons of kind (i).

There are various ways to express the required stability. One way is to require that belief stabilizes at some time:

(17.16)   For each X-proposition A true in @ and each ranking function κ, A alternates between being believed and not being believed only finitely many times and ends up being believed in the process of conditioning κ by all pieces of evidence actually possible in @*.

Only then can A turn out to be determinately true or determinately false in @*; in the case of infinite vacillation there seems to be no detectable truth about A.

This is indeed a scenario that has been closely studied in formal learning theory (cf. Kelly 1996, 2008). Its setting is much simpler, in a way. There are no reasons, continuously in flux and to be weighed; at each stage of data acquisition there is only one preferred hypothesis among all the hypotheses that are not rejected by the data obtained up to this stage. The problem it answers is this: Provided there is detectable truth at all, i.e., that the preferred hypothesis will stabilize after an arbitrary, but finite, amount of data, which inductive or learning strategies are guaranteed to hit upon the true hypothesis generating the data, whatever it is, after finite time, and which inductive strategies do so most efficiently? Efficiency is not our problem. However, (17.16) shares an analogous presupposition, and declares it to hold a priori. Thus, finite detectability again ties together internal truth and rational ranking functions. (See also Haas (2011, ch. 4) who pursues similar ideas within the framework of belief revision theory, though only regarding empirical significance and not regarding truth.)

However, I have not formally studied which kinds of ranking functions conform to (17.16); this is not an easy matter. I prefer a variant of (17.16), stated not in terms of belief and finite oscillation, but in terms of reasons and their stability. Let us define for any frames U or possibility spaces W:

Definition 17.17: Let $\mathcal{A}$ be an algebra over W, $A, B, C \in \mathcal{A}$, $w \in W$, and κ be a ranking function over $\mathcal{A}$. Then B is a *w-stable reason for A within C* w.r.t. κ iff $w \in A \cap B \cap C$, $B \cap C \neq \emptyset \neq \bar{B} \cap C$, and B is a reason for A w.r.t. κ conditional on any $D \in A$ with $w \in D \subseteq C$ and $B \cap D \neq \emptyset \neq \bar{B} \cap D$. And B is an *ultimately w-stable reason for A* w.r.t. iff B is a *w*-stable reason for A within some C w.r.t. κ.

This definition gives a formal version of the 'good' reasons of kind (i). The result of our discussion was that truth is to be distinguished by those 'good' reasons:

(17.18a)   *The Stable Reason-Truth Connection*: Let $X \in U^*$ be an empirical variable and A be an empirical X-proposition with $@^* \in A$. Then for any rational ranking function κ for U^* there is a $U^* - \{X\}$-proposition B that is an ultimately $@^*$-stable reason for A w.r.t. κ. (That B is true in $@^*$ is part of Definition 17.17.)

In view of (17.16), we might think of strengthening the notion of B being an ultimately stable reason for A, such that the range C within which B is an $@^*$-stable reason is reached after finite evidence. Thus strengthened, (17.18a) is very similar to (17.16). However, its formal study would be as difficult as that of (17.16). This is why I did not add the finiteness clause in (17.18a).

Without that clause (17.18a) may be straightforwardly simplified. If the ultimately stable reason B for A is a true atomic Y-proposition ($Y \neq X$), then it must also be a reason for A given $@^*[U^* - \{X, Y\}]$, which is the strictest true proposition within the range of stability of B. Conversely, if B is a reason for A, given $@^*[U^* - \{X, Y\}]$, then B is an ultimately stable reason for A, namely within $@^*[U^* - \{X, Y\}]$. Where B is true, but not an atomic proposition, then its being an ultimately stable reason for A still requires that it is a reason for A, given $@^*[U^* - \{X, Y\}]$ for some $Y \neq X$. Then, however, $B \cap @^*[U^* - \{X, Y\}]$ is also a reason for A, given $@^*[U^* - \{X, Y\}]$, and it is a true Y-proposition. To summarize, what we have just shown is that we might have stated (17.18a) in the following equivalent, but simpler way:

(17.18b)   *The Stable Reason-Truth Connection*: Let $X \in U^*$ be an empirical variable and A be an empirical X-proposition with $@^* \in A$. Then for any rational ranking function κ for U^* there is a variable $Y \in U^*$ and a Y-proposition B with $@^* \in B$ such that B is a reason for A, given $@^*[U^* - \{X, Y\}]$.

(17.18) postulates true reasons for A given true conditions. (17.8) tells us that in this case there must exist true unconditional reasons for A as well. Thus we have

Assertion 17.19: The Stable Reason-Truth Connection (17.18) entails the Basic Reason-Truth Connection (17.13).

By contrast, the Total Reason-Truth Connection (17.15) and the Stable Reason-Truth Connection (17.18b) are of incomparable strength. Consider a toy frame consisting of three binary variables that are possibly realized, respectively, by A, B, and C. And look at the following ranking function:

κ	$B \cap C$	$B \cap \bar{C}$	$\bar{B} \cap C$	$\bar{B} \cap \bar{C}$
A	0	1	1	0
$\bar{A}$	1	2	2	0

Then it is easily verified that each conjunction of two of A, B, and C is a reason for the third, though neither B given C nor C given B is a reason for A. Hence, (17.15) does not entail (17.18b). Conversely, if you look at the table

κ	$B \cap C$	$B \cap \bar{C}$	$\bar{B} \cap C$	$\bar{B} \cap \bar{C}$
A	0	2	2	0
$\bar{A}$	0	1	1	0

you find that each of A, B, and C is a reason for the second given the third, excessively satisfying (17.18b). Still, $B \cap C$ is not a reason for A, and so (17.15) is not entailed. I have not studied whether more positive assertions are available under suitable conditions.

These are poor beginnings, no more. They show, however, that the theoretical machinery developed in this book provides the means to partially extricate the internal or pragmatic notion of truth from its notorious obscurity. And perhaps these means are more suitable than those available before, because ranking theory best captures the dynamic context that must be considered for an adequate treatment of pragmatic truth. By displaying the connections between the various principles introduced I have shown – this was indeed my goal – that these means generate some potential for rigorous theorizing. If the beginnings are reasonable, this potential should be developed much further.

17.4 Principles of Causality

There is one direction for further development, which is all too tempting and which I want to take up in this final section. If causes are, as extensively argued in Chapter 14, a specific kind of conditional reasons, then the principles concerning the rational a priori structure of reasons stated in the previous two sections should entail a priori principles concerning the structure of causes. We have to explore what is entailed. This need not, and in fact will not, be the classical principle of causality, though we will get close to it. After all, this is a matter of proof, not of positing.

In a way, the principles of rationality and the principles of causality to be derived go hand in hand. However, I have taken care here to separate them. The previous sections (should) look exclusively at epistemic rationality and its relation to truth, which as such have nothing to do with causal matters. Likewise, none of the arguments I advanced in those sections referred in any way to causation. Whatever their implications for causation, they presuppose, of course, the theory developed in Chapter 14, and only in this section are we to make use of this presupposition.

Traditionally, there are two principles of causality, not always clearly distinguished. The one commonly referred to as *the general law of causality* asserts: *equal causes, equal effects*. This does not say how many causal relations there are, maybe very few or even none. It only says that causal relations are generalizable in the form of causal laws.

There may be singular causal constellations occurring only once in this world, but in the space of possibilities this is excluded by the principle. Of course, the general law is indeterminate as long as we do not fix what "equal" means. The general law may well be equated with the principle of the uniformity of nature, the defeasible or unrevisable apriority of which I have thoroughly discussed in Section 12.6. So I need not further dwell upon it.

The other principle, here simply called *the principle of causality*, asserts: each event, or in our terminology, *each fact has a cause* (and that relation is generalizable according to the first principle). This principle is the object of my interest in this section.

These principles have been assumed for centuries, more or less unquestioned, and perhaps in some variations – for instance, with accompanying assumptions about the size of cause and effect, which appear unintelligible today (see, however, Yablo (1992) for an attempt to make sense of the claim that the cause has to be greater than the effect). Kant saw that these principles cannot simply be taken for granted, and famously turned them into synthetic principles a priori provable by a transcendental deduction. His argument was a blatant failure according to Strawson (1966, Sect. III.4), but one may also take a more sympathetic view (cf., e.g., Thöle, forthcoming). I assume that the Kant exegetes have not come to an agreement.

The advent of quantum mechanics was disastrous for the principle of causality. After that, and before the serious conception of probabilistic theories of causation, the principle simply seemed to be empirically refutable (cf. Schlick 1925, pp. 373ff.) and actually refuted. Whatever appeared a priori about it, it must have been an illusion. Since then, the range of opinions has diversified, and lacks conviction. Kantians continued on the Kantian project, something rejected by the empiricist mainstream because it had a poor notion of the a priori. Evolutionary epistemology tried to turn the principle of causality into an evolutionary a posteriori rule, which may well turn out to be false even if we are born with it (cf. Lorenz 1941). Maybe the principle is even analytic. A nice example is Davidson (1969). He concluded (though in Davidson (1985) he withdrew that conclusion) that an event is individuated by the set of its causes and effects. This entails that there can be at most one event without causes and effects – almost the principle of causality. (Davidson does not mention or discuss this consequence; presumably it appeared too cheap to him.) Maybe scientists must believe – or at least not disbelieve – in the principle of causality, perhaps not in the a priori sense, but as a motivation for their demanding business (cf., e.g., Reichenbach 1951, Ch. 6, or Pap 1955, pp. 138f.). Popper (1934/89, p. 33 and pp. 195f.) turned it into a research maxim: never give up searching for causes (even if there is no guarantee that you will find them). And many, of course, still accept the apparent lessons of quantum mechanics.

When I look at the contemporary discussion, I find the situation strangely acquiescent; the principles of causality simply no longer play a prominent role. At least I can say this with some confidence for the mainstream of causal theorizing. Looking at the many important books that have so ingeniously advanced the theory of causation in the last 20 years, I do not find any deeper attempt at coming to grips with those principles. This section would already serve its purpose, if it were able to revive this discussion.

Perhaps the impression was that the principle of causality has been superseded by Reichenbach's common cause principle, which indeed plays a central role in the contemporary literature and is vigorously discussed there (cf. also my Sections 14.8 and 14.10). Presumably it is also more fashionable because of its probabilistic nature. Indeed it looks very similar to the principle of causality. Its claim that each correlation has a causal origin (either by being causal in itself or by springing from a common cause) resembles the claim that each fact has a cause.

However, I see a sharp difference between the two principles. Reichenbach's common cause principle has some (cor)relation as a premise, and then claims it to be of causal origin. This is why I was able to argue in Section 14.8 that the common cause principle is analytic, i.e., derivable from my explication of causation. This was true at least of my frame-relative sense of causation, but it extended to the absolute sense in the universal frame (cf. Section 14.9). By contrast, the principle of causality has no such premise; it unconditionally claims some causal relations to exist. Therefore it cannot be analytic and simply derive from some definitions. So we have to study the principle of causality itself; the common cause principle is not an adequate substitute.

Once more, this study will have to move within the ill-defined universal frame U^*. We certainly cannot expect the existence of causal relations, as claimed by the variants of the principle of causality we will consider, to be guaranteed within restricted frames. And we have to assume U^* to be somehow temporally ordered. However, in view of Section 14.10 a weak temporal order $\leq$ will do.

We may first observe some immediate causal consequences of Section 17.2; I already suggested them in the paragraph after (17.7). We may begin by stating

(17.20)   *The Very Weak Principle of Causality*: For any variable $X \in U^*$ and any empirical X-proposition A, A has some direct cause or some direct effect in some possible world $w \in W^*$ w.r.t. any rational ranking function κ for U^*.

If we neglect the difference between the holding of (17.20) for any or only for some X-proposition A – a difference that disappears when X is binary – we may equivalently express (17.20) as claiming that for each empirical variable X there is another empirical variable Y, such that one directly causally depends on the other (cf. Definition 14.24). And then we can verify

Assertion 17.21: The Special Coherence Principle (17.6) and the Very Weak Principle of Causality are equivalent (within the universal frame U^*).

This is an immediate consequence of Theorem 7.10. (Cf. also Spohn 1983b, sect. 6.3, my first attempt to address these matters.)

Similarly, we may state the following stronger principle:

(17.22) *The Unity of the Causal Nexus*: The universal causal graph $\langle U^*, \to^* \rangle$ is connected (in the sense of Definition 7.20g).

And then we may conclude:

Assertion 17.23: The General Coherence Principle (17.7) and the Unity of the Causal Nexus (17.22) are equivalent (within the universal frame U^*).

So, as expected, the unity of science (that I claimed to be captured by the General Coherence Principle) and the unity of the causal nexus go hand in hand. Note that (17.20) and (17.22) speak about causal dependence in the absolute, not in the frame-relative sense (according to which it may also mean causal join in the absolute sense; cf. 14.42).

However, as indicated by the label, (17.20) is very weak; it only guarantees that each atomic empirical proposition has some causal connections in some possible world, not necessarily in the actual one. We would certainly like to establish something stronger, namely:

(17.24) *The Weak Principle of Causality*: For any variable $X \in U^*$ and any empirical X-fact A in @*, A has some direct cause or some direct effect in @* w.r.t. any rational ranking function κ for U^*.

This is still not the traditional principle of causality that claims the existence of actual causes, not only actual causes or effects. However, this is as far as we can get; as I will explain below, I presently see no way of going beyond (17.24).

How might we establish (17.24)? Well, (17.24) is strikingly similar to the Stable Reason-Truth Connection (17.18). This is why I took some care to introduce and defend (17.18) on purely epistemological grounds. However, there is a crucial gap between the two principles. (17.24) postulates for each A some direct cause or effect B. So each is a reason for the other, given the rest of the past of the later one. By contrast, (17.18) postulates for each A some atomic fact B, such that each is a reason for the other, given the rest of the world, past *and* future. This gap is a real one. (17.18) and (17.24) are logically independent.

To illustrate this gap, consider again the toy frame consisting of three binary variables that are possibly realized, first by A, then by B, and finally by C. And look at the following ranking function:

κ	$B \cap C$	$B \cap \bar{C}$	$\bar{B} \cap C$	$\bar{B} \cap \bar{C}$
A	0	1	1	1
$\bar{A}$	1	2	2	0

Here, A is a direct cause of B, and B is a direct cause of C, and hence a reason for C given A. However, neither B nor C is a reason for A given the other. This situation satisfies (17.24), but violates (17.18). Conversely, look at the table:

κ	$B \cap C$	$B \cap \bar{C}$	$\bar{B} \cap C$	$\bar{B} \cap \bar{C}$
A	0	0	1	0
$\bar{A}$	0	0	0	0

Here, A is a reason for B given C and B a reason for C given A, and vice versa. So, (17.18) is satisfied. Also B is a direct cause of C. However, neither B nor C is a direct effect of A. Hence, A has no direct effects and thus violates (17.24). Once more we see that the behavior of the conditional reason relation is quite complex.

The situation is not desperate, though. Additional assumptions help to close the gap. Many assumptions might do, and one might try to find assumptions that are as weak as possible. Here I am content with presenting at least one that is simple, but strong enough to immediately entail the equivalence of the two principles under consideration. I am thinking of our old friend, condition TIDC (15.12), that required the temporal immediacy of direct causation and already helped us to objectify causal relations in Sections 15.4 and 15.5. So, here is the final theorem of this book:

Assertion 17.25: Given condition TIDC, the Stable Reason-Truth Connection (17.18) and the Weak Principle of Causality (17.24) are equivalent (within the universal frame U^*).

Proof: Since $\le$ is a weak temporal order, any ranking function κ for U^* generates some chain graph $\langle U^*, \to^*, \leftrightarrow^* \rangle$. Thus for each $X \in U^*$:

(a) $X \perp_\kappa \{\le X, \ne X\} - bd(X) \mid bd(X)$ (cf. 14.53a).

If $tip(X)$ denotes the set of variables temporally immediately preceding X, TIDC entails that $bd(X) \subseteq tip(X)$. Therefore, (a) entails:

(b) $X \perp_\kappa \{\le X, \ne X\} - tip(X) \mid tip(X)$.

Since this holds for all simultaneous X, we indeed have, with the help of *intersection*, the graphoid property (7.8f):

(c) $\{\approx X\} \perp_\kappa \{< X\} - tip(X) \mid tip(X)$.

Since this holds for all stages $\{\approx X\}$ or $tip(X)$, *contraction*, the graphoid property (7.8e), finally allows us to infer:

(d) $\{> X\} \perp_\kappa \{< X\} \mid \{\approx X\}$.

In other words, TIDC has the consequence that the stages of U^* form a Markov chain, in which the present renders independent the past and the future. Finally, (d) immediately entails that for each $X, Y \in U^*$ with $X \le Y$ and each X-proposition A and Y-proposition B with $@ \in A$, B A is a reason for B (and vice versa) given $^{@*}[U^* - \{X, Y\}]$ if and only if A is a reason for B given $^{@*}[\le Y, \ne X, Y]$. This is so, because (d) says that $^{@*}[> Y]$ is an independent conjunct that does not make a difference if added to the condition.

Relative to a well-defined frame U, (17.25) could be declared to be a theorem. In the given context, however, I have to admit that the argument for (17.25) is not entirely cogent. For instance, TIDC and the proof presuppose that the temporal order of the frame is discrete; but this is a doubtful assumption about the universal frame U^*. So, there is certainly room for improvement.

The proof also shows that TIDC is in effect quite a strong premise. This may be a motive for seeking for weaker assumptions. On the other hand, TIDC is natural, without doubt. And it is at least interesting that the very condition that was crucial for the objectification of causation becomes important again in the present context. Recall at this point that (15.11) provided another possible entanglement of issues of objectification with principles of causality, indeed even with the traditional principle. However, I am not prepared to further discuss or assess the deeper significance of this entanglement.

Let me only summarize what all this means for the status of the Weak Principle of Causality (17.24). I claimed that the Stable Reason-Truth Connection (17.18) is unrevisably a priori. So, the status of (17.24) depends on the quality of the derivation, i.e., of the additional premise TIDC. It is tempting to take TIDC as analytic. However, in my discussion of (14.3) in Section 14.4 I argued that TIDC had better not be made part of the concept of causation, at least in my frame-relative sense. Whether this attitude should change with respect to causation taken in the absolute sense is not so clear. In Sections 15.4–15.5, I argued instead that TIDC is required for objectification. If so, the Weak Principle of Causality would be a consequence of objectification. Maybe it is unrevisably a priori that objectification is possible. This is a good question, which I did not discuss. In this case, the Weak Principle of Causality would also be unrevisably a priori.

Is there hope of strengthening the Weak Principle of Causality to the traditional principle? Well, the argument given so far suggests that this would be possible if we could strengthen the Stable Reason-Truth Connection to the effect that there are not just some true, ultimately stable reasons for each fact, but that indeed some such reasons are already provided by the past of each fact. However, I cannot see why we should assume anything like that for epistemological reasons. In any case, considerations of the kind presented in Sections 17.2 and 17.3 do not justify such stronger assumptions. This is why I am not optimistic about the traditional principle. Indeed, as the German word for reality, "Wirklichkeit", suggests, for something to be real, it must stand in some causal relations, even if these are provided only by effects and not by causes. So we might perhaps be content with the Weak Principle of Causality (17.24); it is miraculous enough that it should be demonstrable as an unrevisably a priori principle (if TIDC is so).

Let me return to my ongoing comparison of probabilities and ranks for a final time. Most of my considerations in this chapter – certainly those of Section 17.2 and this section – could have also been presented in probabilistic terms. However, this would presumably have diminished their compelling force. Moreover, the transferability is much less clear with respect to the ideas of Section 17.3 (and hence also with respect to (17.25)). After all, internal or pragmatic truth is related to belief and believability, and that relation translates badly into probability. So, this seems to be a further field where ranking theory is philosophically more fruitful than probability theory.

I confess I am not very sure of what I have done in the last three sections. Apparently, I was after some deep first principles of philosophy. If I should have succeeded, and in particular succeeded in reviving the rational, normative source of the a priori, then the inquiry should be carefully elaborated and extended. The present study was a mere beginning and not yet satisfactory at all, though it suggested various ways of doing better. Maybe, though, the verdict is that I have lost myself in wild speculations – too wild, indeed. In that case, I think that after so many pages of hard analytic efforts and sober, often pedantic, argument it should be legitimate to spend a few pages with another genre. At least, the speculations may be stimulating. Sometimes, though, it appears to me as if I have merely reshuffled the rich conceptual machinery developed in this book in order to produce platitudes. If so, I myself would have fallen victim to my slogan at the end of Section 6.4: that depth and triviality are close neighbors in philosophy. Then, however, one might say that one kind of progress in philosophy just consists in the transformation of the profound into the trivial. In any case, this is a good point to bring this overly long book to an end.

Bibliography

Aczél, Janos (1966), *Lectures on Functional Equations and Their Applications*, New York: Academic Press.

Adams, Ernest W. (1975), *The Logic of Conditionals*, Dordrecht: Reidel.

Albert, Hans (1991), *Traktat über die kritische Vernunft*, Tübingen: Mohr.

Alchourrón, Carlos E., Peter Gärdenfors, David Makinson (1985), "On the Logic of Theory Change: Partial Meet Functions for Contraction and Revision", *Journal of Symbolic Logic* 50, 510–530.

Anderson, John (1938), "The Problem of Causality", *Australasian Journal of Philosophy* 2, 127–142.

Anderson, Terence, David A. Schum, William Twining (2005), *Analysis of Evidence (Law in Context)*, Cambridge: Cambridge University Press, 2nd edn.

Andersson, Steen A., David Madigan, Michael D. Perlman (1997), "A Characterization of Markov Equivalence Classes for Acyclic Digraphs", *Annals of Statistics* 25, 505–541.

Arló-Costa, Horacio (2000), "Qualitative and Probabilistic Models of Full Belief", in: S. R. Buss, P. Hájek, P. Pudlák (eds.), *Logic Colloquium '98. Lecture Notes in Logic 13*, Natick, MA: A. K. Peters, pp. 25–43.

—— (2001a), "Bayesian Epistemology and Epistemic Conditionals: On the Status of the Export-Import Laws", *Journal of Philosophy* 98, 555–598.

—— (2001b), "Hypothetical Revision and Matter-of-Fact Supposition", *Journal of Applied Non-classical Logics* 11, 203–229.

——, Richmond H. Thomason (2001), "Iterative Probability Kinematics", *Journal of Philosophical Logic* 30, 479–524.

Armstrong, David M. (1968), *A Materialist Theory of the Mind*, London: Routledge & Kegan Paul.

Aronson, Jerrold L. (1971), "On the Grammar of 'Cause'", *Synthese* 22, 414–430.

Audi, Robert (2003), *Epistemology. A Contemporary Introduction to the Theory of Knowledge*, New York: Routledge, 2nd edn.

Austin, John L. (1962), *Sense and Sensibilia*, Oxford: Clarendon Press.

Ayer, Alfred J. (1940), *The Foundations of Empirical Knowledge*, London: Macmillan.

Bacon, Francis (1620), *Novum Organum*.

Bartelborth, Thomas (1996), *Begründungsstrategien. Ein Weg durch die analytische Erkenntnistheorie*, Berlin: Akademie Verlag.

Barwise, Jon, John Perry (1983), *Situations and Attitudes*, Cambridge, Mass.: MIT Press.

Baumgartner, Michael (2008a), "The Causal Chain Problem", *Erkenntnis* 69, 201–226.

—— (2008b), "Regularity Theories Reassessed", *Philosophia*, 36, 327–354.

—— (2009), "Uncovering Deterministic Causal Structures: A Boolean Approach", *Synthese* 170, 71–96.

——, Gerd Graßhoff (2004), *Kausalität und kausales Schließen. Eine Einführung mit interaktiven Übungen*, Bern: Bern Studies in the History and Philosophy of Science.

Beauchamp, Tom L., Alexander Rosenberg (1981), *Hume and the Problem of Causation*, Oxford University Press.

Beebee, Helen (2006), *Hume on Causation*, London: Routledge.

Benkewitz, Wolfgang (1999), "Belief Justification and Perception", *Erkenntnis* 50, 193–208.

—— (2011), *Wahrnehmen, Glauben und Gegenstände. Eine historisch-systematische Untersuchung*, Heidelberg: Synchron Wissenschaftsverlag der Autoren.

Bennett, Jonathan (1976), *Linguistic Behavior*, Cambridge: Cambridge University Press (reissued Indianapolis: Hackett, 1990).

Bennett, Jonathan (1988), *Events and Their Names*, Oxford: Clarendon Press.

—— (2003), *A Philosophical Guide to Conditionals*, Oxford: Oxford University Press.

Berkeley, George (1713), *Three Dialogues Between Hylas and Philonous*.

Besag, Julian (1974), "Spatial Interaction and the Statistical Analysis of Lattice Systems", *Journal of the Royal Statistical Society, Series B*, 36, 302–339.

Binkley, Robert W. (1968), "The Surprise Examination in Modal Logic", *Journal of Philosophy* 65, 127–136.

Blackburn, Simon (1980), "Opinions and Chances", in: D. H. Mellor (ed.), *Prospects for Pragmatism*, Cambridge: Cambridge University Press, pp. 175–196; also in: Blackburn (1993), pp. 75–93.

—— (1993), *Essays in Quasi-Realism*, Oxford: Oxford University Press.

Block, Ned (1990), "Inverted Earth", *Philosophical Perspectives* 4, 53–79.

—— (1994), "Consciousness", in: S. Guttenplan (ed.), *A Companion to the Philosophy of Mind*, Oxford, pp. 210–219.

——, Owen Flanagan, Güven Güzeldere (eds.) (1997), *The Nature of Consciousness: Philosophical Debates*, Cambridge, MA: MIT Press.

Bochman, Alexander (1999), "Contraction of Epistemic States: A General Theory", in: M.-A. Williams, H. Rott (eds.), *Frontiers in Belief Revision*, Dordrecht: Kluwer, pp. 195–220.

Bolker, Ethan D. (1966), "Functions Resembling Quotients of Measures", *Transactions of the American Mathematical Society* 124, 292–312.

BonJour, Laurence (1985), *The Structure of Empirical Knowledge*, Cambridge, Mass.: Harvard University Press.

—— (2007), "Epistemological Problems of Perception", *Stanford Encyclopedia of Philosophy*, http://plato.stanford.edu/entries/perception-episprob/

Booth, Richard, Thomas Meyer (2006), "Admissible and Restrained Revision", *Journal of Artificial Intelligence Research* 26, 127–151.

Boutilier, Craig (1993), "Revision Sequences and Nested Conditionals", in: R. Bajcsy (ed.), *IJCAI 93 – Proceedings of the 13th International Joint Conference on Artificial Intelligence*, San Mateo, CA: Morgan Kaufmann, pp. 519–525.

Bovens, Luc, Stephan Hartmann (2004), *Bayesian Epistemology*, Oxford: Oxford University Press.

Boynton, Robert M. (1979), *Human Color Vision*, New York: Holt, Rinehart and Winston.

Brafman, Ronen I., Moshe Tennenholtz (2000), "An Axiomatic Treatment of Three Qualitative Decision Criteria", *Journal of the Association of Computing Machinery* 47, 452–482.

Brandom, Robert (1994), *Making It Explicit*, Cambridge, Mass.: Harvard University Press.

Breiman, Leo (1968), *Probability*, Reading, Mass.: Addison-Wesley.

Bunzl, Martin (1979), "Causal Overdetermination", *Journal of Philosophy* 76, 134–150.

Burge, Tyler (1979), "Individualism and the Mental", in: P. A. French, T. E. Uehling jr., H. K. Wettstein (eds.), *Midwest Studies in Philosophy, Vol. IV: Metaphysics*, Minneapolis: University of Minnesota Press, pp. 73–121.

Burge, Tyler (1986), "Individualism and Psychology", *Philosophical Review* 95, 3–45.

—— (1997), "Two Kinds of Consciousness", in: Block et al. (1997), pp. 427–434.

Canfield, John, Keith Lehrer (1961), "A Note on Prediction and Induction", *Philosophy of Science* 28, 204–211.

Cantwell, John (1997), "On the Logic of Small Changes in Hypertheories", *Theoria* 63, 54–89.

Carnap, Rudolf (1928), *Der logische Aufbau der Welt*, Hamburg: Felix Meiner.

—— (1936/37), "Testability and Meaning", *Philosophy of Science* 3 (1936), 419–471, and 4 (1937), 1–40.

—— (1947), *Meaning and Necessity*, University of Chicago Press, Ill.: 2nd edn. 1956.

—— (1950, 1962), *The Logical Foundations of Probability*, University of Chicago Press, Ill.: 2nd edn. 1962.

—— (1952), *The Continuum of Inductive Methods*, University of Chicago Press, Ill.

—— (1956), "The Methodological Character of Theoretical Concepts", in: H. Feigl, M. Scriven (eds.), *Minnesota Studies in the Philosophy of Science, Vol. I*, Minneapolis: University of Minnesota Press, pp. 38–76.

—— (1971a), "Inductive Logic and Rational Decisions", in: Carnap, Jeffrey (1971), pp. 5–31.

—— (1971b), "A Basic System of Inductive Logic, Part I", in: Carnap, Jeffrey (1971), pp. 33–165.

—— (1980), "A Basic System of Inductive Logic, Part II", in: Jeffrey (1980), pp. 7–155.

——, Richard C. Jeffrey (eds.), *Studies in Inductive Logic and Probability, Vol. I*, Berkeley: University of California Press.

Cartwright, Nancy (1979), "Causal Laws and Effective Strategies", *Noûs* 13: 419–437; also in Cartwright (1983), pp. 21–43.

—— (1983), *How the Laws of Physics Lie*, Oxford: Clarendon Press.

—— (1989), *Nature's Capacities and Their Measurement*, Oxford: Clarendon Press.

—— (1993), "Marks and Probabilities: Two Ways to Find Causal Structure", in: F. Stadler (ed.), *Scientific Philosophy: Origins and Development. Vienna Circle Institute Yearbook* 1, Dordrecht: Kluwer, pp. 113–119.

—— (1999), *The Dappled World*, Oxford: Clarendon Press.

—— (2007), *Hunting Causes and Using Them. Approaches in Philosophy and Economics*, Cambridge: Cambridge University Press.

Chalmers, David J. (1996), *The Conscious Mind*, Oxford: Oxford University Press.

—— (2006), "The Foundations of Two-Dimensional Semantics", in: García-Carpintero, Macia (2006), pp. 55–140.

Chisholm, Roderick M. (1957), *Perceiving. A Philosophical Study*, Ithaca, N.Y.: Cornell University Press.

Coates, Paul (2000), "Deviant Causal Chains and Hallucinations: A Problem for the Anti-Causalist", *Philosophical Quarterly* 50, 320–331.

Cohen, L. Jonathan (1970), *The Implications of Induction*, London: Methuen.

—— (1977), *The Probable and the Provable*, Oxford: Oxford University Press.

—— (1980), "Some Historical Remarks on the Baconian Conception of Probability", *Journal of the History of Ideas* 41, 219–231.

Collingwood, Robin G. (1940), *An Essay on Metaphysics*, Oxford: Oxford University Press.

Collins, John (2000), "Preemptive Prevention", *Journal of Philosophy* 97, 223–234; also in: Collins et al. (2004), pp. 107–117.

——, Ned Hall, Laurie A. Paul (eds.) (2004), *Causation and Counterfactuals*, Cambridge, Mass.: MIT Press.

Cowell, Robert G., A. Philip Dawid, Steffen L. Lauritzen, David J. Spiegelhalter (1999), *Probabilistic Networks and Expert Systems*, Berlin: Springer.

Cox, Richard T. (1946), "Probability, Frequency, and Reasonable Expectation", *American Journal of Physics* 14, 1–13.

—— (1961), *The Algebra of Probable Inference*, Baltimore Johns Hopkins University Press.

Darroch, John N., Steffen L. Lauritzen, Terry P. Speed (1980), "Markov Fields and Log-Linear Interaction Models for Contingency Tables", *Annals of Statistics* 8, 522–529.

Darwiche, Adnan, Judea Pearl (1997), "On the Logic of Iterated Belief Revision", *Artificial Intelligence* 89, 1–29.

Davidson, Donald (1969), "The Individuation of Events", in N. Rescher (ed.), *Essays in Honor of Carl G. Hempel*, Dordrecht: Reidel, pp. 216–234.

—— (1971), "Agency", in: R. Binkley, R. Bronaugh, A. Marras (eds.), *Agent, Action, and Reason*, Toronto: University of Toronto Press, pp. 3–25.

—— (1985), "Reply to Quine on Events", in: E. LePore, B. McLaughlin (eds.), *Actions and Events: Perspectives on the Philosophy of Donald Davidson*, Oxford: Blackwell, pp. 172–176.

Davies, Martin, Daniel Stoljar (eds.) (2004), *The Two-Dimensional Framework and its Applications: Metaphysics, Language, and Mind*, Special Issue of: *Philosophical Studies* 118, Nos. 1–2.

Dawid, A. Philip (1979), "Conditional Independence in Statistical Theory", *Journal of the Royal Statistical Society, Series B*, 41, 1–31.

de Finetti, Bruno (1937), "La Prévision: Ses Lois Logiques, Ses Sources Subjectives", *Annales de l'Institut Henri Poincaré* 7. Engl. translation: "Foresight: Its Logical Laws, Its Subjective Sources", in: H. E. Kyburg jr., H. E. Smokler (eds.), *Studies in Subjective Probability*, New York: John Wiley & Sons 1964, pp. 93–158.

Dennett, Daniel C. (1988), "Quining Qualia", in: A. Marcel, E. Bisiach (eds.), *Consciousness and Contemporary Science*, Oxford: Oxford University Press, pp. 42–77.

Domotor, Zoltan (1969), *Probabilistic Relational Structures and Their Application*, Technical Report No. 144, Institute for the Mathematical Studies in the Social Sciences, Stanford University, Stanford.

Douven, Igor, Timothy Williamson (2006), "Generalizing the Lottery Paradox", *British Journal for the Philosophy of Science* 57, 755–779.

Dowe, Phil (1992), "Wesley Salmon's Process Theory of Causality and the Conserved Quantity Theory", *Philosophy of Science* 59, 195–216.

—— (2000), *Physical Causation*, Cambridge: Cambridge University Press.

Doyle, Jon (1979), "A Truth Maintenance System", *Artificial Intelligence* 12, 231–272.

—— (1992), "Reason Maintenance and Belief Revision. Foundation vs. Coherence Theories", in: P. Gärdenfors (ed.), *Belief Revision*, Cambridge: Cambridge University Press, pp. 29–51.

Dretske, Fred (1970), "Epistemic Operators", *Journal of Philosophy* 67, 1007–1023.

—— (1977), "Referring to Events", *Midwest Studies in Philosophy* 2, 90–99.

Dubois, Didier, Henri Prade (1988), *Possibility Theory: An Approach to Computerized Processing of Uncertainty*, New York: Plenum Press.

——, —— (1998), "Possibility Theory: Qualitative and Quantitative Aspects", in: Gabbay, Smets (1998), pp. 169–226.

Dummett, Michael (1964), "Bringing About the Past", *Philosophical Review* 73, 338–359.

Earman, John (1992), *Bayes or Bust? A Critical Examination of Bayesian Confirmation Theory*, Cambridge, Mass.: MIT Press.

——, John Roberts (1999), "*Ceteris Paribus*, There is No Problem of Provisos", *Synthese* 118, 439–478.

——, John Roberts, Sheldon Smith (2002), "*Ceteris Paribus* Lost", *Erkenntnis* 57, 281–301.

Edgington, Dorothy (1995), "On Conditionals", *Mind* 104, 235–327.

Eells, Ellery, Elliott Sober (1983), "Probabilistic Causality and the Question of Transitivity", *Philosophy of Science* 50, 35–57.

Ehrlich, Philip (ed.) (1994), *Real Numbers, Generalization of the Reals, and Theories of Continua*, Dordrecht: Kluwer.

Ellis, Brian (1976), "Epistemic Foundations of Logic", *Journal of Philosophical Logic* 5, 187–204.

—— (1979), *Rational Belief Systems*, Oxford: Blackwell.

—— (1990), *Truth and Objectivity*, Oxford: Blackwell.

Ernst, Gerhard (2007), *Einführung in die Erkenntnistheorie*, Darmstadt: Wissenschaftliche Buchgesellschaft.

Esfeld, Michael (2001), *Holism in Philosophy of Mind and Philosophy of Physics*, Dordrecht: Kluwer.

Fair, David (1979), "Causation and the Flow of Energy", *Erkenntnis* 14, 219–250.

Feigl, Herbert, Grover Maxwell (eds.) (1962), *Scientific Explanation, Space, and Time. Minnesota Studies in the Philosophy of Science, Vol. III*, Minneapolis: University of Minnesota Press.

Feldman, Richard (1985), "Reliability and Justification", *Monist* 68, 159–174.

Fermé, Eduardo, Hans Rott (2004), "Revision by Comparison", *Artificial Intelligence* 157, 5–47.

Festa, Roberto (1997), "Analogy and Exchangeability in Predictive Inferences", *Erkenntnis* 45, 229–252.

Feyerabend, Paul (1962), "Explanation, Reduction, and Empiricism", in: Feigl, Maxwell (1962), pp. 28–97.

Field, Hartry (1978), "A Note on Jeffrey Conditionalization", *Philosophy of Science* 45, 361–367.

—— (1996), "The A Prioricity of Logic", *Proceedings of the Aristotelian Society, New Series* 96, 359–379.

Fine, Terrence L. (1973), *Theories of Probability*, New York: Academic Press.

Fishburn, Peter C. (1964), *Decision and Value Theory*, New York: Wiley.

—— (1970), *Utility Theory for Decision Making*, New York: Wiley.

Fitch, Frederic B. (1963), "A Logical Analysis of Some Value Concepts", *Journal of Symbolic Logic* 28, 135–142.

Fitelson, Branden (2001), *Studies in Bayesian Confirmation Theory*, Ph.D. Dissertation, Department of Philosophy, University of Wisconsin, Madison.

Fodor, Jerry A. (1975), *The Language of Thought*, Hassocks: Harvester.

—— (1987), *Psychosemantics. The Problem of Meaning in the Philosophy of Mind*, Cambridge, Mass.: MIT Press.

—— (1991), "You Can Fool Some of the People All the Time, Every Else Being Equal: Hedged Laws and Psychological Explanations", *Mind* 100, 19–34.

—— (1998), *Concepts: Where Cognitive Science Went Wrong*, Oxford: Clarendon Press.

Freund, John E. (1965), "Puzzle or Paradox?" *American Statistician* 19 (4), 29–44.

Friedman, Michael (2001), *Dynamics of Reason*, Standford: CSLI Publications.

Frydenberg, Morten (1990), "The Chain Graph Markov Property", *Scandinavian Journal of Statistics* 17, 333–353.

Fuhrmann, André (1988), *Relevant Logics, Modal Logics, and Theory Change*, Ph.D. Thesis, Australian National University, Canberra.

—— (1991), "Theory Contraction Through Base Contraction", *Journal of Philosophical Logic* 20, 175–203.

—— (1997), *An Essay on Contraction*, Stanford: CSLI Publications.

——, Sven Ove Hansson (1994), "A Survey of Multiple Contractions", *Journal of Logic, Language, and Information* 3, 39–76.

Gabbay, Dov M., Christopher J. Hogger, John A. Robinson (eds.) (1994), *Handbook of Logic in Artificial Intelligence and Logic Programming, Vol. 3: Nonmonotonic Reasoning and Uncertain Reasoning*, Oxford: Oxford University Press.

——, Philippe Smets (eds.) (1998–2000), *Handbook of Defeasible Reasoning and Uncertainty Management Systems*, Vol. 1–5, Dordrecht: Kluwer.

Gaifman, Haim (1988), "A Theory of Higher Order Probabilities", in: B. Skyrms, W. L. Harper (eds.), *Causation, Chance, and Credence*, Dordrecht: Kluwer, pp. 191–219.

Garber, Daniel (1980), "Field and Jeffrey Conditionalization", *Philosophy of Science* 47, 142–145.

García-Carpintero, Manuel, Josep Macia (2006), *Two-Dimensional Semantics*, Oxford: Clarendon Press.

Gärdenfors, Peter (1978), "Conditionals and Changes of Belief", in: I. Niiniluoto, R. Tuomela (eds.), *The Logic and Epistemology of Scientific Change* (= *Acta Philosophica Fennica* 30), Amsterdam: North-Holland, pp. 381–404.

—— (1980), "A Pragmatic Approach to Explanations", *Philosophy of Science* 47, 405–423.

—— (1981), "An Epistemic Approach to Conditionals", *American Philosophical Quarterly* 18, 203–211.

—— (1986), "Belief Revisions and the Ramsey Text for Conditionals", *The Philosophical Review* 95, 81–93.

—— (1988), *Knowledge in Flux*, Cambridge, Mass.: MIT Press.

—— (1990), "The Dynamics of Belief Systems: Foundations vs. Coherence Theories", *Revue Internationale de Philosophie* 172, 24–46.

——, Hans Rott (1995), "Belief Revision", in: D. M. Gabbay, C. J. Hogger, J. A. Robinson (eds.), *Handbook of Logic in Artificial Intelligence and Logic Programming, Vol. 4, Epistemic and Temporal Reasoning*, Oxford: Oxford University Press, pp. 35–132.

Gasking, Douglas (1955), "Causation and Recipes", *Mind* 64, 479–487.

Geiger, Dan, Azaria Paz, Judea Pearl (1988), "Axioms and Algorithms for Inferences Involving Probabilistic Independence", Technical Report R-119, Cognitive Systems Laboratory, University of California, Los Angeles, also in: *Information and Computation* 91 (1991), 128–141.

——, Judea Pearl (1988), "Logical and Algorithmic Properties of Conditional Independence and Qualitative Independence", Technical Report R-97-IIL, Cognitive Systems Laboratory, University of California at Los Angeles, 1988, published in: *The Annals of Statistics* 21 (1993), 2001–2021.

Giang, Phan Hong, Prakash P. Shenoy (2000), "A Qualitative Linear Utility Theory for Spohn's Theory of Epistemic Beliefs", in: C. Boutilier, M. Goldszmidt (eds.), *Uncertainity in Artificial Intelligence, Vol. 16*, San Francisco, Morgan Kaufmann, pp. 220–229.

——, —— (2005), "Two Axiomatic Approaches to Decision Making Using Possibility Theory", *European Journal of Operational Research* 162, 450–467.

Gilboa, Itzhak (1987), "Expected Utility with Purely Subjective Non-Additive Probabilities", *Journal of Mathematical Economics* 16, 65–88.

Glennan, Stuart S. (1996), "Mechanisms and the Nature of Causation", *Erkenntnis* 44, 49–71.

Glymour, Clark (2004), "Critical Notice on: James Woodward, Making Things Happen", *British Journal for the Philosophy of Science* 55, 779–790.

——, Richard Scheines, Peter Spirtes, Kevin Kelly (1987), *Discovering Causal Structure. Artificial Intelligence, Philosophy of Science, and Statistical Modeling*, San Diego: Academic Press.

Goldman, Alvin I. (1967), "A Causal Theory of Knowing", *Journal of Philosophy* 64, 357–372.

—— (1979), "What is Justified Belief?", in: G. Pappas (ed.), *Justification and Knowledge*, Dordrecht: Reidel, pp. 1–23.

Goldstein, Matthew (1983), "The Prevision of a Prevision", *Journal of the American Statistical Association* 78, 817–819.

Goldszmidt, Moisés, Judea Pearl (1992a), "Reasoning With Qualitative Probabilities Can be Tractable", in: D. Dubois, M.P. Wellman, B. D'Ambrosio, P. Smets (eds.), *Proceedings of the 8th Conference on Uncertainty in Artificial Intelligence*, San Mateo, CA: Morgan Kaufmann, pp. 112–120.

——, —— (1992b), "Rank-Based Systems: A Simple Approach to Belief Revision, Belief Update, and Reasoning About Evidence and Actions", in: B. Nebel, C. Rich, W. Swartout (eds.), *Proceedings of the Third International Conference on Knowledge Representation and Reasoning*, San Mateo, CA: Morgan Kaufmann, pp. 661–672.

——, —— (1996), "Qualitative Probabilities for Default Reasoning, Belief Revision, and Causal Modeling", *Artificial Intelligence* 84, 57–112.

Good, Irving J. (1961–63), "A Causal Calculus", *British Journal for the Philosophy of Science* 11, 305–318; 12, 43–51; and 13, 88; also in I. J. Good, *Good Thinking: The Foundations of Probability and Its Applications*, Minneapolis, University of Minnesota Press, 1983.

—— (1962), "Subjective Probability as the measure of a Non-measurable Set", in: E. Nagel, P. Suppes, A. Tarski (eds.), *Logic, Methodology and Philosophy of Science*, Stanford: Stanford University Press, pp. 319–329.

—— (1980), "A Further Comment on Probabilistic Causality: Mending the Chain", *Pacific Philosophical Quarterly* 61, 452–454.

Goodman, Nelson (1946), "A Query on Confirmation", *Journal of Philosophy* 43, 383–385.

—— (1955), *Fact, Fiction, And Forecast*, Cambridge, Mass.: Harvard University Press.

Granger, Clive W. J. (1969), "Investigating Causal Relations by Econometric Models and Cross-Spectral Methods", *Econometrica* 37, 424–438.

—— (1980), "Testing for Causality. A Personal Viewpoint", *Journal of Economic Dynamics and Control* 2, 329–352.

Grenda, Vytautas (2008), *Humean and Antihumean Arguments in Contemporary Theories of Causation*, Ph.D. Dissertation, Vilnius University, Lithuania.

Grice, H. Paul (1957), "Meaning", *Philosophical Review* 66, 377–388.

—— (1961), "The Causal Theory of Perception", *Proceedings of the Aristotelian Society, Suppl. Vol.* 35, 121–168.

—— (1991), *The Conception of Value*, Oxford: Oxford University Press.

Haack, Susan (1993), *Evidence and Inquiry. Towards Reconstruction in Epistemology*, Oxford: Blackwell.

Haas, Gordian (2005), *Revision und Rechtfertigung. Eine Theorie der Theorieänderung*, Heidelberg: Synchron Wissenschaftsverlag der Autoren.

—— (2011), *Minimal Verificationism*, unpublished Habilitationsschrift, Universität Bayreuth.

Haas-Spohn, Ulrike (1995), *Versteckte Indexikalität und subjektive Bedeutung*, Berlin: Akademie-Verlag.

——, Wolfgang Spohn (2001), "Concepts Are Beliefs About Essences", in: A. Newen, U. Nortmann, R. Stuhlmann-Laeisz (eds.), *Building on Frege. New Essays on Sense, Content, and Concept*, Stanford: CSLI Publications, pp. 287–316; also in: Spohn (2008a), pp. 305–328.

Haavelmo, Trygve M. (1943), "The Statistical Implications of a System of Simultaneous Equations", *Econometrica* 11, 1–12.

Hacking, Ian (1967), "Slightly More Realistic Personal Probability", *Philosophy of Science* 34, 311–325.

—— (1975), *The Emergence of Probability*, Cambridge: Cambridge University Press.

Haenni, Rolf (2005), "Towards a Unifying Theory of Logical and Probabilistic Reasoning", in: F. B. Cozman, R. Nau, T. Seidenfeld (eds.), *Proceedings of the 4th International Symposium on Imprecise Probabilities and Their Applications*, Pittsburgh, pp. 193–202.

—— (2009), "Non-Additive Degrees of Belief", in: F. Huber, C. Schmidt-Petri (eds.), *Degrees of Belief*, Dordrecht: Springer, pp. 121–159.

Hájek, Alan (2003a), "What Conditional Probability Could not Be", *Synthese* 137, 273–323.

—— (2003b), "Conditional Probability is the Very Guide of Life", in: H. E. Kyburg jr., M. Thalos (eds.), *Probability is the Very Guide of Life*, Chicago: Open Court, pp. 183–203.

Hall, Ned (2000), "Causation and the Price of Transitivity", *Journal of Philosophy* 97, 198–222; also in: Collins et al. (2004), pp. 181–203.

—— (2004), "Two Concepts of Causation", in: Collins et al. (2004), pp. 225–276.

Halpern, Joseph Y. (2003), *Reasoning about Uncertainty*, Cambridge, Mass.: MIT Press.

——, Judea Pearl (2005a), "Causes and Explanations: A Structural-Model Approach, Part I: Causes", *British Journal for the Philosophy of Science* 56, 843–887.

——, Judea Pearl (2005b), "Causes and Explanations: A Structural-Model Approach, Part II: Explanations", *British Journal for the Philosophy of Science* 56, 889–911.

Hammersley, John M., Peter Clifford (1971), "Markov Fields on Finite Graphs and Lattices", unpublished manuscript.

Hansson, Sven Ove (1993), "Reversing the Levi Identity", *Journal of Philosophical Logic* 22, 637–669.

—— (ed.) (1997), *Special Issue on Non-Prioritized Belief Revision. Theoria* 63, 1–134.

—— (1999), *A Textbook of Belief Dynamics. Theory Change and Database Updating*, Dordrecht: Kluwer.

Hardin, Clyde L. (1988), *Color for Philosophers. Unweaving the Rainbow*, Indianapolis: Hackett.

Harper, William L. (1976), "Rational Belief Change, Popper Functions and Counterfactuals", in: W. L. Harper, C. A. Hooker (eds.), *Foundations of Probability Theory, Statistical Inference, and Statistical Theories of Science, Vol. I*, Dordrecht: Reidel, pp. 73–115.

—— (1977), "Rational Conceptual Change", in: F. Suppe, P. D. Asquith (eds.), *PSA 1976, Vol. 2*, East Lansing: Philosophy of Science Association, pp. 462–494.

Hart, Herbert L. A., Tony Honoré (1959), *Causation in the Law*, 2nd edn., Clarendon Press, Oxford 1985.

Hausman, Daniel M., James Woodward (1999), "Independence, Invariance, and the Causal Markov Condition", *British Journal for the Philosophy of Science* 50, 521–583.

Hempel, Carl G. (1945), "Studies in the Logic of Confirmation", *Mind* 54, 1–26 + 97–120; reprinted in Hempel (1965), ch. 1.

Hempel, Carl G. (1948), "Studies in the Logic of Explanation", *Philosophy of Science* 15, 135–147; reprinted in Hempel (1965), ch. 10.

—— (1952), *Fundamentals of Concept Formation in Empirical Science*, in: *International Encyclopedia of Unified Science*, Vol. II, No. 7, Chicago: University of Chicago Press.

—— (1961/62), "Rational Action", *Proceedings and Addresses of the American Philosophical Association* 35, 5–23.

—— (1962), "Deductive-Nomological vs. Statistical Explanation", in: H. Feigl, G. Maxwell (eds.), *Minnesota Studies in the Philosophy of Science, Vol. III, Scientific Explanation, Space, and Time*, Minneapolis: University of Minnesota Press, pp. 98–169.

—— (1965), *Aspects of Scientific Explanation and Other Essays in the Philosophy of Science*, New York: Free Press.

—— (1973), "The Meaning of Theoretical Terms: A Critique of the Standard Empiricist Construal", in: P. Suppes, L. Henkin, A. Joja, G. C. Moisil (eds.), *Logic, Methodology and Philosophy of Science IV*, Amsterdam: North-Holland, pp. 367–378.

—— (1988), "Provisoes: A Problem Concerning the Inferential Function of Scientific Theories", *Erkenntnis* 28, 147–164.

——, Paul Oppenheim (1948), "Studies in the Logic of Explanation", *Philosophy of Science* 15, 135–175; reprinted in Hempel (1965, ch. 10).

Hesslow, Germund (1976), "Two Notes on the Probabilistic Approach to Causality", *Philosophy of Science* 43, 290–292.

Hilbert, D. R. (1987), *Color and Color Perception. A Study in Anthropocentric Realism*, Stanford: CSLI Lecture Notes No. 9.

Hild, Matthias (1997), "A Representation Theorem for Iterated Contraction", unpublished manuscript.

—— (1998a), "Auto-Epistemology and Updating", *Philosophical Studies* 92, 321–361.

—— (1998b), "The Coherence Argument against Conditionalization", *Synthese* 115, 229–258.

—— (2001), *Introduction to Induction: On the First Principles of Reasoning*, unpublished manuscript, CalTech, Pasadena.

——, Wolfgang Spohn (2008), "The Measurement of Ranks and the Laws of Iterated Contraction", *Artificial Intelligence* 172, 1195–1218.

Hintikka, Jaakko (1962), *Knowledge and Belief*, Ithaca, N.Y.: Cornell University Press.

—— (1966), "A Two-Dimensional Continuum of Inductive Methods", in: J. Hintikka, P. Suppes (eds.), *Aspects of Inductive Logic*, Amsterdam: North-Holland, pp. 113–132.

——, Ilkka Niiniluoto (1976), "An Axiomatic Foundation for the Logic of Inductive Generalization", in: M. Przelecki, K. Szaniawski, R. Wójcicki (eds.), *Formal Methods in the Methodology of Empirical Sciences*, Dordrecht: Reidel, pp. 57–81.

Hitchcock, Christopher (1996), "The Role of Contrast in Causal and Explanatory Claims", *Synthese* 107, 395–419.

—— (2001), "The Intransitivity of Causation Revealed in Equations and Graphs", *Journal of Philosophy* 98, 273–299.

Hohenadel, Stefan A. (forthcoming), *Efficient Epistemic Updates in Rank-based Belief Networks*, Ph.D. thesis, University of Konstanz, 2012.

Howard, Ronald A., James E. Matheson (1981), "Influence Diagrams", in: R. A. Howard, J. E. Matheson (eds.), *Readings on the Principles and Applications of Decision Analysis, Vol. 2*, Menlo Park, Ca.: Strategic Decisions Group, pp. 719–762.

Howson, Colin, Peter Urbach (1989), *Scientific Reasoning. The Bayesian Approach*, La Salle, Ill.: Open Court.

Huber, Franz (2006), "Ranking Functions and Rankings on Languages", *Artificial Intelligence* 170, 462–471.

—— (2007), "The Consistency Argument for Ranking Functions", *Studia Logica* 86, 299–329.

Humburg, Jürgen (1971), "The Principle of Instantial Relevance", in: Carnap, Jeffrey (1971), pp. 225–233.

Hume, David (1739), *A Treatise of Human Nature, Book I, Of the Understanding*. (Page numbers refer to the edition of L. A. Selby-Bigge, Oxford: Clarendon Press 1896.)

—— (1748), *An Inquiry Concerning Human Understanding*.

Hunter, Daniel (1990), "Parallel Belief Revision", in: Shachter et al. (1990), pp. 241–251.

—— (1991a), "Graphoids and Natural Conditional Functions", *International Journal of Approximate Reasoning* 5, 489–504.

—— (1991b), "Maximum Entropy Updating and Conditionalization", in: W. Spohn, B. C. van Fraassen, B. Skyrms (eds.), *Existence and Explanation. Essays in Honor of Karel Lambert*, Dordrecht: Kluwer, pp. 45–57.

Isham, Valerie (1981), "An Introduction to Spatial Point Processes and Markov Random Fields", *International Statistical Rview* 49, 21–43.

Jackson, Frank (1998), *From Metaphysics to Ethics*, Oxford: Clarendon Press.

—— (2005), "The Case for A Priori Physicalism", in: C. Nimtz, A. Beckermann (eds.), *Philosophy – Science – Scientific Philosophy. Main Lectures and Colloquia of GAP.5, Fifth International Congress of the Society for Analytical Philosophy, Bielefeld, 22–26 September 2003*, Paderborn: Mentis, pp. 251–265.

Jaffray, Jean-Yves (1989), "Linear Utility Theory for Belief Functions", *Operations Research Letters* 8, 107–112.

Jeffrey, Richard C. (1965/1983), *The Logic of Decision*, Chicago: University of Chicago Press, 2nd edn. 1983.

—— (1971), "Probability Measures and Integrals", in: Carnap, Jeffrey (1971), pp. 167–223.

—— (ed.) (1980), *Studies in Inductive Logic and Probability, Vol. II*, Berkeley: University of California Press.

Jensen, Finn V. (1996), *An Introduction to Bayesian Networks*, London: UCL Press.

—— (2001), *Bayesian Networks and Decision Graphs*, Berlin: Springer.

Joyce, James M. (1998), "A Nonpragmatic Vindication of Probabilism", *Philosophy of Science* 65, 575–603.

—— (1999), *The Foundations of Causal Decision Theory*, Cambridge: Cambridge University Press.

Kahneman, David, Amos Tversky (1979), "Prospect Theory: An Analysis of Decision under Risk", *Econometrica* 47, 263–292.

Kaila, Eino (1941), "Über den physikalischen Realitätsbegriff", *Acta Philosophica Fennica* 3, 33–34.

Kant, Immanuel (1781/87), *Kritik der reinen Vernunft*; english translation: *Critique of Pure Reason*, London: Macmillan 1929.

Kaplan, David (1977), "Demonstratives. An Essay on the Semantics, Logic, Metaphysics, and Epistemology of Demonstratives and Other Indexicals", in: J. Almog, J. Perry, H. Wettstein (eds.), *Themes from Kaplan*, Oxford: Oxford University Press, 1989, pp. 481–563.

Keisler, H. Jerome (1976), *Foundations of Infinitesimal Calculus*, Boston: Prindle, Weber & Schmidt.

Kelly, Kevin T. (1996), *The Logic of Reliable Inquiry*, Oxford: Oxford University Press.

—— (1999), "Iterated Belief Revision, Reliability, and Inductive Amnesia", *Erkenntnis* 50, 11–58.

—— (2000), "The Logic of Success", *British Journal for the Philosophy of Science* 51, 639–666.

—— (2004), "Justification as Truth-Finding Efficiency: How Ockham's Razor Works", *Minds and Machines* 14, 485–505.

Kemmerling, Andreas (1998), "Eine Handvoll Bemerkungen zur begrifflichen Unübersichtlichkeit von 'Bewusstsein' ", in: F. Esken, H.-D. Heckmann (eds.), *Bewusstsein und Repräsentation*, Paderborn: Mentis, pp. 55–71.

—— (2006), "Kripke's Principle of Disquotation and the Epistemology of Belief Ascription", *Facta Philosophica* 8, 119–143.

Kiiveri, Harri, Terri P. Speed, John B. Carlin (1984), "Recursive Causal Models", *Journal of the Australian Mathematical Society, Series A,* 36, 30–52.

Kim, Jaegwon (1973), "Causation, Nomic Subsumption, and the Concept of an Event", *Journal of Philosophy* 70, 217–236.

Kim, Jin H., Judea Pearl (1983), "A Computational Model for Causal and Diagnostic Reasoning in Inference Systems", in: A. Bundy (ed.), *Proceedings of the Eighth International Joint Conference on Artificial Intelligence*, San Mateo: Morgan Kaufmann, pp. 190–193.

Kistler, Max (1998), "Reducing Causality to Transmission", *Erkenntnis* 48, 1–24.

Kitcher, Philip (1985), "Two Approaches to Explanation", *Journal of Philosophy* 82, 632–639.

Klein, Peter (1999), "Human Knowledge and the Infinite Regress of Reasons", *Philosophical Perspectives* 13, 297–332.

Köhler, Eckehart (2004), "Physical Intuition as Inductive Support", in: F. Stadler (ed.), *Induction and Deduction in the Sciences*, Dordrecht: Kluwer, pp. 151–167.

Krantz, David H., R. Duncan Luce, Patrick Suppes, Amos Tversky (1971), *Foundations of Measurement, Vol. I*, New York: Academic Press.

Kripke, Saul A. (1972), "Naming and Necessity", in: D. Davidson, G. Harman (eds.), *Semantics of Natual Language,* Dordrecht: Reidel, pp. 253–355 and 763–769; ext. edn: Oxford: Blackwell 1980.

—— (1982), *Wittgenstein on Rules and Private Language*, Oxford: Blackwell.

Krüger, Lorenz, et al. (1987), *The Probabilistic Revolution.* Vol. 1: *Ideas in History,* Vol. 2: *Ideas in the Sciences*, Cambridge, Mass.: MIT Press.

Kuhn, Thomas S. (1962), *The Structure of Scientific Revolutions*, Chicago: University of Chicago Press, 2nd edn. 1970.

Kuipers, Theo A. F. (1978), *Studies in Inductive Logic and Rational Expectation*, Dordrecht: Reidel.

Künne, Wolfgang, Albert Newen, Martin Anduschus (eds.) (1997), *Direct Reference, Indexicality and Propositional Attitudes*, Stanford: CSLI Publications.

Kuokkanen, Martti (ed.) (1994), *Idealization VII: Structuralism, Idealization and Approximation. Poznan Studies in the Philosophy of the Sciences and the Humanities 42*, Amsterdam: Rodopi.

Kusser, Anna, Wolfgang Spohn (1992), "The Utility of Pleasure is a Pain for Decision Theory", *Journal of Philosophy* 89, 10–29.

Kyburg, Henry E. jr. (1961), *Probability and the Logic of Rational Belief*, Middletown, Conn.: Wesleyan University Press.

—— (1963), "A Further Note on Rationality and Consistency", *Journal of Philosophy* 60, 463–465.

—— (1970), "Conjunctivitis", in: M Swain (ed.), *Induction, Acceptance, and Rational Belief*, Dordrecht: Reidel, pp. 55–82.

Lange, Marc (2000), *Natural Laws in Scientific Practice*, Oxford: Oxford University Press.

—— (2002), "Who's Afraid of *Ceteris Paribus* Laws? Or: How I Learned to Stop Worrying and Love Them", *Erkenntnis* 57, 407–423.

Lauritzen, Steffen L. (1979), *Lectures on Contingency Tables*, University of Copenhagen, Institute of Mathematical Statistics, 2nd edn. Aalborg University Press 1982.

——, A. Philip Dawid, B. N. Larsen, H.-G. Leimer (1990), "Independence Properties of Directed Markov Fields", *Networks* 20, 491–505.

——, Thomas S. Richardson (2002), "Chain Graph Models and Their Causal Interpretations", *Journal of the Royal Statistical Society, Series B*, 64, 321–361 (with discussions).

——, David J. Spiegelhalter (1988), "Local Computations with Probabilities on Graphical Structures and Their Application to Expert Systems", *Journal of the Royal Statistical Society, Series B*, 50, 157–224 (with discussions).

——, Nanny Wermuth (1989), "Graphical Models for Associations Between Variables, Some of Which are Qualitative and Some Quantitative", *Annals of Statistics* 17, 31–57.

Lehman, R. Sherman (1955), "On Confirmation and Rational Betting", *Journal of Symbolic Logic* 20, 251–262.

Lehrer, K. (1971), "Induction and Conceptual Change", *Synthese* 23, 206–225.

—— (1974), "Truth, Evidence, and Inference", *American Philosophical Quarterly* 11, 79–92.

Lehrer, Keith (1990/2000), *Theory of Knowledge*, Boulder, Colorado: Westview Press, 2nd edn 2000.

—— (2003), "Coherence, Circularity and Consistency: Lehrer Replies", in Olsson (2003), pp. 309–356.

Leitgeb, Hannes (2004), *Inference on the Low Level. An Investigation into Deduction, Nonmonotonic Reasoning, and the Philosophy of Cognition*, Dordrecht: Kluwer.

—— (2010), "Reducing Belief Simpliciter to Degrees of Belief", unpublished manuscript, Munich.

Levi, Isaac (1967a), *Gambling With Truth. An Essay on Induction and the Aims of Science*, New York: Knopf.

—— (1967b), "Probability Kinematics", *British Journal for the Philosophy of Science* 18, 197–209.

—— (1977), "Subjunctives, Dispositions, and Chances", *Synthese* 34, 423–455.

—— (1984), *Decisions and Revisions. Philosophical Essays on Knowledge and Value*, Cambridge: Cambridge University Press.

—— (1986), *Hard Choices. Decision Making Under Unresolved Conflict*, Cambridge: Cambridge University Press.

—— (1991), *The Fixation of Belief and its Undoing*, Cambridge: Cambridge University Press.

—— (1996), *For the Sake of Argument*, Cambridge: Cambridge University Press.

—— (2004), *Mild Contraction: Evaluating Loss of Information Due to Loss of Belief*, Oxford: Oxford University Press.

—— (2006), "Replies", in: E. Olsson (ed.), *Knowledge and Inquiry. Essays on the Pragmatism of Isaac Levi*, Cambridge: Cambridge University Press, pp. 327–380.

Lewis, David K. (1969), *Convention: A Philosophical Study*, Cambridge, Mass.: Harvard University Press.

—— (1973a), *Counterfactuals*, Oxford: Blackwell.

Lewis, David K. (1973b), "Causation", *Journal of Philosophy* 70, 556–567.

—— (1976), "Probabilities of Conditionals and Conditional Probabilities", *Philosophical Review* 85, 297–315.

—— (1979a), "Attitudes *De Dicto* and *De Se*", *Philosophical Review* 88, 513–543.

—— (1979b), "Counterfactual Dependence and Time's Arrow", *Noûs* 13, 455–476.

—— (1980), "A Subjectivist's Guide to Objective Chance", in: Jeffrey (1980), pp. 263–293.

—— (1986a), *On the Plurality of Worlds*, Oxford: Blackwell.

—— (1986b), *Philosophical Papers, Vol. II*, Oxford: Oxford University Press.

—— (1994), "Humean Supervenience Debugged", *Mind* 103, 473–490.

—— (1996), "Elusive Knowledge", *Australasian Journal of Philosophy* 74, 549–567.

—— (1997), "Finkish Dispositions", *Philosophical Quarterly* 47, 143–158.

—— (2000/04), "Causation as Influence", *Journal of Philosophy* 97 (2000), 182–197; extended version in: Collins et al. (2004), pp. 75–106 (references are to the extended version).

Lipton, Peter (1991, 2004), *Inference to the Best Explanation*, London: Routledge, 1st edn. 1991, 2nd edn. 2004.

Loève, Michel (1978), *Probability Theory*, Heidelberg: Springer, 4th edn. in two volumes.

Lorenz, Konrad (1941), "Kants Lehre vom Apriorischen im Lichte gegenwärtiger Biologie", *Blätter für Deutsche Philosophie* 15, 94–125.

Luce, R. Duncan, Howard Raiffa (1957), *Games and Decisions*, New York: Wiley.

Machamer, Peter, Lindley Darden, Carl F. Craver (2000), "Thinking About Mechanisms", *Philosophy of Science* 67, 1–25.

Mackie, John L. (1965), "Causes and Conditions", *American Philosophical Quarterly* 2, 245–264.

—— (1974), *The Cement of the Universe*, Oxford: Clarendon Press.

Madsen, Anders L., Finn V. Jensen (1999), "Lazy Propagation: A Junction Tree Inference Algorithm Based on Lazy Evaluation", *Artificial Intelligence* 113, 203–245.

Maher, Patrick (1993), *Betting on Theories*, Cambridge: Cambridge University Press.

—— (2002), "Joyce's Argument for Probabilism", *Philosophy of Science* 69, 73–81.

Makinson, David (1997a), "Screened Revision", *Theoria* 63, 14–23.

—— (1997b), "On the Force of Some Apparent Counterexamples to Recovery", in: E. G. Valdés et al. (eds.), *Normative Systems in Legal and Moral Theory. Festschrift for Carlos Alchourrón and Eugenio Bulygin*, Berlin: Duncker & Humblot, pp. 475–481.

Malzkorn, Wolfgang (2000), "Realism, Functionalism and the Conditional Analysis of Dispositions", *Philosophical Quarterly* 50, 452–469.

Margolis, Eric, Stephen Laurence (eds.) (1999), *Concepts. Core Readings*, Cambridge, Mass.: MIT Press.

Martel, Iain (2003), "Indeterminism and the Causal Markov Condition", Working Paper Series *Philosophy and Probability* No. 4, Philosophy, Probability, and Modeling Research Group, University of Konstanz.

Martin, Charles B. (1994), "Dispositions and Conditionals", *Philosophical Quaterly* 44, 1–8.

Matús, Fero (1988), *Independence and Radon Projection on Compact Groups* (in Slovak), Ph.D. Thesis, Prague.

—— (1994), "Stochastic Independence, Algebraic Independence and Abstract Connectedness", *Theoretical Computer Science A* 134, 445–471.

—— (1999), "Conditional Independences Among Four Random Variables III: Final Conclusion", *Combinatorics, Probability and Computing* 8, 269–276.

McDermott, Michael (1995), "Redundant Causation", *British Journal for the Philosophy of Science* 46, 523–544.

McGee, Vann (1994), "Learning the Impossible", in: E. Eells, B. Skyrms (eds.), *Probability and Conditionals. Belief Revision and Rational Decision*, Cambridge: Cambridge University Press, pp. 179–199.

McGrew, Timothy (2003), "Confirmation, Heuristics, and Explanatory Reasoning", *British Journal for the Philosophy of Science* 54, 553–567.

Meek, Christopher (1995), "Strong Completeness and Faithfulness in Bayesian Networks", in: P. Besnard, S. Hanks (eds.), *Uncertainty in Artificial Intelligence 11*, San Mateo, CA: Morgan Kaufmann, pp. 411–418.

——, Clark Glymour (1994), "Conditioning and Intervening", *British Journal for the Philosophy of Science* 45, 1001–1021.

Menzies, Peter, Huw Price (1993), "Causation as a Secondary Quality", *British Journal for the Philosophy of Science* 44, 187–203.

Merin, Arthur (1996), *Die Relevanz der Relevanz. Fallstudie zur Semantik der englischen Konjunktion "but"*, Habilitationsschrift, University of Stuttgart 1996. Distributed as *Arbeitsberichte des SFB 340*, Nr. 142, Universities of Stuttgart and Tübingen, 1999.

—— (1997), "If All Our Arguments Had To Be Conclusive, There Would Be Few of Them", *Arbeitsberichte des SFB 340*, Nr. 101, Universities of Stuttgart and Tübingen. (Online at www.semanticsarchive.net)

—— (1999), "Information, Relevance and Social Decisionmaking: Some Principles and Results of Decision-Theoretic Semantics", in: L. S. Moss, J. Ginzburg, M. De Rijke (eds.), *Logic, Language, and Computation, Vol. 2*, Stanford, CA: CSLI Publications, pp. 179–221. (Online at www.semanticsarchive.net)

—— (2002), "Unique Evaluative and Inductive Characterizations of Boolean Complementation", *Forschungsberichte der DFG-Forschergruppe Logik in der Philosophie*, No. 76, University of Konstanz.

—— (2003a), "Replacing Horn Scales by Act-Based Relevance Orderings to Keep Negation and Numerals Meaningful", *Forschungsberichte der DFG-Forschergruppe Logik in der Philosophie*, No. 110, University of Konstanz. (Online at www.semanticsarchive.net)

—— (2003b), "Presuppositions and Practical Reason: A Study in Decision-Theoretic Semantics", *Forschungsberichte der DFG-Forschergruppe Logik in der Philosophie*, No. 114, University of Konstanz. (Online at www.semanticsarchive.net)

—— (2005), "Proportion Quantifier Interpretations of Indefinites and Endocentric Relevance Relations", *Belgian Journal of Linguistics* 19, 147–186. (Online at www.semanticsarchive.net)

—— (2006a), "Multilinear Semantics for Double-Jointed Coordinate Constructions", Ms. University of Konstanz. (Online at www.semanticsarchive.net)

—— (2006b), *Decision Theory of Rhetoric*, Monograph manuscript, University of Konstanz.

—— (2008), "Relevance and Reasons in Probability and Epistemic Ranking Theory: A Study in Cognitive Ergonomy", *Forschungsberichte der DFG-Forschergruppe Logik in der Philosophie*, No. 130, University of Konstanz.

Mill, John Stuart (1843), *A System of Logic, Ratiocinative and Inductive*, London: John W. Parker.

Miller, David (1966), "A Paradox of Information", *The British Journal for the Philosophy of Science* 17, 59–61.

Mosteller, Frederick (1965), *Fifty Challenging Problems in Probability with Solutions*, Reading, Mass.: Addison-Wesley.

Mott, Peter (1992), "Fodor and Cetris Paribus Laws", *Mind* 101, 335–346.

Müller, Olaf (2003), *Wirklichkeit ohne Illusionen, Band 1: Hilary Putnam und der Abschied vom Skeptizismus oder Warum die Welt keine Computersimulation ist, Band 2: Metaphysik und semantische Stabilität oder Was es heißt, nach höheren Wirklichkeiten zu fragen*, Paderborn: Mentis.

Mumford, Stephen (1998), *Dispositions*, Oxford: Oxford University Press.

—— (2001), "Realism and the Conditional Analysis of Dispositions: Reply to Malzkorn", *The Philosophical Quarterly* 51, 375–378.

Nayak, Abhaya C. (1994), "Iterated Belief Change Based on Epistemic Entrenchment", *Erkenntnis* 41, 353–390.

——, Randy Goebel, Mehmet A. Orgun (2007), "Iterated Belief Change from First Principles", in: M. Veloso (ed.), *IJCAI-07. Proceedings of the 20th International Joint Conference of Arificial Intelligence*, Menlo Park, CA: AAAI Press, pp. 2568–2573.

Neapolitan, Richard E. (1990), *Probabilistic Reasoning in Expert Systems: Theory and Algorithms*, New York: Wiley.

Nida-Rümelin, Martine (1993), *Farben und phänomenales Wissen*, Wien: VWGÖ.

—— (1996), "Pseudonormal Vision: An Actual Case of Qualia Inversion?", *Philosophical Studies* 82, 145–157.

—— (1997), "The Character of Color Terms: A Phenomenalist View", in: Künne et al. (1997), pp. 381–402.

Niiniluoto, Ilkka (1972), "Inductive Systematization: Definition and a Critical Survey", *Synthese* 25, 25–81.

—— (2004), "Truth-Seeking by Abduction", in Stadler (2004), pp. 57–82.

Nowak, Leszek (1992), "The Idealizational Approach to Science: A Survey", in: J. Brzezinski, L. Nowak (eds.), *Idealization III: Approximation and Truth. Poznan Studies in the Philosophy of the Sciences and the Humanities 25*, Amsterdam: Rodopi, pp. 9–63.

Oddie, Graham (2001), "Truthlikeness", *The Stanford Encyclopedia of Philosophy (Fall 2001 Edition)*, E. N. Zalta (ed.), http://plato.stanford.edu/archives/fall2001/entries/truthlikeness

Okasha, Samir (2000), "Van Fraassen's Critique of Inference to the Best Explanation", *Studies in History and Philosophy of Science* 31, 691–710.

Olsson, Erik J. (1998), "Competing for Acceptance: Lehrer's Rule and the Paradoxes of Justification", *Theoria* 64, 34–54.

—— (ed.) (2003), *The Epistemology of Keith Lehrer*, Dordrecht: Kluwer.

—— (2005), *Against Coherence: Truth, Probability, and Justification*, Oxford: Clarendon Press.

Pap, Arthur (1955), *Analytische Erkenntnistheorie*, Wien: Springer.

Papineau, David (1985), "Probabilities and Causes", *Journal of Philosophy* 82, 57–74.

Peacocke, Christopher (1992), *A Study of Concepts*, Cambridge, Mass.: MIT Press.

Pearl, Judea (1982), "Reverend Bayes on Inference Engines: A Distributional Hierarchical Approach", in: D. L. Waltz (ed.), *National Conference on Artificial Intelligence, Pittsburgh*, Menlo Park, CA: AAAI Press, pp. 133–136.

—— (1988), *Probabilistic Reasoning in Intelligent Systems: Networks of Plausible Inference*, San Mateo, CA: Morgan Kaufmann.

—— (1990), "System Z: A Natural Ordering of Defaults with Tractable Applications to Nonmonotonic Reasoning", in: R. Parikh (ed.), *Theoretical Aspects of Reasoning about Knowledge. Proceedings of the Third Conference*, San Mateo, CA: Morgan Kaufmann, pp. 637–649.

—— (2000), *Causality. Models, Reasoning, and Inference*, Cambridge: Cambridge University Press.

——, Azaria Paz (1985), "Graphoids: A Graph-Based Logic for Reasoning About Relevant Relations", Technical Report R-53-L, Cognitive Systems Laboratory, University of California, Los Angeles. Short version in: B. du Boulay, D. Hogg, L. Steels (eds.), *Advances in Artificial Intelligence 2*, Amsterdam: North-Holland 1987, pp. 357–363.

Pears, David (1990), *Hume's System. An Examination of the First Book of his Treatise*, Oxford: Oxford University Press.

Peirce, Charles S. (1877), "The Fixation of Belief", *Popular Science Monthly* 12, 1–15; reprinted in: M. Fisch (ed.), *Writings of Charles S. Peirce, Vol. 1*, Bloomington: University of Indiana Press 1982.

Perler, Dominik, Ulrich Rudolph (2000), *Occasionalismus. Theorien der Kausalität im arabisch-islamischen und im europäischen Denken*, Göttingen: Vandenhoeck & Ruprecht.

Piantanida, Tom P. (1974), "A Replacement Model of X-linked Recessive Colour Vision Defects", *Annals of Human Genetics* 37, 393–404.

Pietroski, Paul, Georges Rey (1995), "When Other Things Aren't Equal: Saving *Ceteris Paribus* Laws from Vacuity", *British Journal for the Philosophy of Science* 46, 81–110.

Plantinga, Alvin (1993a), *Warrant: The Current Debate*, Oxford: Oxford University Press.

—— (1993b), *Warrant and Proper Function*, Oxford: Oxford University Press.

Pollock, John L. (1986), *Contemporary Theories of Knowledge*, Savage: Rowman & Littlefield.

—— (1990), *Nomic Probability and the Foundations of Induction*, Oxford: Oxford University Press.

—— (1995), *Cognitive Carpentry: A Blueprint for How to Build a Person*, Cambridge, Mass.: MIT Press.

——, Joseph Cruz (1999), *Contemporary Theories of Knowledge*, Lanham, Md: Rowman & Littlefield, 2nd edn.

——, Anthony S. Gillies (2000), "Belief Revision and Epistemology", *Synthese* 122, 69–92.

Popper, Karl R. (1934/89), *Logik der Forschung*, 9th edn., Tübingen: Mohr, 1989; engl. translation: *The Logic of Scientific Discovery*, London: Hutchinson, 1959.

—— (1938), "A Set of Independent Axioms for Probability", *Mind* 47, 275–277.

—— (1955), "Two Autonomous Axiom Systems for the Calculus of Probabilities", *British Journal for the Philosophy of Science* 6, 51–57.

Price, Huw (1991), "Agency and Probabilistic Causality", *British Journal for the Philosophy of Science* 42, 157–176.

Prior, Elizabeth W., Robert Pargetter, Frank Jackson (1982), "Three Theses About Dispositions", *American Philosophical Quarterly* 19, 251–257.

Psillos, Statis (2004), "Inference to the Best Explanation and Bayesianism", in Stadler (2004), pp. 83–91.

Putnam, Hilary (1962), "The Analytic and the Synthetic", in: Feigl, Maxwell (1962), pp. 358–397.

—— (1975), "The Meaning of 'Meaning' ", in: H. Putnam, *Philosophical Papers, vol. 2: Mind, Language and Reality*, Cambridge: Cambridge University Press, pp. 215–271; and in: K. Gunderson (ed.), *Minnesota Studies in the Philosophy of Science, Vol. VII, Language, Mind and Knowledge*, Minneapolis: University of Minnesota Press, pp. 131–193.

—— (1978), *Meaning and the Moral Sciences*, London: Routledge & Kegan Paul.

—— (1980), "Models and Reality", *Journal of Symbolic Logic* 45, 464–482; also in: H. Putnam (1983a), pp. 1–25.

—— (1981), *Reason, Truth and History*, Cambridge: Cambridge University Press.

—— (1983a), *Philosophical Papers, Vol. 3, Realism and Reason*, Cambridge: Cambridge University Press.

Putnam, Hilary (1983b), "Why There Isn't a Ready-Made World", in: H. Putnam (1983a), pp. 205–228.

Quine, Willard V. O. (1951), "Two Dogmas of Empiricism", *Philosophical Review* 60, 20–43; also in: W. V. O. Quine, *From a Logical Point of View*, Cambridge, Mass., 1953.

—— (1952), "The Problem of Simplifying Truth Functions", *American Mathematical Monthly* 59, 521–531.

—— (1960), *Word and Object*, Cambridge, Mass.: MIT Press.

—— (1969), "Epistemology Naturalized", in: W. V. O. Quine, *Ontological Relativity and Other Essays*, New York: Columbia University Press, pp. 69–90.

—— (1985), "Events and Reification", in E. LePore and B. McLaughlin (eds.), *Actions and Events: Perspectives on the Philosophy of Donald Davidson*, Oxford: Blackwell, pp. 162–171.

—— (1986), "Reply to Morton White", in: L. E. Hahn, P. A. Schilpp (eds.), *The Philosophy of W. V. Quine*, La Salle, Ill.: Open Court, pp. 663–665.

—— (1995), *From Stimulus to Science*, Cambridge, Mass.: Harvard University Press.

Rabinowicz, Wlodek (1995), "Global Belief Revision Based on Similarities between Worlds", in: S. O. Hansson, W. Rabinowicz (eds.), *Logic for a Change. Essays Dedicated to Sten Lindström on the Occasion of His Fiftieth Birthday*, Uppsala Prints and Preprints in Philosophy 1995: 9, Department of Philosophy, Uppsala University, pp. 80–105.

—— (1996), "Stable Revision, or Is Preservation Worth Preserving?", in: A. Fuhrmann, H. Rott (eds.), *Logic, Action, and Information. Essays on Logic in Philosophy and Artificial Intelligence*, Berlin: de Gruyter, pp. 101–128.

Railton, Peter (1978), "A Deductive-Nomological Model of Probabilistic Explanation", *Philosophy of Science* 45, 206–226.

Ramsey, Frank Plumpton (1928), "Universals of Law and of Fact", in: Ramsey (1978), pp. 128–132.

—— (1929), "General Propositions and Causality", in: Ramsey (1978), pp. 133–151.

—— (1978), *Foundations. Essays in Philosophy, Logic, Mathematics and Economics*, ed. by D. H. Mellor, London: Routledge & Kegan Paul.

Reichenbach, Hans (1951), *The Rise of Scientific Philosophy*, Berkeley: University of California Press.

—— (1956), *The Direction of Time*, Los Angeles: University of California Press.

Rényi, Alfred (1955), "On a New Axiomatic Method of Probability", *Acta Mathematica Academiae Scientiarium Hungaricae* 6, 285–335.

—— (1962), *Wahrscheinlichkeitsrechnung*, Berlin: VEB Deutscher Verlag der Wissenschaften.

Rescher, Nicholas (1964), *Hypothetical Reasoning*, Amsterdam: North-Holland.

—— (1973), *The Coherence Theory of Truth*, Oxford: Oxford University Press.

—— (1976), *Plausible Reasoning*, Assen: Van Gorcum.

—— (1985), "Truth as Ideal Coherence", *Review of Metaphysics* 38, 795–806.

Richardson, Thomas S. (1998), "Chain Graphs and Symmetric Associations", in: M. I. Jordan (ed.), *Learning in Graphical Models*, Cambridge, Mass.: MIT Press, pp. 231–259.

——, Peter Spirtes (2002), "Ancestral Graph Markov Models", *Annals of Statistics* 30, 962–1030.

——, Peter Spirtes (2003), "Causal Inference Via Ancestral Graph Models", in: P. J. Green, N. L. Hort, S. Richardson (eds.), *Highly Structured Stochastic Systems*, Oxford: Oxford University Press, pp. 83–105.

Rockafellar, R. Tyrrell (1970), *Convex Analysis*, Princeton: Princeton University Press.

Rosefeldt, Tobias (2000), *Das logische Ich. Kant über den Gehalt des Begriffes von sich selbst*, Berlin: Philo.

Rosenthal, David M. (1986), "Two Concepts of Consciousness", *Philosophical Studies* 49, 329–359.

—— (1997), "A Theory of Consciousness", in: Block et al. (1997), pp. 729–753.

Rott, Hans (1986), "Ifs, Though, and Because", *Erkenntnis* 25, 345–370.

—— (1989), "Conditionals and Theory Change: Revisions, Expansions and Additions", *Synthese* 81, 91–113.

—— (1991a), *Reduktion und Revision. Aspekte des nichtmonotonen Theorienwandels*, Frankfurt a.M.: Peter Lang.

—— (1991b), "Two Methods of Constructing Contractions and Revisions of Knowledge Systems", *Journal of Philosophical Logic* 20, 149–173.

—— (1999), "Coherence and Conservatism in the Dynamics of Belief. Part I: Finding the Right Framework", *Erkenntnis* 50, 387–412.

—— (2001), *Change, Choice and Inference: A Study of Belief Revision and Nonmonotonic Reasoning*, Oxford: Oxford University Press.

—— (2003), "Coherence and Conservatism in the Dynamics of Belief. Part II: Iterated Belief Change Without Dispositional Coherence", *Journal of Logic and Computation*, 13, 111–145.

—— (2007), "Bounded Revision: Two-Dimensional Belief Change Between Conservatism and Moderation", in: T. Rønnow-Rasmussen et al. (eds.), *Hommage à Wlodek. Philosophical Papers Dedicated to Wlodek Rabinowicz*, Lund: http://www.fil.lu.se/hommageawlodek

—— (2009), "Shifting Priorities: Simple Representations for Twenty Seven Iterated Theory Change Operators", in: D. Makinson, J. Malinowski, H. Wansing (eds.), *Towards Mathematical Philosophy*, Dordrecht: Springer, pp. 269–296.

Ryle, Gilbert (1949), *The Concept of Mind*, London: Hutchinson.

Sahlin, Nils-Eric (1991), "Obtained by a Reliable Process and Always Leading to Success", *Theoria* 57, 132–149.

—— (1997), " 'He Is No Good For My Work'. On the Philosophical Relations Between Ramsey and Wittgenstein", *Poznan Studies in the Philosophy of Science and the Humanities* 51, 61–84.

Salmon, Wesley C. (1966), *The Foundations of Scientific Inference*, Pittsburgh: University of Pittsburgh Press.

—— (1970), "Statistical Explanation", in R. G. Colodny (ed.), *The Nature and Function of Scientific Theories*, Pittsburgh: University of Pittsburgh Press, pp. 173–231.

—— (1975), "Confirmation and Relevance", in: G. Maxwell, R. M. Anderson (eds.), *Minnesota Studies in the Philosophy of Science, Vol. VI*, Minneapolis: University of Minnesota Press, pp. 3–36.

—— (1978), "Why Ask, 'Why?'? An Inquiry Concerning Scientific Explanation", *Proceedings and Addresses of the American Philosophical Association* 51, 683–705.

—— (1980), "Probabilistic Causality", *Pacific Philosophical Quarterly* 61, 50–74.

—— (1984), *Scientific Explanation and the Causal Structure of the World*, Princeton: Princeton University Press.

—— (1988), "Intuitions – Good and Not-So-Good", in: B. Skyrms, W. L. Harper (eds.), *Causation, Chance, and Credence*, Dordrecht: Kluwer, pp. 51–71.

—— (1994), "Causality Without Counterfactuals", *Philosophy of Science* 61, 297–312.

Salmon, Wesley C. (2001), "Explanation and Confirmation: A Bayesian Critique of Inference to the Best Explanation", in: G. Hon, S. S. Rankover (eds.), *Explanation: Theoretical Approaches and Applications*, Dordrecht: Kluwer, pp. 61–91.

Samuelson, Paul. A. (1938), "A Note on the Pure Theory of Consumers' Behaviour", *Economica* 5, 61–71.

—— (1947), *Foundations of Economic Analysis*, Cambridge, Mass.: Harvard University Press.

Sarin, Rakesh K., Peter P. Wakker (1992), "A Simple Axiomatization of Nonadditive Expected Utility", *Econometrica* 60, 1255–1272.

Sartwell, Crispin (1991), "Knowledge is Merely True Belief", *American Philosophical Quarterly* 28, 157–165.

—— (1992), "Why Knowledge is Merely True Belief", *Journal of Philosophy* 89, 167–180.

Savage, Leonard J. (1954), *The Foundations of Statistics*, New York: Wiley, 2nd edn.: Dover 1972.

Schaffer, Jonathan (2000), "Trumping Preemption", *Journal of Philosophy* 97, 165–181; also in: Collins et al. (2004), pp. 59–73.

Scheines, Richard (2005), "Causation", in: M. C. Horowitz (eds.), *New Dictionary of the History of Ideas, Vol. 1*, Framington Hills, MI: Thomson Gale, pp. 280–289.

Schiffer, Stephen (1991), "Ceteris Paribus Laws", *Mind* 100, 1–17.

Schlick, Moritz (1925), *Allgemeine Erkenntnislehre*, Berlin: Springer.

—— (1931), "Die Kausalität in der gegenwärtigen Physik", *Die Naturwissenschaften* 19, 145–162.

—— (1934), "Über das Fundament der Erkenntnis", *Erkenntnis* 4, 79–99.

Schmeidler, David (1989), "Subjective Probability and Expected Utility Without Additivity", *Econometrica* 57, 571–587.

Schneeweiss, Winfried G. (1989), *Boolean Functions with Engineering Applications and Computer Programs*, Berlin: Springer.

Scholz; Oliver R. (1999), *Verstehen und Rationalität. Untersuchungen zu den Grundlagen von Hermeneutik und Sprachphilosophie*, Frankfurt a.M.: Klostermann.

Schopenhauer, Arthur (1847), *Über die vierfache Wurzel des Satzes vom zureichenden Grunde*.

Schum, David A. (2001), "Species of Abductive Reasoning in Fact Investigation in Law", *Cardozo Law Review* 22, 1645–1681.

Schurz, Gerhard (2001), "What is 'Normal'? An Evolution-Theoretic Foundation of Normic Laws and Their Relation to Statistical Normality", *Philosophy of Science* 28, 476–497.

—— (2002), "*Ceteris Paribus* Laws: Classification and Deconstruction", *Erkenntnis* 57, 351–372.

—— (2011), "Abductive Belief Revision in Science", in E. Olsson, S. Enquist (eds.), *Belief Revision Meets Philosophy of Science*, New York: Springer, pp. 77–104.

Segerberg, Krister (1998), "Irrevocable Belief Revision in Dynamic Doxastic Logic", *Notre Dame Journal of Formal Logic* 39, 287–306.

Seidenfeld, Teddy (2001), "Remarks on the Theory of Conditional Probability: Some Issues of Finite Versus Countable Additivity", in: V. F. Hendricks, S. A. Pedersen, K. F. Jørgensen (eds.), *Probability Theory – Philosophy, Recent History and Relations to Science*, Dordrecht: Kluwer, pp. 167–178.

Sellars, Wilfried (1963), *Science, Perception and Reality*, Atascadero, Ca.: Ridgeview.

Sen, Amartya K. (1970), *Collective Choice and Social Welfare*, San Francisco: Holden-Day.

Shachter, Ross D. Tod, S. Levitt, Laveen N. Kanal, John F. Lemmer (eds.) (1990), *Uncertainty in Artificial Intelligence 4*, Amsterdam: North-Holland.

Shackle, George L. S. (1949), *Expectation in Economics*, Cambridge: Cambridge University Press.

—— (1961), *Decision, Order, and Time in Human Affairs*, Cambridge: Cambridge University Press, 2nd edn. 1969.

Shafer, Glenn (1976), *A Mathematical Theory of Evidence*, Princeton: Princeton University Press.

—— (1978), "Non-Additive Probabilities in the Work of Bernoulli and Lambert", *Archive for History of Exact Sciences* 19, 309–370.

—— (1985), "Conditional Probability", *International Statistical Review* 53, 261–277.

——, Prakash P. Shenoy (1990), "Probability Propagation", *Annals of Mathematics and Artificial Intelligence* 2, 327–352.

Shenoy, Prakash P. (1991), "On Spohn's Rule for Revision of Beliefs", *International Journal of Approximate Reasoning* 5, 149–181.

Shimony, Abner (1970), "Scientific Inference", in: R. G. Colodny (ed.), *The Nature and Function of Scientific Theories*, Pittsburgh: University of Pittsburgh Press, pp. 79–172.

Shoemaker, Sydney (1990), "First Person Access", in: J. E. Tomberlin (ed.), *Philosophical Perspectives, Vol. 4, Action Theory and Philosophy of Mind*, Atascadero, CA: Ridgeview, pp. 187–214; also in: Sidney Shoemaker, *The First-Person Perspective and Other Essays*, Cambridge: Cambridge University Press, 1996, ch. 3.

Shore, John E., Rodney W. Johnson (1980), "Axiomatic Derivation of the Principle of Maximum Entropy and the Principle of Minimum Cross-Entropy", *IEEE Transactions in Information Theory IT-26*, 1, 26–37.

Silverberg, Arnold (1996), "Psychological Laws and Non-Monotonic Logic", *Erkenntnis* 44, 199–224.

Skyrms, Brian (1980), *Causal Necessity. A Pragmatic Investigation of the Necessity of Laws*, New Haven: Yale University Press.

—— (1984), *Pragmatics and Empiricism*, Yale University Press.

—— (1990), *The Dynamics of Rational Deliberation*, Cambridge, Mass.: MIT Press.

—— (1991), "Carnapian Inductive Logic for Markov Chains", *Erkenntnis* 35, 439–460.

Smith, Martin (2010), "A Generalized Lottery Paradox for Infinite Probability Spaces", *The British Journal for the Philosophy of Science* 61, 821–831.

Sober, Elliott (1982), "Dispositions and Subjunctive Conditionals, or, Dormative Virtues Are No Laughing Matter", *Philosophical Review* 91, 591–596.

Speed, Terri P. (1979), "A Note on Nearest-Neighbor Gibbs and Markov Distributions Over Graphs", *Sankhya Series A* 41, 184–197.

Spirtes, Peter, Clark Glymour, Richard Scheines (1993), *Causation, Prediction, and Search*, Berlin: Springer; 2nd ext edn 2000.

Spohn, Wolfgang (1976/78), *Grundlagen der Entscheidungstheorie*, Ph.D. Thesis, University of Munich 1976, published: Kronberg/Ts.: Scriptor 1978, out of print, pdf-version at: http://www.uni-konstanz.de/FuF/Philo/Philosophie/philosophie/files/ge.buch.gesamt.pdf.

—— (1977), "Where Luce and Krantz Do Really Generalize Savage's Decision Model", *Erkenntnis* 11, 113–134.

—— (1980), "Stochastic Independence, Causal Independence, and Shieldability", *Journal of Philosophical Logic* 9, 73–99.

—— (1983a), "Deterministic and Probabilistic Reasons and Causes", in: *Erkenntnis* 19, i.e., in: C. G. Hempel, H. Putnam, W. K. Essler (eds.), *Methodology, Epistemology, and Philosophy of Science. Essays in Honour of Wolfgang Stegmüller on the Occasion of his 60th Birthday*, Dordrecht: Reidel, pp. 371–396.

Spohn, Wolfgang (1983b), *Eine Theorie der Kausalität*, unpublished Habilitationsschrift, Universität München, pdf-version at: http://www.uni-konstanz.de/FuF/Philo/Philosophie/philosophie/files/habilitation.pdf (page references are to the electronic version).

—— (1986), "The Representation of Popper Measures", *Topoi* 5, 69–74.

—— (1988), "Ordinal Conditional Functions. A Dynamic Theory of Epistemic States", in: W. L. Harper, B. Skyrms (eds.), *Causation in Decision, Belief Change, and Statistics, vol. II*, Dordrecht: Kluwer, pp. 105–134; also in: Spohn (2008a, ch. 1).

—— (1990), "Direct and Indirect Causes", *Topoi* 9, 125–145; also in: Spohn (2008a, ch. 2).

—— (1991), "A Reason for Explanation: Explanations Provide Stable Reasons", in: W. Spohn, B. C. van Fraassen, B. Skyrms (eds.), *Existence and Explanation*, Dordrecht: Kluwer, pp. 165–196; also in: Spohn (2008a, ch. 9).

—— (1993a), "Causal Laws are Objectifications of Inductive Schemes", in: J.-P. Dubucs (ed.), *Philosophy of Probability*, Dordrecht: Kluwer, pp. 223–252; also in: Spohn (2008a, ch. 5).

—— (1993b), "Wie kann die Theorie der Rationalität normativ und empirisch zugleich sein?", in: L. Eckensberger, U. Gähde (eds.), *Ethik und Empirie. Zum Zusammenspiel von begrifflicher Analyse und erfahrungswissenschaftlicher Forschung in der Ethik*, Frankfurt a.M.: Suhrkamp, pp. 151–196.

—— (1994a), "On Reichenbach's Principle of the Common Cause", in: W. C. Salmon, G. Wolters (eds.), *Logic, Language, and the Structure of Scientific Theories*, Pittsburgh: University of Pittsburgh Press, pp. 215–239.

—— (1994b), "On the Properties of Conditional Independence", in: P. Humphreys (ed.), *Patrick Suppes: Scientific Philosopher. Vol. 1: Probability and Probabilistic Causality*, Dordrecht: Kluwer, pp. 173–194.

—— (1997a), "Begründungen a priori – oder: Ein frischer Blick auf Dispositionssprädikate", in W. Lenzen (ed.), *Das weite Spektrum der Analytischen Philosophie. Festschrift für Franz von Kutschera*, Berlin: de Gruyter, pp. 323–345; engl. translation, "A Priori Reasons: A Fresh Look at Disposition Predicates", in Spohn (2008a, ch. 12).

—— (1997b), "The Character of Color Predicates: A Materialist View", in: Künne et al. (1997), pp. 351–379; also in: Spohn (2008a, ch. 13).

—— (1997/98), "How to Understand the Foundations of Empirical Belief in a Coherentist Way", *Proceedings of the Aristotelian Society, New Series* 98, 23–40; also in: Spohn (2008a, ch. 11).

—— (1998), "The Intentional versus the Propositional Conception of the Objects of Belief", in: C. Martinez, U. Rivas, L. Villegas-Forero (eds.), *Truth in Perspective. Recent Issues in Logic, Representation and Ontology*, Aldershot: Ashgate, pp. 271–291.

—— (1999a), "Two Coherence Principles", *Erkenntnis* 50, 155–175.

—— (1999b), "Ranking Functions, AGM Style", in B. Hansson, S. Halldén, N.-E. Sahlin, W. Rabinowicz (eds.), *Internet-Festschrift for Peter Gärdenfors*, http://www.lucs.lu.se/spinning/

—— (2000a), "Über die Struktur theoretischer Gründe", in: J. Mittelstraß (ed.), *Die Zukunft des Wissens. Akten des 18. Deutschen Kongresses für Philosophie*, Berlin: Akademie Verlag, pp. 163–176.

—— (2000b), "Wo stehen wir heute mit dem Problem der Induktion?", in: R. Enskat (ed.), *Erfahrung und Urteilskraft*, Würzburg: Königshausen & Naumann, pp. 151–164.

—— (2001a), "Vier Begründungsbegriffe", in: T. Grundmann (ed.), *Erkenntnistheorie. Positionen zwischen Tradition und Gegenwart*, Paderborn: Mentis, pp. 33–52.

—— (2001b), "Bayesian Nets Are All There Is To Causal Dependence", in: M. C. Galavotti, P. Suppes, D. Costantini (eds.), *Stochastic Dependence and Causality*, Stanford: CSLI Publications, pp. 157–172; also in: Spohn (2008a, ch. 4).

—— (2002a), "A Brief Comparison of Pollock's Defeasible Reasoning and Ranking Functions", *Synthese* 131, 39–56.

—— (2002b), "The Many Facets of the Theory of Rationality", *Croatian Journal of Philosophy* 2, 247–262.

—— (2002), "Laws, *Ceteris Paribus* Conditions, and the Dynamics of Belief", *Erkenntnis* 57, 373–394; also in: Spohn (2008a, ch. 6).

—— (2003), "Lehrer Meets Ranking Theory", in Olsson (2003), pp. 129–142.

—— (2005a), "Enumerative Induction and Lawlikeness", *Philosophy of Science* 72, 164–187; also in: Spohn (2008a, ch. 7).

—— (2005b), "Anmerkungen zum Begriff des Bewusstseins", in: G. Wolters, M. Carrier (Hg.), *Homo Sapiens und Homo Faber. Festschrift für Jürgen Mittelstraß*, Berlin: de Gruyter, pp. 239–251.

—— (2006a), "Isaac Levi's Potentially Surprising Epistemological Picture", in: E. Olsson (ed.), *Knowledge and Inquiry. Essays on the Pragmatism of Isaac Levi*, Cambridge: Cambridge University Press, pp. 125–142.

—— (2006b), "Causation: An Alternative", *British Journal for the Philosophy of Science* 57, 93–119; also in: Spohn (2008a, ch. 3).

—— (2007), "The Core of Free Will", in: P. K. Machamer, G. Wolters (eds.), *Thinking About Causes. From Greek Philosophy to Modern Physics*, Pittsburgh: University of Pittsburgh Press, pp. 297–309.

—— (2008a), *Causation, Coherence, and Concepts. A Collection of Essays*, Dordrecht: Springer.

—— (2008b), "Two-Dimensional Truth", *Studia Philosophica Estonia* 1, 194–207.

—— (2010a), "Chance and Necessity: From Humean Supervenience to Humean Projection", in: E. Eells, J. Fetzer (eds.), *The Place of Probability in Science*, Dordrecht: Springer, pp. 101–131; also in: Spohn (2008a, ch. 8).

—— (2010b), "The Structural Model and the Ranking Theoretic Approach to Causation: A Comparison", in: R. Dechter, H. Geffner, J. Y. Halpern (eds.), *Heuristics, Probability and Causality. A Tribute to Judea Pearl*, San Mateo, CA: Kauffmann, pp. 493–508.

—— (2011), "Normativity is the Key to the Difference Between the Human and the Social Sciences", in D. Dieks et al. (eds.), *Explanation, Prediction, and Confirmation*, Dordecht: Springer, pp. 241–251.

Stadler, Friedrich (ed.) (2004), *Induction and Deduction in the Sciences. Vienna Circle Institute Yearbook 11*, Dordrecht: Kluwer.

Stalnaker, Robert C. (1968), "A Theory of Conditionals", in: N. Rescher (ed.), *Studies in Logical Theory*, Oxford: Blackwell, pp. 98–112.

—— (1976), "Propositions", in: A. F. Mackay, D. D. Merrill (eds.), *Issues in the Philosophy of Language*, New Haven: Yale University Press, pp. 79–91.

—— (1978), "Assertion", in: P. Cole (ed.), *Syntax and Semantics, Vol. 9: Pragmatics*. New York: Academic Press, pp. 315–332.

—— (1999), *Context and Content*, Oxford: Oxford University Press.

——, Richmond Thomason (1970), "A Semantic Analysis of Conditional Logic", *Theoria* 36, 23–42.

Steel, Daniel (2005), "Indeterminism and the Causal Markov Condition", *British Journal for the Philosophy of Science* 56, 3–26.

Stegmüller, Wolfgang (1970), *Probleme und Resultate der Wissenschaftstheorie und Analytischen Philosophie, Vol II, Theorie und Erfahrung, 1. Halbband*, Berlin: Springer.

Stegmüller, Wolfgang (1973), *Probleme und Resultate der Wissenschaftstheorie und Analytischen Philosophie, Vol IV, Personelle und Statistische Wahrscheinlichkeit, 1. Halbband*, Berlin: Springer.

Stoer, Josef, Christoph Witzgall (1970), *Convexity and Optimization in Finite Dimensions I*, Berlin et al.: Springer.

Strawson, Peter F. (1966), *The Bounds of Sense*, London: Methuen.

Stuart, Alan, J. Keith Ord (1991), *Kendall's Advanced Theory of Statistics, Vol. 2, Classical Inference and Relationship*, London: Edward Arnold, 5th edn.

Studeny, Milan (1989), "Multiinformation and the Problem of Characterization of Conditional Independence Relations", *Problems of Control and Information Theory* 18, 3–16.

—— (1992), "Conditional Independence Relations Have No Finite Complete Characterization", in: S. Kubik, J. A. Visek (eds.), *Information Theory, Statistical Decision Functions, and Random Processes. Transactions of the 11th Prague Conference 1990, Vol. B*, Dordrecht: Kluwer, pp. 377–396.

—— (1995), "Conditional Independence and Natural Conditional Functions", *International Journal of Approximate Reasoning* 12, 43–68.

—— (2005), *Probabilistic Conditional Independence Structures*, Berlin: Springer.

——, Remco R. Bouckaert (1998), "On Chain Graph Models for Description of Conditional Independence Structures", *Annals of Statistics* 26, 1434–1495.

Suppe, Frederick (ed.) (1977), *The Structure of Scientific Theories*, Urbana: University of Illinois Press, 2nd edn.

Suppes, Patrick (1970), *A Probabilistic Theory of Causality*, Amsterdam: North-Holland.

—— (1984), *Probabilistic Metaphysics*, Oxford: Blackwell.

Tan, Sek-Wah, Judea Pearl (1994), "Qualitative Decision Theory", in: B. Hayes-Roth, R. Korf (eds.), *Proceedings of the Twelfth National Conference on Artificial Intelligence, Vol. 2*, Menlo Park, CA: AAAI Press, pp. 928–933.

Teller, Paul (1976), "Conditionalization, Observation and Change of Preference", in W. L. Harper, C. A. Hooker (eds.), *Foundations of Probability Theory, Statistical Inference and Statistical Theories of Science*, Dordrecht: Reidel, pp. 205–259.

Terillus, Antonius (1669), *Fundamentum Totius Theologiae seu Tractatus de Conscientia Probabili*, Leodii/Louvain.

Thöle, Bernhard (forthcoming), "Kant's Justification of the Principle of Causality", in: O. Höffe, R. Pippin (eds.), *Kant, Critique of Pure Reason. Contemporary German Perspectives*. Cambridge: Cambridge University Press.

Toulmin, Stephen (1953), *The Philosophy of Science*, London: Hutchinson.

van de Geer, John P. (1971), *Introduction to Multivariate Analysis for the Social Sciences*, San Francisco: Freeman.

van Fraassen, Bas C. (1980a), *The Scientific Image*, Oxford: Oxford University Press.

—— (1980b), "Critical Notice on: Brian Ellis, Rational Belief Systems", *Canadian Journal of Philosophy* 10, 497–511.

—— (1983), "Calibration: A Frequency Justification for Personal Probability", in: R. Cohen, L. Laudan (eds.), *Physics, Philosophy, and Psychoanalysis*, Dordrecht: Reidel, pp. 295–319.

—— (1984), "Belief and the Will", *Journal of Philosophy* 81, 235–256.

—— (1989), *Laws and Symmetry*, Oxford: Oxford University Press.

—— (1995a), "Belief and the Problem of Ulysses and the Sirens", *Philosophical Studies* 77, 7–37.

—— (1995b), "Fine-Grained Opinion, Probability, and the Logic of Full Belief", *Journal of Philosophical Logic* 24, 349–377.

Vendler, Zeno (1967), "Causal Relations", *Journal of Philosophy*, 64, 704–713.

Verma, Thomas S., Judea Pearl (1990), "Causal Networks: Semantics and Expressiveness", in: Shachter et al. (1990), pp. 69–76.

——, Judea Pearl (1991), "Equivalence and Synthesis of Causal Models", in: P. P. Bonissone, M. Henrion, L. N. Kanal, J. F. Lemmer (eds.), *Uncertainty in Artificial Intelligence* 6, Amsterdam: Elsevier, pp. 255–268.

Villanueva, Enrique (ed.) (1998), *Concepts. Philosophical Issues 9*, Atascadero: Ridgeview.

von Kutschera, Franz (1982), *Grundfragen der Erkenntnistheorie*, Berlin: de Gruyter.

—— (1994), "Zwischen Skepsis und Relativismus", in: G. Meggle, U. Wessels (eds.), *Analyomen 1. Perspectives in Analytical Philosophy*, Berlin: de Gruyter, pp. 207–224.

von Neumann, John, Oskar Morgenstern (1944), *Theory of Games and Economic Behavior*, Princeton: Princeton University Press, 2nd edn. 1947.

von Wright, Georg Henrik (1971), *Explanation and Understanding*, Ithaca: Cornell University Press.

—— (1974), *Causality and Determinism*, New York: Columbia University Press.

Wagner, Carl G. (1992), "Generalized Probability Kinematics", *Erkenntnis* 36, 245–257.

Wakker, Peter P. (2005), "Decision-Foundations for Properties of Nonadditive Measures: General State Spaces or General Outcome Spaces", *Games and Economic Behavior* 50, 107–125.

—— (2010), *Prospect Theory for Risk and Ambiguity*, Cambridge: Cambridge University Press.

Walley, Peter (1991), *Statistical Reasoning with Imprecise Probabilities*, London: Chapman & Hall.

Watkins, John W. N. (1957/58), "Farewell to the Paradigm-Case Argument", *Analysis* 18, 25–33.

Weirich, Paul (1983), "Conditional Probabilities and Probabilities Given Knowledge of a Condition", *Philosophy of Science* 50, 82–95.

Wermuth, Nanny (1980), "Liinear Recursive Equations, Covariance Selection, and Path Analysis", *Journal of the Ameican Statistical Association* 75, 963–997.

——, Steffen L. Lauritzen (1983), "Graphical and Recursive Models for Contingency Tables", *Biometrika* 70, 537–552.

Weydert, Emil (1996), "System J – Revision Entailment: Default Reasoning Through Ranking Measure Updates", in: D. Gabbay, H. J. Ohlbach (eds.), *Practical Reasoning. International Conference on Formal and Applied Practical Reasoning 96*, Berlin: Springer, pp. 121–135.

—— (2003), "System JLZ – Rational Default Reasoning by Minimal Ranking Constructions", *Journal of Applied Logic* 1, 273–308.

Williams, Mary-Ann (1995), "Iterated Theory Base Change: A Computational Model", in: C. S. Mellish (ed.), *IJCAI '95 – Proceedings of the 14th International Joint Conference on Artificial Intelligence, Vol. 2*, San Mateo: Morgan Kaufmann, pp. 1541–1547.

Williamson, Jon (2005), *Bayesian Nets and Causality. Philosophical and Computational Foundations*, Oxford: Oxford University Press.

Wittgenstein, Ludwig (1922), *Tractatus logico-philosophicus*, London: Routledge & Kegan Paul.

—— (1953), *Philosophical Investigations*, Oxford: Blackwell.

Woodward, James (2002), "There Is No Such Thing as a *Ceteris Paribus* Law", *Erkenntnis* 57, 303–328.

—— (2003), *Making Things Happen. A Theory of Causal Explanation*, Oxford: Oxford University Press.

Wright, Sewell (1921), "Correlation and Causation", *Journal of Agricultural Research* 20, 557–585.

Wright, Sewell (1934), "The Method of Path Coefficients", *Annals of Mathematical Statistics* 5, 161–215.

Yablo, Stephen (1992), "Cause and Essence", *Synthese* 93, 403–449.

Yule, G. Udny (1911), *An Introduction to the Theory of Statistics*, London: Griffen.

Zhang, Jiji (2006), *Causal Inference and Reasoning in Causally Insufficient Systems*, Ph.D. dissertation, Department of Philosophy, Carnegie Mellon University.

Name Index

Subject Index

Italic numbers refer to definitions or explanations of the relevant terms.

9 780198 705857